THE GREAT
PALEOLITHIC WAR

THE GREAT PALEOLITHIC WAR

*How Science Forged an Understanding
of America's Ice Age Past*

DAVID J. MELTZER

THE UNIVERSITY OF CHICAGO PRESS | CHICAGO AND LONDON

David J. Meltzer is the Henderson-Morrison Professor of Prehistory at Southern Methodist University, and a member of the National Academy of Sciences. He is the author of *Folsom* and *First Peoples in a New World*. He lives in Dallas.

The University of Chicago Press, Chicago 60637
The University of Chicago Press, Ltd., London
© 2015 by The University of Chicago
All rights reserved. Published 2015.
Printed in the United States of America

24 23 22 21 20 19 18 17 16 15   1 2 3 4 5

ISBN-13: 978-0-226-29322-6   (cloth)
ISBN-13: 978-0-226-29336-3   (e-book)
DOI: 10.7208/chicago/9780226293363.001.0001

Library of Congress Cataloging-in-Publication Data

Meltzer, David J., author.
    The great Paleolithic war : how science forged an understanding of America's ice age past / David J. Meltzer.
        pages cm
    Includes bibliographical references and index.
    ISBN 978-0-226-29322-6 (cloth : alkaline paper) — ISBN 978-0-226-29336-3 (ebook)   1. Archaeology—United States—History.   2. Paleoanthropology—United States—History.   3. Natural history—United States—History.   I. Title.
    CC101.U6M45 2015
    930.1'2—dc23

                                                                        2015006581

♾ This paper meets the requirements of ANSI/NISO Z39.48-1992 (Permanence of Paper).

For Suzanne,
who has been there since the beginning of this lengthy project,
and was mostly sure I'd see it through to the end.

CONTENTS

ROSTER OF INDIVIDUALS

THOSE WHOSE names follow participated in one degree or another in the long dispute over human antiquity in North America. This is not a comprehensive list, omitting as it does mention of those participating before the dispute got started, as well as the anthropologists, archaeologists, and others who throughout the long years of controversy occasionally commented from the sidelines.

Each entry provides the name, birth and death dates, education, and a brief identifying description of the occupation and affiliation of the individual at the time he or she was involved in the controversy, as well as the site or sites in which they worked or with which they were associated. Some, like Holmes and Hrdlička, were involved at virtually all the sites, and hence no particular ones are specified. Chapter 10 discusses in greater depth the relative status of different individuals connected with the controversy; however, a quick snapshot of the standing of individuals can be garnered from whether they were elected as fellows of the American Association for the Advancement of Science (AAAS Fellow), received a star in the *American Men of Science (AMS)*, or were elected to the National Academy of Sciences (NAS). These honors are noted in the entries that follow, along with the year of AAAS or NAS election or, in the case of the *AMS* star, the volume number.

1. **ABBOTT,** Charles C. (June 4, 1843–July 27, 1919), physician, naturalist, popular author and archaeologist, Trenton gravels. Highest earned academic degree: MD 1865, University of Pennsylvania. Honors: AAAS Fellow 1883.
2. **BABBITT,** Frances (January 24, 1824–July 6, 1891), Minnesota schoolteacher, Little Falls. Highest earned academic degree: None. Honors: AAAS Fellow 1887.
3. **BALDWIN,** Charles C. (December 2, 1834–February 2, 1895), judge, Western Reserve Historical Society. Highest earned academic degree: JD 1857 Harvard University. Honors: AAAS Fellow 1891.
4. **BARBOUR,** Erwin (April 5, 1856–May 10, 1947), paleontologist, Nebraska state geologist, director of the Nebraska State Museum, Long's Hill. Highest earned academic degree: PhD 1887, Yale University. Honors: AAAS Fellow 1898.
5. **BLACKMAN,** Elmer E. (1865–September 13, 1942), archaeologist, Nebraska State Historical Society, Long's Hill. Highest earned academic degree: None. Honors: None.
6. **BRINTON,** Daniel G. (May 13, 1837–July 31, 1899), anthropologist, University of Pennsylvania, contributing editor to *Science*. Highest earned academic degree: None. Honors: AAAS Fellow 1885.

7. **BROWN,** Barnum (February 12, 1873–February 5, 1963), paleontologist, American Museum of Natural History, Folsom. Highest earned academic degree: BS 1908, University of Kansas. Honors: AAAS Fellow 1931.

8. **BRYAN,** Kirk (July 22, 1888–August 20, 1950), geologist, Harvard University, Folsom. Highest earned academic degree: PhD 1920, Yale University. Honors: AAAS Fellow 1925; AMS 5.

9. **CALVIN,** Samuel (February 2, 1840–April 17, 1911), geologist, Iowa Geological Survey, Lansing. Highest earned academic degree: PhD 1888, Lenox College. Honors: AAAS Fellow 1889; AMS 1.

10. **CARR,** Lucien (December 15, 1829–January 27, 1915), archaeologist and administrative assistant, Peabody Museum, Harvard University, Trenton. Highest earned academic degree: AB 1846, St. Louis. Honors: AAAS Fellow 1877.

11. **CHAMBERLIN,** Rollin (October 20, 1881–March 6, 1948), geologist (son of Thomas), University of Chicago, Vero. Highest earned academic degree: PhD 1907, University of Chicago. Honors: AAAS Fellow 1909; AMS 3; NAS 1940.

12. **CHAMBERLIN,** Thomas Chrowder (September 25, 1843–November 15, 1928), glacial geologist, University of Chicago, chief of United States Geological Survey Glacial Division, founder and editor *Journal of Geology*. Highest earned academic degree: AM 1869, Beloit College. Honors: AAAS Fellow 1877; AMS 1; NAS 1903.

13. **CLAYPOLE,** Edward W. (June 1, 1835–August 17, 1901), geologist, Buchtel College, Ohio (now University of Akron), editorial board of *American Geologist*. Highest earned academic degree: DS, 1888, London. Honors: AAAS Fellow 1882, Fellow Geological Society of London 1879.

14. **COOK,** Harold J. (July 31, 1887–September 29, 1962), paleontologist, honorary curator, Colorado Museum of Natural History, Snake Creek, Lone Wolf Creek, Frederick, and Folsom. Highest earned academic degree: None. Honors: AAAS Fellow 1911.

15. **COPE,** Edward Drinker (July 28, 1840–April 12, 1897), paleontologist, University of Pennsylvania. Highest earned academic degree: None. Honors: AAAS Fellow 1875; NAS 1872.

16. **CRESSON,** Hilborne (1848?–September 6, 1894), physician and archaeologist, Field Assistant Peabody Museum, Harvard University, Claymont, Medora, Holly Oak pendant. Highest earned academic degree: None. Honors: None.

17. **DALL,** William H. (August 21, 1845–March 27, 1927), naturalist, invertebrate paleontologist, United States Geological Survey. Highest earned academic degree: None. Honors: AAAS 1874; AMS 1; NAS 1897.

18. **FIGGINS,** Jesse D. (August 17, 1867–June 10, 1944), naturalist, artist, and administrator, director of the Colorado Museum of Natural History, Lone Wolf Creek, Frederick, and Folsom. Highest earned academic degree: None. Honors: AAAS Fellow 1932.

19. **FOWKE,** Gerard (June 25, 1855–March 5, 1933), archaeologist and field assistant, Bureau of American Ethnology, Lansing. Highest earned academic degree: None. Honors: None.

20. **GIDLEY,** James W. (January 7, 1866–September 26, 1931), paleontologist, United States National Museum, Smithsonian Institution, Melbourne. Highest earned academic degree: MS 1901, Princeton University. Honors: None.

21. **GILDER,** Robert F. (October 6, 1856–March 7, 1940), artist, printer, and journalist, *Omaha World Herald*, Long's Hill. Highest earned academic degree: None. Honors: None.

22. **GODDARD,** Pliny E. (November 24, 1869–July 12, 1928), linguist, American Museum of Natural History, New York. Highest earned academic degree: PhD 1904, University of California. Honors: AAAS Fellow 1912; AMS 3.

23. **GREGORY,** William K. (May 19, 1876–December 29, 1970), paleontologist and physical anthropologist, American Museum of Natural History, *Hesperopithecus*. Highest earned academic degree: PhD 1910, Columbia University. Honors: AAAS Fellow 1925; AMS 3; NAS 1927.

24. **HAY,** Oliver P. (May 22, 1846–November 2, 1930), paleontologist, Carnegie Institution of Washington, Vero, Melbourne, Lone Wolf Creek, Frederick, Folsom. Highest earned academic degree: PhD 1884, University of Indiana. Honors: AAAS Fellow 1901; AMS 2.

25. **HAYNES,** Henry W. (September 20, 1831–February 16, 1912), archaeologist, Boston Society of Natural History. Highest earned academic degree: AM 1851, Harvard University. Honors: AAAS Fellow 1884.

26. **HOLMES,** William H. (December 1, 1846–April 20, 1933), archaeologist, artist, geologist, Bureau of American Ethnology and United States National Museum, Smithsonian Institution. Highest earned academic degree: AB 1870, McNeely Normal College. Honors: AAAS Fellow 1883; AMS 1; NAS 1905.

27. **HRDLIČKA,** Aleš (March 29, 1869–September 5, 1943), physical anthropologist, United States National Museum, Smithsonian Institution, founder and editor *American Journal of Physical Anthropology*. Highest earned academic degree: MD 1892, Eclectic Medical College New York. Honors: AAAS Fellow 1897; AMS 1; NAS 1921.

28. **INGALLS,** Albert G. (January 16, 1888–August 13, 1958), editor *Scientific American*, New York. Highest earned academic degree: BS 1914, Cornell University. Honors: None.

29. **JENKS,** Albert E. (November 28, 1869–June 6, 1953), anthropologist, Folsom, Minnesota Man, Brown's Valley Man, Sauk Valley Man. University of Minnesota. Highest earned academic degree: PhD 1899, University of Minnesota. Honors: AAAS Fellow 1902.

30. **KIDDER,** Alfred V. (October 29, 1885–June 11, 1963), archaeologist, Carnegie Institution of Washington, Folsom. Highest earned academic degree: PhD 1914, Harvard University. Honors: AAAS Fellow 1929; AMS 3; NAS 1936.

31. **KNAPP,** George (unknown), geologist, New Jersey Geological Survey, Trenton. Highest earned academic degree: BS, University of Wisconsin. Honors: None.

32. **KUMMEL,** Henry B. (May 25, 1867–October 23, 1945), geologist, New Jersey Geological Survey, Trenton. Highest earned academic degree: PhD 1895, University of Chicago. Honors: AMS 1.

33. **LESLEY,** J. Peter (September 17, 1819–June 1, 1903), geologist, University of Pennsylvania, director of Pennsylvania Geological Survey. Highest earned academic degree: DD 1844, Princeton Theological Seminary. Honors: AAAS Fellow 1874; NAS 1863.

34. **LEVERETT,** Frank (March 10, 1859–November 15, 1943), geologist, United States Geological Survey, Ohio paleoliths to Minnesota Man. Highest earned academic degree: BS 1885, Iowa Agricultural College. Honors: AAAS Fellow 1891; AMS 1; NAS 1929.

35. **LEWIS,** Henry C. (November 16, 1853–June 21, 1888), geologist, Geological Survey of Pennsylvania, Trenton. Highest earned academic degree: MA 1876, University of Pennsylvania. Honors: AAAS Fellow 1880.

36. **LOOMIS,** Frederick B. (November 22, 1873–July 28, 1937), geologist and paleontologist, Amherst College, Melbourne. Highest earned academic degree: PhD 1899, University of Munich. Honors: AAAS Fellow 1913; AMS 4.

37. **MACCURDY,** George (April 17, 1863–November 15, 1947), archaeologist, Yale University, Vero. Highest earned academic degree: PhD 1905, Yale University. Honors: AAAS Fellow 1900; AMS 4.

38. **MASON,** Otis (April 10, 1838–November 5, 1908), ethnologist, United States National Museum, Smithsonian Institution. Highest earned academic degree: PhD 1875, Columbian [George Washington University]. Honors: AAAS Fellow 1877; AMS 1.

39. **MATTHEW,** William D. (February 19, 1871–September 24, 1930), vertebrate paleontologist, American Museum of Natural History and University of California, *Hesperopithecus* and Snake Creek. Highest earned academic degree: PhD 1895, Columbia University. Honors: AAAS Fellow 1906; AMS 3; NAS 1919. (Matthew's NAS election was subsequently rescinded when it was learned he held Canadian citizenship.)

40. **MCGEE,** William John (WJ) (April 17, 1853–September 4, 1912), geologist and anthropologist, United States Geological Survey and Bureau of American Ethnology, Trenton. Highest earned academic degree: None. Honors: AAAS Fellow 1882; AMS 1.

41. **MCGUIRE,** Joseph D. (November 26, 1842–September 6, 1916), attorney and archaeologist, Bureau of American Ethnology and United States National Museum. Highest earned academic degree: None. Honors: AAAS Fellow 1891.

42. **MERCER,** Henry C. (June 24, 1856–March 9, 1930), archaeologist, curator University of Pennsylvania. Highest earned academic degree: BA 1879, Harvard University. Honors: AAAS Fellow 1893.

43. **MERRIAM,** John C. (October 20, 1869–October 30, 1945), paleontologist and administrator at University of California, later president of Carnegie Institution of Washington, Rancho La Brea, Los Angeles. Highest earned academic degree: PhD 1893, University of Munich. Honors: AAAS Fellow 1905; AMS 2; NAS 1918.

44. **MILLS,** William C. (January 2, 1860–January 17, 1928), archaeologist, Ohio State University, New Comerstown. Highest earned academic degree: MS 1902, Ohio State University. Honors: None.

45. **MOOREHEAD,** Warren K. (March 10, 1866–January 5, 1939), archaeologist, Phillips Academy, editor *Archaeologist*. Highest earned academic degree: None. Honors: AAAS Fellow 1890.

46. **NELSON,** Nels (April 9, 1875–March 5, 1964), archaeologist, American Museum of Natural History. Highest earned academic degree: ML 1908, University of California. Honors: AAAS Fellow 1925; AMS 3.

47. **OSBORN,** Henry Fairfield (August 8, 1857–November 6, 1935), paleontologist and director, American Museum of Natural History, Long's Hill, *Hesperopithecus*, and Snake Creek. Highest earned academic degree: DSc 1880, Princeton University. Honors: AAAS Fellow 1883; AMS 1; NAS 1900.

48. **OWEN,** Luella (September 8, 1852–May 31, 1932), geologist, Lansing. Highest earned academic degree: None. Honors: AAAS Fellow 1911.

49. **PEET,** Stephen D. (December 2, 1831–May 24, 1914), editor *American Antiquarian*. Highest earned academic degree: AB 1851, Beloit College. Honors: AAAS Fellow 1881.

50. **POWELL,** John Wesley (March 24, 1834–September 23, 1902), anthropologist, director United States Geological Survey, founder and director Bureau of American Ethnology. Highest earned academic degree: None. Honors: AAAS Fellow 1875; NAS 1880.

51. **PUTNAM,** Frederic Ward (April 16, 1839–August 14, 1915), archaeologist and administrator, director Peabody Museum of Archaeology and Ethnology (Harvard University) and Peabody Professor of American archaeology. Highest earned academic degree: None. Honors: AAAS Fellow 1874; AMS 1; NAS 1885.

52. **RAU,** Charles (1826–July 25, 1887), archaeologist, United States National Museum, Smithsonian Institution. Highest earned academic degree: None. Honors: None.

53. **ROBERTS,** Frank H. H. (August 11, 1897–February 23, 1966), archaeologist, Bureau of American Ethnology, Folsom, Lindenmeier. Highest earned academic degree: PhD 1927, Harvard University. Honors: AAAS Fellow 1931; AMS 5.

54. **ROMER,** Alfred (December 28, 1894–November 5, 1973), vertebrate paleontologist, University of Chicago, Pleistocene faunal sequence. Highest earned academic degree: PhD 1921, Columbia University. Honors: NAS 1944.

55. **RUSSELL,** Frank (August 6, 1868–November 7, 1903), physical anthropologist, Peabody Museum, Harvard University, Trenton. Highest earned academic degree: PhD 1898, Harvard University. Honors. AAAS Fellow 1897.

56. **SALISBURY,** Rollin D. (August 17, 1858–August 15, 1922), geologist, University of Chicago, glacial geology, Trenton. Highest earned academic degree: AM 1884, Beloit College. Honors: AAAS Fellow 1890; AMS 1.

57. **SELLARDS,** Elias H. (May 2, 1875–February 11, 1961), geologist, Florida state geologist, and (later) Texas state geologist, Vero. Highest earned academic degree: PhD 1903, Yale University. Honors: AMS 5.

58. **SHALER,** Nathaniel S. (February 20, 1841–April 10, 1906), geologist, Harvard University, Trenton. Highest earned academic degree: BS 1862, Harvard University. Honors: AMS 1.

59. **SHIMEK,** Bohumil (June 25, 1861–January 30, 1937), malacologist and sedimentologist, University of Iowa, loess, Lansing, and Long's Hill. Highest earned academic degree: CE 1883, University of Iowa. Honors: AAAS Fellow 1904.

60. **SHULER,** Ellis (October 15, 1881–January 2, 1954), paleontologist, Southern Methodist University, Lagow Pit. Highest earned academic degree: PhD 1915, Harvard University. Honors: AAAS Fellow 1931.

61. **SPIER,** Leslie (December 13, 1893–December 3, 1961), archaeologist, American Museum of Natural History, University of Oklahoma, Trenton, Frederick. Highest earned academic degree: PhD 1920, Columbia University. Honors: AAAS Fellow 1931; AMS 5; NAS 1946.

62. **STOCK,** Chester (January 28, 1892–December 7, 1950), vertebrate paleontologist, California Institute of Technology, Los Angeles, Clovis. Highest earned academic degree: PhD 1918, University of California. Honors: AAAS Fellow 1921; AMS 6; NAS 1948.

63. **TODD,** James E. (February 11, 1846?–1922), geologist, South Dakota Geological Survey, United States Geological Survey, Lansing. Highest earned academic degree: AM 1870, Yale University. Honors: AAAS Fellow 1886; AMS 1.

64. **UPHAM,** Warren (March 8, 1850–January 29, 1934), geologist and archaeologist, Minnesota Historical Society, editor *American Geologist*, Little Falls to Lansing. Highest earned academic degree: AB 1871, Dartmouth College. Honors: AAAS Fellow 1880; AMS 1.

65. **VAUGHAN,** T. Wayland (September 20, 1870–January 16, 1952), geologist, United States Geological Survey, Vero. Highest earned academic degree: PhD 1903, Harvard University. Honors: AAAS Fellow 1906; AMS 1; NAS 1921.

66. **VOLK,** Ernst (August 25, 1845–September 17, 1919), archaeologist, field assistant Peabody Museum, Harvard University, Trenton. Highest earned academic degree: None. Honors: None.

67. **WARD,** Henry B. (March 4, 1865–November 30, 1945), anatomist, physician, and dean of the Medical College, University of Nebraska, Long's Hill. Highest earned academic degree: PhD 1892, Harvard University. Honors: AAAS Fellow 1899; AMS 3.

68. **WHITE,** Israel C. (November 1, 1848–November 25, 1927), geologist, West Virginia University, director West Virginia Geological and Economic Survey. Highest earned academic degree: PhD 1880, University of Arkansas. Honors: AAAS Fellow 1882; AMS 1.

69. **WHITNEY,** Josiah D. (November 23, 1819–August 15, 1896), geologist, Harvard University, and California Geological Survey, Calaveras. Highest earned academic degree: BA 1839, Yale University. Honors: NAS 1863, resigned 1874.

70. **WILLISTON,** Samuel W. (July 10, 1852–August 30, 1918), paleontologist and physician, University of Chicago, 12 Mile Creek, Lansing. Highest earned academic degree: MD 1880, Yale University. Honors: AAAS Fellow 1902; NAS 1915.

71. **WILSON,** Thomas (July 18, 1832–May 4, 1902), archaeologist, United States National Museum, Smithsonian Institution. Highest earned academic degree: None. Honors: AAAS Fellow 1888.

72. **WINCHELL,** Newton (December 17, 1839–May 2, 1914), geologist, Minnesota state geologist, editor *American Geologist*, Little Falls, Lansing, Kansas. Highest earned academic degree: AM 1869, University of Michigan. Honors: AAAS Fellow 1874; AMS 1.

73. **WISSLER,** Clark (September 18, 1870–August 25, 1947), anthropologist, American Museum of Natural History, Trenton. Highest earned academic degree: PhD 1901, Columbia University. Honors: AAAS Fellow 1906; AMS 2; NAS 1929.

74. **WRIGHT,** George F. (January 22, 1838–April 20, 1921), geologist and minister, chair of the Harmony of Science and Revelation, Oberlin College, Trenton, Newcomerstown, Nampa, Lansing. Highest earned academic degree: AM 1862, Oberlin College. Honors: AAAS Fellow 1882.

75. **YOUMANS,** William J. (October 14, 1838–April 10, 1901), editor *Popular Science Monthly*. Highest earned academic degree: MD 1863, New York University. Honors: AAAS Fellow 1889.

CHAPTER ONE

A Study in Controversy

FLORIDA STATE GEOLOGIST Elias Sellards had watched for over a year as excavations in the rich fossil beds at Vero produced bones of mammoth, mastodon, sloth, tapir, horse, and other extinct Pleistocene animals. Then in October 1915 they yielded a nearly complete human skeleton; two more were found in the summer of 1916. Here at last were human traces in "undoubted Pleistocene deposits," prime evidence, he was sure, to resolve the long-standing dispute over human antiquity in America. Sellards fired off letters to the scientific Goliaths of his day, inviting them to Vero to examine the evidence: glacial geologist Thomas Chamberlin (of the University of Chicago and the United States Geological Survey [USGS] Glacial Division), vertebrate paleontologists Oliver Hay (Carnegie Institution of Washington), Samuel Williston (University of Chicago), and—tempting fate—the Smithsonian Institution's archaeologist William Henry Holmes and its physical anthropologist Aleš Hrdlička.[1]

Hrdlička was happy to come down but warned Sellards that "the occurrence of human remains in ancient strata while a great incentive . . . is, *per se*, not of course as yet a proof of the antiquity of the [human] remains."[2] Privately, Hrdlička sneered that "Mr. S. [Sellards] is so cocksure of his 'discovery' . . . that the case inspires rather suspicion than confidence."[3] Still, it was important to the "morale of the profession," as Chamberlin put it, that such finds be subject to critical examination.[4] And so they were.

Hrdlička went to Vero in late October 1916, and was joined there by geologists Rollin Chamberlin (also of the University of Chicago, substituting for his father Thomas) and T. Wayland Vaughan of the USGS, Yale University archaeologist George MacCurdy, and Hay. They met Sellards, examined the Vero deposits, and haggled over whether the human remains and those of the extinct mammals were associated. Hrdlička had little to say, though Sellards guessed he would find objections.[5] Sellards was right.

In Washington a week later Hrdlička happily assured Holmes he would able to cast "the age and nature of the skeletal remains . . . in the true light."[6] Of course he would: Hrdlička was rarely plagued by doubts. "When you came back to Hrdlička," a longtime colleague observed (with little hint of affection), "he was always there, just where the Lord created him, on the rock of ultimate Hrdličkian knowledge."[7]

Not everyone shared Hrdlička's version of Vero's "true light." There was not then nor in the years to follow consensus on the age of its human remains.[8] They were either from the Early or Middle Pleistocene (according to Hay and Sellards), the Historic era (Hrdlička and Holmes), or the hundreds of thousands of years in between (MacCurdy and others). Even so, Holmes claimed the critics "called the matter into question so decidedly, that the world will not be in haste to accept [Sellards's] radical views."[9]

The irreconcilability of interpretations badly strained relations among the participants and dissolved rapidly into interdisciplinary bickering among archaeologists, vertebrate paleontologists, and geologists. In 1918 Holmes insisted he would not "stand in the way of legitimate conclusions" in geology or paleontology, but imperiously dismissed Sellards's Vero work (and Hay's efforts to bolster its antiquity) as "illegitimate determinations [that] have been insinuating themselves into the sacred confines of science and history." He made no apology for his criticism, and in fact returned to that theme seven years later, publicly denouncing the Vero evidence as "not only inadequate but dangerous to the cause of science."[10]

Sellards, new to the fierce controversy over human antiquity of which Holmes and Hrdlička were hardened veterans, was hurt, embittered, and deeply humiliated.[11] He soon left Florida for Texas, carrying with him the scars of Vero. More than a decade later and several years after a Pleistocene human antiquity was demonstrated at Folsom, New Mexico in 1927, Sellards gathered the nerve to ask Holmes his opinion about human antiquity in America in general and Vero in particular. Holmes, then 83 years old, dodged the question, saying "no trace whatever" remained of his earlier antagonism and he had "dropped the matter entirely."[12]

That was not acceptable. Sellards did not want indifference: he wanted vindication. Over the next three decades, and long after his own retirement in 1945, Sellards continued to press the case for Vero and obsessively sought sites testifying to a great human antiquity. More "stones to heap on Hrdlička's grave," his colleagues said as they watched and knowingly nodded their heads.[13] In 1952, at the age of 77 and suffering with an abdominal hernia long overdue for surgery, Sellards returned to Vero to collect charcoal or bone suitable for the newly invented technique of radiocarbon dating. He would prove Vero was just as old as he had said it was. His longtime field assistant Glen Evans accompanied him, having left Texas with careful instructions from Sellards's physician about what to do if the hernia suddenly bulged. It did, and Sellards collapsed unconscious at the excavation. Evans propped him under the shade of a tree, treated him as best he could, and made plans to transport him to a hospital. But the moment Sellards regained consciousness he insisted on continuing to excavate.[14] Nearly four decades after Vero, Sellards still desperately wanted to show that he and not his long-dead critics had been right all along.[15]

Sellards was not the only casualty in the long and bitter dispute over human antiquity in America, though perhaps he was one of the more obsessed. And just as this controversy profoundly influenced the careers of the scores who participated in it, so too it forever changed North American archaeology and helped set the discipline's course into the twentieth century.

1.1 BEGINNING AND ENDING

It took hardly any time at all. Only a few years after the discovery in Europe in the late 1850s that humanity had roots predating history and the Biblical chronicles and reaching deep into the Pleistocene came the suggestion that North American prehistory might be just as old.[16] And why not? There seemed to be an "exact synchronism [of geological strata] between Europe and America," and so by extension there ought to be a "parallelism as to the antiquity of man."[17] That triggered an eager search for traces of the people who may have occupied North America in the recesses of the Ice Age.

It quickly became obvious, however, that North America's archaeological record was not Europe's. Here caves and river valleys were not producing rich layers of primitive stone artifacts intermingled with the massive bones of extinct Ice Age animals. But perhaps there were other indications: in the 1870s Charles Abbott began finding stone artifacts along the banks of the Delaware River near his Trenton, New Jersey home. He pronounced these alike in form and evolutionary "grade" to those of the European Paleolithic (Stone Age), and reasoned that if these artifacts were similar in form they must be comparable in age. Certainly, the artifacts were distinct from those of the Lenni Lenape, the historically known tribe of the region, and were found in "gravels" that could be Pleistocene in age. Abbott was sure that "had the Delaware River been a European stream the implements found in its valley would have been accepted at once as evidence of the so-called *Paleolithic man*."[18]

Abbott's apparent discovery of an "American Paleolithic" triggered a cascade of claims, and within the decade scattered reports of paleoliths came in from the East, Midwest, and Great Basin. Their precise age was difficult—impossible, really—to pin down.[19] Yet, the specimens so readily mimicked European specimens of undeniable antiquity that it was thought they must be as old. "Comparison," Thomas Wilson famously asserted, was "as good a rule of evidence in archaeology as in law."[20]

By 1889 the American Paleolithic was accepted fact. The idea that the earliest North Americans were here thousands, if not tens of thousands, of years ago when the continent's northern latitudes lay shrouded in glacial ice was triumphantly paraded in symposia, feature articles, and books.[21] European savants praised Abbott as America's Boucher de Perthes, whose Somme River valley collections had tipped the balance on the question of the European Paleolithic.[22] The only lingering issue was how much earlier the first peoples may have arrived in the New World. Some speculated America's prehistoric roots might reach into the Pliocene or even the Miocene.[23]

Yet, scarcely a year later the American Paleolithic was under withering fire, led by William Henry Holmes of the Smithsonian's Bureau of Ethnology. His studies of stone tool manufacturing debris at the Piney Branch quarry site in Washington, DC, revealed a fatal flaw in the assumption that form corresponded to age. As he saw it, artifact production transformed rounded cobbles into long leaf-shaped bifaces through "successive degrees of elaboration." Drawing on biologist Ernst Haeckel's then-popular refrain that "ontogeny recapitulates phylogeny," Holmes argued that if a stone tool was discarded or rejected early on in the manufacturing process (stone tool ontogeny), it would naturally resemble the "rude" and ancient stone tools of Paleolithic Europe (early stages of stone tool phylogeny), but of course that meant resemblance alone had no "chronologic[al] significance whatever."[24]

Holmes used artifact form not to infer antiquity but to deny it. Demonstrating the Pleistocene age of archaeological remains, he insisted, was entirely a matter of geology. But then Holmes cynically assumed that any artifact found within unconsolidated, demonstrably glacial-age deposits was "an adventitious inclusion," effectively eliminating the possibility of a Pleistocene antiquity. And he was neither modest nor lacking in ambition. He declared his intent to "revolutionize" American archaeology, and deemed his Piney Branch study "one of the most important periods of [my] labors in the field of science, and one of the most important in the history of American archaeological research." He even insinuated that the reason American and European paleoliths looked alike was that many European specimens were themselves manufacturing rejects that prehistorians there had failed to recognize as such. Perhaps, as one onlooker tartly put it, "Boucher de Perthes may turn out to have been the Dr. Abbott of France."[25]

Abbott and Holmes were both notoriously stubborn and uncompromising, and what started as a difference of opinion grew quickly into mistrust and then raced on to mutual loathing. They angrily debated one another on archaeological matters and called on their allies in geology to testify that the deposits in which those supposed paleoliths were found were indeed Pleistocene in age (George Frederick Wright in Abbott's corner), or most assuredly were not (Thomas Chamberlin seconding Holmes). The controversy that exploded that spring of 1890 ultimately went unresolved for nearly four decades, as archaeologists and nonarchaeologists alike sought evidence of a Pleistocene human antiquity.

The question being asked was straightforward enough: had people arrived in North America in Ice Age times? But easily asked was not easily answered. Throughout the dispute critics admitted the possibility that people were here during the Pleistocene: the great variety of Native American languages, cultures, and physical appearance certainly suggested as much.[26] Of course, that was circumstantial evidence, and it rested heavily on the assumption that the "arrivals in America were of a single or homogeneous stock" and that vast time was required to produce that much diversity among its descendants. Until that assumption was demonstrated, Holmes argued, there was no reason to suppose it true, and equally plausible reasons to doubt it: one could as readily argue that early Americans descended from many different groups, for whom the passage through the "new and constantly changing conditions" of the New World further "greatly accelerated differentiation," rendering Native American diversity moot as a measure of antiquity.[27]

Unlike scientific controversies in which a phenomenon is known to exist and dispute centers on, say, its causes or consequences,[28] in this instance there was no guarantee the phenomenon even existed. Demonstrating that it did proved enormously challenging and at times extraordinarily complicated, for that required developing a deeper understanding of the archaeological record, laying secure chronological foundations (with the necessary, if not always welcome, help of geologists and vertebrate paleontologists), and building conceptual frameworks for making sense of the evidence. Looking back a half-century later, Emil Haury thought the Folsom discovery in 1927 finally brought the human antiquity controversy to an end because it provided "unequivocal evidence of man and extinct animals."[29] That it did. But it was only possible to recognize Folsom for what it was because the decades of dispute leading up to that moment had closed critical conceptual gaps in archaeological, geological, and paleontological knowledge. Were it otherwise, the controversy would have been over in 1896 at the 12 Mile Creek site in Kansas (a site that gained little purchase at the time but in retrospect proved to be Pleistocene in age).[30] Having the right site matters, of course, but even more so the ability to recognize it as such.

As it was, the effort to resolve the question ultimately involved dozens of sites and claims as well as scores of participants haggling over what those meant, and was a shape-shifting affair in its empirical content. It began in the 1870s with supposed American Paleolithic tools, but when those failed to prove the case attention turned to "primitive" human skeletal remains ostensibly reminiscent of earlier fossil humans such as *Pithecanthropus* or *Neanderthals*. When these too were deemed inadequate to the task, they were followed by claims of skeletal remains of anatomically modern humans from deposits with the bones of extinct Pleistocene animals. In the 1920s attention shifted once again, this time to sites in which stone artifacts were found in apparent association with those Ice Age animals. It was a process of adaptive response to what was at times intense scientific selective pressure in this environment of controversy.

Each of these types of evidence came with its own suite of analytical problems and interpretive baggage, though common to all were questions of a specimen's context or its association with geological time markers. These were questions often rendered difficult by the happenstance nature of discoveries, the degree of trust one placed on the finders' ability to report accurately what they had seen of the specimen while it was in situ, and the lack of agreed-upon rules for reading and transforming archaeological field data into evidence.

Then too there were methodological and theoretical concerns, which participants approached from very different perspectives. For some the Native Americans "were here and must be recognized in every theory, must be a factor in every conclusion." Unabashed uniformitarians they were, and they worked from the known (ethnographic present) to the unknown (archaeological past), an approach that demanded continuity between present and past and its corollary, a shallow past.[31] Yet, those who sought a Pleistocene human presence, possibly of a distant unrelated people, "knew little about modern tribes and cared less."[32] Native Americans were merely the latest in a parade of races that had migrated to the continent, making analogies to Indians and their quarry sites irrelevant. And what did a deep antiquity, or a shallow one for that matter, say about cultural evolutionary notions of historical progress? If American Indian prehistory began, as prehistory did in Europe, in Pleistocene times, why had Native Americans not fully transcended Lewis Henry Morgan's "savagery" stage and achieved civilization?[33]

There were differences too in how and to what degree participants sought to harmonize the archaeological and human fossil records of the Old World and the

New. Although the questions being asked and the evidence being examined were unique to these shores, the story of global human prehistory was not one of parallel and separate tracks in the two hemispheres, but a biological and cultural skein that began in the Old World and ended in the New. The more or less constant tugging at one end of the evolutionary thread, such as the dispute in Europe over whether Neanderthals were our immediate ancestors and when modern humans first appeared, was felt at the other. So it was that the 1912 discovery of the Piltdown fossil, which seemingly affirmed that modern-looking humans existed in the Early Pleistocene, had repercussions in distant Vero, Florida. Divergent views of human evolution, the rate and degree of anatomical change over time (and its corollary, the question "What should a Pleistocene human look like?"), and indeed varying views of the one or more mechanisms of evolution itself drove controversy.

That was likewise true of artifacts, except for them expectations and tolerances were broader, insofar as cultural evolutionary sequences were seen as more variable than those in the human fossil record (there were far more structural constraints on cranial variation and rates of anatomical change than on artifact change). American Museum of Natural History (AMNH) archaeologist Nels Nelson vigorously objected to the Colorado Museum of Natural History's Jesse Figgins and Harold Cook's claims that metates and arrowheads at the Frederick, Oklahoma, site could be 365,000 years old: if that were so, he growled, "we shall have to revise our entire world view regarding the origin, the development, and the spread of human culture."[34] He was not ready to do that on principle, but what exactly was the principle? That was not so evident.

That such questions were raised in the complete absence of any comprehensive or guiding theory in archaeology made efforts to find answers all the more challenging, with the result that with each discovery arguments flared anew.

Yet, this controversy was about more than just archaeological and related anthropological issues. Throughout, efforts to determine the relative or absolute age of artifacts and human fossil remains were hopelessly entangled in disputes not of archaeology's making. There was general, though not universal, agreement that a site's antiquity had to be determined independently by geological methods, associating artifacts or skeletal remains with glacial-age deposits—such as glacial gravels or loess—or with the bones of extinct fauna thought to be Pleistocene in age. There were no easy answers from that quarter, either, for at the outset of this controversy glacial geologists were themselves grappling with fundamental questions in their own field, such as when the Pleistocene ended, how many separate glacial episodes there had been over the Ice Age, and even how to recognize a Pleistocene-age deposit beyond the limits of the continental ice margin. For their part, paleontologists had only an incomplete listing of extinct vertebrate taxa, and it was not known whether those had disappeared during the Pleistocene or if they had survived into the Recent period (the latter was suspected to be the case for many taxa).

As a result there were wide-ranging disagreements over archaeological and non-archaeological evidence and issues. Joint visits to the sites in question, necessary at a time when antiquity had to be assessed from circumstances on (and in) the ground, mostly served to widen rather than narrow the gap between antagonists. There was scant agreement about what the empirical record meant, and even less about the methods to interpret it, even as attention shifted from stone tools to human skeletal remains to artifacts associated with extinct fauna. "Facts are facts," Harold Cook had assured John Merriam as Cook was trying to convince Merriam of the antiquity of several sites, but Cook was wrong.[35] Facts were not just facts: they were theory-laden and "controversy-laden" observations about the empirical realm.[36] Until the end of this dispute, the "facts" were never viewed in quite the same way by all who saw them.[37]

Incommensurability led to impasse, as "neither side show[ed] an inclination to recede from the advanced position it had taken."[38] Impasse triggered fierce intradisciplinary and interdisciplinary border wars in which no one was quite sure what belonged to whom. These sparked harsh exchanges in print and led to fiery encounters at meetings, especially at the American Association for the Advancement of Science (AAAS), which yearly brought the participants together from across the disciplines.

Widening the sphere of controversy further, these differences played out amid struggles between amateurs (as most everyone was), an emerging professional class (to which many aspired), and those jockeying for elite status within the professional community (as only a few could be). Whatever the "official myth of democratic equality" in science, Martin Rudwick observes, all of its practitioners know deeply and viscerally that not everyone is created equal. Some are more equal than others. Such inequality is often most visible and matters most during episodes of controversy, when the stakes are highest.[39] Inevitably, there were some who sought to delegitimize others, and fierce disputes broke out over who among them was entitled to write and speak to the public on behalf of science.[40]

The controversy took place among institutions (and newly created specialized journals) vying to establish their preeminence and centrality and in a climate of intense competition for patronage and support. There were few fair fights: the desperate need for funding badly hobbled most, while those in research bureaus within the federal government (its own role in science then rapidly expanding) flexed considerable financial muscle.[41] As this was happening, the ground beneath the discipline shifted from its longtime base in local scientific societies and museums to government research bureaus and finally to the burgeoning university system. With these transitions the theoretical disposition and direction of archaeology and anthropology changed, along with its membership and sociopolitics.[42]

Exacerbating matters was the fact that archaeological sites are inherently contested spaces: they are unique, fixed spots on a landscape. Their factual claims cannot be independently replicated by others.[43] The sites in this dispute were usually located on what someone considered their home turf, at a time when there was still overt territoriality and tension between eastern and western scientific institutions, and between federal and state governments. Friction was often unavoidable between "locals" who found and interpreted the sites and those from distant museums, federal agencies, or universities who swooped in (not always by

invitation) to reinterpret their meaning. Nor was that friction unique to archaeology: the latter decades of the nineteenth century contained recurring strains
between state and federal geological surveys over glacial mapping in the states,
which at times bore directly on the dispute over the number of glacial epochs.[44]

Finally, and perhaps most important of all, this controversy took place at a
time when archaeology, as with many turn-of-the-century disciplines, was self-
consciously attempting to define its boundaries and create an intellectual identity
while others were threatening to breach its borders.[45]

As a result, battles over the content of science spilled over into the conduct of
science, and from there into raw personal, institutional, disciplinary, regional, and
political conflicts. It created a toxic atmosphere of distrust and suspicion between
advocates and critics. One of those moments, a Thomas Chamberlin-orchestrated
venomous assault on Wright's *Man and the Glacial Period*, even sparked an angry
wildfire that blazed into the halls of Congress. In such a wide open field there
were few rules of engagement; the controversy grew heated and bitter and left lasting wounds, like the ones an aged Sellards sought to heal.

The result was one of the longest-running feuds American archaeology has ever
experienced, one that at times was so vicious that the purposeful search for early
sites was "most harmfully discouraged" and many were "actually frightened away
from participating."[46] Nelson, sufficiently fearless that he thought little of plung-
ing into the remote Gobi Desert for long stretches of archaeological fieldwork,
nonetheless advised others in the midst of this controversy to do what he did, "lie
low for the present."[47]

Yet, for all its ambiguity and acrimony and after stubbornly defying resolution
for nearly forty years, the controversy over human antiquity in North America
suddenly evaporated at Folsom in the fall of 1927. Folsom was not the long-sought
evidence of an American Paleolithic or a human presence deep in the Pleistocene
(or at least not by Old World standards). It was something very different. But
then, after decades of dispute, nothing ended up looking quite the way it was
expected it would at the beginning.

In its time, the problem of human antiquity engaged and defined as few others
did the emerging discipline of American archaeology. It did so not just because
some of the best and brightest of several generations of archaeologists—as well as
glacial geologists, physical anthropologists, and vertebrate paleontologists—were
attracted to it, though they were. Nor was it because the participants thought
this was a significant problem, though they did. Rather, this problem loomed so
large because it cut so deep into American archaeology's conceptual core, forcing the
nascent field to confront haziness in its theories, methods, and evidence, all while
past time and everything that flowed from it vital to understanding the prehistory
of North America was held hostage.

Until it was released, American archaeologists labored under what Alfred
Kroeber called a "flat past," where time and space seemingly collapsed in on one
another[48] and where they had to "stuff a tremendous lot of cultural events into an
ever-shrinking chronological container."[49] As James Snead observes, even after
Alfred Kidder, Kroeber, and Nelson in the second decade of the twentieth century
developed and applied new methods to detect chronological change (seriation and

stratigraphy), the past remained "merely an older version of the present."[50] That would not change until Folsom.

In retrospect it is significant but hardly surprising that Kidder, the greatest archaeologist of his generation, instantly responded when Frank Roberts called from Folsom urging him to come see for himself a projectile point that had just been found in situ between the ribs of an apparently extinct bison. By his own admission, Kidder had long "dodged the issue of origins and comforted [himself] by working in the satisfactorily clear atmosphere of the late periods." But he was acutely aware of the stakes of the game and that resolution of the human antiquity question would have repercussions across American archaeology and beyond.[51] Kidder knew nothing of the Folsom site when Roberts reached him, but as he drove to Folsom that early September day in 1927 he was absolutely certain American archaeology had been waiting decades for this moment to arrive.

1.2 A POWERFUL LENS

The demonstration of a Pleistocene human presence in America inspired a complete rethinking of the colonization of the New World. And because of the lessons learned at Folsom (not least, how to find more sites like it), within a dozen years an entirely new understanding of early North American prehistory was crafted and became the foundation of the next half-century of research into what came to be called the "Paleoindian period." That was to be expected, of course, but was hardly the only result of consequence.

Folsom also revealed a vast chasm in North American prehistory. American archaeologists found themselves staring into a gap many thousands (possibly tens of thousands) of years wide between the Late Pleistocene and the Late Prehistoric, with no idea what happened in between.[52] Filling in that gap gave the discipline a mission and method that guided work over the next half-century and forced a rethinking of core assumptions about the relationship of archaeology to anthropology: the use of ethnohistory and ethnographic analogy were never the same afterward.[53] So too, arguably, the relationship between archaeologists and Native Americans, a change with repercussions that flared up at century's end.

With the demonstration of a Late Pleistocene antiquity, the New World found itself with much-needed "chronological elbow room." No longer, as Kidder observed with palpable relief, did it appear that New World civilizations had no chronological on-ramp, an idea that had fueled hyperdiffusionist claims that these civilizations had not arisen on their own but had had outside (Old World) help.[54] No longer, Franz Boas announced, did linguists have to worry about revising their views about the stability of language types and of fundamental grammatical forms.[55] And no longer, Kroeber declared, did one have to conjure complex "racial" migration scenarios to account for the physical and cultural diversity of Native Americans present and past, when gradual in situ population divergence was a feasible alternative.[56]

In the end this controversy and its consequences revolutionized American archaeology, though scarcely in ways Holmes earlier imagined. Disciplinary revolutions are not always the result of controversy, nor, for that matter, do controversies inevitably provoke disciplinary revolutions—but that is where the smart money bets.[57] Scientific controversies in general can be pivotal (as certainly this one in particular was) in a discipline's historical trajectory.[58]

A controversy as long, complex, and bitter as this offers an especially powerful lens for examining late nineteenth- and early twentieth-century American archaeology. This is so because controversy occurs when consensus over an issue breaks down and yet "substantial parts of the scientific community see some merit on both sides of a public disagreement."[59] The disagreement is deemed worthy of being taken seriously, is debated publicly (allowing others to judge the merits of the case or join), involves sustained argument and counterargument, and is held to be determinable by scientific means.[60]

As such, controversy strips away the veneer and exposes—as tranquil times rarely do— the internal workings of a field (and the sometimes spectacularly boorish behavior of its practitioners[61]), the process by which new knowledge is created, the relative status of participants, and the intellectual and social context in which these developments occur.[62] Controversies can range freely and spread rapidly with no a priori limits "as to where it will stop in its questioning of entrenched beliefs, concepts, methods, modes of interpretation, data, criteria of relevance, norms of formulation, acceptance and rejection of hypotheses, and other components of the scientific enterprise."[63] Controversy, as Marcelo Dascal and Victor Boantza put it, "allows for the eventual conciliation of opposite views in the construction of a new theory and even for a new methodology. It thus contributes to the development of knowledge by paving the way for innovation." More simply, it forges the keys to its own solution, one to which both sides contribute even if, Rudwick adds, "one of them fails to admit it."[64]

This is assuming, of course, that a controversy ends in resolution, though as Ernan McMullin observes, not all of them do: some are *abandoned* when the community as a whole is unable to resolve matters, sees no hope in doing so, and loses interest; others might end as a result of nonepistemic pressures, such as a loss of research funds or from political pressure, in which case the controversy has closure but no solution.[65] The human antiquity controversy began as a woefully underdetermined problem but was *resolved* despite years of false starts and dead ends. Progress was made by rejecting different sites and claims, and by trial and error that made clear what needed to be learned. There was nothing teleological about the process.

Not surprisingly, historians of science find controversies to be fertile ground. Although few have shown a great deal of interest in the history of archaeology (as Matthew Goodrum notes, that is slowly changing[66]), there have been studies of historical controversies that involve archaeology and anthropology (the examples come almost entirely from the Old World), as well as studies of disputes in the nearby fields of geology and natural history (biology).[67] Some of these works, such as the several richly detailed volumes exploring nineteenth-century British geology, each devoted to one of the major disputes in which the British geologist Roderick Murchison was deeply involved, are valuable exemplars (though the unwavering spotlight on Murchison prompted one of the authors to ask, if somewhat self-consciously, whether readers had already "had enough of Murchison and his quarrels").[68]

Although long recognized as a seminal episode in American archaeology, the controversy over human antiquity in North America receives only passing mention in broader histories of the field (and, for that matter, in histories of glacial geology), or in biographies of its major figures. Even the best of those works—such as those by Curtis Hinsley, Frank Spencer, and Bruce Trigger—review a few key individuals in the controversy and gloss the central issues, sites, and outcome before moving on. In so doing, they acknowledge how this episode fits into the overall evolution of the field, or in the work of an individual or institution, but while granting its larger importance do not explore it in detail. More particulars emerge in article-length histories devoted to this episode, but their coverage is necessarily limited and often uneven, and they do not plumb the complex analytical depths of an episode that spanned decades, involved scores, and ranged widely across multiple disciplines.[69]

I hasten to add that I tar myself with this same broad brush: my earlier writing on the topic has focused on a few key participants, on specific moments, or on overviews in this decades-long controversy. These were often written for the purpose of providing a deeper historical context and a broader understanding of the current controversies over the origin and antiquity of the first Americans (the details have changed, but the broad questions we archaeologists ask have remained much the same).[70] And none of them approaches the extent or depth of coverage in this book, and indeed each is essentially superseded by it. More importantly, this is history solely for history's sake. The goal is to understand *why* the evidence was seen in the manner in which it was seen given the knowledge available at the time.

Important as this question—along with the controversy and its resolution—is for our understanding of the prehistory of North America and for American archaeology writ large, it has not received the historical attention it deserves for the transformative impact it had. It is principally for this reason, and as a contribution to the broader intellectual history of North American archaeology, that I undertook the study presented here.

There is more to it than that, however. I would not necessarily privilege the significance of the human antiquity controversy in North America in the grand scheme of the history of science. But this dispute is unusual in one respect that arguably makes its study of wider historical interest and import.

Common to most scientific controversies is that the core of the dispute is intradisciplinary: rich as the Murchison volumes are in revealing the social processes that helped shape geological knowledge in Victorian England, those controversies were driven primarily by geological questions that were asked by geologists and resolved by geological methods. The human antiquity controversy was different: there was no archaeological answer to this archaeological question. The dispute that began in 1890 over whether artifact form (later, human skeletal morphology) scaled with antiquity led to no resolution at all, but merely exposed the challenge of reading time from stone tools or anatomy. Demonstrating antiquity had to be and was ceded, grudgingly in many quarters, to geology and paleontology.

Central precepts in these disciplines, however, were themselves unsettled. Contemporary glacial geologists were at odds with one another as to how supposed Pleistocene-age deposits—most especially glacial gravels or loess—were to be recognized in the field, particularly when those deposits were located beyond the margins of North America's terminal moraines (where, as it happened, virtually all of the purported Pleistocene-age archaeological sites were located). One had to distinguish whether gravels had been deposited by a glacier or meltwater rivers draining the ice sheet, whether that took place during a glacial period or in close temporal proximity, or if those deposits were a result of later reworking and redeposition of glacial materials or simply recent river flow. That in turn raised the question, about which there was much debate but little agreement, as to when the Late Glacial became the Postglacial. Loess presented an additional challenge, as initially it was still unclear if it was water laid or windblown, the two mechanisms implying very different temporal relationships to glaciation. In the "aqueous" theory, loess came with the summer melting of northern glaciers; but if loess was aeolian in origin, its age was free to vary independently of glacial episodes, for all that was needed were winds to blow the sediment.

But even if it could be established that a gravel or loess was glacially derived, it still had to be determined when that occurred within what was still an unsettled Pleistocene sequence. At the very outset of the human antiquity controversy, as noted, it was believed there had been a single advance of glacial ice during the Pleistocene: the number of advances crept up to two by century's end largely as a result of fieldwork in the Midwest (an idea stoutly resisted by many glacial geologists who worked in eastern North America, where there seemed to be just one terminal moraine). The number rose to six a decade or so later, though by the late 1920s it was back down to four. Yo-yoing along with the changing relative sequence were the estimated ages for those periods of ice advance and retreat, and by extension the potential antiquity of any archaeological materials that might be related to those separate glacial episodes.

Yet, even as those issues were being resolved it became apparent that the chronological resolution afforded by dating artifacts or skeletal remains by the geological deposits in which they were found was too coarse to be of archaeological use. With that, attention shifted to associated extinct vertebrates for age control, though at that time (the second decade of the twentieth century) the chronological and taxonomic details of the evolution and extinction of Pleistocene vertebrate faunas were still poorly known. There was a reasonably secure listing of extinct *genera*, such as mammoth, mastodon, horse, and camel (among others). It proved far more challenging to identify extinct *species* within surviving genera—bison, for example, which were still extant in western North America yet were found in potentially ancient archaeological sites. Even if a taxon was identified as extinct, opinions were sharply divided over whether that animal had survived past the end of the Pleistocene (whenever that geological moment was) or whether it had disappeared earlier, and if so how much earlier.

Resolving these questions ought to have been a strictly intramural affair, one for geologists and paleontologists to hammer out on their own. After all, on its face it mattered little to archaeologists how to sort glacial moraines of different ages (or even whether there were moraines of different ages), or if the snails found in the loess were terrestrial or aquatic, or the taxonomic details of extinct species.

But it did matter. The debates over those issues became thoroughly entangled in the human antiquity controversy for several reasons, not least because it was disputes over the age of archaeological sites that often brought those issues to the fore. As well, most of the major figures in glacial geology and vertebrate paleontology in the late nineteenth and early twentieth centuries were themselves deeply involved in the archaeological controversy. In fact, by the turn of the century and in the decades that followed most of the sites were investigated and advocated by geologists and paleontologists, not archaeologists. And, finally, because when the results of one discipline came into conflict with another, neither backed down. Hrdlička deemed it "scarcely safe for the geologist or the paleontologist to assume that the problem of human antiquity is his problem." He thanked them "for every genuine help they can give anthropology," but warned that "they should not clog our hands."[71] The erstwhile paleontologist Figgins drew a more caustic line in the sand: he declared anthropologists were "mostly jokes . . . too ignorantly prejudiced to give intelligence to the diseased meanderings of their warped imaginations."[72] This was not just an intradisciplinary controversy over human antiquity, but an interdisciplinary one as well.[73]

These interdisciplinary disputes in North America compounded the challenges facing archaeology, but they also spurred efforts to resolve the broader issues within archaeology, glacial geology, and vertebrate paleontology and with them to determine the age of purported Pleistocene sites. The net result was that new and essential intellectual tools were forged in all these disciplines and their rules of evidence were thoroughly revised and recast, ultimately making possible the resolution of this controversy.

Fully exploring the many dimensions of the controversy over human antiquity in North America demands that attention be paid to more than just the archaeological and closely related anthropological matters at hand. It will require probing in some detail the history of turn-of-the-century glacial geology and vertebrate paleontology, and the broader intellectual and institutional context within which this controversy took place. As this controversy was thoroughly interdisciplinary and wide ranging in scope, so too its historical investigation must be.

Finally, attention to the history of this controversy is warranted for still another reason. Seeking great antiquity is a recurring theme in archaeology, and not just in America: worldwide there is a perennial search for the oldest sites or the oldest evidence of some aspect of human behavior. Inevitably, just a few decades after Folsom and the discovery of the still older site of Clovis that soon followed, a multidecade controversy flared again in America in the later twentieth century, this time over whether there was a pre-Clovis presence in the New World.[74] The epistemic continuity of the antiquity issue so tightly links the modern discipline with its deeper past that actions and events in the earlier dispute have been used as ammunition in the more recent one (an unfair burden to place on the history of a field, as I have argued).[75]

A detailed comparative study of the two controversies is not my goal. Because the more recent dispute is still fresh and the rhetorical clamor has yet to fully subside, we lack the comfortable historical perspective that comes with exploring decades- or centuries-old events; of discussing individuals who are not one's contemporaries, friends, and colleagues; or of knowing how to critically assess one's own position on a controversy, especially after having played a part in it.[76] The distance afforded by the study of this earlier episode provides the potential for insight into the structure and character of this recurring controversy, and perhaps a deeper understanding of the contemporary uses of history. In this way, research on the history of the controversy over human antiquity sheds light on the nature of the recent dispute.

All well and good for archaeology and archaeologists. But more than that, doing so is, arguably, especially useful at a time when there is tension and mistrust between archaeologists and Native Americans, the latter questioning the substance and reliability of archaeological knowledge and calling on them (us) to "produce reliable histories of their . . . disciplines, and clearly articulate their fundamental doctrines so that we can see the various trees of thought that represent the forests in which we labor."[77] Heeding that call helps close the gap between the near and deeper past, thus closing the more general gap between archaeologists and their own history, as well as the descendant communities of the people they study, which surely benefits all.[78]

1.3 APPROACHING THE INQUIRY

This book is a biography of an intellectual controversy: the interdisciplinary effort to establish a Pleistocene human antiquity in North America. As such, it seeks to understand how and why controversy developed, spread, and stubbornly defied resolution for so long; why it became so embittered; and why it ended as quickly as it did when it did.

Accordingly, the focus is not just on the empirical, methodological, and theoretical issues that were in dispute, though of course these weigh heavily throughout, for by the end empirical and conceptual knowledge was gained (albeit through many fits and starts) and a Pleistocene-age site *was* found and recognized for what it was.[79] Nor is the focus on any one person, site, institution, or discipline, but instead on the community of archaeologists, anthropologists (physical and cultural), geologists, and paleontologists who coalesced around the problem and sought to resolve it. After David Hull and others, I will pay particular attention to the degree of "cooperation and competition, camaraderie and animosity, as well as allegiances and alliances" that emerged among them; to the relative scientific status of individual participants, knowledge of which is vital to understanding the reception accorded their views, for within the hierarchy of a scientific community not everyone's judgment is deemed equal, or even relevant, which in this instance was true regardless of the size of their role in the controversy; and, finally, to the larger institutional, political, and social context within which this controversy played out and that constrained or enabled the "scientific action"[80] (or possibly did both).

My approach in this study is twofold: to provide a detailed historical narrative of the controversy, and to explore the broader themes, processes, and context of the dispute and its resolution. In taking this approach I am following the lead of recent scholarship in the history of science, the explicit goal of which—familiar enough to anthropologists—is history as "thick description." As Rudwick put it, invoking Clifford Geertz, "The trick is not to get yourself into some inner correspondence of spirit with your informants" but rather "to figure out what the devil they think they thought they were up to."[81]

The means to that end are to assemble the historical details to provide a sequential description of the processes and events as the participants saw, experienced, and reacted to them, so as to analyze how and why evidence, ideas, and circumstances evolved. Rudwick insists that to do this, to understand "science in the making," the narrative must "rigorously and self-consciously avoid hindsight." He purges his extremely detailed recounting of the Devonian controversy of any hint of where the actions or events might be heading. If the information became available in September, he argues, it cannot be allowed to enter the discussion of what was happening the previous June.[82] Although I admire the purity of his nonretrospective approach (and the great self-discipline required to maintain it), I will not follow it here.

To be sure, we should tell history the way it happened, but that goal is far easier to state than to do and, more important, can hinder understanding.[83] If the participants are confused about what they are trying to explain, how they should explain it, and which explanations will succeed or fail, telling the story strictly from their point of view will hardly be less confusing. Sometimes clarification based on the knowledge of later events is necessary to make sense of the problems faced in earlier times.[84] For that matter, understanding history even on its own terms presupposes a knowledge of the present that is at once unavoidable and illuminating.[85] It could hardly be any other way when the author is an archaeologist who works on Late Pleistocene–age sites in North America, has grappled with the challenge of assessing artifact context and association in ambiguous stratigraphic circumstances, and has been involved in several of the controversies that still swirl around our efforts to understand America's earliest peoples.[86]

The key, as George Stocking and others remind us, is that historical studies must respect the integrity of the past and judge the work done in terms of the ideas of the period rather than against our modern standards.[87] Thus, the goal ought to be, as Oldroyd has argued, "to attempt to recreate a scientist's earlier thinking, while being fully conscious of the fact that it may all too easily be distorted by modern perspective."[88] By bringing a modern perspective to bear, by critically analyzing how previous archaeologists saw problems (some of which are still around), attempted solutions, and got those solutions wrong, we gain a deeper understanding of history. Errors, or at least nontrivial ones, reveal as much about the context in which they were made as do the moves that turned out correctly.[89] Although the outcome of the human antiquity controversy is well known, knowing the result cannot explain how and why it was achieved in the manner it was, which is what this book is about.

1.4 THE DATA OF HISTORY

How "thick" a thick description can be depends on the depth of the historical data.[90] Fortunately, this controversy is rich in source materials—especially during moments of heated debate—and these make it possible to develop a relatively fine-grained narrative. The historical sources include both published and unpublished materials, the latter consisting primarily of correspondence, but also diaries, memoirs, field notebooks, and early manuscript drafts. A few words on the nature, extent, and potential biases of these data are appropriate.

To identify the relevant sources, and with that knowledge build the list of participants and sites, and the issues and debates that took place over the decades, I began with the problem and its well-known core contributors (Charles Abbott, William Henry Holmes, Aleš Hrdlička, and Harvard University's Frederick Ward Putnam) and worked outward from there. That search had a snowballing effect that led to the increasingly larger network of participants, a listing of which is provided in the Roster of Individuals (with caveats noted therein). Naturally, individual participants had different roles: some hogged center stage, others stood near the wings. A few were active over much of the span of controversy; most others made only brief appearances. Where and how much time an individual spent on stage or the nature of his role was not necessarily reflective of that person's importance; even those peripherally involved could and did have a profound impact on the course of the controversy and its resolution.

The participants proved to be a remarkably prolific bunch, regularly churning out articles, books, and commentary. Few published remarks went unchallenged, and few finds passed without discussion and debate. No surprise there. Throughout the decades of this controversy the participants were hothouse close, so much so that as University of Nebraska paleontologist Erwin Barbour fumed over Hrdlička's seeming about-face regarding the antiquity of the Long's Hill human crania, Barbour's old friend Samuel Williston clucked sympathetically but could not resist a "very broad smile." He had warned Barbour that "there were certain men in the United States to whom a Pleistocene man was a red flag before a very erratic bull." Williston could have even told him "precisely what the report upon your discoveries would have been before the 'investigators' visited the place."[91]

Yet, close is not close-knit, and with that level of familiarity occasionally came brutal contempt: Chamberlin thundered at Warren Upham that his repeated claims about the geological context and Pleistocene age of the Lansing skeleton were "nothing short of reprehensible," and to another correspondent he wrote that they fully justified his having fired Upham from the USGS. Abbott could hardly write Holmes's name without calling him a liar.[92] Whether "amateur" or not (whatever that term meant at the time it was used), advocates and critics of a deep human antiquity were ambitious, strong-minded, even more strongly opinionated and convinced of their inerrancy, and not shy about saying so. As Hull observed, "Neither humility nor egalitarianism has ever characterized scientists, and no one has ever given any good reasons why they should."[93]

All that is to the good from the vantage of historical inquiry, for as a result of all this activity and antipathy there are more than 800 articles and books bearing on the issues related to the controversy that appeared over the course of this multi-decade dispute, written by archaeologists and nonarchaeologists alike. Although a wealthy trove of historical evidence, these publications are not without bias. As the literature in the history of science makes abundantly clear, publications must be read carefully. Then (as now) they are laundered for public consumption: in their advocacy of evidence or argument they seek to convey the impression that they began with indisputable fact, proceeded by impeccable logic, and ended with incontrovertible conclusions.[94] They are also usually written with a civility that belies their underlying antipathy: even at the most heated moments of this controversy there was a striking contrast between the public and private discourse. It was rare when someone (like Abbott or WJ McGee, or Holmes in his later years) breached the boundaries of decorum. That is not hypocrisy; it is just the way science works. Shrillness is to be avoided, lest it suggest the evidence is not on your side.[95]

At the extreme, Peter Medawar has argued that there "is no use looking to scientific 'papers' for they do not merely conceal but actively misrepresent the reasoning that goes into the work they describe."[96] Frederick Holmes challenges this position, replying that the publication itself is an integral part of the sense-making process and is at some level a record of past events; the scientist is expected to present the best arguments and evidence in support of the claims, regardless of the one or more routes that led to them.[97] Holmes adds, further, that Medawar's point is not altogether pertinent because how a conclusion was arrived at is not as important to the intended audience (fellow scientists) as the conclusion itself.

I would not disagree, but when a controversy is at its boiling point—when publications are written with gloves up and bristling with polemics—is precisely the time when *how* a conclusion was arrived at gets the most intense scrutiny. Thus, it is especially important to probe that which is unlikely to appear publicly: the false starts, blind alleys, doubts, biases, and speculations, not to mention behind-the-scenes alliances, plotting of responses, leveraging of advantages, and other less-than-scrupulous activities.

Making sense of science and scientific controversy thus requires not just listening to what the participants claim about their work or ideas in publications after the fact, but watching what they do as they do it. Listening through a keyhole, as Medawar says, if it could be so arranged.[98] That is impossible a century after the fact, of course, but the next best thing to being there is reading over the shoulders of those who were, which can be accomplished by examining their correspondence and other more private materials.[99] These were written for smaller and more sympathetic audiences (one person, usually), and thus have far more potential to reveal thoughts, motives, intentions, doubts, guesses, opinions, and subtle influences or constraints (financial, political, social) that might otherwise be invisible.[100] These can provide a measure of what individuals wanted to believe as opposed to what they could reasonably defend, and thus can be especially useful in revealing conceptual ambiguity (by showing where and what sorts of inferential leaps individuals were willing to risk and why) and larger issues and pressures.

It is again fortunate for the historical study of this controversy that those who were involved knew it was important (or at least considered *themselves* important), with the result that many of them deposited their correspondence and other un-published materials in archives. Some of these archival sources have been used in previous examinations of the participants in this dispute, but in the past the focus has been primarily on a few major players, such as Holmes, Hrdlička, and Putnam, whose papers are found in archives in Washington, DC, and Boston.[101] The systematic search that was undertaken for this study revealed a wealth of untapped archival holdings for many more participants who worked in less vis-ible institutions than the Smithsonian or Harvard, and in disciplines outside of archaeology and anthropology.

Most of the participants were habitual letter writers who kept in contact pri-marily by this means. Widespread telephone use and the subsequent decline in correspondence occurred only at the tail end of the controversy.[102] Thus, there is a well-marked trail of nearly 3,500 letters through this controversy—in many instances, both sides of the correspondence—that are revealing in their tone and spontaneity, and are rich in content and detail. Some of what was written might uncharitably be viewed as "gossip" or "politics," but I will not be snobbish about it: interpersonal dynamics are important in understanding human behavior.[103] Moreover, by virtue of the speed with which the mail moved (a letter sent from Trenton one day could be delivered in Boston the next), the correspondence pro-vides a record of events while they were still fresh.

To be sure, unpublished materials are not without bias, though in contrast to publications their private candor at least makes their biases more readily apparent. Personal memoirs are especially notable in this regard: usually written years after events, they suffer from the lapse of time, the (selective) loss of memory, and the all-too-human tendency to present a tamed version of one's past transgressions (Holmes's unpublished twenty-volume autobiography and scrapbook, for exam-ple, was obviously compiled with posterity and perhaps a sympathetic biographer in mind).[104]

1.5 THE SCOPE AND STRUCTURE OF CONTROVERSY

Between the participants' unpublished correspondence and their often rapid-fire responses to one anothers' publications, a narrative can be constructed on a scale of days, weeks, and months, which makes it possible to investigate this swirling and bitter controversy as it played out for the practitioners. It also makes it pos-sible to better define the structure and path of the controversy.

No historical episode begins or ends on cue, though this one arguably comes close: the controversy began in earnest in the spring of 1890 when Holmes launched the first salvo of his attack on the American Paleolithic, and it was resolved in the fall of 1927 at Folsom. But I have expanded the scope of my inquiry to cover the period from 1862 to 1941. General interest in the origin and antiquity of Native Americans of course goes back even further than 1862 (to 1492, if one wishes), but the problem in its modern form began only after 1859, when Europeans first broke through the Mosaic barrier and glimpsed a human past that reached deep into the Ice Age.[105] The impact of that breakthrough was felt almost immediately in North America. The Smithsonian's Secretary, Joseph Henry, reprinted in his *Annual Report* articles from Europe that described the new-found evidence of antiquity. Under his direction, a *Circular* was issued in 1862 aimed at military officers, missionaries, Indian agents, and travelers and residents of "Indian country," with a call to search North America for evidence in kind and rivaling in age that found in Europe.[106] It launched the search for a Pleistocene human antiquity in North America.

That search ended at Folsom, yet stopping the historical investigation in 1927 does not fully reveal how much was learned in the course of controversy, or just how profound were the changes resolution wrought. These become conspicuous in the rapid strides made in understanding the origins, antiquity, and adaptations of North America's Paleoindians that came in the next dozen or so years up to World War II (when archaeological research effectively ceased). Examining that period reveals how new knowledge could be and was rapidly created in the wake of resolving the antiquity issue, and starkly highlights what had not been known and could not be answered—or even asked—in the previous decades of controversy.

Although there was more or less constant attention to the problem between 1862 and 1941, there were periods of relative quiet, punctuated by episodes of intense commotion. "Hot spots," Oldroyd calls those moments, when "heated discussion occurs; correspondence flows; there is an increased intensity of experimentation, field observation, and publication; and there may be some behind-the-scenes skullduggery or acrimonious public debate."[107] Hot spots can be identified by looking for spikes in activity and interaction, as seen, for example, in publications.

Plotting the annual frequency of publications related to this controversy (Figure 1.1) reveals multiple peaks.[108] These are the hot spots, and their pattern a sort of topographic profile of activity, with periods of relative quiet (one or a few publications) separating episodes of frenetic activity (say, fifteen to twenty publications in a year). The spike of approximately one hundred publications in 1892 and 1893 appeared in the heat of what contemporaries called the "Great Paleolithic War," and is obviously unusual but of telling historical import. Although it might be tempting to assume as much, the publication peaks do not necessarily mark periods of dispute, nor the valleys times of consensus; although the former is mostly true, the latter is mostly not.

The hot spots are episodic but display a somewhat regular pattern, lasting two to three years at a time, with peaks occurring at roughly five-year intervals. I hasten to add, however, their "wavelength" does not reflect any underlying oscillating processes within the scientific community but rather the chance timing of site discoveries, for it is those that most often propelled inquiry and triggered disputes.

That there were hot spots as opposed to a constant "burn" provides a structure to historical investigation and—along with the broader shifts in the nature of the evidence—are used to organize the chapters that follow. In broad outline the book comprises three parts. Starting from the inside out, the central and largest part consists of chapters 3 through 8, which provide an in-depth narrative spanning the period from 1872 to 1928. These detail the slow rise and precipitous fall of the American Paleolithic (chapters 3 to 5), then the successive searches for

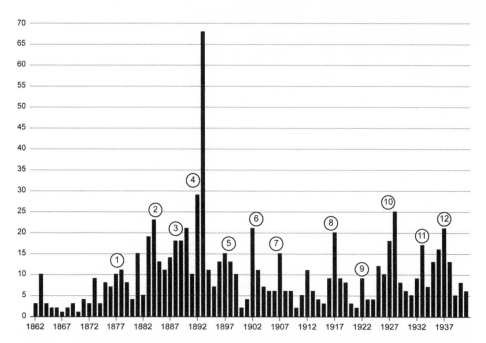

Figure 1.1 Histogram showing the "hot spots" in the controversy over human antiquity in North America and its resolution, as marked by the annual frequency of publications (vertical axis) in archaeology, glacial geology, linguistics, physical anthropology, and vertebrate paleontology that appeared from 1862 to 1941 ($n = 809$; the data for the figure are drawn mainly, though not solely, from the "Printed sources: Primary" portion of the Bibliography). These data provide a fine-grained illustration of the episodic spikes of publishing activity. These spikes were most often related to discoveries of specific sites and ensuing disputes, but they were also related to conceptual breakthroughs and other intellectual pivot points, such as the resolution at Folsom in 1927. Key to select hot spots in the figure: (1) initial discovery of apparent Paleoliths in the Trenton Gravel, NJ (as discussed in chapter 3); (2) report of paleoliths at Little Falls, MN (chapter 4); (3) discovery of paleoliths at Newcomerstown and other Ohio sites, and symposia on the American Paleolithic (chapter 4); (4) the 1892–1893 "Great Paleolithic War" (chapter 5); (5) Trenton site field symposium (chapter 5); (6) discovery of human skeletal remains at Lansing, KS (chapter 6); (7) discovery of human skeletal remains, Long's Hill, NE (chapter 6); (8) discovery of human skeletal remains, Vero, FL (chapter 7); (9) discovery of *Hesperopithecus haroldcooki* at Snake Creek, NE (chapter 8); (10) discoveries of bison remains with projectile points, Lone Wolf Creek, TX, and Folsom, NM, and of artifacts with mammoth and other remains in Frederick, OK (chapter 8); (11) discoveries of artifacts with mammoth remains at Clovis, NM, and Dent, CO (chapter 9); (12) symposia and syntheses on North American Paleoindians (chapter 9).

other evidence of a deep human antiquity that ran from "primitive" humans in glacial-age loess (chapter 6) to anatomically modern human skeletal remains purportedly in association with extinct Pleistocene fauna (chapter 7) and finally to artifacts—albeit ones not playing to Paleolithic type—found in association with extinct Pleistocene fauna (chapter 8).

These chapters do not aim to provide a strict year-by-year chronicle of the controversy. The periods covered by each are not equal, nor does one always pick up where the previous one left off. Between some there are temporal gaps (not necessarily large), and between others there is overlap (not necessarily significant). Structuring the volume in this way makes it possible to adhere to the narrative

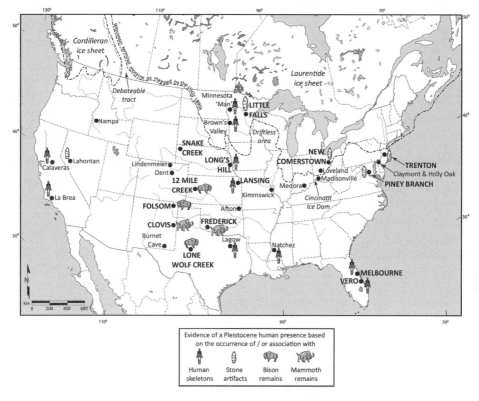

Figure 1.2 Map of archaeological sites and significant glacial geological features (shown in italics) discussed in the text. Sites in uppercase lettering figured prominently in the controversy over human antiquity and its ultimate resolution; they are discussed at greater length in the text. Where included, the figures of artifacts, human skeletons, and bison and mammoth provide a key to the evidence on which a particular claim for a Pleistocene human presence was based. The position of the Wisconsin terminal moraine as it was known in the late nineteenth century is based on Chamberlin 1883b, Chamberlin and Salisbury 1885, and Leverett 1899 (see also Figure 4.1). The position of the "Debatable tract" (now known as the "ice-free corridor," which opened as the Cordilleran and Laurentide ice sheets retreated) is based on Upham 1895.

without having to grant equal time to periods of inactivity. As Rudwick observes, the primary chronology a narrative should punctiliously observe is the sequence of events rather than time as measured by the calendar on the wall.[109] As this is a history of the controversy over human antiquity, the focus is on those sites central to the controversy—the ones that sparked the heated battles that raged within and across disciplinary lines (for example, Trenton)—as opposed to those where (to continue a martial metaphor) only minor skirmishes took place, or ones that became significant only in retrospect (for example, Kimmswick). A roster of sites is provided in map form in Figure 1.2. The same guiding principle holds true for the roster of individuals: the focus is on those involved in the controversy, with more attention to those who figured more prominently—which, as noted (§1.4), may be independent of the time they actually spent on stage.

The six narrative chapters are sandwiched between two chapters that speak to the before and after of the controversy. Chapter 2 sets the stage with a brief history of the establishment of human antiquity in Europe, its catalyzing impact in North America, and the initial if largely inchoate efforts to develop comparable evidence here in America. On the narrative's far side, chapter 9 explores the wide-ranging and rapid gains in knowledge of the archaeology of Pleistocene peoples that came in the wake of the resolution of the controversy at Folsom.

Throughout, attention is on the empirical and conceptual matters at hand (archaeological, geological, and anthropological), but also on the broader contextual elements and dimensions of controversy (iterated earlier). These epistemic and non-epistemic threads are pulled together in chapter 10, the third and final part of the book, which steps back from the narrative altogether to assess and synthesize along thematic lines how and why the controversy developed, why it so stubbornly defied resolution for so long, why it ended as quickly and decisively as it did when it did, and what that meant for American archaeology. Readers who want to see the big picture in advance, or see it without all the narrative details, should feel free to jump ahead to chapter 10, which provides an overview of the human antiquity controversy. If one only reads a single chapter in this volume, chapter 10 is the one to read.

CHAPTER TWO

Setting the Stage

IN THE WINTER OF 1861 Joseph Henry, the Smithsonian Institution's first secretary, was besieged by frequent "perplexing" interruptions and "embarrassments" brought to his doorstep by the outbreak of the Civil War.[1] Determined to maintain a semblance of normality, he devoted much of his *Annual Report* to proudly recounting the Institution's scientific successes for the year. Prominent among them was its role in awakening "a new interest in the study of the remains of the ancient inhabitants of this continent."[2]

Although a physicist by training, Henry well understood that the "nature of man is interesting to the students in every branch of knowledge," and he could see that "no part of the operations of this Institution has been more generally popular" than those related to archaeology.[3] Under his direction the Smithsonian had published to great acclaim Ephraim Squier and Edwin Davis's *Ancient Monuments of the Mississippi Valley* (1848) and Samuel Haven's *Archaeology of the United States* (1856), and was fast becoming the repository for the world's largest and most valuable Native American ethnographic and archaeological collections, the latter primarily from the mounds and earthworks in the Ohio and Mississippi valleys.[4] *Ancient Monuments* had spoken of such remains as having "no inconsiderable antiquity,"[5] but when those words were written few imagined that meant their antiquity went beyond the comfortable confines of Biblical and written history, the foundation on which the understanding of humanity's past had long rested.

But a new dimension of archaeological time opened in 1859 following the discoveries in England and France of stone artifacts definitively associated with the bones of extinct Ice Age animals. The earliest traces of humanity were abruptly plunged deep into the Pleistocene, perhaps hundreds of thousands or maybe (by some estimates) millions of years.[6] Obviously, written history was irrelevant to that distant time "when man shared the possession of Europe with the Mammoth, the Cave bear, the Woolly-haired rhinoceros, and other extinct animals" and used "rude yet venerable weapons" of stone.[7] Human prehistory from the time of the Stone Age or Paleolithic would be knowable only through their artifacts and bones.

Henry had keenly followed the developments in Europe and introduced those discoveries to an American audience by reprinting in his *Annual Report* Adolph Morlot's "General Views on Archaeology," a lengthy article that summarized the latest finds from the "bygone days of our race previous to the beginning of history."[8] Morlot's article generated such excitement that Henry included in the following year's *Annual Report* several more articles on the European evidence of human antiquity. But he wanted to do more than just educate his American audience on developments in Europe: he wanted to investigate the possibility that there was evidence of a deep human antiquity on this continent as well.

It had long been suspected, primarily from the "philological and physiological" diversity of Native Americans, that their origins might extend back, as Haven put it, "among the enigmas of immemorial time."[9] There had been finds in the preceding decades of artifacts and human remains that hinted at a deep antiquity, most especially by Albert Koch in Missouri and Montroville Dickeson in Mississippi, both of whom reported finding human remains alongside fossils of extinct animals. At the time and until the early 1860s, when the outcome of events in England and France demanded they be reconsidered, these and other finds were easily dismissed as fortuitous associations, testimony to the recent nature of animal extinctions, or object lessons on the speed with which human remains could be deeply buried and fossilized.[10] Until news of the European discoveries reached these shores, few took those finds seriously, let alone believed the continent had been peopled in Pleistocene times. That Native Americans were unmentioned in the Bible was easily explained by geography, not time: they lived far from where Biblical history took place, rather than long before it happened. Only Josiah Nott and George Gliddon, the blasphemous authors of *Types of Mankind* (1854) who proclaimed the world's races were separately created species, dared claim a great antiquity for Native Americans, and few took them seriously.[11]

But by 1862 Henry was ready to revisit those earlier finds and, more important, encourage the search for evidence to demonstrate a deep human antiquity on this continent. That year he directed the Smithsonian's George Gibbs to prepare a set of *Instructions for Archaeological Investigations in the United States*.[12] These were broadcast to the Smithsonian's legions of correspondents (a group scattered across the country that numbered in the thousands and submitted observations and specimens[13]), as well as to military officers, missionaries, Indian agents, and travelers and residents of Indian country. Although they had the principal (if unstated) goal of bolstering the Smithsonian's collections of artifacts and human skeletal remains,[14] included within the *Instructions* was the call to search for specific kinds of remains, most notably ones "found under conditions which connect archaeology with geology." It was this kind of evidence that was necessary to "take us back to a very remote period in aboriginal history." This might include artifacts or human skeletal remains found with the bones of extinct animals or under geological circumstances—in glacial-age deposits, say—suggestive of considerable antiquity.

In such cases Gibbs urged that "the utmost care . . . be taken to ascertain with absolute certainty the true [chronological] relations of these objects."[15] Morlot had warned how easy it was to mistake an accidental association of artifacts and the bones of an extinct animal for a genuine one.[16] Only by exercising great care was it possible to resolve the problem of "the origin of the inhabitants of this continent."[17] Solving that problem, Henry believed, would allow comparison of the "analogous stages of the mental development of the primitive inhabitants of this country and those of Europe," and perhaps even answer broader questions of the origin and characteristics of the human races. These were, of course, questions that in the midst of the Civil War could "scarcely be discussed . . . with that dispassionate logic and strictness of induction which is necessary to the establish-ment of truth."[18]

Over the next decade the archaeological collections of the Smithsonian grew rapidly, so much so that Henry began contemplating an "exhaustive work on American archaeology," one that would illustrate "whatever is now known, or can be ascertained by special investigation, of the antiquities of North America."[19] To insure the effort was truly comprehensive, Henry prepared a new set of instruc-tions in 1878, this time with the help of anthropologist Otis Mason. The result-ing *Circular in Reference to American Archaeology* was no less ambitious than the *Instructions* from sixteen years earlier. Yet it differed in one important respect: it said not a word about extinct animals, Pleistocene deposits, or associated artifacts. By saying nothing about human antiquity, the *Circular* said a great deal, for the future of American archaeology as Henry saw it in 1878 was in ancient mounds and pueblos and was not to be found in Haven's enigma of immemorial time.[20] Henry had good reasons in 1878 for giving up on the search for evidence of a deep human antiquity. Yet, even as he was abandoning hope Charles Abbott, a onetime physician and now full-time naturalist and collector, was finding in the Delaware valley what appeared to be traces of Pleistocene-age Paleolithic artifacts.

2.1 ESTABLISHING THE PARAMETERS

Ideas about the origin and antiquity of Native Americans had been discussed and debated almost since the moment Columbus set foot in the Caribbean, and by the nineteenth century much had been written about those issues but little resolved. In the first decades of the century the matter was complicated by the question of the relationship between Indians and the so-called Mound Builders, not least whether they represented separate races, migrations, or both, and if so what might be inferred of their relative antiquity and historical relatedness. These are issues I explore in considerable detail elsewhere[21] but need not do so here as well, for by the 1850s scholarly opinion was leaning increasingly toward the view that Indians and Mound Builders were historically related and descendants of the same popu-lation or populations that came out of Asia.[22] The ancestors of the latter could, in principle, be traced back to the original pair at Creation.

Traditional views of the origin and antiquity of Native Americans were based in large part on inferences drawn from native languages. A vast number of lan-guages were spoken in the Americas—more so even than in Asia and Africa, it appeared. On the assumption that these had diverged from a single ancestral tongue, Thomas Jefferson reasoned that "an immense course of time" must have elapsed since ancestral Native Americans had begun to separate from "their com-mon stock." He suggested their antiquity was "perhaps not less than many people give to the age of the earth,"[23] an inference so heretical as to allow his political enemies to call him a "howling atheist."[24]

Well he might be so accused. Although it was understood that not all peoples were mentioned in the Old Testament, it did not mean any existed earlier than the 6,000 years allotted humanity by the Mosaic chronicles. Even as Europeans spoke of a Stone Age or used the term "prehistoric" in the early nineteenth century,[25] these terms were merely convenient referents to peoples whose existence may have escaped notice—but not the confines of the modern world as depicted in Genesis or historical accounts.

Philosophers and scientists could and did argue over whether the first five days of Genesis should be interpreted literally or allegorically. If they were not days in the usual sense—and by the early nineteenth century few thought they were—the separation of heaven from earth, light from dark, sea from land, and the creation of fish, birds, "cattle and creeping things" might be much older than anyone imagined.[26] But not humans. The last and most recent creation, firmly anchored in genealogies of Genesis from Adam on downward, happened no earlier than 6,000 years ago, on an earth and amid plants and animals that looked just as they do today.[27] On this point the Bible was explicit.

But strange fossils, brought to light in increasing numbers by exploration worldwide and the mining that fueled the Industrial Revolution, had begun to reveal a far older, premodern earth on which a vanished zoo of exotic animals had roamed. Among the most spectacular were the fossil elephants, most especially the mastodon, the massive bones and teeth of which were recovered by French troops at Big Bone Lick, Kentucky. French anatomist and paleontologist Georges Cuvier demonstrated in 1796 that this animal, with its elephant-like tusks and hippo-like teeth, was its own species distinct from living elephants (or hippos) and not simply some curious hybrid. Nor was it alive: he had it on the good authority of American Indians—who, being "nomadic peoples who move ceaselessly around the continent in all directions," were in a position to know—that these "creatures no longer existed."[28] The Christian doctrines of plenitude (the idea that God's world was created in a state of fullness) and providence (that God guarded over all creatures) forbade the idea that any species would have been allowed to go extinct.[29] The mastodon, Cuvier showed, had done just that.

But Cuvier did more than just demonstrate that extinction had occurred: he also used the anatomical and geological evidence to set up an argument about earth history. He realized that the fossil bones of mastodons and the soon-discovered mammoths, rhinoceros, and giant sloths

> are buried almost everywhere in roughly similar beds; they are often mixed there with some other animals likewise fairly similar to those of today. These beds are generally unconsolidated, whether sandy or silty, and always more or less close to the surface.[30]

Because the extinct species in the "unconsolidated and superficial layers of the earth" were often related to living genera ("fairly similar to those of today"), or found with shells of invertebrates still extant, it was an easy step to the conclusion that this suite of now-extinct animals "prove the existence of a world previous to ours, destroyed by some kind of catastrophe."[31] This was only the beginning for Cuvier. He realized those particular fossils were not truly ancient, at least in relative terms, for there were more primitive fossil animals older still, whose remains were found below "regular stony [consolidated] beds, and covered by regular marine strata."[32] Thus, there were animals in near-surface unconsolidated layers that were similar (the same genus) but not identical (different species) to those of the modern world; below them were animals even more distinct. Cuvier concluded the earth must have been inhabited at different times by different suites of extinct animals, each destroyed as its era came to an end.

It was important that one species—humans—was decidedly absent from those earlier worlds, including the most recent. That was to be expected. Even this most recent period was beyond the range of the Mosaic chronicles. The Bible made no mention of mastodons or mammoths. From there it was but an easy step to the conclusion that their extinction neatly demarcated the boundary between the premodern and modern worlds and became a chronological barrier beyond which human antiquity would not extend.[33]

A hint of the character of that premodern world was in the "unconsolidated" gravels and sands that produced those fossils, which in the early decades of the nineteenth century were widely attributed to flooding. *Whose* flood was a matter of debate. Oxford geologist William Buckland attributed these diluvial gravels to "a transient deluge, affecting universally, simultaneously, and at no very distant period, the entire surface of our planet." He supposed the flood was Noah's.[34] These gravels lacked remains of the humans who presumably perished in the Biblical flood, and in retrospect Buckland's Cambridge counterpart Adam Sedgwick admitted that should have given pause. Still, Buckland found their absence not so hard to explain: in Noah's time humans lived only in the "Asiatic" region.[35] Their remains did not belong in floodwater deposits in England and northern Europe.

Few agreed with Buckland that these deposits represented the debris of Noah's flood (the French considered such efforts at "scriptural geology" a peculiarly English aberration[36]), but most were willing to accept the idea that they marked a sudden and drastic event in geologically recent time and involved some manner of deluge. More difficult to accept, however, was how water could account for the scratched bedrock or large boulder "erratics" (rocks from distant bedrock sources) associated with the diluvium.[37]

The answer came from the Swiss Alps. In the 1830s Swiss geologist Jean de Charpentier began to record the telltale signs left by alpine glaciers: as a massive tongue of ice moved, it carried along boulders plucked from high peaks and transported them dozens of kilometers down valleys and along the way smoothed and striated the bedrock it overrode. When the glacier paused, the debris it had carried and bulldozed down-valley was piled in a fronting berm—a moraine—of gravel and sand. Charpentier realized these same signs occurred in now ice-free mountain valleys, sure evidence that glaciers had once been there as well. That they no were longer present he attributed to changes in the elevation of the local mountains rather than changes in climate.[38] That explanation might work in the Alps, but it left unexplained the striated bedrock, erratics, and extensive moraines that occurred in areas where there were no nearby mountains or evidence of uplift, which by the 1830s was known to include much of the low-lying plains of northern Europe and northern North America.

In the summer of 1836 Charpentier shared his observations of alpine glaciers with Louis Agassiz, then professor of natural history (and an expert in fossil fish) at the University of Neuchatel, Switzerland. The ambitious Agassiz quickly made the inferential leap Charpentier had not: because the erratics, striated bedrock, gravelly tills, and moraines at high altitude were the same features (only smaller) as those at lower altitudes in higher latitudes (on the low-lying plains of the British Isles and northern Europe), and because the ones at high altitude were left behind by glaciers, then the ones at high latitude—Buckland's diluvium and associated features—by analogy must have been deposited by long since vanished glaciers:

> As the glaciers are continually pressed forward, and often in hot summers melted back . . . terminal moraines, or curvilinear ridges of gravel and boulders, remain upon the rocks formerly covered by the ice. Thus we can discover, by the polished surfaces and the moraines, the extent to which the glaciers have heretofore existed, which is much beyond the limits they now occupy in the Alpine valleys.[39]

From the size and extent of their traces, these glaciers must have been of almost unfathomable size and existed during a time of global cold. "Since I saw the glaciers," Agassiz crowed, "I am quite of a snowy humor, and will have the whole surface of the earth covered with ice."[40]

Agassiz's Ice Age—and at the outset it was thought to be a singular phenomenon—was soon linked with English geologist Charles Lyell's Pleistocene, a geological period independently defined by the relative percentage of fossil shells to modern ones, and which seemed to coincide perfectly with the timing of global glaciation. It is significant that no one was looking for human remains in Pleistocene deposits, nor did anyone expect to find any. Whatever its age in absolute years, the Pleistocene was old relative to humans.[41] As Lyell put it as late as 1853:

> The comparatively modern introduction of the human race is proved by the absence of the remains of man and his works not only from all strata containing a certain portion of fossil shells of extinct species, but even from a large part of the newest strata, in which all fossil individuals are referable to species still living.[42]

On an earth that by the mid-1800s was becoming almost inconceivably old, humanity's last refuge in the search for ultimate design and its own uniqueness and divinity lie in the affirmation of the Old Testament. Here, fortunately, the Bible and geology seemed to agree. Humans were indeed the last creation or the last in a progressive series of creations, so it seemed on good scientific and scriptural authority.[43]

Naturally, when artifacts or human bones appeared in Pleistocene deposits or alongside the bones of extinct animals, as they did occasionally in the first half of the nineteenth century, the associations were questioned and the finds rejected or ignored.[44] At least, until 1858.

That summer William Pengelly, a part-time geologist of the Torquay Natural History Society, discovered a cave in southwestern England that contained fossil bones. He and paleontologist Hugh Falconer began excavating in Brixham Cave with the goal of resolving the sequence of Pleistocene to post-Pleistocene faunal change in England. These were geologists looking to solve geological problems.[45] The "mud, stalagmites, and breccias of caves, where the signs of *successive* deposition are wanting," are challenging settings in which to establish the stratigraphic control necessary to determining a succession of faunas.[46] To insure the trustworthiness of the results from Brixham Cave, the Geological Society of London, which was funding the excavation, appointed a committee of Britain's top geologists—which included Lyell, Joseph Prestwich, and vertebrate paleontologist Richard Owen—to oversee the research.[47]

That excavation was excruciatingly slow, as it proceeded meticulously layer by layer (unusual for its time). But on July 29, 1858 a flint artifact was found beneath a limestone (stalagmite) layer approximately 7.5 centimeters thick, sealed along with bones of several extinct mammals. By mid-August seven more artifacts had been found beneath that impermeable limestone layer, in a loamy sediment ("cave earth") that proved to contain the bones of woolly mammoth, woolly rhinoceros, cave bear, and cave hyena.[48] By the time excavations were complete the following year, the number of artifacts had more than doubled. Because the site was dug with such great care, there was little purchase to be gained criticizing the excavation methods. Still, Brixham Cave was a cave and vulnerable to mixing, and thus it faced the inevitable skepticism about whether the suggestive association between the artifacts and the extinct mammal bones was fortuitous, or proof of their coexistence in the deep past.[49] In response to a paper by Pengelly on Brixham Cave delivered in the fall of 1858 to the Geology Section of the British Association for the Advancement of Science (BAAS), Owen, then president of the Section, granted that "he was glad that means had been taken for the careful exploration of this cave, but it would be premature to raise any hypothesis until the whole of the facts were before them."[50]

Falconer therefore turned his attention to insuring the authenticity of the Brixham Cave specimens by examining other localities to see if its evidence could be duplicated. In late 1858, he visited a number of the French caves that had yielded stone tools and stopped in at the home of Jacques Boucher de Perthes,[51] a customs official in Abbeville in northwestern France who had since the 1830s been collecting in the Somme River valley the remains of extinct fauna and, apparently, associated stone tools. Many knew of his finds through his two-volume *Antiquités Celtiques et Antédiluviennes* (1847 and 1857). Few believed what they read. A substantial number of the "artifacts" he illustrated were clearly not made by humans, and what genuine evidence he had was embedded in arcane catastrophist flood theories that had long since been rejected.[52] As John Evans later gently put it:

> The announcement by M. Boucher de Perthes, of his having discovered these flint implements . . . was, however, accompanied by an account of the finding of many other forms of flint of a much more questionable character, and by the enunciation of theories which by many have been considered as founded upon too small a basis of ascertained facts.[53]

Boucher de Perthes was perceived as little more than a provincial amateur using outdated theories and presenting spurious data; some considered him "almost a madman."[54] Charles Darwin was hardly alone in looking at Boucher de Perthes's work and thinking many of the artifacts were naturally flaked, and that "the whole was rubbish."[55] Still, Boucher de Perthes's specimens had one important advantage: they came from deep, well-stratified, undisturbed alluvial deposits of Pleistocene sand and gravel and thus were less susceptible to the problems of stratigraphic mixing that plagued cave sites. Boucher de Perthes well understood the importance of what he had:

> Diluvial deposits do not present . . . like the bone caves, an inconcealable cavern, open to all who come, and which from century to century served as a sanctuary and then as a tomb to so many diverse beings. . . . In the diluvial formations, on the contrary, each period is clearly divided. The horizontally superimposed layers, these strata of different shades and materials, show us in capital letter the history of the past: the great convulsions of nature seem to be delineated there by the finger of God.[56]

All of which explains why Boucher de Perthes's evidence became compelling once the illegitimate artifacts were discarded and the genuine ones divorced from his arcane theories.

After his visit to Abbeville, Falconer wrote Prestwich, who was well versed in the Quaternary alluvial stratigraphy of northwestern France, suggesting he do likewise.[57] Prestwich in turn invited the antiquary John Evans to join him and in late April of 1859 they traveled to France, witnessed and photographed the removal of a hand ax in situ from a locality in Amiens, collected a few more hand axes from the workmen, then returned to England.[58]

At a meeting of the Royal Society soon after their return, Prestwich read a paper on the stratigraphy of the Somme valley glacial gravels and Evans gave an extemporaneous talk on the artificial character of the flint implements, in which he assured his audience the specimens from the "drift" were altogether distinct in material, form, and workmanship from the stone tools "so frequently found in this country and on the Continent, and are usually considered to be the work of the primitive, or as for convenience sake I will call them, the Celtic inhabitants of this part of Europe." Rude as the latter might appear, Evans added, "they seem to point to a higher degree of civilization than those [found in the drift] I am now considering."[59] Listening to them that evening, Evans recalled

> were a good many geological nobs . . . Sir C. Lyell, Murchison, Huxley, Morris, Dr. Perry, Faraday, Wheatstone, Babbage, etc. so [we] had a distinguished audience. Our assertions as to the findings of the weapons seemed to be believed.[60]

It was indeed a distinguished audience, which included not only England's geological elite (Lyell, Thomas Henry Huxley, Roderick Murchison) but also some of the greatest and most influential scientists of the age. The question of a deep human antiquity was out of the closet, and the favorable reception the audience accorded Prestwich and Evans immeasurably helped the cause.[61] Prestwich and Evans's papers, coupled with the evidence from Brixham Cave and the testimony of Falconer, was enough to prompt a stream of cross-channel visitors and Boucher de Perthes's fellow countrymen to examine his collections and the sites in the Somme valley.[62] Lyell's pilgrimage to Abbeville was by far the most symbolic, because he had long defended the belief of the recency of human antiquity and had challenged all claims of great human antiquity. Yet, he too returned from France converted:

> I am fully prepared to corroborate the conclusions which have been recently laid before the Royal Society by Mr. Prestwich. . . . I believe the antiquity of the Abbeville and Amiens flint instruments to be great indeed if compared to the times of history and tradition.[63]

Humans had seen Agassiz's glaciers and preyed on Cuvier's fauna.

Lyell announced his conversion in September 1859 at a meeting of the BAAS, where he was then president of the Geological Section. Because he was the premier geologist of the nineteenth century, his announcement carried enormous weight. Where just months before the idea of a deep human antiquity was a dubious claim of provincial amateurs, it was suddenly almost a universally accepted fact.[64] As archaeologist William Boyd Dawkins put it, people were no longer insisting "it was not true" or that "it was contrary to religion" but that "it was all known before."[65] It was, as a contemporary saw it, a "great and sudden revolution."[66] To be sure, there were matters to be clarified, and a brief debate played out over the details of the geological context and the precise age of the associated fauna, but not over the authenticity of a deep human antiquity.[67] There were a small number who attacked the latter, but as Gruber observes, "the very shrillness of their tone betrayed the weakness of their position."[68]

Out of that discovery and confirmation of a deep human antiquity, prehistoric archaeology found itself in possession of a vast dimension in time and the need for methods to fill it. Archeologists began to build a chronology and reconstruct human prehistory from stone artifacts and fossil bones, written history having been rendered utterly irrelevant to the understanding of the earliest humans. In the wake of the events of 1858–1859, and having learned what to look for in the archaeological record, dozens of sites yielding "primitive" stone tools and the bones of extinct animals were soon reported across Europe, providing further support of a deep human antiquity. So much was learned so quickly that by 1863 Lyell had enough material for a wide-ranging synthesis, *Geological Evidences of the Antiquity of Man*.[69] Two years later in *Prehistoric Times*, John Lubbock gave the human presence in the "Drift epoch" a name: he called it the *Paleolithic*—Old Stone Age.

The Paleolithic extended into the deep past: hundreds of thousands of years, perhaps even millions.[70] However theologically disconcerting such a staggering antiquity might be, those rude artifacts were vivid testimony of the savage depths from which human society had climbed:

> To the student who views Human Civilization as in the main an upward development, a more fitting starting point could scarcely be offered than

this wide and well-marked progress from an earlier and lower, to a later and higher, stage of the history of human art.[71]

Not surprisingly, a deep human antiquity was soon entangled in the theory of evolution by natural selection that Darwin had proposed independently in *The Origin of Species*. (Lyell made his first public mention of Darwin's views in his 1859 mea culpa address to the BAAS.)[72] The *Origin*, of course, had no more to say about the evolution of the human species from our animal forebears than the understated one-liner "Light will be thrown on the origin of man and his history."[73] But everyone knew what *that* meant: a shared ancestry with other primates, and pre-*sapien* ancestors deep in time. Although the two ideas developed independently, one could scarcely accept Darwin's views of human evolution without the vast chronology prehistoric archaeology provided (the reverse, of course, was not true[74]). As Darwin himself later insisted: "The high antiquity of man has recently been demonstrated by the labors of a host of eminent men, beginning with M. Boucher de Perthes; and this is the indispensable basis for understanding his origin."[75] And if "there is any truth whatever in the doctrine of evolution as applied to man," Darwin's compatriot Alfred Wallace insisted, "they *must* have had ancestors who *must* have existed in the Pliocene period, if not earlier."[76] Why not in the Pliocene of America?

2.2 BRINGING THE PALEOLITHIC TO AMERICA

Why not, indeed? Joseph Henry went to work. Searching European journals and soliciting help from his network of Old World correspondents, he reprinted in the Smithsonian's *Annual Reports* in the decade of the 1860s more than a dozen papers by European and American archaeologists and geologists. Henry liberally distributed his *Annual Reports*, so they were read not just within the small scientific community but also across an extensive network of local scientific societies and individuals who otherwise had limited access to scientific information, but who might send specimens and information about discoveries to the Smithsonian. (Henry's motives were not entirely altruistic.)[77]

Through Henry's efforts Americans learned about Paleolithic implements, artifacts in Danish *kjoekkenmoedding* (shell middens) and in Swiss lake dwellings, and the latest in archaeological theory and method. Morlot lectured his American audience on the importance of observing artifact association, context, and the value of superposition for establishing "more surely than could be done in any other manner the chronological succession of different ages."[78] Gibbs repeated the message in his *Instructions*.

Issued in early 1862, and updated within a year to add the collection of data on language and culture, the *Instructions* were patterned on the European plan. Gibbs sought information and collections of skeletal remains (insofar as these could be acquired "without offence to the living"—a well-intentioned principle, however often unmet in practice[79]), material culture, and specimens of art, both from contemporary groups, and especially those "found under conditions which connect archaeology and geology." He had specific targets in mind for the latter:

1. The contents of shell beds of ancient date found on the seacoasts and bays, often deeply covered with soil and overgrown with trees. . . . The examination of these collections in Denmark and other countries of northern Europe has led to the discovery of remains belonging to a period when a people having no other implements than those of stone or bone occupied the coast prior to the settlement there of the present race. It is possible that a similar investigation in America may carry us back to a very remote period in aboriginal history.
2. Human remains or implements of human manufacture, bones of animals bearing the marks of tools or of subjugation to fire, found in caves beneath deposits of earth, and more especially of stalagmite or stony material formed by droppings from the roof.
3. Spear and arrow heads, or other weapons, and evidences of fire discovered in connection with bones of extinct animals, such as the mammoth, fossil elephant, &c., among superficial deposits, such as salt licks, &c.
4. Implements of the same description found in the deposits of sand and gravel, or other like material, exposed in bluffs or steep banks, such as have recently attracted the attention of European geologists.[80]

In all these cases, he added, "the utmost care should be taken to ascertain with absolute certainty the true relations [context and association] of these objects." Gibbs had read Morlot and other Europeans carefully. He knew what had been necessary there to establish a deep human antiquity, and it seemed right to begin the search in America with the same approach and even in the same kinds of settings. After all, the New and Old Worlds seemed to have had identical geological histories.

By 1825, Philadelphia naturalist Richard Harlan and others had built on Cuvier's seminal work on Pleistocene mammals and identified and described ten additional extinct North American species.[81] By the 1860s that number had grown larger, though by then differences between the European and American Pleistocene fauna were emerging: here there seemed to be fewer carnivores, and none of the fauna appeared in unstratified drift (the oldest of Pleistocene deposits). Still, the extinct Pleistocene mammals of America, as in Europe, provided a relatively clean chronological marker: the "larger part" of them went extinct at the end of the Pleistocene, and only a few seemed to linger into the Recent period.[82]

Agassiz's Ice Age arrived in North America in advance of the 1840 publication of his *Etudes sur les Glaciers*, and before he himself made landfall in Boston. (In 1847 he took a position at Harvard University's newly created Lawrence Scientific School.) By 1839, his faraway "genius" had already convinced American geologist and paleontologist Timothy Abbott Conrad that there had been a temperature plunge at the commencement of the "Diluvial epoch." The cold created an immense body of glacial ice that scratched, grooved, and "polished [the] surface of the rocks in Western New York" and transported granite boulders great distances south from their source.[83]

Naturally, when *Etudes sur les Glaciers* made its American debut it cast a "flood of light" by which geologists like Amherst College's Edward Hitchcock could "look upon our smoothed and striated rocks, our accumulations of gravel, and the *tout ensemble* of diluvial phenomena, with new eyes."[84] Agassiz's theory neatly accounted for a whole host of drift-related phenomena, although it left unexplained the exact linkage between glaciers and those surface features: Hitchcock was certain that glacial ice and water formed the drift; he was uncertain which "exerted the greatest influence."[85] In this he was not alone.

Into the 1860s, North American geologists still entertained the possibility that drift had formed from massive flooding triggered by earthquakes, or perhaps had been deposited by icebergs floating over vast, submerged portions of northern North America (an idea Hitchcock long before showed was at odds with itself, since icebergs required cold temperatures, submergence warm temperatures).[86] Still, by then most were convinced it was glaciers moving on land that afforded "the best and fullest explanation of the phenomena over the general surface of the continents," one that encountered "the fewest difficulties"—however difficult it was to establish a cause for cooler temperatures and glaciers themselves.[87]

By the 1860s it was no longer possible to speak, as Hitchcock had twenty years earlier, of the "exact identity of the facts in relation to drift on the two continents."[88] The many extant European glaciers testified to an episode of glaciation that, however attenuated, continued to the present, making European glaciation not the well-defined period that it was North America.[89] Still, even though geologists quibbled over such differences, they believed in the "general synchronism between the post-tertiary or quaternary system of Europe and the United States."[90] The geological column in the 1860s showed the Cenozoic (the last part of the geological record) divided into two periods: the Tertiary, consisting of the Eocene, Miocene, and Pliocene epochs (populated by only extinct species), and the post-Tertiary Pleistocene, in which nearly all terrestrial species were extinct but a fraction of the invertebrate species were still living. The Cenozoic was then followed by the most recent era, the "Age of man."[91]

From this point of geological harmony it was but an easy step to believe human prehistory was similar on both continents, as John Foster supposed in his 1870 Presidential Address to the American Association for the Advancement of Science (AAAS):

> With the evidence before us that both hemispheres have been subjected to the same dynamic causes, and peopled by the same races of animals, often identical in species, is it not philosophical to infer that here we shall be able to detect the traces of man and his works, reaching back to as high an antiquity as on the European continent?[92]

Given such similarities, and as human remains were found in the European drift, it seemed only a matter of time before they were uncovered in like Pleistocene circumstances in America. Or earlier. Foster thought the traces of humanity on each continent might reach back to the Miocene, a not-unreasonable suggestion in the post-Darwinian light, when the Pleistocene was viewed as a single and discrete episode, and little was known of when and where humanity had first emerged (or what those geological periods meant on an absolute time scale).[93]

Perhaps. But Gibbs had already seen a daunting challenge to demonstrating a deep human antiquity in America. In every excavation, he urged would-be collectors to take the "utmost care . . . to ascertain with absolute certainty the true relations of these objects." The type, depth, and thickness of strata should be observed, along with the depth and circumstances of all artifacts and bone (which were to be kept separate by strata). These cautions were necessary so as not to "confound ancient with modern utensils" [artifacts], which, as Gibbs and everyone else in America knew, was very easy to do.[94]

2.3 RUDE AMERICANS?

Establishing human antiquity in Europe opened a vast gap in prehistory: having got humans back into the Pliocene, it was necessary, as Stocking put it, to get them back out "without introducing the hand of God."[95] What had transpired over that intervening span was little known, testament to the vast "chasm which separates the flint folks from ourselves."[96] They had the outlines of artifact evolution: the rude drift implements anchored the series, followed by the ground and polished stone artifacts of the Neolithic and then the metal tools of the Bronze and Iron Ages. The sequence was confirmed by the artifacts themselves, their stratigraphic position, and their context: no rude implements, for example, were ever "discovered either in a tumulus, or associated with remains of the later Stone Age."[97]

To paint the fuller picture of prehistory, to conjure the images of the people who fashioned those primitive tools, built houses alongside lakes and created vast middens of shell debris, archaeological eyes turned to cultural data on contemporary "savages."[98] The new breed of prehistorians, most of whom were good uniformitarians (as might be expected, as they were usually geologists), took to heart the catechism "the present is the key to the past."[99] As Cuvier used comparative anatomy to reconstruct extinct animals, so too comparative ethnology could be used to reconstruct extinct cultures. Ethnology was the "thread of induction in the labyrinth of the past."[100] The thread was tied at both ends to artifacts:

> If we wish clearly to understand the antiquities of Europe, we must compare them with the rude implements and weapons still . . . used by savage races in other parts of the world. In fact, the Van Diemaner [Tasmanian] and South American are to the antiquary what the opossum and the sloth are to the geologist.[101]

For Old World prehistorians, stone tool use among the "savage races" of America filled gaps in time and helped reconstitute the lives of the earliest humans. Contemporaneity in space was converted into succession in time, by rearranging the cultural forms existing in the Victorian present along an axis from simple to complex, savage to civilized, rude to refined, and ancient to modern.[102] Julia Wedgwood, Darwin's niece, saw "something dreary in the indefinite lengthening of a savage and blood-stained past."[103] However dreary or theologically unnerving a deep human antiquity might be, those rude artifacts were also vivid testimony of the savage depths from which humanity had climbed, and for many they became a triumphant demonstration of cultural evolutionary progress.[104]

The situation was vastly more complicated in America. Here the past seemingly survived, "a hoary hermit, to the very verge of the newest creations of nature and the latest institutions of man." Stone Age artifacts, including ones reminiscent of those from the European drift, lay on the surface, as "the peoples that made and used them have not yet entirely disappeared."[105] Stone tools of the "savage races" of America were essential to European efforts to fill in the details of their evolutionary sequence, but it was virtually impossible to do that in America: here, obviously, primitive artifacts were not necessarily old.

Matters were complicated still further by Squier and Davis's discovery of such artifacts in the mounds and earthworks of the Ohio valley.[106] Whether constructed by a lost race of Mound Builders (as Squier and Davis supposed), or built by ancestors of the American Indian (the more accepted notion by the latter nineteenth century), the mounds and their implements were obviously older than any in use by contemporary Native Americans. Though the mounds might be old—between 2,000 and 4,000 years old, some supposed—they were still a far cry from Pleistocene in age.[107] No one would "dream of assigning" these earthworks "to a period bearing any relation to that of the Drift Folk of Abbeville or Hoxne."[108]

As for America's drift deposits, none had yielded any rude artifacts. Even if they had, the mere fact that they resembled European ones was not obviously testimony of comparable antiquity. It might mean only that "the early condition of man is everywhere essentially the same, while the rude implements which are obtained from them indicate a similarity of wants and an identity of mental characteristics by which these wants are supplied." Patterns similar by virtue of "common wants and instincts of humanity," it was understood, revealed little of common age or shared ancestry.[109]

Worst of all, Charles Rau and Daniel Wilson realized after reading ethnographic accounts of stone tool manufacture and examining prehistoric stone quarries (at Flint Ridge, Ohio, among other places) that rude forms might just be "rudely-blocked flints fresh from the quarry." They might be no more than the discarded debris of people "already highly skilled in the art of chipping flint."[110]

All this meant, of course, that the mere presence of primitive artifacts was not proof of a deep human antiquity, because by themselves they could not be assigned any particular time period.[111] There was no escaping the realization "that what is called the stone age is not a period of absolute time, but a stage of civilization, long past in one portion of the earth, while existing at present in another."[112] In general, relative rudeness in stone tools might reflect evolutionary grade and chronological age. In Europe it was both: Paleolithic artifacts were old *and* the remains of a past evolutionary stage. That much was proved by stratigraphy (rude artifacts were found deep in Pleistocene deposits) and context (no rude Paleolithic tools were found in Neolithic sites). In America, relative rudeness said something about evolutionary grade, but the message was perplexing (if not upside down), and it said nothing—so far, at least—about age.

Even so, if not all rude tools were old, could some of them be? Even if there were no primitive tools in the drift, was it possible America's earliest inhabitants reached this continent in the Pleistocene?

2.4 LOOKING ANEW

In the wake of the discovery of a deep human antiquity in Europe, scientists and scholars in the 1860s in America took another look at claims that had been made in prior decades but dismissed for a variety of reasons, not least the lack of a conceptual context in which such claims could be accepted (and the strong theological reasons to deny them).

Out of the discipline's closet came Albert Koch's mastodon. Two decades earlier, Koch, a German immigrant explorer and part-time museum operator, found fossils of the extinct *Mastodon giganteus* (syn. *Mammut americanum*) mired in muck in Gasconade County, Missouri, where it had been "stoned and buried by Indians." The bones, some burned, lay in an ash layer intermingled with burned wood, large pieces of skin, broken spears, and rocks weighing two to twenty-five pounds.[113] This unexpected discovery puzzled Koch's readers, including the Benjamin Sillimans (senior and junior), who published his report in the *American Journal of Science*. Over the next two years Koch excavated more mastodon remains at a spring locality along the Mississippi River (at what is now the Kimmswick site, though he found no artifacts there) and then in western Missouri along the Pomme de Terre River.[114] At the latter he uncovered, beneath the femur of what he called the "Missouri Leviathan" (another mastodon),

> an arrowhead of rose-colored flint, resembling those used by the American Indians, but of a larger size. This was the only arrowhead immediately with the skeleton; but in the same strata at a distance of five or six feet, in a horizontal direction, four more arrowheads were found.[115]

Such claims convinced few of Koch's contemporaries, who found them "deficient in responsibility."[116] His reports were occasionally cited but rarely believed.

In the late 1850s Koch again pleaded his mastodon case at the newly founded St. Louis Academy of Science, but his claims were once more rejected by critics still convinced the artifacts and mastodon bones on the Pomme de Terre were fortuitously associated.[117] Koch lived long enough to learn of the discoveries in Europe of human artifacts and extinct Pleistocene fauna (he died in 1867), and though the idea of the contemporaneity of humans and mastodons was "certainly much less improbable" than it had been, it still did not seem as though his proof was satisfactory.[118] Even those who later defended the legitimacy of the association admitted that because the Missouri artifacts were "Indian" in manufacture (Koch himself said so), it meant the humans who were contemporary with the extinct fauna were "more advanced . . . than [their] savage brothers of the European drift period, a circumstance which favors the view that the extinct large mammalia ceased to exist at a *later* epoch in America than in Europe."[119]

A similar find (and fate) was experienced by Montroville Dickeson. In October of 1846 at the Philadelphia Academy of Natural Sciences he exhibited fossil bones he had collected from a "tenacious blue clay" underlying the "drift" (likely, loess) east of Natchez, Mississippi. Among the fossils were two extinct ground sloths, *Megalonyx jeffersoni* and *Mylodon harlani* (syn. *Glossotherium* [or *Paramylodon*] *harlani*); *Mastodon giganteus;* and "an Ursus, a Bos, two species of Cervus, one or two species of Equus." Most important, there was also a "fossil *os innominatum* [pelvic bone] of the human subject." It had been recovered two feet below "the skeletons of the megalonyx and other extinct genera of the quadrupeds."[120] The human bone, Dickeson argued, "is strictly in the fossil state." He insisted it could not have tumbled into the position in which it was found, as it occurred in deposits below the ground sloth (*Megalonyx*) remains and in strata that showed no "displacing power" (evidence of disturbance).[121] He closed his announcement by promising to return to Natchez to "further elucidat[e] this important question."

As it happens, Lyell was headed that way too. The English geologist was on his second tour of the United States and swung through the southern states, where he joined Dickeson on a visit to Natchez.[122] There Lyell learned that the human pelvis had not been dug out "in the presence of a geologist, or any practiced observer," and grew suspicious. Perhaps it had merely fallen into the ravine from "some old Indian grave near the top" of the cliff, where Dickeson then collected it along with the older bones of the Pleistocene fauna. That would make the human pelvis no more than 2,000 years old, far younger than the 100,000 years it would have to be were it actually found in situ at the cliff base. Because Lyell in the 1840s still staunchly resisted the idea of a deep human antiquity, that explanation suited him just fine.[123] Otherwise, the Natchez pelvis implied a human presence "before the last great revolution in the physical geography of this continent," coexistent "with one group of terrestrial mammalia and, having survived [their] extinction . . . seen another group of quadrupeds succeed and replace it."[124]

Some twenty years later, in the wake of the establishment of human antiquity, Lyell confessed that his earlier assessment of Natchez was driven by a strong bias "as to the antecedent improbability of the contemporaneous entombment of man and mastodon than any geologist would now be justified in entertaining." Even so, in 1863 he was still reluctant to concede Natchez had any great antiquity, certainly not comparable to anything in Europe. As long as there was only this isolated case, he argued, and without the testimony of a geologist who saw the bones in place, "it is allowable to suspend our judgment as to the high antiquity of the fossil.[125] Others agreed.

So it went with all finds that had been made in prior decades: a human skeleton found in New Orleans estimated to be 57,600 years old; human jaws, teeth, and bones from Florida cemented in a "conglomerate of rotten coral-reef limestone and shells" that Agassiz thought were at least 10,000 years old;[126] a "fragment of Indian pottery" found with the bones, teeth, and tusk of mastodon, horse, and deer in "Post-Pleiocene" [*sic*] peat and sand deposits along the Ashley River in South Carolina;[127] hearths in "ancient valley alluvium" near Scioto, Ohio;[128] deeply buried stone artifacts along the Cache La Poudre River, Colorado, and in California's "auriferous" gravels;[129] copper spearheads, older than those of the mound builders, found in the "later drift" of Canada;[130] basketry from below mastodon bones on the island of Petite Anse, Louisiana;[131] and most spectacular (and infamous) of all, a human skull from a mine shaft 153 feet below surface (and underneath multiple beds of lava, tuff, and gravel) in Calaveras County, California.[132] Finds of deeply buried artifacts and features—whether stone, copper, basketry, or pottery—might appear old by virtue of their great depth or association with extinct fauna, but their age was problematic, given the "almost marvelous" abundance of "flint and stone implements in the virgin soils of the New World." Because all these allegedly ancient artifacts so closely resembled those in use by modern Indians, was it not more probable, Wilson reasoned, that they simply fell from the surface in the process of excavation or erosion? All "carry with them evidence inconsistent with any geological antiquity." Besides, Foster added, there was no "recurrence of finds in the same deposit . . . which [in Europe] was so conclusive as to the existence of man as contemporary with the great pachyderms."[133]

Least convincing of all was the Calaveras skull, which in Josiah Whitney's opinion—and because he was the California state geologist, with a joint appointment at Harvard University, his opinion ought to be authoritative—was Tertiary in age. However, it had been found under questionable circumstances, and when the skull was sent to Harvard anatomist Jeffries Wyman, he declared that it showed no signs of belonging "to man inferior to the present Indian inhabitants of California."[134] The Smithsonian's Gibbs agreed with Wyman but suggested to Joseph Henry that the Smithsonian invest the small sum of one hundred dollars "to excavate the loose soil, break up the floor of the cavern, and ascertain whether there be not others beneath & with what they are associated."[135] Henry declined, perhaps because by then the air was rife with rumors that the find was little more than a miners' hoax played on the stuffy "Easterner" (Whitney), and the discovery was widely viewed with "extreme distrust."[136] Many even doubted the skull belonged to an Indian. As Bret Harte ended his poetic spoof "To the Pliocene skull":

Which my name is Bowers, and my crust was busted
Falling down a shaft in Calaveras County
But I'd take it kindly
if you'd send the pieces
Home to old Missouri.[137]

In the end, nearly all those weighing the evidence for a deep human antiquity arrived at the same cheerless conclusion.[138] However predisposed they were in principle to accept a deep human antiquity in America, in every instance "where we descend below the alluvium" (into Pleistocene strata), Whittlesey grumbled, "we are thus far met by conflicting testimony as to the facts."[139] That was hardly a firm base to American prehistory: "No solid and enduring scientific fabric can be reared on doubtful premises," Newberry cautioned.[140] On the whole, it seemed to Lubbock that though the idea of a deep human antiquity "is certainly much less improbable than it was some years ago, there does not as yet appear to be any satisfactory proof that man co-existed in America with the mammoth and mastodon."[141]

Even if humans and the extinct fauna were contemporaries, that still might not make them very old, and certainly not older than the "antique flint hatchets of St. Acheul."[142] Most of the evidence "adduced for the antiquity of the American man," Wilson concluded, "has a singularly modern aspect."[143]

2.5 WHERE TO NEXT?

Joseph Henry had had great hopes for an archaeology of the deep human past. Here was the means by "which it will be practicable to reconstruct by analogy and strict deduction the history of the past in its relation to the present."[144] By the 1870s his hopes had faded. He learned, as others had, that American prehistory was not going to emerge readily by European approaches or fit its interpretive schemes. The tantalizing archaeological parallels between the Old World and the New were proving misleading: shell heaps and rude implements hardly meant here what they did over there. The complexities of a continent of living "Stone Age" peoples saw to that. However similar the rude artifacts of Europe and America, they occurred, as Rau observed, "under totally different circumstances."[145] American artifacts were readily classifiable by function but not, as in Europe, by age.[146] Gibbs's 1862 *Instructions* had become irrelevant.

In early 1878, just a few months before he died, Henry prepared with Mason a new *Circular in Reference to American Archaeology,* which, as noted, sought a broader range of archaeological evidence but no longer evidence of the archaeology of human antiquity.[147] As he shepherded the *Circular* through the press, Henry took time to provide popular science writer John Short with material and illustrations for his forthcoming book *North Americans of Antiquity*. It was a book mostly about the Mound Builders, although Short recited—it was now virtually obligatory—the usual claims for a great human antiquity. He dismissed all of these, however, as not "truly scientific."[148] Nothing new here.

Yet, one passage in Short's book provoked a sharp reaction from Frederick Ward Putnam, curator at Harvard's Peabody Museum of Archaeology and Ethnology (to whom Short presented his book in December of 1879). Short placed the Indian on the scale of progress "midway between" the Paleolithic and Neolithic, just as earlier writers had. To Putnam this was "a confused statement and does not express the true condition of things." He then added, "We certainly have a period of rude stone implements—Paleolithic period, and a later one of polished stone & finely made flint impl[ements] etc."[149]

That would have been news to Short, and to Henry and Mason, too. In fact, it may have surprised many at the time. Putnam knew better than to claim an American Paleolithic solely on the basis of rude artifacts or their similarity to European Paleoliths. That would require showing, as in Europe, that rude artifacts were indeed different from and never associated with those that came later, and that they had been found in geological settings like the European drift that testified to their great age. By the time he read Short's book, Putnam believed that Charles Conrad Abbott—geologist Timothy Abbott Conrad's nephew—had done just that.[150]

CHAPTER THREE

Establishing the American Paleolithic, 1872–1881

IN 1872, CHARLES ABBOTT, a naturalist and collector in Trenton, New Jersey, took the first tentative steps toward establishing an American Paleolithic. Over the next ten years his archaeological collecting and writing intensified, particularly as he came under the wing of Harvard University's Frederic Ward Putnam, who provided financial aid, moral support, and scientific respectability at a time when Abbott badly needed all three. Buoyed by that patronage, Abbott laid the foundation for an American Paleolithic.

Abbott showed that objects he was plucking from gravels in the Delaware River valley near his home were the result of human manufacture, unlike those used by the historic Lenni Lenape who had once occupied the region. Abbott's objects were rarely found in burials or sites alongside any Indian artifacts. These were not traces of ancestral Native Americans, either. Instead, they must have been made by an unrelated people. Judging by their similarity to European paleoliths and how deep in gravel deposits—possibly glacial in age—he was finding his specimens, Abbott concluded these were from an American Paleolithic race, one comparable in age and evolutionary grade and likely descended from European Paleolithic peoples.

Those were striking conclusions, and they caught the attention of many others besides Putnam in the small archaeological and geological communities, several of whom proved important in aiding the cause. Chief among them was the Reverend George Frederick Wright, an Andover, Massachusetts theologian, and an up-and-coming figure in glacial geology who seized on the challenge of ascertaining the age of Abbott's finds. It was no easy task. Trenton was south of the southernmost limit of glacial advance by some sixty miles, and in the Delaware valley at Trenton there were multiple gravel layers. Which of these were deposited by, and thus the same age as, the distant glacier? Which were deposited in more recent times as the Delaware River brought to Trenton gravels dislodged from a long-abandoned glacial moraine? And how did Abbott's paleoliths—found in the uppermost layer of the stratigraphic stack—fit into that sequence and relate to the broader history of North American glaciation?

By the late 1870s it was becoming evident—or at least had risen to the level of discussion and the first stirrings of debate—that glacial history may not have been the singular event Agassiz had once supposed (§2.1). There were theoretical reasons for thinking so, based on James Croll's idea that changes in the earth's orbit led to variation in the amount of solar radiation reaching the surface, which alternately enhanced or inhibited glacial growth. And there was empirical evidence, in the thick peat bed found between two glacial drift deposits across the Midwest.[1] The upper of the drift deposits marked what Thomas Chamberlin, then Wisconsin state geologist, soon dubbed the "Kettle Moraine," which traced a sinuous path through the Great Lakes region and appeared far less eroded and weathered than the terminal moraine previously mapped well south of that region. Chamberlin suspected the Kettle Moraine represented a second, later episode of glaciation, an idea that proved a hard sell in New England, whose geologists were skeptical of Croll and saw no such separation of glacial features on their landscape.

Wright, whose geological experience to that point was entirely in the glacial fields of New England, sided with his geological brethren. He also had deeper reasons to resist the call for multiple glacial episodes, reasons rooted in a personal theology that was willing to compromise on some aspects of science and evolution, yet was stubbornly resolute on others. As a result, the age of Abbott's paleoliths landed in a quiet tug-of-war between Wright and Philadelphia geologist Henry Lewis over whether the gravels in which they were found were glacial in age, and if so what that age might be. Wright won the round, but this was only the harbinger of a much fiercer battle to come when Chamberlin made his way east.

That was still a few years away. So far as Abbott was concerned, these were problems for the geologists to resolve, though to him the outcome mattered little. Most geologists who examined the ground at Trenton considered the gravel in which he had found his paleoliths to be glacial in age. That and their typological similarity to paleoliths from the "river drift" of Europe (and their distinctiveness from Lenape artifacts) made for a compelling case that the American Paleolithic was glacial in age.

In January of 1881, less than a decade after Abbott first suggested there might be an American Paleolithic, his Trenton evidence took center stage at a meeting of the Boston Society of Natural History (BSNH). The city's scientific elite—Putnam, Wright, and archaeologist Henry W. Haynes among them—rose to bear witness to Abbott's discoveries and testify to the legitimacy and antiquity of the American Paleolithic. In less than ten years Abbott went from Trenton to Boston, and in so doing traversed the even greater distance from the periphery of archaeology to its center. Abbott's professional triumph unfortunately did little to quell his chronic financial troubles. In those years there were ample opportunities to contribute to archaeology but few chances to benefit from it, and Abbott's personal situation remained precarious.[2] It was as though he never left Trenton—which, in fact, he had not. Still, it was Abbott who took the elusive idea of a deep human antiquity, so recently abandoned by Joseph Henry and Otis Mason (§2.5), and brought it back. It was Abbott, along with Putnam, Haynes, and Wright, who saw that the future of American archaeology appeared to be deep in its geological past.

3.1 CHARLES ABBOTT BUILDS THE FOUNDATION

Charles Conrad Abbott formally trained as a physician at the University of Pennsylvania, strongly influenced there by the anatomist and vertebrate paleontologist Joseph Leidy, who introduced Abbott to the activities of the Academy of Natural Sciences of Philadelphia (ANSP). By inclination, interest, and temperament, Abbott was a natural historian and archaeologist; his first papers on those topics were presented at the ANSP while he was still in his teens.[3] Although he possessed an MD degree, Abbott had neither the caring attitude for nor an interest in earning a living as a physician.[4] But he needed to make a living. The third of three sons, he had lost the primogeniture race and had to work at farming and various odd jobs, all the while attempting to support his family writing on natural history and archaeology.[5]

There was little difference in the manner in which Abbott approached those topics, for he was first and foremost an observer of the birds in the woods and meadows, the crustaceans and fish in creeks and rivers, and the artifacts found eroding from the banks of the Delaware valley and its tributaries. His natural history was not in the spirit of Darwin; or in new fields then emerging, such as ecology, biogeography, and evolution;[6] or in studies that sought to understand human history within the theoretical frame of evolution by natural selection or cultural evolution. European archaeologists like John Lubbock, Gabriel de Mortillet, and Augustus Henry Lane Fox Pitt-Rivers endeavored to link prehistory to an evolutionary natural history.[7] Abbott was aware of those intellectual developments, but his work was largely untouched by such larger theoretical currents.

Rather, as I have detailed elsewhere,[8] his approach hearkened back to an earlier time and tradition, that of the gentlemanly naturalist who observed and catalogued pieces of nature. His natural history aimed at producing a lyrical, firsthand description intended to illustrate nature's charms and intricacies, all the better if it involved a species new to science. Over the years he reported on glimpses of the mind of birds (their feelings, possible thoughts, and communications were perennially favorite topics of his), and the traces of voices in fishes. His archaeology was scarcely different: a typological description of unique artifacts—such as specimens he designated "turtle-backs" because of their shape—and their chronological place. The odd specimens he found, whether from the drift or elsewhere, were no different from the odd behavior he observed among birds or fish. They were all objects to be examined and described. Understanding how those pieces might fit into a broader scheme of human or natural evolutionary history was not Abbott's interest nor care. His theoretical framework was an expedient one, in which principles like the "theory of the gradual progress of mankind" were used to accommodate his facts to the European sequence of human prehistory.

Abbott's earliest publications in archaeology and natural history appear in the ANSP's *Proceedings*, and after 1870 he began submitting his work to the newly founded *American Naturalist*. That journal had been created by a group of Louis Agassiz's former students who left Harvard in rebellion against their mentor over employment rules he had imposed, and because of his staunch anti-Darwinian stance. Frederic Ward Putnam, then in exile at the Peabody Academy of Science in Salem, Massachusetts was one of those mavericks.[9]

Putnam and Abbott's acquaintance by way of natural history soon led to a realization of their common interest in archaeology, and Abbott began sending Putnam artifacts he had collected around his family home perched on the bluffs overlooking Watson's and Crosswicks Creeks, tributaries of the Delaware River.[10] Putnam encouraged Abbott and over the next five years published a half-dozen of Abbott's archaeological papers, in which Abbott established the foundation of his understanding of the archaeology of the Delaware valley.

Abbott well recognized that the Stone Age in the Delaware valley included artifacts that could be ascribed to the Lenni Lenape who occupied New Jersey in prehistory,[11] for they were "a people who had at no time a knowledge of metals."[12] Hence, the term "Stone Age" by itself did not necessarily connote great antiquity (§2.3). Yet, among his collection was a small percentage of artifacts, such as "turtle-back celts," that appeared more crudely made than those found in sites that could be ascribed to Indians.[13] Were these specimens actually artifacts? Abbott was certain they were: several were found together, and "each proves the human origin of the other, and that collectively they show that they are true drift implements,"[14] ones that clearly resembled implements from England, France, and Danish *kjokkenmoddings*.[15] Or so Abbott thought on the basis of illustrations in volumes such as Sven Nilsson's *Primitive Inhabitants of Scandinavia* and John Lubbock's *Prehistoric Times*. He was confident his specimens were similar and sent a number of them to Lubbock himself. (Lubbock was largely noncommittal in reply.)[16]

As for what accounted for those "ruder" forms, Abbott supposed there were either "many execrable workmen among their tool makers; or [their] age . . . far exceeds that of the finely wrought relics." Both rude and more finely made types were found on the surface, yet Abbott could "scarcely imagine that a people who could fashion the latter, would deign to utilize the former."[17] With that, he made the inferential leap Charles Rau and Daniel Wilson had warned against (§2.3). The comparative crudeness of specimens must reflect their antiquity:

> Take a series of whatever class of relics you may, there is always a gradation from poor (primitive) to good (elaborate), which is an indication, we believe, of a lapse of years from very ancient to more modern times, from a palaeolithic to a neolithic age.[18]

Abbott did not ignore Rau and Wilson's warnings altogether. He considered the possibility that these apparent paleoliths were merely "roughly-outlined pieces intended for future more finished work,"[19] or possibly were not unfinished but rather were themselves "cones" (cores), from which flakes for arrowheads were detached.[20] Yet, he did not think either possibility likely because

> the present general character and dimensions of the great bulk of them renders additional chipping impracticable. The larger "rude implements" would,

if further chipped away, form axes and lance-heads of much smaller dimensions than the great majority of those that we now find.[21]

But if that was so, why had geologist Theodore Comstock found in Yellowstone two specimens that were "evidently identical" to one of Abbott's own turtlebacks?[22] Comstock deemed the Wyoming specimens "merely the rejected pieces which have been spoiled during the process of manufacturing more perfect implements, or, in some cases, perhaps they were pieces from which smaller arrow-heads have been chipped."[23] He added that there were other reasons for doubting the remote antiquity of these specimens, notably that the same types of rude implements were still in use among western tribes and routinely found in grave sites of recent age. Yet, whatever the antiquity of Comstock's Yellowstone specimens, Abbott insisted his Trenton specimens were different, neither rejected nor "spoiled":

> The fact that the same forms occur in New Jersey, associated with others of scarcely more definite shape, and *not associated* with "smaller arrow-heads," is evidence, I claim, of their being finished implements. Again, if "failures," is it probable that there would be that uniformity in shape and size, which obtains among them? Thirdly, their outline suggests no other form of implement, such as we know.[24]

Abbott insisted that at Trenton the "rude implements" occasionally occurred near the surface with Indian implements, yet were "never found in [Indian] graves, or in any graves that we have examined," which ought to have been the case were they made and used in recent times.[25] For every "Paleolithic implement found upon the surface, a hundred are quite deeply imbedded in the soil, and in the underlying gravels."[26] Of course, that raised the question of why any of the ostensibly older implements were near the surface at all, but Abbott had an answer for that.[27] Geological processes and the collecting activities of later Indians who perhaps saw the specimens as "venerable relics of a departed people" could account for their occurrence. Naturally, the rude artifacts invariably exhibited, Abbott supposed, "a greater degree of weathering" indicative of their antiquity.[28]

Likewise, there were certain types of artifacts seemingly unique to each, much as certain fossils marked specific geological time periods. Turtlebacks in Abbott's experience were never associated with Indian tools, and seemed, as he later put it, "the most primitive form of the chipped implements of the gravels" (Figure 3.1).[29] In turn, certain "common Indian relics" never occurred deep in the soil alongside turtlebacks. That seemed good evidence that the rude implements were not just Indian artifacts that had tumbled from the surface, for if they were one would also expect other "Indian [stone] relics" as well as their pottery to occur in the same layer of gravels, but they did not.[30]

And, of course, of greatest significance was the apparent similarity of these rude implements to those of Paleolithic Europe. Jeffries Wyman, curator of the Peabody Museum of Archaeology and Ethnology at Harvard University and the recipient of more than 800 specimens from Abbott, agreed:

Figure 3.1 One of the first "rude," ostensibly American Paleolithic implements reported by Charles Abbott (from Abbott 1877a: Figure 1). Abbott found it in the "undisturbed gravel of the bluff facing the Delaware, at a depth of six feet from the surface," and put particular emphasis on the term "undisturbed," "inasmuch as in all cases of the finding of these chipped implements, on the face of a bluff, it is necessary . . . to determine that such specimens occur in the gravel as it exists when first exposed, and not in a talus that may have formed at the base of the bluff." He was quite certain this and other paleoliths he was finding were not from talus debris (Abbott 1877a:32–33).

There are also several implements which, as Dr. Abbott states, very closely resemble the celts of the drift period of Europe, especially those found at St. Acheul, two or three of which, except for their material, could hardly be distinguished from them.[31]

Distinct though these artifacts were, Abbott was not yet ready to declare that the rude implements represented a different people, for he saw in the series "an unbroken line of development in the manufacture of tools" from primitive turtle-backs, hatchets, and spears leading up to the more modern stone weapons.[32] The makers of rude implements must have been ancestors of the Indians, and in "a period of immense duration [since] the puzzling red man passed from the Paleolithic to a Neolithic condition." This was in keeping with what he understood to be the "theory of the gradual progress of mankind," as had been demonstrated in Europe, though Abbott recognized that "the older Stone age of the American race or races does not date back as far into prehistoric ages as is probable in other continents."[33]

Yet, that did not necessarily mean they were recent in age. In 1873 Abbott spotted three artifacts in the river drift of the Delaware valley. All "were taken from this gravel at great depth, and all *beneath undisturbed layers of fine sand*"— none resembled ordinary "Indian relics."[34] Abbott concluded that the "true drift implements [must have been] fashioned and used by a people far antedating the people who subsequently occupied this same territory . . . and we must admit the antiquity of American man to be greater than the advent of the so-called 'Indian'."[35] Abbott's assessment of their geological context, however, was based only on seeing them exposed on the eroding bluff-face of the Delaware valley.

The more drift implements he found, the more Abbott became convinced there was no continuity in technology between them and the more elaborate artifacts of Indians. The former were simply too different in form and finish, raw material, surface weathering, and artifact type.[36] Abbott continued to accept the general principle of the "progressive" and "gradual improvement" of artifact form over time—the "nearer the surface, [the] finer the finish"[37] is how he put it—but he came to reject the idea that the makers of the drift implements and the Lenni Lenape were related. He saw no evidence of "an unchecked development [or] gradual merging" of the two, believing instead these were traces of distinct peoples.[38]

If by Abbott's reckoning the Indian artifacts were as much as 3,500 to 4,000 years old, then by their form and stratigraphic position the drift implements must be a great deal older. Abbott supposed they might even date from a time "not long after the close of the last glacial epoch."[39] As to when that was, Abbott looked to James Croll and James Geikie, who, on the basis of calculations of astronomically driven changes in incoming solar radiation, put the end of the glacial period at 80,000 years ago.[40] If the drift implements were being made then, the Delaware valley had a very deep human antiquity indeed.

Its first occupants had also apparently been its longest, for their artifacts were not only deep in the drift but also sometimes found close to the surface, indicating they had lasted until the arrival of Indians. That in turn suggested what might have become of the makers of the drift implements: Indians, Abbott believed, were "a usurping people," and he thought it likely that on their arrival they drove off the last of the drift implement makers.[41] But driven to where? Abbott looked at North America's native peoples for clues to who might have had ancient ancestors in the drift. He found who he was looking for in the Arctic, although he got there via Paleolithic France.[42]

It seemed to Abbott his Delaware valley implements exactly reproduced those of the "Reindeer Period" of the French caves, the artifacts of which Edouard Lartet and Henry Christy illustrated in their recently published *Reliquiae Aquitanicae*.[43] Lartet and Christy thought the artifacts from this period were similar to ones still in use among the Eskimo, and there were likewise "many circumstances illustrative of the resemblance between the conditions and habits of the modern Esquimaux and these cave-dwellers of France." (It is hardly surprising they drew that connection, as the Eskimo hunted caribou [reindeer] as well.)[44] Abbott connected the dots:

> If, therefore, the rude implements of the Delaware Valley gravels resemble those of the caves of France, and [if] the French troglodytes were identical (?) with the Eskimo, it is fair to presume that the first human beings that dwelt along the shores of the Delaware were really the same people as the present inhabitants of Arctic America.[45]

Some, such as Augustus Grote of the Buffalo Society of Natural Sciences, complained that linking such widely scattered groups could not be done by "the character of the stone implements" because there is "likely to be a similarity in implements between different races, at the same stage of culture."[46] Still, he and others agreed with Abbott's bottom line that the Eskimos are the "existing representatives of the man of the American Glacial epoch."[47]

3.2 FREDERIC WARD PUTNAM COMES ABOARD[48]

In 1874 Putnam left Salem for Cambridge to replace Wyman as curator at Harvard's Peabody Museum.[49] It was a position he had long coveted but not one into which he was warmly received, as Curtis Hinsley has shown. Although Putnam was endorsed for the position by Asa Gray, Harvard's preeminent botanist, there were questions about his competence. Harvard's President Charles Eliot doubted Putnam had the "literary sufficiency" for the task, given he had gone through the Lawrence Scientific School and not the traditional Harvard College curriculum. Besides, Putnam's participation in the revolt against Agassiz meant he had left Harvard without a degree. (More than a dozen years after he had been elected to the National Academy of Sciences, and the same year he was president of the AAAS, Putnam was still having to justify his lack of academic credentials.)[50]

Compounding Putnam's challenge, Harvard had accepted the gift to establish the Peabody Museum only reluctantly: "I have always been of the opinion," Harvard's then-president wrote, "that when a generous man, like Mr. Peabody, proposes a great gift, we should accept it on his own terms, not ours." They did, but it meant Harvard did little to support (or even notice) the museum (into the mid-1880s Harvard made no mention of the Peabody Museum in its own *Annual Report*), and little support came from the city's wealthy donors—the critical lifeblood of any museum—who were far more interested in classical civilizations of the Old World than in what North American archaeology had to offer.[51]

One of the few positive notes for Putnam: when the museum was established in 1866, George Peabody refrained from putting specific conditions on his bequest, save for one: "In the event of the discovery in America of human remains or implements of an earlier geological period than the present, especial attention should be given to their study, and their comparison with those found in other countries."[52] Perhaps with Abbott's help and collections Putnam might fulfill Peabody's bequest and make his own mark. Given what was arrayed against him, Putnam would need to tread carefully, but that suited him just fine: he was by nature cautious. In 1875 Putnam appointed Abbott as a field assistant in archaeology at the Peabody Museum. Although the position accorded more respect than salary, Abbott was glad enough for the respect. But he could have used the salary: "Got a letter from Putnam. What could I not do, had I more money?"[53]

On the first of September 1876, on their way to the Centennial Exposition in Philadelphia, Putnam and his family detrained in Trenton to visit Abbott and see the source for the artifacts he had been sending the Peabody Museum.[54] Abbott, "too excited for [his] own good," met the train,[55] and soon he and Putnam were searching the bluffs of the Delaware below Trenton's Riverview Cemetery for drift implements. Three were found, and Abbott was "as positive [about their] paleolithicity as Putnam was cautious."[56] Then it was on to Philadelphia the next day, where they toured the Exposition and participated through the week in its archaeological convention.

Abbott quickly had grew restless and bored with this "one subject [archaeology] that has been uppermost for so many days." Partly it was listening to others speak in "glittering generalities" about archaeology on a wide canvas, while he could reply only of what he knew and was interested in: the archaeology of the Delaware valley.[57] Even so, he took comfort in being "the only man living [who] knows anything about it."[58] Then too Abbott could see where he stood. Those like Putnam who could speak to a range of archaeological issues made their living in the professional ranks, but for Abbott archaeology was "not in any sense professional."[59] He saw all too clearly his livelihood had to "come from the farm and can never come from any scientific activity,"[60] and he felt trapped:

> The lack of capital makes me unequal to carrying out the business of a farm, especially as the expenses are considerable, due to losses at the outset, and the fact that I am not physically able to do what is known as farm work. . . . As to my scientific tastes, they must be curbed, as they do not bring in any revenue. It is a hard lot, but I cannot change it. It is grin and bear it to the end.[61]

Putnam provided financial help, but it was meager fare: between June 1876 and January 1879 Abbott received $583.25 as recompense for his collecting and collections, and a brief stay at Harvard organizing the collections. Putnam may have been impressed by what Abbott was doing, but when it came to dispersing the Peabody Museum's meager funds Abbott was a low priority.[62] It was his own fault. He had created a buyer's market. When Putnam moved to Harvard he brought the substantial collection of artifacts Abbott had sent to Salem, instantly providing the Peabody Museum with a large assortment of Delaware valley material culture at no cost. Given that this area was well represented in the museum's holdings, it made more sense for Putnam to divert his scarce funds to a collector like Edward Palmer, who was ranging across the Southwest picking up Pueblo ethnographic and archaeological materials, which arguably had greater display value.[63] Better these than another case of Trenton's stone tools, for however old those might be (and Putnam still was treading carefully) they were hardly attractive to museum visitors.

Putnam's support of Abbott did have the strategic benefit (from Putnam's perspective) of insuring Abbott's loyalty, and on occasion Abbott had to apologize to other institutions for being unable to send specimens because he was on a "retaining fee" from the Peabody Museum.[64] As the years went by Abbott came to resent that relationship: the Peabody Museum "squeezed the juice from the lemon, and now throws the skin away. I suppose I am only the skin, now, to them."[65] Nonetheless, Abbott was reluctant to sever his ties to the Peabody, for it gave his efforts a measure of scientific respectability. In turn, Putnam needed to keep Abbott somewhat close, as he was a ready source of archaeological collections and perhaps might one day make a discovery that fulfilled the Peabody's bequest.

Yet, as Hinsley observes, their professional marriage was ill-fated from the start. Putnam, ever-conscious of his new status as curator, was anxious to make no mistakes. If Abbott's Trenton paleoliths were indeed the traces of the earliest Americans, Putnam could bask in the glory. But he also had much to lose. At Harvard he was responsible for the scientific reputation of a discipline as well as a museum, and he was keen to be appointed to the long-vacant Peabody Professorship in anthropology. All of that would be jeopardized if work he endorsed proved to be scientifically unsound.[66] Inevitably, Abbott chafed: "I sent to Putnam what I believe to be four Paleolithic implements. I think Putnam is skeptical still in the subject, but I am not."[67] Just as Putnam had to be cautious, Abbott never could be: not about the meaning and significance of the artifacts from the Delaware gravels.[68]

3.3 FIRMING UP THE STRUCTURE

Fearing the irrepressible and ambitious Abbott might embarrass him, Putnam sought to control Abbott, who was hardly the submissive type. Putnam had the upper hand as Abbott wrote up his Peabody Museum collections for publication, and no doubt owing to his influence the first of Abbott's two reports published by the museum (in 1877) was given the very prudent title "On the Discovery of *Supposed* Paleolithic Implements from the Glacial Drift in the Valley of the Delaware River, Near Trenton, New Jersey."[69] In those reports, Abbott addressed a lineup of tough questions about what he now routinely called "paleoliths." He did so partly to calm Putnam's jitters, but also to convince that wider community of archaeologists who spoke in "glittering generalities" of the importance of Delaware valley archaeology.

In these reports he once again explicitly rejected the possibility that these rude implements were natural, observing that naturally flaked stones lacked sharp points, edges, or other attributes or patterns characteristic of bona fide artifacts. They may appear artifact-like in shape but lack the "finishing" (he had in mind the multiple flaking) of genuine artifacts. Were they a product of nature, they should occur in far larger numbers than the relatively few Abbott collected. Finally, were they naturally flaked, they would not "admit of classification into a primitive form," as turtleback celts, hatchets, spears, and scrapers.[70] Harvard geologist M. E. Wadsworth, though he had no particular expertise in the matter, agreed, and Abbott was duly pleased to record his approval.[71]

So too, he dismissed the possibility that they were unfinished. For by then Abbott believed he knew what unfinished paleoliths looked like: he had collected several in the Delaware valley. They possessed the "same general character of weathering and of chipping" as finished drift implements and were easily spotted as "broken specimens" or manufacturing "failures" of the same origin and age. Even in their unfinished state they exhibited the "peculiarities" that "mark the differences between the Paleolithic and Neolithic forms."[72] Coincidentally, just as Abbott's Peabody Museum reports were appearing a geologist with the Hayden Survey was exploring the Yellowstone region.[73] Like Comstock before him, William Henry Holmes came upon scores of obsidian artifacts he identified as "imperfect, as if broken or unfinished."[74] Holmes did not consider their implications for Abbott's paleoliths. Not yet, anyway.

In Abbott's typological approach, age was an inherent and observable attribute of artifacts, and he had no difficulty recognizing those attributes. On a deep enough time scale he was right: truly ancient artifacts were recognizable as such— that much was clear from the European archaeological record. But Trenton was not the Somme valley, and stone tools were still being made in America. Abbott was aware that others might not see his drift implements as he saw them, which could in turn expose him to the criticism that his drift implements were merely recent artifacts that had tumbled into older, deeper strata.[75] Convinced as he was of the uniqueness of these specimens, he paid little heed to that concern.

On the other hand, that concern was very much on Putnam's mind from his visit to Trenton, where he could see that Abbott was collecting artifacts from an actively eroding valley wall (Figure 3.2). Abbott would on occasion dig into a bluff face to get a fresh exposure, but he was not then nor later inclined to excavate down from the upland surface behind the talus slope to determine if artifacts occurred in situ within the gravel. Nonetheless, Abbott insisted that his finds were coming from undisturbed drift deposits and these were Pleistocene in age "inasmuch as the drift deposits throughout New Jersey, are thus ascribed to the action of ice."[76] That was not sufficient for Putnam. On his visit he questioned Abbott's uncle, geologist Timothy Conrad (§2.2), about Delaware valley glacial geology. Evidently Conrad had not given the subject—or his nephew's efforts— much thought. In fact, Abbott had not found any geologists "to take much interest" in his work. Putnam realized skilled geologists needed to be called in.[77]

When Abbott visited Cambridge in November of 1876, Putnam arranged for him to meet with Nathaniel Shaler, a member of Harvard's geology faculty. Shaler listened with interest to Abbott's ideas about the age of the gravel and the significance of his apparent Pleistocene-age artifacts. Abbott was unimpressed: Shaler seemed "erratic and in no way convincing."[78] Still, plans were made for Shaler to examine the sections, and two months later Abbott guided him to the gravel bluffs below the Riverview Cemetery, where he and Putnam had found drift implements apparently in situ six months earlier. This time, Abbott and Shaler found two artifacts in the gravels. More important, Shaler declared the gravel was glacial drift, very slightly modified by subsequent water action. That night, after a long conversation with Shaler, Abbott privately recorded his triumph: "*I have discovered glacial man in America.*"[79]

Indeed, in his report Shaler went even further. Granting there was considerable "sliding action" of the slopes of the valley where Abbott had found his paleoliths,

Figure 3.2 Charles Abbott searching for artifacts on the banks of the Delaware River, NJ. Date and photographer unknown. Courtesy of the Peabody Museum of Archaeology and Ethnology, Harvard University, 2004.29.1271 (digital file no. 98970060).

Shaler was convinced the specimens could not have slid down from the surface, as they were "really mingled" with the drift in which they had been found. Supposing the age of the specimens predated the deposition of the drift, he concluded that "if the remains are really those of man, they prove the existence of interglacial man on this part of our shore."[80]

Still reluctant to venture out on that limb, Putnam sought a second opinion, asking another Harvard geologist, Raphael Pumpelly, to visit Trenton. Pumpelly did so in April of 1877, but departed far less confident than Shaler had been on the central question of whether the paleoliths had been found in situ. He strongly advised Abbott that when collecting an artifact he should

> determine that such specimens occur in the gravel as it exists when first exposed, and not in a talus that may have formed at the base of a bluff . . . for in a talus, it will be readily seen, that a chipped implement might have very recently fallen from the surface, and now be buried several feet [down] from the face of the bluff.[81]

Pumpelly might have been leery of the eroding bluffs on which Abbott was finding his paleoliths, but Abbott saw "no cause yet, to materially modify my views." He was certain the drift specimens had come from "undisturbed gravel of the bluff facing the Delaware."[82] Nonetheless, Abbott played along and duly considered the circumstances of each find and concluded "no such displacement evidently had taken place."[83] After all, both the gravel and the talus yielded "only the rudest forms of chipped implements" and none of the "ordinary Indian relics."[84] That was a telling point, for

> if these rude forms were of identical origin with common Indian relics [that is, both having tumbled down from the surface], then rude and elaborate alike; jasper, quartz, porphyry and slate together; axes, spears, pottery, and ornaments, all of which are found upon the surface, should have gradually gotten to those depths. Any disturbance that would bury one, would inhume [all]. Such, however, is not the case.[85]

On to the nub of the matter then: if the artifacts were the same age as the gravel, how old was the gravel? George Cook, the New Jersey state geologist, thought they were immediately post-Pleistocene in age.[86] That would make the Trenton paleoliths older than any other human traces found in North America (save Whitney's Calaveras skull), but Abbott was not buying the assessment. Instead, he steadfastly maintained that the gravels were glacial in origin, and thus "man dwelt at the foot of the glacier, or at least wandered over the open sea, during the accumulation of this mass of sand and gravel." He believed this in spite of the geologists' assessments, because the archaeology showed that artifacts "occur at different depths in the gravel" and hence they must have been "made, used, and lost" as the deposit formed. These could not be strata that had been extensively reworked subsequent to their original deposition by the glacier.[87] As to why the gravels appeared water-laid, Abbott had a ready answer: they had been deposited not on the ground surface but in shallow glacial meltwater.[88] The artifacts "demonstrate the presence of inter-glacial man upon the Atlantic coast of our continent; a point in geological time so distant that we are scarcely able to realize it."[89]

That was old. Very old, but hardly surprising to Abbott. After all, "like indications of such vast antiquity are not wanting elsewhere," especially in places like Brixham Cave, where William Pengelly had reported traces of "inter-glacial, if not pre-glacial" human remains.[90] And Yale paleontologist O. C. Marsh had just announced in his August 1877 vice-presidential address to the AAAS that there seemed reason "to place the first appearance of Man in this country in the Pliocene." Not to be outdone, Marsh's bitter rival Edward Cope claimed several months later to have found "numerous flakes with arrow and spear heads of obsidian" in a lake basin in Oregon associated with a Pliocene age fauna.[91] Why not Trenton too?

3.4 THE TRENTON PALEOLITHS GO PUBLIC

Putnam was now in: "From a visit to the locality, with Dr. Abbott, I see no reason to doubt the general conclusion he has reached in regard to the existence of man in glacial times on the Atlantic coast of North America."[92] Even Daniel Wilson was impressed. In his address as chair of the anthropology subsection of the AAAS, delivered just days after Marsh's talk and a few months after the first of Abbott's Peabody Museum reports appeared, Wilson again warned that "isolated and dubious" examples of Paleolithic artifacts must "ever be received with caution . . . on a continent where the indigenous population have not even now ceased to manufacture tools and weapons of flint and stone." Yet, he thought Abbott's work was different: it was the result of "systematic research, based on the scientific analogies of European archaeology." This was the European model, the proper model, as Wilson saw it: identify the tool-bearing drift by its geological characteristics, rather than identify the tools of the drift independent of geological evidence. Abbott had seemingly done just that.[93]

Many others thought so, including William John McGee (WJ, without periods, as he insisted), a young blacksmith turned geologist from Farley, Iowa. Attending his first AAAS meeting the following summer (August 1878), McGee presented the results of his survey of the near-surface (Pleistocene) formations of the northeast corner of the state. In what McGee identified as interglacial deposits were pieces of wood that had been "accurately squared," the prow of a canoe, and a "discoidal stone" artifact."[94] McGee soft-pedaled the finds:

> It is not by any means intended that these instances shall be looked upon as evidence of the contemporaneity of man with the ancient forests whose remains we find so abundantly in this stratum; but a geologist does not perform his whole duty who neglects to record data because they do not happen to verify other observations or strengthen existing generalizations.[95]

Nonetheless, he expected to one day prove such traces were valid, as artifacts had already been found in "the precise equivalent of this stratum" in Europe and by Abbott in the "unmodified glacial drift of New Jersey."[96] McGee expressed surprise there were "still doubts as to the pre-glacial age of man on this side of the Atlantic, notwithstanding the seemingly conclusive evidence afforded . . . by Dr. C.C. Abbott."[97]

All this was just talk; seeing was believing, and soon a stream of visitors descended on Trenton. It started in September of 1878, when Lucien Carr, the Peabody Museum's assistant curator, and Harvard geologist Josiah Whitney (of Calaveras skull fame [§2.4]) arrived from Cambridge. Whitney pronounced Abbott's artifacts Paleolithic and the gravels in which they were found "part and parcel of the glacial conditions once existing." Abbott thought Whitney provided "an authoritative decision on the subject." Authoritative, perhaps, but not final. On the train back to Cambridge later that day, Whitney had second thoughts and concluded the gravels must be *Pliocene* in age, thereby making its paleoliths the same age as the Calaveras skull.[98]

Of course, Whitney had spent less than a day at Trenton; Shaler's and Pumpelly's visits were about as brief. Their "hasty conclusions," as well as those of New Jersey State Geologist Cook, rankled Henry Carville Lewis of the ANSP. Unbeknownst to Abbott, Lewis had been painstakingly mapping the geology around Philadelphia and up the Delaware valley to Trenton since the spring of 1877. On June 2, 1879 they encountered one another:[99]

> [Lewis] from a geological standpoint, working upward, to the present; and I [Abbott] pushing my researches backward, from the historical point, met upon common ground, and each in total ignorance of each other's labors, until our respective studies brought us face to face."[100]

Over the weeks that followed they visited sections together and explained their results to each other. Lewis had identified in the Delaware valley nine distinct stratigraphic units, the upper four of which were Quaternary (Pleistocene) in age. From oldest to youngest, these were the Philadelphia Red Gravel, the Philadelphia Brick Clay, the Trenton Gravel, and an unnamed recent alluvial mud. Lewis believed the Red Gravel and Brick Clay to be glacially derived. These formations were not, he stressed, part of the glacial moraine. That was some sixty miles north of Trenton (crossing the Delaware River near Belvidere, New Jersey). Rather, the Red Gravel was glacial outwash and the Brick Clay from ponding of glacial meltwater. Lewis tied these local episodes into the broader stratigraphic framework of Yale's James Dwight Dana, one of America's most influential geologists, who divided the Quaternary into three periods: the Glacial, Champlain, and Recent.[101] The Glacial period was an era of uplift and high elevation that led to cold climates that in turn triggered snow and glacial ice (in this, Dana favored Lyell's as opposed to Croll's explanation for glaciation). The Champlain period was one of subsidence that brought on warmer climates and glacial melting and retreat. Most of the melting took place during the earlier "Diluvian" phase of the Champlain and produced extensive sand and gravel deposits, the Great Lakes, and in the retreat of Niagara Gorge the possibility of calculating the time elapsed since the ice pulled back. Assuming a retreat of one foot a year, Dana put the Champlain onset at 31,000 years ago.[102] In the Delaware valley the Champlain was expressed as the Red Gravel and the overlying Brick Clay, making these units a by-product of glacial retreat.[103] Neither of these formations—so Lewis had it on Abbott's authority—had ever yielded any paleoliths. That meant these were not strictly glacial in age, let alone preglacial. Lewis seemed relieved to report this because, according to Lyell's then-current (eleventh) edition of *Principles of Geology*, that would have made them 200,000 years old.[104]

Abbott had found paleoliths in the Trenton Gravel, and had assured Lewis they always and only occurred in that formation, indicating they "belonged to and were of coeval deposition with the [Trenton] gravel.[105] The Trenton Gravel, in Lewis's view, was a "true river gravel," deposited in a channel that had incised into the Brick Clay and underlying Red Gravel. That obviously indicated its deposition was substantially later, perhaps even "postglacial or at least very late glacial" in age.[106] Having marched to the brink of proclaiming the Trenton Gravel and its enclosed paleoliths recent in age, Lewis abruptly backed away (no doubt to Abbott's relief). The Trenton Gravel might be a true river gravel, but its cobbles and boulders seemed far too large to have been the work of any modern river. None, even in flood stage, were capable of carrying such massive debris. But a river fed by a vast glacier could. Lewis deemed it "difficult to assign any other cause [for the Trenton Gravel] than that of a melting glacier."

Yet, how could the Trenton Gravel have been carried by glacial meltwater, if the stratigraphically much older Red Gravel and Brick Clay had also been deposited by outflow from a melting glacier? Lewis could surmise only that there must have been a previously unnoticed "second and more recent glacier in the Delaware valley, the melting of which [gave] rise to the Trenton Gravel."[107] The hypothesis of a second glacier had the added virtue of explaining other "facts observed," including Abbott's having found walrus, reindeer, and mastodon bones in the Trenton Gravel, all indicative of conditions far colder than at present.[108] But it also meant the Trenton paleoliths did not have a "vast antiquity" on a geological time scale. Still, Lewis quickly added, "it does not necessarily follow that [they are] by any means recent."[109]

3.5 SUBDIVIDING THE GLACIAL EPOCH

That Lewis suspected there had been a second glacial period was not altogether novel—at least not outside the confines of Delaware valley geology. By the mid-1860s Scottish geologists had begun to suspect that glaciation was not the one-time event Agassiz originally supposed (§2.1). Archibald Geikie, director of the Geological Survey of Scotland, had glimpsed within glacial tills an organic layer that hinted at a warming period amid what was otherwise glacial cold. Archibald's younger brother James (also of the Survey) expanded that observation in a series of influential papers in the early 1870s, culminating in his 1874 volume *The Great Ice Age and Its Relation to the Antiquity of Man*, in which he collated evidence from Scotland, northern Europe, and North America that glacial episodes had repeatedly been interrupted by episodes of milder climate.[110] That implied, of course, an underlying climatic driver, and for that Geikie relied on Scottish astronomer and geologist James Croll's astronomical theory of Ice Ages first proposed in 1864. (By the late 1860s Croll was also in the Scottish Geological Survey.)[111]

Croll attributed glaciation to changes in the earth's orbital eccentricity and the precession of its equinoxes, which would have altered the amount and intensity of incoming solar radiation and thus the heat received at the earth's surface. Reducing the amount of heat in the winter, Croll believed, would favor greater snow accumulation, which in turn should increase albedo, further the buildup of snow and ice, and ultimately trigger glacial growth. On the basis of his calculations of the earth's orbital parameters over past time, Croll concluded that glacial and interglacial cycles must have occurred at regular seesaw intervals in the northern and southern hemispheres (cooling in one hemisphere, reversed in the other).[112] Croll's theory was problematic in its details but among his contemporaries it proved influential, owing in no small measure to its championing by James Geikie, who, by the late 1870s, confidently announced the glacial epoch "was not one continuous Age of Ice, but consisted rather of a series of alternate cold (glacial) and warm or genial (interglacial) periods."[113]

That seemed to be the case in North America as well.[114] Charles Whittlesey of the Ohio Geological Survey had spotted in the 1860s in the Upper Midwest an organic, peaty deposit (he called it the "forest bed") sandwiched between glacial outwash deposits. In the early 1870s, Edward Orton and Newton Winchell (of the Ohio and Minnesota Geological Surveys, respectively) suggested that the forest bed represented an interglacial period, Orton correlating it with the interglacial in Europe.[115] Some of the most extensive work on the forest bed was done by McGee in Iowa in the late 1870s. McGee mapped its extent and concluded that it formed during a time when there was "a mild climate and favorable conditions to the production of vegetable and animal life."[116] This formation of peat and organic remains occurred throughout Iowa but also, McGee reported, "at many points in Illinois, all through southwestern Indiana, and at many localities in Wisconsin, throughout northeastern Iowa, in Canada, and in many other places." Wherever it was found it appeared to rest "upon true glacial Drift" and was in turn "overlaid by true glacial Drift." Therefore, McGee concluded, it had to be interglacial in age.[117]

The interglacial must have been "a period of immense duration," longer than even Croll himself granted, for McGee reasoned it had to have taken a considerable amount of time for the initial glacier to retreat, for plant communities to colonize the deglaciated plains, for animals to follow in turn, and for the thick mat of peat and organic remains to accumulate.[118] Of course, if that was true there ought to be evidence of more than one terminal moraine, and the separate moraines should be of very different ages (as expressed, for example, in different degrees of erosion and weathering).

Thomas Chamberlin, then the state geologist of Wisconsin, had in fact identified a distinctive "Kettle Moraine" looping across his state. Using his own surveys and the reports of others, he was able to trace it northwest into Minnesota, and east by southeast into Michigan, Indiana, and Ohio, where, for lack of fieldwork, it disappeared from sight.[119] In these areas the Kettle Moraine was situated north (sometimes at a great distance) of the southernmost extent of glacial advance. Thus, Chamberlin reasoned it must have formed *after the retreat of the glacier had commenced, and marks a certain stage of its subsequent history.*[120] He was quite certain that stage was not merely a temporary halt in the retreat of the same, single glacier, but rather the terminus from a subsequent glacial advance. The Kettle Moraine constituted a "definite historical datum line" in the glacial epoch, marking a "secondary period of glaciation, with an interval of deglaciation between it and the epoch of extreme advance."[121]

The idea that there had been more than one glacial period was a harder sell in New England. The Kettle Moraine had headed in that direction, but it disappeared en route as a separate and distinctive feature. Yale's Dana was particularly skeptical that there had been a second glacial episode, not just for want of evidence, but also because he rejected Croll's theory underpinning the idea of multiple glacial periods.[122]

In Trenton Lewis had begun with the assumption borrowed from Dana that the Pleistocene was but a single event. But in the face of the anomaly posed by the Trenton Gravel, Lewis had tentatively abandoned that assumption, perhaps partly inspired by the two-stage sequence emerging in the Midwest. If the Trenton Gravel was from melting that followed a second glacial episode, that would make it a geologically more recent phenomena, but the enclosed artifacts would still be older than any other known in America.[123] Beyond that, he would not venture: "The actual age of the Trenton Gravel, and the consequent antiquity of man in the Delaware, cannot be determined by geological data alone." The only route to determining an absolute age Lewis could imagine, and he deemed it "a most unsatisfactory one," was the amount of erosion evident in the deposits. That coarse measure took the Trenton Gravel back only "a very few thousand years at the most," and was just an approximation that required corroboration "by a much more complete series of observations than have yet been possible."[124]

3.6 ABBOTT'S *PRIMITIVE INDUSTRY*

No matter. Abbott had heard enough to confirm his paleoliths were related to a glacial event, albeit a second such episode, and Lewis, the best of the local geologists, had said so. That was all the incentive he needed. In January of 1880, following a suggestion Putnam had made several years earlier, Abbott began work on a book on Delaware valley archaeology.[125] By late February, he had a publisher. He wrote steadily through the summer, and by September the draft of *Primitive Industry* was finished. It was not an easy time financially for Abbott, and were it not for his father's "substantial encouragement" and help meeting expenses the undertaking at times might have been very tough indeed.[126] He hoped the book would prove critically successful and financially rewarding.

Well it might be both, because the *Primitive Industry* that took shape on Abbott's desk was a transparent imitation of John Evans's acclaimed *The Ancient Stone Implements, Weapons, and Ornaments of Great Britain* (1872). The similarities began with Abbott's first sentence, a near-verbatim copy of Evans's; ran through virtually identical chapter headings and contents (for example, "Stone Axes," "Scrapers," "Plummets," "Spearpoints and Arrowheads"); and likewise ended with chapters on Paleolithic implements and the antiquity of the gravels in which these were found.[127] What differences existed between the two were largely a result of geography and the very disparate archaeological records of Great Britain and the Delaware valley. No matter how hard Abbott tried to mimic Evans, and he tried very hard, those differences were inescapable.

Even so, Abbott was confident that a "systematic examination of the surface geology of New Jersey . . . shows as abundant and unmistakable evidence of the transition from a true Paleolithic to a Neolithic condition, as is exhibited in the traces of human handiwork found in the valley of any European river."[128] To back up that claim, he carefully reiterated and embellished the arguments from his Peabody Museum reports on the genuineness of his paleoliths, their differences in technology and material from the more recent Indian artifacts, their position in undisturbed gravels (as now attested by multiple authorities), and especially their similarity to European drift implements, as noticed even by eminent French archaeologist Gabriel de Mortillet, to whom Abbott had sent several Trenton specimens.[129] (Abbott had sent Trenton paleoliths to Evans as well, but if he received a reply it is lost; to another American correspondent, however, Evans confessed to "not being satisfied as to the[ir] true Paleolithic character."[130])

The relatively great age and primitiveness of the paleoliths, Abbott stressed, marked them as the products of "an earlier and ruder race" who lived at a time when a "less wary fauna" was hunted. Otherwise, their weaponry would have needed to be more sophisticated, like the bow and arrow of the Indians. Not coincidentally, Abbott suggested that New Jersey in glacial times would have been much like Greenland and arctic North America, where Eskimos now lived. And because Eskimos until recently used rude stone implements they were likely "traces of the same people."[131]

Although this presumed relationship of American Paleolithic groups to the modern Eskimo had "been frequently questioned," Abbott paid these criticisms little mind.[132] He admitted, however, that there were complications. William Boyd Dawkins had just announced in his *Early Man in Britain and His Place in the Tertiary Period* that while Eskimos might trace their ancestry to Paleolithic Europe, they were descendants of the "Cave-men" and not the earlier, lower (on the cultural scale), and unrelated race of "River-drift men."[133] This was troublesome because up to that moment Abbott had not needed to be concerned about the finer divisions of the European Paleolithic sequence. His concern was showing the overall similarity between his paleoliths and those of Europe, not whether the latter were from the cave or drift periods.

If there were two wholly separate and unrelated Paleolithic races in Europe, as Dawkins argued, and if he was correct that only the Cave men (and not the River-drift men) were the ancestors of the Eskimo, then could Abbott's Paleolithic people be related to the European River-drift men and to the Eskimo? With some manipulation, they could, and Abbott was not one to shy from a bit of manipulation when the circumstances demanded it.

He began by dividing his American Paleolithic into two (previously unspecified) stages: the earlier were a group of River-drift men, and the later a people who fashioned "more specialized argillite implements." Although the later people did not actually inhabit caves (easily explained by the general scarcity of such features in the Delaware valley), they otherwise displayed "the same improvement in the patterns and finish of stone implements" and all the other distinguishing features of European cave life.[134] They were therefore the chronological and historical equivalent of Europe's Cave men.

Moreover, because the American Paleolithic groups on the one hand, and the argillite spearpoint makers on the other, were related to one another as ancestor

**Table 3.1 Abbott's view of the alignment of prehistoric sequences
of America and Europe**

America	Europe
Paleolithic implements in Trenton Gravel	Flint implements in Pleistocene River drift
Argillite implements of more specialized patterns in alluvial deposits and surface	More specialized flint implements of the caves
Jasper and quartz implements of North American Indian—polished stone	Neolithic stone implements of highest grade—polished stone
	Bronze

Note: As depicted in Abbot's *Primitive Industry* (Abbott 1881b:517).

to descendant, and because the descendants of those who made argillite implements were in turn the ancestors of the Eskimo, "then does it not follow that the
River-drift and Cave-men of Europe, supposing the relationship of the latter to
the Eskimo to be correct, bear the same close relationship to each other, as do the
American representatives of these earliest of peoples?"[135] Abbott was implying,
none too subtly, that Dawkins had failed to see, or otherwise misjudged, the historical and evolutionary connections between the River-drift and Cave periods of
the Paleolithic. It seemed to Abbott quite clear that "the sequence of events, and
the advance of culture, have been practically synchronous in the two continents"
(Table 3.1).[136]

Although the relative archaeological sequence on the two continents might be
the same, Abbott appreciated that the earliest American paleoliths might not be as
old as those in Europe. Moreover, he was swayed by Lewis's recent work, confessing in *Primitive Industry* to having been "seriously misled" by earlier geologists.
Nonetheless, Abbott was quick to add that humans could and likely did occupy
the region much longer than the minimum amount of time required for the deposition of the Trenton Gravel.[137]

In the face of geologists haggling among themselves as to which glacial event (a
first or a possible second occurrence) produced the Trenton Gravel, Abbott staked
his ground:

> Personally, I can but express an opinion on the archaeological significance
> of the traces of man found with these gravel deposits, and this is in nowise
> effected by the age and origin of the containing beds. Whatever age the ge-
> ologists may assign to them, be it inter- or post-glacial, these traces of man
> must possess a very great antiquity.[138]

Still, he hedged his bets: he invited Lewis to summarize his assessment of the geology for the final chapter of *Primitive Industry*. Lewis accepted, but did not write
the chapter on his own. By then George Frederick Wright had become involved.

3.7 THE SOUND OF THE APPLAUSE

Wright was an Oberlin-educated Congregational clergyman with an interest in natural history and geology. In the 1860s he accepted a parish on the edge of the Green Mountains of Vermont, then in 1872 moved to Andover, Massachusetts. In both places he spent his leisure time exploring the countryside. Though largely self-taught, he began to acquire a reputation in glacial geology, producing well-regarded studies of New England's kames and moraines. These works brought him to the attention of Boston's scientific community, which centered around the BSNH and included many of the Harvard science faculty as well as other members of the city's educated upper echelon.[139]

One of those was Henry W. Haynes, a Harvard-trained attorney and onetime professor of Latin and Greek at the University of Vermont who spent the 1870s remaking himself as an archaeologist. Haynes traveled extensively across Europe and Egypt to examine archaeological sites and in the fall of 1880 was planning his next trip, this time to visit Abbott. Wright was invited along. Coincidentally, Dawkins was in Boston that fall of 1880 as part of an American lecture tour promoting his *Early Man in Britain*. When Haynes told him of the planned excursion to Trenton, Dawkins invited himself along.[140]

The trio arrived in Trenton on November 17, 1880, and spent three days with Abbott and Lewis "relic hunting" in the gravels.[141] Several paleoliths were found and by visit's end Abbott was elated:

> *Dawkins gives as his opinion that I have made, unquestionably, the discovery of Paleolithic man in the Delaware Valley. He enthusiastically endorses my position, and I need no greater authority to express my opinion.*[142]

Dawkins's enthusiasm quickly waned, however. He went on to Philadelphia immediately afterward and at the end of a day at the ANSP wrote Haynes a "most uncomfortable letter," in which he told of learning that Abbott had tried to pass off an artifact as coming from the Trenton Gravel that had, in fact, been found in Arkansas. That knowledge

> render[s] it impossible for me to believe anything on his [Abbott's] unconfirmed testimony. . . . Under the circumstances I am impelled to fall back on the probabilities that the implements found by you & myself on the face of the bluff really are derived from the sand or the gravel, and the possibility of those which were uncovered before me in the ballast pit having been *in situ*. I feel very much hurt at having to write this, for many reasons.[143]

In England two months later, Dawkins again wrote Haynes of his uneasiness. The Trenton paleoliths appeared to be genuine, but by then he was skeptical about their context, even the ones he and Haynes had found: "I feel strongly inclined to believe that Pal[eolithic] Implements have been found in that gravel, but I do not feel inclined to declare that I saw them dug out from undisturbed strata. I can merely say that the strata seemed to me undisturbed. I wish I could say more, for was not Abbott kindness itself to us?"[144] Dawkins kept his misgivings private, however.[145]

Two months after that Trenton visit the Delaware valley paleoliths were the subject of a special symposium on "The Paleolithic Implements of the Valley of Delaware," held at the BSNH (where, not coincidentally, Putnam was vice president and Wright and Shaler were among the councilors). Putnam had provided a warm-up act two weeks earlier, displaying at the Society paleoliths from the gravels of France, alongside the now "pretty well known paleolithic implements from the glacial gravel at Trenton." However, the symposium on January 19, 1881 featuring Abbott, Haynes, and Wright was the main event, and it proved to be the Society's best-attended meeting of the year (outdrawing the ever-popular annual meeting).

It was also one of the more unusual meetings of the year. Rarely was an entire evening at the BSNH devoted to a single subject, and almost never did its auditorium take on the air of a revival tent, with powerful members of Boston's scientific elite coming forward to bear witness to what they had seen at Trenton. The testimony began immediately after Abbott finished outlining the history of his decade's work at Trenton. First Haynes, then Carr and Putnam testified they too had removed paleoliths from the deep in the Trenton Gravel with their own hands. There was no doubt in their minds the paleoliths were not only genuine, but genuinely like those of Europe. Carr had submitted several of Abbott's paleoliths to "leading archaeologists in London, Paris and Copenhagen [none were named], all of whom unhesitatingly confirmed" their status as paleoliths.[146] Haynes cited his own good authority, believing his years abroad looking at collections provided "opportunities for this kind of study [that] have been unusually great."[147]

If that style of witnessing was unusual, it was also understandable. Archaeology, unlike geology or zoology (the disciplines that dominated the BSNH), was a discipline without a body of theory or method and few established ground rules or procedures. In the circumstances, the evaluation of a discovery and its transformation into evidence depended largely on the integrity of the individual making the claim and, more important, on the testimony of those who bore witness. Wright understood this point well when he emphasized how Abbott's discoveries had been independently confirmed by Lewis and corroborated by many others, then used the fact to praise Abbott: "Nothing but the trustworthiness and accuracy of the discoverer could have secured this discriminating accuracy."[148]

Haynes noticed that some of the Trenton paleoliths were ruder than those of Europe, but he attributed that difference to their being fashioned of argillite, as opposed to the more easily chipped European flint. That distinction notwithstanding, the types of tools were the same on both continents: America had hand axes like those of St. Acheul and "entirely unlike the ordinary Indian axe,"[149] leaving no doubt in Haynes's mind that "the rude argillite objects found in the gravels of the Delaware river, at Trenton, New Jersey, are true Paleolithic implements."[150]

The testimony that night in Boston, as even a cautious Putnam agreed, served to "clear away all doubts as to the importance and reliability of Dr. Abbott's discoveries and investigations."[151] The work had been certified, and could no longer go unacknowledged by those archaeologists who spoke only in "glittering generalities." Abbott's paleoliths were no longer a local Trenton story. Haynes offered one caveat, however: though he thought it "absolutely and incontestably established" that the paleoliths had come from the gravel beds, "there needed to be more discussion and determination of the true geological character of the gravels of the Delaware valley."[152] Wright was happy to oblige.

After Haynes and Dawkins had departed Trenton the previous November, Wright stayed behind and spent two weeks with Lewis examining the gravel and checking Lewis's mapping of the stratigraphy. Wright was impressed, agreeing with Lewis that the Trenton Gravel was separate from the Philadelphia Red Gravel and Brick Clay by "a period of vast physical changes, if not vast time."[153] And he agreed that the Trenton Gravel was laid down at the close of the glacial period. But was there a second glacial stage represented, as Lewis supposed? Here Wright and Lewis parted ways.

Wright, a staunch advocate of the unity of the glacial period, supposed the Trenton Gravel was deposited when glacial ice from the first and, in his mind only, glacial advance (the Champlain) had retreated north of the Catskills, but while "the still swollen stream" draining it was carrying debris from moraines higher in the valley.[154] He must have been persuasive. When Lewis finished a draft of his chapter for *Primitive Industry* he sent it to Wright to "make any alteration in text or sense by adding or omitting sentences without hesitation," and then forward the manuscript on to the publisher. Lewis did not ask to examine any of Wright's proposed changes.[155] Wright jumped at the offer.

Although Lewis and Wright agreed the Trenton Gravel was postglacial, the question still came down to postglacial relative to *which* ice retreat, and how old in years that might be. Since his initial reports on the Delaware valley, Lewis had heard from Chamberlin, who was encouraged by the possibility that his Kettle Moraine and second glacial epoch might be extended into eastern North America. Chamberlin agreed the Philadelphia Red Gravel and Brick Clay were from the first and more extensive glacial period, and the Trenton Gravel from ice retreat of the second. Lewis duly reported Chamberlin's statements in his chapter, though through Wright's influence or editing the idea of a second glacial episode became "an open question."[156] As Lewis (or Wright) summarized the matter:

> The Trenton Gravel, whether made by long continued floods which followed a first or second glacial epoch, whether separated from all true glacial action or the result of the glacier's final melting, is truly a post-glacial deposit, but still a phenomenon of essentially glacial times—times more nearly related to the Great Ice Age than to the present.[157]

Lewis retreated on another front as well. He had been previously unwilling to say anything more than that the Trenton Gravel might be several thousand years old, and he did that only reluctantly, believing there were no geological methods by which one could make more precise estimates. (Lewis was not alone: Dawkins thought attempts to put actual numbers on geological time in the Delaware valley were "so obviously futile that it was not a little strange to find them seriously made."[158]) In *Primitive Industry*, Lewis repeated his earlier warning about the absence of methods for measuring geological time in years, but then proceeded to delineate several ways to do just that.[159]

Lewis's abandonment of his previous position was so abrupt that this portion of his chapter can plausibly be attributed to Wright's heavy editorial hand (if not authorship). For the chronological methods discussed, deriving ages from the retreat of Niagara and St. Anthony's waterfalls and the accumulation of sediment and peat in glacial lakes and kettle-holes, were methods Wright himself advocated as useful for determining the "date of the earliest evidence of man's appearance at Trenton."[160] Yet, no matter the method, the result was invariably the same for Wright: the glacial period ended recently, and human antiquity was therefore recent as well. To understand why he reached that conclusion requires knowing the theological presuppositions Wright brought to his research in geology and human prehistory.

3.8 THE CREED OF GEORGE FREDERICK WRIGHT

Wright was profoundly religious, having assented to the creed of his Congregational Church at the age of twelve, where he was soon regularly reading the weekly sermon. He came from a family that ardently opposed slavery, and when he reached college age (in 1855) he followed his older siblings to Oberlin College. In the decades before the Civil War, Oberlin was a hotbed of abolitionist activity, a main station on the Underground Railroad, and it matriculated—as few colleges did then—"colored students on the same terms as those granted to white." It was a progressive institution in many other ways as well, a "seething pot of religious, social, education, and political reforms," which every winter scattered its students through the upper Midwest to teach at rural schools and spread antislavery sentiment.[161]

When the Civil War broke out in 1861 Wright joined Ohio's Seventh Regiment of Volunteers. He was not long for the war, however, having contracted pneumonia in camp, and he was permanently mustered out before his regiment left to fight. He returned to his graduate work in Oberlin's Theological Seminary, but like those of his classmates who served, Wright's views of life and of his religious, social, and political duties were forever influenced by the war.[162] That influence was manifest in a progressive philosophy that celebrated humanity's great moral purpose and potential, tempered by the staunch belief that its moral purpose was divinely created and inspired.[163] The source of that belief was the evangelical Christianity of his youth and Oberlin's seminary, but it was not a strict fundamentalism. Late in life Wright vividly recalled an early encounter on a New Hampshire lake shore with an encampment of Millerites who had gathered to wait for the Second Coming. He lamented the inevitable loss of faith that followed when the event failed to occur, and thankfully remembered the theological moderation of his own church.[164]

That moderate upbringing put him in good stead when, on graduation from the seminary in 1862 he accepted his first parsonage in Bakersfield, Vermont. There he used his evangelistic methods, supplemented by "more systematic and prolonged efforts in a variety of directions," which for Wright meant reading widely in philosophy and natural history. Darwin's *On the Origin of Species* was on his list, as was Lyell's just-published *Geological Evidences of the Antiquity of Man*.[165] These were books "pressing for attention," and like many contemporary churchmen Wright sought to accommodate the new science without undercutting his theology.[166] He found his solution in Mills's levels of causation: final cause, the ultimate explanation, came not from science but instead "from the assumed wisdom and goodness of God." "The man of science," he wrote in the 1870s, "does not *prove* divine goodness and wisdom; he *studies* their manifestations."[167] Wright insisted this approach was no hindrance but instead

> leaves range enough to satisfy the roving propensities of any reasonable explorer in the realms of natural science. It allows Darwin and his opponents to fight out on purely scientific grounds the battle between their theories. . . . Let them open the book of nature and read, *only they must not be allowed to read too much between the lines*.[168]

This was an attractive accommodation to many, and his saying so got Wright noticed in both theological and scientific circles—the latter including Harvard botanist Asa Gray, himself a Darwinian, albeit a theologically rooted one. Hearing of the young preacher, Gray requested a meeting, and they soon forged a strong friendship built around crafting a theologically infused interpretation of Darwin's evolutionary theory. Wright became a confidante and collaborator of Gray's.[169]

It was in this intellectual context that Wright's geology emerged. Over his years at Bakersfield he spent a great deal of time traveling to attend circular conferences of preachers in this widely scattered district and to work off the "blue Mondays" that followed his busy Sundays. On his rounds he paid close attention to the topography and geology of Vermont's Green Mountains, partly as grist for his sermons: "The interpretation of such natural phenomena by teachers and pastors to their pupils and parishioners should be regarded as a bounden duty."[170] But it was also because he was genuinely curious about the phenomena he was seeing. Although he knew relatively little of geology at that time, his observations of what proved to be glacial deposits made him something of a local authority, and Dartmouth's geologist Charles Hitchcock began to consult with him on geological matters.[171]

Such attention would turn any head, even a pious one, and soon Wright sought wider audiences (theological and geological) and a larger salary. In 1872 he moved to the Free Church in Andover, Massachusetts, the home of the Andover Theological Seminary, then a major force in orthodox circles and the home of Edwards A. Park, editor of *Bibliotheca Sacra*, "the main scholarly expounder of New England theology."[172] Joining this local association of Congregational ministers, Wright was "at once plunged into the midst of theological and scientific discussions."[173] Wright's erudition impressed Park who, recognizing the new minister's scientific interest, asked him to prepare for *Bibliotheca Sacra* a series of articles "for and against Darwinism and showing the bearing of that theory upon the doctrine of design in nature, and upon theological opinions in general." Wright did, then expanded his essays into a volume, *Studies in Science and Religion*, which he dedicated to Asa Gray.[174]

In those works Wright took the position, approved by Gray, that a mechanism such as natural selection could account only for the preservation of varieties and not their origin. Questions of ultimate cause were answerable only in the evident design of nature, a "design either wrought into the original plan, or added by way of increment."[175] Darwinism, in his opinion, was little different from Calvinism, both sharing the idea of a universe set in motion by God, in which sin is ever present and benevolence never guaranteed, and where some subjects are consigned to endless "punishment" (Wright's awkward metaphor for natural selection). The only difference between the two was that "Darwinism has all the unlovely characteristics of hyper-Calvinism without any of the redeeming remedial features."[176]

Where Darwinism and theology collided, of course, was in accounting for people, and there Wright drew the line against evolution by natural selection. He saw good reasons to do so. Following Alfred R. Wallace's lead, he observed that humans in the "savage state" (modern or fossil) had brains much larger than necessary, thereby precluding the possibility that those brains resulted from natural selection, which preserves only variations that are "of positive service at the time of their occurrence" (the implication being that savages have brain power they do not use and hence would not have evolved). Natural selection also failed to explain humans' moral sense, which in Wright's view completely separated humans from all other animals.[177] Finally, there was no fossil evidence linking "the lowest limit in existing men" and the apes.[178] Neanderthals, as Thomas Huxley had observed nearly two decades earlier, did not "take us appreciably nearer to that lower pithecoid form."[179] The most ancient humans were more than mere animals, for the "inventive genius displayed in the rudest flint implement stamps [humans] as a new creation," at least in spiritual terms. "Man is man even in the savage state," Wright proclaimed,[180] for the "size and office of [our] brain" could only have been miraculously bestowed. Wright not only believed in miracles, he saw no conflict between them and science: their occasional occurrence did not necessarily involve disruption in the ordinary uniformity of nature. These were merely God's incremental additions to ultimate design. Or as Gray put it, "now and then, the Deity puts his hand directly to the work."[181]

While Wright envisioned human origins as abrupt and divinely inspired, he did not subscribe to traditional Biblical chronologies that restricted the Creation to just 6,000 years ago. His rationale was straightforward. The Bible deals with eternal truths and was not intended to reveal and explain the course of nature. Its purpose was "to teach us how to go to heaven, and not how the heavens go." The traditional Biblical chronologies, the work of Archbishop Ussher and others, were simply "inferences of particular scholars from very uncertain data." Of past attempts to accommodate those chronologies and geology, Wright thought it difficult to tell which had been most distorted, the rocks or the sacred records.[182]

Still, although he would not force human antiquity into Ussher's short chronology, Wright was none too anxious to have humanity's origins too far back in time and beyond reach of his theology. That unwillingness to relinquish the theological belief that humans were a distinct and geologically recent creation garnered some empirical support in the late 1870s and early 1880s. The European Paleolithic and, perhaps, the American as well reached back to the Pleistocene. But in both, Paleolithic peoples were apparently humans little different from those of today, as Wallace had noted. Deprive the Eskimo of his dog, Wright added, and he would be in "the condition of Paleolithic man in Europe."[183]

Moreover, Wright supposed paleoliths were found only in the very latest Pleistocene deposits. (European archaeologists would have disagreed.) In fact, it appeared to him that paleoliths were found in the gravels and sands left by the "floods marking the close of the glacial period," floods so vast they were suggestive of those of Noah himself. Wright claimed he was speaking allegorically, but he was not averse to readers taking him literally, for he himself thought there was just enough of a possibility these were events that prompted the tales of Noah's flood that he was hedging his bets. (It was a gambit he used throughout his career.)[184]

Finally, playing his Darwinian trump card, Wright argued that the glacial period (and thus the appearance of humans) had to be recent, for "the changes in animals and plants since the glacial period have been so slight, that if the epoch be set too far back eternity itself would scarcely suffice to account for all the divergences which have arisen." Any estimates that shortened the glacial period, Wright triumphantly concluded, should be "as grateful to the Darwinian as to the theologian."[185]

Naturally, Wright could not accept Croll's hypothesis that the glacial period ended as much as 80,000 years ago (§3.1), let alone the corollary that there were multiple glacial events that were themselves of such duration that the span of the entire glacial period would be immense. Fortunately, he saw no need to be "at the mercy" of Croll's astronomical calculations.[186] The similarity of Pleistocene and modern species attested to the recency of the glacial period. So too did the absence of "incontestable evidence . . . in America of any glaciation previous to that of *the* glacial period." He dismissed Chamberlin's Kettle Moraine as merely a line where the ice stalled in its retreat.[187] Wright made no mention of McGee's interglacial peat and forest bed, or of the immense amount of time McGee judged was necessary to its accumulation.

There was also positive evidence for a shorter glacial chronology based on the upstream retreat of Niagara Falls and the Falls of St. Anthony on the Mississippi River in Minneapolis. The channels that drained each of these rivers were incised since glacial times, and because the rate at which incision and the headward retreat of the falls could be determined from historical records (the position of each set of falls a century or so earlier was known), the total distance retreated divided by the rate of retreat yielded a measure of the total time elapsed since the glacial period. In both instances the figure was at or less than 10,000 years.[188] Such estimates, Wright thought, were corroborated by the relatively small amount of erosion of glacial features reported by Chamberlin for the Kettle Moraine (which, of course, Chamberlin suspected was little eroded because it was from the more recent of two glacial episodes), and Wright's own estimate of the infilling time for a glacial kettle-hole in Andover. From its present shape and depth, Wright supposed it had originally been twenty-four feet deeper when it was created "about the close of the Glacial period." If Croll was correct, then the sediment in the kettle had been accumulating for the past 80,000 years; this implied an unacceptably slow rate of filling (an inch every 1,000 years). If, on the other hand, the glacial period ended 10,000 years ago, that was a rate of an inch per one hundred years. Such a rate was "as slow as our imagination can well comprehend" but neatly matched calculations made by Boucher de Perthes of peat accumulation over Roman pottery.[189]

Altogether, it seemed to Wright, the glacial period ended 10,000 years ago, and believing the retreat of the ice was rapid he supposed the Kettle Moraine and the other more northerly moraines would have been deposited "no more than a few hundred years" later.[190] The end of the glacial period, and with it the advent of humanity in North America, was therefore relatively recent, perhaps even reaching down to within sight of the historic period. By the hard numbers provided by Wright and others for the close of the glacial period, it is no surprise Lewis's (Wright's) chapter in *Primitive Industry* concluded that even if "the Trenton Gravel is of glacial age, it is not necessary to make it more than ten thousand years old."[191]

Abbott was not so sure. Ten thousand years seemed hardly enough to accommodate the "several momentous facts in American archaeology," not to mention making it "more probable" that evidence of Paleolithic people would not be found in interior North America because they would not have been able to "wander far inland [before being] met by southern tribes, who drove them northward, exterminated or absorbed them." On the other hand, if humans arrived in preglacial times they would have had ample time to disperse into the continental interior, and—believing that—Abbott was sure their traces would "ultimately be found in the once-glaciated areas of our continent."[192]

3.9 SEEKING HIS JUST REWARD

Scarcely a decade after its traces were first noticed, all was going well for the American Paleolithic—less so for Abbott himself, whose archaeological triumph had little effect on his personal troubles. The scientific community valued his discoveries: even John Wesley Powell of the Smithsonian Institution's newly created Bureau of Ethnology took favorable notice of them in *Science*. Yet that value did not translate into financial security or professional standing. Abbott badly needed the former and desperately wanted the latter. (Praise was not universal: Stephen Peet, editor of the *American Antiquarian*, considered Abbott's evidence inconclusive as to age and used his journal to needle Abbott.)[193]

Abbott was elected an AAAS fellow in 1883 (an election likely orchestrated by Putnam), but it was not enough. He had glimpsed what a professional life might be one memorable evening at Harvard spent in learned conversation with Putnam and others about archaeology and natural history. It starkly highlighted how much the poorer his life was (financially and intellectually) in Trenton, a "purely commercial and fanatical neighborhood."[194] Yet, he was not even successful there. Through the late 1870s and into the 1880s Abbott was "bothered all the time by money matters."[195] In the winter of 1881–1882 he had to accept a low-level clerkship in a Trenton bank.[196]

It was not a position to which he aspired: "I wish to heaven I could be satisfied to be a clerk and a nobody, but I cannot."[197] His days at the bank were dull and even bankers' hours left little time for writing or searching for paleoliths, though plenty of time for feeling sorry for himself: "Up town and at bank until 3 P.M., which seems to wholly use up the day; as I cannot get up early enough to write before breakfast, and am too lazy after I get home."[198] There were many times over the years when he felt like giving up archaeology, and contemplated writing his "farewell address" to the field.[199] When that dark mood struck he sent Putnam mournful pleas:

> Forced out of the ranks of scientific workers, of course you will all very soon forget me, but I have one request to make. Please do not erase my name from the list of recipients of your Annual Reports. It will be a pleasure for me to yearly note your progress.[200]

Even when Abbott's enthusiasm returned, and it usually did, he could not escape the fact that archaeology was "wholly unproductive of pecuniary reward."[201]

Worse, for all that Abbott was sure he had done for Putnam, Putnam had done little to ease Abbott's financial precariousness. Abbott grumbled that Putnam did "nothing but try to cheer with hopes of better times and such stuff," and Abbott put no faith in such hopes.[202] In his unhappiness, he convinced himself Putnam and the Peabody Museum had strung him along all those years and were ready to throw him overboard now that they had his collection.[203]

Abbott's real problem with Putnam was less insidious. He needed attention and assurance, and this Putnam provided but it was never enough: "He [Putnam] cannot find time to answer my brief questions on matters that used to be of importance to him."[204] Yet, Putnam also insisted Abbott not stray in search of archaeological affection. When the Smithsonian Secretary Spencer Baird wrote requesting a donation of Trenton paleoliths, Putnam urged Abbott to do "nothing archaeologically in connection with any other museum" than the Peabody:[205]

> He [Putnam] has much to say about "hope," and "looking forward" &c and that the time may come when the fact that I have stuck to the museum, will result in the museum "sticking to me. . . ." I do not expect to be able to accomplish any work of importance, while tied down to my present position, but I will take Putnam's advice and be an enthusiastic "hoper."[206]

Abbott sent nothing to Baird (two years later Abbott asked Baird for a copy of the Smithsonian's latest *Contributions to Knowledge* volume but was refused outright; Abbott was certain it was because he had not sent Baird any paleoliths and was not about to: "I'll see him dammed first").[207]

And so Abbott continued to hope his archaeological and (especially) his personal fortunes would improve in the coming years. There was good reason to suspect so—at least in regard to the archaeology—for there were already scattered reports being published of Paleolithic sites outside the Delaware valley. In the Schuylkill and Potomac drainages, both of which were thought to be similar geologically to the Delaware, paleoliths were reported "under circumstances pointing to a remote antiquity, although none have occurred at as great depths as at Trenton, N.J."[208] In the coming decade, those reports increased sharply in number, spawned a substantial literature testifying to the prescience of Abbott's vision, and eventually led to a job for Abbott himself.

Yet, so too the difference of opinion over the unity or diversity of the glacial period—which in the late 1870s had only just emerged and was still relatively low key (in part for lack of sufficient empirical evidence)—escalated in the coming decade, particularly after Wright gained temporary employment in the newly created USGS Glacial Division headed by Thomas Chamberlin. Their differences over North American glacial history had significant repercussions for the debate over human antiquity.

CHAPTER FOUR

The American Paleolithic Comes of Age, 1882–1889

SCARCELY A YEAR AFTER he first visited Trenton, George Frederick Wright was invited to the Chair of New Testament Language and Literature at his alma mater, the Oberlin Theological Seminary. Boston friends worried he would be unable to pursue his geological work in this scientific wilderness and would receive little encouragement at home and scarce notice elsewhere.[1] But Oberlin gave Wright the chance to return to his Christian roots, literally and intellectually, and provided new territory in which to search for evidence of a deep human antiquity.[2] Wright moved there in the fall of 1881.

Oberlin's President Charles Fairchild, always "on the lookout for fields of labor in which the various professors might distinguish themselves," knew of Wright's interests in glacial geology and introduced him to Charles C. Baldwin, presiding judge on the Eighth Circuit Court of Ohio and a driving force behind Cleveland's Western Reserve Historical Society. In short order the Society funded Wright to extend his "explorations of the glacial boundary across Ohio and the states farther west."[3]

He did not restrict his efforts to glacial geology. Soon after moving to Ohio Wright announced that although no paleoliths had yet been found, "they may be confidently looked for" in the state's glacial gravels. He urged collectors to search for paleoliths, confident that once they became familiar with their rude form they would "doubtless find them in abundance," and certain these artifacts were worth far more than the "whole mass of Indian implements which now fill our museums." He pointed out where to look, too: along the gravel terraces of the valleys (like the Little Miami) that flowed south from the glacial region and emptied into the Ohio River.[4]

By mid-decade Ohio began to yield paleoliths, as Wright had predicted. It was hardly unique in this respect, for throughout the 1880s paleoliths and other artifacts suggestive of a deep human antiquity—including apparent Pleistocene art objects such as the Nampa Image and the Holly Oak pendant—were reported from the eastern seaboard to the desert West. Naturally, the paleoliths resembled those from Trenton—that was often how they had been identified. Some were found in geological circumstances that suggested a comparable antiquity to Trenton, and others hinted at an older human presence, perhaps dating to the previous interglacial or even the prior glacial period.

Assuming, that is, there had been multiple glacial periods. The 1880s also saw the difference of opinion regarding the number and timing of glacial episodes become more sharply defined. The majority of midwestern geologists, led by Thomas Chamberlin—who in mid-decade became the newly appointed chief of the USGS Glacial Division—considered the question resolved in favor of multiple glacial episodes. Over that same period Wright's glacial studies expanded into the midwestern states, first under the auspices of the Western Reserve Historical Society, then while he was an assistant for the USGS Glacial Division. Although he had been hired by Chamberlin, Wright remained steadfast in his devotion to the idea that there had been but one glacial advance. Almost immediately their relationship soured over geological matters, such as Wright's claim that glacial ice had overridden the Ohio River at Cincinnati and created an ice dam that formed a lake so vast and deep that its waters were still 300 feet deep at present-day Pittsburgh 500 miles upstream. Chamberlin strenuously disagreed, and their relationship began to deteriorate, sped along by Wright's attempt to fashion himself as a public spokesman for glacial geology—a role Chamberlin deemed him utterly unfit to fill.

Yet, even as the geological situation became increasingly contentious, an archaeological consensus was emerging. By decade's end, the AAAS, the BSNH, and the Anthropological Society of Washington (ASW) all sponsored symposia or published papers testifying to a deep human antiquity.[5] The question was no longer whether humans had reached the Americas during the latter part of the Pleistocene, as at Trenton, but just how much farther back in time they may have arrived. An American Paleolithic was almost universally accepted. It was a remarkably smooth coming of age, better than could have been hoped for a decade earlier by even the most optimistic champions of a deep human antiquity. In his vice presidential address to Section H (then comprised of anthropology and psychology) of the AAAS in 1888, Charles Abbott could hardly keep from gloating. When he was hired the following year as the first curator of archaeology at the University Museum of the University of Pennsylvania, it seemed fitting and proper.

The same year Abbott secured his long-coveted position (§3.9), Wright published *The Ice Age in North America and Its Bearing upon the Antiquity of Man*, a volume aimed at explaining North America's glacial and human history to a public audience, and archaeologist William Henry Holmes began work at the Piney Branch site, a prehistoric stone quarry in Washington, DC. Both events signaled the presence of deep fault lines that soon surfaced within the geological and archaeological communities.

4.1 THE PALEOLITHIC COMES IN QUARTZ

Wright's prediction that paleoliths would be found in the glacial drift of Ohio was not, after all, so very bold. In the years since the American Paleolithic debut or, perhaps as important, in the years since illustrations of Abbott's paleoliths appeared in print, new discoveries had been made from the mid-Atlantic region to New England (§3.9). That all were "true" paleoliths seemed obvious: they were identical in form and material (all the same "peculiar variety of quartzite"), were "rude and primitive in the extreme," and one "perfectly" resembled a paleolith illustrated by Abbott.[6] There were doubts about some finds, but that hardly mattered. Traces of an American Paleolithic were spreading and by the early 1880s had crossed the Mississippi River, where a local schoolteacher, Frances Eliza Babbitt, "succeeded in shifting the interest with regard to glacial man from New Jersey to Minnesota."[7]

No mean feat, that, but then Babbitt was no ordinary schoolteacher. Unlike her peers she published in major journals, presented at national meetings, and played an active role in archaeological discussions of the day. That Babbitt did so at a time when women (let alone ones in their mid-fifties) were exceedingly rare in archaeology, and in the sciences generally, makes her brief prominence all the more remarkable.[8] Unfortunately, little is preserved of what motivated her, just that she had long been an enthusiastic student of geology and archaeology.[9]

The vehicle that carried Babbitt from the schoolhouse to the notice of the American scientific community was a discovery initially made several years earlier in Little Falls, Minnesota by State Geologist Newton Winchell. He had traveled through the year before Babbitt arrived in town, and in his *Annual Report* for 1877 he reported finding nonwaterworn quartz chips in water-laid sands a few feet below the surface of a Mississippi River terrace. Winchell thought the artifacts were preglacial in age, because the terrace in which they were found "was spread out by that flood stage of the Mississippi River that existed during the prevalence of the ice period, or resulted from the dissolution of the glacial winter." He was certain the flakes were artificial or at least not natural. Frederic Ward Putnam, who was sent the specimens for a second opinion, pronounced them "formed by the hand of man."[10] But with Minnesota's geology still to be mapped, and a state legislature demanding results, Winchell had to leave the matter there.

Aware of Winchell's finds, Babbitt began collecting in the locality during the winter of 1878–1879. She started in the same deposit as Winchell, but in the coming summer she spotted a rich vein of artifacts near the base of a wagon track notched into the terrace. Here was a find to rival Winchell's. He had called his finds Paleolithic, though Babbitt pointedly observed "the term is often loosely applied to stone remains in a sense quite irrespective of their ancientness" (true enough, but hardly fair to Winchell, who believed that specimens "embraced within the actual drift" were by definition Paleolithic[11]). Babbitt emphasized that her finds were more abundant (about 1,200 specimens), from deeper and possibly older deposits, and still in situ in an "ancient manufactory," making their context "not open to discussion." There was no mistaking these artifacts with those of Mound Builders, or even the quartz "pounded up and strewn about the shores by boys and other idlers" (Babbitt's schoolteacher voice was occasionally heard in her writing).

Babbitt declared the Little Falls quartzes Paleolithic in character and "because they are imbedded, somewhat deeply so indeed, in a drift deposit which is certainly glacial, whatever its other and relative characteristics may yet prove to be."[12] That the terrace appeared to postdate the first of Winchell's glacial epochs but predated the close of the second "destined [Little Falls to be] the object of national study." But sometimes destiny needs a helping hand: after presenting her results at a Minnesota Historical Society meeting (in February 1880), she sent a copy to the *American Antiquarian* to insure a broader audience.

Back home, however, the county historian sounded a sour note. He had visited the site and seen that Babbitt's artifacts were thoroughly mixed with other broken rocks of "no special design." If the latter were not artifacts, were Babbitt's? Given these circumstances, he doubted it.[13] Stephen Peet, editor of the *American Antiquarian*, had second thoughts as well. In a barely anonymous letter in a subsequent issue he declared the Little Falls artifacts to be redeposited surface finds and unlikely to have been "the work of man." If "paleoliths had to depend on such a shallow foundation as was furnished by these alleged discoveries," he scolded, "the matter would better be dropped."[14]

Babbitt realized she needed to fortify the case and sought help from higher authorities. To her dismay Putnam was noncommittal, suggesting a batch of specimens she had sent him "*might* be artificial—possibly." Babbitt tried again, and carefully selected about a hundred specimens to send Henry Haynes, to see if they would pass his authoritative eye.[15] Haynes judged many to be natural forms and others debris "struck off in the work," but he had no doubt some were tools that displayed "marks of use" and the signs of having come from a "workshop of early man."[16] Unlike Putnam he was willing to stick his neck out: at Babbitt's request he sent a statement to add to her presentation at the AAAS annual meeting in August of 1883.

In that presentation Babbitt spoke confidently of the geology and archaeology of Little Falls.[17] She reported the latest thinking of specialists like Chamberlin and others on the geological history and age of the moraine belts of the upper Midwest and related it with the help of geologist Warren Upham (one of Winchell's assistants[18]) to the Little Falls terrace. She fully commanded the archaeological issues too, rejecting a series of alternatives—that the Little Falls quartzes were natural, recent, or the same age as Winchell's—before concluding that the "many particular observations and deductions" forced her to the "working hypothesis" (her term) that Little Falls was an American Paleolithic workshop and cache.[19]

The workshop was on a till surface beneath twelve to fifteen feet of modified drift that formed the terrace; Winchell's specimens were three to four feet from the terrace surface. Babbitt was certain that her quartzes had not slipped down from Winchell's deposit and that there must have been an interval between their deposition, long enough to pile up the overburden of modified drift.[20] Upham's check of the stratigraphy indicated that Winchell's and Babbitt's artifacts were from different levels of what he identified as a glacial floodplain formed during "the retreat of the ice of the last glacial epoch." But, as Babbitt was quick to add, hers came from the bottom of that unit.[21] They were postglacial—very early postglacial.

The Little Falls quartzes were small and found in clusters and included finished and unfinished tools, cores, and "flake blocks" ("preforms," in modern terms) "convenient for transportation and handling, which appear to be designed to be flaked into knives and other implements."[22] Babbitt displayed at the meeting specimens she declared "unquestionably the work of man," but she knew the most compelling testimony that these were artifacts was not their inherent attributes (which, as Haynes admitted, were not always unambiguous) but their context, having been found in clusters that formed coherent assemblages.

Putnam stood after Babbitt's paper to liken Little Falls to Abbott's Trenton Gravel and to proclaim its importance. Heady praise, but it did not last. In the "animated discussion" that immediately followed, Putnam was challenged by Peet. Only a few of the Little Falls specimens showed the slightest signs of human modification, and those that did appeared "accidental rather than artificial." Peet insisted that there was "very little support for their authenticity among the attendees at the AAAS meeting."[23] That was not true, but it left the unmistakable impression that Peet's opinion represented the majority. But Peet could only control what was said in his own journal, and its limited distribution limited his influence.

Babbitt's allies had greater reach. In *Science* Babbitt's work was given top billing in the report on papers delivered in the anthropology session. Parts or all of her presentation appeared there as well as in the *American Naturalist* and the AAAS *Proceedings*. This unprecedented coverage was almost certainly owed to Putnam's efforts (he had editorial positions or influence at all three).[24] An editorial accompanying Babbitt's paper in *Science* praised the Little Falls work for having confirmed Abbott's "theory" that "man existed on this continent during at least a portion of the glacial epoch." Nonetheless, it conceded that there had been "a lively discussion between experts, as to whether these quartz specimens are actual relics of human industry. Thus far, at best, the glacial workman is known only by his chips."[25]

Abbott, clerking in a Trenton bank (§3.9) but still keeping a protective watch over his Trenton claims, would have none of that. Calling his hard-won evidence a theory hit too close to home. Although the *Science* editorial may not have intended for the term to be taken in a pejorative fashion as mere speculation, that was exactly the way Abbott read it. After all, theory in that sense was how Peet used it when speaking of Abbott in the *American Antiquarian*.[26] If Abbott could not very well respond to Peet in his journal, he could at least correct the *Science* editors. He fired off a letter, asking rhetorically what better evidence of deep human antiquity was needed if those chips were of artificial origin: "Are not shavings and sawdust as good evidence of men working in wood?" Abbott did not linger defending Little Falls. He moved quickly to shore up Trenton's ramparts, insisting there was no doubt "overshadowing the existence of man in the Delaware valley as long ago as the close of the glacial period." This was not mere theory, he snarled, it was "a fact susceptible of actual demonstration."[27]

Abbott's sputtering notwithstanding, Babbitt got the message. Even otherwise sympathetic Otis Mason had cautioned against accepting Little Falls too quickly, at least not until after more thorough geological and archaeological study.[28] Babbitt left the geology to Upham; the authenticity of the artifacts she tackled at the AAAS meetings the following year. She arrived in Philadelphia armed with over three dozen specimens. Granted, she argued, these were not typical paleoliths, but there were reasons for that: they were made of quartz, which by virtue of its angular fracture planes was not easily fashioned into ideal forms. For that matter, the angles and points that naturally occurred on quartz meant it did not need much modification to make it suitable for use. It was therefore unreasonable to dismiss the Little Falls quartzes merely because they did not conform to paleoliths made of flint or because they were not obviously tools. The best way to appreciate their human origin, Babbitt concluded, was not by examining individual specimens, any one of which might appear as the "capricious product of natural agencies," but by examining the whole series.[29] There was no fallback this time on their context. Instead, Babbitt relied on the attributes collectively visible, emphasizing their distinctive fracture patterns, traces ("bruises") from their being socketed in horn or wood, and abrasion and fractures resulting from use wear.

Peet was in the audience, but if he had objections he had no chance to voice them. Immediately after Babbitt finished, Abbott leapt to his feet to describe and display several Trenton paleoliths (including ones of quartz). He was keen to avert the type of discussion that had taken place the year before that might do collateral damage to his claims, and to reassert Trenton's importance. It worked. Neither Peet nor anyone in the audience raised objections, and the meeting continued without a repeat of the previous year's debate. Or so *Science* reported. In the *American Antiquarian* Peet insisted the assembled archaeologists still remained "divided in opinion as to the human origin of these specimens."[30]

4.2 LEST TRENTON BE FORGOTTEN

Abbott had other good reasons to jump to his feet. Also in the audience were members of the British Association for the Advancement of Science (BAAS), which for the first time in its history had convened in North America (in Montreal). Anthropologist Edward B. Tylor, then president of the BAAS anthropology section, was among the attendees who traveled on to Philadelphia and had spoken to the AAAS audience on "Some Aspects of American Anthropology." In his address Tylor wondered, as many had before him, how to account for the rich diversity of Native American languages from the limited number presumably spoken by the first arrivals. Unless we can prove "an antiquity sufficiently remote to allow time for the strange diversity of tongues to have occurred," he wrote, "our perplexity is great."[31]

Abbott urgently wanted Tylor to know there was no need to be perplexed: Trenton provided the antiquity he sought. Abbott must have been convincing. After the AAAS meetings Tylor joined him for a day searching for paleoliths on the banks of the Delaware River. Abbott was honored, though Tylor later confessed privately he "did not take" to Abbott's site or its "so-called" paleoliths.[32] Even so, Tylor's visit was one of the few highlights in a continuing low period in Abbott's life (§3.9).

Worsening matters, in the summer of 1884 rumors were once again (§3.7) circulating that something was amiss at Trenton.[33] University of Pennsylvania paleontologist Edward Cope wrote Putnam that a local teacher claimed he had accompanied Abbott in the field and seen him "*bury* the paleolithic 'finds'." Cope urged him to look into the matter:

> My prepossessions have always been in favor of Abbott, but there is no doubt he has a lively imagination, & the charges appear to be—some of them— well founded. You should look them up privately first & when you, as grand jury, find an indictment, then bring him to face his accusers.[34]

Putnam made inquiries that satisfied him that Abbott was innocent of the charges. Even so, for someone as cautious and professionally insecure as Putnam (§3.2), it was another warning sign.

Abbott sensed a change in Putnam's behavior, and in September of 1884, just a couple of weeks after Tylor's visit, he penned an angrier-than-usual letter to Putnam complaining about his relationship to the Peabody Museum. Putnam read the letter, burned it, then scolded Abbott, telling him to write only "when in [his] right mind." An equally infuriated Abbott refused to be the "enthusiastic local archaeologist" Putnam dangled on a string.[35] All Putnam ever cared about, Abbott was sure, "was to get specimens, but the collector could go to the devil afterwards."[36]

Abbott thought it was time to abandon Putnam and "re-unite with Philadelphians," but that required patching up the "bitterness of past experiences" with his onetime teacher Joseph Leidy of the Academy of Natural Sciences (what caused that bitterness occurred in connection with Dawkins's 1880 visit—perhaps Abbott had learned what Dawkins was told that day in Philadelphia about Abbott's trustworthiness [§3.7]).[37] The fence-mending continued over the next year or so but without much success. During that time, however, Abbott's personal circumstances began to improve and so too his outlook.

Despite toiling in misery as a bank clerk, Abbott had found time to publish two books in natural history (*A Naturalist's Rambles about Home* and *Upland and Meadow*[38]). They had been reviewed favorably by critics and were financial successes—so much so that in 1886 Abbott was able to quit his job at the bank. Abbott's archaeological efforts were gaining positive notice as well. In his latest *Annual Report* the ever-forgiving Putnam offered heady praise of Abbott for "having worked out the problem of the antiquity of man on the Atlantic coast" and proclaimed the Abbott Collection as one of the most instructive in the Peabody Museum. Following a "long cherished" plan of Abbott himself, Putnam had the collection arranged "to show the successive periods of man's occupation of the Delaware Valley" from the paleoliths of the glacial period to those of the New Jersey Indians, "superposed in time, arranged in sequence side by side, and in such numbers that the cumulative evidence, always asked for by the archaeologist, is amply presented."[39] A review of the exhibit in *Science* declared the Abbott Collection "invaluable, unique and of extreme importance to all who wish to study the Stone Age" and praised Abbott for his "industry and sharp-sightedness."[40]

Given the echoes in *Science* of Putnam's own description and tribute to the Abbott collection, Putnam likely had a hand in that anonymous review. Regardless, he exploited the promotional windfall shamelessly, reprinting it in its entirety in his *Annual Report*.[41] Here, for the president and board of Harvard College, the museum's trustees, and potential donors among the Boston Brahmin (many who ignored or, worse, disdained, the Peabody's preoccupation with the archaeology of the "low civilization" of the New World), was independent testimony of the valuable research being carried out under his direction. The oldest human remains in America may not equal the high art of classical antiquity that so attracted Boston's philanthropists, but they could at least rival the best of European prehistory.[42]

Abbott returned the compliment with a generous biographical sketch of Putnam for *Popular Science Monthly*.[43] Putnam's own circumstances had improved as well, when in January of 1887 he was finally awarded the prize he had long sought but had long been denied: the position of Peabody Professor of American Archaeology and Ethnology.[44] For the moment Abbott and Putnam were both doing very well and could afford to be cordial, even generous, toward one another. Abbott extended his magnanimity to several new discoveries that had come partly from Putnam's investments in other archaeological fields, two of which made Wright's prediction that paleoliths would eventually be found in Ohio come true.

Putnam had been funding physician Charles L. Metz of Madisonville, Ohio, who was excavating local mound sites on his behalf. Metz had the trained eye Wright thought was needed to spot a paleolith—and did while excavating a cistern. Madisonville was in the Little Miami valley, and, as Wright had foreseen, the paleolith was at the base of an eight-foot section of loess, just atop a glacial gravel. Metz sent the specimen to Putnam, who put it on display at the November 4, 1885, general meeting of the BSNH. The artifact, as Putnam described it, was "about the same size and shape of one made of the same material found by Dr. Abbott in the Trenton, N.J. gravel."[45]

Wright learned of the discovery only after the fact, but when Metz found another paleolith thirty feet below the surface in a gravel terrace at nearby Loveland (also in the Little Miami valley) in the spring of 1887, he immediately alerted Wright. In mid-November Wright visited both sites and came away convinced both paleoliths came from "true glacial" formations. These were, so far as Wright was concerned, among the most important archaeological discoveries made in America, ranking on a par with those of Abbott himself. In a fit of rhetorical excess he proclaimed it now possible to "speak with confidence of *pre*glacial man in Ohio."[46] He did not mean that in the strict geological sense humans were present in America before the glacial period. He meant only that humans came into an area in advance of the actual physical arrival of a glacier. But that rhetorical flourish could (and ultimately did) box Wright into the awkward corner of implying a far greater human antiquity than he was willing to admit.

4.3 THE AMERICAN PALEOLITHIC COMES TOGETHER

Times were good for advocates of the American Paleolithic: at the 1887 AAAS meetings Frances Babbitt was elected a fellow of the association and Abbott vice president of Section H. Later that same year he too was the subject of a flattering biographical sketch in *Popular Science Monthly*.[47] Others were clambering aboard the American Paleolithic bandwagon. USGS geologist WJ McGee and Smithsonian archaeologist Thomas Wilson met with Abbott at the AAAS meetings and afterward stopped in to visit Trenton. According to Abbott, they left

more than ever impressed with what they saw to-day, and the finds were very satisfactory. They were pleased with the whole trip, and McGee will report at Washington to an effect that will astonish a certain clique; if I am not greatly mistaken.[48]

Abbott, ever sensitive to perceived slights, wanted the Washington archaeologists to believe him, not because any had publicly criticized his Trenton work but because he saw a "clique" there with an attitude and approach to archaeology that he perceived as threatening. He had Cyrus Thomas's Moundbuilder Survey in mind, which Abbott complained tended "to modernize everything in connection with the Indian." Whereas Thomas was concentrating on the mounds, Abbott feared that effort (or, more specifically, the approach behind the effort) might "modernize" his Paleolithic as well. Thomas, as it happens, was indeed skeptical about the American Paleolithic, warning his field assistant Gerard Fowke to "hold onto your views despite of [Thomas] Wilson or anyone else, and for goodness sake don't let him [make] you crazy on the subject of Paleolithic implements."[49]

Abbott did not know that, of course, but he was right that Wilson and McGee were impressed. For his part, Wilson, the newly appointed curator of prehistoric anthropology at the United States National Museum (USNM) (the position opened on the death of Charles Rau), was instantly struck by the similarity of the Trenton material to paleoliths he had seen during seven years spent in Europe as a U.S. consul.[50] Trenton, for him, was America's Abbeville, and though he believed Abbott's finds testified to an American Paleolithic (as, for that matter, did the paleoliths from Little Falls, Loveland, and so on), he was concerned by their spotty and widely scattered distribution. Were there an American Paleolithic occupation, evidence ought to be found elsewhere and in more abundance. Filling the gaps in the distribution of the finds, he believed, was necessary to "establish a general occupation, and a general occupation must be established before the scientific world would accept the fact [of an American Paleolithic] as proved."[51]

To fill in those gaps, Wilson—occasionally accompanied by McGee and archaeologist William Henry Holmes—began field investigations in Washington, DC, particularly along Piney Branch Creek and Rock Creek, where "rude implements were found everywhere in profusion." He duly reported those discoveries at a meeting of the ASW in early November.[52] But he was not satisfied to stop there. He convinced Smithsonian Secretary Samuel Langley to issue a call to the Smithsonian's correspondents for information on "rude or unfinished implements of the paleolithic type."

The *Circular Concerning the Department of Antiquities* appeared in January of 1888, almost ten years to the day after Joseph Henry issued his *Circular in Reference to American Archaeology,* in which he was convinced there was no information to be gained from a public solicitation on the matter of human antiquity in the New World (§2.5).[53] Wilson, who wrote this latest *Circular,* thought differently, and he called for reports on how many, of what material, and in what circumstances paleoliths had been found, and especially where they were found relative to the glacial line (Wilson solicited Wright's help in obtaining a detailed map of the glacial boundary[54]).

Lest there be any confusion as to exactly what was sought, Wilson included illustrations of paleoliths from around the country, including Trenton. His experience with European paleoliths had convinced him that illustrations would suffice to help to identify specimens because, after all, their morphology was diagnostic of their age and evolutionary grade. "Comparison," the onetime attorney argued, "is as good a rule of evidence in archaeology as in law." Geological evidence testifying to their antiquity was useful, Wilson supposed, yet proof was unnecessary in every instance. Demonstrating that paleoliths were found "either in the river gravels [or] in the caves" and did not occur where Neolithic artifacts were found would be sufficient. For that "distinction between the two epochs" having been

> established on a firm basis, one not to be denied or doubted, then I have no
> doubt that implements may be found on the surface which by comparison
> [alone] can be identified as being Paleolithic and as belonging to that epoch.[55]

Geological evidence in that case became unnecessary.[56]

Perhaps. But when McGee followed Wilson's paper at the ASW with one of his own ten days later, he focused on the geological context and age of the Trenton Gravel. This deposit, in his view, was one of two in the Delaware valley. The older was the same age as the Columbia gravel he had mapped in the Potomac drainage (corresponding to Henry Lewis's Philadelphia Red Gravel), whereas the younger Trenton Gravel was formed of "morainic material" redistributed by water and ice action from the floods that occurred during the melting of the glacier from the "second or last glacial period."[57] McGee had accepted that there were multiple glacial episodes because he had traced the peat and forest bed across Iowa a decade earlier (§3.5). For that matter, he had long believed Trenton provided "seemingly conclusive evidence" of an early human presence.[58] Nothing in the intervening years had changed his opinions. In fact, he himself had extended the range of Pleistocene humans into the Great Basin. In October of 1882, while searching for fossils in Walker River Canyon in the basin of Pleistocene pluvial Lake Lahontan, he spotted an obsidian "spear-head" twenty-five feet below the modern surface in the "upper lacustral clays" that marked the second and more recent high stand of (Quaternary) Lake Lahontan. The artifact was "associated in such a manner with the bones of an elephant, or mastodon, as to leave no doubt as to their having been buried at approximately the same time."[59]

In late 1887, as Wilson began to expand the sample and range of paleoliths, and McGee to fortify their geological context and age, Wright headed to Boston to deliver a series of lectures at the Lowell Institute. At his urging Putnam arranged a special "paleolithic meeting" at the BSNH to coincide. Abbott came up from Trenton, Upham joined them from Minneapolis, and Haynes participated as well. For the assembly Putnam gathered paleoliths to put on display: representing Europe were specimens from the Abbeville and St. Acheul gravels and the cave of Le Moustier, courtesy of French prehistorian Gabriel de Mortillet, and a paleolith from Milford Hill England, a gift of John Evans. The American specimens were mostly from Abbott's Trenton Gravel, but on display as well were several of Babbitt's Little Falls quartzes and Metz's Little Miami valley paleoliths.

Together publicly for the first time, this "remarkable group of authentic specimens of these implements of early man" revealed just how similar Paleolithic artifacts were in the Old and New Worlds. Putnam drove the point hard: though the artifacts differed considerably in the type of stone used in their manufacture (the European paleoliths were fashioned of flint, not the argillite, chert, or quartz of American paleoliths), they were nonetheless "identical in shape [and] often agreeing in size and in minute points of structure." Presented that way, there could be no doubt these specimens were artifacts and not products of nature. Putnam goaded his audience all the same: "If there are any persons present who may doubt the artificial character of the specimens, I can only say open your eyes and be convinced."[60] That was hardly necessary. Putnam was preaching to the choir. Besides, questions about paleoliths seemed beyond such remedial matters. Their uncanny similarity, Putnam asserted, was testimony that all were made by "man in [the] early period of his existence" on both sides of the Atlantic, when "he was living under conditions of climate and environment which must have been very near alike on both continents, and when such animals as the mammoth and mastodon, with others now extinct, were his contemporaries."[61]

Putnam left unmentioned the age of the American Paleolithic, but others did not. Abbott stood next to declare with more confidence than he had previously been able to muster that the Trenton Gravel was definitely glacial in age. His confidence was rooted in the work and authority of McGee, who had sent Abbott an abstract of his ASW talk, which Abbott duly read.[62]

In turn, Wright took the floor to argue that the Trenton, Madisonville, and Loveland terraces all dated to or even before the close of the glacial period. However, this was only a minimum estimate for human antiquity, as people must have arrived before then and the glacial period may have been a long time in closing. Perhaps, Wright believed, we can "speak with confidence of *inter*glacial man in Ohio." (He was again boxing himself in [§4.2]: by "interglacial" he meant a presence within the full glacial period and not between two separate glacial events, as was becoming the understood meaning of that term.) By his Niagara Falls recession chronology (§3.7), the close of the glacial period was 10,000 years ago.[63]

These conclusions struck Abbott, who must have recalled Wright's earlier insistence that the Trenton Gravel was postglacial and no more than 10,000 years old. He had to be sure he had heard correctly:

> Did I, or did I not understand you to say that the Trenton Gravel was deposited *before* the Niagara Gorge was formed; and so was older than that phenomenon? I thought you said so at the December Bost. Soc. Nat. Hist. meeting. . . . I want to be sure of the matter. . . . All other evidence, collectively considered, requires more than 10,000 years and I hope I am right; for otherwise it conflicts with "Indian matters" a good deal.[64]

Abbott had heard correctly. For that matter, the age Wright assigned was about the same as the one Upham had worked out for the drift containing the Little Falls quartzes, and by virtue of cross-country stratigraphic correlation he deemed the Little Falls artifacts slightly later than those in Ohio or New Jersey. Upham, of course, assumed the Little Falls quartzes were genuine artifacts, quietly ignoring a sour note sounded earlier that year by Winchell himself, who had begun to publicly express his doubts about the human origin of Babbitt's specimens.[65]

Putnam declared the meeting a great success. Questions remained, of course. There was the puzzling disappearance of the Paleolithic population: had they gone extinct along with the mammoth and mastodon, or moved north and become Eskimos? Putnam hoped to find their skulls to resolve whether they were dolichocephalic like the Eskimo. And there remained questions about whether the Paleolithic was one race in Europe and America, and if the American Paleolithic was everywhere the same age.[66]

That question was on his mind because it seemed that the eastern North American sites (along with McGee's Lake Lahontan spearpoint) dated to the end of the glacial period, whereas far western sites were much older. The "auriferous gravels" of the Sierra Nevada range in California had yielded, along with the Calaveras skull, stone mortars, polished stone artifacts, and shell beads in deposits occasionally capped by lava flows that Whitney estimated to date to the Early Pleistocene or possibly the Pliocene. Putnam thought it odd that West Coast groups were fashioning what appeared to be Neolithic artifacts earlier than those making Paleolithic tools had even appeared in eastern North America. Perhaps conditions on the Pacific coast favored rapid cultural advance, or its geological evidence had been "misunderstood." Of one thing Putnam felt confident: the artifacts were found in the auriferous gravels in the circumstances Whitney had described. Others, like Daniel Brinton in his vice presidential address to the AAAS the year before, were far more skeptical, dismissing the auriferous gravel evidence as a "violent anachronism."[67]

These unresolved issues notwithstanding, it seemed to Putnam that the discoveries of Abbott, Babbitt, and Metz had been "so thoroughly *correlated* by the careful geological observations" of Wright, Upham, and McGee that a "steady advance" of progress had been made toward understanding the origins of humanity in America.[68]

McGee thought so too, and knowing a bandwagon when he saw one he jumped on. He decided that winter to publish his ASW address in *Popular Science Monthly*, selecting that journal because, as he smugly lectured its editor, William Youmans, its "standards in geology, archaeology, and cognate branches of knowledge have not always been so high as the student of these subjects . . . could desire." McGee would rectify that.[69] In the course of writing his paper he asked for and received several of Wright's publications on the Ohio paleoliths. That prompted him to expand his "little article on the Trenton Gravels" into a sweeping overview of the Paleolithic in America. In late May of 1888, he sent Youmans his finished paper with the assurance that the issues were "fully up to date both as regards geology and archaeology" and that "every sentence has been carefully weighed and . . . that every utterance is thoroughly reliable."[70]

McGee tallied what was by then a long list of sites testifying to a Pleistocene human presence in North America, arguing that even excluding all doubtful cases (among them, he thought, the Little Falls quartzes and his Lake Lahontan spearpoint), there still remained "a fairly consistent body of testimony indicating the existence of a widely distributed human population upon the North American continent during the later ice epoch." The best evidence was Trenton, where paleoliths were "abundantly embedded" within the Trenton Gravel, though not quite as abundantly as McGee imagined (in his enthusiasm, he put Trenton's count of paleoliths at over 25,000; the actual number was closer to 400, as Abbott immediately pointed out[71]). Ultimately, the "testimony is cumulative, parts of it are unimpeachable, and the proof of the existence of glacial man seems conclusive," at least in regard to the second (later) glacial period. McGee was less confident that humans were here during the first (earlier) glacial period and was altogether unswayed by claims of an even earlier arrival.[72]

Because he believed the older glacial formations had by then been thoroughly searched yet had failed to yield any trace of a human presence (excluding Koch's anomalous Missouri discoveries [§2.4]), it was reasonable to conclude that humans were absent before the second glacial episode. That absence of evidence could be treated as evidence of absence. It was only in the later glacial period that evidence of a Paleolithic was certain, proving both the age and grade of the American "autochthons."

On close examination McGee saw differences in the "degree of development" among the paleoliths, which ranged from the "rudest" turtlebacks to the finest chipped flints, polished stone, and even copper spearheads (the latter recovered by Winchell from the same stratum as Babbitt's quartz chips). It was too soon to tell if the anomalous artifacts were merely "adventitious inclusions" (redeposited from younger levels) or represented variation within the Paleolithic period. For that matter, it was also premature to correlate the archaeological and geological sequences of America and Europe, except in the coarsest of terms.[73]

Nonetheless, McGee was certain North America at the time of the American Paleolithic was habitable, though it must have been cold and damp at sites like Little Falls, Loveland, Madisonville, and Trenton, which were in close proximity to glaciers or the turbulent ice-laden rivers that drained them. Living as they were in conditions approaching those of the Eskimo, McGee surmised they were likewise hunters who pursued now-extinct fauna. He himself had found remains of several extinct faunal species associated with his spearpoint. Abbott had recovered a human wisdom tooth in undisturbed gravel at Trenton "within a dozen rods" (roughly 200 feet) of a mastodon tusk, proving (at least to Abbott) "the *indisputable* fact, that remains of the mastodon and man have been found associated in no uncertain way."[74] Putnam reported a similar find in a Massachusetts peat bog that had produced a human skull and a mastodon skull in the same stratigraphic position (both resting on the blue clay at the base of the peat), although separated from one another horizontally by some eighteen feet.[75] For McGee the first people had a Paleolithic culture, lived on the ice front, and arrived around the time of the second ice advance—and well after the first.

But on May 25, 1888, the day before McGee mailed Youmans his manuscript, a discovery was made near Claymont, Delaware, by Hilborne Cresson, one of Putnam's newly appointed field assistants, that potentially expanded human antiquity in North America many times over. Word of that discovery did not leak out for a few months, and that was deliberate. Putnam privately alerted Abbott and Wright to the details of the find but urged them to "please keep the matter perfectly quiet." There were reasons for secrecy, he assured them, but did not elaborate, saying only that he wanted to "keep our paleolithic and glacial work in our Museum circle." He promised the discovery would be made public at the AAAS meeting that August.[76]

4.4 ABBOTT TAKES CENTER STAGE

Meanwhile Abbott was preparing his Section H vice-presidential address for that meeting. It was the most prestigious stage on which he had ever spoken, and he had been thinking about the talk all spring. Wright was consulted on the fine points of Trenton geology, while questions of what to include and what to omit were floated by Putnam. Abbott assured Putnam "I do not want, in the address, to be otherwise than *very moderate* so as to avoid discussion." But both Abbott and Putnam knew better than to think he would be able to do that. This was the same Abbott who had announced just three months earlier he would make his address "a Peet annihilator."[77]

In Abbott's mind Peet had it coming. Since the late 1870s (§3.9) Peet used his *American Antiquarian* to criticize Abbott's lack of excavation, his ignorance of glacial geology, and especially his reliance on surface collections—a serious flaw, Peet argued, given "the stone age is still in existence" in America and the surface was littered with material of indeterminate age. He mocked Abbott's equating cultural stages (Paleolithic/Neolithic) with geological periods, especially because Abbott's ostensibly Paleolithic material was apparently younger than Whitney's Neolithic artifacts. (Peet was sure one or, more likely, both claims were flawed.) Peet especially doubted that American paleoliths were like those of Europe, but even if they were he pointed out the oddity that specimens said to be similar archaeologically appeared by geological evidence to be an order of magnitude different in age. Were Abbott's paleoliths "accidental fractures" of nature or "rude because they were unfinished?" For a decade Peet had kept up a drumbeat that implied in barely subtle terms that Abbott was an amateur whose work

> lacked the careful supervision of a scientific man who understands the points at issue, when [the specimens] are taken out of their so-called matrix. European archaeologists do not rely upon such haphazard discoveries and why should we?[78]

Harsh words, but now that Abbott had the bully pulpit of the largest scientific society in America he intended to use it. Peet might think him an amateur, but he let his AAAS audience know that on his side were authorities like Upham, whose "careful examinations" at Little Falls proved the great antiquity of Babbitt's "unquestionably worked" quartzes. Like Wright, who had prophesized that there would be paleoliths in Ohio that had proved to be found under exactly the conditions he had anticipated.[79] And like McGee, who had detailed the geology of the Delaware valley, demonstrating beyond doubt that the paleolith-bearing Trenton Gravel was glacial in age. There had been much "needless comment . . . from presumably learned sources . . . and lame attempts to belittle the discovery by those who should know better." Yet all those discoveries (and Cresson's, to which Abbott alluded) were only some of the Paleolithic finds on record. Why, Abbott asked, should the bare mention of the American Paleolithic "still excite a sneer?"

> There will probably always be over-cautious folk who will only accept *cum grano salis*, the Man of the Tertiary, however elegantly he may be plead for; but no one willing to accept other testimony than his or her own eyes— often the most treacherous of guides—can in fairness turn their backs, when we speak of that primitive chipper of flinty rock, who, with no other weapon . . . held at bay, the savage beasts of primeval times.[80]

Trenton, Abbott insisted, was especially significant, as most "competent archaeologists" well recognized. The site had spawned an important literature, excluding the "inanities of the ignorant," and helped addressed fundamental issues of the American Paleolithic. There one could precisely define a paleolith: it was a coarsely chipped mass of flinty rock on which a distinctly designed cutting edge was formed, which (Peet's assertions notwithstanding) showed weathering on its surface indicative of great age and was often but not always found in deposits of glacial or river drift.[81] There was no confusing these with naturally flaked stones, as Haynes and Wadsworth had testified (§3.3, §3.7).

At Trenton, furthermore, there was a cultural sequence: at the base were the rudest forms of paleoliths made of local argillite. These were overlain by well-advanced specialized implements, still made of argillite but of more distinctive types (Abbott speculated the advance was due to climate and environmental change or perhaps the acquisition of language). Atop all were Indian artifacts, pottery, polished stone, and jasper arrowheads. Rude artifacts might be found on the surface, he admitted, but jasper arrowheads were never found in the gravels. That fact, Abbott insisted, was the stratigraphic linchpin: it proved argillite implements predated those of jasper, as every "competent" (there was that word again) investigator realized.[82] Abbott gloated that he never saw argillite implements in any Indian workshops. That was negative evidence, but evidence nonetheless that the rude forms were not simply manufacturing refuse.[83] Otherwise, how could it have aligned in so logical a cultural sequence?

The manner in which these several prehistoric groups related to one another remained a puzzle. Abbott admitted he had abandoned his earlier suspicion that "*Homo paleolithicus*" were ancestral Eskimo, after discovering several skulls in the Trenton Gravel that were morphologically unlike those of the Indian yet failed to show any similarity to those of the Eskimo. A setback to be sure for both Putnam and Abbott, but Abbott's belief in a "pre-Indian man" was hardly challenged, only his hope that the skulls were Eskimo. Perhaps, he speculated, it was later argillite groups who were Eskimo ancestors.[84]

Abbott knew that sufficient positive evidence to clear away *all doubt* (his emphasis) would probably never be forthcoming. Still, he hoped "the 'Doubting Thomases' will be fewer by the year 2000." Doubting Thomases or no, the probability of an American Paleolithic seemed to him to "hug the bounds of certainty."[85] There was a time, he announced, "when, to all appearances, American archaeology would have to be squeezed into the cramped quarters of ten thousand years; but we are pretty sure of twenty or even thirty thousand now." Archaeology now had "time enough to spare."[86] It was about to get more.

4.5 PUSHING THE ANTIQUITY ENVELOPE

In his address Abbott supposed Cresson's discoveries would "excite discussion," but there was no discussion at all. That was because Cresson's report, read by Putnam, was vague about details: he said only that the finds were from "a reddish gravel." He failed to mention that was Lewis's Philadelphia Red Gravel (§3.4), the unit beneath the Trenton Gravel that was deposited in full glacial times or during a previous glacial episode if McGee was right about it correlating with his Columbia formation.[87] Those crucial details omitted, Cresson's paper was scarcely noticed in the meeting summary published in *Science*.

Putnam's reluctance to detail the stratigraphy was likely due in part to the fact that the locality had not been examined by anyone other than Cresson. After the meeting, Putnam asked Wright to examine the geology at Claymont (he turned to Wright and not Henry Lewis, who had defined the formation, as Lewis had passed away that summer. Lewis's death was not a loss as far as Abbott was concerned; he had come to believe Lewis was suppressing evidence of paleoliths).[88]

Cresson had first come across the Claymont locality in July of 1887 while following a work-gang excavating a trench for a railroad line. From a perch atop the berm he spotted an artifact, scrambled down the slope to extract the specimen, then quickly climbed out just ahead of several tons of cascading gravel (and a tongue lashing from the gang boss). The object appeared to be a paleolith, so Cresson sent it to Putnam and began to watch the locality. In May of 1888 another specimen was found by Cresson in the same gravel formation a short distance from the original find.[89]

These were not the first traces of Pleistocene remains reported by Cresson, an erstwhile artist and archaeologist. In the late 1880s he announced several spectacular discoveries, virtually all ostensibly made in the 1860s. For reasons he never fully explained, Cresson reported these only long after the original sites had been destroyed. Still, his testimony was believed (for a time) by Putnam, as Cresson appeared to have credentials and credibility: he had spent the latter part of the 1870s in France pursuing studies at the École des Beaux Arts and the École d'Anthropologie. That apparently included "careful study of gravel deposits in the valley of the Somme, where [he] found implements of the river drift men in place," an experience that led him to archaeology and ultimately Putnam's employ.[90] Once he started working for Putnam, Cresson began to report paleoliths: first from an "undisturbed aqueous deposit" of gravel along the White River near Medora, Indiana (in a geological situation Wright presumed was akin to that at Trenton), then at Claymont.

Wary of the potential antiquity of the Claymont finds, Wright spent a day on-site with Cresson examining the stratigraphy.[91] As Wright saw the situation, Cresson had indeed plucked the paleoliths from the Philadelphia Red Gravel, and to witness the fact he had Cresson photograph the section (no matter that the specimens were no longer in place). Like Lewis, Wright believed the Red Gravel marked the full glacial. The Claymont paleoliths were therefore older than Abbott's, but how much older depended on "the interpretation of the general facts bearing on the question of the duality of the glacial epoch." Pointing out the problem was easy enough; finding a solution Wright could abide was not.[92]

A week after visiting the site, Wright was in Washington, DC, where he conferred with McGee about the Claymont geology. It was not a very satisfying meeting for either. McGee identified the Red Gravel/Brick Clay and his Columbia as the "aqueo-glacial margin" from the first glacial advance. Its age was "long anterior to that of the last ice-invasion," long enough that the two were separated by an interglacial "three, five, or ten times as long as the post-glacial interval."[93] Even by Wright's conservative estimate of postglacial time (10,000 years), that would make the Columbia Formation 30,000 to 100,000 years old. The Claymont paleoliths, by McGee's reckoning, would be very old indeed.

Putnam and Abbott were prepared to accept a great age for the Claymont paleoliths.[94] Wright was not. He said little while in Washington, only suggesting that McGee visit Claymont himself.[95] Wright made his objections public the next month (December 1888) at the BSNH, where Putnam had organized a meeting to showcase the Claymont discovery. There, Wright publicly chided McGee for postulating two glacial epochs on "the basis of facts outside of [the glacial] region." Wright insisted there was no evidence to divide the glacial period; it was one prolonged epoch, with a mild intervening period. McGee's Columbia gravels were likely just the far-flung debris from that single advance. After all, Wright confessed, he and Lewis had themselves underestimated just how far glacial debris that they called the "fringe" had extended.[96]

After Wright returned to Oberlin in January of 1889, he wrote McGee to float his reinterpretation of the origin and age of the Columbia Formation and tried to appeal to his charitable side: "It will greatly relieve the archaeologists if we can get along without the two glacial periods this side of the Alleghenies."[97] McGee was not feeling charitable:

> I trust you will allow me to say that the deposits which I have ascribed to the Columbia Formation on the Delaware and Susquehanna rivers seem to me quite distinct from the "fringe" of Professor Lewis and yourself. At any rate, these deposits may safely be estimated at from five to ten times the antiquity of the intermorainal deposits.[98]

The American Paleolithic had just tumbled headlong into what became an explosive dispute over North American glacial history.

4.6 THOMAS CHAMBERLIN AND THE QUESTION OF GLACIAL HISTORY

Archaeologists had little intellectual capital invested in the number and timing of glacial events, as McGee well knew. The only one who would truly be relieved

Figure 4.1 Antagonists over the question of the glacial history of North America: George Frederick Wright (*left*), professor of the Harmony of Science and Revelation at Oberlin College, and Thomas C. Chamberlin (*right*), longtime chief of the Glacial Division of the USGS and professor of geology at the University of Chicago. Photographs courtesy of the Oberlin College Archives (*left*) and Smithsonian Institution Archives, Image MAH-41213E (*right*).

was Wright, who clung fiercely to the idea of a single glacial period, which at the BSNH the month before he had proclaimed "more and more evident."[99] Others were seeing it differently. In the winter of 1888–1889 the tide of geological opinion was running against Wright, and it would take an extraordinary effort to hold it back, let alone convince others to join him—or at least not abandon him. Making his task much more difficult was the formidable Thomas Chamberlin, who was by then actively antagonistic toward Wright (Figure 4.1).

Chamberlin was born in a log cabin atop the Shelbyville moraine in Illinois, then moved to a farm outside Beloit, Wisconsin. He graduated from Beloit College in 1866 with a basic exposure to geology and the natural sciences but in a curriculum dominated by the Latin and Greek classics ("intellectual murder," he later called it).[100] Following graduation, he attended the University of Michigan for a year of study in geology under Alexander Winchell, then returned to Wisconsin to teach high school. This was followed by a stint at the State Normal School at Whitewater. In the spring of 1873 he became a member of the Beloit faculty and the Wisconsin Geological Survey, assigned the unpromising southeastern region of the state. It was an area with few mineral resources but one deeply buried in glacial drift. Chamberlin became a glacial geologist by legislative default.[101] He was extremely hardworking and indisputably talented, and his geological star rose. In three years he was chief geologist of Wisconsin, a position he held until 1881, when he joined the USGS. In two years he was named chief of the newly created Glacial Division, where he stayed for the next several decades, despite moving in 1892 to the University of Chicago. Over most of this time he unofficially presided over glacial studies in America.[102]

At the outset Chamberlin and Wright were not so very different. Chamberlin was the son of a circuit-riding minister and raised in an orthodox abolitionist home, and like Wright early on sought to balance geology and theology. While at Beloit he spoke at a local church on the harmony of Genesis and geology (aptly taking seven Sundays to complete the sermon).[103] Unlike Wright, Chamberlin never sought to accommodate geology to Genesis. He abhorred any a priori constraints on scientific inquiry, the kinds of constraints that Wright, who viewed science through the lens of his faith (§3.8), willingly accepted.[104]

Their worldviews diverged in the decade after the Civil War, in which neither fought (Chamberlin was still a Beloit student) but which deeply influenced each, albeit in different ways. Wright's war was a profoundly moral tale about human sin and God's goodness, and it fired his faith and reaffirmed his belief in theological and social reform as the road to human salvation (§3.8).[105] The lesson Chamberlin learned was more pragmatic. The "larger vision that came with the wider interests and experiences of the war, visions of that which was national rather than personal, entered into the new mental attitude," irreversibly expanded the scope and role of the federal government in the sciences, especially geology.[106] The war showed him, as it did John Wesley Powell (who became USGS director in 1881), that geology should not be the strictly local (if not insular) enterprise it had been before the war, dictated by individual states' needs to assess their mineral resources and agricultural potential.[107] The centralization of scientific power in federal research surveys and agencies, and the demands to map the resources and outline the geological history of the newly unified nation, gave Chamberlin license and resources to attack glacial geological questions on a larger scale than could have been previously imagined.

Chamberlin did so following a strict scientific code, one rooted in an almost religious philosophy in which science provided the key to unlocking a pantheistic universe.[108] Chamberlin expressed that philosophy in the notion that nature's organization manifest "a supreme universal organism whose power is in itself and pervades itself and its own operating agency."[109] The study of science thus became an inquiry into what he called "secular theology"—it was the study of the truth. As he exhorted the University of Michigan's graduating class of 1888:

> Investigative study calls into continuous exercise [of] certain noble activities and attitudes of the mind; to love the truth supremely, to seek the truth assiduously, to scrutinize evidence rigorously, to withhold judgment when evidence is insufficient, to look upon all sides equally, to judge impartially and to make conscientious corrections for personal bias.[110]

The truth would set him free, but to attain it the scientist must begin with an "undeserved acceptance of things as they are, of complete devotion to finding out what they really are and how they have come to be as they are." Such a sacred trust required scrupulous devotion. As he put it:

The swerving of the mind from absolute rectitude in any of its activities falls under ethical condemnation. Falsity in intellectual action is intellectual immorality. Narrow and loose habits of thought, prejudiced attitudes toward evidence, bias from previous opinions and feelings, shallowness and superficiality in observation and carelessness and reasoning are appropriate subjects of moral reproof.[111]

Chamberlin demanded strict adherence to the scientific method, or at least his revamped version of it: the method of multiple working hypotheses that, not coincidentally, he first articulated in 1889 at the end of a prolonged and unsuccessful effort to instruct Wright in the proper canons of science.

Chamberlin saw science as existing at the public's behest and for its benefit and improvement, and he put that principle into practice. As president of the University of Wisconsin (1887–1892), he pioneered university extension courses, farmers' institutes, and summer schools for science teachers. Yet, for the public to gain the most from those results, they had to be packaged in an intelligible fashion and in a way that honestly and accurately depicted the state of knowledge in the field. When it came to ideas that were weak or flawed or where conclusions were in doubt or dispute, Chamberlin thought it best not to release them at all. If publication of disputed material was unavoidable, it had to include a frank disclaimer explaining what was in doubt or dispute and why. Otherwise, the reputation of science suffered.

As democratic as Chamberlin's ideals were, he was nonetheless an elitist when it came to science. He did not believe everyone was entitled to speak on its behalf—only those who had fully mastered its intricacies and adhered to his strict moral code and extraordinarily high standards. It was a foregone conclusion that those around him occasionally failed. It was just as inevitable he did not tolerate those failings. On many an occasion Chamberlin advised a colleague to give a paper to the world "through the kitchen stove." Even as a young man he was autocratic, literal, moralistic, self-righteous, and utterly humorless.[112]

4.7 THE KETTLE MORAINE MOVES EAST

Chamberlin's days on the Wisconsin Geological Survey had convinced him that the Kettle Moraine marked a second glacial period in North America (§3.5, §3.8), and he strongly suspected it extended across the continent. He could not prove as much working within confines of Wisconsin's borders, especially since geologists in New England had yet to report an analogous feature or, like James Dana, seemed unwilling to admit one might exist (§3.5). Chamberlin himself had not had the chance to examine those other areas to see whether one did, but he soon got that opportunity. In March of 1881 Powell was sworn in as director of the USGS. Two months later he appointed Chamberlin to a position as geologist with the USGS, and his directive to Chamberlin was as broad as it was ambitious:

The terminal moraine that enters the United States on the north border of the Territory of Dakota and stretches thence southward and eastward in

sinuous course probably to the Atlantic will be the subject of your investigations for the coming year. You will also study such collateral deposits as may
be intimately associated with its history.[113]

Over the summer and fall of 1881, and again in the spring of 1882, Chamberlin mapped glacial features west into the Dakota Territory, then east to the
Hudson valley. By the time he finished he no longer felt compelled to soft-pedal
the claim for two glacial periods, as he had previously (§3.5). He had earlier believed caution was appropriate because the "presumption" of multiple glaciation
was not backed by evidence, and the available evidence might reflect only "temporary oscillations" of the ice front or the debris of subglacial streams. His USGS
fieldwork convinced him the Kettle Moraine was no local phenomenon, but part
of something that stretched "across the whole of the glaciated area and belong[ed]
to a system of glacial movements which differ in many important respects from
the earlier ones."[114]

He marshaled that evidence in his "Preliminary Paper on the Terminal Moraine of the Second Glacial Epoch," which was submitted and published within a
year of his USGS appointment. Though done with astonishing speed, it was not
hasty or ill-considered. After a decade of work Chamberlin was well aware of the
potential pitfalls challenging even the most diligent efforts to interpret glacial
features: "Most formations betray their origin in their salient characteristics, but
those of the Quaternary age are apparently capable of diverse interpretations if
their general nature alone is considered."[115] The self-disciplined Chamberlin, however, believed himself unusually diligent, and by and large he was right.

In his "Preliminary Paper" Chamberlin confidently set forth how he understood the telltale but often subtle features that marked glacial phenomena, to
insure the ground rules were known for "any who may be inclined to differ" with
him.[116] It was a slightly veiled warning to those who might lack what he saw as the
critical ability to spot those features, or who failed to abide by the rigorous code
of scientific conduct necessary to bar the "personal predisposition" to see what one
wanted to see that sometimes crept into such studies. He was particularly explicit
about what he meant by a "terminal moraine," for it was this feature that lay at
the heart of his argument. In general, any accumulation of debris along the outer
margin of a glacier was a terminal moraine, but the key was to distinguish those
moraines that marked the farthest limit of an ice advance and thus had the most
historical significance, as opposed to ones that formed during a temporary halt of
a retreating glacier. Chamberlin reserved "terminal moraine" for the former and
called the latter "peripheral moraines." The Kettle Moraine at its fullest extent,
he believed, was a true terminal moraine, and much younger than (and therefore historically unrelated to) the terminal moraine further south that marked the
maximum extent of Pleistocene glaciation in North America.[117]

With "fatherly affection for [his] mentally first born," Chamberlin began tracing the Kettle Moraine out from the Green Bay glacial lobe that produced it, then
followed its trail into the Midwest, across New York, and ending in New England.
Along the way he identified contemporaneous moraines and the geography of the
glacial lobes that produced them. He then returned to Wisconsin and reversed

direction, describing the moraines of Minnesota and the Dakotas, then predicted their extension into the "British Possessions." In all, he reported on over a dozen moraines and interlobate moraines, and he reconstructed from striations, boulder trains, and other features the direction of the ice flow that produced them.

Each of the separately identified moraines shared distinctive features (kettles, subordinate ridges, knobby drift, a distinctive texture, and so on), and many abutted one another. This made it relatively easy to tie all the segments together as a single, historically related terminal moraine, albeit a broad, irregular, and occasionally discontinuous one. The Kettle Moraine was not, Chamberlin stressed, a single and simple continuous ridge (which was "too often" the conception of a terminal moraine). Yet, given that its many pieces formed a single unit that stretched from Cape Cod into the British possessions, it had to have resulted from some "widely prevalent combination of conditions," such as a concurrent ice advance.[118]

The path the Kettle Moraine traced across northern North America proved to be largely (but not entirely) separate from and well to the north of the line that marked the previously mapped maximum limit of ice advance. The Kettle Moraine was also more massive and better defined, and it appeared far less weathered. All that, Chamberlin argued, meant that the two marked the maximum advance of historically separate glacial events, the Kettle Moraine being the younger.[119] The case for a second glacial epoch, so far as Chamberlin was concerned, was grounded on substantial evidence, and he confidently proclaimed its "general correctness." Most of his peers thought so as well.[120]

Still, though the occurrence of a second glacial epoch seemed obvious, there were places where the two terminal moraines seemed to converge into one.[121] Teasing those apart and building the case for separate glacial events was a much more complicated task than it was where the two moraines were hundreds of miles apart. A glance at Chamberlin's map of glacial features in North America (Figure 4.2) instantly shows the knottiest area of overlap: from eastern Ohio east to Cape Cod, Massachusetts.

What prevented Chamberlin from working out the details in the northeastern states was not the complexity of the geology but the constraints of politics. Glacial features in these areas were under investigation by others, notably the respective state surveys, and the USGS as a federal agency had to tread carefully. In those years it was in the perennially awkward position of having to balance the needs of its own continental-scale investigations against the proprietary interests of the state geological surveys. That balancing act was complicated by the fact that its investigations were not requested by either Congress or the states and instead were often initiated by Powell to resolve problems of interest to him or to fit his available personnel, both of which sometimes collided with what individual states perceived as their domain.[122] The USGS, of course, had deeper financial pockets and could generally investigate whatever it wished (insofar as it could be justified to Congress). As one Survey member later put it: "Practically it has carte blanche to carry on geological investigations over the whole territory of the United States, and in every branch of scientific work directly related to geology."[123] It could and often did go wherever it wished, insofar as the problems it was examining such as the history of continental glaciation, could "never be comprehended, stated,

Figure 4.2   Thomas C. Chamberlin's "General Map of the Terminal Moraine of the Second Glacial Epoch," published in 1883. This map was the culmination of a decade of investigations of what he initially termed the Kettle Moraine, first recognized in Wisconsin, then subsequently traced east and west. By the time he made this map he was certain the Kettle Moraine marked the second of two separate glacial epochs, depicting it as "moraine" on this map (and shown as dark, looping bands). The traces of the first glacial epoch are evident in the "drift-bearing area" to the south of the Kettle Moraine. The terminus of the two separate glacial advances roughly coincided between eastern Ohio and New England. This became the root of the disagreement between glacial geologists of the eastern and western states. The former (George F. Wright most especially) interpreted that overlap as indicative of a single advance rather than two asynchronous features that happened to terminate in the same approximate position (from Chamberlin 1883b: Plate XXVIII).

studied or solved in any single state . . . unless one had *carte blanche* for his movements."[124]

And the USGS did not always know (or care?) what the state surveys were doing and vice versa. There was no institutionalized cooperation. As a result, relations between federal and state surveys varied, and in some states they were never better than strained. In Minnesota and Alabama, State Geologists Newton Winchell and Eugene Smith (respectively) vigorously opposed any USGS incursion, fearing it might make their work appear unnecessary in the eyes of their state legislators and lead to a loss of funding.[125] That attitude was anathema to Chamberlin, who viewed such petty concerns as demeaning to the greater good of searching for scientific truth, a goal that certainly transcended mere political boundaries, especially in the wake of the centralization of science that followed the Civil War.

Much as Chamberlin stood by that principle, he was not tone deaf to the political realities and professional courtesies that demanded compromise. The Pennsylvania Geological Survey had just begun plotting its glacial features. It was obviously not the right moment for Chamberlin to map his own way across Pennsylvania, and so he stopped his own eastward survey at its border. Peering across the state line as the two separable moraines merged into one, he could see that the most complicated and potentially vulnerable evidence for two glacial epochs would be in the hands of others.[126]

Chamberlin was willing to wait for their results, but fearing that a single moraine might be mapped in an area in which he strongly suspected that two existed, he could not resist announcing his views of the glacial geology of Pennsylvania, in hopes they might guide the state geologists in their mapping and interpretations. Chamberlin offered the hypothesis (strategically, in Dana's *American Journal of Science* [§3.5]) that the more singular, beltlike moraine that crossed from Pennsylvania to Cape Cod might appear morphologically distinct from the Kettle Moraine (marked by large conspicuous loops), but that it was nonetheless part of the same system from the second glacial advance. And the reason two spatially separate terminal moraines did not occur in that region was that the second ice advance bulldozed away traces of the first. Such a hypothesis was corroborated by the absence of a more interior moraine, at least until further work proved otherwise.[127] Chamberlin knew who was mapping the Pennsylvania moraine and offered his suggestions to them, supposing these "may not be without service to the . . . workers in this somewhat new field," and hoping their "final conclusions are unformed and opinions still plastic."[128] He evidently did not know those workers—Henry C. Lewis and George Frederick Wright—very well.

4.8 MAPPING THE PENNSYLVANIA MORAINE

Lewis was commissioned by the Pennsylvania Geological Survey in 1881 to trace the glacial moraine across the state. The fieldwork spanned two years, and he was joined by Wright some of that time.[129] Wright planned to follow that with his own survey of the Ohio moraine, sponsored by the Western Reserve Historical Society. Both were well aware of Chamberlin's ideas about a second glacial period. Lewis had even briefly flirted with the possibility of multiple glaciation in explaining the Delaware valley sequence, only to downplay it in the chapter he wrote and Wright edited for Abbott's *Primitive Industry* (§3.7–§3.8).

While he was working on his "Preliminary Paper" Chamberlin kept in contact with Lewis, who promised to keep him apprised of his and Wright's Pennsylvania results. Chamberlin awaited these "special interest," though nothing reached him in time to include in his "Preliminary Paper."[130] Still, Chamberlin offered the hypothesis that only "a part of the moraine traced by the Pennsylvania geologists [Lewis and Wright] belongs to the earlier epoch" and the correctness of that assignment to the earlier epoch was "an open question."[131] Describing it as such, Chamberlin—intentionally or not—echoed Lewis's (or Wright's) statement the year before in *Primitive Industry* (§3.7).[132]

If Chamberlin hoped Lewis and Wright would corroborate the evidence for a second glacial period, he was bound to be disappointed. In mapping the southern limit of glaciation across Pennsylvania in 1881 and 1882, Lewis and Wright found but a single line of drift hills marking what they saw as the true terminal moraine. Overall, that line was "slightly undulating, almost everywhere sharply defined, and marked by a continuous line of drift hills." There were features on either side of that line that complicated this otherwise straightforward picture. First, there were moraines north of what they saw as the "true" terminal moraine. They identified these as deposits left by ice retreat and deemed them "more recent" than the terminal moraine itself. "More recent" was a matter of degree: they were younger, but Lewis and Wright still thought them part of the same glacial episode that produced the terminal moraine.[133] As Lewis wrote in his official report, the Kettle Moraine might also be example of one of these "moraines of recession," as he called them, laid down at "halting places or oscillations of the glacier in its northward retreat."[134]

Second, south of the terminal moraine there were "occasional transported or striated boulders and fragmentary patches of till." This outer "fringe," as Lewis and Wright named it, occurred over areas that otherwise seemed to lack bedrock striations, glacial erosion, or other evidence of having been overrun by ice. The fringe was discontinuous: it was largely nonexistent (or, they admitted, might have been missed) in eastern Pennsylvania, and in the western portion of the state it was generally close to the terminal moraine and never more than five miles beyond it.[135] Wright subsequently suggested the fringe resulted from debris falling off the crest of a glacier that had extended beyond its base.[136]

Like their moraines of recession, the fringe was considered by them part of the same single episode of advance and retreat, and not a remnant of a separate glacial event. That explanation worked well for Pennsylvania, but Lewis had by then read Chamberlin's "Preliminary Paper" and was aware that the gap between the fringe and the terminal moraine, so narrow in Pennsylvania, widened out to sixty or seventy miles farther west. That being the case, it was difficult to explain how that fringe could have been deposited: that was too great a distance for it to have been from an ice overhang. Lewis rejected the possibilities that the fringe had been misidentified or was nonglacial in origin. Nor did he consider that it might represent the remains of a much older glacial advance unrelated to the moraine he was mapping. In the end, he offered no explanation of its origin; he merely observed it was becoming a feature "destined to play an important part in glacial geology." Wright agreed.[137]

In transmitting Lewis's final report, Pennsylvania State Geologist J. Peter Lesley provided a lengthy introduction in which he identified several instances of apparent glacial phenomena south of the terminal moraine Lewis mapped, notably striae and polishing (but not fringe deposits), which called into question just how well defined their terminal limit was.[138] Lewis bristled that Lesley and others who claimed to see glacial traces south of their terminal moraine had mistaken fluvial action, slickensides, and other products of nonglacial weathering and erosion as signs of glaciation. He himself had no difficulty tracing the "truly terminal character of the Pennsylvania moraine."[139] The moraine, in Lewis's view, was a single, largely uncomplicated phenomenon, and as he hardened on that point the last vestiges of his earlier infatuation with multiple glaciation vanished.

Wright never even flirted with multiple glaciation, for it implied a deeper geological and human history than even he, who was not uncomfortable with the progress of science, could accommodate with his religious convictions. Like Lewis, he acknowledged Chamberlin's "interior line of moraine hills" but then dismissed it as marking only a temporary interlude in ice retreat.[140] He carried that opinion with him when he entered Ohio the summer of 1882. He met the moraine at the Pennsylvania state line and then on horseback and by foot traced it through Ohio, zigzagging across its path to insure he could see where traces of glaciation ceased. Even with the usual problems of following an irregular line across rural country, Wright had no difficulty finding a sharply defined southern edge: beyond it, till and erratic boulders were absent, there were no scratched stones or striated bedrock, valleys were steep and narrow, crop production was significantly lower, and surface soils were shallow and heavily weathered.[141] He was sure he had "fixed with certainty" the glacial boundary in both space and time. The more heavily weathered unglaciated portion of Ohio south of the moraine contrasted so sharply with the "surprisingly small" amount of erosion and channel deepening Wright observed north of it that glaciation must have occurred at a "later date . . . than that which is advocated by some."[142]

In tracing the moraine, one observation stood out above all others: it appeared that in the vicinity of Cincinnati the glacier had crossed the Ohio River into northern Kentucky, where Wright spotted what he believed was glacial till on high terraces about five miles south of the river. The till consisted of erratic boulders of a distinctive metamorphic conglomerate that contained red jasper pebbles, the source of which was an outcrop on the northern shore of Lake Superior, some 600 miles distant (one of those erratic boulders was destined to serve as the headstone on Wright's grave [§7.8]).[143] Wright suspected that where the glacier overrode Cincinnati and crossed the Ohio River it was 600 feet thick and fifty miles wide and had no subglacial channel. He was sure an obstruction that massive must have dammed the Ohio River, creating behind it a lake so vast and deep that Pittsburgh, 500 miles upstream, would have been under 300 feet of water.[144]

Wright named it the "Cincinnati Ice Dam" and was inordinately proud of it (decades later he fondly remembered "large audiences gathered" to hear his report on it).[145] He first aired the idea in the spring of 1883, and it caught the attention of West Virginia University geologist I. C. White. That summer at the AAAS meeting White announced that Wright's ice dam helped resolve a long-standing anomaly. White had observed a 130-mile stretch of high terraces along the Monongahela River, the highest of which was apparently at a uniform elevation (1,065 feet above sea level), showed unmistakable signs of "rubbish usually transported by streams," and away from the main channel had thick deposits of silt and clay indicative of quiet backwaters. Their origins had puzzled White; the Cincinnati Ice Dam provided an answer. The uppermost silt and clay units marked the settling of sediments from the Monongahela and its tributaries when they entered the lake behind the dam, and the lower terraces were cut during subsequent "spasmodic lowering of the dam." In light of that evidence, White declared "Wright's hypothesis as proved beyond a reasonable doubt."[146]

Pennsylvania's Lesley thought so too. White's paper (at least according to Wright) "brought [Lesley] to his feet to express his delight that at last evidence had been produced to establish such an obstruction to the drainage of the Ohio."[147]

Lesley had also puzzled over those same terraces and had offered—but then withdrew for lack of evidence—the hypothesis of regional submergence. He had been waiting ever since for the "means by which [the terraces] could have been produced. Mr. Wright's ice dam at Cincinnati furnished the means."[148]

Chamberlin was at that AAAS meeting but made no comment. He had two papers to present, one on the outer border of the drift, the other on the inner terminal moraine of the second glacial advance. If Wright was in Chamberlin's audience, and he likely was, he could not have failed to see the gap between their views. Chamberlin's papers were about two distinct moraines across North America. The older of the two, Chamberlin observed, was not nearly so distinct as the younger, having "suffered fully twice as much" erosion.[149] He had seen the "vanishing edge" of the older moraine in Ohio west of the Scioto River valley and must have been troubled by Wright's claim of a single terminal moraine across the entire state.[150] Wright, in fact, made no mention of the possibility of a second moraine, and while that might be explained by the fact that his zigzag course had taken him only twenty miles or so on either side of the moraine, Chamberlin would have seen that portions of the line Wright mapped overlapped the Kettle Moraine. For that matter, Wright recorded areas in Ohio where fringe was present, but he "did not think it best" to draw that broken and discontinuous line across the state, leaving it to others to determine why the fringe was present in some places, absent in others. The fringe soon provided Chamberlin a wedge to pry apart Wright's work, but he made no comment at the time, as he was caught up instead in a discussion of the origin of the Great Lakes that followed his paper.[151]

Warren Upham, who was present and spoke of evidence from Minnesota for two distinct glacial periods,[152] simply assumed Wright was mapping the earlier of the two moraines, and he hoped Wright would have the opportunity to continue tracing the "moraines of the ice-sheet, both at its extreme limit and in the more massive moraine accumulations formed during the latest epoch of glaciation."[153] Wright was about to get the opportunity from Chamberlin himself.

4.9 AN UNEASY ASSOCIATION

Chamberlin and Wright first made contact in December of 1881, when Chamberlin asked Wright for his papers on glacial geology. He watched as Wright helped Lewis on the Pennsylvania survey, and then as Wright had traced the moraine across Ohio from 1882 to 1884. He had heard Wright (and perhaps met him for the first time) at the August 1883 AAAS meeting. By then he was likely aware of Wright's resistance to the idea of multiple glaciation.

In 1884, however, Chamberlin also needed field assistants for his USGS Glacial Division. As was common procedure in those years, part-time staff, from skilled geologists to common laborers, annually constituted nearly half the USGS workforce. Hiring seasonal field workers was a budget-stretching and patronage gambit on Powell's part that enabled him to spread USGS largess across various states, especially ones with powerful congressmen. Sometimes this was done at the insistence of congressmen, but for Powell always with an eye toward winning allies for his Survey.[154] The hiring was left to the discretion of the division heads, and for his part Chamberlin sought college teachers for the temporary slots: many could be deployed in the summer, enabling him to get as much accomplished in a summer at lower cost than he could over "the entire year [with] one fourth the number."[155]

Wright requested one of those temporary positions, and on the recommendation of Upham and Haynes he was hired in the spring of 1884.[156] In hiring such part-time help Chamberlin risked employing individuals whose abilities and expertise varied, who might have a different perspective on geological problems, and who might aspire to be more than just temporary data gatherers. Wright was one of those risks, and though properly pious Wright made little secret of his almost impious scientific ambition. He frankly hoped his original geological and archaeological investigations would "enlarge somewhat the boundary of human knowledge" and fashioned himself a key player in those fields. His USGS appointment affirmed that vision.[157]

Chamberlin was just as ambitious, but also autocratic and intellectually intolerant. He knew the risk in hiring Wright and told him he was well aware Wright was "known to entertain views inharmonious, in some respects, with those who were to become [his] colleagues."[158] He hired him anyway, judging Wright a competent field geologist, hoping his energy could be harnessed for good use and that perhaps in time Wright might realize the flaws in his thinking.[159] He put him to the test straightaway, assigning Wright the task of "tracing the border of the drift [from Ohio] west to the Miss. river & in the further investigation of the effect of ice-interrupted drainage in the Ohio Valley." He particularly wanted Wright to investigate the Cincinnati Ice Dam thoroughly, and "attack both from the affirmative & negative sides." He assured Wright that "the option of its solution should rest with you."[160]

Chamberlin's resolve to leave responsibility to Wright was challenged almost immediately. Two months later in the *American Naturalist*, Wright summarized his ice dam theory, boasting of its "strong" confirmations by White, Lesley, and G. H. Squier in Kentucky. He called for much wider observations all along the Ohio River and its tributaries above Cincinnati, where "there should be found numerous facts, explicable only by this theory."[161]

Chamberlin was in the field at the time but soon came across Wright's paper. After reading it, he promptly wrote to again urge Wright to consider the alternative hypothesis that the terraces of the upper Ohio River that "have been the subject of so much strained theorizing" (Wright's strained theorizing, obviously) were "beyond question" merely glacial drainage terraces and hence had their origins in meltwater from the north, not lakes to the south. Besides, Chamberlin added, the "terraces do not belong to the same epoch as the Cincinnati ice blockage but are later."[162] In case Wright failed to get the message, Chamberlin reminded him that

I was at one time preparing to discuss these points in print, but since you are to investigate the subject I deem it more courteous to submit my views to you. If these brief suggestions are not sufficiently suggestive . . . I will endeavor to find time for a more explicit statement.[163]

And lest Wright miss that veiled threat, Chamberlin wrote two weeks later, making his restrictions more explicit and official:

> The approach of the Associations [the 1884 AAAS meeting] reminds me of a rule of the survey that I may not have stated to you, viz: In cases in which members of the survey desire to present to scientific bodies or to publish any results of survey work, permission to do so is to be obtained from the officer to whom the party makes report [Chamberlin, in this case].[164]

Just in case that somehow was not clear, Chamberlin wrote again a week later, and this time he took the gloves off. He informed Wright that Powell had approved the proposed budget for the Glacial Division, except for the funds for Wright's investigations on the upper Ohio. "In view of this . . . it would not be wise to discuss the subject [of the ice dam] at [AAAS] or elsewhere until the future course of the work shall be determined."[165] Wright's fate with the USGS was none too subtly linked to his submission to Chamberlin's rules. Wright's AAAS paper steered clear of the ice dam.

If Wright was checked, it was only temporary. The following spring (April 1885), he spoke about his USGS-sponsored investigations at the BSNH's monthly meeting, in which he highlighted his Cincinnati Ice Dam. Chamberlin was not there to hear it but learned of it soon enough. Ten days later a notice of the talk was published, and almost immediately an angry Chamberlin fired off a letter. Publications must be submitted through proper channels, he warned Wright, not merely as a matter of form but "as a wise precaution against immature and premature publication." Wright's public comments on the ice dam, by virtue of his employment, came with the apparent endorsement of the USGS. This was "peculiarly unfortunate" because he had failed to give due weight to "opposing opinions [and] evidence thought to be unfavorable to the views advanced."

That was the rub for Chamberlin: he favored open discussion in science. After all, he pointedly reminded Wright, he hired Wright despite his "inharmonious" views. But this spirit of openness demanded that all sides of a controversial issue be weighed carefully. Wright failed to do that, let alone consult "the very large mass of evidence" the Survey had amassed on this issue which, Chamberlin believed, ought to have led Wright to a very different conclusion about the terraces of the upper Ohio.[166] Yet, just as Chamberlin was reprimanding Wright, Powell was grateful Wright was on the USGS rolls, for the Survey was under increasingly intense congressional scrutiny.

4.10 HARD TIMES FOR THE USGS

The 1884–1885 budget that emerged from the appropriations committee provided the USGS with $386,000, an increase of more than 60 percent over the previous fiscal year. But when the budget was approved in July of 1884, two items were tacked on that struck at the USGS. One was a directive to the Joint Committee on Public Printing to reduce the cost and number of items printed, among the most expensive of which were USGS *Annual Reports*, *Bulletins*, and *Monographs*.

The prior year these had cost nearly $150,000; that figure was expected to balloon to more than $200,000 the coming fiscal year.[167]

The other item was the establishment of a Joint Congressional Commission, chaired by Iowa Senator W. B. Allison, that aimed to sort out the responsibilities of the USGS, the Coast and Geodetic Survey, the Signal Service, and the Hydrographic Office of the Navy Department. These several agencies too often appeared to duplicate each other's efforts, especially topographic mapping. The Allison Commission included Representative Hilary Herbert of Alabama, a former Confederate Colonel and staunch opponent of federal involvement in science and other matters he believed were rightfully those of the states, as well as Theodore Lyman, a graduate of Harvard's Lawrence Scientific School and a self-made millionaire who lived in a Horatio Alger world in which fortunes waited anyone with initiative and integrity.[168] He was also Alexander Agassiz's brother-in-law and shared his antipathy toward the USGS. Lyman called on the National Academy of Sciences to consider the proper role of the federal government in science. Through the summer and fall of 1884 they looked into the problem while USGS crews, Wright now among them, went in to the field.[169]

Despite Powell's well-regarded testimony before the Allison Commission,[170] by the spring of 1885 the USGS was under ever-more-intense scrutiny. Democrat Grover Cleveland had been elected as the U.S. president and came into office having sworn to reduce a federal bureaucracy bloated by years of unchecked expansion under successive Republican administrations. Cleveland was particularly scornful of science, and—sensing the mood—non-USGS geologists and others with grudges toward the USGS spread rumors of scandal and corruption in the Survey. Newspapers reported rumors of Powell resigning and publicly floated names of possible successor.[171]

One of those stories appeared in the Cincinnati *Commercial Gazette* in April of 1885, just as Chamberlin was reprimanding Wright. That story went so far as to claim USGS members were forced to believe in evolution against their will, and it gave Powell a golden opportunity for rebuttal. Not only was scientific fitness and not apostasy the main criterion of appointment, he was pleased to say he even had two clergy on the USGS staff. However much the Reverend George Frederick Wright was a thorn in Chamberlin's side as a field assistant for the USGS, Wright was a potent symbol of Powell's generosity and intellectual open-mindedness.[172]

Seeing which way the political winds were blowing, Chamberlin backed off on his criticism of Wright. He insisted it was not his intent to "embarrass [Wright] or in any way trammel [his] independence of investigation." Still, he reserved the right to make suggestions and promised that when Wright's report was written it "will become my duty to review your matter."[173] Indeed, anticipating what might be in Wright's report, Chamberlin and Grove Karl Gilbert (then head of the USGS Appalachian Division) spent much of November of 1885 examining the terraces of the Monongahela, Alleghany, Beaver, and Ohio Rivers for what they might reveal of "the distinctness of the glacial epochs and the discrimination of the character and conditions of the formation of the terraces."[174] Chamberlin let Wright know he was doing so, insisting it was "an old plan of ours." Still, he was coy about it. Wright would have to wait until the results were published in the USGS *Annual Report* to learn what they had found.[175]

In 1885 Wright was on the USGS payroll for nearly two months, at a time when criticism of the USGS was building. Agassiz was stirring up matters behind the scenes but finally showed his hand in a September New York *Evening Post* editorial. He took particular aim at Powell's large discretionary fund and unsupervised spending: government science budgets needed to be "so complete and detailed as to invite a fair and open criticism."[176] When Congress reconvened in December, Allison reopened the Allison Commission hearings. Alabama's Herbert was primed, having heard complaints from Eugene Smith, his state geologist, who had suffered the Survey's sting. (The USGS's WJ McGee had condescendingly reinterpreted one of Smith's reports in accordance with one of his own pet hypotheses, then sent it back with the suggestion that Smith test it in the field.)[177] Herbert also read into the commission report a long letter from Agassiz that excoriated the wastefulness and extravagance of the USGS and insisted that work, especially in fields such as mining and vertebrate paleontology, could and should be done in the private sector. Agassiz, who made his fortune in copper, had an obvious conflict of interest.[178]

Powell fired back, first in testimony before the Allison Commission in late December and early January, then in a February letter responding to Agassiz's charges directly, a letter roughly twenty-five times longer than the one that provoked it.[179] The Survey, he insisted, was operating and attacking problems at a scale well beyond the means of any individual, society, or museum, Agassiz's included. If private enterprise was so effective, why had the work of the USGS not already been done? Scarcely 5 percent of the geological work represented in the Survey's library, Powell observed triumphantly, was the result of private investigation. A "hundred millionaires could not do the work in scientific research now done by the General Government."[180] The Allison Commission ended fifteen months of hearings in February 1886, and the USGS awaited a decision on its fate and that of its budget.

Herbert, too impatient to wait and perhaps sensing the outcome would not be to his liking, made a preemptive strike. In April he introduced a bill (rumored to have been drafted by Agassiz) to severely curtail the Survey's operations, virtually excise its paleontology program, discontinue its *Monographs* and *Bulletins*, and sell off most of its laboratory equipment and other property. On its heels he also publicized the rumor that Powell employed sixty-nine per diem workers around the country, all of whom had private positions and were thus double-dipping at government expense. But Herbert's facts were wrong and he had gone too far. The move backfired.

The scientific community, even those not wholly disposed toward the Survey, rallied behind it. Powell himself attacked Agassiz as a jealous hypocrite whose only motive was to improve his museum at Harvard. Senator Allison almost immediately issued a disclaimer that the members of the commission had not yet finished their report, and the majority of its members repudiated Herbert's move, partly after hearing from the many districts (and from many on the infamous list of sixty-nine double-dippers) where Powell's Survey had so strategically dispersed funds and conducted research in the preceding years.[181]

When the Allison Commission issued its final report in June of 1886 it was a near-total vindication of the USGS.[182] The majority agreed that the Survey was important for the greater scientific good, that it was well administered, and that its scientific direction was its own (Powell's) business.[183] In so doing it affirmed the principle that scientific research was within the purview of the federal government. Herbert made a last-ditch effort to trim the USGS appropriation but failed. In late June the Survey received $467,700 for assistants and expenses, almost $82,000 more than the previous fiscal year.[184]

Chamberlin happily reported these results to Wright, his tone undoubtedly reflecting the fact that with the USGS now sufficiently secure on Capitol Hill it no longer needed to keep token clergy on its rolls. Chamberlin began to cut Wright's ties to the Survey.[185] Wright's work in fiscal year 1885–1886 amounted to only twenty-four days in the field and an equal amount of time completing the report of his fieldwork.[186]

At the same time, Chamberlin's direct involvement with glacial studies was drawing to a close. In two long papers in mid-decade, published with the sixth and seventh USGS *Annual Reports*, he summarized his research on the Driftless Area and on glacial erosion. By then, he took for granted not only that there were two major glacial epochs, but also that the latter was divisible into several stages or subepochs of glaciation and deglaciation.[187] The older of the two major glacial epochs was distinguished by its "attenuated border" (which Chamberlin ascribed to its "feeble tendencies" to aggregate in moraines), whereas the younger was well marked by strong glacial action, erosion, and immense moraines.[188]

Both moraines were visible around the 10,000 square mile Driftless Area of central Wisconsin, an area that, Chamberlin realized, would have been a prime setting for an ice dam. Ice had advanced toward the region, split and skirted around it, then rejoined and extended south another 200 miles. While puzzling over the possible water-laid origin of the loess that covered the western margin of the Driftless Area, Chamberlin considered, then rejected, the hypothesis that an ice dam had backed up either glacial meltwaters or the Mississippi River, which ran through the Driftless Area.[189] It was a lesson well learned in regard to the loess (and from it Chamberlin later moved toward the idea that loess was aeolian in origin [§6.7]), but Chamberlin also saw its implications for Wright's Cincinnati Ice Dam.

There were lessons, too, in the patterns of glacial erosion across the United States: ice scoring was more pronounced and abundant in areas north of the later (second) moraines than south of them. Chamberlin believed this was not merely because those areas would have been subject to recurrent glaciation, or even because they were fresher surfaces (although that played a role). Rather, it was that the younger ice advance was more "vigorous."[190] That the older event was more feeble and less visible on the ground provided Chamberlin with a ready explanation for Wright's fringe.

By mid-decade, Chamberlin was well armed for any dispute in glacial geology, and the substantial body of work he produced in the years since joining the USGS propelled him to the forefront of that field. In August 1885 he was elected AAAS

Section E (geology and geography) vice president. He was the first USGS official so honored. Chamberlin used his vice presidential address the following year to take stock of accomplishments in glacial geology. These included a nearly complete map of the southern limit of the glacial formations in North America; the demonstration that most of this vast tract was buried in ice (as opposed to its debris having been left by waterborne icebergs or some other agency); the evidence of at least two glacial epochs and the identification of their respective traces; the detailing of the origins and distribution of moraines, tills (subglacial debris), loess, and striae across much of North America; and, finally, the "working hypotheses" necessary for the tracing out of glacial features that have "become rich beyond the limits of convenient statement."[191] He did not mention any ice dams.

4.11 WRANGLING OVER THE GLACIAL BOUNDARY

Altogether, it was a long list of achievements, one for which Chamberlin could fairly claim the lion's share. There was, of course, plenty of work left to do, such as more precisely correlating different glacial features across the landscape and teasing apart the remains of the two ice advances in regions (like western Pennsylvania) where they apparently overlapped. But with his AAAS address Chamberlin signaled these were details he would leave to others: he himself was moving on to a much larger problem. However much the study of glacial deposits had advanced—and Chamberlin believed it had advanced a great deal—knowledge of the cause or causes of continental glaciation remained limited to "the same old stock of hypotheses, but all badly damaged by the deluge of recent facts." Even his onetime attraction to Croll's "fascinating hypothesis" had faded in the face of the "asymmetry of the ice distribution in latitude and longitude and its disparity in elevation" (the fact that glaciers did not occur across the northern hemisphere simultaneously).[192] Here was a scientific problem sufficiently challenging to warrant Chamberlin's attention. Within a few years he offered his first ideas on the question, which over the next several decades evolved into an inquiry in the early history of the earth's atmosphere, which in turn lead him late in life toward seeking a theory of the history of the earth and planets.[193]

There was another, far more pragmatic reason Chamberlin left the mopping up of North American glacial studies to others. By the time of his AAAS address Chamberlin had accepted the presidency of the University of Wisconsin; he took office in the latter part of 1887. He still kept his USGS title and nominal duties but turned the field studies and daily operations over to others, notably Frank Leverett, his onetime student and by then USGS assistant. Nevertheless, since Chamberlin retained official charge of the Glacial Division he continued to direct the activities of all the division geologists, including those of Wright, who had a report to publish on the surveys he'd done while in USGS employ.[194]

Wright's report was slated to appear as a USGS *Bulletin*, and Chamberlin insisted it cover only new data and discussion. He warned Wright he would find objectionable any reiteration of previously published matter, unless it was directly relevant.[195] In this, of course, Chamberlin held all the cards. He would judge the adequacy of Wright's report and reject it if he pleased. The opportunity to do so came soon enough. Wright submitted a draft in June of 1886, admitting he would not be surprised if Chamberlin "should express dissent" from some of his views. He was right. As the manuscript went back and forth between them, Chamberlin grew increasingly frustrated with Wright's intransigence, and at one point he was ready to give up, telling Wright that after he weighed Chamberlin's latest comments,

> all the good likely to come from [our] deliberations and discussion will probably have been reached whatever may be the result and the facts may as well go freely to the public.[196]

But he reconsidered, and a year later was correcting more pages, which now included a discussion of the Ohio paleoliths. Wright took this material almost verbatim from a paper he had published in the Ohio Archaeological and Historical Society *Proceedings*, including the rhetorical flourish that the evidence proved a "preglacial" human presence (§4.2). Chamberlin thought this discussion added "interesting and valuable material" to the report, but flatly rejected the idea that paleoliths were preglacial in age:

> They merely indicate that man was present at the time these deposits were laid down, and that, if I interpret correctly, was during the later glacial epoch, which I am inclined to regard as being nearer our own day than preglacial times. In short, I do not think this evidence carries man half way back to preglacial times.[197]

When Wright's *The Glacial Boundary in Western Pennsylvania, Ohio, Kentucky, Indiana, and Illinois* finally appeared in 1890, it did not appear alone. Chamberlin took the extraordinary step of prefacing it with a twenty-five-page Introduction (nearly a quarter of the entire *Bulletin*) "reviewing certain questions of wide interest and debatable interpretation."[198] The way Chamberlin saw it, there was plenty of interpretation to debate, for Wright had not conceded any ground on their differences.

The southern limit of glaciation as Wright traced it between the Alleghenies and the Mississippi River consisted largely of fringe, with a few "marginal accumulations worthy to be called a moraine."[199] The occasional moraines or morainelike features present along the line as well as those well north of the fringe margin were duly noted, but Wright said not a word about the age of the fringe relative to those better defined moraines. And in saying nothing Wright said everything: from his silence one could easily gain the impression that fringe and moraine were both part of the same glacial event. Chamberlin might forgive Wright's disagreeing with him, but he could not forgive Wright's implying no disagreement existed. That, to Chamberlin, was scientifically immoral (§4.6).

Even worse, Wright openly flaunted his Cincinnati Ice Dam. Assuming that the presence of ice straddling the Ohio was sufficient proof the river was dammed, Wright then turned to the upstream terraces to determine whether they were better explained by an ice dam or alternate hypotheses—regional subsidence or possibly drainages from an earlier glacial epoch (Chamberlin's idea). The latter he readily dismissed: the upper terraces were 300 to 400 feet higher than the lower ones, and if both were glacial in age, as he believed they were, that implied a vast interval of time between them. But such was nullified by the "vegetable and animal remains of recent species found in a very fresh state of preservation" in both sets of terraces, and by the spatial congruence of the deposits. It seemed highly improbable to Wright that two glacial events widely separated in time would have overlapping deposits. Moreover, Wright was not swayed by the peat deposits found between tills that were also said to indicate two widely separated glacial events: peats could develop in only "a few hundred years."[200] Besides, if these were drainage channels, how could they account for the thick slack-water deposits of fine clay that "extend across the valley, covering everything to a uniform depth?"[201] The hypothesis that the upper terraces formed during an earlier glacial epoch failed. The other alternative, general regional subsidence, fared no better, as it implied so many other enormous changes that it seemed "an inordinate cause to introduce to account for so small a result." Thus, he concluded that

> both the theory of continental depression and that of such an enormous period between two nearly conterminous fields of successive glaciation . . . must give place to the simpler theory that the ice . . . temporarily obstructed the channel, forming thus an ice dam.[202]

So it seemed to Wright and West Virginia's White (§4.8), who continued to endorse the ice dam as accounting "for all the phenomena of surface geology . . . along the rivers west from the Alleghenies."[203]

This was more than Chamberlin could bear. In his Introduction to the *Bulletin* he duly acknowledged Wright's efforts at tracing the limits of the outer moraine in Pennsylvania and Ohio, for extending that line to the Mississippi River, and "the obviously great value of the data" presented. It was appropriate he accept the report. However, he deemed it necessary to call attention to certain points of "divergent interpretation" lest "confusion or misapprehension" be introduced into the literature, particularly to the general reader who knew nothing of the debate and might otherwise be misled.[204]

Chamberlin pounced first on Wright's fringe. As he saw it in eastern North America the southernmost edge of ice advance was marked by a distinctive ridged accumulation or terminal moraine, whereas in the Mississippi valley "the drift ends in a thin margin, unmarked by specific terminal ridging." The two were not contemporaneous. Rather, the distinctive terminal moraine of the Appalachians and Alleghanies was related to the later (second or kettle) moraine in the Mississippi valley, whereas the outer, thinner drift margin in the Mississippi valley continued east to form the line of "thinly dispersed erratics" in western Pennsylvania. The ice sheets marked by these "attenuated borders" (Chamberlin believed there were more than one) were more ancient than the drift that terminated in the more easily recognized ridged or terminal moraines. In the circumstances, there was no reason to accept the idea that the ice "border is a historical unity" (that is, was from the same single advance).[205]

If Chamberlin's hypothesis was correct, it had to explain Wright's (and Lewis's) observation that the surface immediately north of the fringe showed no evidence of glacial striae or other erosion. Chamberlin easily disposed of that problem, having already set the stage several years earlier in his study of glacial erosion. Here he merely reiterated his observation that an ice sheet might extend over large areas "without forcefully abrading the surface" or "notably plowing up material at its edge."[206] And that fact bore directly on Wright's ice dam.

Chamberlin conceded that "bowldery drift" south of the Ohio River indicated that an ice sheet had made it over to the Kentucky side, although he credited that observation to earlier work by others, not Wright. Yet, the mere presence of ice on the Kentucky side was not proof the river was dammed. The Kentucky drift was so "trivial" that the ice was hardly likely to have formed an obstruction capable of backing up a lake that reached Pittsburgh. Such a dam would have had to withstand pressure nearly ten times greater than that of contemporary Alpine glacial dams, which were themselves largely impermanent and riddled with tunnels, frequently leaked, and did not have to hold back a body of flowing water anything like the silt-and-glacial-meltwater-laden Ohio River. Even supposing the Cincinnati Ice Dam held, the lake water would have soon lifted the ice and breached the dam. "Several conditions favored the stream," he wryly observed, "in its struggle with the supposed encroaching glacier."[207]

Chamberlin was not even willing to grant temporary or partial damming. Unlike White and Lesley, he did not need a dam to create terraces 200 to 300 miles upstream because "intervening agencies" could have formed those. After examining those terraces with Gilbert in the fall of 1885, Chamberlin doubted they were lacustrine in origin, for in contrast to what White claimed, the terraces sloped with the present streams, "precisely as the river terraces do today, and at gradients not greatly different." Further, they were composed of fluvial gravels (and not just clays), were cut into rock (which, if a lake was the erosive agent, would have required an immense amount of time), and were connected to abandoned channels and oxbows, unmistakable proof of fluvial action.[208]

They were, however, ultimately glacial in origin, though not in a way Wright or White supposed. Chamberlin argued that the terraces formed from drainage during the earlier glacial epoch. In addition, below these were a set of "moraine-headed terraces" that were products of streams from the glacier that existed during the second glacial epoch. The terraces thus not only failed to support Wright's ice dam hypothesis, they in fact testified to two glacial periods.[209] As to Wright's claim that the well-preserved "vegetable remains" belied two such long-separated glacial events, Chamberlin merely observed that

> the whole tenor of the study of fossil remains is to emphasize the fact that the preservation has been extraordinarily perfect under certain conditions, while under others the completeness of the destruction has been equally phenomenal. It appears to me that the rate of decay and of lithification [fossilization] of buried vegetal material is a problem rather than a proof. Very fresh-looking wood is found in the *Tertiary* deposits of the lower Mississippi Valley.[210]

Chamberlin's assault began on general principles and ended in the terraces of the upper Ohio River. Wright was put in his place and suffered a very public humiliation. Chamberlin had said at the outset that Wright was competent to collect data, but anyone who read his Introduction could see he hardly meant what he said, let alone that he deemed Wright competent to interpret what those data meant.

4.12 SYNTHESIS AND ANTITHESIS

Chamberlin could easily resolve his disagreements with Wright's *Bulletin*, which had to meet Chamberlin's scientific standards and editorial approval. Chamberlin could hardly control what Wright might publish once released from these official restraints. Chamberlin knew it. Wright unquestionably knew it. Even before the type for his *Bulletin* was set, Wright was wrapping up a book manuscript. He would not kowtow to Chamberlin in it, he assured James Todd. Todd, like Wright, was on the faculty of a small college (Tabor, in Iowa), was a summer USGS employee, and had also run afoul of Chamberlin:

> I am not surprised at your experience with Professor Chamberlin. He is not patient with adverse theories. He has written to me some very crusty letters; but fortunately I was not so far under his thumb that he could altogether control me.[211]

Wright's book aimed to present Ice Age geology and archaeology to the public, and it was a brazen move on his part. Having faithfully discharged his USGS obligations, he would now vault over Chamberlin's head to address a public audience, one that Chamberlin considered it his self-appointed duty to protect from immoral science (§4.6). Knowing full well that doing so would provoke Chamberlin and possibly delay Wright receiving permission to use some of his USGS results in the new book, he sent his request to Powell as well as Chamberlin. Three days later, with Wright's assurance that due credit would be given the Survey for work done under their auspices, Powell granted his approval.[212]

Two weeks went by without response from Chamberlin. Wright sent a second letter, offering to "submit any portion of the MS. which may be necessary to your criticism," but letting him know Powell had already approved the project. Wright even happily chirped that his book would "serve a good purpose in bringing before the public the great services which investigations of this sort are rendering them."[213]

That did it. Two weeks later a seething Chamberlin answered. Yes, he had both of Wright's letters but had delayed responding because he was in doubt as to what to say and was reluctant to say it. But because Wright insisted, Chamberlin told him he was free to use material from his *Bulletin* as it was a matter of public record. But presenting a book to the general unscientific reader, Chamberlin sternly lectured, was altogether premature. "Leading and important truths" relating to glacial formations were not yet known, and in regard to the "antiquity of man there cannot even be a critical and specific *statement of the problem*" until the chronological relations of glacial and nonglacial deposits were determined.

Any work that promised conclusions would only confuse matters later. The only justifiable publication was one that "consists of a very careful, conscientious, critical and appreciative exposition of the varying views held by competent workers, together with a sharp discrimination between that which is demonstrative and that which is but believed." That sort of book could be written only by one with "determinate knowledge."[214]

Chamberlin left it unsaid but it did not matter: he obviously thought Wright incapable of such a book. Frankly, Chamberlin snarled, he could "only look upon such a publication as you propose as being, in the present state of investigation, premature and unfortunate both for science and for the public." Given these circumstances it would be better if they ended their official association:

> I could not obviously write you so frankly were it not for our official relations, but so long as you remain a member of the Survey . . . I in some measure partake of the responsibility for your publication. This responsibility I am not willing to assume, and as the relationship has ceased to be active . . . I think it will free us both from embarrassment and give you perfect freedom to follow your judgment if the relationship shall cease.

Strong words, but Chamberlin knew he was fighting a losing battle. Wright's book would appear whether he cooperated or not. Still, he could not resist a parting insult:

> What I have written, I have written with very great reluctance, but I am impelled to be consistent with my conception of scientific and educational ethics and with the canons of practice which have withheld me from a popular utilization of work whose extent would probably justify me, if any one, in attempting to secure popular returns.[215]

Here, perhaps, was what was most galling. It was not that Wright aspired to write for the public, or that Chamberlin thought him incompetent to the task, but that in doing so Wright was usurping a privilege Chamberlin considered justly his.[216]

Chamberlin wrote Powell the same day, enclosing a copy of his letter to Wright. It was a harsh letter, he admitted, but Chamberlin the scientific moralist saw no alternative "consistent with what seems to me to be due science and to the public." The experiment in hiring Wright had failed:

> I now realize that it was a mistake to accept Professor Wright as a member of the Survey, but at the time I so recommended I thought that he would come to an appreciation of our more critical methods and their necessary results, and that a valuable worker rather than a seeker for public fame would be developed. But he has failed to appreciate the force of evidence that is practically demonstrated and there seems to me little hope of his adoption of thoroughly scientific methods, or of arriving at trustworthy conclusions.

Powell agreed.[217]

Chamberlin's letter to Wright did not have its desired effect, though by now he hardly expected it would. Wright offered his regrets that Chamberlin did not approve of his plan of publication. But that was all he offered. Chamberlin's suggestion that Wright resign from the Survey was ignored, while he tried once again to justify his course:

> I hope, however, and believe, that such a presentation as I can now make will advance knowledge and prevent the spread of misconceptions; and, at the same time, that it will not prejudice, but rather prepare the way for, the more thorough treatment of the subject which by general consent will at last fall to your own hands.

Besides, he assured Chamberlin, his book was needed because "Professor Putnam and others are going to publish pretty soon facts which will create new interest in these geological discussions," and someone (he obviously believed it might as well be him) must assume responsibility for "clearing the ground." Wright neglected to mention that Putnam was unhappy about his book as well, as it "cuts in upon a little work of mine [Putnam's] now in progress . . . [that] will consist entirely of a discussion of the implements found in the gravels." (Putnam's book never appeared.) Wright was assuming tasks both Chamberlin and Putnam thought were duly theirs.[218]

Chamberlin did not bother to reply. There is no reason to think he was anymore convinced by Wright's arguments the second time around. The breach between them was final: Wright's ambitions had collided head on with Chamberlin's imperiousness, and it triggered a mutual loathing that reverberated for decades. Chamberlin's animosity proved savage and unforgiving. Wright's antipathy was tempered with fear, but it solidified into a hardheaded refusal to surrender any ground on issues that divided them. His stubbornness and Chamberlin's malevolence guaranteed they collided each time there was occasion for their differences to surface, and given the aspirations of each, those occasions came often. Their malice bound them together in a relationship as tight as any shared by kindred souls.

4.13 WRIGHT'S *ICE AGE IN NORTH AMERICA*

Wright's *The Ice Age in North America and Its Bearing upon the Antiquity of Man* had its origins in lectures he gave at the Lowell Institute in Boston in November and December of 1887 (§4.3). Lowell lectures addressed issues in philosophy, religion, natural history, and the arts and sciences, and though the Lowells themselves opposed evolution they brought in scientific lecturers of all stripes, albeit often selected ones with religious leanings. The lectureship was a preeminent honor that provided its speakers with a prominent Boston stage, a packed house, an almost certain guarantee of publication, and a substantial honorarium ($1,000).[219]

Wright's friend and scientific mentor Asa Gray (§3.8), himself a prior Lowell lecturer, had suggested the Lowell trustees invite Wright to deliver talks on North American glacial geology. It was a triumphant moment. Six years earlier Wright had left Boston, worried his exodus would mire him in scientific backwaters. Now he was returning as a respected glacial geologist to speak where he had found his scientific calling, and in the city where his first paper had been given just over a decade earlier.

From Wright's vantage the honor was well earned. In the intervening years he could boast of having "traced the southern boundary [of the glacier] myself" from Long Island to the Mississippi River, and even spent two brief seasons (1887 and 1888) in the Dakota Territory and the Northwest. Few others, he believed, had such extensive experience.[220] He enjoyed another edge too, having spent August of 1886 camped in Glacier Bay, Alaska, measuring the movement of the Muir glacier and exploring its surroundings. It was fieldwork favorably viewed even in certain USGS quarters. McGee, for one, thought the Muir study set new standards: "The time has now come when researches in superficial geology generally, and in the glacial drift particularly cannot well be prosecuted further without observations on living glaciers."[221] Chamberlin's reaction to the trip was to deny Wright's request for USGS funds to finance the venture, because doing so would have bestowed official sanction.[222]

Before he left for Alaska, Wright and his backers (notably Judge Baldwin, whose son accompanied Wright) tested public interest in his glacial research and sought to secure additional funding from the New York *Tribune* and the *Century Magazine* in return for letters from the glacier's edge. Those efforts proved unsuccessful, but no matter: the Lowell lectures presented him with a golden opportunity to "meet a widely felt want" he perceived on the part of the public.[223] Judging by the response, he was right. His lectures were well attended, and Wright was invited to speak at Harvard (where he dined with Alexander Agassiz). Afterward, he was pursued by several publishers seeking the rights to his Lowell lectures. It was an exultant moment for Wright, and in the circumstances this was a book he believed would be "a rather important one" and he would not fail to publish it, Chamberlin notwithstanding.[224]

Like Abbott's *Primitive Industry* (§3.6), Wright's *Ice Age in North America and Its Bearings upon the Antiquity of Man* transparently imitated a European model, in this case James Geikie's *The Great Ice Age and Its Relation to the Antiquity of Man*. Like Geikie, Wright first detailed the causes, consequences, and traces of glaciation, then devoted the last few chapters to the evidence of Paleolithic humans. But their books differed in one very obvious respect. As early as 1877 Geikie had heartily endorsed the idea of multiple glaciation in America (§3.5). He did so despite the fact that the physical evidence of multiple glacial advances on this continent was then relatively meager and consisted of little more than the preliminary suggestions of Orton and a few others that the "forest bed" of peat layered between glacial outwash deposits in Ohio and Minnesota might represent an American interglacial.[225] Twelve years later, in the face of much more evidence and arguments in favor of multiple glaciation, and under Chamberlin's menacing glare, Wright still found the idea unpalatable. His book was intended to be "a pretty complete digest of all these investigations" into North American glacial geology, but he could not help but spell out in it his objections to the idea of multiple glaciation.

Ice Age was Wright unshackled. He had assured Chamberlin he would give full attention to alternative hypotheses so as to "prevent the spread of misconceptions," but Chamberlin doubted that would amount to more than window dressing.[226] He was right. Considerable effort, Wright claimed in *Ice Age*, had "been directed towards [the] verification or disproval" of the Cincinnati Ice Dam, and he insisted a theory of such wide significance should not be too hastily accepted. But the only arguments he mounted against it were feeble: a few minor errors of interpretation, which he conceded; the flimsy idea of regional subsidence, which he wheeled out and easily toppled; and, the claim that the terraces were formed during an earlier glacial epoch, which he argued implied a completely unacceptable lapse of time. Those duly disposed of, he was pleased to claim that the theory of an ice dam "so naturally accounts for so complicated a set of facts" as to be "well-nigh proved." Wright was "confident that close reflection upon the evidence already presented will be sufficient to produce conviction in most minds." It certainly had for other authorities, and the emphasis was on authority: like White, "whose long experience and careful work upon the Pennsylvania Geological Survey has made his name a synonym for accuracy of observation and skill in drawing conclusions," and Lesley, "for so many years the organizing mind of the Pennsylvania Geological Survey."[227] Chamberlin's rejection of the Cincinnati Ice Dam went unmentioned.

Elsewhere, however, it suited Wright to cite Chamberlin's objections to Croll's theory of the cause of glaciation, for which neither of them cared. For Wright there was too little known of the effects of changing orbital parameters on earth temperatures, too many discrepancies between the glacial records of the northern and southern hemispheres and within North America, and too little direct geological evidence of earlier glacial periods.[228] Wright also had other, deeper reasons to dispute Croll's projected series of glacial and interglacial events occurring between 70,000 and 240,000 years ago, "in which geologists and archaeologists are invited to distribute their remarkable discoveries concerning glacial man."[229] Human antiquity of that magnitude was too great to concede, and it did not square with his evidence and belief that the single glacial period had ended a scant 10,000 years earlier (§3.8, §4.3).[230]

In rejecting Croll's theory, Wright insisted he did not mean to disregard the possibility of there having been two glacial epochs, admitting that most of his calculations of the timing of the end of the ice age "relate to the chronology of what President Chamberlin calls 'the second glacial epoch.'" He even conceded that "the majority of authorities of most weight" sided with Chamberlin on the number of glacial epochs.[231] Yet still Wright attacked ("with diffidence," he insisted) the idea of multiple glaciation. The interglacial forest bed may have been widespread, he granted, but that did not mean it was necessarily contemporaneous across all those areas. It could represent forests that grew near the ice front that were buried by local oscillations of the ice front. More to the point, the vegetation in these beds (peat and coniferous trees) was akin to what he had seen in Glacier Bay, which to his mind bespoke a cold, ice-marginal setting, not a warm interglacial. Such vegetation could grow rapidly and did not require (or indicate) a long lapse of time.

The fact that the southernmost moraine was more weathered and oxidized than the Kettle Moraine was just as easily explained as debris picked up earlier and transported farther than the fresher (recessional) moraines further north. Similarly, glacial scouring ought to be less further south, not because the area was earlier overrun by ice, as Chamberlin and others supposed, but because glacial erosion "would necessarily grow fainter as the [southern] boundary was approached."[232] Finally, Chamberlin himself had said there were no cases where tills of two glacial periods were directly superposed, though Wright neglected to add that Chamberlin had made that comment a decade earlier, before he had examined moraines outside Wisconsin.[233]

Wright admitted that when he originally set out to map the terminal moraine, he expected to easily trace a distinct line across North America. But as he moved west he had had to abandon that hope and be content with "finding marginal deposits more evenly spread over the country, ending, in some cases, in an extremely attenuated border." That attenuation was readily explained because a glacier may not always leave distinct moraines; that depended on the bedrock over which the ice moved, how long it was in position, whether subsequent melting obliterated traces of the glacial deposit, and even variation in the snowfall feeding the glacier. As a result, a glacial moraine might appear broken, scattered, or even invisible in places, its debris possibly transported "far beyond the extreme limits reached by the ice itself."[234]

Having been over nearly the entire southern moraine, Wright (and, he implied, few others) was especially aware of the challenge of separating deposits that "belong to the extreme and true terminal moraine" from that which was merely fringe.[235] In his opinion, what Chamberlin, McGee, and others mapped as a distinct terminal moraine was fringe, or possibly temporary pauses in ice recession.[236] After all, there are great irregularities in the advance and retreat of ice from a "combination of causes."[237] Left behind by that to-and-fro movement were less-conspicuous piles of debris that marked the temporary position of ice at that point in space and time. Such features should not be confused with the terminal moraine that marked the southernmost extent of the one great ice sheet in North America.

While he was writing *Ice Age*, Wright had floated his defense of a single glacial period past McGee, in the hope that McGee might agree. Wright even offered McGee a few pages in the book to explain his "belief in the connection of the Columbia with a preceding glacial period." McGee declined, both to the offer to write a few pages and to the idea that the Columbia deposits were anything like Wright's fringe. It seemed to McGee, in fact, that the two were quite distinct.[238]

What, then, to make of the interior moraine line that Chamberlin had explicitly designated as a terminal moraine of the second Glacial epoch? Wright decided to follow George Cook of the New Jersey Geological Survey and call those "moraines of retrocession."[239] The term implied far more ephemeral features, and this Wright was willing to accept, as it seemed only "proper to hesitate before recognizing the theory of two distinct glacial epochs in America as an established doctrine to be taught." The facts in support of the theory of distinct epochs he

deemed just as explicable "on the theory of but one epoch with the natural oscillations accompanying the retreat of so vast an ice-front.[240]

That cleared the way for Wright to conclude the glacial period was a single
event that ended relatively recently, about 10,000 years ago, just as humans appeared in the Americas. Such a conclusion required "a considerable extension of
man's antiquity as usually estimated," but Wright assured readers of *Ice Age* there
was no reason why such an extension "should seriously disturb the religious faith
of any believer in the inspiration of the Bible."[241] By foreshortening the close of
the glacial period to within theologically acceptable limits, Wright had seen to
that. In doing so, he had to work against a strong current of opinion in geology—
for which he would ultimately pay a price.

For the moment, though, he retrieved the Claymont paleolith from its deep
chronological abyss. No matter that Cresson had plucked the specimen from the
Philadelphia Red Gravel, a glacial formation that by McGee's estimate made this
paleolith "three to ten times as remote" in antiquity as the ones from all other
American Paleolithic sites (§4.5). Wright doubted McGee's estimates and, secure
in the knowledge that there was but a single glacial period, he assigned Claymont
an age that was only a few "thousands of years older than the deposits at Trenton,"
still safely within the one (and only) glacial period.[242]

Ice Age appeared in June of 1889. Within months its initial edition of 1,500
had sold out.[243] Wright's colleagues treated it well: Abbott thought it "knocks
spots out of any scientific publication issued in this country in the last decade."
Winchell was "delighted with the book in every respect." Anonymous reviews in
the *American Geologist* (almost certainly written by Winchell) and the *American
Journal of Science* (likely written by James Dana) praised Wright's experience and
expertise, though both acknowledged that some of Wright's conclusions would
"find opponents" on certain issues for which there remained differences of opinion among geologists.[244]

The *Science* reviewer, Harvard geologist William M. Davis, was less effusive.
The Cincinnati Ice Dam was an open question despite Wright's "paternal fondness," and he warned that readers ought not to take Wright's word on the number
of glacial episodes or even accept his synopsis of the views of others. Indeed, Davis
urged readers to "follow up on his [Wright's] footnotes" and "consult the original
essays." Davis wanted everyone to know contradictory evidence and information
had been omitted from *Ice Age*. So too, it seemed to Davis, Wright had not been
altogether charitable in his treatment of others' ideas, and he admonished Wright
for his stale citations, for centering the discussion on his own work (there were
others, Davis dryly observed, who had published about terminal moraines), and
for not fully considering alternative arguments. The review opened with Davis
wondering whether it was premature to write such a book for a public audience.
He did not think so: teachers and intelligent readers needed one. But *Ice Age*
served that purpose only if its readers carefully noted "the expressions of doubt as
well as the expressions of fact." To insure the point was made, Davis closed his
review with a final warning to keep an open mind about the issues Wright himself
had presented as settled.[245]

Here, expressed in Davis's diplomatic tones, were Chamberlin's dark fears about what *Ice Age* might become. Now that the book was out and those fears were realized, Chamberlin seethed. An innocent comment from Leverett about "moraines of recession" (§4.8) triggered a sharp outburst:

> A moraine of recession is an incident of retreat. There are moraines that mark advance. I urge this because Professor Wright in his recent work is still beclouding the public mind and belittling the facts of history on this basis. It is time that this crudity of conception and this lack of knowledge should be antagonized and rebuked, especially as it assumes the function of public instructor.[246]

But that was said privately. Publicly there was nothing from Chamberlin about the book, only stony silence. One could choose (as a few did) to interpret his silence as approval. Their surprise came later.

4.14 THE BANDWAGON ROLLS

Wright's book appeared at a low point (but not the lowest) in his relationship with Chamberlin, but it was the high point for studies of the American Paleolithic. The subject was by the late 1880s highly fashionable in archaeological circles, and the discussion of human antiquity in *Ice Age* was hardly criticized at all, doubtless because Wright was not saying anything others had not said already. In fact, *Ice Age* was one of nearly a dozen favorable treatments of the American Paleolithic that appeared in the last years of the decade. Abbott, of course, had devoted his August 1888 AAAS vice presidential address to the subject of human antiquity (§4.4). He followed that in December of 1889 with a long article in *Popular Science Monthly* in which he further explored the idea that in the tripartite scheme of North American prehistory, the Paleolithic might be ancestral to both the later argillite groups and the Eskimo (one following the melting glaciers to the north, whereas the others, not "so enamored of an Arctic life," remained in the Delaware valley, only to be run off by arriving Algonquins).[247]

At the BSNH in December of 1888 Putnam hosted a symposium on "Paleolithic Man in Eastern and Central North America." McGee's "Paleolithic Man in America" appeared in *Popular Science Monthly* (November 1888), and he and Wilson participated in an April 1889 symposium on "The Aboriginal History of the Potomac Tidewater Region," organized by the ASW.[248] Not all of these papers were sounding the same notes, but they were close enough to be reasonably harmonious. All the while, returns to the Smithsonian's *Circular* (§4.3) continued to come into Wilson's office at the USNM, amounting to literally hundreds more reports of paleoliths from around the country.

There were occasionally jarring sounds. Putnam urged Wright not to make the "blunders which McGee made" in his *Popular Science Monthly* summary. "That McGee should cast a doubt upon Miss. Babbitt's specimens" Putnam thought "very strange. He evidently knew nothing about them."[249] Or so it seemed to Putnam. Yet, by now there were statements critical of Little Falls in print (tellingly, in Peet's *American Antiquarian* [§4.1]), and when Winchell—who authored one of those statements—sent a photograph of the Little Falls quartzes to Wright to use in *Ice Age*, it came with a strong warning. Winchell had examined Babbitt's collection and came away "impaired [in his] faith in the human origin of any of them," including those in the photograph. Worse, Winchell went to the locality with Babbitt, only to be convinced "her imagination had been substituted for science, and had played a large part in prompting her writings."[250] He pointedly reminded Wright that Upham and Putnam's endorsements of Babbitt's claims carried little weight, as Upham was speaking only to the geological situation (not whether the specimens were artifacts) and Putnam had never seen the locality or the entire collection.

Winchell's letter must have bothered Wright, for shortly thereafter he asked Haynes for his opinion of the Little Falls quartzes. Haynes, of course, had examined many of the specimens a half-dozen years earlier (§4.1). He shared Winchell's opinion of Babbitt (she "had no knowledge of archaeology") but nonetheless firmly believed "her fanciful speculations" could not wholly obscure the fact that the Little Falls collection contained "well-marked examples" of paleoliths. Wright published Haynes's letter in *Ice Age* but made no mention of Winchell's doubts nor used the photograph Winchell provided.[251] Babbitt devoted her next (and last) AAAS paper the summer of 1889 to yet another defense of the authenticity of the Little Falls quartzes, though by now it had come down to little more than a weary plea that "the artificial character of the specimens in hand is not established upon my own personal *ipse dixit* [authority] nor that of any other individual, but upon the unanimous verdict of qualified scientists of international repute."[252] This time, she did not bother to bring the artifacts.

There were other sour notes, some of which marked deeper philosophical differences. In *Ice Age*, Wright cautioned that because almost all of the specimens collected by or reported to Wilson were surface finds, "there is no means of determining their age except from their general weathered appearance, as implements of the same forms have been made and used all through the Stone age." Wilson would have none of that. At the ASW symposium in April of 1889, he outlined the evidence for the Paleolithic in the District of Columbia, which consisted principally of surface finds. He granted that surface material could not furnish "complete proof of the *antiquity* of the paleolithic period" but nonetheless established "the *existence* of a paleolithic period" simply through their morphology and through comparison with like implements in river gravels elsewhere in the United States and other countries. At odds here was the role of geology in establishing archaeological claims. Wilson, and to a considerable extent Abbott as well, assumed that artifact form was sufficient to make the case for the antiquity of paleoliths. Wright, Haynes, and even Putnam were reluctant to go that far.[253]

For the time being, however, the American Paleolithic was secure. Better, at decade's end it seemed to be expanding, with several discoveries coming in late 1889, each of which was destined for controversy. The first came from New Comerstown, Ohio, where on the banks of the Tuscarawas River, W. C. Mills spotted a possible paleolith some fifteen feet below surface. He compared it with several thousand specimens he had previously picked up from the surface and dug from

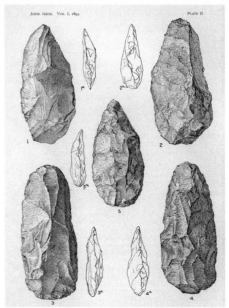

Figure 4.3   George F. Wright's 1890 illustration (*left*) of the New Comerstown specimen alongside a paleolith from Amiens, France, that John Evans had given to Wright's mentor, Asa Gray, who in turn had given it to Wright (from Wright 1890c: Plate C). In response, William H. Holmes crafted a composite illustration (*right*) of the supposed New Comerstown paleolith alongside "four ordinary rejects of the blade-maker." He left it to the reader to separate the "'implement' of glacial age and Paleolithic type from the modern rejects" (from Holmes 1893d: Plate II).

mounds, and when it proved unlike any of those he sent the specimen to Wright, explaining that he suspected the gravels containing the specimen were "similar to those in Trenton." Wright was instantly struck by the specimen's resemblance to a paleolith from Amiens, France that John Evans had sent Asa Gray, who in turn gave it to Wright (Figure 4.3). He grilled Mills for more details on the find: Mills assured him that he had removed it from fresh gravels and it had not tumbled down from the surface above. These were "facts," Mills insisted.[254]

Wright, Baldwin, and others subsequently visited the locality, bringing with them paleoliths from France, Egypt, and Trenton. Mills was delighted to see his was "exactly like" one from the Somme valley. Wright, for his part, confirmed that the New Comerstown paleolith had come from an undisturbed stratum within the glacial terrace. Like Cresson at Claymont, Mills was instructed to take a photograph of the discovery site. Granted, the specimen was long out of the ground, but Wright believed a picture of the section was better than no record at all. Together they planned papers on the find for an upcoming meeting of the Western Reserve Historical Society. However, with one eye firmly on his public audience Wright sent the first report of the New Comerstown discovery to the *New York Nation* newspaper.[255]

About the same time that Mills spotted the New Comerstown paleolith, Wright received word that a miniature clay image of a woman, barely an inch and a half high, had been found during the drilling of a deep water well in Nampa,

Idaho. The Nampa Image, it came to be called, and it was reportedly found several hundred feet below the surface under alternating layers of clay, sand, and even fifteen feet of lava. Its context bespoke a deep antiquity, perhaps comparable to that of the Calaveras skull and the artifacts from the auriferous gravels in California, which Wright (unlike Whitney) deemed Pleistocene in age.

Wright sent a letter to the BSNH in November of 1889 announcing the discovery, then arrived there on the first day of 1890 to discuss the find in detail. He brought with him testimony of its discovery from Charles Adams, president of the Union Pacific Railroad and a devoted reader of *Ice Age*. While on a stopover in Boise City, Adams heard about the Nampa Image, contacted the driller who found it (M. A. Kurtz of Nampa), and recorded his deposition of how and where it was found, then wrote Wright. Wright in turn contacted Kurtz.

Wright's probing questions to Kurtz and Alexander Duffes, a "prominent citizen" of the town who was there when the image was found, convinced him the specimen could not have fallen or been thrown into the well by a bystander, that it was not inserted into the pump at night, that it was not made by the well-drilling crew, and that it was found 300 feet below the surface (a character witness testified on behalf of Kurtz and Duffes's integrity). Wright was naturally concerned the whole thing might be an elaborate hoax, but he assured his BSNH audience that

> the whole appearance of Mr. Kurtz's letters show him to be a genuine man. There was no sensational publication in the papers, nor has there been any suggestion of mercenary motives. There were no archaeologists or scientific men on the ground to be humbugged.[256]

In one letter Kurtz mentioned that dirt from the well had been piled up and schoolchildren used it as a playground, but Wright accorded that comment no particular significance.

Wright, in turn, wrote Samuel Emmons of the USGS for information on the geology of the Nampa region. Emmons was hesitant to guess an age of the deposits from which the image was supposedly derived, but he offered they were likely "of far greater antiquity than any deposits in which human implements have hitherto been discovered." Wright's Oberlin colleague, geologist Albert A. Wright (no relation), examined the specimen but saw "no satisfactory marks of the tooth of time." Wright must not have found that answer to his liking, for he turned over the specimen to Frank Jewett, an Oberlin chemist, who observed an iron oxide accumulation that led him to conclude the Nampa Image had considerable antiquity. That was the answer Wright wanted, and he "attach[ed] much weight" to Jewett's observation and accepted "without further question the genuineness of the image."[257] Putnam and Haynes did too.

Another piece of apparent Pleistocene-age art surfaced late in 1889, when Cresson showed Putnam and Wilson a whelk shell on which was engraved the image of a mammoth. (Oddly, Cresson did not mention the find to Wright, with whom he was in regular correspondence.) True to form (§4.5), Cresson claimed the artifact had been discovered decades earlier near Holly Oak, Delaware, but that he had been unable to obtain it until then. The Holly Oak pendant's bearing on the human antiquity issue was obvious: here was proof humans had seen an extinct Pleistocene animal. At Cresson's request, Putnam announced the find at the February 5, 1890, meeting of the BSNH.

It was a decidedly unenthusiastic presentation. Putnam talked for a few minutes on the shell, read a brief letter from Cresson about its discovery, then "spoke at length on early man of both sides of the American continent and of the relationship of the modern Indians to the early peoples of America."[258] It was not simply that Putnam had Abbott's just published *Popular Science Monthly* paper on Paleolithic descendants on his mind that evening; he had doubts about the pendant's authenticity.[259]

Unlike others who may have downplayed the pendant's chronological significance on the grounds that some of the Pleistocene megafauna survived until recent times, Putnam believed such an association indicated a Pleistocene antiquity, just not in this case.[260] He was troubled by the striking resemblance of the Holly Oak mammoth and the one engraved on the La Madeleine tusk found in France by Eduoard Lartet in 1864. Putnam made a composite photograph of the two, then sent Cresson a copy. Cresson, straight-faced, replied it was "strange how faithfully all the copies represent the mammoth."[261] Strange, indeed. Putnam knew better. The Holly Oak pendant soon sank from sight.[262]

By decade's end Abbott was lionized at home and abroad as America's Boucher de Perthes, and his Trenton Gravel was accorded the same pride of place in American archaeology as the Somme gravels of Europe (§2.1).[263] His personal circumstances were finally about to turn as well. Although he could barely afford to attend the August 1889 AAAS meeting (he sent a plea like "drowning man catching at a straw" to Putnam, who offered to help pay his expenses[264]), just four months later he was hired as the first curator of archaeology at the University of Pennsylvania University Museum. Abbott's financial troubles instantly vanished with the salary of $1,000 a year.[265] Putnam had helped engineer the hiring, though he also candidly explained to William Pepper, president of the University of Pennsylvania, that

> [Abbott] is a singular man, and one who has to be treated differently from common mortals, and as I understand him and thoroughly appreciate him I have not put him under any obligations that he did not prefer himself.

Knowing Abbott's tendency to become unhappy when left to his own devices, he suggested to Pepper it was best to hire Abbott as a full-time curator, not a part-time one, otherwise Abbott would "get into an unhappy state of mind by hopes deferred."[266] To his (and Abbott's) delight, Pepper followed Putnam's advice, and it represented more than mere employment—it was testimony that Abbott was right:

> Dr. Abbott was for years placed in a very unpleasant position by the non-belief of many persons in the great discovery which he made showing that man existed in the Delaware Valley at a time preceding the deposition of the

Trenton gravel. . . . Now the scientific world gives him the full credit he so richly deserves.[267]

It appeared that the worst of Abbott's troubles were finally behind him. The Peabody Museum's Board formally thanked Abbott for his service and announced that the nearly 30,000 specimens he had donated would thereafter be known as the Abbott Collection.[268]

Putnam, however, was not about to let his access to the rich archaeological ground of the Delaware valley slip away. Even before Abbott was officially gone he hired Ernst Volk as his replacement. Volk, a Trenton bachelor who lived with his elderly mother, was quiet, unambitious, and devoutly loyal to Putnam, almost to the point of obeisance. He was, in temperament, everything Abbott was not. Abbott dutifully piloted Volk around to some key gravel exposures, then "stood back and let him hunt." Within a few minutes, Volk had made "a splendid find." He will work out, Abbott assured Putnam, and "you can rely on the unswerving accuracy of his statements."[269]

4.15 LOOKING TO THE FUTURE OF THE PAST

The future of the past certainly looked bright. Yet, secure as it seemed, changes were coming—and not just in archaeology, but in glacial geology as well. As the decade ended, Chamberlin showed no signs of easing his campaign against Wright, though for a time he remained discreet about it. In October of 1889 he delivered to the Society of Western Naturalists the first version of a paper describing what he called the "method of multiple working hypothesis." A later iteration published in *Science* gained lasting fame, its notice reaching far outside geological circles.[270] Wright was not mentioned by name, but he was on Chamberlin's mind when he wrote the paper. In the preamble Chamberlin condemned what he termed the "method of the ruling theory," in which a premature explanation or working hypothesis became a tentative theory and then, as "parental affections cluster about his intellectual offspring," an overriding idea:

> There is an unconscious selection and magnifying of phenomena that fall into harmony with the theory and support it, and an unconscious neglect of those that fail of coincidence. The mind lingers with pleasure upon the facts that fall happily into the embrace of the theory, and feels a natural coldness toward those that seem refractory.[271]

On occasion, perhaps, "truth may be brought forth by an investigator dominated by a false ruling idea," as errors might stimulate investigations on the part of others, but that was a mixed blessing. "Dust and chaff are mingled with the grain in what should be a winnowing process."

In this portrayal of improper scientific methods and their consequences were unmistakable echoes of Chamberlin's lectures to Wright over much of the previous decade. He did not have to search hard for illustrations of a ruling theory in action. Wright's idea of a Cincinnati Ice Dam was one; his insistence on a single glacial epoch was another. Chamberlin had repeatedly urged Wright to consider alternatives to his pet theories, or consider multiple working hypotheses, and let "their mutual conflicts whet the discriminative edge of each."[272] But just as repeatedly, Wright ignored the urgings or paid them only lip service:

> On this supposition [of an ice dam], so natural in itself, all the facts are easily explainable. So confident have I become in the reality of this dam that I have not hesitated to use it as a means of putting myself in the line of discovering other facts which are the natural consequence of this.[273]

Which, of course, was just what Chamberlin most mistrusted:

> The search for facts, the observation of phenomena and their interpretation, are all dominated by affection for the favored theory until it appears to its author or its advocate to have been overwhelmingly established.[274]

That, in Chamberlin's eyes, was utterly unscientific.

At the first annual meeting of the Geological Society of America (GSA) in December of 1889, Chamberlin returned to the question of the time separating the glacial periods.[275] He marshaled evidence from the lower Mississippi, upper Ohio and Allegheny, Susquehanna, and Delaware Rivers that showed in each instance there were high glacially derived terraces (from an earlier glacial period) that had subsequently been deeply cut and then partially filled by debris from a later glacial period. Although the height of the older terraces and the depths of the later trenches obviously varied, in all cases they bespoke long intervals between the two episodes of glaciation.[276]

Wright was not present, but White was in the audience and immediately stood to protest "the facts presented by President Chamberlin from the valley of the Ohio," which had "always been" interpreted differently by other geologists who studied the region (among whom, it went without saying, were Wright and himself). White insisted the Ohio and Monongahela terraces were better explained as submergence behind Wright's ice dam, and "without recourse to a 'second glacial epoch,' where the evidence of neither a 'first' nor a 'second' ever existed."[277] Besides, he asked rhetorically, "how are we to discriminate" glacial material laid down directly by ice or indirectly by meltwater?

That, Chamberlin retorted, was what geologists were *supposed* to know how to do, and they ought to be able to distinguish material deposited by running water and formed in "static or horizontal waters." Having done just that with Gilbert in the very region where White himself was presumably expert, Chamberlin replied with dripping condescension that the differences were clear and sharp, and he left no doubt that the terraces of the upper Ohio and its tributaries were formed by running (not standing) water.[278] White retreated, admitted the terraces were probably fluvial, and tried to cut his losses by insisting they were covered by lacustrine deposits (his thirty-foot clay unit) *after* their formation. It was a fatal error. The terrace features he had just surrendered to Chamberlin were essential to Wright's ice dam theory. Chamberlin did not even deign to reply, but McGee stood to vouch for the correctness of Chamberlin's interpretations and their importance to "the complex subject of Pleistocene history."[279]

If Wright read the subsequently published proceedings, he could hardly have failed to grasp the significance of McGee endorsing Chamberlin. This was not the same McGee who had earlier explored the geology of the upper Ohio valley with Wright's report as his guide, nor, for that matter, the McGee who just the year before had been a staunch supporter of the American Paleolithic and to whom in January of 1889 Wright had offered space to contribute to his *Ice Age*.[280]

This newly invented McGee was now publicly allied with Chamberlin and was abandoning his earlier support of the American Paleolithic. That much became evident in the summer of 1889. Wright and Cresson had urged McGee to see the Claymont locality, and on his way north to the AAAS meetings in August McGee visited the site. With him was Bureau of Ethnology archaeologist William Henry Holmes. They spent a day at the site and just as they were about to leave, Holmes, who had been a couple of hundred feet away, called Cresson and McGee over to see something:

> Holmes handed me a specimen which he had found in undisturbed gravel. It was very dirty and had to be taken home and washed. Holmes & McGee think it is not artificial but I think it is. . . . It is a large turtle back quartz of the (Babbitt) Minnesota type.[281]

Cresson insisted McGee examine precisely where the specimen came from and photograph it in place. McGee did, declaring its context "undoubtedly undisturbed" Columbia deposits, but he simultaneously claimed the specimen was natural. McGee, Cresson snarled, "is totally unfit to judge whether a stone has been artificially fractured," and sounded the alarm: "McGee & Holmes I think are determined to kick against any implements that are found in Columbia deposits."

Anxious not to have them gain the advantage, Cresson quickly shipped the find to Trenton for Abbott to carry to the AAAS meetings. Abbott, Putnam, and Wright were peppered with letters from Cresson, imploring them to examine the specimen for themselves, and if it proved to be an artifact to

> kick up a dust on it, and if we win, the government will be forced to acknowledge that their officials have found chipped implements in Columbia deposits. McGee, however, won't do this—he will dig hard, & so will Holmes, who stands in with him.

"Wade in Putnam," Cresson exhorted from the sidelines, "and give early man in the Delaware Valley a boom! . . . explode a ton of dynamite!" What "a great triumph for our Museum."[282]

Putnam did no wading. Instead it was Abbott who had a long, unsettling talk with McGee, whose take on the Holmes specimen was that if it was in place in the Columbia gravels then it was not artificial, and if it was artificial then it must not have been in place. Abbott found this "curious," if not inconsistent, but he may not have realized he had just gotten a preview of McGee's forthcoming paper, which outlined his newly formulated principles regarding questions of human antiquity. The specimen found by Holmes failed on Principle 3: "It is a fair presumption that any unusual object found within, or apparently within, an unconsolidated deposit is an adventitious inclusion." Abbott was not buying it. He showed the specimen to Dana and a few others. They pronounced it "unquestionably the work of man."[283]

McGee was changing his stripes. Cresson asked Wright afterward how he and McGee had gotten along, but evidently they had not spoken about the specimen or much at all. Wright must have noticed the cooling: when in October of 1889 McGee sent Wright photographs of the "supposed paleolith" found by Holmes, he no longer greeted him as he previously had with the warm and familiar "My dear Professor Wright." Now it was the icier "Dear Sir."[284] Soon their correspondence ceased altogether, and McGee became Wright's most vicious critic. He did so with Chamberlin's blessing and armed with archaeological evidence from Holmes's work on the same Washington, DC, hillside where Wilson had found his proof of the American Paleolithic.

In September of 1889, just a few weeks after he returned from his visit to Claymont with McGee, Holmes began excavating at the Piney Branch quartzite quarries a few miles north of the White House. Holmes had first visited the spot two years earlier as Wilson collected paleoliths from its surface (§4.3), and he had been at the Anthropological Society of Washington's symposium in April 1889 when Samuel Proudfit dismissed Wilson's specimens as little more than the "debris of Indian workshops." Putnam, the evening's discussant, remarked that Piney Branch reminded him of quarry sites in New England where he had seen scattered in the debris great numbers of "rude implements in various stages of manufacture." They occurred there, he presumed, since

> every perfectly chipped implement of the knife, spear, or arrow kind had to pass through all the stages noticed in these rude forms before the perfect implement was obtained, and as on many village sites which I have explored I have found . . . rude implements of paleolithic forms associated with the highest type of chipped implements, with ornaments, and with pottery, *I have been led to the belief that form alone can tells us but little of the time when an implement was made.* It seems to me that all we can go by in this country is the fact that up to this time the implements found in the Eastern gravels are all of rude forms, very closely corresponding with those actually found in similar gravels in the Old World.[285]

Putnam, ironically, was among the first to articulate what became a central refrain of critics in the Great Paleolithic War.

By November of that year, Holmes was publicly echoing Putnam's concern at a meeting of the ASW, announcing, on the basis of his work at Piney Branch, that artifact form "had no chronologic[al] significance whatsoever." With this realization came the inevitable corollary: without geological evidence to confirm its antiquity, the American Paleolithic was essentially adrift in time.[286] Several months later Holmes published his first paper on Piney Branch in the *American Anthropologist*. Lucien Carr read it but was unruffled. As he explained to Henry Henshaw, the journal's editor:

> In regard to Holmes find . . . [it does not] affect the general question of the antiquity of man except as far as this particular find was used to prove it. There are others & if human evidence can prove it how are you going to get around the Trenton Gravel & California cement where implements of human workmanship have been found? The only way to discredit them is to say "the boy lied."[287]

Two months passed before Abbott heard about Holmes's paper, and he was hardly so nonchalant. He dashed off a letter to Henshaw: "Will you please get for me a copy of Holmes (?) paper on the Paleolithic!!! finds near Washington?"[288] Abbott found out soon enough what Holmes had to say about the American Paleolithic.

CHAPTER FIVE

The Great Paleolithic War, 1890–1897

IN THE SPRING OF 1890 Charles Abbott got his first look at William Henry Holmes's paper on the Piney Branch quarry. It seemed innocuous enough. Holmes had studied the sequence of stone tool manufacture at the site with an eye on the by-products of the process, most especially the manufacturing failures, the specimens that were unfinished, broken, or otherwise jettisoned. He observed that Delaware valley paleoliths were similar to the "failure shapes" at Piney Branch, but their geological occurrence "seem[s] to make them safe indices of the steps of progress."[1] Yet, even a careless reader could see that Piney Branch's implications struck at the conceptual core of the American Paleolithic. It would be only a matter of time before Holmes trained his sights on those claims.[2] Abbott had cause for concern (§4.15), and it came sooner than expected. Henry Henshaw sent Holmes's paper to Abbott, then followed up with an invitation for Abbott to Washington, preferably immediately. Why immediately? Henshaw had just visited Holmes at Piney Branch, and it seemed to him that Abbott would be

> particularly interested in the matter since you have done so much work at Trenton, and a visit here just now could not fail to prove instructive. Mr. Holmes would be very glad to see you, and we will all do what we can to make your visit pleasant and instructive.[3]

"Instructive," Henshaw said. Twice. Abbott was not looking for instruction. Puzzled and wary, he went to Washington.[4] He saw that Holmes might have rightly interpreted the process of stone tool manufacture at Piney Branch, but Abbott thought Piney Branch irrelevant to "the question of man's antiquity in America."[5] If Abbott failed to see the implications of Holmes's work, Holmes wondered about Abbott's archaeological acumen:

> Having brought to light evidence of importance at Wash[ington] bearing upon the nature of rude forms of flaked stone, I decided to ask Dr. A. as I had asked others to pay the open trenches a visit. On parting [Abbott] said, "I have learned more arch[aeology] in three hours than ever before in three months." This I was content to think was a pleasant compliment but from my subsequent studies and increased wisdom I concluded that he probably meant what he said.[6]

Their initial exchange rapidly escalated into intense debate. American Paleo-lithic proponents like Abbott and George Frederick Wright viewed the similarity between paleoliths and the Piney Branch quarry debris as merely coincidental. Holmes retorted that American and European paleoliths might look alike because the latter were themselves quarry rejects, and he set out on a scorched-earth march through the sites of the American Paleolithic—Abbott's included—to prove it.

The growing dispute over the American Paleolithic was resolving itself as a geological issue, but geology provided little guidance. Questions of whether the paleoliths were found in situ were complicated by the absence of protocols for recognizing stratigraphic units and depositional contexts. There were questions too about the age of the artifact-enclosing formations and whether these were primary or secondary deposits. These questions in turn got entangled in the debate over how to recognize Pleistocene-age formations and the number and timing of the glacial periods. Thomas Chamberlin and Wright's behind-the-scenes disagree-ments (§4.9, §4.11) erupted into full view in the early 1890s.

The catalyst was the 1892 publication of Wright's *Man and the Glacial Period*. It was much like his *The Ice Age in North America and Its Bearing upon the Antiq-uity of Man*, but this latest book was mercilessly and publicly savaged by critics. The explosion over *Man and the Glacial Period* was only nominally about artifacts and their geological contexts, but it rapidly escalated into a wide-ranging battle over theory, method, and the interpretation of evidence; forced discussion of the role and worth of amateurs; and rekindled long-simmering resentment (§4.10) over the perceived heavy-handedness of the USGS. As the Panic of 1893 came to grip the nation's economy, it triggered a new wave of attacks on federal science in general and the USGS in particular.

By the August 1893 AAAS meeting, the talk was fiery and positions on both sides were hardened beyond compromise. Proponents and critics spoke past one another, then they gave up talking altogether. Following a few years of relative quiet, the principals separately visited Trenton in June and July of 1897 for brief fieldwork stints, then met to confront each other at the AAAS meetings that Au-gust. Once more, their arguments were irreconcilable. Coincidentally, the AAAS was meeting jointly with the BAAS, and John Evans himself was present. At the meeting he examined a set of the Trenton paleoliths but found them unconvinc-ing. It was a devastating blow. Still, by decade's end contemporaries called the American Paleolithic battle a draw. The actions of the participants suggested otherwise.

5.1 THE BUREAU OF ETHNOLOGY TAKES THE FIELD

The American Paleolithic attracted considerable attention in the archaeological community of the 1870s and 1880s. Yet, with the exception of a brief public state-ment by John Wesley Powell and some private remarks by Cyrus Thomas (§3.9, §4.3), the Bureau of Ethnology, the nation's largest, most powerful, and most visible institution conducting archaeological research, had up until 1889 scarcely noticed the American Paleolithic. When it did, the character and tone of the discussion radically changed.

Bureau archaeologists brought to the debate a particular vision of anthropology and human history and had the ambition, funding, and intellectual talent to pull it off. Like the USGS it was an institution doing the work of a scientific society but with the power of the federal government behind it.[7] By the late nineteenth century, government research bureaus and scientific agencies—concentrated in Washington as a by-product of the economic and political centralization bred by the Civil War—had become the most powerful players in American science.[8] On Capitol Hill, and ripe for the plucking, was funding at unprecedented levels in any one of dozens of scientific fields in which the government was trying its hand. The opportunities to pull those levers were there for the boldest, most ambitious, determined, shrewd, and politically savvy scientist-politicians of the generation—especially those who had formed connections with key legislators during the Civil War. In this regard, few could match Powell, who for a time had more sweeping power and financial resources than any scientist in the nation.[9]

The Bureau of Ethnology was founded under Powell as an almost unnoticed rider on the 1879 bill that consolidated the four competing geological surveys of the American West into the single USGS.[10] The pretext for the Bureau's founding was the publishing of ethnographic material left over from those earlier surveys, and for that an appropriation of $20,000 was allotted for "completing and preparing for publication the *Contributions to North American Ethnology*."[11] With that appropriation Powell created the Bureau of Ethnology (hereafter BAE, in accordance with its name change in 1894 to the Bureau of American Ethnology, which aimed to more accurately reflect its scope and venue).

Created as it was without formal congressional authorization, the BAE was at the mercy of the annual funding cycle and occasionally had to justify its existence to cost-conscious members of Congress. But bureaucratic inertia was on Powell's side: once created, federal agencies are extraordinarily difficult to dislodge. His ability to manipulate generations of congressmen (making good use of his lavish *Annual Reports* as well as "a little amiable nepotism") insured that the BAE thrived.[12] His initial appropriation increased annually in $5,000 increments, plateauing at $40,000 in the early 1880s and never falling below that for the next decade or so.[13] In contrast, Frederic Ward Putnam at the Peabody Museum, the only other player in anthropology who had a significant institutional program, reported in 1890 that his museum's endowment yielded less than $2,500, and donations the previous decade averaged scarcely more than $3,000 per year. "We can realize," Putnam acknowledged with a dollop of self-pity, "how little we could have accomplished, had it not been for the generosity of friends." He lamented that until there were more donations to the permanent fund, "we shall be entirely dependent on continuous aid of this kind for the means of keeping the Museum in the advanced position it now holds."[14] Almost to the day Putnam was bemoaning his financial straits, Powell sent his yearly budget request to Capitol Hill. He asked for $50,000 and got it.

Once the BAE was established, Powell devoted only a small part of its budget to the *Contributions*. Instead, most funds went to support new research.[15] For Powell saw the BAE as a vehicle "to organize anthropological research in America" around human biology, material culture, institutions, language, and philosophy. Unifying these disparate fields would be a holistic anthropology "of final, positive knowledge to explain and justify the wide disparities in human conditions, past, present and future."[16] Its theoretical centerpiece was Lewis Henry Morgan's cultural evolutionary vision of a universal, progressive history of humanity—Morgan being at the time "the presiding genius of Bureau thought."[17] The human mind, Powell believed (echoing Morgan), was the motor of change and evolution. Darwinian selection applied only in the natural world. ("Animals . . . were adapted to the environment; man, through his activities, adapted the environment to himself.")[18] The empirical evidence of the progressive evolution of the mind was naturally elusive in certain areas (language and philosophy, among them), but obvious in others (technology). But in Powell's view those realms were inextricably linked:

> All the grand classes of human activities are inter-related in such a manner that one presupposed another, and no one can exist without all the others. Arts are impossible without institutions, languages, opinions, and reasoning; and in like manner every one is developed by aid of the others. If, then, all the grand classes of human activities are interdependent, any great change in one must effect corresponding changes in the others. The five classes of activities must progress together.

Here, then, was a role for archaeology: it provided insight into the origin and development "of arts and industries," the minds that created them, and glimpses of the nonmaterial realms of human existence.[19]

Reinventing the discipline as he was from first principles, finding trained anthropologists or archaeologists (of which few were available) mattered far less to Powell than finding talent. He put this theory into administrative practice, hiring individuals who by experience or training included artists (Holmes), entomologists (Thomas), geologists (WJ McGee), lawyers (Lester Ward), and ornithologists (Henshaw). During the years in which Powell was simultaneously director of the BAE and USGS (1881–1894), their administrative and scientific staffs were often interchangeable, and little heed was paid to bureaucratic fences between them.[20] Employees moved with ease among offices and fields of study.[21] Sometimes it was thought they moved with too much ease:

> Major Powell called about him [at the BAE] his former [USGS] assistants, and thus we have the singular spectacle presented of explorations among American antiquities, conducted by geologists, ornithologists, entomologists, and ethnologists, without the aid of experienced archaeologists! When it is remembered how exacting are the requirements of science, and how its most minute departments have become the life-work of trained *specialists*, it may well be questioned whether the genius of man is capable of passing successfully from one to another of these fields of research.[22]

Such a question never occurred to Powell. He carried a decidedly populist chip on his shoulder and surrounded himself with those who also earned their scientific credentials climbing the mountains of the American West and mapping its sun-baked plains. He had little regard for academic formalities and was unimpressed, even disdainful, of USGS geophysicists and geochemists who possessed doctorates from European universities. Allegiance was not to a specific discipline but to science as a whole and to Powell himself who, as even his allies admitted, "likes strong personal loyalty from those nearest to him."[23]

If human history was the unbroken progress from savagery to barbarism to civilization up Morgan's evolutionary ladder, then turning that ladder upside down revealed an archaeological methodology. As paleontologists and geologists had done with such stunning success, archaeologists should work from the known (ethnographic present) back into the unknown (archaeological past). For Powell, a confirmed uniformitarian, the present was the key to the past.[24]

BAE archaeologists adopted the spirit, if not the details, of Powell's grand vision. For them, the natives "were here and must be recognized by every theory, must be a factor in every general conclusion." Their material remains were a bridge from present to past: as A. V. Kidder later described it, the "effort was directed toward identification of ancient sites with modern tribes."[25] That was possible because it was assumed the prehistoric record evinced "continuity of the pre-Columbian population of North America, subject to known evolutionary laws [progress], as against the cataclysmic theories postulating intrusive or extinct races." The latter included "lost races" of Mound Builders or an American Paleolithic race comparable in age and evolutionary grade to that of Europe.[26]

Thomas Wilson had surmised those who made the paleoliths at Trenton and Madisonville were "a race of people, if not different in type & appearance, at least different in their industries and civilizations from those heretofore known as aborigines."[27] But to accept such a lost race who were apparently non-Indian and left no succeeding generations would rupture the BAE's methodological link from present to past. The BAE's approach demanded continuity between present and past. In merging the archaeological and ethnographic records,

> there would then be no more blind groping by archaeologists for the thread to lead them out of the mysterious labyrinth. The chain which links together the historic and prehistoric ages of our continent would be complete; the thousand and one wild theories and romances would be permanently disposed of; and the relations of all lines of investigations to one another being known, they would aid in the solution of many of the problems which hitherto have seemed involved in complete obscurity.[28]

As it was generally assumed Native Americans once settled scarcely moved, that seemed reasonable.[29]

At the same time, there were theoretical reasons to deny an ancient Mound Builder or Paleolithic race. For such antiquity mocked the BAE's grand vision of evolutionary progress: if Native Americans personified a primitive stage in that sequence, and if their prehistory began, as it did in Europe, deep in Pleistocene times, why had Indians not transcended savagery and achieved civilization? Why were they, Powell wondered, "just passing into barbarism when the good queen sold her jewels"?[30] That said, Powell was not averse to the idea that people may have reached the Americas in later Pleistocene times, but on one point he was resolute: "Nowhere is any great break found, nowhere is a higher culture interpolated, and nowhere do we find evidence of peoples other than the North American Indians and their ancestors."[31]

Predisposed as the BAE was to reject the idea of separate races, Mound Builder or Paleolithic, and significant as those questions were in American archaeology, it is no surprise the Bureau took on both. Their approach in each instance was the same: demonstrating there was no separate Mound Builder (Paleolithic) race only required demonstrating that their artifacts and Indian artifacts were one and the same.[32] Through the 1880s Thomas did just that in the BAE's Mound Survey. As it wound down, Powell looked to move the BAE's archaeological research in another equally high profile direction. There was little doubt what direction that should be. On June 30, 1889 he appointed Holmes to take charge of the "archaeologic fieldwork of the Bureau" and resolve the vexing question of the American Paleolithic.[33]

5.2 WILLIAM HENRY HOLMES AND THE LESSONS OF PINEY BRANCH

Holmes was an obvious choice.[34] As he fondly mused, he "was born on the same day with the [Smithsonian] Institution . . . and have come to regard myself as an original predestined member of the family." As a young man in the Midwest, he showed more interest in sketching and painting than working the family farm. He earned a teaching certificate from the McNeely Normal School in Ohio, but he had grander ambition. Wanting to study art, he left the farm and teaching behind, heading to Washington in the spring of 1871.[35] There he met Joseph Henry's daughter, who told him of the natural history collections on display at the Smithsonian. Sketching mounted birds there one day, he was spotted by one of the scientific staff, who gave him piecework illustrating specimens. Within the year he was hired as an artist for Ferdinand Hayden's *Geological Survey of the Territories* and found himself en route to Yellowstone.

Over the next decade on Hayden's survey, Holmes learned to produce his trademark geologically correct panoramas of western landscapes, and he had his first exposure to archaeology in exploring Anasazi ruins in the San Juan valley of New Mexico and Arizona. His ability to decipher geological formations soon won him promotion as a geologist, and in that role he joined the USGS when it was formed in 1879. Though employed as a geologist for the next decade, Holmes was increasingly drawn to "the study of primitive art in all its branches."[36] Under Powell he freely pursued that interest and was encouraged to do so: Powell put him in charge of illustrating BAE reports and gave him access to artifacts streaming in from BAE projects. Holmes was appointed (in 1882) as honorary curator of aboriginal pottery at the U.S. National Museum (USNM), and he produced descriptive and analytical studies of shell, textile, and ceramic artifacts.[37] In 1889 Holmes was assigned full time to the BAE. Predestined or not, with the exception of three years at Chicago's Field Museum (1894–1897), he spent nearly six decades in Washington's museums and research bureaus. He ultimately succeeded Powell as BAE head and served stints as curator of anthropology at the USNM and curator and director of the National Gallery of Art.[38]

Holmes thoroughly absorbed Powell's progressive evolutionary perspective.[39] Still, his was not the BAE's evolution unvarnished. Holmes had learned enough about artifacts and material culture to realize Morgan's evolutionary stages worked best on a grand scale and were not precisely defined classes of empirical reality: the difficulty of wholly separating the phenomena of one stage from another, and the unwillingness of artifact forms and other attributes to follow strictly progressive lines, attested to that.[40] Stages were analytical tools, not facts of human history. Holmes learned to use stage concepts ("savagery") and terms like "culture grade" as handy didactic devices, but they carried none of the empirical baggage they did for Morgan or Powell.

Moreover, Holmes believed people should be considered part of the natural world. Attempting to rectify this notion with Morgan led Holmes to posit a prehuman and presavage stage to incorporate "modification by environment" and aspects of biological evolution in human development.[41] Holmes's most striking appropriation from biological evolution was his use of Ernst Haeckel's then-popular notion that ontogeny recapitulates phylogeny.[42] It was a cornerstone of his attack on the American Paleolithic. What made it so was that it was part of a general theory of technology and technological change.

His interest in and talent for analyzing material culture and technology was attributed to his having "learned how to think as the Indian thinks," but that was somewhat misleading. During his career, and despite many years spent out west, Holmes had "singularly few contacts with the living Indians." John Swanton, who wrote Holmes's biographical memoir for the National Academy of Sciences, thought his work "lacked the control which might have been supplied by direct observations of native artisans."[43]

That was of little concern to Holmes, who felt that if "savages learned it others can learn it." He compensated for his lack of ethnographic experience by carefully reading others' accounts, by studying artifacts with the skilled eyes of an artist, and experimentally replicating ceramic vessels and stone tools. (The latter came to an abrupt end when he permanently disabled his left arm trying to flake an especially large boulder.)[44] Holmes, it was said, could make Indian arrows out of "a beer-bottle, a piece of cannel coal, or anything that has a shell-like fracture"[45] (Figure 5.1). That was a skill he had picked up over the years, and though it had undeniable parlor appeal it was more than that. Artifacts had been conceived in static terms as particular forms or functional classes (turtlebacks, arrowheads) or as chronological markers (paleoliths), and only in the most general sense as revealing change. (Paleolithic proponents recognized differences, but they did not see evolution except in categorical terms.)[46] Holmes looked beyond material forms and products to discern underlying processes of manufacture, technology, and evolution.

Holmes probed for the first time—and systematically across artifact classes (of varying materials)—the manner and stages in which prehistoric artisans fabricated their tools. In so doing, he showed how one apparently distinct and immutable artifact type might be related to another by virtue of being different stages of

Figure 5.1 William H. Holmes knapping stone tools at the Piney Branch quarry, circa 1890. Holmes included this image in his *Random Records of a Lifetime in Art and Science* and labeled the photograph "Holmes in an ocean of the paleoliths of Abbott, Putnam, Wilson and the rest of the early enthusiasts of American antiquities. All are merely refuse of Indian implement making." Courtesy of the Smithsonian Institution Libraries, Washington, DC.

the same process of manufacture. This was a novel approach to material culture,[47] for it showed that artifact form was merely the crystallized moment of an underlying dynamic of manufacture and use. Form or morphology was therefore not a straightforward indication of age, as previously descriptive classifications (like Abbott's) had assumed. Holmes saw variation within artifact classes and sought to understand its archaeological meaning, as well as the meaning of the classes themselves.

He did so by interpreting the archaeological past in ethnographic terms. Unlike American Paleolithic proponents who essentially ignored Samuel Haven's observation that "the flint utensils of the Age of Stone lie upon the surface of the ground . . . [and t]he peoples that made and used them have not yet entirely disappeared," Holmes embraced the point.[48] He showed that by viewing artifacts in technological terms and by unveiling the processes, underlying types, and their variability, one could link the present with the past and give meaning to the past.

Holmes's approach to the archaeological record was thus very different from that of American Paleolithic proponents: it was evolutionary, anthropological, and uniformitarian. It sought to bridge the gap between the archaeological record and the processes that created it and close the gaps between theory, method, and data. His was, Holmes insisted, a *scientific* approach, and he was at pains to distinguish it from the approach of amateur natural historians, among whom Holmes came to number proponents of the American Paleolithic.[49]

Holmes arrived at the American Paleolithic in 1889 with a clear charge and the tools to investigate it in a way that left little doubt of the outcome. To accomplish his task, he began fieldwork on the quartzite cobble-strewn terraces along Piney Branch Creek. He deliberately chose the locality following both Wilson's announcement that paleoliths had been found there and Proudfit's rejoinder, echoed by Putnam (§4.15), that the Piney Branch "paleoliths" were merely "the resultant debris of Indian workshops [and not attributable] to paleolithic man."[50] The remarks likely reminded Holmes of a lesson he had learned in 1878 in Yellowstone, Wyoming (§3.3), where the ground was littered with worked obsidian nodules: "It occurred to [me] . . . that the various Indian tribes of the neighboring valleys had probably visited this locality for the purpose of procuring material for arrow points and other implements." He had found artifacts among the debris, but "nearly all . . . are imperfect, as if broken or unfinished."[51]

At that time he did not accord any particular significance to the observation. It became the leitmotif of his work at Piney Branch and emerged in his first paper on the site, which was delivered in mid-November of 1889 to the Anthropological Society of Washington and subsequently published in the *American Anthropologist* in January of 1890. Holmes began by claiming he came to Piney Branch "without preconceived notions of what the results should be." He was being disingenuous: he knew exactly what the issues were and how to resolve them; otherwise, he scarcely would have bothered to go there.[52]

He did, however, take a very different approach to the site from those who had been there before. For one, he excavated trenches upward of eighty feet long and nearly ten feet at the deepest. In them Holmes examined the quarry pits, the associated debris, and the underlying cobble deposits being worked. That provided, as no surface collection could, more complete evidence of the archaeological context of the artifacts on the site. Paleolithic proponents had frequently been chided by Stephen Peet for collecting materials from the surface, which made it difficult to assign an age on geological grounds (§4.4). But then Abbott, Wilson, and others were convinced antiquity could be assigned to artifacts found on the surface by their form alone (§4.3).

Holmes addressed that too by examining nearly 2,000 Piney Branch specimens that represented all stages of artifact production, from whole and unworked quartzite boulders to nearly finished but broken and discarded points. He stressed the material came from "but one period of work, and that by one race, whose clever artists had in mind, so far as this shop was concerned, but one ideal." Holmes was confident he could test whether Proudfit's suspicions were correct and whether, more generally, form indicated age.[53] As Holmes saw it, the process of artifact manufacture involved three stages that transformed a cobble through "successive degrees of elaboration" into long leaf-shaped blades—which he envisioned as the "ideal" product of the process. During the first stage the cobble was flaked by percussion on one side, thereby producing a "typical turtle-back." It was then turned over and the other side flaked to create a rough bifacial preform, and finally the specimen was thinned, trimmed to achieve a symmetrical form, and then pointed. The results of the process were "neat, but withal rude, blades," which then "graduated from the school of direct or free-hand percussion" and were transported to locations of residence or use where final finishing took place. Because finishing did not occur at the quarry, and the rude forms were pressed only into occasional "emergency" service, they would not be expected to show signs of use.[54] Of course, not all efforts at readying those blades for transport off the quarry were successful: artifacts broke while being flaked, or were discarded because they were too thick, had humps that could not be removed, or were otherwise flawed. The production process "left many failures by the way." Everything at a quarry site like Piney Branch was "mere waste" or a "failure" of the manufacturing process, and that included what Abbott called turtlebacks, of which Holmes had collected "fully 1000" at Piney Branch:

> It causes me almost a pang of regret at having been forced to the conclusion that the familiar turtle-back or one-faced stone . . . together with all similar rude shapes, must, so far as this site is concerned, be dropped wholly and forever from the category of [finished] implements.[55]

Holmes said *almost* a pang of regret, but he did not mean it. He was sure he was correct, for he believed that what he saw at Piney Branch "was representative of a class, and will serve in a measure as a key to all."[56] He was not speaking only of other quarry sites in the Potomac valley. He was convinced all stone tool production followed the same narrow and well-defined pathway in which "every implement made from a bowlder or similar bit of rock *must* pass through the same or much the same stages of development, whether shaped to-day, yesterday, or a million years ago; whether in the hands of the civilized, the barbarian, or the savage man."[57]

As to who had produced all those rude forms, the answer from Piney Branch was obvious: it had to have been "our Indians," among whom mining and quarrying activities were well known and on whose village sites in the area Holmes had observed the finished quartzite artifacts that had been quarried at Piney Branch (or other like sources). Holmes saw the "quartzite art of the [Potomac] valley as a unit" in which rude, unfinished forms were restricted to workshop sites, whereas the villages had the "specialized and minute forms" of artifacts. He granted there might be temporal variation within the artifact assemblages, for the "American Indian did not always occupy one plane," but stone artifacts seemed "*not very sensitive to cultural changes.*"[58]

Anticipating skeptics who might wonder why, if workshops were used by Indians, they lacked pottery (assumed to be diagnostic of Indian occupations), Holmes declared there was "little reason" for any miner to carry finished tools and especially ceramic vessels into a quarry pit. To those who might observe that turtlebacks were sometimes found outside workshops, Holmes replied that any place in which unfinished forms were found was, by definition, a workshop. Under no archaeological circumstances would Holmes permit a turtleback to represent anything other than early-stage manufacturing refuse of Indian groups: it had "no ethnic or chronologic significance whatever."[59] The Piney Branch case made, he fired his first shot across the bow of the American Paleolithic:

> It may be said with much truth that the archaeologist who studies flaked stones of *any country* without having made himself familiar with the functions and character of such a workshop is liable to make serious blunders.[60]

"Serious blunders," like mistaking unfinished artifacts and quarry debris for Paleolithic artifacts, blunders that seemed to Holmes to respect no boundaries.

Yet, having gone that far, Holmes went no further. The rude implements of Paleolithic Europe and perhaps even Trenton seemed remarkably like the Piney Branch quarry forms, but he hedged on claiming they were necessarily recent in age as well. After all, there seemed to be independent geological testimony that those artifacts might have considerable antiquity. Instead of rejecting them as ancient stone tools, he drew from their similarity with the Piney Branch specimens the general Haeckelian lesson (Figure 5.2):

> As in biology the growth of the individual epitomizes the successive stages through which the species passed, so in art the flaked stone tool of the highest type advances through stages of manufacture each step of which illustrates a period of human progress in culture.[61]

These were the lessons Holmes had for Abbott on his Piney Branch visit in the spring of 1890. Whether or how much of their discussion touched on Trenton is lost to history, although it almost certainly came up. Judging by the record that survives, their exchanges that day were reasonably cordial.

But as Holmes continued to think about these issues his views changed, and later that year at the AAAS meetings he discussed his Piney Branch study and its implications.[62] Reporting on the Section H (anthropology) presentations in the *American Naturalist*, Wilson reprinted an Indianapolis newspaper's rendition of Holmes's conclusion:

> The rude forms of chipped stones are not tools at all, as has been taught by archaeologists for half a century, and so the "rough stone age" and the "smooth stone age" of the District of Columbia and all the rest of the world are knocked into smithereens. This is not Prof. Holmes' exact language, but that it just what he meant.

There was more: the newspaper cheerily reported that Holmes's paper "knocks some dear old theories in the head, and particularly sets the French archaeologists in the background." Holmes evidently had not said so directly, but that could be drawn from his view of the manufacturing process. Putnam listened to Holmes's paper that afternoon and agreed "the many divisions based on the shape of the implements is [*sic*] artificial and of no value." Nonetheless, Putnam was not ced-

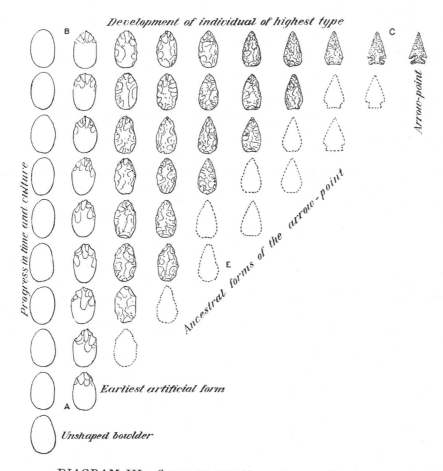

DIAGRAM III.—STEPS IN THE EVOLUTION OF SPECIES.

Figure 5.2   William H. Holmes's view of stone tool manufacture and evolution, from a lecture he gave at the World's Columbian Exposition in 1893 (from Holmes 1894: Diagram III). He labeled the diagram "Steps in the evolution of species" and drew on the then-popular notion in biology that ontogeny recapitulates phylogeny, such that in species' evolution the ancestral adult forms become descendant juvenile stages, and thus the life history of an individual "replays" in condensed stages the life history of the species. Translated by Holmes into archaeological terms, that meant that the early stages in the process of stone tool manufacture (*top line, left to right*) mimicked the "ancestral forms" made early on in human cultural evolution (*lower lines*), "although the steps in the former are all of one time, and the latter represent all times" (Holmes 1894:137). Hence, it was no surprise to him that early-stage artifact failures found at the Piney Branch quarry—and, he supposed, Trenton, New Comerstown, and other American Paleolithic sites—resembled the ancient stone tools of Paleolithic Europe, but that did not mean they were the same age. It meant age had to be determined by geological evidence and not on the basis of morphology of the specimen—as Abbott and other American Paleolithic proponents often did.

ing any ground on the American Paleolithic. Nor was Abbott, who smirked that the whole discussion was "quite amusing." Mason did not think so: "The views of Mr. Holmes . . . are accepted in this country almost universally."[63] Mason may have exaggerated Holmes's influence at that moment, but Abbott could scarcely afford to be amused. He had to respond.

5.3 ABBOTT RETURNS FIRE

That fall Abbott used his inaugural *Annual Report* to the University Museum to reply. He was willing to grant that Holmes might be right, but only about Piney Branch. Abbott would not countenance generalizations beyond that, particularly in regard to the American Paleolithic, for in his mind "the inferences drawn [by Holmes] are too sweeping, and have not necessarily the bearing upon the question of man's antiquity in America which he [Holmes] practically claims."[64]

Abbott had his reasons for drawing the line: he thought the Piney Branch discards were not at all identical to Trenton's "true Paleolithic implements." Attributes of stone artifacts that Holmes took as a sign of manufacturing failure (such as humps that could not be removed) were often found on Delaware valley paleoliths that had been as "elaborately worked as any arrow-point," and were even present on more refined Indian artifacts. Further, argillite paleoliths did not form a graded series with, for example, argillite arrow points, drills, and scrapers—testimony that the former were not merely the early stages of the latter. There was a wide gap between the two, far wider than American Paleolithic critics could close. Finally, it seemed to Abbott that Holmes was laying too much stress on the artifacts and not enough on their context. The conditions under which paleoliths were found in the Delaware valley were wholly different from what Holmes saw at Piney Branch. At Trenton the specimens were in a distinct horizon, and "no verbal jugglery" (he had McGee's paper at the AAAS the summer before in mind [§4.15]) could make them an adventitious inclusion.[65] The evidence of antiquity, so far as Abbott was concerned,

> must be the same here as in Europe, and only when we find the geological and archaeological conditions in accord, i.e. rude implements in undisturbed deposits, can we assert that such evidence has been found.[66]

Abbott stood firmly by Trenton on the strength of its geological context, not just the resemblance of its artifacts to European paleoliths. The gauntlet was thrown: if Holmes aimed to reject Trenton as an American Paleolithic site, he would have to go to Trenton to prove it.

It was a bold challenge, and Holmes willingly accepted. But Abbott's brave words were not just for Holmes. They were also aimed at the University Archaeological Association, his governing council, in the event that Holmes had raised doubts in their minds about Abbott's scientific credibility. It was a strategic move: Joseph Leidy, the association's president, had concerns about Abbott, his onetime student, and had opposed his appointment (§3.7, §4.2).[67] Abbott did not want those concerns to escalate, particularly as the results of his first field season had not gone well, and he had burned his bridges (again) with Putnam (this time over Abbott's hiring Ernst Volk away from the Peabody Museum in late 1890, without first asking for Volk's release).[68]

Aiming for greater rewards than had come from the Delaware valley, Abbott planned "extensive explorations" of the region's caves and rock shelters. If the Delaware was like the Somme, as a stream of distinguished foreign archaeologists had assured him, then surely there were archaeologically rich caves to match those of Paleolithic Europe. Abbott selected several promising ones, marshaled start-up funds, then devoted part of the summer to "cave-hunting." But eastern Pennsylvania was not the Dordogne region of France. Abbott found few caves and nothing of archaeological interest. It was not a promising start, though Abbott sugarcoated it as best he could, assuring the council that negative results were valuable "principally in experience" (his), even if they had not produced additions to the museum collection.[69]

The following spring he requested funds to continue the work, but first the council asked for a detailed proposal with plans, maps, methods, probable results, and costs.[70] They were not keen to continue contributing merely to his experience. Abbott proposed a change in strategy: he would float the Delaware River and make "an exhaustive examination, both geological and archaeological" of two islands and a long stretch of the river bank. To insure the council material for exhibit would result, he promised to explore several Indian village sites along the way. He proposed to then join Henry Mercer, who was inspecting quarries and exploring other sites across eastern Pennsylvania on behalf of the museum.[71] A key part of his plan was to revisit his old grounds and "demonstrate beyond all cavil" that paleoliths were found in the Trenton Gravel. The council approved the proposal.[72]

Notwithstanding his acknowledgment of the importance of geology in fixing the context and age of paleoliths, Abbott still did not see that "impracticable and difficult" excavations were necessary, and once again he put little effort into his fieldwork, mostly spending time as a "mere on-looker" watching Mercer work. (Mercer urged Abbott to do more on his own so as to have more material to go into his report.)[73] Yet, when Mercer submitted a report on his summer's activities, Abbott could afford onlooking no longer, for Mercer had come onto a jasper quarry north of Limeport and was suddenly sounding a great deal like Holmes.[74]

Mercer had collected at Limeport unmistakable turtlebacks and other artifacts at various stages in the manufacturing process that resembled the specimens Holmes had found at Piney Branch and even, Mercer reluctantly admitted, Abbott's supposed paleoliths. But Limeport was not a Pleistocene site: not by its geology, other artifacts, or the stone used to fashion its turtlebacks, which were made of jasper. Mercer had it on Abbott's authority that jasper was used only by Indians and not by the Delaware valley's Paleolithic peoples (§3.3, §3.6). If these jasper turtlebacks were "blocked-out or unfinished implements," did that mean Holmes was right about the independence of form and age? Mercer dodged the question, suggesting that perhaps Neolithic groups had a "method of implement manufacture" that along the way mimicked paleoliths just as, he supposed, some paleoliths might be unfinished. Either way Abbott, who had insisted rude implements ought not to be fashioned of jasper or appear on Indian sites, stood to lose. And Abbott was not keen to lose, not in front of his governing board.[75]

A summary of Mercer's activities had to be included in Abbott's official report to the council, and though he extracted long passages that Mercer had written on other sites, Limeport was mentioned only in passing, not by name, and nary a word of Mercer's troublesome discussion of jasper turtlebacks and paleoliths.[76] Abbott was not about to allow Mercer's skepticism to seep in. Privately, a snarling Abbott told him why: "You seem to think that I have been or could be mistaken in these matters. *I Can't.*"[77]

Officially, Abbott now admitted that rudely fashioned stone implements "*may be*" found in some Indian sites, though that had not been proved. And he maintained that an artifact's "mineralogical identity" made it possible to unambiguously separate paleoliths from Indian artifacts; that was beyond the shadow of a doubt.[78] But Abbott was being disingenuous here.[79] The isomorphism of artifact form and raw material was not as absolute as he had supposed. Mercer had shown that. So too had Edward Paschall, a collector in southeast Pennsylvania, who a year earlier had sent Abbott several unambiguous paleoliths from the Point Pleasant argillite quarry, a site Paschall was certain was an Indian village occupied as recently as historic times. He warned Abbott to pay more attention to these unfinished argillite forms.[80] Abbott had not heeded the advice, nor thought he had to: "Why should not 'doubters' *prove* the falsity, if such there is?"[81] The doubters, Holmes foremost among them, aimed to do just that.

5.4 THE GATHERING STORM

While Abbott was floating downstream in 1891, Holmes was at quarry sites in the Tidewater region and Midwest, with an eye toward understanding how artifacts were prepared for transport away from a quarry and the relationship between artifacts and raw material availability.[82] Holmes also began looking at the geology of American Paleolithic sites with the help of Frank Leverett of the USGS Glacial Division (§4.11).

At Chamberlin's behest, Leverett had been attempting to correlate midwestern glacial deposits with distant eastern formations such as Lewis's Philadelphia Red Gravel and McGee's Columbia (§3.4, §4.5).[83] Leverett was skeptical about the possibilities and cautioned Putnam and others at an 1890 meeting of the BSNH that any conclusions about the absolute or relative age of paleolith-bearing deposits were at best premature, and thus more analysis had to be done before any "decision concerning the antiquity of man can be reached by geologists."[84] Putnam urged Leverett to go to the Little Miami valley, where he was convinced earthworks sat atop "older drift" and were blanketed by silt from the younger drift, and where Charles Metz had recovered paleoliths in what Wright identified as glacial deposits (§4.2).

In the field in Ohio in 1891 Leverett did just that and concluded that the "implement-bearing" gravel terrace at Loveland was indeed glacial in age, having formed behind a moraine that dated from the "maximum advance" of the second glacial stage. He said as much at the AAAS meeting that August,[85] and that caught Holmes's attention. If Leverett was right about the antiquity of the Loveland terrace, then Holmes needed to take a closer look. Leverett was headed to Ohio after the meetings; Holmes joined him.

In early September they were in Loveland, where they met with Metz and spent several days at the gravel pits where he had found his paleoliths. While Leverett examined the geology, Holmes searched for artifacts. Finding none, he wondered (as he later put it) "whether there had not been some mistake." Were the specimens really artifacts? If so, was it possible they had slumped into older deposits from the surface above? An examination of the specimens convinced Holmes only one was genuine. Yet, Holmes noticed (and Leverett confirmed) that the specimen had dark sediment adhering to it, quite unlike the lighter, fine-grained matrix of the glacial terrace and very much like the surface stratum. That being the case, Holmes suspected the artifact had recently been on the surface but in the excavation for the railroad gravel had fallen into the pit and been covered by debris, making it appear as though it was in situ twenty-five feet below the surface. For Holmes this testified to the importance of having the geological context of finds confirmed by several observers and not rely on a single individual's say-so—especially if that individual was "not a professional student of geological phenomena."[86]

Holmes drew another lesson from Loveland. If that specimen was in situ, it implied people were inhabiting an area "overrun by [glacial] torrents." That seemed improbable, and the power of those torrents and the size of the terrace gravels raised the possibility that this "artifact" was flaked naturally. Holmes spotted in a nearby railroad cut many specimens that were so well flaked and shaped that they appeared artificial, but all were covered by glacial striations that obviated a human origin.[87] So much for Loveland. He reached the same conclusion at Madisonville. There was no evidence that the specimen was a paleolith, let alone in primary context in glacial gravels. It was, instead, a "typical reject of the modern blade-maker," and as at Loveland it must have tumbled from the surface into the deeper gravels.[88]

Holmes intensified his examination of the supposed American Paleolithic in the summer of 1892, starting with an excursion to Little Falls with Newton Winchell, then going to Trenton, where he and WJ McGee spent two days combing over the site, at times in the company of Abbott, Volk, Stewart Culin (who was on Abbott's governing board), and Chamberlin's colleague, glacial geologist Rollin Salisbury.[89]

Salisbury had been borrowed from the USGS in 1891 by the New Jersey Geological Survey to map the Pleistocene formations of the state. One of his first efforts was to determine whether there was evidence that had "escaped observation" of a glacial stage that predated the mapped terminal moraine (which Chamberlin suspected was equivalent to the Kettle Moraine [§4.7]). Salisbury soon found, some distance south of the terminal moraine, glacial deposits that had no "genetic connection with the [previously mapped] moraine," whether in material, degree of weathering, or oxidation. These were neither water-deposited nor an instance of Wright's "fringe" (§4.8). Instead, they came from an earlier glacial advance, though how much earlier was uncertain: perhaps an interval several times as long as that which had elapsed since the end of the last glacial period. And much as Chamberlin had a decade earlier (§4.7), Salisbury from the opposite border peered across the state line into Pennsylvania, where Wright and Lewis had not seen more than a single glacial advance: Salisbury could not help but remark that he could see drift deposits there in "similar situations" as those in New Jersey.[90]

In the summer of 1892 Salisbury examined the Trenton Gravel. It appeared to him to be part of a valley train of debris deposited by outflow streams from the glacial front. Wright had suggested (§3.7) the gravel was contemporaneous with the ice front about sixty miles to the north. Salisbury was skeptical: the gravels were not continuous from Trenton to the ice front (contra Wright), suggesting the formation had been deposited long after the Ice Age ended. Regardless, it was necessary to carefully determine the age of any specific deposit in which artifacts were found.[91] Holmes joined Salisbury in Trenton that July, and they along with McGee examined the walls of a long, wide, and deep sewer trench through the Trenton Gravel freshly dug by the city. No paleoliths were found. Holmes arranged for BAE assistant William Dinwiddie to spend a month "in the trench with the workmen" watching for artifacts. Still no paleoliths were found.[92]

After Trenton, McGee and Holmes headed upriver and examined the Point Pleasant argillite quarries and saw what Mercer and Paschall had seen: "An Indian workshop with abundant material . . . including several specimens corresponding to the typical turtle-backs hitherto supposed to occur [only] *in situ* in the Trenton Gravel."[93] Abbott was with them, and in writing of the visit later Holmes could not hide his glee that on seeing the Point Pleasant sites Abbott was "entirely at a loss to explain the occurrence." The explanation seemed to Holmes "at once apparent to any one not utterly blinded by the prevailing misconceptions."[94]

A few days after Holmes's party left Trenton, George Frederick Wright, Albert Wright, and Judge C. C. Baldwin arrived to spend the day. They were escorted to the sites by Abbott and Volk. Despite temperatures over 100°F, Abbott deemed it "a most delightful day," quite in contrast to his cheerless diary entry after Holmes's visit.[95] But Abbott's mood worsened only a few weeks later when he realized his curatorial position was "slipping away." For that he blamed Daniel Brinton and Culin, declaring both of them "liars." Unfortunately for Abbott, however, both were on his governing board and Culin, for one, had a low opinion of Abbott: he doubted Abbott's claims about the Trenton Gravel and thought his methods were "loose" and did not warrant the "official approval of respectable institutions." On the other hand, Culin thought Mercer could be saved, if Holmes was willing to "set him right."[96] It was clear by August of 1892 that Culin did not expect Abbott to last at the University Museum.[97]

Abbott must have understood the situation, for once again he made overtures to Putnam to take him back. Putnam, not wanting to meddle in another institution's affairs and happy with his arrangement with Volk (and perhaps enjoying not having the mercurial Abbott as an employee), urged Abbott to stick it out; he could offer nothing at the Peabody. It was up to Abbott to be on his best behavior, and his Trenton Gravel would have to withstand critical scrutiny.[98] No easy tasks, those. Especially not with the very public cannonade of the American Paleolithic that began weeks later at the AAAS meetings.

5.5 THE PRELIMINARY SKIRMISH

Abbott did not attend the AAAS meetings in Rochester that summer, but Holmes was there to deliver his address as Section H vice president along with two papers on the American Paleolithic, prompting a pained cry from Putnam, who accused him of "annihilating Paleolithic man."[99] Putnam had reason to worry: in his first paper Holmes pronounced "a very large percentage" of Paleolithic claims to be "defective or erroneous," as in most cases it had not been demonstrated the alleged paleoliths were finished or found in secure glacial contexts, let alone recovered by geologically competent observers. He showed that Babbitt's Little Falls quartzes were all "failures left by arrow-makers," and none were in primary context. Mercer followed Holmes and testified to finding turtlebacks in argillite quarries upstream of Trenton. Although these were recent in age, he was not willing to concede that the supposed Paleolithic turtlebacks from Trenton were as well. But he came close.[100]

If Abbott and Babbitt's paleoliths were having a bad time of it, so too were Wright's glacial interpretations. Wright was in Rochester for the GSA's summer meeting, which immediately preceded the AAAS meeting, and he spoke on "The Extra-morainic Drift of the Susquehana Valley." The subsequent report of the session in the *GSA Bulletin* made the laconic observation that after Wright finished, "Professor R. D. Salisbury and Mr. WJ McGee remarked upon the matter of the paper, challenging the observations and inferences of the author." Indeed they did, for Wright had once again raised his "fringe" and insisted it was merely an extension of a single terminal moraine. McGee and Salisbury objected, as Chamberlin had before (§4.11), that the fringe belonged to an entirely different and earlier ice advance. Salisbury repeated that objection when Albert Wright invoked the fringe to explain the very same features in New Jersey that Salisbury had just assigned to an earlier glacial event.[101] In the opinion of George Frederick Wright's allies, the critics "came out second best at Rochester on the moraine question."[102] Salisbury thought otherwise, but regardless of who won that skirmish the meeting marked the start of all-out conflict, for at Rochester a conscious decision was made to launch an attack against the American Paleolithic and the unity of the glacial period.

Organizing the attack was Chamberlin himself, who arrived too late to witness Salisbury and Wright squabble but sat quietly through the remainder of the meetings. It was not for lack of something to say. He was plotting. Chamberlin at that moment was creating the Department of Geology at the newly founded University of Chicago. He had just hired Salisbury and appointed Holmes (they knew one another through the USGS) as nonresident professor of archaeological geology. He also planned to launch a new journal under the department's auspices.[103] As he listened to the papers in Section E (geology) and Section H, it occurred to him that several were ready-made for the inaugural issues of the *Journal of Geology*. As he explained to Salisbury:

> I have been thinking that it would be a fine idea for you and Whitson to prepare a special paper for our Jan[uary] number [of the *Journal*] on the Trenton Gravel. Then Holmes a paper, on his search for implements. Then perhaps a note from you on the geology of the [other] localities Holmes discusses & then I will discuss the general conditions of the time & their bearing on the subject, much as I talked to Holmes & you at Rochester. If you are ready, we could make a strong combination. It was with a view to something of the kind that should bring out the full strength of our side

with everybodies [*sic*] work in its proper place & [alongside] the others, that I did not discuss the subject [publicly] at Rochester.[104]

Ultimately, the first two numbers of the *Journal of Geology* looked very much as Chamberlin envisioned, though by the time they appeared in early 1893 he decided not to publish the review he had intended to write. For by then he had already "prodded Prof. Wright so severely . . . that I would seem to be a prosecutor if I carried out my original purpose."[105] Chamberlin was shedding no crocodile tears; he was being honest. Wright had indeed been prodded severely, and not just by Chamberlin, for the debate over the American Paleolithic and the unity of the glacial period exploded in the fall of 1892.

For most proponents and, indeed, for most critics, there was no particular reason to link the American Paleolithic (an archaeological issue) with the unity or diversity of the glacial period (a glacial geological issue). Proving people were in America in glacial times did not require knowing how many glacial advances there were, although such knowledge could certainly help narrow the age of that occupation, if it were shown that artifacts were associated with a particular advance (as, for example, Cresson's Claymont paleoliths [§4.5]). Similarly, resolving the number and timing of glacial events was strictly a geological problem and would not be settled by archaeological evidence. Most archaeologists and geologists had no intellectual capital invested in the internal debates of the others' field, except for Wright, who was willing to put humans into the Pleistocene, but only within the chronological confines of a single glacial period. Because Wright perceived these archaeological and geological issues as inextricably linked, because he was vocal and public in his pronouncements on the subjects, and because he had garnered powerful foes, his linking of the archaeological and geological issues led to battles erupting on both fronts simultaneously.

That both archaeological and geological issues were linked likely did not appreciably change the content of the debate in each of the respective fields. However, it had the effect of expanding the debate and bringing more (and more powerful) players into the arena, which in turn raised the stakes and the volume, and made the controversy considerably more visible and assuredly more acrimonious than it might otherwise have been. It also added a new dimension to the controversy, for as archaeologists and geologists aligned themselves it was obvious one side was comprised largely of BAE and the USGS scientists and the other side was not. This fanned long-simmering resentment toward government science and scientists (§4.10). The Great Paleolithic War began as a relatively straightforward discipline-specific empirical dispute but quickly escalated far beyond that.

5.6 THE GREAT PALEOLITHIC WAR

The spark that ignited the Great Paleolithic War was the publication in September of 1892 of Wright's *Man and the Glacial Period*. Like his earlier *Ice Age*, this book had its origins in Lowell Institute lectures (§4.13). Wright was invited back in March of 1892 on the strength of his success five years earlier and the book's favorable reception that came from that appearance. Once again, his Lowell lectures were successful, and as before he planned a book from them. In just months a manuscript was at D. Appleton and Company for publication in its International Scientific Series, which included books by Darwin, Huxley, Lubbock, and Tylor—fitting company, as far was Wright was concerned.

By Wright's own admission, *Man and the Glacial Period* covered much the same ground as *Ice Age*, but it was offered to the public as a condensed, better balanced, and international version, with new material from his recent field excursions on both sides of the Atlantic and the latest archaeological discoveries here and abroad.[106] Wright had no reason to suppose he was on shaky ground. In fact, so confident was he in the American Paleolithic that he opened his discussion of Pleistocene-age human remains not in Europe but in America, where "the problem is so much simpler."[107] The usual suspects from *Ice Age* were there: Trenton, Loveland, Madison, Little Falls, and Claymont, now joined by Nampa and New Comerstown. Claymont and Nampa, of course, suggested a considerably greater antiquity than the other localities (§4.14), but no more than that allotted by the single glacial period, with the Nampa Image deposited at the very outset of that period, Claymont when the ice was at its maximum, and the remainder as the ice retreated but had not yet vanished from the landscape.[108] To this summary Henry Haynes added an appendix assessing the evidence for the even earlier "Tertiary Man," notably the discovery of apparent artifacts (eoliths) in Miocene and Pliocene deposits in Europe and California.

Man and the Glacial Period was mistitled: Baldwin, to whom the volume was dedicated, gently complained to Wright that there was much "more glacial period & less man" than he expected.[109] He was right: nine of the ten chapters tilted toward geology. Yet these provided Wright the opportunity to herald his geological triumphs—his recent observations of Alaska's Muir Glacier, the mapping of the Pennsylvania and Ohio terminal moraine, the detection of the fringe, and of course the discovery of the Cincinnati Ice Dam, all of which were insistently embedded within the idea that there had been but a single glacial period.

As in *Ice Age*, he admitted there were differences of opinion, but with the same great enthusiasm for his views and tepid consideration of others, for which Davis previously criticized him publicly and Chamberlin seethed privately (§4.13). Wright again cited White and Lesley's testimony on behalf of an ice dam and toppled the same straw-alternatives he had easily disposed of in *Ice Age*. Chamberlin's arguments about the terraces on which White was hoisted two years earlier (§4.15) were misstated, then rejected. Chamberlin, Wright insisted, made too "unnecessary [a] demand upon the forces of nature, when the facts are so easily accounted for by the simple supposition of the dam at Cincinnati."[110]

So too the "much-mooted" question of whether there were several glacial epochs. Granted, there were features that to "Chamberlin and many others" seemed to indicate different ice advances: moraines showing differences in their weathering and oxidation, peat layers sandwiched between till deposits, variations in the amount of glacial erosion, and the like. Yet, Wright thought none of these required the long-term disappearance of glacial ice, just the temporary oscillations of a single ice front. He had little difficulty accepting the idea that local glacier retreat over a few centuries was sufficient to account for twenty feet of "vegetable accumulations" that would then be buried when the ice readvanced.[111] Wright admitted some hypothesized multiple glacial periods, but when it came to selecting which alternative was more acceptable, he invoked his own brand of Occam's Razor:

> The introduction of a whole Glacial period to account for such limited phenomena is a violation of the well-known law of parsimony, which requires us in our explanations of phenomena to be content with the least cause which is sufficient to produce them.[112]

That "law of parsimony" assumes, however, that alternative causes (hypotheses) were equally sufficient explanations of the phenomena at hand. Wright thought that so, and therefore he accepted the idea that there had been only a single glacial event. He believed his position could be maintained "without forfeiting one's right to the respect of his fellow geologists."[113] Wright was either very optimistic, very naïve, or both. Although styling himself a public spokesman for glacial geology, he was utterly blind to the depth of the animosity toward him and his ideas by the USGS geologists. Wright even counted himself among their number, listing himself on the title page as an "Assistant on the United States Geological Survey."

Among the many unpardonable sins of *Man and the Glacial Period* in Chamberlin's eyes, this was among the worst. Chamberlin had heard that summer that Wright was coming out with another book and was determined to distance it and its author from the USGS. Only, he discovered Wright's USGS ties had never been officially severed. After checking with Powell to make sure the USGS no longer needed token clergy on its rolls (§4.10), Chamberlin formally requested Wright's dismissal on the grounds of incompetence, intransigence, and unwillingness to consider alternative views in publications. Within a week Wright was told his USGS services were dispensed with effective immediately. Over the eight years of his appointment to the temporary force of the USGS, Wright had worked a total of 137 days, most in separate stints lasting no more than three weeks. His active service had ended by 1888.[114] In Chamberlin's eyes Wright's dismissal in July of 1892 was just a long overdue formality.

Wright intended *Man and the Glacial Period* to give a clear view of the "present state of progress." Even sympathetic readers could see otherwise. Abbott grumbled the book was just "a re-hash of his older one." But Abbott's uncharitable opinion was offered in the privacy of his diary. Unfortunately for Wright most of the published reviews were far more severe.[115] In the months following its appearance the book was reviewed more than a dozen times and in most of the major scientific and literary journals of the day. The majority of the reviews were written by just three individuals: Salisbury (who wrote four), McGee (two), and Upham (two). (Other reviews were published by Brinton, Chamberlin, Holmes, Peet and Winchell.[116]) Multiple reviews by a single author were possible, as most were published anonymously and journal editors could not check whether the same person had already reviewed the volume elsewhere. Moreover, since most reviews were volunteered, the individuals fastest to offer had first pick of the most prominent venues. Both sides were well aware of that: within days of receiving his copy, Upham wrote Wright asking if he knew if there were reviewers "at work on your book for the *Am. Geologist* and *Am. Jour. of Sci.*" or perhaps *Popular Science Monthly*. Upham ultimately reviewed the book for the *American Geologist* and *Popular Science Monthly*, but it was too little too late. Chamberlin had already organized the critics' assault, writing a review himself and enlisting Salisbury, McGee, and (later) geologist John Branner (of Stanford University) to do likewise. Like McGee, they were all happy to comply:

> I shall carry Wright's book with me for light reading in lieu of "Peck's Sun" and "Texas Siftings", and with the view of putting myself in a position to adequately review the document in accordance with your suggestion. I shall advise you later as to the media selected. Rejoicing over the tone of righteous indignation expressed in your letter.[117]

As the critics were faster and more determined to make their views known, the inevitable result was that the majority of the reviews of *Man and the Glacial Period* were highly negative, some downright ugly.

Salisbury, still smarting over his exchange with Wright at Rochester, launched the first blast scarcely a month after the book appeared. Published anonymously in late October in the *Chicago Tribune*, it opened with barely disguised contempt: "To every reader who is in a position to judge [its] merits . . . the book will be disappointing. To every reader who is not in a position so to judge the book will be misleading." Salisbury then reeled off a litany of complaints: there were factual errors on "almost every page," even on matters as fundamental as the parts of a glacier (Wright had blundered in describing the névé). Worse, the book "partially or wholly ignored the work of most American glacialists"; presented theories as accepted (a single glacial epoch) while ignoring contradictory evidence; and offered interpretations like the Cincinnati Ice Dam that were "wholly wrong." This was a theologian's book, Salisbury sneered, not written by an author with "the scientific habit of mind or the scientific mode of expression." It was said Salisbury could be "very gruff" and insulting. So he could.[118]

He also proved to be persistent, hardly pausing with the *Tribune*. That was, after all, only his local newspaper, and he was determined to attack Wright on a more prominent stage. Even as his *Tribune* review was working its way into type, Salisbury petitioned the editor of the *Nation* to review *Man and the Glacial Period*. Upham had already asked, but the *Nation*'s editor, W. P. Garrison, chose Salisbury. A few weeks later, Garrison received Salisbury's review and must have had second thoughts about his choice: "I have your notice of Wright," he told Salisbury, "and will try to let justice be executed, though the heavens fall" (Garrison's idea of justice was not Wright's: he later declined Wright's request to respond to Salisbury on the grounds that he did not want controversy in the journal). Salisbury's review in the *Nation* did not appear until year's end, but there was nothing in it that he had not already said in virtually identical language in his *Tribune* review. His review in *Nature* that December plowed the same ground.[119]

All of those reviews were anonymous, but Salisbury hardly tried to hide his identity. Anyone who knew who wrote one knew who wrote the other two.

Upham wrote immediately after seeing the "envious and vituperative" *Tribune* review and accused Salisbury of authoring it. The University of Chicago's anthropologist Frederick Starr soon confirmed Upham's suspicion, though Wright also could see that the reviews had been "inspired" by Chamberlin. Upham assured Wright his forthcoming review in *Popular Science Monthly* would offset Salisbury's "malignant notice" and urged him not to "pine away in melancholy," nor fall in to "like sins" in responding. He counseled Wright to keep a "friendly attitude toward even such detractors, remembering the commands of the Bible and the proverb 'Noblesse oblige'."[120]

It was sincere advice, but at the moment Upham himself could not offer much more than that: his own reviews were still months from appearing. And because Salisbury's reviews were publicly anonymous Wright could not respond to him directly. For that, he would have to wait until Salisbury's signed review appeared in the *American Geologist* appeared a few months later. (Winchell, editor of the *Geologist*, would not let Salisbury hide behind the cloak of anonymity.) Until then, however, Wright had enough to deal with at the hands of critics who signed their reviews.[121]

Just days after Salisbury's *Tribune* review appeared, Brinton's review appeared in *Science*. Brinton the anthropologist was far more generous in his assessment of Wright as a glacial geologist, numbering him "among the first in this country" and praising the "scientific manner" in which Wright marshaled the geological evidence. There he checked himself: such "unreserved praise" could not be extended to Wright's archaeology. Brinton rejected each of Wright's American Paleolithic cases. Better evidence than any as yet brought forward would be necessary to establish a Pleistocene human presence in America.[122] A week later Brinton revisited the issue in his weekly *Science* column on anthropology:

> For two or three years past there has been in the air—I mean the air which archaeologists breathe—a low but menacing sound, threatening some dear theories and tall structures, built, if not on sand, at least on gravels offering a scarcely more secure foundation.[123]

That sound was Holmes's demonstration that many of the purported paleoliths were "never real implements at all," a point Brinton found convincing. If these objects occurred in gravels, it was only because this was where the objects were quarried. It had "even been hinted," Brinton said, that the artifacts from the Trenton Gravel "of our own land, may have to lose their laurels."[124]

Watching from the sidelines, Abbott could stand no more, even if it meant publicly attacking a member of his governing council. The following week in *Science* he denounced "the geologists at Washington; and those [like Brinton] that look upon them as little gods." If the critics were correct, why were only rude argillite forms found in gravel deposits and not any other Indian artifacts? If these were quarry rejects, where were the "chips resulting from their fashioning?" And why were signs of use critical to the issue: "Does the spear or arrow point show signs of use?" Most insulting of all, who were these self-styled "expert geologists" to lecture him about the Trenton Gravel or judge that formation's age when they cannot agree among themselves? Abbott declared he too knew "a smattering of gravel-ology. . . . [and] when I find gravel stratified and unstratified, I know and assert the difference." Those who had not spent as much time as he had examining the Trenton Gravel could not claim greater knowledge of its geology. These "geological jugglers," he complained, were not going to stop until they proved "the Indians bought the Delaware Valley from William Penn."[125]

In the same issue of *Science* Wright also took the stand in Abbott's and his own defense. The key was experience. Abbott was "not a professional geologist," but Wright was happy to grant him a mantle of expertise by virtue of his long hours in the field. Likewise, Metz and W. C. Mills each had extensive field experience, and all were "well qualified to judge the undisturbed character of the gravel" in which they had found implements.[126] That Wright pronounced them geologically competent by virtue of the hours each had spent searching for artifacts was intended to preempt the critics. Wright knew Holmes and McGee had not found paleoliths at Trenton and believed they considered that significant. He admitted he had not found artifacts at Trenton either, but what of it? He had not spent much time there, but that was no reason "to doubt the abundant testimony of others who have." After all, "if we are limited to believing only what we ourselves have seen, our knowledge will be unduly circumscribed." What Wright could not see was that Holmes's criticisms of Trenton did not rest on whether he (Holmes) himself had found any artifacts (though he had pointed that out), but that the objects and their context was misinterpreted.[127]

Holmes's article on Trenton had yet to appear, so Wright was swinging blindly hoping to land a few early punches. Criticisms of the American Paleolithic had so far been relatively disjointed. (Brinton, by his own admission, was mostly reporting what was "in the air.") It was easy for Wright to dismiss the criticisms as "offhand opinion," and at that juncture he was mostly correct. He did admit one case had been questioned more thoroughly—he had heard Holmes's AAAS presentation on Little Falls—but was quite willing to jettison that one site, certain that doing so "would not disturb confidence" in Trenton or the Ohio sites. Besides, he was sure that a thorough discussion would "dispel the uncertainty that may exist."[128]

Although he would not admit it publicly, Wright sensed another weakness in the American Paleolithic case. He was not alone: Putnam thought Holmes and McGee had raised legitimate questions about the geological context of the Trenton finds and urged Abbott to "do all you can" to show the character of the gravel that yielded the paleoliths.[129] Continuing to rely on Abbott was not the wisest strategy, however. He was distrusted by critics, and even a few American Paleolithic proponents had begun to perceive him as a liability: Haynes thought the principal objections to an American Paleolithic were "doubts in regard to Abbott's truthfulness & from Wilson's foolish & ignorant claims that all rudely chipped implements are paleoliths." Wright agreed.[130] Abbott was well aware such things were being said about him, telling Wright, "Come [to Trenton] by all means, and I won't salt the ground in advance."[131]

Abbott could joke, but such doubts had been raised before, and that hardly helped his reputation or cause. It also put Wright in the awkward position of having to defend a case he himself had not made and to rely on the testimony of an individual perceived as untrustworthy. It was better to fortify the case himself. Yet, as he began making plans to undertake more work at Trenton, hoping to find paleoliths in place or clarify their geological context Chamberlin struck.[132]

The venue was the *Dial*,[133] a biweekly magazine of literary criticism and discussion published in Chicago. In it, Chamberlin's most severe rebuke of Wright's *Ice Age*, made privately three years earlier (§4.12), now became very public in his attack on *Man and the Glacial Period*. Wright had assumed the mantle of writing about glacial geology and human antiquity for the general public, and in so doing had laid his work open to "the test of a critical examination," for in Chamberlin's mind:

> No one is entitled to speak on behalf of science who does not really command it. No one can be trusted to lead the public who does not himself know the way accurately. The first question is, naturally, how trustworthy is this work?

In Chamberlin's eyes, *Man and the Glacial Period* was not. Beyond making a litany of factual errors and misleading statements regarding glacial geology, Wright had failed to answer three essential questions regarding the American Paleolithic. Were paleoliths finished artifacts? Were they deposited at the same times as the formations in which they occur? And, were the formations "properly placed in the glacial series?" Chamberlin the geologist then lectured Wright the theologian on Holmes the archaeologist's "epoch-marking investigations," which "practically demonstrated" that paleoliths were merely "flakings, failures, and rejects." Any reader, it seemed to Chamberlin, could see for himself that there was "an entire absence of any marks of use" on the paleoliths illustrated in Wright's book.[134]

Of course, that was hardly a topic in which Chamberlin was an authority, and the criteria that then existed for "marks of use" separating finished from unfinished artifacts were at best in dispute (as Abbott had just insisted) and hardly apparent from a photograph. Unless, of course, one conceived of "marks of use" in a different manner than Holmes (and therefore Chamberlin) did. Holmes distinguished finished from unfinished on the basis of shape. "Marks of use" were understood not as wear, as Abbott conceived it, but as extensive flaking and specialized forms that would be visible in a photograph. Because Holmes had yet to explain this in detail, Chamberlin's parroting of his criteria undoubtedly made little sense to American Paleolithic proponents. (Wright ignored the point in his response to Chamberlin.)[135]

Chamberlin was no more willing to grant expert status to the discoverers of the alleged paleoliths, summarily dismissing all claims on the grounds that none rested "on expert geological testimony." Wright had bestowed geological expertise on Abbott, Metz, and Mills, but Chamberlin would have none of that. No matter how diligent or experienced, they were not experts:

> Common testimony is worth something when it is conscientious, as it is, doubtless, in the most of these cases; but when an author assumes to teach

the people on behalf of science, he ought to tell them what is science, as distinguished from what rests merely on inexpert testimony.[136]

Inexpert testimony like Wright's. Chamberlin declared Wright unfit to undertake so basic a task as tracing a terrace formation, and he rejected Wright's interpretation of the glacial formations at Trenton and Claymont as confused, misconceived, and disingenuous, especially in what he saw as Wright's deliberate slighting of the work of others who disagreed with him. Chamberlin never directly answered his opening rhetorical question of whether Wright was in command of the science of which he wrote, but there could be no mistaking what he thought.

Chamberlin's damning review ended on a more serious charge than gross incompetence. He accused Wright of misrepresenting his ties to the USGS in order to give *Man and the Glacial Period* the gloss of official approval. Although Wright proclaimed himself an "Assistant on the United States Geological Survey," Chamberlin declared that Wright's relationship with the Survey was tenuous at best, and had essentially been severed seven years before when Wright proved "seriously defective in discrimination, breadth, and correctness of interpretation." Wright might insist he was still on USGS rolls (even if in name only) at the time his book appeared, but that was a moot point so far as Chamberlin was concerned (though he contacted Wright's publisher to find out when the book was going to press to determine if Wright had time to revise his title page[137]). Besides, Wright's distortion of the work of Survey geologists was testimony enough that he had no real connection to them.

Wright did not subscribe to the *Dial*, but he was not given a chance to miss Chamberlin's review. Starr sent a copy from Chicago, thinking Wright should know what was being written. That was about all the help he could offer: "I am quite unable to do anything here for you. 'Justice' is a convenient term, is it not?" Baldwin too sent a copy but was more bemused than troubled and urged Wright to look at the bright side: "T. C. is advertising your book and quite successfully. . . . Did your publishers hire him to do it? He is likely to do you quite amount of good."[138] Wright could hardly afford to be cheery, especially after James Dana, editor of the *American Journal of Science* and sympathetic to Wright's glacial views, quietly warned (in an admitted breach of his own editorial policy not to divulge "journal secrets") that Chamberlin was also planning to respond to Wright's just-published article on the "Unity of the Glacial Period" (§5.12). In pained reply Wright poured out his unhappiness, and pleaded for help:

> Professor Chamberlin is not in good humor. He has just published in the *Dial* a very abusive and misleading review of my last book in which he charges me without reason of taking unwarranted advantage of my relation to the survey to get a reputation that I do not deserve. He also challenges the genuineness of all the American palaeoliths, and speaks very contemptuously of my knowledge of the Delaware region. Professor Salisbury and Mr Magee [*sic*] took my criticisms of their views at the last geological society

[GSA] meeting in very high dudgeon [*sic*] and through their influence on the publication committee refuse[d] to publish my paper. . . . They do this on the plea that the society wants to avoid controversial matter. But they had published Salisbury['s] misleading report and my conflicting observations were couched in most respectful language. All I ask of you now is to protect me from abuse by Professor Chamberlin in his reply to my article.

Dana assured him "nothing incourteous [*sic*] will be published in the *Jour. Sci.*"[139] Unfortunately for Wright, Dana's help only extended to the *American Journal of Science.*

Wright sent the *Dial* a reply, but it and a response by Chamberlin in January of 1893 hardly brought the two of them closer, although Wright tried. After defending his work and time at the Survey and his geological interpretations, he called for courtesy and candor, that "men of science may not fall into the habits of controversy which are supposed to prevail to too great an extent in ecclesiastical circles."[140]

Chamberlin was not buying that argument, but at least the central issues each saw came into focus, and in so doing the substance of their disagreements over glacial history and archaeology fell away, revealing that for Chamberlin this dispute was, at its core, about credentials. Wright insisted his time in the field qualified him to speak on glacial history. Chamberlin rejected that claim on the grounds that Wright had exaggerated his experience and qualifications ("it bothers my limited arithmetical ability to figure in 'eight years' [of fieldwork] between 1880 and 1884"), but mostly because Wright was unable to see things that "any discriminating geologist can see for himself." As proof, Chamberlin reprinted nearly in its entirety his letter of three years earlier (§4.13) in which he lectured Wright, then putting together *Ice Age*, that it would be "premature and unfortunate" to write a book for the general public.

But Chamberlin made a tactical blunder: he left out the final paragraph of his letter, in which he insisted that if anyone was qualified or "justified" to write a book for the public, he himself was the one. Wright seized on the opening. Chamberlin's relentless attack was not about differing views on human antiquity and glacial history: it was instead being driven by the basest of motives, jealousy. That infuriated Wright. Who was Chamberlin to decide who should write for the public? Wright refused to "frustrate [his] literary plans" for the sake of Chamberlin's ego. Besides, Chamberlin had already put his ideas on glacial history out in the public in several publications. Wright was merely keeping an alternative interpretation of the facts before the public. All in the interests of truth, of course.[141]

Leverett, watching from the sidelines and trying very hard to think the best of Wright (it was a professional friendship Wright certainly needed and wanted to preserve), thought Wright's parting salvo a cheap shot:

It seems to me that in imputing jealousy . . . you bring a charge that scarcely anyone that knows [Chamberlin] will consider fair. . . . He is in my opinion much too large a man to attempt to frustrate your literary plans for the sake of personal gain.[142]

Having been declared incompetent by the chief of the USGS Glacial Division might have been seen as a badge of honor in some circles of the geological community, but it was mostly deeply humiliating. Still, it was no surprise. Wright knew Chamberlin's opinion of him, though he might have hoped he would keep that opinion to himself as he had three years earlier. The next round of reviews were more of a surprise and much more malicious. But then the reviewer, WJ McGee, saw Wright's book as a chance to demonstrate loyalty to a new-found cause.

5.7 THE "BETINSELED CHARLATAN" AFFAIR

A few years earlier McGee had been an enthusiastic American Paleolithic proponent, as well as a cordial correspondent of Wright's (§4.3, §4.5, §4.14). His departure from the fold began about the time Holmes started his work at Piney Branch. Whether he underwent that conversion as a result of being compelled by the sheer force of Holmes's logic, or simply because (in Swanton's words) he "desperately feared to depart from the 'party line,'" is not particularly relevant, though given McGee's ambitions and intense desire for status the latter was surely an important factor.

What mattered was that his renunciation of his Paleolithic faith was swift and total, and like all who have sacrificed strongly (or at least very publicly) held beliefs, he became more zealous than those who had not suffered through the difficulty of rejecting previously cherished views. As one contemporary put it as a prelude to rubbing McGee's nose in some of his former writings: "The zeal of the proselyte is proverbial, and [so is] the readiness with which he forgets his former faith."[143] Yet, it is best not to put too fine a point on this: McGee certainly appeared fervent in his prior belief in an American Paleolithic, but how strongly he held to it is uncertain, because at the time that belief was a handy vehicle to advance his scientific career. Fueled as he was by blind ambition, that belief served him well. As the tide changed on the Paleolithic, McGee saw which way it was running and rowed accordingly. When he was enlisted by Chamberlin to review Wright's work, McGee took to the task with enthusiasm, if not hypocrisy:

> It seems to me that the sooner geologists (other than myself—I have long been awake) awaken to the necessity of stamping Wright as a pretender incompetent to observe, read or reason, and devoid of sound moral sense, the better it will be for the world.[144]

Converts are often greeted with suspicion and face (or feel) the need to silence doubts about their conversion and faith to the new cause. What better way to prove themselves than by eviscerating onetime friends and allies? American Paleolithic proponents might have hoped for "a more tolerant spirit from one who has so recently seen fit to change his own faith."[145] Yet, what changed was only McGee's faith he could advance his status by attacking rather than promoting the American Paleolithic.

McGee began his assault in *Science*, suggesting changes to Brinton's "otherwise excellent review." McGee took exception to Brinton's having identified Wright as one of the foremost glacial geologist in the country. Wright knew nothing of the "New Geology" emerging under the "discriminating genius of Chamberlin." Its core was the recognition of multiple glacial advances, a fact accepted by all the "leading authorities." Because Wright failed to accept that premise, McGee argued, his geological interpretations were obsolete, his archaeological claims had no "time-basis," and the whole was "an offense to science." Why, McGee asked, was a theologian speaking on behalf of a science in which his competence had been tested and found wanting? The world, McGee decided, "would be wiser if the book were not written."[146]

Wright had just mailed to the *Dial* his reply to Chamberlin, which ended with the hope that "courtesy and candor may prevail."[147] It was too late for that. Perhaps, Wright replied to McGee, he should be thankful for his frankness. It gave full public expression to matters that might otherwise linger sub rosa. Wright then patiently explained that the errors McGee and others singled out were more apparent than real. He claimed his USGS position had ended a month after *Man and the Glacial Period* was published (that was not exactly true, but his book was in production by the date of his termination). And, yes, he did take a very different view of glacial history, as he had just explained in the November *American Journal of Science* (§5.12). Yet glacial geology, Wright insisted,

> is not an exact science. There is no infallible court of appeal for the settlement of theories. Observers and students of the fact may widely differ for a long time in their conclusions without discredit to either party. I can only ask for freedom of opinion and freedom of utterance.[148]

In responding to his critics, Wright again called for civility. Again, the call fell on deaf ears.

McGee had no desire to be civil. "If my strictures are just," McGee told *Science*'s editor, "controversy is impossible; if they are unjust, I am a libelist." McGee did not think he was a libelist, and he was not. Not yet, anyway. If McGee had regrets over his comment in *Science*, it was only that he found the case "difficult to treat" because providing more detailed criticism "unduly dignifies it."[149]

Even so, he found a way to do that and humiliate Wright simultaneously. He reviewed Wright's book in the *American Anthropologist* in early 1893 alongside a small, privately printed booklet by Francis Doughty,[150] a numismatist from New York, who had allegedly found on Tertiary pebbles in New England carvings of bird-headed men, anthropoid apes, elephants, dolphins, ships, pyramids, various human races (in color-coded pebbles), and the sphinx. McGee made his intent explicit: neither of these works had scientific merit, and the only difference was that one of the authors was merely foolish (Doughty), the other a charlatan.

It was a savage ad hominem attack that quickly brushed by Doughty's pamphlet ("a bundle of absurdities worthy of notice only because it is representative of the vain imaginings so prevalent among unscientific collectors"), then moved on to a lengthy assault on Wright's book. The usual flaws were highlighted (McGee was very good at parroting Chamberlin and Salisbury), with Wright's chapters and interpretations variously labeled puerile, superficial, warped, distorted, misleading, absurdly fallacious, crude, egotistic, unscientific, and an "offense to the nostrils." It was possible, McGee supposed, that Wright was merely incompetent, his "distortions" arising out of ignorance of the New Geology and the "brilliant researches" of Chamberlin and others. But McGee insisted Wright knew better and deliberately sought to deceive readers so as to privilege his own incompetent work above all others. What better evidence of dishonesty, McGee argued, than *Man and the Glacial Period*'s title page:

> The imposing list of titles which the author appends to his name conveys the impression that he is a geologist rather than a theologian, which is misleading; that he is a professor of geology, which is not true; and that he is an "assistant on the United States Geological Survey," which is sheer mendacity and theft of reputation."[151]

Doughty's pamphlet and Wright's book were dismissed as the work of "harpies," but Doughty's was destined for little notice. Wright's, on the other hand, was issued by a reputable publisher and thus "its maleficence is multiplied." McGee deemed the one laughable, the other dangerous:

> Doughty is a simple hearted-quack whose bread pills but tickle the fancy of weakling dupes; Wright is a betinseled charlatan whose potions are poison. Would that science might be well rid of such harpies, especially the latter![152]

Appended to McGee's *American Anthropologist* review was a reprint of his *Science* review, but he did not stop there. He struck again in the *Literary Northwest* with a review no less vicious than its predecessors. "It will of course strike you as a piece of destructive criticism," he admitted to the editor, but with no small dissembling McGee justified it by observing it was "no more severe than notices already published." He failed to mention that the other severe notices were his.[153] No matter. He easily rationalized his antagonism, and even urged Chamberlin to be more aggressive:

> It really seems to me that you and Professor Salisbury as well as a good many other geologists, err in dealing with this pest [Wright]. You are assuredly in a position to declare his utter incompetence as an observer and utter lack of veracity as a writer; it would seem to me your best course is to state these disagreeable facts in unqualified terms whenever occasion arises, as I have never hesitated to do in his presence and in publications. Of course there are those who would deplore what they mistakenly conceive to be acrimonious controversy, and you will find the pretender's dupes (and they are many as legion, more the marvel) to rise in their petty spleen and smite your feet; but you would only evade and rise above real controversy by justly characterizing the man and his methods; for when they are justly characterized he can make no adequate response, save by a libel suit, which he dare not enter.[154]

Although McGee could justify his actions, his reviews triggered swift and strong condemnation throughout the scientific community. Letters streamed into Oberlin, even from government scientists, several of whom sought to disassociate themselves from McGee's abusive review. USGS paleontologist William Dall insisted there was not

> in the Survey or Bureau of Ethnology a single person who approves of [McGee's] publications, or sympathizes in their method or spirit. I have heard it emphatically disavowed & condemned by Mr. W. H. Holmes and in fact by every one whom I have heard discuss it.[155]

Perhaps so, though the week before Holmes wrote Branner to say, with neither irony or disapproval, "We are having a pretty warm time over this question and McGee has stirred up the animals at a great rate."[156]

Holmes soon changed his tune. Wanting to borrow the New Comerstown artifact to illustrate it for an article he was preparing on the Ohio finds, he wrote Wright to request the specimen. That prompted a flurry of letters between them and Baldwin (who as president of the Western Reserve Historical Society where the artifact was housed had to approve the loan). Baldwin angrily asked why Holmes deserved the loan and questioned his culpability in McGee's reviews. "In law," Judge Baldwin decreed, "each conspirator [is] liable for more violent acts of others in furtherance of a common purpose." That seemed sufficient grounds to deny Holmes's request. Baldwin's letter, forwarded to Holmes by Wright, elicited a diffident response. Holmes insisted he too had been harshly treated, though by whom he did not say, and he did his best to distance himself from McGee:

> As to associations here we work as individuals and the one person who alone has chosen to be over severe . . . did so with the approval of none and without the knowledge of nearly all.

Baldwin would hear none of that nor, for that matter, would Wright: "There seems to me hardly any comparison between the abuse which has been heaped upon you and that which has been poured out so liberally upon myself." Baldwin wanted to know why Holmes, as a member of the editorial board of the *American Anthropologist*, had not reined McGee in. Wright, shell-shocked by the ugliness of the attacks against him, privately thanked Holmes for his courtesy. He should not have bothered. Holmes illustrated the New Comerstown specimen but made it clear the illustration was not from the original specimen. And in a statement as true as it was unfair, he blamed his inability to examine the actual artifact on Wright (who had no control over the specimen) and not Baldwin (who did): "I desired to have a new drawing direct from the specimens, but a request looking to that end, made to Professor Wright, met with no response."[157]

If within official Washington there was some scrambling to distance oneself from McGee, outside Washington the reactions were markedly more clamorous. Starr considered McGee's reviews inexcusable and despicable. An appalled Dana condemned them as a "disgrace to American science." Winchell thought the attack a "murderous onslaught." Abbott, showing not the slightest self-awareness, counseled Wright to ignore his critics and let them convict themselves on their own "villainy." Wright's many correspondents clucked sympathetically and admired his forbearance in facing up to such vitriol.[158]

There was, however, an unmistakable theme to those letters of encouragement. Although they were deeply sympathetic to the abuse Wright was suffering, and they universally deplored McGee's review (and those of Chamberlin and Salisbury), most were silent in regard to the substantive criticisms made of *Man and the Glacial Period*. Wright was certain Dana and "many others agree with me." Yet, after expressing how terribly he had been treated, several even gently admitted that questions about the American Paleolithic and the unity or diversity of the glacial period were not yet resolved, and that perhaps Wright might even be wrong. Winchell privately agreed with some of Salisbury's criticisms, whereas Dall thought the "whole question of prehistoric man seems to be one where skepticism is a duty until overcome by irrefutable proof."[159] Bad as they felt for Wright, they were not going to defend weaknesses in his position. That realization must have been deeply disappointing to Wright. Worse, several who agreed with him on substantive matters chose to lie low. Warren Moorehead and Upham did so because they were young or vulnerable or both (Upham's position at the USGS was tenuous, and his relationship with Chamberlin fraught). Others, neither young nor particularly vulnerable (such as Dana and Wilson) feared getting publicly embroiled in the controversy.[160]

Wright's supporters agreed that McGee's review, although more malicious than Chamberlin's and Salisbury's, in substance followed theirs. Their near-simultaneous appearance and the "sameness of tone" of the critics' reviews bespoke a "sameness of origin." It all smacked of a conspiracy (that was Baldwin's legal opinion), and there seemed little doubt who was behind it: "The whole lot have . . . been hatched in the Geological Survey, and Chamberlin has been the incubator."[161]

Dana further attributed the violence of the reaction to Wright's having dared criticize Chamberlin directly in the *American Journal of Science* on the unity or diversity of the glacial epoch (§5.12). Dana was likely wrong about the immediate cause of the firestorm: the reviews of Wright's book preceded the appearance of Wright's article. And Chamberlin had long been privately vilifying Wright. It was only a matter of time and trigger before he went public: *Man and the Glacial Period* gave him both. Unquestionably, however, Dana and Wright's other defenders were on target when they considered Chamberlin the leader of the assault in tone and in substance.[162]

To Wright's defenders, the conspirators' intent was self-evident: it was an attempt by sanctioned government science to advance itself at the expense of university and local practitioners, the attack symbolizing the arrogance and abusiveness of heavy-handed government scientists. As they saw it, the "American amateur geologist can claim to have attained the respectable age of manhood, and to have a right to be heard." The reaction to Wright was proof enough for Winchell that "our 'official' geologists will brook no criticism of their work."[163] This was more than a defense of one man as a scholar and scientist: this was a defense of a common cause on behalf of all the non-USGS geologists of the country and their right to fully participate in American science without fear of censure from "official science."[164]

Those on the other side saw it from a very different vantage. Wright "did not fairly represent the present state of scientific opinion on these two questions [glacial history and human antiquity] in a book which especially professed to set forth the present status of the problems with which it dealt." Others' views and evidence had not been adequately presented in *Man and the Glacial Period*, thereby misleading the public as to the state of knowledge and debate. As far as critics were concerned, Wright's defense and defenders were only diverting attention from substantive issues by questioning the motives and right of the critics to criticize.[165] Obviously, there was little common ground here.

As it happened, this very public fight played out at a time when the USGS was once again particularly vulnerable to attack on Capitol Hill. In 1892 the national economy took a turn for the worse: the first budget deficit in two decades was projected and the Panic of 1893 was on the horizon. Looking to cut expenses, the Fifty-Second Congress again targeted the Survey budget. Old enemies (Herbert of Alabama [§4.10]) and new ones (Senator Edward Wolcott from Colorado) re-ignited opposition to the Survey's seemingly unending and expensive topographic mapping, its dearth of economically valuable results, its wasteful expenditures on "abstract" sciences like paleontology, and especially Powell's secretive and cavalier use of the USGS budget. The Survey appropriation, which had peaked at nearly $800,000 in the late 1880s, was slashed to $376,000 in the summer of 1892.[166]

Powell, badly shaken by the congressional action, and in great pain from his amputated arm (lost in the Civil War), was absent from Washington much of the fall of 1892. When he returned that winter he published a vigorous defense of the Survey in *Science*, and he watched with some satisfaction as yet another congressional attack on the Survey fizzled.[167] Still, at this moment in its history the Survey badly needed friends, not enemies. Chamberlin, Salisbury, and especially McGee had made enemies. For a time, those enemies defended Wright by attacking the Survey. That, at least, is how Powell saw it, although his antagonists denied it. It is difficult to shake the judgment that both were partly right.

5.8 MOUNTING A DEFENSE

Leading Wright's defense were Baldwin, Edward Claypole, Winchell, and William Youmans. Baldwin arranged a reprint of McGee's *American Anthropologist* review in its entirety, but annotated with his own acidic commentary on McGee's errors and misrepresentations plus a vigorous defense of Wright against McGee's accusations of plagiarism and deception. Baldwin raised the ante as well, tying McGee's malicious reviews to his official USGS position:

> Who is Mr. McGee? How did he come to be in a leading position on the U.S. Survey? Is it any part of his business to use the weight of his position—in the service and pay of this great nation—in branding high minded, learned Christian gentlemen as thieves, robbers, poisoners . . . betinseled charlatans and harpies?[168]

Baldwin stopped short of saying publicly what he suspected privately, that there had been a conspiracy by Survey geologists to silence Wright. But he came awfully close.

McGee (being McGee) was so perversely flattered and unabashedly ravenous for attention that he asked Baldwin for a hundred copies for his own distribution. (He received them, but only after he assured Baldwin that Otis Mason, then president of the Anthropological Society of Washington, was absolved of any responsibility for his *American Anthropologist* review.)[169] Even with McGee's help Baldwin's self-published tract was too limited in circulation to fully redress the wrongs felt by Wright's allies. As Haynes put it, "it is not possible to reach people by a pamphlet, as you can by a popular magazine."[170]

Haynes was correct, but fortunately for Wright's allies a raft of new journals had appeared in the preceding decade that quickened the pace and opportunity for discussion and debate (*Science* magazine, for example, appeared weekly) and provided venues in which to place a spirited defense of Wright and express pent-up resentment toward the Survey. Firmly on Wright's side in their editorials and content were the *American Geologist*, which over much of its time (it was founded in 1888) had been the unofficial organ of opposition to the USGS, and *Popular Science Monthly*, published by D. Appleton and Company, which also published *Ice Age* and *Man and the Glacial Period*.[171] On the other side were the *American Anthropologist* and Chamberlin's *Journal of Geology*. The other outlets, such as the *American Journal of Science*, *Science*, and the *Proceedings of the American Association for the Advancement of Science*, were more neutral territory, not securely controlled by either faction. Of course, each had audiences of different size, scale (whether local or national), and interest.

Winchell was appalled by the sudden "murderous onslaught" on Wright and was determined to recenter the discussion and allow Wright the opportunity to be heard, whether "right or wrong." Claypole suggested Winchell publish a symposium devoted to *Man and the Glacial Period* in the *American Geologist*, with papers by government and nongovernment geologists who took a positive—or at least dispassionate—view of the evidence for human antiquity.[172]

As the editors of the *American Geologist* gathered the papers for that symposium, Chamberlin, with an astonishingly poor grasp of the politics of American geology at that moment (or perhaps not caring), mounted a takeover bid of the *American Geologist*. He sent a circular letter to its editorial board of (all of whom were part owners) proposing a consolidation with the *Journal of Geology*. By the terms offered, the *American Geologist* would within the year be absorbed by the *Journal of Geology*, its assets split by its owners and the University of Chicago, and (the intellectual bottom line) the newly enlarged *Journal of Geology* would be under the editorial control of the geology faculty at the University of Chicago. A few of the editors from the *American Geologist* might be asked to serve on the editorial board of the *Journal of Geology*, but that was entirely at Chamberlin's discretion.[173] All of this was part of Chamberlin's plan to elevate the scientific standards of geology as he defined them. That he sought to establish publication standards was unusual. This was a time when peer review was rare; the common custom was to accept papers that came across the transom. In those days, editors were often more concerned with finding papers to fill the pages of a journal, usually one in which they had a financial interest, and could not be too finicky about the quality of the work.[174]

Chamberlin seemed mildly surprised when the proprietors of the *American Geologist* declined his proposition. What he had not realized was the deep resentment toward him and the swelling pride nonfederal geologists took in their independence from the USGS, which he so obviously represented. *American Geologist* was the self-styled "organ of the opposition." Keeping it viable and independent was not being done for economic reward (the journal lost money). It was nothing less than a crusade on behalf of intellectual autonomy. As Winchell put it, subsuming the *American Geologist* under the *Journal of Geology* would "be the end of independent geological work and opinion, so far as its expression goes, in America for a long time."[175] That independent voice lasted for another two decades.

The *American Geologist* proceeded with its planned symposium, which appeared in March of 1893. The ten authors included Wright (who helped select the contributors), defenders of the unity of the glacial period, proponents of the Cincinnati Ice Dam, and advocates of an American Paleolithic, with a couple of skeptics in the mix.[176] What was said was mostly a rehash of known views; its importance was more the symbolic value of the gesture. The *American Geologist* editors were making the point that Wright's book demanded serious attention and a nonincendiary discussion.[177] Chamberlin thought otherwise:

> The mischief of the whole matter is that real scientific questions that ought to be treated with delicacy and discretion are forced into abominable controversial relations that cast a cloud over the whole subject.[178]

Noticeably absent from the contributors to the symposium were Abbott, Holmes, and McGee. All were invited, but Abbott brusquely refused ("I say *NO!*") and McGee declined, citing his already published reviews, which he shamelessly described as "in no way controversial." He and Holmes thought they could do better responding "after the symposium is printed" if they so chose. (It is not known if Salisbury was invited to participate, but he took the opportunity to respond afterward.)[179]

Over at *Popular Science Monthly*, Youmans had already begun his defense of Wright in late 1892 with a flattering biographical sketch that described Wright as one of the "foremost . . . authorities in geology and the antiquity of man." Wright's ever-vigilant critics were not buying: Branner promptly sent a six-page rebuttal, describing Wright as little more than "a talkative amateur and a sensationalist" and his book merely "trash." Youmans was not pleased: "Your letter about Prof. Wright . . . I did not think it well to publish. Perhaps what you say of him as a geologist is fully warranted by the facts, although the opinion is not shared by a large number of able specialists in that department."[180] There was the rub: the government geologists were so overweening that they failed to see there were geologists in the scientific community who might have opinions different from theirs.

Youmans deeply resented their attitude and sought a "man with a little ginger in his composition" who would not be reluctant to use *Popular Science Monthly* to "show up" the Survey geologists.[181] On Wright's recommendation, Claypole accepted the task, opening with a very simple question: why had Wright's book triggered any controversy at all? Little of it was sensational or new, and certainly Wright was well qualified to write such a volume. Granted, there were mistakes, but these hardly warranted the storm of "one-sided, persistent, and personal attacks" made on the book and its author.[182]

Claypole was quick to insist that he had no objection to criticism but that these were condescending, shallow, and hypocritical. (Dismissing the Nampa Image on the hearsay of an anonymous statement from a "well-known" government official, he argued, was weaker and feebler than the weakest and feeblest of the cases Wright brought forward.) Worse, those critics oozed the infallibility and omniscience of which "men in official positions" were often infected. As he had privately written to Wright, "There is something very amusing in the air of infallibility which official connection usually gives." Publicly he wondered:

> By what right does he [Chamberlin] set himself up as a judge of the competence of all other workers who think they *have* found such stones? Who, in his opinion, are experts? Where shall such men be found, and by what touchstone shall they be tried? Is official connection the grand *sine qua non*?[183]

Who was Chamberlin to "protect" the public by denying Wright the opportunity to present his views to the public? American geologists, Claypole growled, "will not be silenced by official insolence or warned off their fields of investigation by 'notices of trespass' from self-appointed owners."[184] Especially amateurs like Wright who had few advantages and whose contributions were at their own expense yet were often singularly important. Claypole reminded readers of Boucher de Perthes and, not missing an opportunity at a jab, observed that even Boucher de Perthes never had to suffer the indignity of being called a charlatan. Claypole was mostly but not entirely correct (§2.1), but if the point was stretched it was well intentioned. Amateurs, in Claypole's mind, were the backbone of science, had much to contribute, and should be made welcome. Even by official science.[185]

What of McGee? Appalled as Claypole was by McGee's reviews, which he considered dishonest, caustic, and "outside the pale of honorable scientific warfare," he was also puzzled. Just a few years earlier McGee had been a true believer. Now McGee described his previous work as "obsolete." Why did he seek to destroy "a faith which once he preached?"[186] Claypole was no less amused by the acolyte falling from grace than he was by the image of the scientist failing to keep pace with the theologian. Why was the "theologian in the van of the evolutionary army with the geologist and the archaeologist lingering in the rear"? The metaphor had a dark side, for those lingering in the rear were conducting an inquisition:

> The onslaught made on Prof. Wright by chiefly official geologists savors too strongly of the old-time, intolerant, theological method of crushing a formidable rival by dint of concerted action or force in default of reason.[187]

Many besides Claypole commented on this apparent role reversal of religion and science, though this was not a struggle between the two. Wright's religious views may have been an integral, if unspoken, part of his own position on glacial and human history, but others shared his scientific views without paying the least attention to his religious beliefs, save in one regard. He was regularly praised by correspondents for, as Haynes put it, honoring the "character of a Christian":

> You have been patient, & long-suffering, & have refrained from hitting back, when you had ample material for a damaging assault upon Chamberlin & his character.[188]

To his defenders, Wright's religious stripes hardly disqualified him from scientific pursuits, nor were they even relevant to this dispute, though several of his critics had tried to make them so. More irony: just a few months earlier Wright had been appointed professor of the Harmony of Science and Revelation.[189]

5.9 COLLATERAL DAMAGE

Claypole conjured a stark image of the arrogance and malevolence of "official science" but carefully aimed it at particular individuals—notably Chamberlin, Holmes, McGee, and Salisbury (Figure 5.3)—so as to avoid blanket censure of government science. It was a fine line to walk to avoid casting aspersions on the Survey as a whole (some distinction, Powell grumbled), but Claypole more or less hewed to that line.

Youmans, on the other hand, danced across it without a second thought. In an editorial accompanying Claypole's piece, he made little effort to distinguish guilty individuals from the government agencies they represented. The attack on Wright was, in his view, an example of how the USGS responded to work by individuals done along lines they did not approve. "Of all the arrogant things in the world official science is perhaps the most arrogant," he intoned, "and of all the obstructive things official science is perhaps the most obstructive."[190] And then:

> How much does the country really want of this kind of thing? In granting an appropriation for the Survey did [Congress] mean to endow a Holy Inquisition or a Sacred Congregation of the Index? We think not.[191]

Claypole and Youmans's pieces gained hearty approval from Wright's defenders. Abbott thought they were grand; Baldwin purchased 150 copies for Wright to distribute. Yet, some had misgivings. Winchell praised Claypole's precise targeting but thought Youmans went too far and had indiscriminately condemned all USGS geologists, many of whom were appalled by the assault on Wright. Winchell, keen to maintain good relations with Survey members on whom he depended to support his journal and possibly his future candidacy as director of the USGS, distanced himself from Youmans's editorial.[192]

Youmans's editorial forced Powell's hand. Powell could ignore Claypole, but he could not let Youmans "draw a lesson therefrom in condemnation of the work of the Geological Survey."[193] Within weeks, he penned a response to *Popular Science Monthly*. Powell opened with a detailed explanation of the history and work

Figure 5.3 An outing to a steatite (soapstone) quarry in Clifton, VA, circa 1894 included several of the major critics of the American Paleolithic. From left to right: Anita Newcomb McGee (a physician and founder of the U.S. Army Nurse Corps), WJ McGee, Margaret Hetzel (the owner of the site), USGS geologist David Day, USNM anthropologist Otis Mason (seated, with walking stick), William H. Holmes (seated, with elbow propped on rock), unidentified man, and Rollin D. Salisbury (with umbrella). Courtesy of the Smithsonian Institution Libraries, Washington, DC.

of the Survey under his direction, painting a glowing picture of its successes and the warm camaraderie between government and nongovernment "professorial" geologists, the latter by his count making up the majority of the geologists working on glacial matters. As Powell saw it through his rose-colored glasses, there was "harmonious co-operation . . . again and again" among the participants.

There were, however, those who did not fit into this Elysian scene, and Wright was one of them. The fault was Wright's: he was incompetent, unwilling, or unable to keep up with new knowledge and changing developments in geology, and he had the temerity but not the qualifications to write for the public. *Ice Age* was bad enough: it ignored the conclusions of his co-workers, denied the accuracy of their observations, and was altogether erroneous and misleading. Powell and others had chosen to ignore its shortcomings on the assumption that the book would sink from sight. But then Wright published *Man and the Glacial Period*, a book that "would do harm." It was not only a misstatement of glacial geology, it failed to appreciate the "new methods" sweeping through archaeology. These were methods anathema to the many collectors in America who were keen to find proof of a deep human antiquity for the rewards it would bring, including, Powell inexplicably suggested, being "decorated with ribbons . . . [and] knighted" (a gaffe Claypole later delighted in lampooning). Powell found it particularly galling that Wright's response was to skirt the substantive issues and cry persecution. Wright deserved the criticisms he had received.[194]

Those criticisms, Powell insisted, came from Wright's "fellow workers" in the field, professorial geologists like himself who were connected with colleges and universities. Only one of the critics was a permanent member of the USGS, and all the criticisms were in scientific journals; not a single one appeared in an official Survey publication. Powell was splitting hairs. To be sure, McGee was the only permanent full-time employee of the Survey, but Chamberlin and Salisbury each had USGS appointments along with their university positions, and Holmes was certainly part of official science, though in the BAE, not the USGS. Having convinced himself his staff was beyond reproach, Powell dismissed Claypole's article with a wave (an "error in every paragraph") and demanded that Youmans retract his editorial, now that Powell had fully explained the facts.[195]

That did not happen, though it is hard to suppose Powell thought it would. Youmans, as was his custom and editorial privilege, declared Powell's reply "a failure." (Powell's reply did not wash with Claypole or Winchell, either, both of whom later responded.) As far as he was concerned, Powell might have cleared himself of culpability, but he failed to refute the charge that this was a concerted attack orchestrated by Survey members. Worse, Powell had not expressed even mild disapproval of McGee and instead had tried to deflect the blame for those harsh reviews on Wright's "fellow workers." But even if Powell had censured McGee it would not have been enough. Youmans wanted McGee's head on a pike:

> The services of Prof. Wright were dispensed with from the Survey—so the director [Powell] tells us—because he failed to distinguish 'overplacement' [reworked material] from original glacial deposit. . . . [but] might we suggest to the director that the writing of so discreditable an article as that which proceeded from the pen of Mr. W J McGee might perhaps be at least as serious a reason for removal from the Survey as even the non-recognition now and then of 'overplacement'?[196]

Which, in the event, is what happened. On June 30, 1893 McGee resigned from the Survey and moved to the BAE. It was likely no coincidence that his resignation came just days after the publication of Youmans's editorial. His critics took credit for it. As Dana put it to Wright: "[McGee's] awful kick at you had a tremendous recoil, breaking abruptly his U.S.G.S. connections, while leaving you unharmed."[197] At the time there was no admission from official Washington that McGee's resignation was linked to his "betinseled charlatan" review, or to his having focused such an unfavorable light on the Survey. That is no surprise. Even so, everyone involved knew that was precisely why McGee was moved. Branner, not without sin in this controversy, admitted as much a decade later:

> [McGee] published . . . a criticism of Prof. Wright that was so violent as to bring down upon him and upon Major Powell . . . opposition that com-

pelled the immediate removal of McGee from the Survey. . . . But Major
Powell never abandoned a friend, so McGee was simply transferred to the
Bureau of Ethnology. . . . And this is how he came to be an ethnologist.[198]

That statement was made at a time when McGee was immersed in yet another,
equally bitter dispute (this time over whether he or Holmes would become Powell's successor), and when old Survey friends like Branner had become enemies. It
is often in such a context that buried truths emerge. In the end, the "betinseled
charlatan" affair proved costly to both McGee and the Survey.

Removing McGee may have deflected some of the criticism aimed at the USGS,
but its detractors continued to fire at Powell and the Survey. Ultimately, he too
resigned, coincidentally (or not) a year to the day after McGee. Powell was not,
however, drummed from office over the Wright matter. He had operated on a
much larger stage, had made more powerful enemies, and in the end lost support
for his position "by antagonizing many members of the Congress, by leading the
Survey in directions that they and others questioned or found inadequate, and by
attempting to interject his own ideas into national policymaking—even lawmak-
ing."[199] The USGS under his successor, Charles Walcott, was very different. Less
controversial, too.

5.10 HOLMES'S MARCH THROUGH THE AMERICAN PALEOLITHIC

Holmes had not participated in the frenzied attack on *Man and the Glacial Period*. His review of the book was brief, focused strictly on the archaeology, and
suggested that because this was not Wright's area of expertise the archaeological
material ought to be omitted in subsequent editions of the volume to give the
"slowly accumulating evidence a few years to overtake the already well developed
theory."[200] Still, Holmes did his part to make Wright's book obsolete. In the winter of 1892–1893 he published a string of papers that forged the link between his
Piney Branch quarry specimens and the artifacts of the American Paleolithic, and
then set off in critical pursuit of each of the major Paleolithic claims.

The foundation for this work was his "Modern Quarry Refuse and the Paleolithic Theory," published in *Science* in late 1892, in which Holmes restated his
argument that all tools passed through a single sequence of manufacture (varying
only along lines determined by differences in raw material types), from the cobble
to the finished arrowhead, spearpoint, knife, or drill (§5.2). Initially crude items
became progressively more refined until they reached the stage at which they were
then taken off-site:

> The final quarry-shop form—and it must be especially noted that there was
> practically but one form—is naturally something beyond or higher than the
> most finished form found entire among the refuse [at the quarry]. . . . A most
> exhaustive examination of the great quarry sites has shown beyond a shadow
> of a doubt that this final form was almost exclusively a leaf-shaped blade.[201]

En route to producing that form, mistakes were made, pieces were broken, and the quarries and workshops became littered with rejected specimens. For every one of the millions of successfully flaked spear or arrow point, knife, or scraper found across America, there must be, he estimated, "as many millions" of manufacturing failures. Those rejects, by their very nature rudely flaked, might resemble stone tools from an occupation that dated to glacial times, made by a people "not yet advanced beyond the primal or Paleolithic stage of culture."[202] But they were not the same.

In an infant science, before an understanding of the "true nature" of quarrying and its debris had been reached, it was understandable, Holmes supposed, that manufacturing rejects were mistaken for Paleolithic artifacts. It was no surprise to him that the archaeological literature and museum collections were rife with such wrong-headed classifications. Times had changed, however, and it was "now conceded by scientific men that this is all wrong," given there was nothing inherent in their rude form to separate them from quarry rejects or to pinpoint their age and evolutionary grade. And in regard to the latter, with rare exceptions none of the American specimens showed the "specialization of form" or "adaptation to definite use" characteristic of European Paleolithic specimens.[203]

As for determining their age, that was altogether beyond the bailiwick of archaeology; that was strictly a matter to be determined by "competent and reputable observers of geologic phenomena." That some supposed paleoliths were found in gravels was immaterial, partly because their geological context had not been securely attested to by skilled geologists, but also because that was precisely the context where one would expect manufacturing debris and rejects because gravel was a prime source of raw material. Indeed, even if a specimen was found in a deposit "ten times as old" as the Trenton Gravel, it would not be proof of an American Paleolithic unless it represented the remains of an "exclusively rude culture."[204]

Holmes insisted all claims be resolved on geological grounds, but he demanded a central role for archaeology. Thus, while condemning American Paleolithic proponents for using form and geological context to determine the age and grade of a specimen, he too used form and geological context to identify age and grade. It came down to which analogy was more appropriate—manufacturing failures or European paleoliths—and whose geological observations (or the observations of whose geologist colleagues) were more trustworthy. Holmes made it clear who his friends and authorities were:

> My explorations have been made with the greatest care and rarely without the aid and advice of some of the foremost geologists and anthropologists of the country. The conclusions reached have been freely discussed, and are generally approved by those familiar with the facts.[205]

Holmes's paper was widely hailed, even by those with no stake in the dispute.[206]American Paleolithic proponents were unimpressed. The very afternoon he received his copy of *Science* with Holmes's paper, Abbott—far too volatile to follow the advice he had given Wright to ignore his critics (§5.7)—quickly responded. He insisted his Trenton paleoliths were found in undisturbed and stratified contexts; that their antiquity was not exclusively a geological issue; and that the onus was on the critics to prove "paleolithic man is an impossibility." And then he turned to verse:

I.
The stones are inspected,
And Holmes cries "rejected,
    They're nothing but Indian chips"
He glanced at the ground,
Truth, fancied he found,
    And homeward to Washington skips.

II.
They got there by chance
He saw at a glance
    And turned up his nose at the series;
"They've no other history,
I've solved the whole mystery,
    And to argue the point only wearies."

III.
But the gravel *is* old,
At least so I'm told;
    "Halt, halt!" cries out W.J.,
"It may be very recent,
And it isn't quite decent,
    For me not to have my own way.

IV.
So dear W.J.
There is no more to say,
    Because you will never agree
That anything's truth
But what issues, forsooth,
    From Holmes or the brain of McGee.[207]

Holmes, never given to humor, was not amused being the butt of Abbott's doggerel.[208] Perhaps not coincidentally, it came on the heels of Abbott's *Annual Report*, which opened by denouncing the "disgraceful articles in pretentious periodicals, written by persons wholly ignorant of the subject."[209] He did not name names, but Holmes knew Abbott had him in mind. He had intended to review Abbott's report for the *Journal of Geology*, but once he read it decided it had "such vile personal insinuations and such a mass of errors, geologic, archaeologic, and otherwise, that it cannot be mentioned save to be damned." Still, he could not ignore it altogether: "I may perhaps write something for 'Science' which is the trough into which the slops are thrown."[210] He did, but like Abbott did not name names. But then he did not need to either, for it was obvious he had Abbott in mind when he declared the American Paleolithic was built on the blunders and misconceptions of "amateurs" who worked and wrote for themselves, had little scientific understanding of stone toolmaking let alone of geological age and context, and could not see the critical distinction between evidence of a glacial human presence (age) and evidence for humans in a Paleolithic stage (evolutionary grade). The American Paleolithic, Holmes announced, was so "unsatisfactory and in such a state of utter chaos that the investigation must practically begin anew."[211] And he was just getting started.

Over the next several months he published sharply critical studies of Wilson's Tidewater paleoliths and Abbott's from Trenton; the Ohio finds at Loveland, Madisonville, and New Comerstown; and Babbitt's Little Falls quartzes.[212] He approached each confessing indifference to the outcome of his investigation, a willingness in principle to accept a Pleistocene human presence, and an insistence that he found controversy and dispute "most distasteful." Just doing his scientific duty, he said.[213]

Perhaps. But it is difficult to square that diffident assertion with the zeal with which he prosecuted these cases, especially Trenton (which by then had become personal).[214] Nor was there much doubt about the outcome. Holmes was convinced each purported American Paleolithic case was flawed. Their specimens were "typical rejects of the modern blade maker" mistakenly identified as Paleolithic. To show how easy it was to mistake the two, Holmes illustrated the New Comerstown specimen alongside several quarry rejects but without identifying which was the supposed paleolith and which was not (Figure 4.3). He challenged his readers to tell them apart.[215]

None of the specimens, Holmes argued, were in primary context in Pleistocene-age deposits. Some had fallen into older deposits, and others were collected from redeposited gravels or recent formations such as talus slopes. None were correctly recognized as occurring in secondary deposits. That was no surprise to Holmes: "Talus deposits form exceedingly treacherous records for they would be chronologist. They are the reef upon which more than one Paleolithic adventurer has been wrecked." To show how easily artifacts on the surface could tumble into deeper deposits, Holmes provided hypothetical illustrations of the consequences of slope collapse at New Comerstown and Trenton.[216]

The American Paleolithic was so flimsy, he argued, that were it not for "supposed analogies with European conditions and phenomena, and by the suggestions of an ideal scheme of culture progress, it would vanish in thin air."[217] Why had so many been so mistaken? He saw several reasons, none charitable. Such errors were the inevitable consequences of work by mere "relic hunters" who "lacked adequate knowledge," were weighed down by "preconceived notions" and hamstrung by the application of "unscientific methods," and whose "prematurely announced and unduly paraded" finds were vestiges of the "old archaeology."[218] Naturally, none had been verified by independent competent observers, as the Pleistocene deposits in those sites "are and always were wholly barren of art [artifacts]."[219] Holmes did not suggest the Trenton paleoliths were modern; he only rejected the claim that any were Pleistocene in age. He took obvious pleasure in returning Abbott's accusation—that he would prove that "Indians bought the Delaware Valley from William Penn" (§5.6)—by suggesting that if William Penn had only paused in his trading with the Delaware and glanced up he might have seen one of their number

> fabricating rude ice age tools, making the clumsy turtle-back, shaping the mysterious paleolith, thus taking that first and most interesting theoretical step in human art and history. Had he looked again a few moments later he might have beheld the same tawny individual deeply absorbed in the task of trimming a long rude spear point of "Eskimo" type from the refractory argillite. If he had again paused . . . he would have [then] seen the familiar redskin carefully finishing his arrow points. . . . Thus in a brief space of time Penn might have gleaned the story of the ages—the history of the turtle-back, the long spear point and their allies—as in a single sheaf.

But Penn had not looked, and so "two hundred years of aboriginal misfortune and Quaker inattention and neglect [Abbott was a Quaker] have resulted in so mixing up the simple evidence of a day's work, that it has taken twenty-five years to collect the scattered fragments, separate and classify them, and to assign them to theoretic places in a scheme of cultural evolution that spans ten thousand years."[220]

Holmes's 1893 papers signal a shift from a strong reliance on the quarry thesis to a greater dependence on geological testimony. Arguments that the alleged paleoliths were historic Native American manufacturing debris worked readily enough when those specimens were found on the surface, as the context could not preclude recent deposition. For artifacts found in purportedly glacial contexts, Holmes (and for that matter McGee, Chamberlin, and Salisbury) argued the objects must have been on the surface in recent times, but that owing to factors like slumping, the uprooting of trees, and rodent burrowing, they had fallen into older strata at greater depths. The attitude of Holmes and his geologist colleagues, deeply resented by American Paleolithic proponents, was that if there was any possibility of intrusion or mixing, then it must have occurred. Where the geological context seemed secure, Holmes changed tactics and cast aspersions on the collector or called into question the age of the deposit; that is, the artifacts might be in situ, but in a unit that was redeposited in postglacial times. Here he even turned Lewis against Abbott, claiming Lewis and Leidy both thought the Trenton Gravel was reworked.[221] Abbott was livid: "Leidy never visited the Trenton Gravel but once . . . [and] Lewis instructed by me, not vice versa."[222]

5.11 POINT/COUNTERPOINT

Holmes's critiques appeared in early 1893 and triggered a running debate with Abbott, Haynes, and Wright (among others) that played out primarily in *Science* alongside exchanges over *Man and the Glacial Period* and the USGS (§5.6, §5.8). Wright immediately challenged Holmes's declaration that the American Paleolithic was "utter chaos," marshaling a defense around the New Comerstown find.

He detailed its discovery and testified to Mills's integrity (his "character and reputation are entirely above suspicion") as well as the specimen's authenticity and its
resemblance to Evans's Amiens paleolith (§4.14). There was "no possibility of any
doubt about the undisturbed character of the gravel." He promised a photograph
showing the terrace at New Comerstown where the specimen was found. All of
this showed the American Paleolithic was neither "chaotic" nor "unsatisfactory"
but "as specific and definite and as worthy to be believed as almost anything any
expert in this country, or any country, can be expected to produce."[223]

Haynes weighed in alongside Wright in *Science*, growling that no competent archaeologist—not to be confused with that "little knot" of Survey geologists (Haynes named no names) who possessed the "half-wisdom half-experience
gives"—would have difficulty distinguishing true paleoliths from unfinished
Neolithic ones. The former were "well-known and perfectly defined," the latter easily apparent. Granted, fewer had been found in America, but there was no doubt
in his mind that "Paleolithic man lived here also." Only the autocratic or envious
could possibly think otherwise, so far as Haynes could see.[224]

Abbott thought Haynes's letter superb, and not wanting to be left out jumped
in with notes to the *Archaeologist* and *Science*, this time speaking to the question of
how to distinguish between finished and unfinished implements.[225] Like Haynes,
he claimed this was not difficult (baldly contradicting his assertion just a few
months earlier that discriminating between the two was "always a difficult and
sometimes an impossible feat"[226]), and argued that Holmes's defining characteristics of rejects (their large size, rough shape, and absence of evidence for hafting)
were found on heavily and undeniably used forms in village sites. The basis for the
"lovely picture" (made by Holmes, of course) of stone tool manufacture from raw
material to finished product was never "reproduced in nature." Then, more verse:

> When Holmes shall drive the fog away
>   That now enwraps the scene,
> And in the light of later day
>   He stands with smile serene,
> And points to how in modern time
>   The red man came equipped
> With every blessing of the clime,
>   From elsewhere newly shipped;
> We can but hope he'll name the date
>   When first upon the strand
> This red man stood with heart elate,
>   And where he chanced to land.
> Then, noble efforts, nobly made,
>   Before he seeks a rest
> Point out how far is *truth* displayed,
>   And just how far he guessed.[227]

Holmes was still not amused—not by Abbott and especially not by Haynes,
whose paper he found particularly offensive. A week later he fired back in *Science*.

Although he insisted that he "strongly deprecate[d] personalities in scientific discussion and hesitate[d] to refer in a critical way to the legitimate work of other investigators," Holmes nonetheless accused Haynes of having so "little knowledge of native art and less of geology" that "he has rarely touched the subject of glacial man without adding to its obscurity." Haynes had committed, in Holmes's eyes, an unpardonable sin, expressed in seven words written in 1881: Haynes spoke of Paleolithic specimens "either from the gravel, or the talus."[228]

That was it—an innocuous statement made by Haynes a dozen years earlier and merely intended to convey the idea that specimens came from different contexts. Haynes was not saying he had no idea or cared whether a specimen came from gravel or talus. But that is how Holmes read it, and if Haynes apparently could not distinguish whether an alleged paleolith came from Pleistocene gravel or recent talus, Holmes considered everything he said on the subject of the American Paleolithic useless. Worse than useless, even: "It would be difficult to find within the whole range of scientific writing lines containing less of science or evincing a greater degree of incompetence . . . than these." Nor did Holmes stop there: he scoured Haynes's papers on the Nile valley Paleolithic (where Haynes had made his reputation) and concluded that none was worthwhile. Haynes possessed "no comprehension of the real problems involved." His observations made "in archaeologic obscurity and geologic darkness amount to naught."[229]

Caustic words, and, as Haynes wryly noted in response, the readers of *Science* could see just how much "Mr. Holmes 'strongly deprecates personalities in scientific discussion.'"[230] Privately, Haynes was taken aback by the "flank movement turned upon me," especially as his *Science* letter and an article he had just published in the *American Antiquarian* on distinguishing Paleolithic and Neolithic artifacts had not been aimed at Holmes:

> Personally I entered into the fight because I could contain myself no longer after reading McGee's article in "The Anthropologist." Evidently Holmes was thoroughly angry at what I wrote in the "American Antiquarian" for I named no names in my letter to "Science" & was thinking about McGee & not especially about Holmes, any more than Chamberlin or Salisbury. It does not displease me that Holmes chose to put the cap on.[231]

Haynes was probably right about the impact of his *American Antiquarian* piece, which made it clear why he was not as troubled by the issue of geological context as was Holmes. Haynes believed any trained archaeologist, especially one who had actually handled European paleoliths, could distinguish a true paleolith from a rudely chipped, unfinished later one. Holmes's "much vaunted" manufacturing rejects, Haynes proclaimed, had not the "slightest resemblance to real Paleolithic implements." Four decades earlier Evans likewise had had to contend with critics who claimed the flint implements from river gravels were merely "wasters." Haynes did not have to remind anyone how that debate turned out. Paleolithic implements were identifiable by their "shape and style, and the method of their use . . . the character of their chipping . . . and clear and unmistakable traces of great antiquity." Finally, revealing the broad chasm separating proponents and critics of the American Paleolithic, Haynes insisted it was not difficult to discriminate between disturbed and undisturbed gravels. "Nothing more is required to do this than a trained habit of observation on the part of the student." A skilled archaeologist, "like Putnam, Carr, or Abbott, is just as competent to determine whether the gravel has been disturbed . . . as another archaeologist, like Holmes, is to assert the contrary."[232] Holmes could not have disagreed more.

Haynes did not care. He cheerily dismissed Holmes's ad hominem attack, then counterpunched in *Science*, condemning Holmes's Trenton critique ("a long and labored article"), which appeared to him little more than an effort by Holmes "to demonstrate that because he has failed to find any evidence of the existence of Paleolithic man in the Trenton Gravel, therefore no such evidence has ever been found by anyone else." What Haynes found even more surprising was Holmes's "astonishing statement . . . that 'most of these so-called gravel implements of Europe are doubtless the rejects of manufacture'." Holmes, it seemed, had a fondness for making startling assertions, but Haynes could not take them seriously.[233]

This time, having attacked Holmes directly, Haynes anticipated a reply with "a good deal of amusement." It never came. By then all of Holmes's site critiques were published, and he let them stand without further comment. Wright and Haynes did not and could not. Holmes's critiques were shaking the faith of the believers: Upham privately confessed to Wright that "Holmes in showing his hypothetical talus sections brings room for doubt at Trenton, New Comerstown, and Little Falls."[234]

Even so, Wright was confident. He privately told Holmes he was "all off [his] base at Trenton." In a surprisingly modest admission, Holmes agreed that might be true: "Your restrictions may be well founded. I try not to hold to anything save for the truth that may be in it. I am probably given to stating things too strongly at first. Time will straighten this all out."[235] Of course, that was the private Holmes admitting doubt. The scrappy public Holmes never did.

After reading Holmes's critiques, Wright conceded in *Science* that it was "probable that we have been mistaken about the character of Miss Babbitt's discoveries at Little Falls." Trenton and New Comerstown, however, he did not concede. Leaving the defense of Trenton to others, he focused on New Comerstown (saying nothing of the other Ohio claims he had previously defended, which now seemed less secure).[236]

Wright had earlier promised a photograph showing the New Comerstown find spot, but Mills, much to his regret, had not taken one. ("Had I known of its importance I could have taken a photo of it in place in the bank. I have been sorry a dozen times that I did not know enough to have a picture taken."[237]) The best Wright could do was a photograph of Mills standing near where it was found, which he duly published—along with more details of the finder, the find, and its context—in *Popular Science Monthly*. It appeared almost at the same time as Holmes's critique, and Holmes, knowing the photograph was coming, wondered why Wright bothered:

Professor Wright has visited and photographed the site and will speedily prepare a plate for publication, for just what purpose, however, it is rather hard to see. . . . A photograph made of the tree after the bird has flown will not help us determining the bird. No more will observations on Mr. Mills' moral character, his education or business reputation, diminish the danger of error.[238]

Wright had an answer:

Doubtless it will strike the reading public rather strangely to have Mr. Holmes speak so slightingly of the value of a photograph showing it as it actually was soon [within 6 months] after the discovery, when he has himself given two fancy sketches, representing an impossible condition of things, to inform us how he thinks it might have been. . . . most readers will prefer to see a photograph, in which there is no danger of the incorporation of fanciful elements.

Moreover, Wright found it odd that Holmes insisted on expert testimony and yet dismissed observations on Mill's character and reputation. After all,

if there is doubt about his moral character, that . . . vitiates the evidence to a high degree. So, also, if there is doubt about his ability to discern the difference between disturbed and undisturbed gravel in such a situation, that would largely vitiate the observations. But Mr. Mills's education and habits of observation are such that his evidence in so clear a case as this is, is as good as that of any expert.[239]

Wright had badly misunderstood Holmes's central point. Amateurs like Mills (or, for that matter, Abbott, Babbitt, and so forth), no matter how broadly educated or well intentioned, lacked necessary specialized knowledge. Polemics aside, the "new geology" had highlighted the complexity of geological features, the problems of redeposition and stratigraphic mixing, and the challenge of identifying the age of glacial units. Interpreting artifact context was hardly the straightforward issue it was assumed to be, and it could not be accomplished by untrained individuals or an image in a photograph.[240]

American Paleolithic proponents refused to back down. Unlike Wright, Haynes was even unwilling to concede Little Falls, though by this point he confessed his motives were not entirely pure:

I suppose the next "Science" will contain one more letter from me apropos of Miss Babbitt's finds. I don't feel inclined to give them up upon Holmes' unsupported assumptions & I asked for the opportunity to give him a final slap. I think I am fully justified in expressing my contempt for him after his very plain speaking of his opinion about me.[241]

Defending Little Falls, however, required some sleight of hand. Responding to Holmes's accusation that he had accepted Babbitt's conclusions, Haynes replied he had only ventured the judgment that the specimens were artificial. It was Babbitt who claimed they came from undisturbed glacial deposits. If what she said was true, then because some of the specimens resembled paleoliths they must be "true Paleolithic implements." Of course, and here Haynes tacked sharply to the right, that was "a question for the geologists to answer." Paleolithic look-alikes could occur on sites of much younger age (especially where the same stone was used), and thus whether a particular specimen was a paleolith depended on the "conditions of its occurrence." Only if the site was deemed of glacial origin was it probable that the implements were Paleolithic. Although conceding a greater role to geology and context than he previously granted, Haynes stood firm that the testimony of firsthand observers like Babbitt was sufficient, Holmes's disparaging remarks about her competence notwithstanding. Indeed, Haynes professed to be glad for the company:

> I had supposed that such crass ignorance as this was confined, in Mr. Holmes' judgment, to myself; but it seems that there are others falling under a like condemnation. How fortunate it is for the rising generation that Mr. Wm. H. Holmes has appeared to set them quite right in regard to the prehistoric archaeology of North America.[242]

With this last volley, the rapid-fire exchanges in *Science* ended. Each side had laid out its views, hardened its position—and utterly failed to make converts on the other side.

For French paleontologist Marcellin Boule reading the debate in Paris, it was all too reminiscent of what European archaeologists had been through some forty years earlier with the establishment of human antiquity on that continent. That puzzled him. Two years earlier he had seen the Abbott paleoliths at the Peabody Museum and those Wilson had amassed at the Smithsonian. He had also visited Trenton and spent a day "archaeo-geologizing" in the company of Abbott and Wilson.[243] As Dawkins had over a decade earlier, Boule could as easily imagine he was in the valley of the Seine as in the Delaware valley, and what he saw of the artifacts was also strikingly familiar: "Ce qui m'a le plus frappé, c'est la similitude, je dirai presque l'identité de forme des instruments américains aver les instruments paléolithique européens."[244] He was sure these were not manufacturing rejects but finished pieces, doubted they had worked their way into glacial-age gravels, and pointed out that even if he had not found any paleoliths at Trenton it did not mean none were there. He had not found paleoliths at Chelles either, and that was a legitimate Paleolithic site.

But then Boule had not realized there was growing suspicion in America about genuineness of European paleoliths. Holmes had wondered aloud if some of them could be manufacturing failures as well, but no American archaeologist was so brazen as to suggest the European Paleolithic was built on a foundation of quarry rejects. Otis Mason was only joking when he said, "Boucher de Perthes may turn out to have been the Dr. Abbott of France.[245] Still, after a firsthand look at European museum collections in late 1892, Mercer and Brinton were convinced Holmes's "quarry rejects and unfinished implements theory" explained away or cast doubt on many of the specimens from French sites. Even so, they had "no doubt but that certain specialized forms are truly 'paleolithic,' and that the stratigraphic evidence is conclusive as to their vast antiquity."[246]

Doubt was easy to dispel where the geological evidence was secure. However, that was not the case in America, where the depositional context of the artifacts was complicated and made more so by the dispute over glacial history simultaneously being fought over the winter of 1892–1893.

5.12 ON THE UNITY OR DIVERSITY OF THE GLACIAL PERIOD

After the Rochester AAAS meeting (§5.5), Upham urged Wright to stick to his guns on the idea of a single glacial period. He felt sure his views were too valuable to be abandoned, especially in light of "the estimates of such vast interglacial epochs as are supposed by McGee and Salisbury." That was the rub for Wright: he was willing to grant humans a Pleistocene antiquity, so long as the Pleistocene did not extend too far back in time. Divided into multiple glacial periods with tens of thousands to hundreds of thousands of years between them, it threatened to increase human antiquity by orders of magnitude. Although Leverett thought that ought to make Wright especially keen to understand the age and history of those deposits to gain a better handle on just how old any associated human remains might be, Wright instead stubbornly resisted and actively disparaged efforts to divide the glacial period into separate episodes.[247]

Wright was not entirely alone in advocating a single glacial period. Dana, like many geologists who had cut their teeth on the Pleistocene geology of New England, saw no evidence of multiple glacial epochs (§4.7). To him it was a "mystery" how two separate ice advances moving comparable distances south, each apparently having "very great thickness, great weight & great transport & abrading power," had left behind only one substantial terminal moraine and south of it a scattering of much lighter glacial debris (Wright's fringe). Dana could better account for that evidence with a single advance, explaining to Wright:

> If, instead of making the moraine lines those of a Second glacial period, we make them the terminal moraine deposited during an epoch of cessation of motion and of gradual retreat in the ice-front for some uncertain distance, we do not encounter the difficulty above mentioned. The ice after such a dropping of its moraine material (gathered from hill tops mainly) would not be able, whenever the time of increase & extension began, to pick up what had been dropped; it would slip on over it; and being increased in thickness by additions to its upper surface, the mass would move on with comparatively little transported material and hence distribute relatively little along the way. This makes one prolonged glacial period, with a time of retrogression or ablation in the course of it.[248]

That was music to Wright's ears, but Dana was reluctant to defend that idea publicly. Nor was Upham much help. In the fall of 1892 he submitted to

Chamberlin his Lake Agassiz monograph. It was seven years in the making, and Upham thought it the "greatest work of my life." He feared doing anything that might elicit Chamberlin's disapproval and make him "liable to hinder" its publication as a USGS *Monograph*. The best Upham could offer Wright were anonymous reviews of *Man and the Glacial Period* that applauded the unity hypothesis.[249] Wright was on his own.

In November of 1892, Wright published "Unity of the Glacial Epoch" in Dana's *American Journal of Science*. (Dana helped there, at least.) Wright knew he was playing a zero-sum game: if advocates of multiple glacial periods won, he lost. He therefore made the tactical decision to not just advocate the unity hypothesis, but also to attack the diversity hypothesis and show that its evidence was readily explained by a single ice advance. He broadly followed the approach in *Man and the Glacial Period*, but the book touched only lightly on this controversy and scarcely mentioned Chamberlin or anyone else. This time around Wright came out swinging, accusing Chamberlin of mistaking inferences for facts, Salisbury of misreading field evidence, and Leverett (one of Wright's few remaining friends in the Survey) of twisting his observations to fit Chamberlin's theories.[250]

It was vintage Wright. Each piece of evidence Chamberlin and others identified in support of multiple stages Wright turned on its head to show it fit just as well with a single advance. To the claim that the earlier advance rarely was manifest in moraine ridges and occurred in what Chamberlin called an attenuated border, Wright responded that the attenuated border was nothing more than his fringe, and a by-product of the same single advance. To the objection that the fringe was often too far from the moraine to have been part of the same depositional episode (§4.13), he claimed Chamberlin was now forced to admit that additional mapping was pushing the two (moraine and fringe) ever closer together.[251] In regard to those glacial deposits with organic peats and plant remains sandwiched between them and supposed to represent long ice-free periods, Wright observed that thick forests can quickly recolonize deglaciated terrain, as the area near Muir Glacier showed. Two or three centuries, he estimated, were all that was needed to grow a thick forest atop glacial deposits, which could then be overridden by subsequent ice advance.[252]

Salisbury's claim that the attenuated border showed greater weathering and oxidation and hence must be older than the terminal moraine Wright dismissed as "partly an illusion" and partly a result of the glacier moving over preglacial terrain that was already covered with oxidized materials. Just how badly Salisbury had been misled by looking at oxidation could be seen in the Delaware valley, where Salisbury had identified a till at Pattenburg and High Bridge whose heavy oxidation seemed to indicate it had to be very old. If this was a till, it should contain "foreign material" carried in by glacial action. None occurred there, according to Wright, a blow "fatal to [Salisbury's] theory."[253]

And then there were the high terraces of the Ohio, which Chamberlin attributed to an earlier glacial advance but Wright saw as deposits from the lake that filled behind his Cincinnati Ice Dam. Leverett had recently attempted to show those deposits were part of a much larger depositional process that extended well beyond the area of Wright's ice-dammed lake, but Wright breezily dismissed Leverett for not having "duly considered the facts."[254]

In the end, Wright argued, the present state of knowledge was such that it was pointless to hold any theory, whether unity or diversity, "with great positiveness." Indeed, "over-confidence on this point at the present time is likely to blind the eyes of the investigator, and to hinder progress."[255] Even if one was not inclined to accept the unity hypothesis as the outright winner, as he did, it could not be rejected either, nor could it be considered anything less than a viable and equal alternative to the hypothesis of multiple glaciation.

Wright knew Chamberlin would object to his paper, but it was not written for him. Wright was aiming at those who already accepted the unity hypothesis (New England geologists like Dana) and who might be looking for reinforcement of their views. He was also playing to the larger audience of geologists who may have had no strong opinion on the matter and might be swayed to his side.

What Wright may not have anticipated was Leverett's reaction. Leverett had greater knowledge of the glacial record in the regions where Wright worked than anyone (Wright included). He was able to evaluate Wright's evidence and arguments in intimate detail and instantly spot any errors in fact or interpretation. Leverett was at first too polite to say anything publicly, but that did not stop him from complaining privately. After *Ice Age* appeared in 1889, Leverett gently criticized Wright, explaining he had "grave doubts" about the Cincinnati Ice Dam.[256] Those doubts he made public at the 1891 AAAS meeting when Leverett, without mentioning Wright's name, observed that the silts said to mark the lake behind the ice dam extended both east and west of Cincinnati. Moreover, those silts rested atop and appeared much later than the local glacial drift, a unit with which they ought to be contemporaneous. Finally, the absence of glacial striations and the meagerness of glacial debris on the south side of the Ohio River did not suggest the ice crossing the river was massive enough or present long enough to have blocked the river's passage. An ice dam as Wright envisioned could not have existed.[257] Wright obviously disagreed, but he and Leverett maintained friendly relations, occasionally joined each other in the field, and corresponded regularly, sometimes in considerable detail about glacial matters.

When the firestorm over *Man and the Glacial Period* began in the fall of 1892, Leverett was "much grieved over the existing state of affairs" between Wright and Chamberlin and hoped to "influence" Chamberlin more favorably toward Wright. That would not be an easy task, as Leverett himself was taken aback by Wright's "Unity" paper. He understood how out of sorts Wright had been from the attacks by McGee, Chamberlin, and Salisbury. Nevertheless he complained to Wright that in the "Unity" paper

> you have displayed a heated feeling and have dwelt somewhat unfairly with your fellow glacialists, myself included. That you have taken an attitude of hostility toward all of us who hold to the duality of the glacial period will lead disinterested parties to wonder what is the matter.[258]

Worse, Wright was being disingenuous: Leverett reminded him of a conversation they had had two years earlier when he showed Wright the evidence that the silts in southeast Indiana could not have come from a slack-water lake behind the proposed ice dam. Leverett also showed him his maps indicating the ice sheet did not touch the Ohio River and thus had not extended far enough to form a dam. Obviously Wright had paid little heed, leaving Leverett feeling "as if you [Wright] had dismissed my ideas with a dictum that smacked of unfairness . . . and I am sorry for it." Leverett, caught between Chamberlin (his mentor and employer) and Wright (a friendly fellow investigator), was anxious to "bring about a better state of feeling," yet he still could not help but feel ill-treated all the same.[259]

Leverett grew angrier two months later when he looked again at Wright's paper and was surprised "even more than when I first read it." After grilling Wright on several substantive points, Leverett got to what especially "stirred [him] up":

> I read what you say of me on pp. 369–370. You do not quote me fairly and it looks as if you wished to convey the impression that I am a careless observer and not aware that something more than microscopical tests are needed. I think that it is largely this apparent endeavor to disparage honest scientific effort that had made several glacialists hostile to you.[260]

It was hard to escape the message: Leverett could easily become one of those hostile glacialists.

Chamberlin already was, and he lost little time in replying. Scarcely a week after Wright's "Unity" paper appeared he asked Dana for space to respond. Dana agreed but assured Wright he had little to worry about. He was convinced Chamberlin had been unduly influenced by his Midwest experience, did not fully appreciate the very different character of eastern glacial processes, and had been so focused on "local movement in local glaciers" that he failed to see the "actual grandeur of the Ice Age in North America."[261]

In early 1893 Dana received Chamberlin's rejoinder and marveled at its "over-correct, overconfident & otherwise unscientific" tone. Still, he published it as the lead article in his March issue. Chamberlin's "The Diversity of the Glacial Period" was pointed, condescending, and utterly self-righteous: "His [Wright's] description of the Ice Age in his two books and elsewhere seem to me to convey an archaic and bedwarfed impression both of the extent and complexity of the period."[262]

An "attenuated border," Chamberlin lectured, was not a fringe. The former was a drift sheet that thins out gradually, the latter a light scatter of debris off a terminal moraine formed by minor, incidental action. There were, in fact, two more fringelike features one might find beyond the drift edge: a bordering tract of gravel and boulders, sometimes miles from the drift edge, which were likely deposited by ice-fronting lakes (proglacial lakes) or glacial floods; and a bordering tract of pebbles and silt, laid down by low-gradient lakes unattached to the ice sheet. All were fringe deposits, in the narrow sense of occurring beyond the ice edge, but each represented a very different mechanism of deposition, and they were hardly equivalent in any meaningful way. To confound them, Chamberlin sneered, "is to push science backward." Worse, calling a feature a fringe without understanding how it was deposited made any effort to link it with nearby or distant terminal moraines misguided. In effect, Wright could hardly claim that a fringe and a terminal moraine were part of the same (single) glacial epoch if

THE GREAT PALEOLITHIC WAR

he failed to show or understand how the fringe formed and whether it was part of the same ice advance represented by the moraine. No matter how close those fringe deposits might end up relative to the moraine (and Chamberlin showed that Wright had systematically ignored or misrepresented his mapping to make it *look* as though he was now forced to move the two together), if they were the result of different ice advances, their proximity was meaningless.[263] Besides, if a fringe in Wright's sense was indeed a normal by-product of glacial action, why was it so proving hard to find? Chamberlin went in for the kill:

> Professor Wright claims to have spent five seasons on the [glacial] border between Pennsylvania and the Mississippi River, and yet he has not found in Western Ohio, Indiana, and Illinois any such moraine-and-"fringe" in any such relationship as he has so often described in Western Pennsylvania. One would think that five seasons work would give a better [result].[264]

(In 1902, when Leverett published his massive report on the glacial deposits of the Erie and Ohio basins he firmly rejected the fringe, observing that Wright mistook older Illinoian drift for Wisconsin age materials. That same year Salisbury's report on the glacial geology of New Jersey came to a similar conclusion. Wright's fringe in Pennsylvania and Ohio was squeezed out of existence from both sides.[265])

Wright's Cincinnati Ice Dam fared no better in Chamberlin's hands. If there was a lake, where was its outlet? Why were lake silts found well outside the lake boundaries and even below the dam that created it? Why were the gravels purportedly laid down by the lake only found on terraces and not across the lake floor? Why were the terraces Wright claimed were the same elevation as the dam (1,000 to 1,100 feet) actually nearly 300 feet higher than its top? The answer was obvious: there was never a lake. Wright had confused river and lake deposits and treated as contemporaneous features widely separated in age. Coupled with the absence of any evidence for a dam itself, "the hypothetical lake may well be dismissed."[266]

The fringe gone and the lake drained, Chamberlin pounced on Wright's objections to multiple glaciation, showing that trees growing near an Alaskan glacier were irrelevant (their growth depended on local conditions, not global climate), and that weathering and oxidization of glacial debris was indeed a viable measure of relative age when in the hands of "an expert competent to eliminate preglacial and englacial factors"—someone like Salisbury, say, whose work Chamberlin used to trump Wright. The Pattenburg and High Bridge deposits Salisbury identified as glacial, which Wright claimed were not because they lacked "foreign material" (that "fatal" blow to Salisbury's theory), indeed had nonlocal material in it. Henry Kummel of the New Jersey Geological Survey had independently collected a range of exotic specimens, testimony that what Salisbury had seen, and Wright denied, was evidence of glacial action distinct from that of the moraine some twelve miles distant.[267]

Chamberlin had Wright thoroughly boxed in, and with a healthy dose of disdain announced that the majority of "critical workers" in America and Europe accepted the evidence for at least two glacial epochs (James Geikie, whose work Wright sought to emulate [§4.13], read Wright's "Unity" paper and declared him "behind in his views"). In the Great Lakes region the current working hypothesis was that there might have been as many as four glacial episodes (in 1890 Chamberlin was already chiding McGee for claiming there were "two and only two" glacial periods). To continue to argue for a single event was "to put the question back where it was two decades ago."[268]

In late 1893 Dana finally weighed in and sought to reconcile Wright and Chamberlin's differences, suggesting that because New England received more precipitation its ice sheet grew larger, moved more slowly, and left a more singular record. In the Midwest, ice pulsed back and forth owing to its irregular and poorer supply of moisture, hence it left complex features that appeared as multiple advances, even though only one had occurred. That would explain why geologists from different parts of the country saw glacial history so very differently.[269] Chamberlin was not compromising in this issue.

Glacial climates may well have been as Dana proposed, but their effects would have been precisely the opposite of what Dana claimed, because the leading edge of a glacier reached further where the moisture supply was poorer. The more important difference was preservation. In New England, earlier glacial episodes were overridden, buried, or otherwise obscured. In the Midwest "a very considerable series of episodes is well displayed." A New England geologist was inadequately prepared to see subtle traces of earlier events. Dana was right about the differences between geologists, but for the wrong reasons: "Those who have studied the formations of one epoch believe in one epoch; those who have studied the formations of more than one epoch, believe in more than one epoch."[270]

Wright's "Unity" paper had also criticized Leverett and Salisbury, and though Chamberlin had defended them well and no reply was needed on their part, Leverett had finally had enough. He publicly chastised Wright for claiming that features that were demonstrably widely separated in time were part of the same fringe-moraine.[271] Salisbury continued his pounding of Wright, but his most detailed response to the "Unity" paper was one in which he said nothing of it or its author. Chamberlin had established a "Studies for Students" section in the *Journal of Geology* to provide "critical and analytical dissertations on topics that advanced students need to pursue in a thorough-going way." These were also intended for working geologists who, Chamberlin suspected, could use the remedial lessons (making the "Studies" "rather delicate work," as he put it). In the inaugural article of the series Salisbury explained at length how to recognize distinct glacial periods. The implications were obvious: there had been more than one, there were criteria (Salisbury listed a dozen) by which these could be recognized, and distinguishing them was comprehensible even to students. Wright was not mentioned by name, but contemporaries following the debate over glacial history would have known that he was one of the principal targets of the remediation.[272]

5.13 SHOWDOWN IN MADISON

By the late spring of 1893 Wright and his allies, and Holmes and Chamberlin and theirs, were congratulating themselves for having emerged victorious from the long winter of battling over glacial history and the American Paleolithic. Haynes praised Wright's final *Science* paper in May as having "a very convincing effect upon the geologists."[273] On the other side, McGee saw signs that the "literary volcano" triggered by Wright's book was now growing feeble:

> The matter is in pretty good shape; neither Wright's defenders nor his own defenses have been able to clear him of the serious charges which have been made by a number of different reviewers, and while some prejudiced persons may lean toward the Scotch verdict, the great majority, so nearly as I can learn, look upon him as not cleared—and this is just as it should be.[274]

Leverett, ever the mediator, told Wright he saw a silver lining to the unpleasantness: "I am convinced that the discussions have been wholesome and profitable to all concerned." Wright could be excused if he did not agree.[275]

As the combatants were brushing themselves off, however, one question that had lingered through the winter and spring remained unanswered: where had Putnam been? It was a good question. He alone among Paleolithic proponents had the status and position to stand up to the critics and could not be so readily dismissed as a mere amateur or collector (as could Abbott, Babbitt, Metz, and Mills) or a part-time dabbler from another field (Wright). Proponents could have used his help, yet Putnam stayed silent. Even Wright's urgent pleas for him to get involved got no more than a perfunctory reply from a secretary that Putnam was too busy preparing for the World's Columbian Exposition in Chicago to write anything. Still, if it was any solace, she added "He [Putnam] says that nothing he has yet read on the subject has shaken his faith in the matter." Hardly a ringing defense, but it was better than nothing. That letter was duly forwarded to the *American Geologist*, where it was published in its special symposium that spring (§5.8).[276]

Putnam was always busy and was especially busy that year, tending to his usual commitments and the upcoming Columbian Exposition. Even so, Holmes and Wright, who were also very busy and likewise involved in Exposition matters, found time to participate. The difference? They had more invested in the outcome of the controversy. Putnam had spoken in favor of the American Paleolithic and believed in it, but he also maintained an escape plan (part of which was to keep Abbott at arm's length) in case the idea failed. Putnam was not about to damage his or his museum's reputation by becoming embroiled in a controversy as nasty as this one (§3.2, §4.2). Fortunately, his frenetic schedule gave him the excuse to avoid getting involved.

Even so, among Putnam's many duties was being permanent secretary of the AAAS, and his attendance could be expected at the annual meeting. A good thing, too, for the upcoming meeting in Madison in August of 1893 marked the first time since the explosion over *Man and the Glacial Period* that all the participants would come together. Haynes hoped Putnam would be there to "lend a hand." He wanted as many of the brethren present as they could muster.[277]

Moorehead, secretary for Section H and responsible for assembling the section program, extended personal invitations to McGee and Wright, flattering McGee that no one could represent those opposed to the "new fad" better than he and

Holmes, and pledging to Wright that the Program Committee would insure that the tone of the papers would be entirely proper and the discussion fair to all. In case those carrots failed, there was also a stick: he let each know the other side was to be well represented. McGee and Wright both attended; Abbott and Holmes did not. Moorehead was right about representation but badly wrong about the ability of the Program Committee to police the participants' behavior.[278]

The opening act took place in Section E with a docket of papers on Pleistocene geology. The sessions attracted both George and Albert Wright, Upham, Leverett, McGee, Chamberlin, and Salisbury but was a relatively quiet affair, the participants' antipathies notwithstanding. Wright argued for fringe deposits and a single glacial epoch; Chamberlin and Salisbury spoke in rebuttal. The dialogue they had had in print played out in person with few changes but one concession on Wright's part. It came subtly and without acknowledgment that it was in response to Chamberlin's "Diversity" paper. Of course, Chamberlin sitting in the audience, knew exactly what Wright was doing and resented his unwillingness to be open about it.[279]

The excitement was in Section H. The anthropologists had set aside a day "for the great debate over the existence of the Paleolithic Man." The members of Section E who were involved joined them for what was by all accounts a very long day: tedious in some spots, lively in others, bitter and personal at the closing. "We were all relieved," Moorehead admitted, "when it came to an end." There were papers by Volk, Mercer, Holmes (read in his absence by McGee), and others, but the main event was the animated and frequently heated discussion that followed papers by Wright and McGee, each summarizing the evidence for human antiquity. No statement went unchallenged: McGee flew into a "terrible rage" when Wright cited his discovery of the Lake Lahontan spearpoint (§4.3). Archaeologists and geologists alike participated to the extent that the discussion became, as Peet described it, like "border warfare, in which neither geologists nor archaeologists are quite sure as to what belongs to them."[280]

The geologists moved to seize territory, with Chamberlin pointedly questioning Paleolithic proponents about whether "any other testimony than that of an expert geologist [can] be relied upon," and why some of the cases previously trumpeted by Wright were now ignored. He knew the answers: the point was to ask the questions. Putnam was finally forced to get involved when Mercer called on him to say what he thought. Putnam's response sparked an extended debate between him and Chamberlin over whether it was possible for an artifact on the surface to fall down the hole of a decaying tree root. Putnam had seen no such thing in his years of excavating in earthworks, but Chamberlin countered, and even a sympathetic Moorehead conceded, that Putnam "lost the point." On it went, from arguments about the minutiae of turtlebacks to discussions about what constituted sufficient evidence to make a Pleistocene case. Volk was heard to mutter, "We should do more digging, and less talking."[281]

But Volk was wrong: more digging was not going to resolve the case. What was needed was resolution of issues relating to evidence and method, and it was these that were being hammered out, albeit painfully, at Madison. None were settled, of course, and there were no winners, however keen some were to declare them. Wright was convinced that "Holmes and his crew . . . made a pretty poor showing every way" and that Chamberlin's view of glacial phenomena had been so roundly rejected "he is going to be pretty thoroughly discredited as an authority."[282] Wright never lost faith in his scientific beliefs, which clouded his perception of reality. Nonetheless, a none-too-subtle shift in the debate was visible by day's end. Any presumption in favor of the American Paleolithic had vanished: the best that proponents could hope for now was to keep the issue sub judice.[283]

The Madison meeting also showed how far apart the sides were, a gap mirrored in reports of the meeting written by McGee and Claypole (for the *American Anthropologist* and *American Geologist*, respectively[284]). Yet, what most lingered in the minds of the participants was what happened when "one of the disputants lost self-control and allowed himself to employ unparliamentary language."[285] McGee had exploded following accusations of a discrepancy between what he and Powell had said about the Nampa Image. It was so harsh an outburst that an old friend who witnessed it became alarmed about McGee's mental well-being and urged him to let up:

> You are irritable and can be led into saying things which you should not say; the change in this respect is incredible, for you used to be the coolest man within the limits of my acquaintance—which is saying a great deal. You are in the first stage of a break-down and you will be a wise man if you will take a rest. You owe it to yourself, to your family and to your friends.[286]

Given that McGee had just been unceremoniously exiled from the USGS and sent to the BAE, it was good advice—but given too late for much good to come of it.

5.14 INTERREGNUM

After the Madison AAAS meeting the participants dispersed and the most ferocious battles of the Great Paleolithic War were over. There was one more public fight coming over paleoliths at the 1897 AAAS meetings (§5.16), but the intervening years were relatively quiet.

In late 1893 Leverett, ever the optimist, tried again to reconcile Chamberlin and Wright, urging each to "drop all personal feeling and simply work for the advancement of truth."[287] But they were long past reconciliation; the best he could hope for was an end to open hostilities. That did not happen either, and Leverett was partly to blame. He had misunderstood Wright to have admitted that the fringe was "not a dependency of the terminal moraine, but was much older." Anticipating Wright's mea culpa at that December's GSA meeting, Chamberlin refrained from critical comments in his own presentation to allow Wright to save face in his. Yet, Wright made no concession. Instead he attacked Chamberlin on a "less essential point" where he thought he had the upper hand. Chamberlin was furious, not least because of Wright's "refusal to state his position on more vital points" in response to a direct question.[288] Chamberlin distrusted Wright ever after. In fairness to Wright, his views had not changed. In fact, at the time of that meeting he had two articles forthcoming on the unity of the glacial period and the Cincinnati Ice Dam.[289]

It was a common enough mistake Leverett made in thinking that Wright had changed his mind. Over the next several years participants, especially on the American Paleolithic side, repeatedly circulated rumors that one or more of their antagonists was about to come over. Abbott claimed to have heard that Salisbury "thinks it very likely I am right," whereas Upham believed "our good brother McGee may now repent of his rashness; and the same is true of the trio, C.[hamberlin], S.[alisbury], and McG.[ee]."[290] It was so much wishful thinking. Few minds were changed, and indeed Chamberlin only grew angrier, blaming Wright and Dana (who "compromises himself with W[right]'s ideas"[291]) for damaging the status of American science:

> I think that American glaciology has lost the leadership that it had ten years ago, and has been seriously put back in the eyes of geologists of the more discriminating school, by the reactionary advocates of the old classification of Professor Dana, and the attempt to force our results into subservience to it. Ten years ago American glaciology was leading the world. We are now distinctly and markedly behind Europe.[292]

Wright continued to push his favorite themes, though his papers were largely repetitive.[293] New editions of *Ice Age* and *Man and the Glacial Period* appeared—to Chamberlin's dismay, the public he wanted to defend were buying Wright's books— each with a new preface responding to critics. The new editions were virtually ignored in the scientific community. Seeking opportunities to enhance his scientific reputation, Wright joined a tourist excursion to Greenland in 1894 to study its ice sheet, but his trip was cut short when his vessel hit an iceberg and sank. Still, he produced a book from that ill-fated trip, though its Greenland content was limited and its discussion of North American glacial history largely recycled.[294] Wright's concerns for his reputation were well founded. Although he fashioned his *Ice Age* after Geikie's *The Great Ice Age and Its Relation to the Antiquity of Man* (§4.13), when Geikie needed chapters on North American glacial history for his new edition he turned not to Wright, whose competence he questioned (§5.12), but to Chamberlin, whose knowledge Geikie deemed "at once extensive and profound."[295]

Chamberlin continued to keep a critical eye on Wright: "I find it hard to decide," he said, "whether Wright is unable to see that the case cited has no essential bearing on the main question, or whether he wishes to create a desired impression without regard to its legitimacy."[296] Chamberlin was in Greenland the summer of 1894 as well, serving as a geologist with the Peary Auxiliary Expedition. Although that expedition had its share of Arctic challenges, it was generally agreed that Chamberlin alone among the scientific staff accomplished what he set out to do.[297] On Chamberlin's recommendation Salisbury went to Greenland in 1895 with the expedition sent to fetch Peary home. Salisbury also published on his Arctic work, but most of his mid-1890s fieldwork was in New Jersey mapping its Pleistocene deposits.[298]

Although he had been eased out as USGS director (§5.9), Powell retained his position at the head of the BAE. He spent the next several years occasionally writing philosophical pieces on anthropology and tending his painful arm. With Powell gone from the USGS, Winchell made a desultory effort to be named his successor and, failing that, petitioned for a radical restructuring of the Survey. Neither happened.[299]

McGee, newly appointed as a BAE ethnologist and its self-proclaimed "ethnologist-in-charge," went to northern Mexico in the fall of 1894 for anthropological fieldwork. He found the "warlike" Seri eking out a living on the California Gulf coast and collected a small amount of linguistic and ethnographic data. He returned in 1895, but thanks to bad luck and leadership (his) the expedition nearly perished in the harsh desert. Although his anthropological results were meager at best, from them McGee produced a 300-page monograph that won him his spurs in late nineteenth century Washington anthropology. Because Powell withdrew as well from administration of the BAE, McGee unofficially took over.[300]

Holmes, detailed since the late spring of 1892 to work on the BAE's exhibits at the World's Columbian Exposition, found himself by the end of 1893 at a cross-road. His previous half-decade of research he immodestly described as "one of the most important periods of [my] labors in the field of science, and one of the most important in the history of American archaeological research." Exaggeration, perhaps, but without question it catapulted him onto center stage in American archaeology, changed the character of the discussion of the American Paleolithic, and culminated in a massive monograph on stone tool technology that won him the 1898 Loubat prize as the best work in anthropology over a five-year period. It also brought him job prospects. When the Exposition ended, stealthy lobbying by Chamberlin led to Holmes's appointment as curator of anthropology in the newly formed Field Columbian Museum (a position a young and chronically underemployed Franz Boas desperately wanted and believed would be his; that Holmes was hired instead embittered Boas, and its aftereffects reverberated over the next three decades).[301] Powell deemed it a fitting reward for Holmes's having provided the "genius and enthusiasm" by which "American archaeology has been revolutionized."[302]

No revolution is bloodless, and as Holmes gained another job in archaeology, Abbott lost the only one he had ever had. It was perhaps inevitable. Placed in a position he had so long coveted, he discovered the reality of museum work was not nearly as appealing as he imagined. He never quite recovered from his dismal start in the summer of 1890 (§5.3). William Pepper, president of the University of Pennsylvania, could not see that his curator had engaged in any substantial museum work (§4.15). In fact, Abbott often spent his days in Trenton instead of Philadelphia. Abbott realized he was not cut out for museum work. (A decade later he found himself back in the museum for a visit and it seemed to him "strange and dream-like." In contrast, "Out-door work in archaeology is beyond comparison, the most delightful of occupations, but the museum and contact with humanity—I shudder at the thought."[303])

The Great Paleolithic War took its toll on Abbott. Brinton and Culin had closely followed Holmes's work (§5.4) and considered his Trenton critique "a very strong case against the Abbott finds" that would "greatly aid in placing the Delaware Valley relics in their proper cases." It also put in sharp relief Abbott's "loose methods," which they could see were detrimental to the reputation of the institution.[304] As Abbott's supervisors they were in a position to do something about this state of affairs. In October of 1893 Abbott learned his curatorship was over.[305] He comforted himself that the position had been "a veritable thorn in my flesh for over two years" and he was "glad to be free" of the museum.[306] A month later he received an invitation to return there to hear a lecture by WJ McGee. The old wounds burst open: "What a pack of asses these people are!"[307]

5.15 RETURNING TO THE FIELD OF BATTLE

Abbott spent the next several years cursing archaeology and archaeologists and insisting he had not the slightest interest in the subject anymore. He said it so loudly and often he almost convinced himself. But his letters and diary carefully recorded any tidbit, published or otherwise, that he thought showed he was right.[308] He urged on Volk, who with Putnam's funding continued work at Trenton, and he harangued Mercer, who now had Abbot's old job and whose support for Abbott's claims was steadily eroding: "Who are you," he snarled at Mercer, "that these questions can only be settled by your decision?"[309] Mercer was not Abbott's sole target, just the most convenient one.

At the AAAS meetings in 1896 Putnam described Volk's excavations at the Lalor Farm in Trenton and assured the audience there was now unambiguous stratigraphic evidence of two distinctive archaeological assemblages—an underlying one dominated by argillite artifacts, and one above by jasper artifacts. Brinton rose immediately to challenge him, and McGee followed, citing the exposures Dinwiddie had examined several years earlier without finding anything, let alone artifacts associated with a strata of ascertainable age (§5.4). Putnam was stopped, but with Volk continuing to excavate in the area McGee was sure more such claims would follow. He decided the critics needed to return to Trenton and urged Holmes to

> visit Trenton during the year, if possible in company with Salisbury, and be ready to come out at Detroit [the venue for the upcoming 1897 AAAS meeting] with a final statement of the whole Trenton question. I am very much in earnest about this, and disposed to say that you must do it, in your own interest and in the interests of American archaeology.[310]

Holmes agreed, and with Salisbury began planning summer fieldwork at Trenton in 1897 in advance of the AAAS symposium that August. Unbeknownst to them Mercer had plans of his own.

Mercer had invited Wright to join him at Volk's excavations at Lalor Farm to see for themselves if the "argillite rests clearly under stratified lines." The work was scheduled for late June of 1897 and was to include several geologists as well. Putnam, wanting to insure credit was given his hard-won Peabody funds that supported Volk's work (the money was from the Duke of Loubat, who had endowed the prize money Holmes won that year), asked the visitors for a joint statement afterward of their findings.[311]

Mercer and Wright were joined by Volk, Abbott, geologist Arthur Hollick (Columbia University), Harrison Allen (Academy of Natural Sciences, Philadelphia), and later by New Jersey State Geologist John Smock and two of his assistants, Henry Kummel and George Knapp. A trench nearly twenty-two feet long and three feet wide was dug, along with several small excavation pits; these produced several hundred artifacts, many in situ.[312] It was "all that was hoped for and more than was anticipated," as far as Abbott was concerned. By the time everyone departed, an exhausted Abbott was certain (as was Volk) that the visitors were convinced the yellow sands in which argillite artifacts were found in situ was deposited by swollen glacial rivers. For Abbott it was of course the "same old story. Old deposits and relics in place," but still:

> I cannot say, looking back on the past four days, that I have enjoyed it. There is too much assumption of extra carefulness, as they call it, which is simply a lot of childish twaddle. . . . They may be very eminent men, but it took me a good deal less time to learn that we had here evidence of man's antiquity.[313]

Putnam heard the good news about the site visit from both Abbott and Volk, and on almost the same day he received a letter from McGee inviting him to jointly organize a session on Delaware valley archaeology at the upcoming AAAS meetings.[314] McGee had already invited Holmes and Salisbury. Putnam was keen to get Mercer, Wright, and Hollick on the program. Yet, just as he was about to send his invitation to Mercer, he learned Mercer planned to send the reports on their fieldwork to *Science* for publication in advance of the AAAS meeting. (Mercer was not planning to go to Detroit but wanted his results to be available to those who were.) Putnam quickly added a postscript to Mercer urging him not to submit his manuscript on the flimsy excuse that if the results were published beforehand they could not "come up at the meeting" in August. In the event that was not convincing, Putnam showed Mercer the fist in the glove:

> While I do not wish to interfere in the least with a report from all who were present last week, yet you must remember that this whole work is under my direction and that I have the right to be consulted and, to a certain extent, to indicate what should be done.[315]

Mercer brushed Putnam's plea aside.

Putnam could not bring himself to tell Mercer the real problem, which, as he explained to Wright, was that he needed something to show the donors who had supported Volk's work and his trustees who had approved it. If the reports went straight to *Science*, Putnam would get only a footnote of credit. But having positive results to announce at the AAAS meetings in a session that he co-organized would "show the importance of the work and help me to continue such investigations, as well as being just to me as the instigator and director of all this work in the valley for so many years." Putnam was not without ego, but even more he was desperate to maintain funding for his programs (§5.1).[316] He might not have pushed so hard had he realized the fieldwork had not gone as strongly his way as Abbott and Volk imagined. Smock, Kummel, and Knapp all left Trenton convinced the yellow sands were a secondary aeolian deposit and "probably not very old."[317] That argillite artifacts were found in situ in the yellow sands therefore meant little. Soon both Kummel and Knapp had manuscripts saying so ready for submission to *Science*.[318]

Meanwhile, Salisbury and Holmes headed to Trenton, unaware of all the activity that had just taken place or Mercer's publication intentions. In mid-July they found out and likewise were not keen to be scooped in *Science*. Like Putnam they wanted control over what information was disseminated from Trenton that summer. On their behalf Chamberlin wrote Kummel (who had received his PhD at the University of Chicago a few years earlier) to explain that any publication before the AAAS meeting would be premature: it was important to have a field conference of all parties at Trenton, followed by discussion in Detroit, before any manuscripts were submitted. This was no less disingenuous than Putnam's excuse. If a field conference had been planned for Trenton, Salisbury and Holmes had failed to mention it. But Kummel got the message: he had "no desire to intrench upon work which has been done prior to this."[319]

Mercer's plans fell apart. Wright acquiesced to Putnam and refused to sign a joint statement or contribute a paper, despite Mercer's increasingly anxious pleading. ("Discuss it afterwards if you wish at Detroit with those who do not know as much about it as we do."[320]) Within a few days, under strong pressure from Salisbury and Chamberlin, both Kummel and Knapp withdrew their papers.

As Kummel understood it, Salisbury was convinced Wright had learned of his and Holmes's plans, had hurried to Trenton in advance to "cut the ground out from under his old antagonists," and duped Kummel and Knapp into helping (neither of whom was convinced they were unwitting pawns). In the circumstances, not only would it "prejudice the case" were they to publish with Mercer beforehand, it would somehow play into Wright's hands (which raises the question of how that could be, as Kummel and Knapp had a very different opinion of the origin of the artifact-bearing deposits than Wright). Salisbury's scenario was pure fiction but it hardly mattered: Knapp and Kummel were or had been Salisbury's assistants, still had to work with him, and knew what kind of person he was. As Knapp explained to Mercer, Salisbury would be "greatly provoked" if he published his paper and, he added ominously, "slow to forget." Kummel regretted having to withdraw and he was not proud of what he had done either, confessing to Mercer that

> my only reason for withdrawing is that I know it will be in my material interest not to cross Prof Salisbury in this matter. I am satisfied that I am in the right, and that his attitude is unjustifiable. Nevertheless, I wish not to have his ill will. I relinquish what I believe to be my rights in favor of my material interests. Low ethical standards? Perhaps so.[321]

Only Hollick was still willing to go along with Mercer, but by the time their two papers reached *Science* with a plea to the editor, James McKeen Cattell, to publish them at once, it was too late. Knowing Mercer was going ahead with his plans, McGee urged Cattell to stall. Puzzled by the delay, Mercer telegraphed Cattell demanding the papers be published or returned. They were returned. Mercer was furious:

> If the rules of the publication of *Science* compel you to supersede earlier manuscripts by later on a subject like this, for the sake of the American Association I have nothing further to say but to regret the fact. If no such rule intervenes I trust to your sense of justice to give Prof. Hollick and myself our rights and to place the Washington party where they belong in this instance . . . as settlers of a settled question.[322]

Realizing his *Science* plan had collapsed, Mercer wrote unhappy letters to Salisbury, Holmes, and Putnam. Putnam tried to sooth him and urged him to send his paper to Detroit to be read. Holmes made the same request but was less sympathetic ("the young men will lose nothing by deferring to . . . the elder"). Salisbury was not sympathetic or apologetic. Mercer pushed back but got a cold reply:

> Everything which I said in my note of recent date seems to have been misunderstood or misinterpreted. . . . I am very sure you do not understand even the elements of the situation or you would not have responded as you do in the letter just received. In view of the response I do not think it wise to continue the correspondence.[323]

Mercer perhaps understood for the first time why Knapp and Kummel feared Salisbury, whose own students described him as a dictator who "loved the feeling of power."[324]

Salisbury and Holmes made their planned July visit to the Lalor Farm, joined by Kummel and Wilson but not Mercer. He declined a last-minute invitation from Holmes, likely seeing it for what it was: a transparent attempt to placate him. Wilson came away convinced the results of the visit were "decidedly favorable to Dr. Abbott's contention." He had seen and heard only what he wanted to see and hear.[325]

5.16 AN END AND A BEGINNING

In August 1897 the players assembled in Detroit for a special session of Section H on human antiquity in the Delaware valley. In a replay of Madison four years earlier (§5.13), the Section E geologists joined them. At issue were two key points: were the artifacts in situ, and did they occur in Pleistocene formations. The essentials of the stratigraphic section were agreed on: the uppermost zone was six to ten inches thick and contained artifacts of flint and jasper and ceramics ("Indian artifacts"); below that was a yellowish sand, some thirty-six inches thick and interspersed with thin, reddish, clayey bands, which contained argillite artifacts along with small and large pebbles (too large to have been carried in by wind). Beneath that was a reddish sand some eight to ten inches thick that all agreed contained no artifacts and that graded downward into a cross-bedded sand and gravel, thence into the Trenton Gravel.[326]

All further agreed the argillite artifacts were found in the yellowish sand. But there was no agreement as to whether they were in situ. Putnam thought so, believing them to be the same age as the sand, whatever that might be—he left that to the geologists. Kummel agreed about the primary context of the artifacts, but like Knapp he considered the sand to be aeolian, dating from a period much later than the Trenton Gravel. (That these deposits were at least postglacial seemed apparent from a lack of soil development.) Salisbury was not convinced the sands were aeolian rather than water-laid. Those different processes implied different ages (fluvial sands had the potential of being at least terminal Pleistocene), but it hardly mattered to Salisbury because he believed the artifacts were redeposited.[327] Wright, naturally, objected to the idea that the sands were aeolian or that the artifacts were in secondary context, and he declared that the clay bands (lamellae) indicated undisturbed fluvial deposits. On that basis he argued the yellow sand must be old and therefore the argillite artifacts also dated to the end of the glacial period, deposited there by people who came to the edge of glacial meltwater rivers to collect stone for toolmaking.

Mercer (whose paper finally appeared after the meeting, as McGee had seen to it) was of several minds. On the one hand he emphasized, as Wright did not, that not only were jasper artifacts in the yellowish sand, there were also jasper, flint, and quartz artifacts of the sort found in the "Indian village" stratum immediately above. Moreover, the frequency of the artifacts seemed to diminish with depth, as one might expect were they "leaking" out of that stratum. Still, Mercer was swayed by the fact that a significant number of the argillite implements were found beneath "well observed and unbroken films of stratification" (the clay lamellae) and that no pottery was in those lower units, and that was enough to nudge him into concluding these were in situ, separated from the later occupation by an unmeasured amount of time. Hollick provided no such ages, but like Wright he saw the yellowish sand and lamellae as representing "successive periods of flood and sedimentation."[328]

Holmes was hardly interested in the argillite. The yellowish sand was a near surface deposit, and all the artifacts found in it, argillite or otherwise, were almost certainly redeposited: "Every bank that crumbled, every grave dug, every palisade planted, every burrow made, every root that penetrated and every storm that raged took part in the work of intermingling." It was pointless to conjure up "shadowy images of other races." More striking was that the focus was on a "superficial deposit" well above the Trenton Gravel and therefore irrelevant to arguments about deep human antiquity. That the conversation was no longer about paleoliths he saw as a concession that there were none. He had said so five years earlier and said it again now: the matter of a Pleistocene human presence at Trenton was "practically closed."[329]

Wilson too had been struck by silence about the Trenton Gravel but spun it in the opposite direction: it was not in question in these studies. Instead, they were aimed at showing the presence (which he himself accepted) of "an earlier human occupation of the territory with a different culture, at least a different industry, from that of the modern Indian. That, he judged, was "entirely in harmony with, and vindicatory of, the position taken by Dr. Abbott" in 1883, of a pre-Indian, post-Paleolithic occupation.[330]

Perhaps, but it was not much of a rope to throw a drowning idea. In the discussion that followed, ironically led by McGee as chair of Section H and Claypole as chair of Section E, even that rope proved hard to grasp. Despite Wright's insistence that the clay lamellae were undisturbed, Claypole conceded Salisbury was probably correct. The presence of artifacts beneath the lamellae was not a sign that they were in primary context. Still, Claypole was quick to insist, after Chamberlin tried to use the failure of the Trenton Gravel to produce artifacts as leverage to dismiss the entire American Paleolithic, that everyone needed to remember that the "discussion relates solely to the New Jersey district." Even so, Claypole himself was starting to harbor doubts about the American Paleolithic.[331]

Putnam had no such doubts but did not want a fight. Instead, he went back to the argillite artifacts. Even if these were redeposited, it required an "immense" length of time for them to be buried. As they were so unlike what was found above, they should be considered an earlier, pre-Indian culture. It was as Abbott said fifteen years earlier: Putnam thought a "grand advance" had been made. Holmes vigorously objected. The notion of a separate culture from the Indian offended his conceptual (and methodological) beliefs (§5.2): "The first step in acquiring a knowledge of the past is to seek to understand the present." The argillite artifacts were indistinguishable from those of the Indians. The two had very different understandings of the significance of stratigraphic separation. For Putnam, it was proof of a gap in time and therefore of different cultures; for Holmes, it was a fortuitous product of geological processes, which might have no temporal or cultural meaning.

McGee, as chair, had the last word that afternoon, and he aimed straight at Trenton's most vulnerable spot. "Fifteen years ago," he reported, "there was hardly an archaeologist who did not regard the Trenton region as affording conclusive evidence of glacial man; to-day the manner in which the evidence has been torn to shreds is apparent to everyone. What I say relates only to glacial man."[332] McGee specifically excluded the argillite occupation from that summary, but then there was no reason to include it. All that arguing about whether there was a pre-Indian culture and whether the sands might be postglacial in age could not disguise the fact that on the most important question, the American Paleolithic, Trenton was failing. The final blow came just two weeks later across Lake Ontario.

Several of the participants—Putnam, Claypole, and McGee among them—went on to Toronto where the BAAS was holding one of its rare meetings in this hemisphere. Four years earlier Haynes had insisted "only a jury of the acknowledged pre-historic archaeologists of the world is competent to pronounce judgment upon this question."[333] Toronto provided that opportunity: John Evans, the dean of English archaeology, whose work—perhaps more so than any other Paleolithic archaeologist—had influenced the search in America, was present.[334] After Putnam and Claypole delivered talks on Trenton and Ohio paleoliths (respectively), Putnam handed Evans a set of Trenton paleoliths. As reported by McGee, Evans declared "clearly and emphatically" that "whatsoever the Trenton material may mean, it has absolutely nothing to do with that which in Europe is called Paleolithic and assigned considerable antiquity."[335] McGee was never without an agenda, but this was not just his wishful spin. Commentators in *Nature* and *Science* duly reported that was indeed Evans's reaction, adding that he considered the specimens to be decidedly Neolithic in form. Putnam tried to change Evans's mind by suggesting the American paleoliths were genuine and appeared different only because they were not made on European flint. Evans rejected the argument.[336] In later years an outraged Abbott was certain Evans was "misled [in that] unfortunate Toronto business." He blamed Putnam for the "awful blunder," but then he always did.[337]

Abbott knew the score: that Evans rejected the American Paleolithic was devastating. Evans's *The Ancient Stone Implements, Weapons and Ornaments of Great Britain* had been a template for his own *Primitive Industry* (§3.6), and its illustrations of paleoliths the authoritative source by which many American specimens were deemed credible by their resemblance. McGee gloated publicly over the outcome, and claimed credit privately:

> In a sort of esoteric fashion I had been preparing the way for a satisfactory verdict all through the meeting, and had cultivated the minds of the president of the section [Evans] and the sectional committee, and indeed of the section generally.

He also saw to it that Anita McGee's summary of the Detroit session was published in *Science* in advance of the publication of any papers from that meeting, in order to "prevent misapprehension on the part of the public . . . which will be fostered by certain influences too long and devious for description."[338]

In the wake of Evans's pronouncement, many proponents started to drift from the American Paleolithic fold, though a core remained faithful. Putnam and Wright agreed Volk's Trenton work should continue "in the hope of finding something under the thicker red layer" (the unit below the argillite-bearing sand) and clarifying the age of the clay bands. In mid-September, another field conference was held at Trenton, but it added nothing to the debate, though Wright confidently assured Abbott the day of his triumph over the clique that tried to dismiss him was "near at hand." It was not. The best that advocates could achieve was a favorable column in the *New York Times* assuring its readers that Evans had been mistaken about the Trenton specimens. But this was not a genie that could be put back in the bottle.[339]

Proponents continued to claim that the earliest North Americans arrived in the Pleistocene. Critics agreed this was possible in principle, but they continued to deny it as fact.[340] Notwithstanding that impasse, on one point the critics undeniably won: by the turn of the century new discoveries of American paleoliths had virtually ceased. Advocates of an American Paleolithic could do little more than recycle the same, now badly battered, specimens.[341]

In the fall of 1897 Putnam returned to Trenton to see Volk's trenches, in the hope that something of importance might appear. He brought with him a young physical anthropologist he had recently hired for the American Museum of

Natural History's Hyde expedition, Aleš Hrdlička. Hrdlička had gotten his first glimpse of the controversy over human antiquity at Detroit and now was seeing Trenton firsthand. They did not spot any artifacts that day, but Putnam was not discouraged. The active search for a deep human past continued there, and with Volk's discovery on December 1, 1899, of a human femur deep in the Trenton Gravel,[342] the character of the evidence and the nature of the controversy took an abrupt turn, and with it Hrdlička moved to center stage.

CHAPTER SIX

Cro-Magnons in Kansas, Neanderthals in Nebraska, 1899–1914

BY THE CLOSE OF THE nineteenth century Ernst Volk had spent nearly a decade in Frederic Ward Putnam's employ, faithfully walking Trenton's drainages, railroad cuts, and newly excavated sewer ditches in search of paleoliths or any other traces of a Pleistocene human presence. He had had considerable success but had not been able to satisfy Putnam's hopes of a discovery that would clinch the case for Trenton's antiquity. Finally, in March of 1899, it looked as if that find had been made. An excited Volk stopped by Charles Abbott's house with a bone fragment that Abbott, trained as a physician, immediately identified as a human tibia. It had been plucked from the Trenton Gravel.[1]

So far as Abbott was concerned, this discovery called for even more than the usual hyperbole: this specimen "forever set at rest [the] antiquity of man in Delaware Valley."[2] He crowed to Putnam that the find was "next in importance to the discovery of the missing link in Java" (*Pithecanthropus* [*Homo*] *erectus*), then commanded him to *do* something. Here at last was compensation for Putnam's "miserable blundering" at the BAAS meetings in Toronto in 1897 (§5.16), about which Abbott was still bitter.[3] Redemption in sight, Abbott began alerting the brethren of the good news. But Putnam was cautious as ever (§3.2), and a good thing too. On closer examination, the tibia proved to be a scapula. Worse, it was from a bovid. A best-case scenario was that it might be an extinct Pleistocene musk ox. The worst case? A farmer's lost cow.[4] Putnam was not wandering out on any limbs for the sake of a cow.

"Damn the cow," Abbott snapped. If "Paleolithic man is never acknowledged, it will be your *fault*, and *inexcusably* so!" As far as the ever-impatient Abbott was concerned, Volk already had enough evidence to substantiate everything Abbott had been saying about human antiquity at Trenton for nearly twenty-five years. The way he saw it, Putnam owed it to Abbott to publish Volk's report of investigations to counter the "lies" of Holmes, McGee, Salisbury, Brinton, and others and close the "mouths of Washington calumniators . . . if anything short of God Almighty can close them."[5] Abbott's frustration at Putnam only mounted over the next decade as each passing year failed to bring Volk's report.

Worse, the Delaware valley's central role in discussions of human antiquity in America was being usurped as attention shifted to sites in the Midwest that were producing very different indications of a deep human antiquity—namely, ancient-looking human skeletal remains in what appeared to be Pleistocene-age loess deposits near Lansing, Kansas and Omaha, Nebraska. The shift from paleoliths to human skeletal remains, and from sites in gravels near the glacial front to ones in loess well away from the terminal moraine, triggered a new round of debate that was at once familiar and not. There were the usual questions about whether the remains were in primary context or redeposited; if the formation in which they were found was a primary deposit or had itself been reworked; and how old both the human remains and the deposits in which they were found might be. Thwarting efforts to answer those questions was the fact that archaeological claims once more became entangled in geological disputes, this time over the origin of loess, whether from wind or glacial meltwater, its age, and whether there was more than one Pleistocene loess formation, and if so how those might be identified.

On the archaeological side, antiquity was again based on morphology, but now it was the form of human skeletons (skulls primarily) rather than stone artifacts. Discussion veered into the unfamiliar terrain of what a Pleistocene-age human skeleton ought to look like. It was a question not easily answered given the relatively meager human fossil record then known, uncertainty over the nature and rate of anatomical change or changes over time, and disagreements (back to geology again) over just how long ago modern-looking humans appeared and whether that was in the Pleistocene.[6]

Some of the participants in this next round were old hands who continued in their roles as the controversy entered the twentieth century. There were several newcomers, foremost among them Aleš Hrdlička, a young medical doctor introduced to the controversy over human antiquity by Putnam, who employed him at the American Museum of Natural History (AMNH) as a physical anthropologist. Starting at the turn of the century and soon aided and abetted by William Henry Holmes, who hired Hrdlička in 1903 to head the Smithsonian Institution's newly formed Division of Physical Anthropology, Hrdlička visited, evaluated, and publicly criticized virtually every newly discovered site with purportedly ancient skeletal remains found in North America.

Hrdlička became a formidable presence in discussions of human antiquity in the Americas, and he remained an intimidating figure well into the twentieth century. Observers then and later attributed Hrdlička's critical stance to his being hired by Holmes, and they even suggested that had he stayed with Putnam the human antiquity dispute in America might have been very different.[7] The matter is more complicated. What is certain is that the foundation for Hrdlička's role in the human antiquity controversy was laid during his first few years in this arena and began, appropriately enough, where the controversy itself started: in Abbott's Trenton Gravel on the banks of the Delaware River.

6.1 HUMAN SKELETAL REMAINS EMERGE FROM THE TRENTON GRAVEL

Push as Abbott did in the spring of 1899, Putnam did not allow Volk to publish the results of his decade of work in the Trenton area. Abbott neither knew nor cared why, he just wanted to see Volk's results in print. When there was no sign of that happening he angrily swore off archaeology, vowed to Putnam he would cease writing to him (by now, Putnam surely wished Abbott was a man of his word), and insisted he would pay no further attention to Volk. That was no easy task, because at the moment Volk was literally excavating outside Abbott's window and having an "awful time" with Abbott watching over him daily, unable to "control his pen nor tongue."[8]

Abbott's shunning of Putnam lasted scarcely three weeks before he was once again railing at him. "Patience," Abbott snapped, "has ceased to be a virtue."[9] But there was good news. The scapula had indeed come from an extinct musk ox, not a lost Holstein,[10] thereby at least underpinning the case for the Trenton Gravel being Pleistocene in age and sparking hope that other bones, perhaps human, might yet be found. Better still, Volk's excavations that summer around Trenton were producing unmistakable human skeletal remains. Abbott, naturally, was convinced those were Pleistocene in age. Putnam less so.[11] As always, Abbott was going too fast for Putnam, Putnam too slow for Abbott.

These were not the first discoveries of human remains from Trenton. Over the previous two decades several complete crania, along with cranial and postcranial fragments, had been found, some in apparently ancient deposits. Those specimens were examined in the fall of 1898 by Peabody Museum physical anthropologist Frank Russell, who deemed them similar anatomically and morphologically to those of "modern Indians, probably of the Lenni Lenape" (the Delaware). Given their questionable geological context, he doubted they had any great antiquity.[12]

Russell's paper irritated Abbott, as would any paper that reached such a conclusion. Abbott saw nothing questionable about their geological context and told Russell so. Putnam was displeased as well. A decade earlier he had declared the specimens unlike the Delaware Indian type but instead "crania of the Paleolithic people of New Jersey." In his 1899 presidential address to the AAAS, Putnam pointedly ignored Russell's paper.[13] He determined that the next time skeletal material was found he would turn it over to someone else for examination. That opportunity came on the first of December.

That morning Volk was checking one of his customary locales, a railroad cut near the Lalor Farm (site of the 1897 dueling visits [§5.15]). Heavy rains had exposed a lens of greenish sand streaked by reddish bands that lay beneath a coarse gravel, which in turn was below the yellow sand that Wright and Abbott believed to be outwash from glacial streams (§5.15). In a small pocket in the greenish sand, Volk spotted two fragments of what proved to be a conjoinable segment of a human long bone shaft. He marked the place, hustled home to fetch his camera, returned, and photographed the bones in place, then rushed home again and hastily wrote Putnam to announce he had found a "human tibia" in Pleistocene deposits.[14]

Volk believed this find to "be the key to all."[15] Putnam was delighted but was taking no chances: "Do not let anything get out about it until we have gone into the matter more carefully. Be sure that there is no possibility of a mistake as to the specimen's being as old as the deposit."[16] If Abbott was one of those whom Putnam wanted to keep in the dark about the discovery, it was too late. Volk, positively "silly about some great 'news,'" had already appeared on Abbott's doorstep and shown him the specimen. Abbott was uncharacteristically diffident, perhaps as a result of his earlier taxonomic blunder: "It is, I am confident, human, but then so was, I thought, the musk-ox."[17]

Volk kept careful watch over the railroad cut, and a week later near where the long bone had appeared he spotted three freshly broken fragments that proved to refit into a portion of a human parietal.[18] Two days later, Putnam had all the specimens in Cambridge. He too thought the long bone fragment a human tibia, and perhaps a very old one. It even looked as though one end was deliberately cut. He peppered Volk with more questions about how the specimens had appeared when found, then boarded a train the next day for New York, carrying the bones.[19] This time, he was not going to make the mistake of having Russell study them. Instead, he took them to the AMNH, where he would entrust the examination to Aleš Hrdlička, whom Putnam had recently hired under the auspices of the AMNH's Hyde Expedition to the American Southwest.

6.2 ALEŠ HRDLIČKA

Hrdlička was born in Czechoslovakia in 1869 and emigrated with his father to New York in 1881. After a few years in night school, a brief stint working in a cigar factory, and having learned English by visiting churches on Sundays, he enrolled in the Eclectic Medical College of the City of New York. He earned a homeopathic medical degree, finishing first in the class of 1892.[20] After graduation, he established a medical practice on the Lower East Side, but having higher aspirations he enrolled at the New York Homeopathic Medical College in the fall of 1893, graduating in the spring of 1894, once again first in his class. After that graduation he took a position at the Middletown (New York) State Homeopathic Hospital for the Insane, where he conducted research on insanity and began a gradual transition to anthropology. The shift was sparked by his growing interest in anthropometry and medico-anthropology. These were fields then in vogue, particularly in Italy and France, and which subscribed to the idea that psychological abnormalities—defined to include insanity, prostitution, and criminal behavior—were correlated with measurable physical attributes, primarily head shape and size.[21]

Hrdlička saw that identifying physical correlates of behavioral abnormality required knowing something of physical normality, and that in turn necessitated representative samples of the range of variation in cranial and somatic attributes. He also realized he himself had insufficient anthropometric training. This realization prompted him to quit his position and sail for France in January of 1896 to study anthropometry, "pathological anthropology," and related courses at the École d'Anthropologie in Paris.[22]

Founded by Paul Broca several decades earlier, the École was at the time of Hrdlička's arrival headed by Léonce Manouvrier. In general, École anthropologists were convinced of the fact of evolutionary change, though they were indifferent (occasionally even hostile) to Darwin's theory of natural selection to account for it. Rather, they attributed evolution to multiple factors, not least a neo-Lamarckian notion that the environment (natural and cultural) influenced morphology and anatomy.[23] A grudging acceptance of evolution notwithstanding, Broca saw human races as fixed and used craniometry to rank them.[24] Manouvrier broke from his predecessor, taking the more enlightened view that criminals were not necessarily anatomically predetermined but were influenced by social factors.[25] Both were confident that anthropometric methods reliably measured and tracked human types.

Hrdlička's own views were an awkward blend of both: he absorbed some of Broca's racism in seeing "natural causes" for the social and economic disadvantages of nonwhites and Eastern European immigrants, manifested by differences in their cranial shapes and sizes.[26] He allied himself in later years with the eugenics movement, though he was not so blatant a racist to argue, as some did (for example, Charles Davenport, Madison Grant, and John Kellogg), that certain groups should be prevented from reproducing. Nonetheless, Hrdlička was opposed to intermarriage, and not just between races but also between different immigrant classes. His was hardly an uncommon prejudice in turn-of-the century American cities, then bursting with large and newly arrived immigrant populations, nor among immigrants of one generation toward those of another, especially if the earlier and later arrivals came from different regions in Europe.[27]

On the other hand, following Manouvrier (and with distant echoes of Lamarck), Hrdlička also granted an evolutionary plasticity to human anatomy and physiology. As he later expressed it:

> In pigmentation, in stature or strength, in form and size of the teeth and jaws, in the dimensions of all other organs, including the brain, and in his functional qualities and effectiveness, including those of the mental powers, [man] is seen to respond to changing conditions; and these reactions are observed to be the more prompt and effective the higher in civilization and refinement is the individual.[28]

Thus, the key to evolutionary change was intelligence and cultural advancement: "The more man is developed intellectually the more there is of this striving for a higher state."[29]

Regardless of the forces behind evolutionary change, Hrdlička's position had two nonnegotiable entailments: first, the long skein of human evolution was dominantly unilinear, a steady anatomical evolutionary progress through the Pleistocene from *Pithecanthropus* to Neanderthals to *Homo sapiens*. This viewpoint hardly changed over the next several decades despite the Piltdown discovery and Marcellin Boule banishing Neanderthals from the human family tree. Hrdlička challenged both.[30] Second, Hrdlička saw the interaction of heredity, adaptation, and environment leading to gradual but ultimately profound anatomical change over time. Fossil humans should slowly dissolve back in time into a mosaic of increasingly primitive features. The degrees of morphological difference became the hands on an evolutionary clock. His approach came to be called "morphological dating," and by it a glacial-age human skeleton had to look primitive.[31] This was in principle the argument Abbott made in regard to paleoliths, but for Hrdlička it worked, "calibrated" as it was with primitive forms (Neanderthals) known to be

Pleistocene in age, and rooted in the assumption that skeletal evolution was irreversible. (It failed for Abbott because modern people could make primitive tools.) Anatomically modern humans could not be Pleistocene in age (at least not until Piltdown came along and asserted as much).[32] Of course, given the differences of opinion regarding human evolution and when the Pleistocene ended, there was plenty of room for disagreement.

Returning from France in mid-1896, Hrdlička joined the Pathological Institute of New York, where he spent nearly three years amassing anthropometric data on over 11,000 individuals, including, he boasted, "all classes of the insane, the epileptic, the idiot and to some extent the criminal," as well as much comparative data on "normal" populations. Owing to chronic difficulties gaining support for his research and other problems, he grew unhappy in his position and sought a way out.[33]

He escaped into anthropology. Although the details are uncertain, at some point in 1897 he met Putnam, and through him he had the opportunity to study a collection of skeletal material from northern Mexico.[34] Hrdlička presented those results at the 1897 AAAS meeting, where he witnessed the intense debate over Trenton. The following year Hrdlička conducted his first fieldwork among Native Americans in Mexico and was subsequently offered a position by Putnam with the Hyde Expedition, and in accepting the offer Hrdlička turned to physical anthropology.[35]

Over the next three years he participated in four expeditions to the Southwest and northern Mexico, where he amassed anthropometric data on several thousand Native Americans and others. On the strength of the stream of publications that issued from it, Holmes, then the newly appointed BAE chief, engineered Hrdlička's hiring in 1903 as the assistant curator-in-charge of the newly formed Division of Physical Anthropology of the United States National Museum (USNM, part of the Smithsonian Institution). Hrdlička remained at its helm until his death forty years later.[36]

Like Holmes, whom Hrdlička came to deeply admire (the feeling was mutual), Hrdlička was inflexible, largely humorless, and rarely plagued by doubts over the correctness of his views, which he was never modest about expressing. To repeat the observation earlier made by John Swanton, "When you came back to Hrdlička he was always there, just where the Lord created him, on the rock of ultimate Hrdličkian knowledge."[37] To his friends and colleagues Hrdlička could be charming and kindly, but he was not widely popular and intimidated friend and foe alike. Only a very few dared joke with him, Harvard University's Ernst Hooton most notably.[38] Fearsome and self-assured though he was, Hrdlička generally avoided public debate, preferring to make his stand in print and correspondence. Ironically, this avowed empiricist also spent decades with psychics and spirit mediums trying to commune with his long-deceased first wife. That he never did failed to convince him the effort was pointless.[39]

6.3 THE TRENTON FEMUR: A PRELIMINARY LOOK

In late 1899 Putnam asked Hrdlička to examine the newly found bones from Trenton. Hrdlička confirmed they were human, though the long bone was a femur rather than a tibia, and he thought it "in no way especially remarkable," not in comparison to Native American femora in the AMNH collections or to Native Americans he had recently examined. As Putnam had suspected, the femur was cut, perhaps intentionally; a portion of the medullary canal had been gouged out. Putnam supposed the bone had been fashioned to use as an implement handle, a modification he had seen before in bones from Ohio mounds. The parietal fragment was similar in color and condition to the femur, but it was so small that comparisons with various American Indian crania were inconclusive. As for the critical question of their antiquity, the bones themselves said nothing. Hrdlička concluded the determination of their age "must be based principally on their location with regard to geological formation."[40]

To resolve that geological question, Putnam turned as he had before to Wright, who reported that the coarse gravel overlying the greenish sand where the femur was found "indubitably belongs to the coarsest of those laid down by the glacial floods at Trenton." The greenish sand was itself a slightly earlier remnant of those floods, perhaps from a time of the year when the flow of glacial meltwater was diminished. Although Wright did not see the femur in situ, he confidently proclaimed that its placement in the greenish sands precluded the possibility of "accidental or fraudulent burial."[41] Hrdlička's and Wright's reports convinced Putnam that this was, indeed, an important find, and he hastened to fortify the case. He instructed Volk to "say nothing about it yet," but to send him at once

> a little of the sand from the spot where the bone was found. This is very important. When you found the bone did it seem to you that it was firmly in the sand? Was there an impression of the bone in the sand when you took it up? What do you mean when you say there was a "wash out"?

Putnam was concerned: "We must not make any blunders about it."[42]

He was right to worry. Within weeks of the discovery, the news had somehow reached the Smithsonian, where it was received "with scorn and contumely." Thomas Wilson, the Smithsonian's longtime American Paleolithic champion, warned Putnam that the criticisms of his colleagues were becoming dangerously personal:

> What evidence, they do say and will say, have we that Volk in his naturally ardent desire to do something, to show something, all this to justify his continuance in employment, should not have planted this bone to the end that he should afterwards find it with his camera. And they enquire—"Who is Volk, anyhow?"[43]

Wilson stated the obvious: the claim rested on the "faith and credit" given Volk, and that depended on the perception of his "character for truth and honesty." There was little Putnam could do about Volk's reputation, but Volk could bolster the case by providing details of the discovery. Putnam suggested Volk write up his field notes in his native German, so as to more fully and precisely express himself.[44]

This extra effort was necessary, and not just to appease the skeptics: Putnam doubted they would change their minds. It was his perennial concern to protect his reputation and those of the several institutions he represented (§3.2). Even more critical, he had to convince his donors their funds had yielded significant scientific returns. Volk's work was funded by Frederick Hyde, then on the AMNH Board of Trustees. Hyde visited Volk in Trenton in 1899 and been impressed by what he had seen, but that was before the femur appeared. Once its identification was confirmed, Putnam telegraphed Hyde, then en route to Egypt, that "his patronage of the research has come to something of particular importance." Lest the message be lost, he spoke at length on the discovery in his annual report to the AMNH Board of Trustees, closing with the transparent plea that "Mr. Volk's employment in archaeological researches in various parts of the upper Delaware valley may be continued."[45]

By the spring of 1900, Putnam was feeling confident. Doubts might be expressed about whether Abbott's paleoliths were artifacts or manufacturing debris and whether they occurred in glacial deposits. But finding skeletal remains surely enhanced the case for a great human antiquity. What better way to bolster the sagging credibility of those Pleistocene tools than to find the bones of the Pleistocene toolmaker? In his mind, "That implements have been found in the glacial deposits at Trenton there is *no question now*, notwithstanding what some folks say."[46]

The skeletal evidence, however, needed to be more fully developed, for as Putnam announced (with a disparaging glance at Russell), the Trenton bones had not been subject to a "thorough comparative study," certainly not one that reached a satisfactory conclusion, which he believed would prove these were a "type of their own" and "unlike those of the recent Indian."[47] Putnam insisted he was not advocating any specific conclusion, only that he was dissatisfied with the methods Russell had used to reach his. Nor was he unfair in saying so: Russell himself admitted he did not have sufficient comparative material for the analysis. Moreover, Volk had recovered additional skeletal remains since Russell's study. Some were from "Indian burial-places" and thus surely of comparatively recent age, but others were of unquestionably greater antiquity, like the femur. Further examination was needed. Hrdlička was Putnam's choice to do it.

6.4 HRDLIČKA FINDS HIS METHOD

Hrdlička spent much of 1901 on the description and analysis of the Trenton skeletal remains, supplemented by correspondence with Abbott and Volk on their geological context.[48] The specimens included the virtually complete Gasometer, Riverview Cemetery, and Burlington skulls; four small cranial fragments found by Abbott from other localities between 1882 and 1895; parts of five skeletons excavated by Volk in the spring of 1899; and the femur and parietal bone found in December of 1899.[49]

In a pattern Hrdlička was to repeat many times over his career, he first identified the historic occupants of the area. They appeared to be the Lenape, though they were not "autochthonous" to the region. Rather, their tradition and ethnohistoric evidence suggested they had arrived recently, perhaps in the last decades of the fourteenth century. Because their language showed no admixture with other languages, he supposed the region was uninhabited at the time of their arrival. Thus, he inferred that skeletal remains in the region ought to display "a fairly uniform physical type,"[50] and for an anatomical baseline against which to compare the Trenton remains he measured several dozen Lenape crania from the Peabody Museum and the AMNH and obtained data on four Lenape skulls from George Dorsey at the Chicago Field Museum.[51]

Hrdlička used that sample to derive their cephalic index, the ratio of skull width to length, expressed as a percentage. If the index was ≤75, the individual was designated "dolichocephalic" (long-headed), and if ≥80 or greater "brachycephalic" (round-headed); values in between were designated as "mesocephalic". The index was considered a useful marker of race, though by 1902 it was suspect in some quarters. Franz Boas in particular had shown that skull length and width were not necessarily related, and that the index was "greatly influenced by causes other than head length and breadth." It might be a "convenient practical expression," but Boas did not think it expressed "any important anatomic relation."[52]

Hrdlička nonetheless found the index useful and determined that most Lenape skulls were dolichocephalic, as was true of crania from related northeastern groups. Most of the skeletal material collected by Volk from recent "Indian burial-places" around Trenton were sufficiently similar that they too could be "safely considered as having belonged to individuals of [the Lenape] tribe." However, there were a few crania in Volk's collection that were markedly brachycephalic. As that form was common only among tribes that lived west and southwest of New Jersey, Hrdlička attributed their presence in Trenton to "a fight at this location."[53]

There were, however, two archaeological specimens from Trenton that fell outside the Lenape anatomical range and had no counterpart among any of the crania of historic Indians from other areas: the brachycephalic Burlington County and Riverview Cemetery skulls. (The Gasometer skull, though unusual in some respects, was within range of the Lenape type.) Russell had also flagged the Burlington and Riverview Cemetery skulls as outliers, though he believed they were "well within the limits of variation" of the Lenape.[54] Hrdlička thought otherwise. The skulls' great breadth but low height put them "totally beyond the scope of individual variation" and made them a type "totally different from that of the Lenape, or any other Indians from the East or elsewhere of which thus far we have any knowledge." He had no doubts on this score, but no ready explanation either. Two hypotheses came to his mind: first, they were crania of Lenape captives or visitors, though if that were the case Hrdlička had little idea where they originated. Alternatively, they were people who preceded the Lenape in the Delaware valley.[55] That greatly pleased the brethren. Volk alerted Wright that Hrdlička's results were "very interesting and in our favor."[56]

Perhaps. But if the Burlington and Riverview Cemetery skulls were from a "pre-Lenape race," then for Hrdlička their antiquity was a matter "wholly one of geology."[57] Given the long and rancorous dispute over Trenton geology, resolving their age would be no easy task. Worse, the geological context of these finds was at best indeterminate. The Burlington skull had changed hands before it reached Abbott, and he was unable to speak to the discoverer. All he knew was that the crania had rolled out of the bank of a creek running through a field. The River-view Cemetery crania was uncovered by one of the gravediggers, and though Volk went over the ground with the man who found it, its ostensible discovery in "apparently undisturbed sand and gravel" could never be accepted at face value, and everyone knew it.[58]

Hrdlička's report appeared in January 1902. After reading it, Abbott was certain no more convincing evidence of a Pleistocene human presence could possibly be found. Putnam agreed.[59] Yet, if either harbored hopes that the critics would be convinced by Hrdlička's report, their hopes went unrealized. The Burlington and Riverview Cemetery skulls were anomalous (in their brachycephaly), but anatomical anomalies were not necessarily evidence of antiquity, and in neither case was there secure geological evidence to back up the claim.

Still, there was the human femur. Little could be said of its anatomical affinities, but its geological context was certainly suggestive. Even Rollin Salisbury, whose compendium on New Jersey glacial geology had just appeared, admitted as much. But he added the caveat that the femur's Pleistocene antiquity would be accepted only if it was recovered in situ and if the stratum overlying the greenish sands was primary, nonreworked fluvio-glacial Trenton Gravel. Salisbury did not accept either premise, and neither did Holmes, who pointedly observed that this single specimen could hardly support the full weight of claims for a glacial antiquity in the Delaware valley.[60]

That was no surprise. The well at Trenton was poisoned. Any evidence emerging here, whether artifacts or skeletal remains, was inevitably viewed with skepticism and suspicion as deep as Abbott's convictions were strong. As Putnam ruefully admitted, too much controversy surrounded the area and those working in it for it to be otherwise.[61] None of the critics of a great human antiquity in the Delaware valley were swayed by the femur or the other skeletal evidence, save one: WJ McGee. Putnam put the femur on display at the AMNH during its hosting of the International Congress of the Americanists in October of 1902. After examining this and the other specimens from Trenton, McGee admitted privately (and even publicly to a reporter) that this was indeed significant testimony:

> I was . . . impressed by certain crania from the gravels . . . discussed by Putnam, Russell, and Hrdlička, and which undoubtedly present an archaic or primitive aspect suggesting high antiquity; I was still more impressed with a bone [the femur] . . . which, as shown by Mr. Volk's oral and written testimony as well as by his photographs, was taken from the undisturbed Pleistocene deposits. . . . This osteologic material . . . [warrants] reopening the Trenton case for Glacial Man, with so strong a presumption in his favor that the burden of proof must fall on the opposition.[62]

It was a startling victory, but not a significant one, for it had little to do with the merits of the Trenton case and everything to do with McGee's established practice of putting ambition above principle. Just as his prior metamorphosis from American Paleolithic supporter to vicious critic came as he sought to ingratiate himself in the inner circle of Washington science (§4.15, §5.7), his latest conversion occurred as he was fighting a desperate, angry, and ultimately unsuccessful battle with Smithsonian authorities to be appointed John Wesley Powell's successor at the BAE, a position he believed was owed and bequeathed. Holmes was picked instead. The details of the fight are not relevant, save in two respects: first, McGee's prior alliance with Holmes had drawn McGee into the American Paleolithic debate, but the bitter ruin of that friendship between 1902 and 1903 turned him against Holmes and his causes. Second, Holmes's appointment as BAE chief led directly to Hrdlička's hiring at the USNM.[63]

6.5 HOLMES GETS HIS MAN

Holmes himself had little experience with skeletal remains, save a brief 1899 foray into the debate over the Calaveras skull. The approach he took was familiar: this ostensibly Tertiary cranium ought to "show or suggest inferior development" in accordance with its great antiquity. The recent discovery of *Pithecanthropus* was testimony on that score. Yet Calaveras, claimed to be half again as old, was by Jeffries Wyman's assessment no different from the modern crania of the Pacific Coast tribes (§2.4), and this, Holmes believed, "speaks very eloquently against [its] extreme antiquity."[64] As always, "an acquaintance with the historic peoples of a region is the best key to [understanding] the prehistoric peoples."[65]

But so too was an acquaintance with skeletal anatomy, and Holmes had long been keen to see work in physical anthropology developed at the Smithsonian, an area largely neglected under Powell. Within just a few weeks of Holmes's appointment as BAE chief, he began lobbying Smithsonian Secretary Samuel Langley to create a Division of Physical Anthropology in the USNM. He assured Langley that the "many and diverse racial elements" of the nation needed to be studied to promote the welfare of the greater nation. The funds for the position, Holmes explained, could come from Thomas Wilson's salary (he had died the year before). In the ensuing decades, as Hrdlička emerged as a fearsome critic of a Pleistocene human antiquity, the irony of having created his position from Wilson's recovered salary was surely not lost on Holmes. Langley approved a search for someone to fill the position.[66]

Holmes did not search very hard. Having seen Hrdlička's report on the Trenton skeletal remains and heard his presentation on the Lansing skeleton (§6.8) at the International Congress of the Americanists in 1902, he recognized a kindred spirit. He could be confident Hrdlička would aggressively pursue questions related to skeletal remains, for Hrdlička was indefatigable and had already published nearly fifty papers on physical anthropology of Native Americans in less than a decade.

Nonetheless, Civil Service rules required that Holmes administer an examination to Hrdlička and the other two applicants. Unfortunately for the latter, Holmes crafted the exam using material drawn from Hrdlička's publications. The other two never stood a chance. As Holmes confessed, "The position was really made for [Hrdlička] as the only satisfactory candidate available at the time." Boas, with whom Hrdlička had been working at the AMNH for several years, mostly agreed. He considered Hrdlička "a man of very great learning in physical anthropology [though] narrow in his points of view." Boas was sure he would "contribute much material in his special line of work."[67]

Hrdlička's Smithsonian successor, T. Dale Stewart, later argued it was no coincidence that Holmes and Hrdlička approached purportedly ancient skeletal remains in a similar way—namely, their insistence that these should be anatomically distinct from historic Indians of the region. He believed that Hrdlička was merely parroting Holmes, and that when Hrdlička came to the Smithsonian he aligned his approach with his new superior (Holmes). Stewart, in fact, inferred that had Hrdlička stayed in Putnam's employ his subsequent role in the human antiquity dispute would have been very different.[68] That seems unlikely. Hrdlička's approach to skeletal remains after he arrived at the Smithsonian was not different from the one he had taken at the AMNH under Putnam. And he had developed his ideas about skeletal variation and morphological change during his time in France under Manouvrier.

What is striking is how quickly Holmes and Hrdlička became close as friends and colleagues. That was due partly to the similarity in their views, but perhaps also because Hrdlička was hired when Holmes was in the midst of the painful successional BAE fight. McGee, for all his faults, had long been one of Holmes's closest friends, and after McGee was driven off Holmes found himself in the awkward position of being first among longtime equals. With Hrdlička there was no prior history, making for a fresh start. It was a fresh start for Hrdlička as well, whose employment until then had been precariously year-to-year on money Putnam might raise. Holmes offered him financial security, lifetime employment, and near-endless opportunities for research.

Individually Holmes and Hrdlička were staunch critics of a deep human antiquity in the Americas, and together they made a formidable pair. As for Putnam, he could never provide Hrdlička secure employment, nor was he ever comfortable with Hrdlička's approach, even disagreeing with its basic premise. As Putnam put it a few years later:

> The fact that "Kansas Man" [Lansing] had a skull form like some of the Indians only means to me that the Indian type in [North America] is older than some have supposed, but [this] has been my view for a long time as you may well know.[69]

Hrdlička began his Smithsonian career on May 1, 1903,[70] arriving there about the same time as "Kansas Man," which had been found deep in a hillside flanking the Missouri River in the eastern corner of the state.

6.6 CRO-MAGNONS IN KANSAS?

The discovery was made in early 1902 by brothers Michael and Joseph Concannon while digging a fruit storage tunnel into the bluffs beneath their parents' home, which sat on a high terrace of the Missouri River outside Lansing[71] (Figure 6.1). The remains, the nearly complete skeleton of an adult male along with a jaw fragment of a child, were shown to a reporter who published a brief note on their

Figure 6.1 The Concannon farmhouse atop the Missouri River terrace. The wood-framed entrance to the fruit cellar where the Lansing human skeletal remains were recovered by Michael and Joseph Concannon is at the base of the bluff. The photograph was taken by Bohumil Shimek around 1903 or 1904 as he was studying the origin of the sediments of the bluffs. Courtesy of the Smithsonian Institution Archives, Image SIA2014-07066.

discovery in the *Kansas City Star*. That caught the eye of M. C. Long, an archaeologist and curator at the public museum in nearby Kansas City. Long visited the locality and found more human bones, which the *Kansas City Star* in a widely distributed account suggested were perhaps as much as 35,000 years old.[72]

That notice in turn attracted the attention of University of Kansas vertebrate paleontologist Samuel Williston, who visited the Concannon farm in mid-July with Long and then announced in *Science* a few weeks later his "firm conviction that the specimen [represents] the oldest reliable human remains hitherto discovered in North America." Williston was quick to add, however, that the remains were not glacial but postglacial in age, dating to the period of the "so-called *Equus* beds, the time of *Elephas, Mastodon*, extinct bisons [*sic*], moose, camels, llamas and peccaries." Williston, who had trained under O. C. Marsh at Yale University, had been conducting research on the vertebrate paleontology and geology of Kansas for the previous dozen years, so he was reasonably familiar with the Pleistocene deposits of that state.[73]

Yet, none of those extinct mammals were found with the human skeletons. Williston's assessment of their age was based on their stratigraphic position relative to the high water mark of the Missouri River. He reasoned the river had fallen at least forty to fifty feet since the deposition of the skeletons and the terrace in which they occurred, and assuming that downcutting occurred in the "period of depression following the glacial epoch" he concluded the remains must be as old. Williston declined to say anything of their anatomical form, saving that for analysis by a "professional ethnologist."[74]

Williston was to join the geology faculty at the University of Chicago that fall, and he wrote his soon-to-be colleague Thomas Chamberlin to describe the geological situation and ask his opinion on the age of the deposits. Chamberlin demurred, not having had a chance to examine the setting, but he gave this warning to Williston:

> As you doubtless know, the great majority of human relics found in deposits of this class are not really in the original formation, but are in secondary modifications of it, usually very late. These secondary deposits are so deceptive as to have misled many geologists. I assume however that you have sufficiently guarded against this error.[75]

Williston wrote again and this time provided Chamberlin a sketch of the location and more details on the stratigraphy. His second letter only made Chamberlin more wary, as did a letter Chamberlin received the same day from Frank Leverett, who was then working upriver in Iowa and questioned Lansing's antiquity on the basis of his own knowledge of the region. Chamberlin replied to Williston again, lecturing him at length on the complexity of river terraces, the deceptiveness of reworked secondary deposits, and the skepticism that ought to be applied to any claims of a deep human antiquity, given the lack of Pleistocene human remains in America (particularly their scarcity in caves, quite in contrast to Paleolithic Europe) and the many claims for antiquity that had previously failed to pass critical muster.[76]

Meanwhile, word of the Lansing discovery reached Warren Upham in Minnesota. Upham made a quick visit to Lansing in August. Long and Williston were with him, along with Winchell and University of Kansas geologist Erasmus Haworth.[77] They collectively agreed the human remains were in or on coarse sediments that lay immediately below a clean and "horizontally stratified water deposit of fine loess," and as Upham announced in *Science* three weeks later, all further agreed that the bones were

> entombed at the beginning of the loess deposition, which would refer it to the Iowan stage of the Glacial period, long after the ice-sheet had receded from Missouri and Kansas, but while it still enveloped northern Iowa and nearly all of Wisconsin and Minnesota.[78]

If indeed the Lansing remains dated to the end of the Iowan Stage, that would make them two to three times older than the Trenton and Little Falls remains (Table 6.1).

Table 6.1 The Pleistocene glacial and interglacial sequence as seen in the first decade of the twentieth century

Chamberlin and Salisbury 1906			Winchell 1903		
Stage name	Episode	Time (units)	Stage name	Episode	Time in years (units)
Later Wisconsin	Sixth glacial	1	Wisconsin	Fifth glacial	8,000 (1)
Unnamed	Fifth interglacial				
Early Wisconsin	Fifth glacial	2.5			
Peorian	Fourth interglacial		Peorian	Fourth interglacial	
Iowan	Fourth glacial	5	Iowan	Fourth glacial	
Sangamon	Third interglacial		Sangamon	Third interglacial	
Illinoian	Third glacial	8	Illinoian	Third glacial	
Yarmouth/ Buchanan	Second interglacial		Yarmouth/ Buchanan	Second interglacial	
Kansan	Second glacial	15	Kansan	Second glacial	160,000 (20)
Aftonian	First interglacial		Aftonian	First interglacial	
Sub-Aftonian or Jerseyan	First glacial	?	pre-Kansan or sub-Aftonian	First glacial	200,000

Note: Gray = Pleistocene glacial; white = interglacial. Sequence from Chamberlin and Salisbury 1906 and Winchell 1903a. This sequence was virtually unchanged from ones offered in Chamberlin (1896d), Leverett (1899), and Leverett (1902), and was used by Chamberlin and allies as the standard through the first two decades of the twentieth century, though with slight modifications by Leverett and Taylor (1915). See also Tables 8.1 and 9.2.

By Upham's reckoning, in part following Chamberlin's estimate of arbitrary units of time elapsed since the termination of each of the major glacial stages, Lansing could possibly be 20,000 to 25,000 years old.[79] A month later in the *American Geologist* he revised that estimate to between 12,000 and 33,000 years old, suspecting Lansing was closer to the former than the latter. That was substantial antiquity in either case, given that he thought the Quaternary was only about 100,000 years in duration.[80] Upham thought the Lansing cranium looked its age. He described it as "long-skulled [dolichocephalic] with beetling eyebrows, low and receding forehead, and projecting jaws." He considered Lansing to be the "contemporary of the Late Paleolithic men of Europe, who in the Solutrian [*sic*] and Magdalenian development of implement-making . . . were far advanced beyond primitive savagery."[81] As the European sequence was then expressed, that would link Lansing to Cro-Magnon, which was already recognized as Pleistocene in age yet was essentially anatomically modern. Nonetheless, Upham used terms ("low-browed," "prognathous") evocative of Neanderthals and *Pithecan-thropus*, suggesting he was partly describing Lansing in terms of what he thought a Pleistocene-age skeleton ought to look like.[82]

In Upham's view the loess in which the bones were found represented the fine fraction of glacial drift brought there during the summer melting of the vast glaciers to the north. Glacial floodwaters swept down the Missouri and Mississippi rivers and formed a "lakelike sheet" that, after evaporating, left behind a blanket of loess. That blanket extended over the uplands, indicating a high water mark at least 200 feet above the present floodplain at Lansing. Upham granted a small role to the wind in carrying sediment out of the valley onto the higher elevations, but he did not need wind to account for the loess: Upham was an epeirogenist, who envisioned the surface of the earth bobbing up and down like a cork in geological time over a range of roughly 500 vertical feet. Interior North America in the late Iowan glacial stage, having sunk under the load of ice, was lower than at present, and thus glacial floodwaters need not have been that deep to flood present upland surfaces and blanket them in loess. That surface was subsequently uplifted when glaciers melted back. In a brief accompanying editorial in the *American Geologist*, Winchell affirmed Upham's conclusions about both the loess and the cranial form.[83]

Haworth was not convinced. He wrote to several others, Chamberlin and Wright among them, urging them to visit the locality while it was still open and easily studied. As he put it, "Neither Dr. Williston nor myself, nor anyone else in Kansas, is an authority on glaciology, we shall have to look to outsiders for a determination of the age of this loess."[84] That Haworth ostensibly agreed with Upham's interpretation yet issued his request *after* Upham's visit is telling: he wanted to hear from more authoritative sources.

Upham feigned similar concern: on his return from Lansing he wrote both Chamberlin and Holmes reporting his results, requesting that they visit the locality "for verification or correction of our view."[85] His actions belied his words. He had already announced in print that the site was Iowan in age. He was looking for verification, not correction. Chamberlin, who had little patience for unwarranted claims of great antiquity and absolutely none for premature publication, had heard enough. He decided to visit Lansing himself, warning both Haworth and Upham that the determination of the age of a loess so close to the valley floor was at best extremely difficult. Deposits in that setting were subject to erosion and redeposition by river movements. The not-so-subtle subtext of his message was obvious: sufficient care had not yet been exercised, and the question of Lansing's antiquity was open.[86] To Leverett he was more blunt:

> I am very sorry that Upham should repeat his incautious expression of opinion. . . . This demonstration of fundamental untrustworthiness . . . justifies me in not committing to him more important work than I have done.[87]

Chamberlin began assembling his troops.

Three weeks later Chamberlin, his son Rollin (also a geologist), Dorsey, BAE field assistant Gerard Fowke, Holmes, Salisbury, and Samuel Calvin visited Lansing in the company of Haworth and Long. (McGee had been invited as well, but at that moment he was in Nova Scotia on a death watch at John Wesley Powell's bedside, angling for Powell's dying blessing to succeed him.) The guest list was picked by Chamberlin, and Calvin's inclusion is noteworthy: as director of the Iowa Geological Survey, Calvin was especially familiar with the Pleistocene deposits and loess of nearby southeast Iowa.[88]

A newspaper account of the visit by Chamberlin's party, as Upham triumphantly announced to Wright, reported that they believed the Lansing skeleton was indeed at the base of the loess and glacial in age. Upham should have known better than to believe a newspaper, given what Chamberlin had already written to him. Chamberlin later remarked that one of the reporters he spoke to had been "conscientious" about expressing his views, whereas "the other man drew on his imagination considerably."[89] It is easy to guess which reporter's story Upham was reading.

Upham learned what Chamberlin and colleagues actually thought when he and Winchell had a brief and unexpected debate with Calvin over matters they considered already settled. Not all loess, Calvin lectured them, was the same age, let alone of Pleistocene antiquity. Nor was it certain the Lansing skeletal remains were in primary loess as opposed to redeposited loess. Redeposition was always a concern, Calvin reminded them, on slope and valley deposits, where the scars of redeposition can "heal up" so cleanly as to leave no traces. The furthest Calvin would go was that the Lansing skeleton was buried in a loess-like deposit, the age of which was "altogether inconclusive."[90]

Upham had already been warned about redeposition by Leverett, but he and Winchell were untroubled.[91] Upham sent Calvin a long, twelve-point reply (with copies to Winchell and Wright), insisting the Lansing skeleton was in primary loess and therefore Iowan in age. Citing the authority of Chamberlin himself, Upham insisted all North American loess was Pleistocene in origin, having originated in an "aqueous" deposit from the melting of the vast Iowan ice sheet. He knew Calvin thought otherwise; perhaps they should have to "agree to disagree."[92]

Calvin was dumbfounded by Upham's reply. He had stood on the Missouri River floodplain at Lansing just weeks earlier and watched great clouds of dust blowing about, forming loess "before our very eyes." Not all loess could possibly be Pleistocene in age. Moreover, "neither the loess of this region nor of any other region ever shows satisfactory evidence of having been deposited in water. . . . There is no true loess of aqueous origin!" Worst of all, Upham's appeal to Chamberlin's authority was either poor scholarship or disingenuous: Upham was citing Chamberlin and Salisbury's Driftless Area publication from two decades earlier. "I am quite sure," Calvin scolded Upham, "Chamberlin and Salisbury would not now write as they did in 1884–5. The world has moved, knowledge has broadened, opinions have been reconstructed and greatly advanced since then."[93] Upham had either missed that advance or chose not to follow along. Calvin forwarded Upham's twelve-point reply to Chamberlin, who shook his head:

I have read the Epistle from Ephesus [Upham] which you enclose. It wipes away the last qualm of my executive conscience relative to the discontinuance of official work under one who with the opportunities presented is still in the mental condition implied by the category of [12] propositions which he submits to you in this Year of Geologic Science, nineteen hundred two.[94]

A few weeks later, Chamberlin himself wrote Upham that "it would be wise to recognize the fact that my views have undergone a gradual change with increasing knowledge of the [loess] formation."[95] Once again, an archaeological question was about to spill over into a geological dispute.

6.7 ON THE ORIGIN AND AGE OF LOESS

Upham had at least accurately portrayed what Chamberlin and Salisbury wrote nearly two decades earlier. They had favored an aqueous origin for several reasons: the loess was thickest near the main river valleys and areas of glacial drift; it showed distinct stratification and coarser texture (stratified beds of sand and clay) along the river valleys; and it contained occasional aquatic snail shells and gravel (the latter too large to have been transported by wind). Still, they had doubts. The aqueous theory failed to explain the topographic difficulties that the "singular distribution of the loess" presented: the loess-capped hills of Mississippi valley seemed far too high to be within reach of lake or floodwaters. Even more awkward, the loess on opposite sides of the Mississippi River was at different elevations, as it ought not to be if left behind by a receding lake or an overflowing river, the waters of which would have formed a horizontal plane. For that matter, Chamberlin and Salisbury were uncertain as to whether the source was a lake, river, or some odd fluvio-lacustrine hybrid, which had just enough energy to keep the clay afloat and deposit the silt, but not enough to have lifted sand into suspension.[96]

To explain the topographic anomaly the loess presented, Chamberlin and Salisbury explored mechanisms that might have affected the surface level of the glacial meltwater, including ice dams, sea level change, and changes in elevation of the earth's surface owing to crustal deformation under the glacial ice load. Ultimately, they concluded that differential crust depression and not sea level change was an important factor in increasing the area that might potentially be flooded by meltwater, and they allowed for an aqueous origin. But they were not the epeirogenists Upham was: far more information on landscape depression was necessary to "constitute a sure groundwork for belief." In the end, although they favored the hypothesis of an aqueous origins, they were uncomfortable about it.[97]

They maintained their aqueous-origins theory into the 1890s, continuing to admit its anomalies, which now included the fact that loess deposits produced more land snails than aquatic ones. But then anomalies to a theory can always be explained away by clever advocates, and these two were very clever: "It may be much more troublesome to account for the presence of even a few aquatic shells in an eolian deposit," Salisbury rationalized, "than for the presence of many land shells in a water deposit."[98]

By the late 1890s their opinions had begun to shift, largely because of the nagging difficulty of explaining, as an exasperated Chamberlin put it, that "special fondness [of loess] for summit heights." No matter how much he tried to conjure a mechanism by which glacial meltwater could reach such heights and yet do so unevenly, the idea that it had was a "severe tax upon belief."[99] Chamberlin, who by then emphatically rejected epeirogenesis,[100] could not conceive of any geological processes that could provide what he needed to make the aqueous theory work. He was approaching the limit of how far he would go in defense of a theory.

Pushing him toward that breaking point was Bohumil Shimek, a malacologist and botanist at the University of Iowa who, beginning in the mid-1890s and with a confidence that grew into arrogance, published a string of papers attacking the aqueous theory. Armed with data gathered in snail-collecting forays in loess from Iowa to Mississippi, he showed the loess was dominated by large numbers of land snails "out of all proportion" to the far fewer numbers of aquatic species. That lopsided relationship was just the reverse of what ought to be true were loess and its snails deposited by water, especially as aquatic snails are more prolific than their terrestrial brethren.[101]

Moreover, the majority of the few aquatic snail species Shimek recovered were air breathers, whose presence could easily be explained by small pools or streams on the loess surface. Fully aquatic, gill-bearing snails were exceedingly rare, and their occasional presence could readily be explained as their having been deposited by birds or small mammals that fed on these species. Shimek spotted tooth marks of small mammals on some shells, and the presence of the fragile opercula in land snails indicated they had not been transported significant distances by any agency. In contrast, the land snail species recovered were virtually all herbivorous and required substantial plant cover, and thus Shimek envisioned a landscape shrouded in vegetation, which trapped wind-blown dust carried from more barren regions. And because most of Shimek's snails were extant species that lived in temperatures not significantly different from those of the present and in relatively dry settings, the climate of loess deposition was relatively moderate, not glacially cold, and sufficiently dry to create perfect conditions at least part of the year for "clouds of dust . . . to be taken up by the winds."[102]

By the late 1890s Chamberlin was familiar with Shimek's results and began to retreat from his half-hearted support of the aqueous theory.[103] He at first backed away slowly, crafting a hypothesis that invoked both water and wind. Chamberlin's hypothesis assumed a very low slope to the land surface and therefore a widespread flood of glacial meltwater, one that left behind vast mudflats of sediment derived from glacial grinding. As the landscape dried and was exposed to the corrosive effects of the wind, the silt fraction was lofted up onto the surrounding terrain.

> The material thus derived would be essentially identical with the glaciofluvial deposition, and thus the hypothesis seeks to account for the glacial element in the constitution of the eolian portion of the loess. The presence of land mollusks in the upland eolian loess finds in this way a ready explanation, while their presence in the lowland loess mingled with aqueous mollusks finds an almost equally obvious elucidation.[104]

Shimek saw an easier explanation: the abundance and type of snails simply reflected their proximity to bodies of water and the amount of plants and shade;

that was true of the present, and it ought to have been true in the Pleistocene.[105] Chamberlin admitted his hypothesis that loess was primarily aqueous and secondarily aeolian was an unhappy, even "Janus-faced" compromise, and Chamberlin was one who preferred more elegant and parsimonious solutions to geological problems. But then the loess was proving an especially obstinate geological problem. At least for Chamberlin.[106]

Not so for Wright. He had long attributed loess to an aqueous origin,[107] and even a trip around the world in 1900–1901—which included a long traverse of Asia, where he had the chance to examine Chinese loess firsthand—failed to convince him otherwise. Still, it briefly appeared that he might change his mind. In a letter to *Science* sent while en route, Wright agreed with German geologist Baron Ferdinand von Richtofen (not to be confused with his nephew, Germany's World War I flying ace the "Red Baron") that the vast loess blanket of northern China was was "a wind deposit without doubt."[108] Supposing Wright might be coming around to his point of view, Shimek had several of his Iowa loess papers waiting for Wright on his return. Wright responded in kind but also quizzed Shimek on why the North American loess could not have been deposited by water. Shimek, thinking he was preaching to the near-converted, laid out the snail evidence and the impossibility of glacial or aquatic agencies being responsible for it. But Wright did not budge: could there be some role for the action of the water? Shimek, replying with a noticeably sharper tone, only granted the Mississippi River the task of carrying sediment downstream, where it accumulated on bars where, at times of low water level, it was then airlifted onto the uplands.[109] The upland loess itself was not deposited by water. On this point Shimek saw no room for discussion. As to its age,

> I have found nothing thus far in our American loess which would appear inconsistent with the glacial origin of the material. At least that seems to be its source. It may have been re-washed or transported, but I believe that all of it here can be [ultimately] traced back to the drift.[110]

Shimek was unable to convince Wright of the aeolian-origin hypothesis. But then it may not have yet been obvious to Shimek or anyone else why it mattered (and mattered deeply) to Wright that the loess had an aqueous origin. Everyone found out soon enough.

After his return, Wright elaborated on loess origins in the *Bulletin of the Geological Society of America*, where he willingly granted that winds blowing off the Gobi Desert contributed to the Chinese loess formation. Yet, he was unconvinced wind alone could account for its widespread accumulation and the associated terraces of sand and gravel that blanketed valleys between the mountain chains. That seemed to him to bespeak deposition by water. But how and from what source? Wright conjured a solution in an enormous inland sea, one that reached from the present Black Sea in the west and lapped up against the Altai Mountains on the east. When that sea dried, it left behind a vast expanse of loess.

That sea had been brought into existence, he supposed, by epeirogenic tilting as Asia and Europe "play[ed] see-saw," the one subsiding as the other rose.[111] That inland sea and the loess it left behind were not very old, judging by apparent references to it in Chinese histories and the relative lack of erosion of its surface. There must have been a rapid, postglacial lowering of the earth's crust that enabled the sea to flood across the sunken landscape of Asia and drown the land in several thousand feet of water. There Wright abruptly stopped. The *Bulletin of the Geological Society of America* was not the place to show all his cards, though readers of *McClure's*, a literary rather than scientific magazine, had already seen them. There Wright announced that the vast inland sea that had recently drowned Asia was none other than Noah's flood, and the loess its sedimentary imprint. It was a bold and reckless stroke for a geologist, though perhaps not for the chair of the Harmony of Science and Revelation at Oberlin (§5.8). Yet, in his attempt to harmonize scripture and geology Wright violated one of geology's sacred canons:

> [The] instability of the earth's crust [shows] that the present cannot be made a measure of the past. Man has certainly witnessed catastrophes by flood which are quite analogous to the one described in Genesis.[112]

Of course, Wright was far less concerned about abandoning Charles Lyell's uniformitarianism than he was with reassuring *McClure's* readers no Biblical canons were being violated. Nothing Genesis said required that the floodwaters were over the tops of the highest mountains (as this flood clearly was not), or that Noah's flood was necessarily a universal phenomenon. The Bible could well accommodate the geological evidence.

It could also accommodate the archaeological evidence: Wright saw a "sharp line of distinction between Paleolithic and Neolithic man," so much so that they appeared unrelated. Perhaps that was more than mere happenstance. Paleolithic peoples had spread out across the globe (reaching America, of course), only to be caught in the "extreme and rapid changes" at the closing of the glacial period that "exterminated man in company with many of the animals accompanying him both in America and in Europe." The "Paleolithic man of science," Wright triumphantly concluded, "may well be the antediluvian man of Genesis."[113] Noah, it seemed, begat the Neolithic.

That announcement confirmed Chamberlin's darkest suspicions about Wright's competence and motives, and within weeks he attacked him anonymously in *Science*. Wright, he growled, deliberately ignored contrary evidence, notably the absence of marine fossils from the supposed sea, the presence of steppe fossils, and the work of Richtofen and others who had demonstrated an aeolian origin for loess. Wright was once again misleading the public, presenting old and hotly disputed claims as new and remarkable. Chamberlin was not buying— not the geology, not the archaeology, and especially not the harmony of Genesis and geology: "One is led to wonder how far respect for the Scriptures is fostered by 'remarkable discoveries' of this sort and by the much trumpeted stage-play that preceded and accompanied them."[114]

If Chamberlin needed another push in the direction of the aeolian-origin theory of loess or another reason to loathe Wright, this episode provided it. Yet, if Wright cared a whit about Chamberlin's approval or anyone else's, he did not show it. He reprised the theme of the geological evidence for the Noachian deluge the following year in *Bibliotheca Sacra*. He was not afraid, as he later put it, to make "some religious use of my [geological] investigations," even though he believed this led to prejudice against him.[115] And there positions stood on the eve of the Lansing discovery.

6.8 LOESS AND THE LANSING MAN

After the initial round of visits to Lansing in the summer of 1902, Holmes arranged for Fowke, then excavating in the "well-known fossil bone-beds of Kimmswick, Missouri,"[116] to dig a series of trenches at Lansing to better expose the stratigraphy. Up to that point, all observations had been done under artificial lighting in the Concannon's tunnel, hardly the best circumstance in which to assess the deposits. Following Fowke's trenching, parties returned to Lansing that fall for another look.[117] Proponents continued to insist the Lansing remains were beneath primary Pleistocene loess of Iowan age. Critics thought the opposite.

But precisely what critics thought was not immediately known to Upham, and it made him anxious. He suspected that Chamberlin, Calvin, and probably Holmes disagreed with him. Anticipating their reaction, he repeatedly urged Wright—who was standing far too quietly on the sidelines for Upham's liking—to visit Lansing and "enter the discussion" with an extended consideration of loess origins.[118] As a preemptive strike Upham published a flurry of papers on Lansing.[119] These contained few surprises and added little to what he had already said in *Science* or the *American Geologist*. But he believed either that no one would notice the absence of new data or that reaching as wide an audience as possible justified the repetition. Chamberlin, as Upham soon learned, noticed the lack of new data and objected to the repetition.

When Upham sent his *American Geologist* paper to Chamberlin, along with a copy of his twelve-point reply to Calvin, he fished for a hint of what Chamberlin's forthcoming paper on Lansing's geology might say. Chamberlin already knew. He had drafted the paper a few weeks after his visit to Lansing and already sent copies to Calvin, Holmes, and Salisbury for comment. Of course, Upham was not a member of Chamberlin's inner circle. He got only a brief, unpromising preview:

> I do not propose to discuss the origin of the loess in my paper on the subject because in my judgment the deposit which overlies the human relics in Lansing is not loess, nor of the Iowan age, and does not require discussion of the question.[120]

That Chamberlin deemed the loess question irrelevant did not bode well.

Upham pressed him on the matter and sent Chamberlin still more of his papers on Lansing, loess, and human antiquity in the New World. The effort backfired. The ever-righteous Chamberlin erupted with a long, caustic reply, excoriating Upham for violating the canons of scientific ethics in a "wholly inexcusable way" by presenting to the public as fact matters that were at the very least debatable. It was a grisly repeat of Chamberlin's attack on Wright of a decade earlier, only this time Chamberlin kept the assault mercifully private. Still, it was just as ferocious. Chamberlin told Upham that his repeated claims of secure evidence of Pleistocene human presence on this continent were "nothing short of reprehensible."[121] Chamberlin apologized for speaking "with vigor" but made no apologies for having spoken. He took the defense of science personally, and Upham was getting no less than what he deserved. That letter permanently ended all but the most formal correspondence between the two.[122]

When Chamberlin's paper on Lansing appeared in the *Journal of Geology*, it duly acknowledged Upham and Winchell's claim that the Lansing skeleton was buried by Iowan loess, but true to his word Chamberlin completely ignored the question of loess origins, opening instead with a lengthy "academic" discussion (his term) of the "science of river action," the understanding of which was "critically involved in the interpretation of the Lansing remains." As this thoroughgoing uniformitarian geologist insisted:

> It is a vital principle of good practice that the agencies and phenomena nearest at hand [fluvial action] be first considered, and, if the case requires, be eliminated, before recourse is had to more remote agencies [Iowan-age loess]. This is peculiarly true when, as in this case, the agencies closest at hand in time have quite certainly swept away the most of a more ancient record in making their own.[123]

The scouring, erosion, reworking, and redeposition of sediments, especially at the mouths of small tributaries where they met large and active rivers, was precisely the topographic situation at Lansing. It resulted in complicated and deceptive geology, as Chamberlin had lectured Williston about two months earlier, and made it doubtful that deposits in that topographic position could be pure loess preserved since Iowan times. Chamberlin, Upham, and Winchell all agreed water was involved in the formation of the Lansing deposit—but for Chamberlin it was as fluvial scour and fill, whereas for Upham and Winchell it led to deposition of loess under glacial meltwater lakes.

Chamberlin freely admitted there was Iowan loess in the region; some upland surfaces were blanketed by it. But the mottled silt in which the Lansing skeleton was buried was not the same as the Iowan loess. Instead, it was highly variable in color, texture, and effervescence, as true loess was not. Nor had it been laid down by a stream, lake, or other "subaqueous" process, for (as opposed to Upham's belief) it lacked lamination, distinct stratification, and size sorting. There was a thin clay band within it, and though undoubtedly water-laid, it was discontinuous and most likely the result of a small puddle of standing water. As Salisbury added, "I have seen thousands of sections of loess, but never one with such a seam of clay." Most revealing about the origins of this deposit was its abundance of shale and limestone fragments.[124] That debris originated in the terrace bedrock on which the site rested; it could not have been brought in aloft or afloat.

For Chamberlin the Lansing loess was no loess at all but a "wash product" of loesslike silt carried from the uplands and blown in by winds from the Missouri bottoms, mixed with detritus eroded from the adjacent slopes and water-laid clay and other debris that washed in at times of high water.

> Very similar deposits seem to have been formed at all ages since the main loess epoch, and are being formed now, and apparently must continue to be formed as long as the general loess mantle remains the chief source of erosion and redeposition, but these deposits generally betray their origins by their secondary characters, as in this case.[125]

Such deposits readily formed as the Missouri River meandered in its course, at times higher and closer to the site, other times lower and more distant. That action would have had consequences for the gradient of the tributary and the slopes flanking the valley, and it led to erosion and redeposition from the valley walls onto the valley floor. The unmistakable geomorphological evidence that the slopes and deposits at the site were "pretty closely adjusted to the Missouri River bottoms which are features of recent origins" could mean only that the deposit in which the skeletal remains were found and buried was also recent. Thus,

> the antiquity of the burial is measured by the time occupied by the Missouri River in lowering its bottoms, two miles more or less in width somewhere from fifteen to twenty-five feet, a very respectable antiquity, but much short of the close of the glacial invasion.[126]

Chamberlin favored this interpretation because it involved a minimum of assumptions and neatly conformed to the "natural order of things."

Still, Chamberlin, champion of multiple working hypotheses (§4.15), duly acknowledged there were alternative, though less probable, explanations. One of those, of course, was Upham and Winchell's, but their alternative failed because the Lansing deposit was not true loess but a secondary deposit, and its age was "that of its derivation, not that of the parent loess."[127] So Chamberlin said publicly. Privately, he sneered that Upham and Winchell's position was "pitiably weak and incompetent,"[128] even insulting:

> You [Leverett] and I and Calvin and the rest have been trying for years to find some reliable means of correlating the Iowan [loess] on the Mississippi with its equivalents a little east of the Mississippi without success, and we know of no way of placing a distant formation in the Iowan horizon, but our northern brethren need only a half day's inspection to do this.[129]

Chamberlin's paper came with its own Greek chorus. Calvin and Salisbury, each of whom was in "perfect accord" with Chamberlin, appended statements of approval.[130] Salisbury also took the opportunity to rebut the one piece of evidence recovered by Williston on which proponents of an aqueous origin had placed so much weight: a unio (clam) shell found in the wall of the original tunnel near the skeleton. There was no known instance in which a unio shell had ever been found in undisturbed loess. Fowke's much more extensive trenching had failed to produce a single unio, only land snails. Not wanting to impugn the field skills of their new colleague at Chicago, Salisbury and Chamberlin took Williston at his word that the unio was in situ when found, and they suggested the shell had been brought in by some agency other than floodwater—humans, perhaps. But if so, that was in recent geological time, "much short of the close of the glacial invasion."[131]

As for the bones themselves, Hrdlička examined the Lansing remains in early October of 1902. He was then en route to Mexico, so he sent his "somatological notes" to Dorsey, who presented them on his behalf at the meeting of the International Congress of the Americanists later that month, where both the Lansing skull and Volk's Trenton femur were on display.[132] Precisely what Hrdlička said about Lansing on that occasion went unrecorded, but press reports from the meeting gave the essence of his remarks, and they were enough to distress Upham. He learned that Hrdlička (and Dorsey) believed Lansing was the same morphological type as the modern Plains Indians and thus, presumably, of no great antiquity. Yet it seemed to Upham, who wrote Hrdlička to clarify the point, that such a similarity was

> not inharmonious, but fully accordant, with its having an antiquity of 10,000 to 15,000 years or more; and I believe, with the late Major Powell, that men have lived in America for longer than that time . . . and they may have attained nearly their present racial character, as to the skeleton, fully 15,000 years ago, the time that seems to me the most reliable estimate for the age of the man at Lansing. Is not this acceptable to you?[133]

Hrdlička, by then in Mexico, did not answer.

The realization the Lansing battle was about to break out on a second front of human morphology and evolution prompted Upham to fire off letters to Chamberlin, Dorsey, Holmes, McGee, Winchell, and Wright (among others) urging Lansing not be "prejudiced by the near affinity with the present Indians."[134] In each he appealed to the "sagacious opinion of Major Powell" that humans arrived from the Old World very long ago, "probably 50,000 or 100,000 years ago," as though Powell's almost-two-decade-old writings were, like Chamberlin and Salisbury's mid-1880s' ideas on loess, forever authoritative. Upham's pleas to Wright to get involved in Lansing became even more urgent.

For his part, Holmes must have been delighted to hear the results of Hrdlička's examination, for he had independently studied the Lansing skull and concluded it "corresponds closely in type with crania of the historic Indians of the general region. It presents no unique features and offers no suggestion of great age or of inferior organization."[135] The antiquity of Lansing had resolved itself as a matter of geology rather than anatomy, and as Holmes followed Chamberlin's interpretation of the geological history of the site he was confident on this score as well. Holmes erred, however, in believing Chamberlin's "mastery of the intricate problems of glacial and post-glacial geology is everywhere acknowledged."[136] Upham sneered at that "worshipful remark" about Chamberlin[137] and felt confident that he and Winchell, not Chamberlin, were right about Lansing's geology. Upham began to craft a response.

Winchell already was at work on his. As president of the GSA he had the bully pulpit of the banquet address at its 1902 December meeting. Although his announced title, "Was Man in America in the Glacial Period?," implied a wide-ranging essay, his focus from the outset was Lansing. To anticipate what he might be up against, Winchell asked Chamberlin for his assessment of the Lansing situation. Chamberlin advised him that

> the deposit at Lansing cannot . . . properly be referred to typical subaqueous deposition, but is the combined product of aqueous and alluvial agencies. In my judgment the deposit is not true loess, for it does not bear the typical characteristics of loess. Nor does the evidence of its topographic relations support the view that it was deposited during the main loess period, that is, the Iowan. In my paper [forthcoming in the *Journal of Geology*] I will set forth explicitly the mode in which I think the deposit was formed, and this I trust will be in your hands before you will need to complete your presidential address.[138]

That Winchell was able to elicit a preview from Chamberlin and not the harsh scolding that Upham received reflects their very different relationship. Winchell was Chamberlin's senior, never his employee, and as director of the Minnesota Geological Survey he held a position of comparable status. Besides, he was giving the GSA's presidential address, and Chamberlin as much as anyone understood the propriety in such matters. As he explained to Holmes, "I thought it fair to say this much beforehand."[139]

Chamberlin knew Winchell well enough to realize there was little hope of influencing his views. Still, he hoped Winchell would not altogether ignore what he had to say. In that he was disappointed. Nowhere in Winchell's address was explicit mention made of Chamberlin's discussion of Lansing, which by then had appeared in print. Yet, any reader could see Winchell was aware of what Chamberlin had written, and in fact his address was laid out in parallel fashion to Chamberlin's *Journal of Geology* article.

However, instead of opening with a lengthy discourse on river action, which Winchell deemed irrelevant to Lansing, he began with a discussion of the "acute" problem of the loess and its origins. The fix was in from the start. Winchell insisted loess was a single formation that "has not been found to be separable, so far as the latest surface deposit is concerned, into different parts as to date or mode of origin."[140] That gambit made the explanation of its origin far less complicated.

Nor was much nuance apparent in how he laid out the possible explanations for loess origins. The way Winchell saw it, accepting the aqueous hypothesis necessarily required rejecting the aeolian hypothesis. Not for Winchell was there the Janus-faced compromise of wind and water that Chamberlin was earlier willing to grant (§6.7). Nor did he make any effort to disguise the hypothesis he preferred. He scarcely bothered to "show the weakness of the eolian hypothesis," preferring instead to cite, in a patently passive-aggressive stab, Chamberlin and Salisbury's 1885 support of the aqueous hypothesis.

As a vigorous advocate of the aqueous hypothesis, Winchell was not inclined to cross-examine it too closely. He was not bothered by the anomalies that had troubled Chamberlin. Winchell claimed all loess throughout North America was or originally had been stratified and water-laid. Saying that the deposit was once stratified but no longer was might avoid a few trouble spots, but it was a weak strap on which to hang a hypothesis. Nonetheless, Winchell believed faint traces of one-time stratification could be seen in many sections in the layered arrangement of snails (testimony of a now vanished surface), the horizontal and vertical fissures along which the loess broke (evidence of a weakness of cohesion along stratification planes), and the "indistinct fine lamination" that appeared on weathered surfaces. Winchell never asked whether such stratigraphic features could occur in an aeolian deposit; he was satisfied supposing they would occur in an aqueous one. But as Shimek later responded, water deposits are often laminated, but not all laminated deposits formed in water.[141]

The striking scarcity of aquatic snails was no more of an obstacle than the lack of stratification: "Many land [snails] may exist in an aquatic formation, but the *existence of a single aquatic fossil species in the loess requires the presence of water*." This was an "extremely rash" statement, in Shimek's view, given the known instances of aquatic snails dropped on loess hills by birds.[142] Even so, it raised the question of why so many land snails were in a water-laid deposit. Winchell supposed the snails had foraged out across the loess floodplain soon after water levels dropped below the surface.

Just as Chamberlin's "academic" discussion of river action set up his assessment of Lansing, Winchell believed his loess discourse cleared the way "for the rational and intelligent" consideration of the site (implying rationality and intelligence had been lacking up to that point).[143] He saw at Lansing several feet of "geest," clay weathered in situ from the limestone bedrock, overlain by a thin and discontinuous clay lens, then Pleistocene loess. The skeleton was in or on the geest (its precise position uncertain) and covered by the loess. Winchell granted that the Lansing deposit was an atypical loess, being a mix of silt and drift, such as he supposed occurred near the margin of the ice sheet in an area covered by stagnant water. In saying it was atypical, however, he was not abandoning his belief that loess was a singular phenomenon. Instead, he was asserting that because loess was a by-product of silt-laden water and glacial debris, the texture of a loess deposit could vary depending on the distance to the ice margin and the relative contribution of each. Although "a product of the miscegenation of ice and water, a hybrid which has no name, but whose parentage is well known,"[144] loess by any other name was still loess, so it must date to the Iowan stage.

Still mimicking Chamberlin's essay, Winchell closed by listing other possible hypotheses to explain the Lansing site deposits and age—but dismissed them all. The rock fragments in the sediment precluded this being an aeolian deposit (Chamberlin would have agreed); and the horizontal stratification precluded the deposit being a result of slope wash of drift and loess (Chamberlin would have disagreed). There was but one conclusion: "Man existed in North America at the time of the Iowan epoch of the ice-sheet." It was a conclusion he believed well fortified by the many other finds made over the years of human artifacts in glacial deposits of North America.[145]

There is no published record of any discussion that followed Winchell's address that evening at the Willard Hotel. Conferees retired afterward to a "smoker," but at some point amid the cigars and liqueurs a furious Chamberlin confronted Winchell:

> I took Professor Winchell sharply to task, in private, for his quotation of my early [1885] views . . . without being fair enough to his hearers and read-ers to indicate that I had abandoned the view which he adopted, so far as the main loess was concerned. He promised to modify his statement in the printed copy, but whether he has done so or not, I have not taken the trouble to see.[146]

Winchell did modify his words, though only by the addition of a footnote that admitted Chamberlin's thinking on the loess had evolved. It was a hollow gesture. He never changed the main body of the text in which Chamberlin's earlier work was proclaimed a "classic" of American glacial geology that "can not lightly be set aside."[147] Chamberlin had been hoisted by his own obsolete petard.

6.9 REMEDIAL LESSONS

Winchell's closing remarks about the wealth of sites testifying to a Pleistocene human antiquity grated on Chamberlin. He had long brooded over what he saw as the appallingly low standards by which geological claims for human antiquity in the Americas were judged and the naiveté of archaeologists in so quickly ac-cepting them. If "Putnam spreads glacial man finally over the American map," he snarled that fall of 1902, "he is merely making a spectacle of himself":

> There is perhaps no help for this, but it is greatly to be regretted. There is in America *not even a reasonable approach* to a good case of human remains in true glacial deposits.[148]

Proponents obviously thought so and further believed that because there were so many apparent glacial-age sites in America—"collateral circumstances and evi-dence," Winchell called it—they gave benefit of the doubt to all claims of a Pleis-tocene human antiquity.[149] If Paleolithic people were at Trenton, then they were likely at Lansing. Chamberlin deemed that nonsense: "No amount of multiplica-tion of *bad cases*," and he believed Trenton a bad case, "makes a *good* case."[150] In thinking thus, he was hardly alone.

That proponents and opponents of a great human antiquity failed to agree on the legitimacy of any claim, or even the criteria for identifying an ancient site, was nothing new. This dispute had boiled for more than a decade without consensus on standards and methods for judging early sites. It was doubtful that consen-sus was possible. Still, most agreed that determining the age of these sites was a geological problem, and as such Chamberlin believed it ought to be amenable to a geological solution. It is assuredly no coincidence that a few weeks after exploding at Upham for proclaiming there was abundant evidence of a Pleistocene human presence in the Americas, Chamberlin decided to speak at the upcoming AAAS meeting on the "Criteria Requisite for the Reference of Relics to the Glacial Age."[151] The meeting was held jointly with that of the GSA, and thus his presentation came just three days after Winchell's banquet address. Chamberlin never named names, but his presentation would have been readily understood by all as a rebuttal aimed at both Upham and Winchell.

Chamberlin's paper, subsequently published in the "Studies for Students" section of the *Journal of Geology* (§5.12), was less a discussion of criteria and more an assessment of the types of geological settings most likely to yield secure evidence of a Pleistocene human presence. Glaciated regions were the best places, where artifacts were embedded in, say, till sheets, moraines, kames, or eskers—assuming those artifacts were in primary context and had not themselves been redeposited. The case would be even more robust if those artifacts were striated by glacial ice. Equally compelling would be remains recovered from interglacial soils, peats, or other formations subsequently buried by later glacial deposits. In all cases, there was always lurking that "bête noire of the incautious glacialist, secondary displacement and rearrangement" of the deposits. But he supposed a "duly trained and fairly skillful" geologist could spot those.[152]

The trouble began when a geologist ventured even a few miles beyond the terminal moraine, where nonglacial deposits readily imitated glacial ones and could contain what might appear to be glacial-age outwash debris but which was instead from postglacial redeposition. There were several variations on this theme, and Chamberlin described the geomorphic processes behind reworking and redeposition, where it would occur (primarily along major valleys), how it would be manifest (differential weathering, rapid and deep burial of relatively young materials via scour and fill, and so on), and why it might not be readily detectable (postglacial sediments can be "indistinguishable in kind and structure from true glacio-fluvial material"). Here, only the "most critical glacialists" would be able to determine if a site was acceptable or not.[153]

Interpreting geological settings became even more challenging as one ventured still farther from the ice front, where even "a careful glacialist would hardly venture to discriminate." Here one encountered troublesome circumstances such as inverted deposits caused by bluff edges slumping (like the Trenton talus [§5.10]) or "recomposed" secondary deposits, as these might appear in the mouths of tributary valleys. In illustrating the latter he did not mention Lansing by name, but in near-recitation of his just-published *Journal of Geology* paper on the site he described precisely what he believed were the geological processes responsible for its deposits: formations in which older sediments were later repackaged in a way that "the new aggregation must necessarily bear a close *general* resemblance to the original loess and may easily be mistaken for it."[154] Beyond that, he said surprisingly little about the difficulties of identifying glacial-age remains in primary loess. (He regretted that omission a few years later, when human remains were found in what appeared to be loess deposits in Nebraska.[155])

Chamberlin made no effort to hide his opinion of how challenging it was to discern antiquity in these more complicated geological circumstances, nor of the skills a geologist needed to see through nature's tricks. He counseled "sagacity, guided by experience and controlled by the most diligent circumspection." Even that might not be enough. Some circumstances—and he said this just before turning to what his audience knew was Lansing—were the "special danger-ground of the unwary and are liable to deceive the very elect." Damning with faint praise, he was. Upham and Winchell had ventured into the geological thicket at Lansing without due diligence and had failed to see the ground properly. They might be good geologists, but they were not capable of surmounting problems that test even the elect.[156]

Winchell and Wright were in his audience, but only Wright challenged Chamberlin afterward. No record exists of what was said then or in the dialogue between Chamberlin, Winchell, and Wright that followed Wright's paper on the Missouri valley loess later that day. Wright did report, "So far as I could see Winchell had the best of it." But then one would scarcely imagine Wright saying anything else.[157] It is a safe bet Chamberlin had a different view of the outcome.

On reading Chamberlin's paper, Upham dismissed it as merely "another case of special pleading and casuistry" and mistakenly supposed Chamberlin wrote it to "retain McGee in his camp."[158] Upham obviously did not know Chamberlin well. It had been years since McGee mattered to Chamberlin (if he ever had), and McGee certainly did not matter at that moment, given Chamberlin's anger at him over the BAE succession fight then unfolding. Chamberlin's loyalty was always foremost to Holmes. More surprisingly, Upham failed to realize he himself was one of Chamberlin's targets.

Nor was it just Upham, Winchell, and Wright that Chamberlin was aiming at. He closed by assessing the reliability of using associated vertebrate fossils to establish a Pleistocene human antiquity. It was an odd topic to bring up in a paper otherwise devoted to surficial geological deposits, especially for one like Chamberlin who had never previously shown interest in fossil remains. Nor was it included to make his discussion comprehensive. After all, he said nothing about criteria by which a "paleolith" could be recognized as a finished tool as opposed to a quarry reject, or about what a Pleistocene-age human crania ought to look like. Rather, Chamberlin was speaking, again without naming names, to a specific fossil claim recently made by one individual. Once more his audience would have filled in the blanks, especially the one about whom the discussion was directed: Williston, Chamberlin's new faculty colleague at the University of Chicago.

Five years earlier, buried deep in one of his Kansas State Geological Survey reports, Williston gave a brief report on a find of "arrowheads associated with the remains of *Bos [Bison] antiquus*" along the banks of 12 Mile Creek in the western part of the state. He believed it evidence of the contemporaneity of "man with the Equus [Pleistocene] fauna."[159] Few paid that report any mind. In October of 1902 Williston tried again, this time at the International Congress of Americanists. There it was noticed. Upham solicited an article on the site for the *American Geologist*. In it Williston reported that two of his experienced field assistants and fossil collectors, Handel Martin and Thomas Overton, had excavated a bison bone bed of approximately ten animals along the banks of the creek over a two-year period (1895 and 1896). The second year after flooding exposed more bones, an "arrowhead" was spotted beneath the right scapula of one of the largest bison skeletons. It was seen in place by both Martin and Overton as well as a bystander.[160]

There was no possibility in Martin's mind—and Williston considered Martin a reliable witness—of there having been an "accidental intrusion" of the point with the bones. It must have been "within the body of the animal at the time of death." The deposit with the bones was buried by twenty feet of "Plains marl," which not far from this locality yielded the remains of "*Elephas* [*Mammuthus*] *primigenius*," marking the fauna of the Pleistocene age *Equus* beds (§6.6).[161]

Upham was thrilled by Williston's report, convinced as he was that the 12 Mile Creek discovery was "fully as old as the 12,000 years ± which I assign for the Lansing skeleton." He immediately wrote to Wright with the news, sure he would not permit "that light to be hidden." Wright hardly ever made mention of the discovery.[162] Chamberlin did in his "Criteria" paper, albeit obliquely. It was, he argued, no easy task to determine the age of fossils from geological formations that were not physically connected with glacial deposits and hence could not be directly correlated with Pleistocene stages. There was no question that specific taxa did mark certain geological periods and not others, but the complication was one of temporal scale:

> General inferences and faunal aspects may be applicable to periods measured by hundreds of thousands or millions of years, and yet be quite too broad to discriminate epochs of a few thousands or even tens of thousands of years.

Stratigraphic ranges of faunas needed to be pinned down much more securely than they were, for it was known that many species, such as the faunas of the *Equus* beds, were extinct. But whether they had gone extinct during the later stages of the glacial period, or after it ended, was not at all certain. Some species, particularly mammoth and mastodon, appeared to be late survivals, possibly lasting into the Holocene. This raised an obvious quandary: if artifacts were associated with an extinct Pleistocene species, did that mean a great antiquity for humans *or* the recent extinction of the fauna? It was for Chamberlin another of those "Janus-faced" questions, the resolution of which required working out the "specific stratigraphic and correlated evidences of a degree of precision and refinement commensurate with the geologic minuteness of distinctions involved in the problem."[163]

It was a reasonable argument to make. *Bison bison* (the living species) was still roaming the Great Plains. Using a nonextinct genus as a chronological marker required knowing a great deal more about the temporal ranges of the now extinct *species* within that *genus*. That information was lacking at 12 Mile Creek because the age and taxonomic identity of the specimen was in flux. Williston had initially identified the specimen as *Bison antiquus*. Frederic Lucas of the USNM subsequently assigned it to *Bison occidentalis*, a species he defined in 1898 (on an Alaska type specimen).[164]

Moreover, Lucas observed that *B. occidentalis* "is the species most nearly resembling the existing bison, with which it was probably for a time contemporaneous."[165] If the modern species occurred in fossil form and the extinct species was contemporaneous with the modern, it would not have been possible to identify a particular occurrence of *B. occidentalis* as Pleistocene in age without independent evidence of the sort Chamberlin called for. Lucas made a "rational effort to stabilize bison taxonomy," but neither he nor anyone at the time was certain *B. occidentalis* had not survived into the Recent.[166] Questions about bison taxonomy and extinction lingered for decades (§8.6, §8.8).

It is little wonder then that Chamberlin raised the concerns he did, or that Wright and virtually everyone else put so little weight on the 12 Mile Creek site. As Matthew Hill observes, those who accepted the site were ones who were already strong supporters of a Pleistocene human presence. Ultimately, the site's significance became evident only later, when more was known of bison taxonomy and evolution and what the form of the "arrowhead" found with it signified. To insist, as subsequent commentators have, that 12 Mile Creek "should have been decisive in determining the time of arrival of humans in the New World" ignores its historical context.[167]

6.10 DRESSED FOR BATTLE, NO ONE TO FIGHT

Winchell understandably did not want to tarnish the dignity of a presidential address with a bare-knuckle brawl with Chamberlin over Lansing. It was sufficient to subtly swipe at Chamberlin's objections while putting his most positive spin on the case. But now that the meeting was over and Chamberlin had raised the stakes by broadening the human antiquity debate to one of general geological interpretation and competence, it was time for him and Upham to face the attack head-on. Upham published in the *American Geologist* a brief reply to Chamberlin's Lansing critique, but it was little more than a reassertion of his belief in the site's antiquity, along with a couple of "decisive" objections (Upham's term) to Chamberlin's arguments.[168] Upham broadcast reprints of his paper as usual and got the usual cool response from Chamberlin, who could only express his "surprise that the evidence at Lansing seems to you 'very fully conclusive for the glacial age of man there,' as your opinion seems to me to rest on an erroneous identification of the formation involved."[169] Upham must have realized the weakness of his rebuttal, for he assured the brethren that Winchell was busy preparing a full exposé of all the "fallacies" in Chamberlin's position (and along the way yet again urged Wright to enter the debate).

Winchell spent the first few months of 1903 crafting his reply.[170] He opened with a scornful rejection of Chamberlin's "academic discussion," but after reading the first several paragraphs a colleague advised him to "cut 'em out." And "for mercy's sake," his friend added, "don't even mention the ancient row over glacial man."[171] Winchell mostly followed that advice but could not resist repeated jabs at Chamberlin's "mere academician's play."[172] That play, Winchell argued, could be boiled down to twenty-three points, a "glittering phantasma of Pleistocene geology," nearly all of which Winchell disagreed with; those he agreed with were trivial or irrelevant. At its core Winchell saw the debate over Lansing as being about just three questions: What was the nature of the deposit immediately overlying the bedrock? Was loess present? And, had the Lansing locale been subject to fluvial erosion since Iowan times?[173]

"Phantasma" or not, much of Winchell's rebuttal aimed at demonstrating that the tributary drainage where the site was located was not, as Chamberlin insisted, subject to repeated erosion and redeposition but was a deeply incised and well-protected preglacial-age valley that had been geologically stable and largely untouched since Iowan times. The proof of that was the several feet of geest ("homogenous, uniform, exceedingly fine, non-effervescing clay") atop the bedrock at the site, which Winchell presumed had weathered in place over a vast span of time. That Chamberlin deemed the geest a "heterogeneous mixture" of bedrock, glacial drift, loess, and slope wash merely showed his examination had been too "hasty and casual" and unduly influenced by "preconceived" ideas.[174]

As for Chamberlin's judgment that the stratum overlying the geest was "incompatible with a typical loess deposit," Winchell could barely disguise his disdain. Of course it was not typical loess: there was no "typical" loess. Lyell's classic definition of loess as he observed it at Vicksburg, Mississippi would hardly fit comparable deposits in, say, Iowa. Yet, these were loess all the same. The differences between them merely reflected their proximity to the glacial front; the more northerly the loess, the more it incorporated elements of glacial till, ultimately grading into glacial drift. Chamberlin himself had endorsed the idea of the transition "horizontal and vertical, of the Iowan till to loess," as had many of Chamberlin's students and colleagues, including H. F. Bain (assistant State Geologist in Iowa), Calvin, McGee, J. E. Todd (South Dakota State Geologist), and J. A. Udden (Augustana College). Because

> the typical idea of loess is subject to great extension and alteration . . . the deposit on the Concannon farm, overlying the human skeleton cannot at present be excluded from the category of typical loess [cf. §6.8].[175]

Obviously, if geest and loess were present at Lansing, and all loess was Iowan in age, that undermined Chamberlin's contention that the absence of (post-Iowan) Wisconsin-age glacio-fluvial deposits meant the site had been scoured and filled in post-Wisconsin (postglacial) times.[176] Indeed, Winchell suggested there had never been Wisconsin-age deposits here that could have been removed by erosion. Todd's mapping of Wisconsin-age terraces of the Missouri valley suggested that these declined in elevation southward from the ice margin and that by the latitude of Lansing had disappeared (a point made earlier by Upham in his reply). "If no Wisconsin terrace ever existed at the place of the burial of the human relics . . . no later Pleistocene record has been superimposed so as to erase or obscure the Iowan record." Thus, the age of the Lansing skeleton, which lay in the "pre loessian geest, or very near the top of such geest," was as old as "the length of time elapsed since the spreading of the Iowan loess."[177]

It all made for a coherent, even compelling explanation, but only if one generously defined what constituted loess, assumed it was water-laid (necessary to insure that pebbles and larger clastic particles could be floated in), and understood all loess to be the same age (Iowan). Calvin's reminding Upham of having watched loess form the summer before as winds lifted sediment off the dry Missouri floodplain would have meant nothing to Winchell. However generous his definition of loess, it was not generous enough to take in last year's dust.[178] Just so, Winchell's rebuttal made little sense to Calvin, Chamberlin, Shimek, and others, who saw in the very same sections a completely different depositional and erosional history.

Winchell asked several colleagues for comments that—still mimicking Chamberlin—he then appended. Wright, as expected, agreed in all particulars. Williston knew better than to walk down that beckoning dark alley. He merely attested to where and how the unio shell had been found. But perhaps Williston was just being polite. That spring in *Popular Science Monthly*, he reiterated his belief that the Lansing humans remains were contemporary with some of the extinct fauna, but then he backed away from suggesting this indicated great antiquity, as the animals may have lived "within comparatively recent times." The final word, he insisted, "must be said by the geologist," and he was content to let them fight that battle. It is telling, however, that Williston described the Lansing deposit as talus (Chamberlin's view) and not loess (Winchell's view).[179]

Todd was not so subtle. He had examined Lansing, and his comment on Winchell's paper sounded several jarring notes. He did not think geest or loess was present at Lansing; he did not believe Winchell's contention that the site had been untouched by erosion since pre-Wisconsin times; and he politely but firmly rejected Winchell's construal of his interpretation of the age and elevation of the Wisconsin-age terraces of the Missouri River as "not in harmony with my conception of their relations."[180]

To Winchell's credit he published Todd's criticisms along with his paper, though they negated nearly everything he had just argued. Yet, Winchell's confidence in doing so cannot distract from the fact that an independent, authoritative observer considered his case deeply flawed. Chamberlin, of course, thought so too, but his response, at least in public, was stony silence. Privately, he dismissed Winchell as no "expert in this field" and one whose ideas had been abandoned "by most experienced and progressive students."[181] He briefly entertained the idea of responding. As he told Leverett:

> Judged from [Winchell's] summary, he has greatly distorted and misinterpreted my views. I have been wondering whether I should give it any attention or not and if so, how. The only thing I have even momentarily considered is a review in which I should point out his misstatements, but I do not know that even that is worth while.[182]

In the end, the imperial Chamberlin deemed a response unworthy of his time.

Holmes and Hrdlička had also been tarred by Winchell's brush, albeit in passing, but they did not respond either. By then, Hrdlička and the Lansing skeletal remains had both arrived at the Smithsonian, and Hrdlička undertook a more detailed examination that fully confirmed his (and Holmes's) earlier conclusion: "Considered anthropologically, all the parts of the skeleton, and the skull in particular, approach closely, in every character of importance, the average skeleton of the present-day Indian of the Central states."[183] The possibility demanded, so far as Hrdlička was concerned, that

> the very far-reaching and far more difficult conclusion that man was physically identical with the Indian of the present time, and that his physical characteristics during all the thousands of years assumed to have passed have undergone absolutely no important modification.[184]

Hrdlička was unwilling to assume that the hands on the morphological clock (§6.2) had stood still.

Chamberlin, Holmes, and even Hrdlička (young as he then was) were all supremely confident in their scientific judgment. As Chamberlin expressed it, "I do not like to use over-positive terms, but I think [my] identification [of Lansing] is firm beyond possible revocation."[185] They stated their position and believed impartial observers would recognize they were correct and their opponents in error, so no further comment was necessary. Besides, why dignify Winchell (or Upham) with a reply? It was better to ignore them, and from that moment Chamberlin, Holmes, and Hrdlička did.

The deafening silence that greeted Winchell's rebuttal effectively ended the debate over the antiquity of the Lansing human remains. Although there were a few more sputters from Upham[186] as well as belated testimony from Wright, by late 1903 Lansing had fallen into that netherworld of claims believed by proponents, rejected by critics, and now dangerous terrain for all others. For there was no mistaking which side had gathered the most formidable scientific talent. One would be foolish to jump into this fight and expect to win against the fearsome chief of the USGS Glacial Division (Chamberlin), the state geologists of Iowa (Calvin) and South Dakota (Todd), and one of the most prominent anthropologists in the country (Holmes).

The next few years saw sporadic flare-ups over the origin and age of loess, though here too Chamberlin and others had little to say, for that debate was past its tipping point and the weight of opinion was rapidly shifting to the aeolian theory. At that moment Wright finally engaged in an effort to save the aqueous theory, and with it Noah's flood. It was his last significant foray into a geological dispute. Chamberlin could not be bothered to reply, but he cheered on Shimek, who carried the banner for the aeolian hypothesis and in the process learned the frustration of arguing with Wright.[187] Abbott, watching from afar, thought it like the "merry old days of earnest work here about, and then too . . . the controversial days that embittered me."[188]

The debate turned on the usual classes of evidence: snails; the elevation, distribution, evenness, and thickness of loess deposits; the lack of evidence for the vast glacial-age lake from which the loess had ostensibly settled; and the question of whether there was one loess formation or more and what their age or ages were. Wright was then in his mid-sixties and taking on a younger (by nearly two dozen years) and very aggressive opponent in Shimek. Moreover, Shimek was a master of the snail evidence that had come from the loess; much of it he had amassed himself, as even Wright acknowledged.

Shimek had one weakness, however: the myopia that infects those expert in a highly specialized realm who think their evidence trumps all. For Shimek the snails "will continue to be the rock which will wreck the arguments of the advocates of the glacio-fluvatile and aqueous theories." Not surprisingly Winchell insisted, "It matters not what fossils are found in the loess. If it be water laid from the bottom to top, as it seems, and shows stratification at all altitudes, as I believe, it was certainly deposited in water. Nothing can abrogate that evidence." So too Leverett, otherwise sympathetic to Shimek's view, publicly chided him that snails were just one piece of evidence on which the outcome of this debate would turn.[189]

Well-armed as he was, Shimek was nonetheless attacking Wright, who by then had perfected his ability to withstand body blows to his theories: years of boxing with Chamberlin had taught him that. Wright was further armored by the certainty his creed provided (§3.8) and, Chamberlin would add, by his ignorance of glacial geology. Wright never conceded defeat, he merely kept shifting his line of defense. On this occasion, Wright deflected Shimek's punches by sidestepping or ignoring the empirical flaws Shimek identified in the aqueous theory, such as the lack of a single aquatic snail.

He instead turned attention toward a "remarkable discovery" that seemed to support the theory: a half-dozen boulders in the Osage River valley near Tuscumbia, Missouri that he guessed (without evidence) were floated into place by glacial meltwater. He used them to pen a "new chapter in the glacial history" by conjuring those boulders into the flotsam of glacial floods that, by his calculations (making assumptions about the size of the ice sheet and the rate of melting), annually released some 500 cubic miles of water. When that water reached the spot where the Missouri River bends at Kansas City, it dammed up, creating a vast lake some 200 feet deep that flooded back up the Missouri valley and its tributaries. As the lake drained each year (doing so, Wright estimated, in ninety-six days), it left behind thick loess deposits.[190]

Loess occurred at elevations higher than 200 feet, but as a staunch epeirogenist (and nonuniformitarian) who believed land level changes occurred "out of all analogy with those which take place now," Wright had no difficulty making land subside so floodwaters could scale the heights. As Shimek observed with no small measure of scorn, the earth's surface "conveniently move[s] down and up through a vertical distance of 300–500 feet to accommodate [such] theories."[191] He was not exaggerating: when Shimek pointed out the obvious problem with Wright's scenario of water piling up behind a dam near Kansas City, namely that it failed to account for the high loess bluffs south of the supposed dam at places like Vicksburg, Wright glibly replied, "We shall be compelled to suppose that the epeirogenic downward movement affected to a considerable extent the whole valley of the Mississippi."[192] The loess-tail was wagging the epeirogenic dog.

The flaws in Wright's reasoning were hard to miss: loess was spread over high elevations, and because loess was derived from water the water must have overtopped the landscape as well. For that to have occurred, either floods were of great magnitude, the land had subsided, or both. And though the floods were sufficient to move boulders of three or four tons, they also left behind fine-grained loess sediments. Yet, there was no empirical evidence for any of these claims. Where,

Shimek wondered, were the traces of flooding, the ancient shorelines from the lake and its drainages, or the coarse-grained deposits left by this rapidly draining lake—and, of course, why were the snails only terrestrial species? An ally of Wright's, Luella Owen, had an answer to the last question: perhaps it was simply impossible to identify the snails properly, and mistakes had been made.[193] Shimek was apoplectic:

> Miss Owen is not familiar with mollusks, and it is to be regretted for her sake that her statement was published, as it is sure to invite the ridicule of all who have even the most rudimentary knowledge of shells.[194]

Owen and Wright discussed the merits of replying to Shimek. Wright was for it. He had never been one to run from a fight. Owen wavered, perhaps from having been singled out for Shimek's "hottest fire" (though, as she explained to Wright, she "adopted the attitude of the early Christians and found some pleasure in being roasted in such a goodly company"). Winchell, whose own ox had been gored when Shimek attacked Winchell's claim that there was no "typical" loess, encouraged Wright and Owen to keep fighting and promised to publish their papers in the *American Geologist*. They began to map a strategy for doing so. Owen counseled Wright to back away from his insistence on heaving epeirogenic movements, and she took a crash course in malacology.[195]

Their replies appeared in the *American Geologist* in the spring of 1905, but little was new. Wright surrendered a small piece of ground, admitting he was not completely excluding wind action in loess formation, though he continued to insist loess bore unmistakable evidence of water deposition and was "clearly connected with the closing stages of the Iowan glacial epoch."[196] Owen reversed course and granted that snails could be identified to the species level with reasonable accuracy. But she still tried to turn Shimek's snails against him, arguing the aqueous theory did not require the presence of freshwater snails, only that the glacial floodwaters receded early enough in the summer for the newly deposited loess to dry, be covered in vegetation, and attract land snails. Because, as she incorrectly supposed, the aeolian theory required summer aridity, it must be incorrect, as the snails' shell color indicated moist conditions. Unfortunately, the experts to whom Owen sent the shells for identification told her the specimen color offered "no decisive evidence" on the question of aeolian or aqueous conditions.[197]

This time Shimek made no reply. By this time none was needed. Not only did most geologists now accept that loess was aeolian in derivation, it was also realized there were multiple loess deposits of varying ages, some like the Iowan (now Peoria) loess, which was Pleistocene in age, but others were undeniably postglacial.[198] Winchell had been right that there was no "typical" loess. His error was assuming that some of those "atypical" loess deposits were loess (they were not), and that those differences merely reflected proximity to the ice sheet. Given the complexity of the loess record, including the fact that not all loess deposits are glaciogenic,[199] it is little wonder there was confusion about loess origins and age or ages. But if these details were not fully understood in the first decade of the twentieth century, certainly the broad parameters of the complexity of the loess record were, save by the ever-intractable Wright.

Although the battle over loess origins was finished so far as the great majority of glacial geologists were concerned, loess remained central to the human antiquity controversy when, a few years later, human skeletal remains emerged in eastern Nebraska from what was again identified as Iowan loess. This time, attention focused primarily on the morphology of the skeletal remains themselves and only secondarily on the age of the loess in which they were found, in large part owing to the growing influence of Hrdlička in the debate over human antiquity. Moreover, the debate unfolded in a very different way and in a very different venue than Lansing had, though it dissipated far more rapidly.

6.11 NEANDERTHALS IN NEBRASKA?

In late October of 1906, readers of the Sunday *Omaha World Herald* awoke to the breathless headline, "Is This the Link? A Sensational Find Near Florence May Startle the Scientific World." A lavishly illustrated multipage article proclaimed that human skeletal remains discovered in a mound atop Long's Hill outside Omaha might end the quest, begun in Darwin's day, to fill the gap "between the lowest type of man and the highest type of monkey."[200]

That newspapers ratchet up stories is hardly news, nor that word of the find first appeared here rather than in a scientific journal. This was an unusual case, however, for amateur archaeologist Robert Gilder, discoverer of the Long's Hill specimens, was a reporter at the *Omaha World Herald*. He both made the news and wrote it up. Through the fall of 1906 and the spring of 1907 he milked the find to burnish his archaeological reputation and sell his newspaper. But his aspirations reached well beyond his Omaha neighbors. Gilder's siblings Joseph and Jeannette published *Putnam's Monthly and the Critic*, a semipopular illustrated magazine, and Gilder granted them exclusive rights to the first national announcement of the discovery, which they were delighted to accept.[201]

Gilder was not entirely to blame for the Long's Hill hype: the local scientific community helped fuel it. In the *Omaha World Herald* R. C. Henry, an anatomist at Creighton University, proclaimed the nearly ten skeletons, with their massive postcranial bones and skulls in which the "frontal eminence is entirely wanting," a race of physical giants but mental pygmies. Erwin Barbour, Nebraska State Museum vertebrate paleontologist and director of the state's Geological Survey, wondered aloud how humans with such small brains could even exist. Perhaps, as a member of Creighton's dental school faculty observed, their success was owed to the "tremendous crushing power" of their jaws. The *Omaha World Herald* conjured up a race of tiny-brained giants living on Long's Hill, with jaws that could crack walnuts and whose crushed skulls spoke darkly of ancient violence.[202]

There were more sober voices. Henry Ward, dean of the Medical College at the University of Nebraska, thought the skulls from a "very primitive" race but wanted a more thorough comparative study before drawing any conclusions. So too did Elmer Blackman, an archaeologist with the Nebraska Historical Society who was familiar with skeletal remains from prehistoric sites across the state. The Long's Hill specimens seemingly differed from those he had seen and might even be older, but beyond that he would not venture.[203] Still, there was incentive to learn more.

Barbour and Ward immediately made plans for a detailed study of the skeletal remains and the site itself. It was not a moment too soon. The discovery had caught the eye of Henry Fairfield Osborn (tipped off by Gilder's brother in New York[204]), zoologist at Columbia University and curator of vertebrate paleontology (and soon to be president) of the AMNH. Osborn had not much experience with fossil humans, but he believed horses and primates had followed similar evolutionary and migratory paths out of their shared Central Asian homeland, and therefore the rich record of fossil horses in Nebraska suggested there ought to be early humans there as well.[205] Osborn quickly caught a train for the 3,000-mile round trip to Omaha in late October of 1906 to examine Gilder's specimens. He was stunned: the skulls were "extremely primitive & small-brained," reminding him of Neanderthals.[206]

But he was only stunned, not stunned senseless. Back in New York he did his due diligence and promptly wrote Hrdlička to ask for anatomical details on the oldest crania found in America. Were any a "small-brained race, with prominent eye-brows and retreating forehead"? Hrdlička had just finished "a revision and study of all the crania found on this continent" and replied immediately. Low foreheads, he warned, were not a sign of antiquity, and skulls of the sort Osborn described were found throughout the continent and might merely reflect individual variation or artificial cranial deformation. Osborn got the same message a few days later from Clark Wissler in the AMNH anthropology department.[207]

Duly cautioned, Osborn promoted the Long's Hill from Neanderthal-like to a form reminiscent of the more advanced Galley Hill type (a specimen found in England in the late 1880s in apparent Pleistocene deposits but possessing more modern anatomical features[208]). He forwarded Wissler's letter to Barbour and suggested that because other North American archaeological crania showed those same "primitive" features, the geological antiquity of the Nebraska specimens "cannot be very great."[209]

Barbour was not interested in Osborn or Wissler's opinions and was glad Osborn's enthusiasm had waned. For if this was as significant a find as it seemed to Barbour, he was keen to have the spotlight shine on his home state and institution and not have it stolen by Osborn, who had long been mining Nebraska's rich paleontological remains for display in New York. Osborn and Barbour had worked for Edward D. Cope and Othniel C. Marsh (§7.1), respectively, and though they themselves were on generally good terms, some of the bitterness and all of the lessons from the long-standing feud of their mentors still lingered, not least the importance of protecting one's fossil territory and establishing priority of discovery. Barbour bluntly told Osborn he was willing to share photographs and descriptions of the Long's Hill specimens but that Osborn needed to "observe our rights in the matter of a scientific announcement."[210]

Barbour did not wait for Osborn to reply. Together, he and Ward quickly penned a note announcing the find in *Science*. Long's Hill, they reported, was a loess formation atop which sat an artificial earthen mound composed of two strata separated by a burned clay layer. Human remains were in both layers. Those in the upper stratum were Indian and those in the lower a "much more primitive type." The latter had thick protruding brow ridges and "feeble" frontal eminences and were brachycephalic. Barbour and Ward likened them to Neanderthals (odd, as most Neanderthals were dolichocephalic), though they stopped well short of claiming the Nebraska specimens were that old or ascribing evolutionary significance to these anatomical patterns. For the moment, they merely observed that the specimens were "inferior" to Mound Builder crania and were perhaps "early representatives of that tribe."[211]

The caution was Ward's, for when left to his own devices Barbour was less reserved. He reported to his old friend Williston that the lower stratum was of "plain unmistakable undisturbed loess." Thus, the skeletal remains within it must have been deposited as the loess was accumulating; there was, he insisted, no evidence that these were intrusive burials. The "Nebraska Loess Man," he proclaimed, was "closer to the Neanderthal skull than anything we know of."[212]

Long's Hill seemed to confirm Lansing was no anomaly, and Williston for one was delighted to hear it. He had never changed his views on Lansing's antiquity. But from his vantage of occupying an office down the hall from Chamberlin, Williston offered Barbour some advice: he should leave portions of Long's Hill unexcavated, so that critics might have a chance to examine it. And there surely would be critics, he warned, for such a find was tantamount to waving "a red rag before a bull," at least in front of certain anthropologists who "are not impartial."[213]

Barbour took heed, but after subsequent work at the site[214] he was convinced, as he told Williston, that the skeletal remains came from "straight undisturbed natural original *LOESS*." Were these bones from a species other than human, he was certain no one "would hesitate for a moment to declare them loess fossils." Perhaps, he inquired, the *Journal of Geology* might be interested in a report on the site? Williston did not reply right away—but did not need to. Barbour had sent a copy of the letter to Chamberlin.[215] Although Barbour's letter arrived without a cover note, Chamberlin knew what was being asked of him ("I assume that you desire a frank statement of the way the matter strikes me"), and took the bait:

> You appear to me to have overlooked one important feature of the case, namely the chronological diversity in the deposition of the loess. While you appear to have scrupulously endeavored to guard yourself against secondary deposits of the certain kind . . . there are upland and summit accumulations of loess in many sections, particularly those near the great rivers, which probably contain representatives of nearly every considerable age since the main loess formation was deposited. Some of these upland deposits are very recent—certainly much younger than the last stage of glaciation.[216]

Chamberlin had no confidence in any claims of great antiquity for an upland loess, unless its specific age was determined, a task Chamberlin admitted was almost impossible with the available data. Any claims for the antiquity of an upland loess—and he left unsaid that Barbour's fell into that category—"tends to awaken [a] lack of confidence in the whole inquiry."

Nonetheless, he would be glad to publish Barbour's statement in the *Journal of Geology*, provided the "critical factors" of the case were sharply delineated and Barbour avoided the "loose and untenable deductions" that in Chamberlin's view were so persistently associated with finds of this kind and so often presented to the public. Indeed, Chamberlin had just heard that Osborn was coming out with a paper on Long's Hill in a popular outlet (*Century Magazine*), and that, he announced,

> seems to me a great pity. In a matter in which so much wretched work has been foisted upon the public without critical scientific assortment, it would seem that a submission of the case to scientific criticism were the part of prudence before exploiting the matter before the public, who have no adequate means of distinguishing the character of the conclusions reached.[217]

If Chamberlin hoped he might sway Barbour's thinking, he only partly succeeded. Barbour did change his mind, but only by deciding not to send his paper to the *Journal of Geology*. With the sure knowledge that Chamberlin was watching, he went ahead with the article he had already prepared for *Putnam's Monthly*. Chamberlin's warning notwithstanding, Barbour was not about to let Osborn steal the glory of the national announcement of Nebraska Man. He only hoped that what he presented would not meet with Chamberlin's "entire disapproval." Barbour was well aware he had to defer to a critic of Chamberlin's stature. Williston had advised as much: "Such criticisms and doubts as those of Professor Chamberlin are valuable and important, because every proposition of the kind must stand the severest tests before being accepted." One could not just preach to the choir.[218]

As it happens, Chamberlin found Osborn's paper in *Century Magazine* far less objectionable than Barbour's in *Putnam's Monthly*. By the time Osborn's article appeared in January of 1907, he had completely retreated from his initial enthusiasm for the specimens' antiquity: "This cranium is of a more recent type by far than that of the Neanderthal man. It may prove to be of more recent type than that typified by the early Neolithic man of Europe."[219]

While Osborn was hauling the Long's Hill specimens up the evolutionary ladder, Barbour was sending them back down. In a burst of publications written in a span of just two months—a second paper in *Science*, two reports in the *Nebraska Geological Survey* series, a piece in *Putnam's Monthly*, and one for *Records of the Past*—Barbour laid out the "harmony of testimony" of the anatomical and geological evidence. The thick brow ridges, low forehead, large parietals, narrow temples, thick skull walls, and small brain capacity all pointed toward the Neanderthal type. It was evident visually: a composite profile of the Long's Hill crania superimposed atop profiles of European, Mound Builder, Neanderthal, and *Pithecanthropus* skulls showed that the Nebraska crania were somewhat "higher in the human scale than Neanderthal man, but lower than the Moundbuilder." He could be even more specific: "They resemble the man of Spy," a later-stage Neanderthal possessing more gracile jaws and a tool kit more Aurignacian than Mousterian.[220] It was telling that human anatomist Ward's study of the Long's Hill skeletons, which immediately preceded vertebrate paleontologist and geologist Barbour's paper in *Putnam's Monthly*, made no such far-reaching claims.[221]

Yet, the remains of this "low and savage" tribe had been found immediately below those of two other races, Mound Builder and Indian. Barbour himself was untroubled by the coincidence of a relatively recent burial mound built atop the apparent remains of Pleistocene humans, though he realized he was vulnerable on the point. Who was to say all the skeletons at Long's Hill were not from the same group?[222] To preempt that criticism, Barbour declared that the lower stratum bones were in undisturbed loess. The deposit in which they were found showed customary loess color and structure, carbonate ("lime") tubes, concretions, and (snail) shells; some of the remains reached a depth of twelve feet below surface, and it was "improbable that a primitive race would dig graves" to that depth; and, most important for this believer in the aqueous origins of loess, the bones were water-worn, disarticulated, and scattered, as they would be if they came to rest in the fine-grained sediments of a glacial flood.[223]

In an effort to save his friend from embarrassment, Williston gently pointed out that most geologists now attributed loess origins to wind.[224] As Barbour's claim essentially demanded a water origin, he ignored Williston. Besides, the condition of the remains suggested water erosion, and that enabled him to link the deposition of the skeletons and loess deposition to the same geological process, thus making them of comparable antiquity. With Chamberlin's lecture on the chronological diversity of loess still ringing in his ears, Barbour explicitly disavowed any intent to estimate the age of the loess or its skeletal remains, but then he promptly went back on his word. If those remains were in undisturbed loess, as he believed, the conclusion was inescapable. This evidence "carries man in America back to glacial times." The Long's Hill loess was atop Kansan drift, and "though as young as the later Wisconsin sheet or younger, it is nevertheless old." In *Putnam's Monthly* he went even further out on that limb: "These bone fragments . . . were doubtless deposited with the loess, the age of which may be safely reckoned at ten thousand to twenty thousand years or more."[225]

That announcement of Barbour's came as a postscript to his *Putnam's Monthly* article, and Gilder thought it might create more of a furor than all the original articles.[226] He was right.

Blackman watched as events unfolded, and he was skeptical of Barbour and Gilder's claims. He had seen like crania in other sites. When Upham came fishing for information, Blackman shared his suspicion that the specimens were no more than 1,000 years old. That suited Upham just fine. He began spreading the word that Long's Hill was "quite modern." Upham rarely met a purportedly ancient site he did not like, but that he did not like this one is perhaps unsurprising. From the tone of his letters to Blackman, envy had much to do with it. His Lansing claims had failed and it would have been a bitter pill to swallow were Barbour's to succeed. Worse, the Nebraska claims had captured the public's attention "chiefly because of the literary talent of the discoverer and of his family and friends."[227] He was untroubled by the memory of Chamberlin chastising *him* about presenting scientific results to the public (§6.8).

Blackman was not ready to dismiss Long's Hill and admitted as much to Upham. Still, he wondered how bones ostensibly that old yet so close to the surface could survive for that length of time, or if their presence in the loess was better explained by gophers having carried bones down from the upper stratum of the mound. He knew that many geologists believed loess was still in the process of being deposited by the wind, and that the Long's Hill specimens could therefore have been recently buried, though as a geologist Barbour "should be able to tell whether the loess had been moved or not." Upham did what he could to encourage Blackman's skepticism, suggesting that the primitive features of the specimens was due to their being "a family of abnormal skull characters, not necessarily very ancient."[228] In arguing against the Nebraska finds, Upham found himself in the awkward position of agreeing with Holmes and Hrdlička.

6.12 HRDLIČKA'S *SKELETAL REMAINS SUGGESTING OR ATTRIBUTED TO EARLY MAN IN NORTH AMERICA*

When Ward backed away from Barbour's interpretations of the Long's Hill specimens it meant Barbour lost the benefit of his expertise in human anatomy. (Barbour's was in large quadrupeds of the Cenozoic.) That was unfortunate, but Ward inadvertently did more to weaken Barbour's position when, in conversation with Holmes at the Smithsonian in early 1907, he explained that no detailed study of the Long's Hill remains had been conducted. At that moment, Holmes had on his desk a just-completed manuscript by Hrdlička on all the purportedly ancient skeletal remains of North America. After talking to Ward, Holmes decided to postpone its publication to give Hrdlička the chance to study the Long's Hill specimens and incorporate them in the volume. Permission from the ever-protective Barbour was secured, and within weeks Hrdlička was on a train to Nebraska.[229] Over two days in January Hrdlička studied the skeletal remains at the State Museum in Lincoln, and then for several hours he, Barbour, Gilder, and Ward examined the frozen ground at Long's Hill (Figure 6.2). Hrdlička assured Holmes that the visit threw "some much needed light on the whole subject of the Nebraska/Loess Man."[230]

It also was a finishing touch to his *Skeletal Remains Suggesting or Attributed to Early Man in North America*, which firmly established Hrdlička's reputation in the human antiquity controversy. At Holmes's direction, Hrdlička had spent the prior year assessing the claims made on behalf of the human remains at Natchez and Calaveras (§2.4), Trenton and Lansing, and several lesser known finds, to which he now added Long's Hill. Hrdlička was rarely given to elaborate discussions of theory or method, but in *Skeletal Remains* he offered a set of criteria he thought needed to be met to assign any of those skeletal remains to a deep geological antiquity. These were (1) indisputable stratigraphic evidence, (2) some degree of fossilization of the bones, and (3) marked serial somatological distinctions in the more important osseous parts.[231]

Easily stated, perhaps, but meeting those criteria was a task "fraught with exceptional difficulties," because traps to the unwary lurked in each.

Figure 6.2 Aleš Hrdlička in the excavation at the Long's Hill site, NE on a cold January day in 1907. He was accompanied to the site by Henry Ward and Erwin Barbour (who likely took the photograph), and after examining the site and the skeletal remains he left unconvinced of its Pleistocene antiquity. Courtesy of the Nebraska State Museum.

The first criterion (stratigraphic integrity) was troublesome, owing to the propensity of humans to bury their dead. To insure remains were not intrusive burials, Hrdlička urged that skeletal finds be photographed in situ and examined in place by competent scientists, particularly geologists familiar with the formations involved. Hrdlička was content (for now) to leave detailed geological matters to the geologist. However, there was one aspect of a site's geology he kept a close eye on: a site's potential to fossilize or chemically alter bone, a condition "very generally regarded as an important indication of antiquity." He had become wary after seeing highly fossilized bones in Osprey, Florida: their resemblance to skeletal remains of the region's native populations suggested no great antiquity. That contradiction was resolved by a trip to the area in February of 1906 where he discovered that the local environment favored "in an extraordinary way the infiltration of the bones and even replacement of their normal constituents, the latter process constituting fossilization."[232] The problem was not just that recent bone could fossilize; it was also that ancient bone might not. Hrdlička suspected the former was more common, and he expected it might one day be possible to distinguish recent from ancient fossilized bone using chemical analyses, but for him that day had yet to arrive.

For others it already had. In the early 1890s, and following the lead of studies in France, Thomas Wilson tested various chemical constituents of the Calaveras (California) skull and Natchez (Mississippi) pelvis, and in the case of the latter he was able to show that the Natchez human pelvis had fluorine amounts like those of the extinct Pleistocene ground sloth (*Mylodon* [*Glossotherium*]) Dickeson found in the same deposit (§2.4). The results indicated the human and the sloth were "substantially of the same antiquity" and this was, Wilson believed, compelling evidence of a Pleistocene human presence. Wilson declared the fluorine test "a valuable discovery, and one which may afford large opportunities for determining the antiquity of man in America, thereby aiding to settle some of those disputed questions about which the dogmatism of certain scientists has had such free rein." He urged its application to other cases, Calaveras included.[233]

Hrdlička had a reprint of Wilson's paper on the fluorine test yet did not cite it, even in discussing Natchez in *Skeletal Remains*. Speculation on that specimen's age, he announced, would be "quite useless." Obviously, in 1907 Hrdlička judged the fluorine test's reliability very differently from the way Wilson had in the 1890s (or Kenneth Oakley in the 1940s when he used it to crack the Piltdown case[234]).

Although Hrdlička doubted the reliability of chemical tests, somatological tests were another matter. He admitted there were complications here too. Large gaps in the evolutionary record made it difficult to discern how anatomically distinct late Pleistocene humans might be. And the movement of populations might result in different morphological types in the same region, even over relatively recent periods. Nonetheless, there was "no such thing as absolute stability in any human structure," which changed over time with heredity and adaptation. Thus, after accounting for local adaptations, ancient skeletal remains should be markedly different from recent ones (§6.2). Certainly the geologically ancient crania of Europe were testimony to that fact. As to how far back modern humans extended, it was at least 7,000 to 8,000 years, according to Egyptian crania similar to their modern counterparts.[235]

In the end, Hrdlička was confident—given a sufficient sample of well-preserved remains to insure one was "outside of the territory of semipathological occurrences, and features of reversion, degeneration, or purely accidental variations limited to individuals or small numbers"—that separating ancient from recent skeletal forms was not an insurmountable problem. One had to be both careful and competent: ancient skeletal material might be fossilized or not, and primitive looking or not. And in judging such remains as ancient, the geological evidence "should be absolutely decisive."[236]

That was hardly the case with Long's Hill. On the basis of his examination of the specimens, site, and excavations,[237] Hrdlička raised more than a half-dozen "serious objections" to these being geologically ancient, not least their context. They came from a burial mound atop a ridge dotted with them, and the supposedly distinctive Indian and loess skeletons were alike in condition and color. Moreover, none of the remains were fossilized, and all exhibited cutmarks indicative of bone cleaning before secondary burial. Such coincidences were not easily explained if the remains were of widely different ages. Many of the bones showed signs of rodent gnawing, testifying to the possibility that animals may have moved them through the mound (as Blackman had suggested). Hrdlička supposed all were the same age.[238]

Most important to Hrdlička, there was nothing about the morphology of the skeletons to suggest a deep antiquity. Comparing them with a large sample of skeletal remains in the region, Hrdlička argued there was nothing anomalous about the Long's Hill specimens. They were not substantially larger than known Plains Indians, their crania were not more massive (Barbour had evidently mistaken another skeletal element for a thick parietal portion), and the low forehead and heavy brow ridges that had so excited Barbour were hardly unusual. That such features were present on only two of the Long's Hill specimens said more about "the principles of heredity" than antiquity.[239]

Thus, Hrdlička arrived at the same place in regard to Long's Hill as he had at Lansing (§6.8, §6.10) and earlier still at Trenton (§6.4).[240] In viewing these three sites, along with the remainder of the North American finds of purported antiquity, Hrdlička concluded that, irrespective of other considerations,

> in every instance where enough of the bone is preserved for comparison the somatological evidence bears witness against the geological antiquity of the remains and for their close affinity to or identity with those of the modern Indian. Under these circumstances but one conclusion is justified, which is that *thus far* on this continent no human bones of undisputed geological antiquity are known.[241]

Hrdlička was careful to say "thus far." He did not want to be accused of precluding any possibilities.

With that conclusion, however, Hrdlička tipped his hand. Throughout *Skeletal Remains* he granted the importance of geology in ascertaining a site's age. Yet, that was mostly window dressing. In each case he relied far more on the anatomical evidence. By the conclusion of *Skeletal Remains* the other two criteria, stratigraphy and fossilization, were hardly seen and had become nameless "other considerations." That a physical anthropologist emphasized skeletal morphology is, of course, fully understandable. What this conclusion signaled, however, was the beginning of a shift that in a few short years resulted in the somatological evidence becoming for Hrdlička not just one of three criteria, but the only one that mattered.

6.13 OVER BEFORE IT BEGAN

Skeletal Remains appeared in December of 1907. The initial reaction from Nebraska was surprisingly positive. Gilder wrote to praise its "thorough and scholarly" weight. He quibbled about only one small matter: when and why the bones were turned over to Barbour at the University. "With this single error," he assured Hrdlička, "I believe your paper could not be improved upon." He wrote again five days later asking for an autographed copy.[242]

Ward thought Hrdlička's discussion "very fair," despite Hrdlička having a conclusion "radically opposed to those which we advanced." Like Gilder, he spotted only minor errors, though one touched an extremely raw nerve. Hrdlička had identified Blackman as a "Professor." Blackman, Ward snapped, had no connection with the University, "no title, no degree," and was little more than an uneducated politico with "overweening self-conceit" who had been an obstacle to their efforts to develop a legitimate archaeology program at the University. Worse, "I foresee distinctly the wide publicity which will be given through the newspapers to your support of his views and your rejection of those advanced by us."[243] Wide and negative publicity had already begun. Blackman, in fact, made sure of it.

Blackman took the liberty of making "the substance of your [Hrdlička's] report clear to the public," and sent him and Upham press clippings, crowing that neither he nor the "best thinkers of the state" had been deceived by the claims for Long's Hill. Blackman shed a few crocodile tears, claiming he was sorry anything should occur that would cause Nebraska scientists "to be held in light esteem by the public at large," but those tears disappeared quickly in the delight he took at gaining revenge for the "severe criticism" he had incurred for having taken the stand he did (Blackman had a martyr's streak), and his pleasure at Barbour's discomfort from having said too much and now having to take it back.[244]

Internecine warfare is not unusual in a small provincial scientific community, particularly one as distant from the scientific mainstream as this was and in which financial support was limited and highly dependent on the whims of a state legislature and private donors. Those legislators and donors were no doubt delighted to see Long's Hill acclaimed in national magazines like *Putnam's Monthly* and the *Century* (far more prominent venues than Barbour's in-house Nebraska Geological Survey reports), and by the possibility that North America's oldest human traces were in their state. There was much state, local, institutional, and personal pride riding on Long's Hill. Now it had collapsed. It was painful, if not humiliating, to have the grandiose claims of the state geologist crushed by a scientist at the Smithsonian Institution, and have that news splashed across newspapers in cities and states countrywide. (Chamberlin read about it in Chicago.) Ward feared what might follow: "We shall lose some popular support . . . and the result will be the further postponement of our plan for the inauguration of a properly manned department in the University."[245] Hard budget times indeed came to pass, but there is no evidence that Hrdlička's *Skeletal Remains* was to blame.

Gilder glibly dismissed the fuss and bad publicity as jealousy and the usual competition among Omaha's newspapers, which never let slip an opportunity to shoot at their competitors. He also admitted that as a reporter, and not as a scientist, he had no need to be correct or truthful. "We of the press," he told Hrdlička, "can make a statement and deny it the next day if necessary, but an honest investigator could not afford so haphazard a method." It hardly bothered him to be haphazard.[246]

When approached by the newspapers for comment on Hrdlička's report, Barbour declined to make a statement. He said nothing to Gilder either, but to Williston he poured out his anger and humiliation at Hrdlička's having blindsided him. Earlier that year, Hrdlička had apparently "told Ward and myself [Barbour] most emphatically and positively that it was by all odds the most antique skull ever found in the western hemisphere. I am quoting his words verbatim."[247] Despite Barbour's obvious pain, Williston could not help but break into a "very broad smile":

Did I not tell you that there were certain men in the United States to whom a Pleistocene man was a red flag before a very erratic bull? I could have furthermore told you precisely what the report upon your discoveries would have been before the "investigators" visited the place."[248]

To make matters worse, another bull soon appeared in the pasture. Shimek went to Long's Hill in the spring of 1907 with Barbour and Gilder, then returned again that summer. Gilder was happy for the attention and sent Shimek snails from the site for identification. Later, however, he and Barbour heard rumors that Shimek had a paper on "Loess and the Nebraska Man," rumors Williston confirmed: Shimek was on the program at the December joint meeting of the AAAS and GSA and was "preparing to 'skin' the whole thing." Williston advised Barbour to be ready to reply to both Hrdlička and Shimek, and he did what he could to bolster Barbour's shaky self-confidence.[249]

> I have no use for Hrdlička—he has no opinion of his own on geology, and about as much knowledge of geology as my youngest boy. Shimek is a botanist, with a theory of his own that he is determined to prove—and that is all there is to it.[250]

As a geologist, Barbour could reasonably assume that the damage done by Hrdlička's report would not reach deeply into his own professional community. That was likely true in a general sense, but key Pleistocene geologists—Chamberlin chief among them—were well aware of Hrdlička's conclusions. Barbour believed Shimek's attack could cause more damage and "convince the best geologists in the country that there is nothing in [the Long's Hill] discovery." To ready his defense he quizzed Gilder about how much time Shimek had been at the site ("comparatively little") and whether Shimek was biased from the outset ("there is not an idea [in the paper] which he did not tell me before he had even seen the mound"). Barbour suspected Shimek's prejudices had poisoned the mind of "his countryman Hrdlička," a suspicion Gilder was quick to confirm, believing there was no question the Czech compatriots were conspiring before Hrdlička's report was even written. (There is no trace of such in their correspondence, and according to Shimek he learned of Hrdlička's paper only after his was completed.)[251]

In any case, Shimek concluded Barbour badly misunderstood or misinterpreted the sediments and stratigraphy. Shimek deemed the stratum that contained most of the supposed loess human fossils a recently disturbed soil rather than undisturbed loess. The typical loess below it in which Barbour found a few deeply buried human bone fragments (a finding Shimek could not duplicate) was pockmarked by gopher holes. Like Blackman and Hrdlička, Shimek thought animal burrowing a more likely explanation for the bone fragments in this deeper level. Shimek failed to see any signs that the bones were waterworn, and he remarked laconically that Barbour evidently still adhered to the old aqueous hypothesis. This time Shimek could not be bothered to debate that issue, as the aeolian hypothesis was now well beyond its "merely conjectural stage." In any case, none of the Long's Hill remains were in loess, let alone glacial in age.[252]

Gilder wrote Shimek to say he saw "nothing at all objectionable in your Loess Man paper," but he urged Barbour to reply. Barbour did nothing. His lack of response to Hrdlička is perhaps understandable, as he likely conceded Hrdlička's authority on human anatomy and morphological variation. His lack of response to Shimek is puzzling because that was in his domain. After his burst of publications on Long's Hill in the fall of 1906, the last of which was completed by year's end and before Hrdlička and Shimek were on the scene, Barbour ceased writing about the site. By all accounts a kindly man, he later explained to Wright that a public fight was not his style: "We have worked in an entirely different spirit."[253] That spirit, however, did not preclude him from privately denouncing Hrdlička and Shimek.

It was left to Gilder to carry the flag for Long's Hill. After excavating in Wyoming in the summer of 1907 with Harlan Smith, then an assistant curator in archaeology at the AMNH and an accomplished field archaeologist, Gilder and Smith visited Long's Hill, where (according to Blackman, who was present) Smith asked "some very pointed questions about the *why* of the situation." Among them, Smith wondered whether any "prominent archaeologist with experience in excavating had examined the situation." Gilder seemed a little shaken, but then Blackman supposed "contact with a well versed scientist will naturally weaken the confidence of a lay-man in his own judgment on scientific subjects."[254]

Gilder was a journalist and not a scientist, but he read the situation well enough: the acceptance of Hrdlička's conclusions (even by Ward), the unwillingness or inability of Barbour to defend the site's antiquity, and Smith's skepticism made apparent whom he needed to ally himself with. "You know," he explained to Shimek, "I am seeing phenomena with your eyes now." He conducted more excavations at Long's Hill and published an update in 1908. Despite his private unhappiness with Hrdlička and Shimek (he continued to complain of them to Barbour), he cited their conclusions without disapproval, sent them samples, and in an article on excavations at another site even included a report by Hrdlička on the recovered skeletal remains. (He had tried to elicit one from Shimek on the snails as well).[255] He hoped he might ultimately convince the critics that Nebraska Man was legitimate and ancient.[256] He never did.

6.14 LANSING TO LONG'S HILL: LOESS TO DUST

Although the Long's Hill case ended on the same failed note Lansing had, it played out very differently. There were no high-profile visits by teams of skeptics and proponents. Those who came—Hrdlička, Osborn, and Shimek—went quickly and on their own. There were no broad discussions of anthropological or geological themes raised by the find, in large part because few were competent to challenge Hrdlička on the former, and on the latter loess origins were by then widely accepted as aeolian (save, of course, by Upham, Winchell, and Wright). For that matter, there was hardly any debate about the particulars of Long's Hill, and for reasons that go beyond Barbour's unwillingness to engage his critics.

For one, the articles on Long's Hill (aside from two brief notes in *Science*) were published in nonscientific magazines such as *Century*, *Putnam's Monthly*, and *Records of the Past*. These were not venues that lent themselves to scientific discourse and indeed rarely accepted responses or published debates. Not even the *American Geologist* included a paper on Long's Hill, perhaps because Barbour never received the warm invitation to publish that Upham sent Williston on hearing of his talk on the 12 Mile Creek locality (§6.9). The one critical reply to Long's Hill (Shimek's) appeared in the *Geological Society of America Bulletin*. In contrast, the Lansing debate played out in the *American Anthropologist, American Geologist, Geological Society of America Bulletin, Journal of Geology,* and *Science.*

Barbour (and Gilder as well) were keen to have the attention that publication in a nationally distributed popular magazine like *Putnam's Monthly* provided. But doing so had its costs. Chamberlin made it clear to Barbour what was required to have a paper on Long's Hill published in the *Journal of Geology.* Barbour did not rise to the challenge but instead detoured around it, and in Chamberlin's eyes made the unpardonable sin of placing his work in front of the public without first subjecting it to scientific review. Chamberlin and most of his colleagues did not approve of that action or see need to respond. And after Hrdlička's and Shimek's works appeared, indicating there was no reason to consider Long's Hill Pleistocene in age, no response was needed. They instead elected to ignore the Long's Hill evidence. One of the chief weapons of a scientific community, David Hull observes, is a conspiracy of silence.[257]

Barbour badly misread that silence, proclaiming to Wright that Long's Hill was a far stronger case than Lansing, because those who were "apt to be most critical [of Lansing] have had nothing to say publicly against the Loess Man."[258] In identifying those apt to be most critical, Barbour could only have been referring to Chamberlin; after all, Hrdlička and Shimek had been critical of both finds. Moreover, Barbour was well aware of Chamberlin's deep skepticism regarding Long's Hill: Chamberlin told Barbour so himself. He merely refrained this time from announcing his objections publicly. If Barbour felt vindicated, he should not have. Chamberlin was simply unwilling to waste his time when Hrdlička and Shimek said all that needed to be said. Chamberlin reengaged in the human antiquity debate a decade later, when there appeared a new class of claims with different geological challenges.

It is curious that proponents of a deep human antiquity failed to come to Barbour's aid. Upham was opposed, as noted, but both Winchell and Wright thought Long's Hill sealed the case for glacial-age human remains in loess, seemingly affirming the evidence from Lansing. Yet their public statements of support were limited and their assistance to Barbour was restricted to a few private gestures.[259] It was only in a talk given a week before his death (in May of 1914) that Winchell responded to Hrdlička and Shimek. With one eye on emerging radical movements in Russia, China, Mexico, and the United States, Winchell denounced Shimek's advocacy of the aeolian hypothesis as "radically anarchistic, revolutionary, and destructive." He waved away Hrdlička's claim that the Lansing and Long's Hill skulls were like those of historic Native Americans. As poor as curation procedures were at the Smithsonian, Winchell smirked, the comparative crania Hrdlička used were likely not from historic Native Americans at all but rather from Pleistocene loess deposits. Naturally they looked similar to those from Lansing and Long's

Hill. When Hrdlička received a reprint of Winchell's last address, he got partway through before scrawling "Awful" across the title page, then filed the paper in his library, the remainder of its pages uncut.[260]

Although most of Lansing's proponents (with the exception of Upham) accepted the Long's Hill evidence and vice versa, virtually all failed to see that the two sites gave conflicting testimony. Found 200 miles apart, both sets of remains were ostensibly in loess and both were estimated to be more than 10,000 years old. Certainly, Pleistocene fauna of that age looked very different from modern forms. Should not one expect the same of humans of that antiquity (a reasonable expectation in those pre-Piltdown years[261])? At the very least they ought to both be similarly primitive. Lansing, however, did not appear as anatomically primitive as the Long's Hill specimens, at least according to their respective proponents.

Only Wright seemed to have noticed that if Long's Hill was accepted it negated the efforts of proponents (Upham, most notably) to argue that human crania were modern tens of thousands of years ago. Wright, long-experienced in accommodating troublesome evidence, reconciled the conflict by suggesting there was considerable variety in artifacts and skeletal remains in glacial times. Both primitive and more advanced forms co-occurred. That solution introduced another problem, for unlike his fellow proponents, Wright had a telescoped view of geological time. He thought that the Iowan period "did not close until about the time that the civilization of Egypt, Babylonia and Western Turkestan had attained a high degree of development, and that the cranial capacity had at that time in those regions reached that of the higher races of the present time." Naturally, Hrdlička's acceptance of the "greatly exaggerated ideas of the antiquity of the glacial epoch" held by others (Chamberlin, but also Upham and Winchell) led him to err in expecting the Lansing and Long's Hill specimens to be significantly different from modern American Indians. There was little reason to demand that a glacial human look "primitive."[262]

Wright saw no contradiction in proclaiming that the Long's Hill specimens resembled Neanderthals. These were not arguments acceptable to Hrdlička had he noticed them, but by then neither he nor anyone else was paying them any mind. Indeed, in his review of the Long's Hill crania Hrdlička had not even deigned to compare them to Neanderthal specimens.[263]

That his discoveries at Long's Hill were being ignored was painful for Gilder, and the pain became more acute as he realized that despite his continued supplications Hrdlička and Shimek never came around to his view. In the fall of 1910 he snapped and wrote a long, rambling, angry screed on the scientific "inaccuracies" of both Hrdlička's and Shimek's criticisms of the "Nebraska Loess Man." He sent Barbour the draft for comment, explaining he knew it would make him enemies, but "they cannot hurt me as I am a newspaper man not an archaeologist." He then added the ominous footnote that should he "drop off suddenly" the paper would be published by his heirs.

He did not drop off, and Barbour—though unwilling himself to engage in a bare-knuckle brawl—urged Gilder to publish his manuscript and even make the comments "a little stronger." Beyond his questionable action in throwing fuel on that fire, Barbour failed Gilder by not catching and correcting the flawed logic of his "proving" Hrdlička incompetent to speak on physical geology (which Hrdlička had largely ignored) or his claiming that Shimek was out of his element in anthropology (which Shimek had barely mentioned). Gilder had not seen the problem either, but by this time he was beyond clear thinking and paranoid as well. He claimed the second volume of the BAE's alphabetically organized *Handbook of American Indians*, set to cover N–Z entries, was purposefully delayed until Hrdlička's analysis was complete "so as to whack Nebraska Loess Man in the 'Ns.'" There was no entry in the *Handbook* for "Nebraska Loess Man," nor mention of it anywhere else in the two volumes.[264]

Gilder's paper appeared in Wright's *Records of the Past* and opened with a new and dramatic retelling of his reaction on receiving Hrdlička's *Skeletal Remains*:

> In reading Dr. Hrdlička's report the writer's attention was at once attracted to the many and persistent "inaccuracies" in its pages and I wrote the doctor calling his attention to one of the most glaring.[265]

Virtually the only truth to that statement is that he had written Hrdlička on that occasion (§6.13). It was only after he fully realized that his claim to archaeological fame had been killed that he bitterly complained of Hrdlička's glaring factual errors, misleading photographs, and colossal failure to spot crucial clues to the antiquity of the Long's Hill skeletons—namely, their brow ridges that proved they were "ape-like . . . to a degree truly remarkable." As for Shimek's gophers having moved human bone into underlying undisturbed loess, Gilder proclaimed that impossible because gophers "invariably carry foreign substances . . . *upward*."[266]

It was a bizarre performance. Barbour was nonetheless elated and delighted in telling Ward "they are very indignant in Washington at Gilder's criticisms, but the jocose part of it is that no one dares to reply," for Gilder now had a great deal more skeletal remains, which the Washington anthropologists admitted they could not easily dismiss.[267] That was fantasy. It was Gilder who told Barbour all this, and by then Gilder was hardly a credible witness on the thoughts or actions of his antagonists.

Neither Hrdlička nor Shimek responded to Gilder's tirade. Shimek was by then working on fossil snails in the Mississippi valley, whereas Hrdlička was plowing new ground, having left in May of 1910 to spend two months in Argentina and Brazil. There, he took on South America's claims of great human antiquity, particularly those of Florentino Ameghino, who had argued that humans emerged on that continent, presenting fossil evidence of their evolution that purportedly reached back to Miocene (*Tetraporthomo argentinus*) and Pliocene times (*Diprothomo platensis* and *Homo pampeus*).[268]

Hrdlička approached the South American claims the same way as he had the North American evidence, with the same result. None withstood his critical scrutiny on either anatomical or geological grounds, and all were decidedly not Pleistocene in age, let alone from the Pliocene or Miocene. In May of 1911, transmitting to Holmes the manuscript that became his *Early Man in South America*, Hrdlička revealed he was aware Gilder's tirade had just appeared, and he anticipated what his newest work was likely to elicit from proponents of a deep human antiquity:

I regret that the evidence here shown is, all through, adverse to the various contentions regarding the presence in South America of early man . . . [and] notwithstanding the utmost temperate way in which the subject has been handled [in this report], I do expect, on the basis of experience concerning others and because of the knowledge of the mental attitude of some of the scientists concerned, that a considerable abuse shall be the reward of the collaborators of this report from certain quarters.[269]

That abuse was coming was hardly news to Holmes. He had been the target of Abbott's slings and arrows for decades. Of course, he had also given as much as he had received.

6.15 TRENTON REDUX?

Abbott had watched from the sidelines as Lansing and then Long's Hill captured the attention of the archaeological community. He was not envious, just disdainful: "Such discoveries only confirm my own of thirty or more years ago, they are not wonderful, and amuse me because so much attention is paid to them."[270] Although he had loudly proclaimed his disgust with archaeology and promised to stay out of it, with those discoveries stealing the limelight it was time to remind everyone of Trenton's rightful place.

It was a task that should have fallen to Putnam. After sponsoring Volk's Trenton fieldwork for a dozen years, in 1902 Putnam asked Volk to write up his results. Two years later Volk completed the task, and Putnam happily accepted his large manuscript.[271] Then all forward movement stopped. Unfortunately, as Putnam explained to Volk and Abbott, he had no funds to publish the massive report at the museum, and his efforts to have it published elsewhere went for naught.[272] Abbott refused to accept that excuse. He was certain Volk's report was being suppressed: "Putnam will keep it smothered (if it is not destroyed) because it is so written that he (Putnam) cannot get the 'glory' of it and Volk only figure as a mere laborer with pick and shovel." That made Putnam no less an enemy of a human antiquity in the Delaware valley than Holmes, whose name Abbott could scarcely speak without calling him a liar.[273] Abbott's anger and paranoia boiled out in a letter to Winchell:

There was a settled purpose in Washington to make naught of my work, and to do this effectually there was no end of libelous assertions and free use of opprobrious terms. . . . A clique at Academy of Natural Sciences at Philadelphia did all they could to belittle my work, and finally Holmes . . . as champion of "aboriginal modernity" has deliberately *lied* on more than one occasion. But do you not see their cunning? They *say* these things, not *write* them, and so escape the penalty of a suit of libel. There never would have been any trouble, if the Washington people had been honest, and none later if Putnam, Boas, and Saville had not suppressed the report of the meeting of the Society of Americanists [*sic*] held in New York at American Museum of Nat. Hist. in Oct. 1903 [which included a presentation by Volk on Trenton].

Perhaps more blame rests on Putnam than any one else. . . . He talks much of the courage of his convictions, but he has no convictions.[274]

Misery loves company, and Abbott regularly tormented Volk that Putnam was betraying him as well. It pained Volk deeply. On the one hand he felt intense filial loyalty to Putnam for having employed him all those years, and yet it was difficult not to notice that of everyone Putnam hired to work under the auspices of the Columbian Exposition a decade earlier, Volk was the only one still without a permanent position or secure future in archaeology.[275] And why was his report, which would provide him long overdue professional recognition, taking so long to appear?

Fearing it never would, Abbott, then in his mid-sixties and deeply resentful of the lack of recognition of his own labors at Trenton, again picked up the torch. He self-published *Archæologia Nova Cæsarea*, a three-volume series that appeared in successive years starting in 1907. To his friends who were certain (or hopeful) that he was through with archaeology, Abbott's reappearance came "like a bolt from a clear sky."[276] The *Archæologia* series provided virtually no new evidence, nor new interpretations of the cultural sequence of the Paleolithic, Argillite, and Historic Indian he had established decades earlier (§3.6, §4.4). He felt the need only to reiterate, not revise, and no reason at all to change his mind.[277] His primary goal was set the record straight and respond to his critics, Holmes chief among them. While working on *Archæologia Nova Cæsarea* he admitted in his diary that some of it would be criticized as being in poor taste, but he easily rationalized the ugly tone: "Holmes has lied and I claim the privilege of saying so."[278]

Say it he did. The opening *Archæologia* volume began with a long preface criticizing Holmes's writings as disingenuous, petty, dishonest, gratuitous, delusive, reprehensible, insanely grotesque, and simply contemptuous. Abbott, with no sense of irony, claimed he was holding himself back. Here and in the next two volumes Abbott made clear the differences between him and Holmes. Holmes was a museum worker, a theorist who hardly dirtied his hands, and one who rarely let facts get in the way of pet ideas. Abbott, on the other hand, followed the facts in the ground wherever they led, and they led inexorably, inevitably, and unerringly to a glacial, even a preglacial, presence in the Delaware valley.[279]

Wright thought Abbott's thrusts at Holmes were "as clear-cut and keen as those of a Damascus blade" and congratulated Abbott for having so well restrained himself.[280] Holmes, on the skewering end, howled at Putnam: "I hope you do not approve of the style of the recent paper by Abbott on the Delaware valley evidence. It bears the earmarks of vulgarity and dishonesty on nearly every page." There was nothing Putnam could do about Abbott. His reply, which Holmes carefully posted for posterity in his autobiographical *Random Records of a Lifetime in Art and Science*, claimed ignorance of Abbott's latest sally and went further still. He essentially disavowed all of Abbott's Trenton work:

I have endeavored to keep all my work there entirely confined to Volk's careful researches of the past twenty years & he is still working for me. When Volk's detailed statements are published, we shall know the facts & until that is done I am perfectly content with the status as I understand it.[281]

That "status," of course, was that Abbott's artifacts from the Trenton Gravel were not accepted by the archaeological community as being Pleistocene in age.

Had Abbott seen that letter he would have exploded, not just because Putnam had seemingly abandoned him and everything he had done and said, but because he believed it was entirely Putnam's fault that Volk's report—which Abbott saw as vindication of *his* work—was not yet published, and likely never would be. By now he knew only too well the archaeological community paid him no mind, but maybe Volk's report would convince them.[282]

Or perhaps Winchell would. He had long lent a sympathetic ear to Abbott and in 1906 had retired to devote himself to the human antiquity cause and Abbott's vindication. Within a few years he boldly reformulated the Trenton sequence to begin in preglacial times and at the same time completely rewrote Holmes's interpretation of the Piney Branch quarries, to make its rude artifacts not manufacturing debris at all but instead artifacts of preglacial age. It took Abbott's breath away.[283]

Breathtaking it was, for although Winchell admitted Piney Branch was well south of the glacial front, he insisted the site was carpeted in transported glacial debris and thus contemporary with the Trenton Gravel. Among the Piney Branch debris, Winchell claimed, were artifacts deposited by glacial meltwater. These were of preglacial age, he reasoned, because they must have been in place before the melting of the ice sheet that rafted the materials south, and because Piney Branch lacked any postglacial or Neolithic artifacts. Finally, the artifacts were coated with "a sort of glacial patina" (that Holmes had overlooked), which was "invariably present" in old gravels of the drift age and never on recent artifacts.[284] Holmes's work was turned on its head. As Winchell reported to Abbott:

> The results of Dr. Holmes' excellent work seem, therefore, in spite of the conclusions of his seemingly careful and candid research, to confirm the conclusions which you [Abbott] reached so many years ago as to the paleoliths of the Trenton Gravel—and on this you are to be congratulated and should be honored by your fellow men.[285]

Abbott was thrilled. Arguments and evidence such as this coming from a geologist of Winchell's standing were surely the death knell to Holmes's position— and redemption for Abbott's nearly forty years of effort. But Abbott had badly overestimated Winchell's influence, and there is little indication that his radical revision of Trenton or Piney Branch, announced in Wright's *Records of the Past*, drew any attention.[286] Even Wright, who steadfastly opposed Holmes and supported Trenton, made no mention of this apparent breakthrough in the fifth edition of his *Ice Age* (released in early 1911).

Abbott continued to rage—at the archaeological community for ignoring him, at Holmes for disputing him, and at Putnam for failing him. By May of 1911, Abbott was approaching seventy and thinking of his own mortality and how he would be remembered by later generations. The thoughts depressed him. The epitaph he had planned for his tombstone, "Discoverer of Paleolithic Man in North America," would not come to pass, not unless the situation changed dramatically.[287]

He was unaware that at that very moment Putnam was correcting page proofs for Volk's report on *The Archaeology of the Delaware Valley*. Scarcely two months later all of Abbott's accumulated bitterness, resentment, and disappointment abruptly vanished when Volk telephoned with the news that his long-awaited volume was about to appear. Abbott, realizing he had wronged Putnam, immediately wrote a heartfelt apology. The long-awaited news naturally

> brought the past most vividly to mind . . . and much that I recalled was present to accuse me of having been hasty, and with genuine grief I felt impelled to admit that I erred in concluding that those whom I supposed were friends had turned aside. Now, the battle is over. Volk has demonstrated my contention as to the Delaware Valley was correct, and as I am and must remain unable to take active part in archaeological research, it would be inexpressibly satisfactory to me, if you were willing to let by-gones be by-gones.[288]

Putnam was willing, though he could hardly resist lecturing Abbott that he had been his own worst enemy by discarding friends who failed to support him and failing to take the "trouble to present the scientific evidence that would secure their unqualified support." Warming to the theme, Putnam poured out his own hurt and unhappiness with Abbott, for failing to appreciate all Putnam had done for him and the archaeology of the Delaware valley (including spending more than $25,000 on his and Volk's research), all of which he did because he believed in the importance of Abbott's work.

> No, my old friend! There never was any sense in your getting angry and writing me such letters as you have. Every one of which I was sure you would regret, but until you assured me, as you now have, I could only keep the sorrow to myself and await the day that has now come. Now you have opened your heart again to your old friend and have shown your regret for the injustice you did him. I am pleased . . . and gladly comply with your wish that by-gones shall be by-gones . . . and remembering only those [years] when we were working together in the ever-broadening study of early man in America.[289]

After nearly a decade of sporadic and unhappy exchanges, Abbott and Putnam reunited over Volk's report and began a flurry of correspondence that ended only with Putnam's death four years later.

Although there was a common thread running through Abbott and Volk's efforts in the Delaware valley, Putnam well knew that for Volk's report to emerge untainted by the poisonous air of Abbott's disputes it had to stand alone. His heavy editorial hand insured the volume contained no reference "to any other persons or their work," save for a listing in a bibliography safely quarantined in the back of the book. As Putnam explained to Wright (but not, understandably, to Abbott), Volk's book was strictly a descriptive account of his field observations that deliberately steered clear of "the old Abbott controversy."[290] Abbott's name was hardly mentioned in the more than 250 pages of laboriously detailed testimony and the sworn affidavits of other discoverers. Volk's readers who had no knowledge of the controversy might have imagined Abbott merely a local resident who occasionally visited Volk's trenches. Far more telling, there was no mention of Trenton Gravel paleoliths. Volk described one artifact as bearing a resemblance to "one of the Paleolithic forms of Europe," but he carefully avoided assigning it a great antiquity on that score. Abbott chose not to notice that omission, limiting his complaints to Volk having intentionally excluded from discussion any artifacts collected from the surface of the river bed.[291]

So far as Volk (and presumably Putnam) was concerned, the evidence of human antiquity in the Delaware valley rested strictly on "chipped quartz, quartzite pebbles broken by fire and fracturing, fragments of a human cranium, and a part of a human femur which had been cut and worked by man." Those came from trenches, railroad lines, or other cuts made into the gravel, having been transported there during the annual meltwater floods that occurred during the "last stages of the glacial epoch."[292] Thus,

> the conclusive evidence so eagerly demanded and with so much difficulty secured asserts the antiquity of man on this continent at least as far back as the time of these glacial deposits in the Delaware Valley.[293]

That is, so long as the geological situation had been properly read. Putnam himself admitted the geology of the region was controversial, but insisted that Volk's work would be successful if it provided archaeological evidence contemporary with geological deposits from "*any* portion of the glacial period and its immediate close."[294] On that point, he appended testimony to Volk's report from several geologists, Wright included, who had visited Trenton. However, none of them had spent more than a few days in the area (save Wright), and all of them had submitted their briefs in 1899, before Salisbury's extensive study of the glacial geology of New Jersey appeared (§5.4).[295]

Pleased as he was by the unexpected appearance of Volk's volume, Abbott still had mixed feelings. He feared it would be Volk's name that was forever linked to Trenton's Pleistocene archaeology and not his, and it would be Volk and Putnam's triumph that brought the controversy over human antiquity to an end. And so a month after receiving Volk's book, Abbott embarked on his last, *Ten Years' Diggings in Lenape land*. He vowed it would not be "disputatious" and only "deal in facts that *I* have accumulated." Being Abbott, he could not help being disputatious: even Volk was on the receiving end of petty and insulting comments.[296]

Putnam knew better than to believe that Volk's book or another from Abbott would settle decades of controversy. Knowing how important it was to keep Abbott's spirits up and dampen his volatile moods, he pointed out that history would surely record Abbott's seminal role, though he gently lowered Abbott's expectations by explaining it might be a very long time before that verdict was read. The Washington archaeologists continued to be skeptical because the results did not agree with their "pet theories." Still, he asked Abbott, "why should we care what such ones may say?" Putnam counseled patience and looked to the future when a new generation of workers would revisit the issue and decide it on its merits.[297]

Advising Abbott not to worry about what others thought of him was futile, and Putnam surely knew it. And by implying he did not care deeply about the outcome or about what others said of Volk's work, Putnam was only fooling himself. After all, Trenton was his legacy too. He was the one who had convinced multiple individuals and institutions to support Volk's efforts for more than a decade, and Abbott's in the years before that.

In the end, Volk's report proved anticlimactic. Its appearance did not force Abbott off the Trenton stage, nor did it vindicate Putnam's long-term financial and intellectual investment in the Delaware valley, for the simple reason that it never grabbed the spotlight.[298] It appeared on the scene with little fanfare and was largely ignored by an archaeological audience skeptical of there being anything new or compelling to be said about Trenton. Volk's enumeration and illustration of artifacts and bones found in the gravels would not resolve the case, not while their context and the geological history of the region remained unsettled.

It is telling that *The Archaeology of the Delaware Valley* was ignored not only by skeptics. Wright, though he believed that Volk's "marvelous" report left no room for doubt about Trenton and though he remained steadfast in his support of a Pleistocene human antiquity at Trenton, made no mention of it in his *Origin and Antiquity of Man* (dedicated to Putnam) in 1912, nor in his sixth and final edition of *Ice Age* (1920). In the latter, his discussion of Trenton remained almost precisely as first presented in 1889; so too his glacial geology. Despite the now-universal acceptance of multiple glacial episodes, he steadfastly remained convinced of the unity of the Pleistocene period, saw no reason to change his opinion that loess formed by water, and even stood by the attenuated border of the ice sheet in Pennsylvania and the Cincinnati Ice Dam. Wright was nothing if not fiercely loyal to his own ideas. But then Wright was convinced by Abbott decades earlier, and as he had not accepted any subsequent criticism of the Delaware valley claims reference to Volk's report was unnecessary.[299]

There the archaeology of the Trenton Gravel might have died a silent and unobserved death, and with it all trace of the American Paleolithic, were it not for another burst of effort by Winchell. In an examination of stone artifacts from Kansas, he seized on the notion that their patination (sheen and color) was proportional to their age, regardless of where or in what context they were found. The syllogism was simple yet flawed: genuine European Paleolithic artifacts were patinated and were Pleistocene in age; artifacts found in America that were patinated were therefore Pleistocene in age. Winchell paid no heed to processes that might cause patination and whether these were time dependent, nor did he consider that not all ancient tools were patinated and more recent ones might be. After all, "the fact that only Paleolithic forms are found with the patina of age proves that the Neolithic forms were developed later."[300] Because his Paleolithic and Neolithic artifacts were identified as such by their patina, he was chasing his own tail.

In *The Weathering of Aboriginal Stone Artifacts No. 1—A Consideration of the Paleoliths of Kansas*, Winchell crafted an elaborate prehistoric sequence based on his patination dating that ran from Early Paleolithic (highly patinated, relatively rude artifacts) through Paleolithic and Early Neolithic to Neolithic (not patinated, less rude). So certain was he of the relationship of patination to time that he identified single artifacts with multiple flake scars displaying different degrees of patination as ones first flaked in Early Paleolithic times, then picked up and successively flaked by later Paleolithic, Early Neolithic, and finally Neolithic peoples.[301] Thus, the same item was used over tens (or possibly hundreds) of thousands of years (given that he correlated the Early Paleolithic with the Aftonian and the Neolithic with the Wisconsin).[302] Best of all, he claimed, this proved Hrdlička wrong. Of course Pleistocene-age skeletal remains resembled those of historic Indians: there was now archaeological proof that glacial-age artifacts resembled those of the historic Indians. This "coincidence" of archaeological and anatomical results was testimony that in body and behavior modern-looking humans were present in glacial times.[303]

Winchell dedicated *The Weathering of Aboriginal Stone Artifacts* to Abbott, whom he considered the "Boucher de Perthes of America" for championing Paleolithic artifacts from the Delaware valley, despite his discoveries being "discredited and discarded by the powerful influences that are localized at Washington."[304] Abbott was overjoyed. Winchell's method would "simply revolutionize" American archaeology and finally conquer the "ultra-asinine Holmesian-pseudo-archaeologic monkeys that have set themselves up as know it alls." If he felt conquered, Holmes hardly showed it: he declared Winchell's volume "the best illustration of utter error & silly foolishness ever issued from the press."[305]

Winchell plowed ahead. He requested samples of Trenton paleoliths and argillite artifacts and soon received more than 1,000. By their patination and relative rudeness he crafted an elaborate chronology of the specimens, which, as Abbott reported, showed he had "post-dated everything rather than ante-dated," thus

> that the Trenton Gravel was "Kansan" and not "Wisconsin" and the interglacial period was high-water mark of flint chipping and not the day of historic Indians and that man originated in pre-Pleistocene times. He said my sequence of events . . . was correct, but it should be pushed back and not brought forward.[306]

Wright, always happy to have affirmation of Trenton's glacial age, was nonetheless none too keen to have it that old. He dryly replied that "there does not seem to be so much difference in the age of the so called Kansan deposits and the Wisconsin."[307]

No matter. Abbott was thrilled, and Winchell was on a roll. In the spring of 1914 he geared up for a season of fieldwork to culminate in Trenton in October, where he planned to accompany Abbott and Volk, visit their sites and collections, and examine the regional geology, all with an eye on gathering the material for his next book, a definitive study of Delaware valley archaeology. But in early May Winchell died. Two days later, a melancholy Abbott tossed in the towel. He boxed up the letters and papers and specimens Winchell had sent him and shipped the lot to Putnam, in hopes it would aid someone with "younger feet" to take up the subject. None did. As usual, he blamed Putnam.[308]

Sending the material to Putnam was auspicious. Six months later a fast-moving fire set by hunters engulfed Three Beeches, Abbott's family home. In forty minutes it reduced the house to ashes. Everyone in the house escaped unharmed, but Abbott lost nearly all his possessions, his family's "accumulated treasures of centuries."[309] The pain was so acute he could barely speak of it in a letter to Wright six weeks later.

But work at Trenton still was not over. In the summer of 1912 Leslie Spier, an engineering student at City College of New York with a growing interest in anthropology, was hired by the AMNH to conduct an archaeological reconnaissance of the Trenton area. Spier and colleagues, working under the direction of curator Clark Wissler, did a surface survey of the region to assess Abbott and Volk's inferences that "'Argillite Man' and 'Paleolithic Man' were racially distinct from the historic [Lenni Lenape] Indian." The work continued into the following summer and led to the conclusion that there were sites with distinctive artifact assemblages, some clearly historic and others not. (The latter lacked pottery and had argillite artifacts.) Spier was sure the patterns were not attributable to contemporary differences in site function.[310]

Resolving whether they represented chronologically distinct occupations required excavation, and in the summer of 1914, alongside the lane to the burned shell of Three Beeches, a series of trenches was excavated. It was cutting-edge fieldwork. Everyone at the AMNH was well aware of Nels Nelson's innovative methods then being employed at San Cristobal Pueblo to craft a ceramic chronology by tracking the stratigraphic distributions of artifacts.[311] Accordingly, Spier's excavations involved carefully troweling four-inch arbitrary levels measured from the well-marked stratigraphic contact between the surface soil (thick with pottery, bone, shell, and copper implements of the historic Indians) into the underlying "yellow loam," in which all agreed the argillite artifacts seemed most common. All the sediment was "sifted." This was one of the earlier occasions of screening systematically employed in an archaeological excavation.[312]

Spier collected, counted, and plotted the density of artifacts by depth and relative to the occurrence of clay lamellae and pebble lenses in the yellow loam. The argillite artifacts had a "unimodal vertical distribution" and were quite separate stratigraphically from artifacts in the surface soil above, therefore almost certainly "culturally distinct."[313] In the months before the fire Abbott had watched with delight: "All I have to do is sit on my north porch, and occasionally walk down to the pits or trenches and say 'I told you so'."[314]

Wissler followed Spier's study with a statistical analysis (means and standard deviations) of Spier's stratigraphic data on both artifacts and pebbles occurring in the yellow loam, concluding all were deposited in "the same series" and at the same time. Wissler inferred this meant "the depositing agent" was geological, not cultural; given the size of the pebbles, that agent was likely water and not wind (wind lacking sufficient competence to carry in the pebbles).[315]

Neither Spier nor Wissler said anything of the Paleolithic or the Trenton Gravel, the absolute age of the yellow loam, or the Pleistocene antiquity of any of the remains they recovered. At least not in print. Privately Wissler told Putnam he thought the sands at Trenton could be linked to a glacial meltwater deposit—no doubt welcome news to the dying Putnam, though since the conclusion was tentative Wissler forbade him from mentioning it. But such concerns hardly mattered to

Abbott. Winchell had assured him that the argillite was, in fact, decomposed basalt and must be of great antiquity. Therefore, Abbott easily spun Spier and Wissler's results about the argillite in the yellow loam into affirmation of the still-greater antiquity of the Paleolithic artifacts in the underlying gravels:

> Now, at last, I am on top. The existence, one-time, of Glacial Man is not a fairy tale as the Washington crew declared, and I feel that my work was not in vain, but who else would have stood up under the tempest of interpretation as I did?[316]

This time, Abbott was not the only one who overreacted.

In the fall of 1915 Holmes heard about the AMNH fieldwork and wrote Henry Kummel, now the New Jersey state geologist, to complain that "explorers . . . entirely unqualified to observe or discuss geological phenomena" were trying to raise the old ghost of Abbott's "argillite culture . . . bugaboo." What, he wondered, did Kummel think of that? Kummel, having innocently ventured into the Trenton controversy nearly twenty years earlier and been badly burned (§5.15), was not making that mistake again. He carefully replied that he himself was not directly involved, but that the geologist at the AMNH assisting the archaeologists was of the opinion that the artifact-bearing deposit (the yellow loam) was not glacial in age but was older than the surface soil. Beyond that, he would not venture to say.[317]

Holmes tried again a couple months later (in early 1916), when Wissler sent him a summary of the Trenton work that he hoped Holmes would place in the *Proceedings of the National Academy of Science*. Wissler's paper threw Holmes for a loop. It was mostly about the methodological achievement of applying "the machinery of frequency curves" to the study of artifact distributions, as well as its implications for stratigraphic work. Holmes, always pleased to develop the science of archaeology, could hardly disapprove of that. On the other hand, in setting the context for the Trenton work, Wissler spoke briefly of the decades of controversy, though he was scrupulously nonjudgmental, as one who was anxious to avoid becoming himself entangled might be.

Nonetheless, even the mere mention of Abbott as the person who had initiated the discussion of Trenton archaeology was too much for Holmes. He wanted censure, not neutrality. He urged Wissler "to leave out any statement that seems to carry with it the approval of the great body of rotten work done on the Trenton site." He feared that Wissler writing about the yellow sands would inevitably lend some credence to Abbott's sequence of "a Paleolithic culture, and an argillite culture," and Holmes could not countenance that "seeming approval of the mass of error that has been read into the gravels at Trenton by Abbott, Putnam, and others." He returned Wissler's paper, telling him that if he wanted it in the *Proceedings* he was "perfectly at liberty to forward it to Editor Wilson."[318] Wissler did not bother.

Another Trenton publication not to Holmes's liking had been suppressed (§5.15).

CHAPTER SEVEN

Dangerous to the Cause of Science, 1915–1925

IN AUGUST OF 1915 Frederic Ward Putnam passed away. Clark Wissler used the occasion to take stock of the state of affairs regarding human antiquity in America. Although surely not meaning to tarnish Putnam's memory, Wissler's epitaph had that unfortunate consequence:

> The progress so far made may be said to stand as a historical memorial to Professor Putnam, who, in spite of long years of almost barren search, still held firmly on to such faint glimpses of man's antiquity as chance threw in his way.[1]

An "almost barren" search: damning with faint praise, that. If Putnam had been unable to see anything more than faint glimpses, just what progress had been made?

Wissler pointed to the large literature based mostly on odd finds of stone or bone, which by their nature were unverifiable, dependent on the "mere assertions of an individual," had little scientific worth, and ultimately left no recourse to the archaeological community "but extreme skepticism." Putnam had failed. Still, Wissler was an equal-opportunity critic. He had no use for Hrdlička's arguments against antiquity, though he was savvy enough not to mention Hrdlička by name ("some writers," he said). The presence of a morphologically distinct skeletal form would be a good argument in favor of antiquity, Wissler argued, but its absence "would not be equally good evidence against it." After all, the Cro-Magnon of Pleistocene Europe bore a resemblance to modern peoples.[2]

But it was not just the scrappy nature of Putnam's evidence or the "fallacy" of Hrdlička's reasoning that were to blame for the controversy and unsettled fate of so many claims to great antiquity. It was "the absence of associated faunistic or other chronological indicators." If there had been remains of extinct fauna of Pleistocene age at Trenton, Lansing, Long's Hill, or any of the other contenders found over the decades, Wissler thought it might have been possible to resolve the antiquity controversy.

If Wissler had read Thomas Chamberlin's warning a dozen years earlier about the danger of using extinct fauna as chronological indicators (§6.9), he showed no sign of it.[3] All that was needed, so far as Wissler was concerned, was a site with human skeletal remains (modern-looking or not) or stone artifacts (rude or not) in direct association with the bones of a now-extinct animal. Of course, that had not worked for Williston at 12 Mile Creek. Chamberlin had seen to that (§6.9). Obviously, Wissler believed, as Chamberlin did not, that however unsettled the chronology of the extinct Pleistocene fauna, it was sufficiently well understood by 1916 that fossils of these animals could serve as diagnostic time markers.

As it happens, the year before Putnam died a nearly complete human skull and additional skeletal elements were recovered in Rancho La Brea, California from an asphalt "chimney" in the famed tar pits in apparent association with the extinct giant bird *Teratornis*.[4] Scarcely three months after Putnam's death, and about the time Wissler was drafting his elegy, human skeletal remains were found on the opposite side of the continent near Vero, Florida. They were spotted by fossil collectors actively mining a wealth of Pleistocene fossils, including those of mammoth, mastodon, horse, ground sloth, and sabertooth cat from a recently excavated drainage canal.[5]

The discoveries at Rancho La Brea had little impact. The *Teratornis* notwithstanding, the bulk of the fauna in that particular chimney was obviously younger, including as it did modern bear, coyote, wolf, skunk, antelope, rabbit, and pocket gopher. Even the horse mandible found near the human skull seemed to be a domesticated form. These were not the extinct species that typified Rancho La Brea, such as the dire wolf, sabertooth cat, and camel. And the human skull itself, according to anthropologist Alfred Kroeber, did not differ markedly from skulls of the Native Californians and had none of the features of the "ancient races" that might place it in the Pleistocene.[6] Either the *Teratornis* survived after the extinction of the other Rancho La Brea Pleistocene fauna, or its bones had been reworked from an older deposit and come to rest alongside the remains of more recent animals. The latter was certainly possible given the considerable potential for movement of bones within the viscous asphalt of the chimney.[7] In the circumstances, University of California vertebrate paleontologist John Merriam declared the age of the human remains measurable in thousands, and not tens of thousands, of years. With that, Rancho La Brea quietly slipped off the stage of contenders of a deep human antiquity.

Not so the Vero find. Shepherding this discovery was Elias Sellards, then Florida's state geologist. He was certain Vero offered "conclusive proof" of a Pleistocene human presence, and in October of 1916 he invited proponents and skeptics alike to come to Florida to see the evidence for themselves.[8]

The Vero site visit signaled the beginning of a decade (1915–1925) in which a half-dozen finds were made of human skeletal remains in purported association with extinct Pleistocene fauna. No longer was the issue the apparent rudeness of the stone artifacts, the morphology of human skeletal remains (though Hrdlička continued to play that card), or the age of the gravel or loess in which they were found. There were the perennial questions of whether those remains were in primary depositional context and whether the association with the chronological markers (in this case, the bones of the extinct fauna) was secure, but there were the added complications of whether the fauna was Pleistocene in age and, if the faunal evidence conflicted with that from glacial geology or anthropology and archaeology, how that conflict might be decided. Resolving a case for a Pleistocene human antiquity where human skeletons and Pleistocene vertebrates were present was not going to be as easy as Wissler supposed.

7.1 OLIVER HAY OFFERS A FAUNAL SOLUTION

In late eighteenth- and early nineteenth-century North America, vertebrate pale-ontological attention focused on near-surface forms that were dominantly Pleis-tocene in age, such as mastodon, mammoth, and ground sloth, which included the specimens Gorges Cuvier used to demonstrate the reality of extinctions (§2.1). After the Civil War, the focus turned to fossils of older Cretaceous reptiles and Tertiary mammals and swirled around the epic feud between Edward D. Cope and Othniel C. Marsh, who competed in the rich fossil beds of the American West for specimens to fill the exhibit halls and laboratories of rapidly sprout-ing museums and universities (a feud that, by virtue of Marsh's appointment as chief paleontologist of the USGS, spilled over into the early 1890s USGS bud-get battles [§4.10]).[9] In the late nineteenth century, Pleistocene mammals were mostly (though not wholly) neglected by Cope and Marsh and many who worked with them, such as Erwin Barbour, Oliver P. Hay, Henry F. Osborn, and Samuel Williston (all of whom had a role in the human antiquity controversy [§6.6, §6.9, §6.11]). By century's end it was known what Pleistocene fauna looked like. Not known was how to differentiate Early from Late Pleistocene taxa, when a particu-lar genus or species went extinct, and whether some had managed to survive into the Recent period. Which, of course, was why in 1903 Chamberlin was reluctant to accept faunal evidence as proof of a Pleistocene antiquity, either for the fossils themselves or for any apparently associated human remains (§6.9).

Ironically, Chamberlin and Holmes lobbied hard for the employment of pale-ontologist Oliver Hay, who in the early twentieth century emerged as the most prominent authority on Pleistocene vertebrates, but also as one of the most vigor-ous proponents of a Pleistocene human presence in the Americas. Yet, unlike other proponents of a deep human antiquity, Hay had the respect of his peers—for a time.

Hay was born the same year as Holmes (1846), but his career had a much longer takeoff. Raised and educated in the Midwest, he taught in several small colleges in the later decades of the nineteenth century, did a year of graduate work at Yale, and ultimately earned a PhD at Indiana University (in 1884). For over a decade (1879–1892) he was on the faculty at Butler College (Indiana), where as a biologist and geologist he helped organize the Indiana Academy of Sciences. After a fossil-collecting trip to western Kansas in 1889–1890, he began to move into paleontology. He slowly made his way into research positions, with a brief stint in the Indiana Geological Survey and then in the Field Museum of Chicago (1895–1897), where his tenure overlapped with that of Holmes (it was likely where they first met), and then on to the AMNH (1900–1907), where he produced a massive study of the fossil turtles of North America.[10] After a brief retirement, he joined the Carnegie Institution of Washington as a research associate in 1912, a slot he held until his second retirement in 1926.[11]

Among those who wrote in support of his Carnegie Institution appointment were John Branner, Merriam, Holmes, and Chamberlin. Holmes was effusive in his praise. In words he may have come to regret, he proclaimed Hay's work on the fauna of the separate stages of the Pleistocene period as vitally important to issues of human antiquity and professed "a very high opinion of Dr. Hay's ability and his qualifications." Chamberlin backed away from the hard line he had taken in 1903. Now that North America's glacial formations were more fully mapped and understood, it was time to link glacial and vertebrate histories in order to identify the relative ages of specific, now-extinct species. It seemed to Chamberlin "especially fortunate" that Hay, with his prior experience and contributions along these lines, was willing to take on the task.[12]

It is no surprise that Chamberlin thought so. Two years earlier in *Science*, Hay had taken to task the imperial Osborn, his former superior at the AMNH, for failing to notice the "great advances" made in North American glacial history, specifically the work of Chamberlin and others in delineating multiple glacial and interglacial deposits. That failure, Hay lectured Osborn, meant "no correct solution of the paleontology of the Pleistocene is possible."[13] Hay knew better. Armed with a detailed knowledge of the fossil locations of extinct genera, he was able to demonstrate that some of those taxa—horses, say—were found only in interglacial (Aftonian) soils below glacial drift, and not in regions glaciated during later Iowan or Wisconsin times, whereas other taxa—mastodon and mammoth, for example—were found in lakes and pond deposits created by the retreat of the Wisconsin ice and had therefore survived even into postglacial times.[14] This was just the sort of hard evidence lacking in 1903 when Chamberlin dismissed vertebrate faunas as ambiguous and unreliable chronological indicators (§6.9). Hay aimed to resolve that ambiguity and unreliability.

At the time they recommended Hay for the Carnegie Institution appointment, neither Chamberlin nor Holmes fully understood another of Hay's qualities, notably just how much he relished using highly visible venues like *Science* to needle those with whom he disagreed. Hay proved to have an extremely sharp tongue, and he was capable of mercilessly skewering opponents, all with a mischievous sense of humor. For some like Osborn, the humor removed some of the sting (in 1918 Osborn warmly praised Hay's knowledge and abilities), but for others—especially Holmes and Hrdlička, who became favorite targets of Hay's barbs—he was not the "useful and lovable" figure his biographer painted him to be.[15]

Within a year of his Carnegie Institution appointment, Hay began to compile the location of Pleistocene vertebrate fossils and published a series of maps showing their distribution relative to the most current maps of glacial features, the results of which led him to define within the Pleistocene a series of successive faunal periods. The youngest was the Wabash, which Hay named for several particularly rich post-Wisconsin marshes, ponds, and lakes in Indiana, the fauna of which included

a species of megalonyx, the American mastodon, at least two species of elephants [mammoths], the giant beaver, one or two extinct genera of peccaries, at least three extinct genera of musk-oxen, and the extinct moose, *Cervalces scotti.*[16]

The faunas that preceded the Wabash were composed of herbivores and carnivores that Hay provisionally divided into two chronologically separate faunal communities:

> The early Pleistocene was characterized by the existence of numerous edentates [sloths], horses, camels, tapirs, and saber-tooth cats, and few bisons, while during the later pre-Wisconsin Pleistocene there were few edentates, few horses, no camels, few saber-tooth cats, but numerous bisons.[17]

The Early Pleistocene faunal unit Hay designated the "Aftonian." The later but still pre-Wisconsin fauna he labeled the "Sangamon." That Hay's faunal periods coincided with the interglacial periods of Chamberlin and others (Table 6.1) is no coincidence. Hay was convinced that one of the most potent factors in determining the community composition of species was not "moisture or heat" but time, which "never ceases to act and its influence is inexorable. A great assemblage of animals is swept away and another is put in the place of the old, and this is doomed itself to disappear in its turn."[18]

In 1914 Hay tackled the taxonomic challenge of differentiating the extinct bison species.[19] That effort established his expertise on that genus, knowledge he later called on in the human antiquity debates of the 1920s (§8.6). But in 1914 human antiquity was scarcely of interest to him; he made only passing mention of 12 Mile Creek and its associated artifact.[20] Within just a few years, however, Hay became far more vocal on the subject of human artifacts associated with extinct taxa—and in fact a passionate advocate for the idea of a Pleistocene human presence in the Americas. And once Hay's mind was made up on a subject, he held tenaciously to his opinions, making it difficult to "move him from his final decision."[21] Hay's stubbornness and his cheeky, confrontational style burst forth in the Vero dispute. By then Hay was already past his seventieth birthday, yet he leapt into the fight with a zeal that belied his age.

7.2 MEN AND MAMMOTH AT VERO

Samuel Williston trained many students at the University of Kansas before he left in 1902 for the University of Chicago (§6.6). Two of them, Barnum Brown and Elias Sellards, began their careers as undergraduates with Williston in the mid-1890s. Decades later they each had significant roles in the human antiquity controversy. Brown's at Folsom in the 1920s was triumphant; Sellards's at Vero a decade earlier was not. The humiliation Sellards suffered there scarred him deeply (§1).

Sellards had not gone looking for a role in the human antiquity controversy. As a student in geology and paleontology he discovered and worked on a rich collection of Permian insects under Williston, earning a BA and MA (in 1899 and 1900). He then followed the path Williston had taken two decades earlier and went to Yale for a PhD in paleontology (1903). After brief stints at Rutgers University and the University of Florida, Sellards was appointed Florida state geologist in 1907.[22] For much of the next decade his research focused on the geology and mineralogy of the state, but in November of 1913 his attention was drawn to Vero's rich fossil deposits by two local collectors, Frank Ayers and Isaac Weills.

Weills and Ayers had been gathering fossil bones from a recently excavated drainage canal near the Florida coast. At one spot the canal sliced through a buried deposit where Weills spotted a variety of vertebrate, invertebrate, and plant fossils that Sellards recognized as being Pleistocene in age. The importance of the locality was amplified in October of 1915 when Ayers found "in place" much of the lower half of a human skeleton, likely that of a female. The bones were in a "cross-bedded river-wash sand," stained brown by organic matter and with inclusions of "fresh-water marl rock." The deposit, which Sellards designated Stratum 2, was sandwiched between an underlying marine shell marl (Stratum 1) and overlying alluvial sediments (Stratum 3) "consisting chiefly of vegetable material intermixed with sand," which graded at the surface into a freshwater marl.[23] In addition to the human remains, Stratum 2 also contained a rich Pleistocene fauna that included *Elephas* [syn. *Mammuthus*] *columbi, Equus, Megalonyx, Mammut, Smilodon,* and *Canis.*

Sellards was not unaware of the controversy over human antiquity in America and the significance of finding human remains among these fossils: he was, after all, a student of Williston. In announcing the Vero find in July of 1916, Sellards insisted that the human and nonhuman bones came from the same stratum ("the section is continuous along the canal bank and the deposits identical in appearance") and that chemical analyses (of phosphoric acid, calcium oxide, "insoluble matter," and iron and aluminum oxides) of the human, *Canis,* and *Megalonyx* bones in Stratum 2 showed that they were similar in the elemental constituents, and that the human remains were "quite as well mineralized as are the associated bones of the Pleistocene animals." He had included in his analysis human bone from a nearby Indian mound to show that it differed in its chemical content and mineralization from the Vero Stratum 2 specimens.[24]

Sellards knew that many claims for a great human antiquity had failed for lack of demonstrable stratigraphic integrity, and he understood Vero was vulnerable on this score. Its human remains were found relatively close to the surface (in some instance, within two feet)—and this would spark the inevitable criticism that the human skeleton was of more recent age and became buried in older deposits. Anticipating that, he observed that neither the stratum in which the bones were found nor the one above it were disturbed or showed any signs of burial. Were it a burial, he argued, "more of the skeleton would have been found." Moreover, human bone of recent age would not have had time to become as "thoroughly mineralized" as the bones of the Pleistocene fauna.[25]

The Stratum 2 remains he regarded as "conclusive proof" of a Pleistocene human presence at Vero. Even better, these were not the only human remains from the site. Ayers had kept a watchful eye on the locality, and four months after his initial discovery he spotted an ulna in loose dirt on the floor of the canal; by its adhering sediment and the degree of mineralization, it appeared to come from Stratum 3. Watching the spot carefully, he soon found again in loose dirt a humerus and then, accompanied by Weills and Sellards, a series of additional human bones culminating in the discovery in situ in Stratum 3 (just above the Stratum 2 stratigraphic contact) a variety of cranial and postcranial elements of an adult male. These remains were scattered along a horizontal distance of some ten feet and, because of the slope of the stratum, a vertical span of several feet.[26]

In that same deposit were bones of a dozen other vertebrates, of which Sellards believed at least half were extinct (such as the armadillo *Chlamytherium* [syn. *Holmesina*] *septentrionalis*). Although fragmentary mammoth and horse remains were also found in the lower portion of Stratum 3, Sellards could not preclude these having been redeposited from Stratum 2, especially as there was evidence that Stratum 3 filled a channel cut into Stratum 2. Excluding the mammoth and horse, he was confident the remainder of the Stratum 3 fauna was associated with the human remains. All were mineralized in the same manner and showed no signs of the wear associated with reworking. As before, Sellards believed "the position of this [human] skeleton and the conditions of preservation are such as to exclude definitely the possibility of its representing a human burial."[27]

Sellards was certain Stratum 2 was deposited during the Pleistocene and Stratum 3, with its extinct fauna, was of "great antiquity" as well. As to how old it was, he observed that after the human skeleton in Stratum 3 was "entombed," an additional "two feet of vegetable material and sand" had accumulated in the valley. Given the sluggish nature of the stream responsible for its deposition, Sellards inferred Stratum 3 accumulated "very slowly" over a long period. Beyond that, he did not venture.[28]

However, he soon learned from Hay that Vero's fauna might be much older than he had imagined. Sellards had sent Hay the Vero turtles for identification, and Hay was of the opinion that those from Stratum 3 were Pleistocene in age— and "not the later part of the Pleistocene," either.[29] That was surprising because, as Sellards reported, "a considerable amount of broken pottery" and "well worked flint arrowheads" also came from Stratum 3.[30] Sellards was struck by the implications of Hay's identifications:

> If that stratum is Pleistocene then it is certain that man in America was making both pottery and arrowpoints in Pleistocene time, and . . . you know, particularly the manufacture of pottery, is a more advanced stage of culture than is supposed to be shown by the Pleistocene of Europe.[31]

Sellards published on the findings, then he made his next move: letters went off to Hrdlička and Osborn, and then to Chamberlin, vertebrate paleontologist William D. Matthew, and Williston, telling each about the site and inviting them to visit. Hrdlička had not seen Sellards's paper, but he immediately responded with a caution: human burials, he told him, "may be made in almost any strata that can be worked, and isolated human remains may be introduced into such strata in many different ways." Human bones in ancient deposits demanded a higher standard of proof of their antiquity than nonhuman remains. Having duly warned Sellards, Hrdlička told him he would be pleased to visit to see the site for himself.[32]

Hrdlička wasted little time showing Sellards's letter to Holmes, who in turn wrote Chamberlin urging him or Rollin Salisbury to visit Vero. "The matter is of considerable importance," Holmes stressed, "not because of any great probability that early man has actually been discovered, but that a mistake by a man in Mr. Sellards's position may be difficult to correct." If allowed to pass unchallenged, Vero would become "another stumbling block to the process of human chronology on this continent."[33]

By the time he received Holmes's letter, Chamberlin had already heard from Sellards, who also told him he planned to send a paper to the *Journal of Geology* on the Pleistocene vertebrates of Florida, including those associated with human remains. Chamberlin assured Holmes that

> papers of this kind, whose implications go so far—usually so far beyond the real evidence of the case—should be subjected to close scrutiny before they are admitted to the columns of a responsible scientific magazine. We have spent a great deal of precious time on points of this sort . . . and are earnestly desirous of establishing a higher standard in this important respect [§5.8].

In replying to Sellards, Chamberlin proposed that there ought to be a committee of visitors to examine the site and that it ought to include a paleontologist, a geologist who was expert in Pleistocene formations, and an anthropologist. He suggested that Salisbury or perhaps his own son Rollin could handle the geology, and that Holmes or Hrdlička could speak to the anthropology. As for the paleontologist, Chamberlin went down the hall to see Williston, who was happy to visit his former student's site, schedule permitting.[34]

Over the next few months, plans came together for a joint visit to Vero.[35] With the planning there was considerable jockeying for position. Sellards assured his correspondents the human bones were not from a burial, owing to the intact and laminated strata above, and were directly associated with the bones of the extinct vertebrates, as some of those bones appeared to have "markings" (cut marks) made by stone tools found nearby.[36] There was also new evidence to report. Further work in the Stratum 2 sands yielded more human remains, primarily lower limb bones along with a few "flints" (flakes), a "small bone implement," a cut mammoth tusk, and additional fossils of extinct horse, tapir, and armadillo as well as those of several extant mammals. As before, the human and nonhuman bones showed the same degree of mineralization and the overlying deposit was laminated and undisturbed, further affirmation that "man was present in America in association with a Pleistocene vertebrate fauna."[37] Sellards announced this latest discovery in *Science* in late October of 1916, at the same moment as the Vero site visit was taking place. Intentionally or not, he had raised the stakes.

Hrdlička, who was busy working the other side of the street, could see Sellards was lobbying hard for his claims and was unhappy about it. Without the slightest hint of irony, he complained to Chamberlin that Sellards was using every possible opportunity to accentuate his "utmost conviction " that Pleistocene man has at last been discovered. This cock-sureness from the very start and repeated in every letter to Prof. Holmes, myself, and others in the Institution make us naturally a little bit more suspicious instead of convinced; and his rush on all sides into print with his discovery seems also a little hasty.[38]

Reading of the Vero find, Charles Abbott thought otherwise: "Got official report of Florida 'antiquity' finds. Same old story as I told up here, only more details. It confirms me absolutely."[39]

7.3 A NONHARMONIC CONVERGENCE

All was set for the site visit. Rollin Chamberlin, Hay (a last-minute replacement for Williston), Hrdlička, Yale University archaeologist George Grant MacCurdy, and USGS geologist T. Wayland Vaughan arrived in Vero the last week of October of 1916. On the eve of their arrival Williston warned Sellards (as he had Barbour a decade earlier [§6.11, §6.13]) of the trouble headed his way:

Messrs Holmes and Hrdlička are to visit you soon I believe. *They are wholly partisans* and will not examine your evidence in a true scientific spirit. Furthermore, as I have urged to Professor Chamberlin they are in no wise competent to pass upon the evidence. They are ethnologists and not geologists or paleontologists and while I would trust fully their opinions on the nature of the human remains I would not give a rap for their opinion as to their age. . . . Furthermore, they assume from the very beginning that you are wrong. I am saying but what is notorious. Do not let the matter rest with what they or their friends say.[40]

But Williston also assured Sellards he could trust Rollin Chamberlin to "find the real truth in the matter," and that he considered Hay "the most competent of the paleontologists to give an opinion."[41]

Holmes did not attend but hoped the visitors would give Vero "a thorough sifting."[42] A sifting can yield very different results, particularly when participants bring different perspectives, their visits do not overlap, and their examinations are conducted largely independently of one another.[43] Their interpretive differences were obvious almost immediately afterward, at least to Thomas Chamberlin, to whom many of the participants confided their opinions. Chamberlin reported to Sellards that Rollin had found Sellards's "observations trustworthy, as he anticipated, and the only essential points of difference will be those of interpretation, and that, of course, is all in the work of a day, with geologists."[44] A few days later Chamberlin heard from Vaughan that the evidence was not decisive as to whether the site was Pleistocene or Recent, and then from Hrdlička (via Holmes) that "the majority of the evidence will be against Sellards's conclusions."[45]

Seeing this divergence of views, Chamberlin proposed to Sellards that the participants prepare their individual reports for publication in the next (January 1917) issue of the *Journal of Geology*. They would have to be quick about it. Chamberlin wanted to release the issue in December of 1916 in time for the holiday meetings of the AAAS and Paleontological Society, "at which discussion may not unnaturally arise."[46] Chamberlin had other reasons for rushing the authors: he wanted "independence of the reports," and with that accelerated schedule there would be no time to circulate the manuscripts among the authors.[47] His gambit worked. Because Hrdlička "had very little to say" while at Vero, Sellards had no idea what Hrdlička thought. He prepared for the worst, anticipating that Hrdlička would "find objections of some kind."[48]

When Sellards received his issue of the *Journal of Geology* he learned just how far Hrdlička's objections went, and indeed how far apart all the participants were in their assessments of the site. Worst of all, he learned how few seemed to agree with him. Opinions split along multiple lines. On the question of whether the human bones were, in either stratum, in situ, MacCurdy, Vaughan, and (most vocally) Hrdlička insisted that the possibility of human burial precluded these remains being the same age as the Pleistocene fauna. Sellards, obviously aware that the case hinged on whether the human bones were in primary context, insisted a later burial was unlikely. Who, he asked, would bury their dead in that "muck bed"? Besides, the overlying strata were intact. Hrdlička anticipated that response: the "muck and sand" disturbed by burial, he observed, "would tend in the course of time . . . to assume the appearance and characteristics of the original deposits."[49]

Sellards and Hay pointed to the human skeletons themselves, which were as "scattered, imperfectly preserved, and frequently broken" as the Pleistocene vertebrates. But Hrdlička had anticipated that argument as well. The human bones in Stratum 2, he learned from Ayers, were found close together and apparently articulated, "the position they would occupy in the body." The Stratum 3 human remains were disassociated, but they were still within a relatively small area and at roughly the same depth below surface, which was "about that of a common Indian burial." Besides, the human bone was in far better condition than the thousands of nonhuman bone fragments from those strata, and this indicated a different depositional history, notwithstanding their similar degree of mineralization. En route home from Vero, Hrdlička detoured to the Florida coast to examine a burial mound in which human bones were enclosed in "concrete," further testimony to the rapidity of mineralization under the right environmental conditions.[50]

On his return to Washington, Hrdlička submitted to Edgar Wherry, a USNM mineralogist, samples of the Vero bone along with comparative material of recent age from other Florida sites (including a specimen from a shell mound near Osprey collected a decade earlier that had first convinced him mineralization was not invariably correlated with antiquity [§6.12]). The object of the analysis, Hrdlička explained to Sellards, was "to ascertain the fluorite contents of the bones and also to show if possible the different grades of fossilization of the different bones in the two lots." The result demonstrated there was nothing to indicate "any great antiquity of the Vero specimens," let alone differences with the several bones "of undoubted Indian derivation."[51] Hrdlička relegated Wherry's report to an appendix of a long critique of Vero he was busy preparing.

Sellards got no help on these points from Rollin Chamberlin, who—though he accepted Sellards's claim that the human skeleton was not a burial—agreed with Hrdlička that the Pleistocene vertebrate remains were far more fragmented than the human bone and had almost certainly been "transported to a greater or lesser extent." Indeed, Chamberlin undermined Sellards's geological interpretation altogether by suggesting that because the fauna had been reworked from Pleistocene deposits on the nearby uplands it mattered little whether the human bone was buried or not, as it was not the same age as the redeposited Pleistocene animal bones. All that could be said of the age of Stratum 2 was that it was older than Stratum 3—which Chamberlin considered Recent—but closer to the Recent than to the age of the upland Pleistocene deposits that "originally housed the old mammalian bones." Sellards likewise had no help from Vaughan, who accepted the possibility of human burial and rejected the evidence by which Sellards concluded that Stratum 2 or Stratum 3 was Pleistocene in age. Vaughan thought judgment on the antiquity of these deposits "should be suspended."[52]

Hay was having none of that. There was neither stratigraphic nor paleontological evidence that the Pleistocene fauna had been washed in: the animal bones did not have the appearance of "transported fossils." There were fragile remains such as a nearly complete carapace of a turtle, which could not have withstood transport of any significant distance. Even the apparently younger forms, such as the deer and fox bones in Stratum 3, were not Recent species (or so Hay supposed, but he was a taxonomic splitter[53]). Had human bones not been found, he added, no one would have questioned the age of the fauna. (Hrdlička, insistent on a higher standard when human remains were involved, would have agreed [§7.2].) The strata were old: the fauna was "essentially that which is found in the Aftonian interglacial beds in Iowa and in the *Equus* beds of the Plains." The only objection Hay could see to Vero being Early to Middle Pleistocene in age was that it "contravenes our present ideas regarding the history of the human race."[54]

Although Hay and Sellards were willing to have Vero contravene human history on this continent, the anthropologists were not. Hrdlička brought samples of the Vero pottery and human remains back to the Smithsonian for more detailed study. He saw nothing to "remind the anthropologist of early man." All the anatomical features were consistent with "recent, more especially Indian, bones." Holmes examined the pottery and likewise concluded it was a type "in common use among the Indian tribes of Florida."[55]

MacCurdy echoed Hrdlička's and Holmes, but then—Chamberlin's efforts notwithstanding—he had compared notes with Hrdlička in advance. MacCurdy further undermined the case for Vero's antiquity by pointing out that the cut bone and tusk from Stratum 2 were natural, not cultural. The "flint" flakes and bone tools were undeniably artifacts (some flakes manufactured of stone from a source a hundred miles distant). Yet, they were so few in number and looked so much like those in the overlying Stratum 3, where stone artifacts and bone tools occurred with greater frequency (some flakes seemed to come from the same "parent block" [core]), that MacCurdy strongly suspected there had been stratigraphic mixing by way of roots or burrowing animals. He doubted any human remains or artifacts were in primary context in Stratum 2.[56]

MacCurdy harbored similar suspicions about the Stratum 3 remains. If its "pottery, bone implements . . . and flint arrowheads" were as old as the Pleistocene vertebrates, then either the Neolithic in America was older than anyone realized, "fossil mammals continued to live on in Florida until a comparatively recent date," or the whole was a mass of redeposited cultural and natural material. Because the site was a "meeting-place of waters," the pottery showed signs of being waterworn, and some of the leaves in Stratum 3 appeared as though "buried only a few years ago," MacCurdy suspected Stratum 3 was badly mixed, making it "hazardous" to attribute any great antiquity to its remains.[57]

The AAAS meetings were held in late December in New York, and as Chamberlin anticipated, the just-published Vero papers in the *Journal of Geology* sparked considerable discussion.[58] Hrdlička presented his views of Vero "quite without gloves." Hay was there too, though according to Hrdlička he came "especially to oppose, but his narrowness of view, credulity and ignorance of the human problems involved was so apparent that I was quite glad he spoke."[59] Over at the Paleontological Society, Sellards declared that Vero showed the "undoubted presence of man in this country during the Pleistocene period," a statement Hay followed with another shot across Hrdlička's bow: "Even though anthropologists are wont to question the association of the human remains with the true Pleistocene forms . . . he considered the fauna as of early Pleistocene age." Matthew, who had privately told Sellards he accepted the Vero evidence, now went public, announcing after Hay finished that "he regarded the opinions of expert collectors of fossil vertebrates as to the validity of this discovery as carrying great weight, more probably, than stratigraphic data."[60] No surprise there: Matthew was himself an "expert collector of fossil vertebrates."

7.4 SPINNING THE MESSAGE

Over the next several months Rollin Chamberlin, Hay, Hrdlička, MacCurdy, and Sellards published a string of papers on Vero (several each, in Sellards's and Hay's cases). This time around, all were well aware of the others' views, and on this next round their papers were not intended just to bolster their own position but to undercut the positions of the others.

Sellards had the most to lose and moved fast. Within weeks he sent the *American Anthropologist* the first of his rejoinders to his critics. The choice of that journal was strategic: as the brunt of the criticism came from anthropologists, where to better confront them than on their own turf? It was an auspicious choice. The journal's editor was the linguist Pliny Earl Goddard, who like others in his discipline harbored the strong suspicion, based on the diversity of Native American languages (§2.1), that humans had been in the Americas for a very long time. As summarized by linguist Edward Sapir that same year:

> If the apparently large number of linguistic stocks recognized in America be assumed to be due merely to such extreme divergence on the soil of America as to make the proof of an original unity of speech impossible, then we must allow a tremendous lapse of time for the development of such divergences, a

lapse of time undoubtedly several times as great as the period that the more conservative archaeologists and paleontologists are willing to allow as necessary for the interpretation of the earliest remains of man in America.[61]

Sapir supposed that even 10,000 years was "a hopelessly inadequate span of time."

Goddard was receptive, even "quite anxious," to have an article on the subject of human antiquity in the *American Anthropologist*, and he thought Sellards's paper was "just the right sort."[62] He thought so despite the fact that Sellards, when confronted with a discrepancy between the geological and anthropological records regarding human antiquity, readily sided with the geology.

Sellards's rebuttal, published in the spring of 1917, was ostensibly a response to all the objections raised in relation to Vero's antiquity, but the majority of its pages focused on Hrdlička's concerns. (Sellards devoted a couple of pages to Chamberlin's reinterpretation of the geology, but scarcely noticed Vaughan and MacCurdy's criticisms.) Knowing the Vero case depended on whether he could prove the skeletal remains were deposited along with the Pleistocene fauna and were not the result of a later burial, Sellards detailed the condition of the human bones. Just like the bones of the extinct vertebrates, the human remains were scattered and isolated, had broken while fresh (and not, as Hrdlička had suggested, when the canal was dug), were often incomplete, and were missing many skeletal parts. In saying all this, Sellards implied that *all* the human remains from Vero were scattered, broken, and so forth, though late in the paper he admitted (albeit only in passing) that he was speaking primarily of the human remains from Stratum 3. (MacCurdy was quick to point out that Sellards ignored the articulated human skeletal elements from Stratum 2.) Sellards argued that fragmentary remains were not to be expected in a burial. Indeed, there was a partially articulated skeleton of an extinct wolf in that same deposit that was more complete and in better condition than any of the human remains. No one suggested it had been buried by its wolf brethren.[63]

Sellards believed all the remains from Stratum 2 and Stratum 3, human and nonhuman alike, were deposited by flowing water. The bones, shell, wood, and other debris were found lying horizontally, as might be expected were they "washed by the stream to their present resting place." However, judging by the presence of the nearly intact turtle carapace, the material had not come from far away, perhaps the nearby banks of the stream. He suspected they were moved while the bone was still "green," and only after deposition began to erode and come apart. That interpretation matched the stratigraphic record, as the remains were found beneath undisturbed deposits of muck and then marl, which were laid down after the flowing water of earlier times slowed and the whole was covered in a quiet pond.[64]

Admitting all the bones were redeposited was risky, for it exposed Vero to the criticism that any association among the species was entirely fortuitous. But it was a risk worth taking, because that criticism had already been expressed, and by crafting an interpretation of how and when the bones came to be associated Sellards could potentially blunt its damaging effects. Insisting there was "no essential difference" in the skeletal completeness or preservation of the human bones and those of the Pleistocene fauna, Sellards argued they must have originally been together in primary context and then were transported together into the same secondary deposit. If all the bones had the same redeposited history, they must be the same age. The only objection Sellards could see to that scenario (aside from Hrdlička's insistence that these were burials) was Rollin Chamberlin's suspicion that the nonhuman bones had eroded out of older upland deposits west of the site. Sellards dismissed that notion: "*The formation from which Dr. Chamberlin would derive the bones is almost if not entirely non-fossiliferous.*" Privately, he was far less generous: "Why Chamberlin should have hit upon such a completely impossible theory as to the source of the fossils is entirely incomprehensible to me."[65]

Sellards insisted the human remains were the same age as the extinct fauna, but he began to retreat from his earlier assertion about the relative ages of the strata in which they were found. He came to realize—from the possibility that Stratum 3 might just be reworked from or a facies of Stratum 2—that there was not necessarily a long time span separating the two, Hay's view of the faunal ages notwithstanding. Yet, whether the strata were the same or different times, all the remains were Pleistocene in age.[66] In the end, it seemed to Sellards that "none of the objections [to Vero's antiquity] are valid." There was only one inescapable conclusion:

> The American race reached its present structural [anatomical] development and attained the cultural stage [Neolithic] indicated by the artifacts of this deposit earlier than has been heretofore supposed, the conclusion is but in accord with other lines of geologic investigations which are constantly bringing to our notice larger time intervals and the earlier origin of types than had been previously assumed.[67]

And why not? In his view the Cro-Magnons of Pleistocene Europe were hardly different from modern humans, and the "much more isolated American race may have persisted through a longer period without sensible change in structure."[68]

Such a conclusion might have seemed inescapable to Sellards, but not to Mac-Curdy, whose paper on Vero immediately followed in the *American Anthropologist*. MacCurdy repeated his earlier observation that many of the specimens Sellards identified as artifacts were not cultural, whereas those that were genuine artifacts were largely confined to Stratum 3. There was but a single "flint spall" from Stratum 2, and because it was made of identical stone to the flakes from Stratum 3, MacCurdy assumed it had moved downward from that overlying layer.[69]

Further, despite their presumed antiquity neither the artifacts from Stratum 3 nor the skeletal remains differed significantly from those of the Florida Indians, even though, MacCurdy pointedly observed, they ought to just as Cro-Magnon differed from the recent inhabitants of the Dordogne, or for that matter from those of Neolithic age. There was "nothing new or unexpected" in the cultural and human remains from Vero, which otherwise pointed to the Recent prehistoric period: "In no respect whatsoever do they resemble the remains from the Pleistocene of any known part of the world."[70] The Vero evidence signaled to the "cautious observer" that the site either lacked stratigraphic integrity, or else (and far less likely) its cultural and biological remains appeared earlier or its Pleistocene fauna persisted later than previously supposed. MacCurdy had little doubt which interpretation was flawed:

> At all events in the face of irreconcilable differences between the combination of anthropological phenomena on the one hand and of paleontological on the other, and until there is forthcoming exact evidence as to the date of disappearance in Florida of the extinct vertebrates in this list, to say one is assured of the accuracy of the conclusion that the human remains and artifacts from Vero are of the Pleistocene period is to base one's conclusion on a forced correlation.

It was better to assume, as MacCurdy understood from estimates by Johns Hopkins paleobotanist Edward Berry, who had studied the Vero plants, that the Vero assemblage was no more than 3,000 to 4,000 years old.[71]

As before, Sellards did little more than acknowledge MacCurdy's anthropologically based criticisms.[72] He was convinced this was a problem that had only a geological solution. Of course, that still meant swaying Rollin Chamberlin, and in the spring of 1917 Chamberlin agreed to revisit Vero—although, as Sellards noted with bemusement, "he expects to be able to locate the original source from which the vertebrate fossils washed!"[73] Chamberlin spent four days at Vero, and in the end, unable to find his hypothesized upland fossil source, he admitted the extinct vertebrates in the Vero deposits "were primary and not secondary." Moreover, Chamberlin agreed that Hrdlička's claim that the bones were from a burial was "entirely without merit."[74] But any relief Sellards felt was merely temporary.

Writing his father from a northbound train "somewhere in Alabama," Rollin summarized his new interpretation of the Vero deposits. The crux of the problem, he now saw, was Hay's pronouncement that the vertebrate fauna was Early to Middle Pleistocene in age. Having a fauna of that antiquity alongside pottery made for "a violent case" of mixing. But if Hay was wrong about the age of the fauna, as Vaughan, Berry, and possibly even Sellards himself suspected, then a "less violent case [of mixing] . . . requires a less violent hypothesis." Moreover, despite his realization that the fauna in Stratum 2 had not been moved far, it was nonetheless apparent that the Vero deposits had been subject to extensive scour and fill. As a result, the top of Stratum 2 was highly eroded, with the erosional pockets filled by Stratum 3. This not only made it difficult in places to separate the two strata, it made it far more likely that the human skeletal and cultural remains, admitted by all to be far more abundant in Stratum 3, had been mixed by geological processes into Stratum 2. Not surprisingly, the spot where human bones were found in Stratum 2 was an area where "scour & fill were at their maximum."[75]

Chamberlin subsequently expanded on this theme in the *Journal of Geology*.[76] In essence, he argued that the "undisturbed specimens of extinct vertebrates" were in primary context in Stratum 2, whereas the human bones and artifacts were originally deposited in Stratum 3, but because of mixing each stratum had contributed elements to the other.[77] Assuming the extinct vertebrates within Stratum 2 were not as old as Hay claimed (even Hay admitted some of the Aftonian fauna lasted to later periods), and assuming the Stratum 3 pottery was Recent (Chamberlin was not averse to using cultural remains as diagnostic time markers), and given that the marine shell marl under all the deposits at Vero contained only living shell species and was at most Late Pleistocene in age, a relative and absolute chronology for Vero readily suggested itself:

> If (1) the testimony of the human relics, particularly that of the pottery, be taken at its apparent paleontological [chronological] value; if (2) the upper creek fill [Stratum 3] . . . be regarded as embracing all the human relics; if (3) the critical extinct vertebrate fossils found in this upper creek fill be regarded as derivative of the lower creek fill [Stratum 2]; and, if (4) the lower creek fill be regarded as contemporaneous with the last living stages of the extinct vertebrates whose fossils it holds as primary inclusions, as Dr. Sellards contends, the whole history becomes consistent physically and paleontologically, and the gist of its lesson is that the Pleistocene fauna lived longer in this genial southern clime that it has been credited with in the more northern latitudes, while the evidence of man's presence here falls into harmony with the general tenor of other evidences which fail to assign him an antiquity beyond the mid-Recent.[78]

As Chamberlin pushed to make Vero younger, Hay dug in to insure it stayed old. Speaking to the Biological Society of Washington in the spring of 1917, he reiterated his belief that the vertebrates from Stratum 2 and Stratum 3, human and nonhuman alike, were in situ, the former Early Pleistocene in age and the latter Middle Pleistocene (Illinoian). He had previously conceded that the only possible flaw in the Vero case was that it contravened known human history, but now he was more aggressive, cheerily dismissing Hrdlička's claim that the Vero individuals were buried ("unless human bones possess some unexplained means of underground dispersal") and rejecting Hrdlička's right to even speak to the case. The problems to be solved at Vero, Hay announced, were the domain "primarily [of] the geologists and paleontologists; only secondarily the anthropologists."

It had to be so, he insisted, because anthropologists had shown themselves unable to interpret their own data, as marked by their rejection of ample "independent evidences that man with a culture much like that of modern Indians existed in America during approximately the Sangamon stage" (the interglacial stage that followed the Illinoian). That evidence included Natchez and 12 Mile Creek, paleoliths in loess, and the Afton site in Oklahoma that Holmes had briefly excavated in 1901 following a report of a co-occurrence of artifacts and Pleistocene fauna, but had dismissed as attributable to recent Indians (Hay thought it Early Pleistocene in age). Hay admitted each of the sites could be individually criticized: "We may not be able to rely absolutely on any one of these reputed finds." Nonetheless, he insisted that together they "produce a probability of man's existence in Pleistocene times." (Hay was evidently unaware that odds are multiplied, not summed.)[79]

Hrdlička heard reports of Hay's talk and, learning that Hay was about to submit the presentation to *Science*, alerted James McKeen Cattell, the journal's editor to expect the paper:

When the article comes, kindly give it your personal attention. As you know we are trying hard, since many years now, to counteract the uncritical spirit in anthropology which would people this continent with modern forms of man long before the existence of the Heidelberg man in Europe; but our road is made hard on the one hand by the inexperienced easily swayed young man [Sellards], and on the other by the biased old man [Hay] belonging usually to some other branch of science, but anxious to impress his opinion on the poor ignorant anthropologist. . . . If his paper found place in some high class journal, much damage has been done.[80]

Hay's paper did not appear in *Science*, but he was only temporarily blocked. A synopsis of the paper soon appeared, and thanks to Goddard a longer version was published the next year in the *American Anthropologist*.[81]

Hrdlička, meanwhile, had worked up his detailed assessment of Vero.[82] Because it was slated as a BAE *Bulletin*, with no page limits or unfriendly editors with whom to contend, he took the opportunity to assess the claims for Pleistocene human antiquity in the Americas that had appeared in the years since his previous syntheses (§6.12, §6.14)[83] and used them as a springboard for a critique of Vero and the "broader anthropological and archaeological problems" it raised. That meant first drawing a disciplinary line in the sand. Not only did Hrdlička categorically reject Hay's demoting anthropology to secondary status, he thought it "scarcely safe for the geologist or the paleontologist to assume that the problem of human antiquity is his problem." Individuals in those disciplines might be helpful in determining the age of geological deposits or associated faunas (given his disciplinary arrogance, it is doubtful he actually believed that), but they were poorly trained to deal with human remains, ignorant as they were of human biological and cultural evolution and especially of the "*human element . . .* of man's conscious activities."[84]

In essence, they paid too little heed to the "universal" penchant of humans to bury their dead. Fossils of plants and animals, which Vero had in abundance, were passive objects that "find their resting places accidentally." Unless moved by some agency after deposition, these constitute "safe evidence of contemporaneity with other similar objects and with the geologic components of the same horizon." That same assumption, Hrdlička insisted, could never be made of human skeletal remains (or even artifacts), because "from the earliest known times man has buried his dead at varying depths, thus introducing his remains into deposits and among other remains with which otherwise they had no relation." How pervasive were burials? Given, say, 4,000 years of pre-Columbian occupation (Hrdlička's estimate), the number of burials on the North American continent might reach 2 billion. With that "vast array of possibilities," it was better in all instances to assume human remains were intrusive and younger than the remains with which they were found, until proved otherwise.[85]

Moreover, intentional burial resulted in well-preserved and relatively complete skeletons. In contrast, a human body abandoned on the surface would lay there for a long time during which it dried, cracked, and weathered and was gnawed on and scattered by scavengers, and "in nearly all cases" wholly or largely destroyed.

Such remains would not enter the geological record in a good state of preservation or in "any degree approximating entirety." That was true of large animal bones as well, despite the fact that those were "on the whole more durable than human bones."[86] Thus, when presented with a case like Vero, one could just play the odds:

> What slight chance, then, can there be of finding in any stratum, but especially in one of slow accumulation, a fairly complete and well-preserved human skeleton of equal age with the deposit? And if one such marvel should happen, what chance would there be of the discovery within a few rods distance, at almost the same depth, and in a distinct geological formation, of a like skeleton? Surely, such a chance would be infinitesimal.[87]

The "scientific" approach was to reject all conceivable alternatives, especially burial, before concluding the human bones were the same age as the deposit or the extinct fauna. And that, Hrdlička peevishly added, should happen *before* publicly announcing the human bones were of great antiquity.

Telegraphing his punches, Hrdlička placed his commentary on human burial immediately after his summary of Sellards's report and just before his own assessment of Vero. Not surprisingly, he was sure the Vero individuals postdated the Pleistocene fauna. Ayers had told him the human remains in Stratum 2 were found in their "natural relations." Because the bones of the lower half of this skeleton were articulated, the upper half—evidently lost when the canal was excavated—must have been as well. That suggested there had once been an entire skeleton here, which could "only" have been a burial, for only burials (as opposed to bodies left on the surface) were complete and articulated.

Hrdlička thought it no coincidence the Stratum 2 human remains were just over two feet below the surface, as he believed "the large majority" of Indian burials reached just two to four feet below surface. With sufficient patience (which, he generalized, "Indians seldom lacked"), the overlying sediments were easily excavated for an interment. There was no other natural means to account for such a well-preserved skeleton: on a landscape as flat as Vero, it was doubtful that sediment-bearing waters would pile up high enough to fully (and rapidly) cover a human body that lay on the surface.[88]

Yet, if all of Hrdlička's sweeping statements were true, how were the broken, disarticulated, and scattered human remains from Stratum 3 to be explained? On initial examination, Hrdlička admitted their condition did not obviously suggest a burial but instead a body abandoned on a surface, followed by trampling and dispersing by animals (though not by water, as Sellards supposed). Then he probed deeper and learned that the area in which these human bones were found was relatively small—far smaller, he believed, than the area that "normally" results from animals dragging a carcass apart (another generalization). Furthermore, none of the human bones showed evidence of animal gnawing or trampling, and the missing skeletal elements were fragile parts that "decompose readily" or were likely lost in their recent exposure in the bank of the drainage canal. Otherwise, the surface weathering of the bones was minor and all were well preserved—testimony, in Hrdlička's view, that the skeleton was buried while the body was still intact. Its "disassociation and fragmentation" doubtless occurred later, "owing to movements, stresses, root action, and other agencies operating on or within the deposits enclosing the body."[89] The human bones at Vero were articulated (or not), scattered (or not), weathered (or not), but all had been intentionally buried. Hrdlička never met a case that could not fit his preconceptions, and as Swanton observed, these preconceptions were so much a part of him that he did not realize he had any.[90]

That Sellards reported both sets of human remains were found beneath deposits that showed no signs of disturbance from later burial gained no purchase with Hrdlička. "In old graves," Hrdlička proclaimed, "except under unusual conditions, all signs of disturbance of the ground are absent or obscured."[91] Stratigraphic healing could occur over a short time, which was all Hrdlička was willing to grant Vero's antiquity, since on morphological grounds the bones were "strictly modern," no different from the type found among the eastern Algonquian or the Sioux.[92] There was nothing to suggest "even remotely an individual more ancient or anthropologically more primitive than the Indian."[93]

For those like Sellards and Hay (and increasingly many European anthropologists[94]) who claimed humans were anatomically modern in Pleistocene times, he replied that "as late as the Aurignacean [sic] period, approximately 15,000 to 25,000 years ago, man had not yet fully reached modern standards in physical development," let alone cultural development (the use of pottery), and "can not possibly be conceived of as having been numerous enough to reach the northeasternmost limits of Asia, from which alone there was a practical way open to the American continent." In the face of such anthropological "realities," it was inconceivable that humans were in the Americas before at least 15,000 years ago.[95]

Hrdlička granted that cultural phenomena changed more rapidly than human anatomy, but that only meant one should expect more heterogeneity in material culture than skeletal morphology, and not that major cultural developments (such as modern types of pottery) outpaced the rise of "modern standards of physical development."[96] Correspondingly, modern pottery and modern skeletal forms could not be geologically ancient (§6.4). Of course, in Hrdlička's hands virtually any human skeleton could be made to fit within the range of modern Native American forms, even if it meant taking remains from Florida and linking them with Algonquian or Siouan physical types. Having handled more skeletal remains than any of his contemporaries, he could do all this without notable opposition: few could meet him on his own ground.[97]

Accordingly, any discussion of the geological age of Stratum 2 and Stratum 3, or of the age of the extinct vertebrates, was "quite irrelevant" to the antiquity of the human bones. Were the human remains indeed as old as the Early Pleistocene, as Hay claimed on the basis of the fauna, it would mean a "new natural history of man, [a] new anthropology." Hrdlička was certain that was unnecessary: "*Sic transit gloria hominis Veroensis.*"[98]

Hrdlička's *Bulletin* was completed in March of 1917, but it did not appear until the following summer. Sellards knew it was coming, and though he could guess its central message he did not learn its details until after he had assembled his own final substantive statement on Vero. That appeared in his *Annual Report* for 1917, which contained responses to his critics (including Rollin Chamberlin, whose latest hypothesis Sellards dismissed as "equally as untenable" as his first[99]), along with Hay's detailed descriptions of the vertebrates, Berry's report on Vero's fossil plants, and R. W. Shufeldt's analysis of the Vero bird remains.[100] Unfortunately for Sellards, his "in-house" contributors were making his task no easier than his outside critics.

Berry saw two alternatives for the Vero flora. It was either the same age as the "Peorian interglacial deposits" of the Mississippi valley or immediately "post-Wisconsin and correspond[s] with what the Scandinavian geologists have named Litorina time" (that is, Recent). In either case, it assuredly meant that the vertebrate remains in those same deposits could not "possibly be of Middle or Early Pleistocene age." Like Chamberlin, Berry blamed Hay for inflating the age of the deposit based on the vertebrates.[101] Hay conceded nothing. Although granting that some geologists (Sellards included) believed there was no "great interval" of time between Stratum 2 and Stratum 3, he insisted the vertebrates proved the deposits were Early and Middle Pleistocene age, and so too the human remains.[102] Sellards tried to strike a middle ground, saying only that "the accepted interpretation of faunas and floras is Pleistocene."[103] He wanted the human remains to break the Pleistocene barrier: how far they went beyond that was of lesser concern.

By late 1917 Vero was caught up between evidentiary and interpretive extremes. Thus, to the question of whether the human skeletal remains in Stratum 2 and Stratum 3 were in situ, Sellards insisted they were based on what Ayers, who had seen the specimens in the ground, had reported to him. Hay and Chamberlin agreed. Yet, in Hrdlička's opinion, according to what Ayers told him about the articulation and spatial patterning of the bones they must have been from a later burial, a point on which MacCurdy and Vaughan concurred. Naturally, Holmes, Hrdlička, and MacCurdy insisted the presence of pottery and projectile points precluded a Pleistocene age, whereas Hay and Sellards, believing humans had attained a "Neolithic" stage of evolution far earlier than archaeologists realized, vigorously disagreed. And, of course, the morphology of the human remains was either relevant to the discussion, as Hrdlička insisted, or not, as Hay and Sellards asserted.

Nor was there consensus on whether the extinct vertebrates in Stratum 2 were in situ. Sellards, Hay, and several of the geologists thought so, but Hrdlička, Mac-Curdy, and even Chamberlin (much to Sellards's dismay) were skeptical. The only point of unanimity was that the extinct vertebrates were Pleistocene, but no one sided with Hay in saying they were *Early* Pleistocene in age. MacCurdy echoed others in arguing the Pleistocene fauna may have lived longer in southern regions, whereas Hrdlička dismissed the antiquity of the vertebrates as irrelevant to the age of the human remains.[104] For Sellards and Hay the Vero skeletal remains and artifacts were Pleistocene in age, though they could not agree on whether Early or Late Pleistocene. For the anthropologists, all the human remains were Recent.

That "bone on bone" conflict, as Thomas Chamberlin put it, left little room for compromise or resolution, with the result that few outside the immediate Vero circle granted a Pleistocene age for those remains.[105] When Vero first appeared, Holmes explained to Chamberlin that "it is not my idea, of course, that we should try to break down all of the evidence of antiquity that may be reported, but rather that we should hold conclusions in abeyance."[106] By late 1917, that had come to pass. The Vero exchanges, Chamberlin observed, "offered no warrant for the public propagation of any decision or consensus, pro or con, respecting the Pleistocene age of the ancient Vero man." Even so, he thought there had been progress that was "wholesome and fruitful . . . however indecisive." The problem had been laid out clearly, the discussion was productive, lines for further work were well indicated, and for the moment all principals were being civil, even joking with one another. In December of 1917, Hay sent Holmes a reprint of his paper from the Florida Geological Survey's ninth *Annual Report*. When Holmes spotted the sentence where Hay pronounced that "men possessing a culture much like that of modern Indians existed in America at least as far back as the Sangamon interglacial stage," he dashed off a note (with one seemingly intentional strikeout):

> I am delighted with your report on the age and stage of culture of the Man of Vero, Florida. It is the most ~~important~~ astonishing announcement respecting the history of man in the world that has ever been made.[107]

That amiable moment lasted no more than a few months, and what followed did not always make for "wholesome progress."[108] Or much progress at all.

7.5 TURF WARS

That individuals working from diverse disciplinary angles with very different evidence came to conflicting interpretations is no surprise. Unfortunately, it was not clear how to assess the relative merits of those various lines of evidence, reconcile the conflicting interpretations, ascertain whether the antiquity of Vero was a matter of geology, paleontology, anthropology, or some combination of all three, or even determine if a consensus on Vero's antiquity was possible. Such problems are not without solution, but only if concern for finding common ground trumps disciplinary loyalty.

In 1919 Thomas Chamberlin tried to bring the warring parties together to reconcile their divergent views. Doing so, he knew, required that each advocate "leave as much leeway as he can . . . for the accommodation of conflicting views." But that meant finding agreement on the relative weight and merit of the different approaches and lines of evidence. Here the disciplinary differences revealed themselves to be about more than just practitioners fighting over whose interpretation was superior; they were also about the validity of the assumptions underlying their respective approaches.[109] What did it suggest of the rates of evolution of human and nonhuman animals if Hay's Aftonian mammals from 450,000 years ago[110] were associated with anatomically modern humans? What of the relative rates of biological as opposed to cultural evolution if those ostensibly ancient humans were associated with modern pottery? Chamberlin supposed that "when a plastic art [artifacts] is set over against the evolution of biotic species, the presumption of the speedier changes lies much in favor of the art."[111] Holmes himself had said that some races experienced "slow development."[112] Perhaps material culture had not changed over many hundreds of thousands of years. How were these discrepancies to be resolved?

Chamberlin saw the burden of reconciliation falling heaviest on the paleontological evidence, given it was the chronological outlier. But following his method of multiple working hypotheses (§4.15) and in fairness to Hay, he insisted all the approaches needed to be tested to see which best explained the Vero evidence with the least "degree of strain." For that to be accomplished, there needed to be a "hospitality to working alternatives" and a "hospitality to revision."[113]

That did not happen. Hay was convinced of the primacy of his methods and results, and thus of Vero's great antiquity, both of which he defended vigorously. He rejected any compromise forced on the vertebrate record. "Our friends the anthropologists," he snarled, were willing to believe the Pleistocene fauna lingered into the Recent period: "The writer holds that the view is wholly wrong."[114] To strike back at Holmes and Hrdlička he began self-publishing a pamphlet series—*Anthropological Scraps*, he called them—that existed for the sole purpose of needling Holmes and Hrdlička. These were clever and droll and maintained an ironic patina of civility ("The scientific writings of Dr. Aleš Hrdlička are always a source of joy, of instruction, and of inspiration"). But his daggers were razor sharp.[115]

Of course, Hay was not solely to blame for the interdisciplinary standoff that resulted. Hrdlička was equally convinced of the infallibility of his views, and he increasingly doubted nonanthropologists could be right or even contribute to the discussions. The fact that most sites were now found and championed by geologists and paleontologists exacerbated tensions. Hrdlička had not always seen matters that way. He had earlier refused to estimate the age of the Burlington County and Riverview Cemetery skulls, conceding such questions to the geologists (§6.4). Yet, over the years he had steadily retreated from that line, partly from an increased confidence in his own abilities and methods, but also because geology did not always provide an answer, let alone one to his liking. By Vero he was also feeling cornered. Rollin Chamberlin was not proving as helpful as his father had once been, and now Thomas Chamberlin questioned Hrdlička's claim that the Vero skeletons were burials, believing that was a geological question and not one for Hrdlička to answer.[116] With Sellards and Hay hammering him from that side of the disciplinary fence as well, it seemed to Hrdlička the dispute had split along strictly disciplinary lines. That suited him just fine. "Our colleagues in collateral branches of science," he announced, were to be "sincerely thanked for every genuine help they can give anthropology," but "they should not clog our hands."[117] Between Hay and Hrdlička, there was little room for compromise.

That the methods used to establish the antiquity of Vero were "subject to challenge," Chamberlin believed, was due to the fact that the "physico-dynamic" criteria geologists used to determine the relative order and age of the stages of the Pleistocene did not apply at a locality so far from the ice margin. Assessing the age of these deposits required knowing the age of vertebrate or archaeological remains or both found under, within, or above them. But how reliable were ages based on those remains?

Hay had spent much of the preceding decade mapping vertebrate fossils relative to the "drift sheets" from North America's "four or five" glacial stages, as well as their position in relation to "river gravels, sands, old soils, and beds of loess" (marking interglacial stages)(§7.1). On that basis he had identified taxa and the relative percentage of extinct species within each glacial or interglacial stage. The percentage increased the farther back in time: thus, roughly 50 percent of Illinoian species and about 90 percent of the Aftonian species were extinct. This pattern, based on faunas from a "thousand or more" localities, gave Hay a two-fold method to determining the age of a fauna, regardless of its proximity to glacial features. One could identify the temporally diagnostic types, calculate the percentage of extinct species in the fossil fauna, or both.

That the Vero Stratum 2 fauna included Aftonian-specific mammoth, horse, sloth, and camel and consisted of approximately 70 percent extinct taxa suggested an Aftonian/Early Pleistocene age. Correspondingly, the roughly 50 percent of extinct forms in Stratum 3 marked a fauna closer in composition to an Illinoian/ Middle Pleistocene age. Hay was aware that percentages could be driven by preservation and collection, though he used that fact to boost rather than reduce the apparent percentage of extinct forms: thus, he deemed the 50 percent of extinct forms in Vero's Stratum 3 an underestimate resulting from those "accidental causes."[118]

Although Hay's fauna seemingly provided a means of extending diagnostic time markers from glaciated to unglaciated areas, the Chamberlins (father and son) as well as others were skeptical about its reliability. One reason expressed by the elder Chamberlin was that "changes of species are slower, and their interpretation more uncertain, than the changes in great ice sheets," not to mention the difficulty at Vero of reconciling such a shallow deposit (about six feet) with such a great lapse of time (450,000 years). His son added that a fauna could outlast a given glacial or interglacial stage (or at least fail to correlate tightly with one), and it was not improbable that at a place like Vero far from the glacial front or fronts the "fauna may have lingered longer . . . than it did at the north, where the advances and retreats of the ice border were putting the fauna under the stress of an oscillating climate." He preferred to rely on the age estimate of the marine shell marl underlying Stratum 2 at Vero, also dated on the basis of its fauna, which in this case was composed almost entirely of recent marine taxa and thus accorded at most a Late Pleistocene antiquity (§7.4). The overlying strata had to be younger.[119]

If the geologists proved the marl was Late Pleistocene, Hay conceded it would "at once end the dispute about the time of the disappearance of the fauna represented at Vero; and vertebrate paleontology will become once more indebted to geology." But he doubted that would happen, was unhappy Rollin Chamberlin seemed "to respect rather lightly the vertebrate fossils," and altogether resented Rollin's suggestion (echoed by others) that the Florida Pleistocene fauna may have lingered into the Recent period. How, Hay asked, could animals "subjected to the extreme climates of glacial and interglacial epochs" last "practically unchanged" from the Aftonian to the Recent? If that were so, why were these species absent from, say, Recent deposits in neighboring Louisiana? Hay answered his own questions: "The extinct animals so often mentioned did not live on up to the Recent epoch."[120]

Humans were undeniably in Florida in more recent times, and on the basis of the pottery found at Vero Holmes, Hrdlička, MacCurdy, and even Rollin

Chamberlin deemed the site to be postglacial, possibly no more than a few thousand years old. Hay expected as much from the anthropologists but was keenly disappointed that Chamberlin had more faith in the apparent age of the archaeological remains than the paleontological ones. Hay insisted the Neolithic antiquity of pottery was unproved: it had merely been assumed, based on sequences from Europe. Who was to say that American prehistory played out in the same manner? Hay reminded all that pottery had been found alongside a mastodon tusk in South Carolina in 1859 (§2.4). And then there was the Nampa Image made of clay (§5.6).[121]

AMNH archaeologist Nels Nelson, one of the few American archaeologists of the time deeply knowledgeable about European prehistory,[122] admitted Hay had it partly right. Finds in America were judged to a certain extent on the basis of European sequences. Still, that was not the only reason he and other anthropologists were unwilling to accept the notion of Early Pleistocene-age pottery at Vero. Rather, it was because all archaeological work in America up to that point had shown that

> pottery-making is of relative late date in culture history. . . . The archaeology of the eastern United States seems particularly clear on this point. Thus, it has been demonstrated over and over again that the lower strata of artificial deposits . . . are devoid of ceramics. . . . [Only] after a time they began making a plain dull-reddish earthenware, and that finally, some time before the arrival of European explorers, they took to ornamenting this ware.[123]

To accept an Early Pleistocene age for Vero demanded the invention of pottery at an unheard-of antiquity and "oblige[d] us to assume that this early culture of Pleistocene times was snuffed out and that after some millenniums . . . a new and lower type of culture became established which only after a very considerable period reached the level of the original culture." Such a sequence, Nelson supposed, was "conceivable, but it is not plausible."[124] Nelson had briefly visited Vero in 1917 while working on the chronology of Florida shell mounds, and he saw nothing to indicate it was any older than "the middle period of shellmound occupation along the Florida East Coast."[125] As Nelson saw it, a solution to Vero required that

> either the anthropologist must surrender not only his present lightly held opinion regarding the antiquity of man in America, but also his rather more firmly fixed notion regarding the order and progress of cultural traits in general, or else the paleontologist must concede us a very much narrower margin of time as having elapsed since the close of the Pleistocene than he has hitherto.[126]

Nelson would not surrender his ground, but then neither would Hay.

Instead, Hay tried a flanking maneuver. If artifacts were alike across broad expanses of space (as MacCurdy and others admitted), could they not also be alike through long spans of time and date to the Early Pleistocene? The artifacts of Australia were similar to those of Mousterian Europe. Perhaps "some races of men stand still in their culture or move so slowly that their progress is imperceptible."[127] Warming to the idea, Hay saw evidence in the fossil record that 20 to 50 percent of Early Pleistocene species "which still exist . . . have undergone no changes that are perceptible." This included "Virginia deer, beaver, brown bear, gray wolf, coyote, and gray fox." There were other taxa (fossil tapirs, horses) that were distinguishable from modern forms, but only with difficulty. The vertebrate divisions of the Pleistocene differed principally from one another and from the Recent "in the successive extinction of animals . . . and not so much in changes in structure." Likewise, he argued that the fossil records of "the nearest relatives of man, the higher apes" (chimps, orangutans, baboons) were virtually identical from Pleistocene times to the present. Why, he then asked rhetorically, "should it be supposed that our [human] ancestors were so retarded in their evolution until well past the middle of the Pleistocene and that this evolution then should have suddenly been quickened?"[128] Hay was unwilling to grant that patterns of human evolution were somehow distinct or unique as opposed to those of other mammals.

In time, Hay's labors on behalf of his vertebrates and of Vero might have served to isolate him, but they did not because he very shrewdly raised the stakes, shifting the discussion away from the specifics of Vero's antiquity to a broader appraisal of the evidence for human antiquity in North America. In doing so, he completely disregarded his earlier advice to other scientists (anthropologists) that they keep out of matters (paleontological) that did not concern them. Instead, he launched an effort to revivify the many previously rejected sites, a process that inevitably became a referendum on the methods of those who had rejected those sites: Holmes and Hrdlička. In doing so, Hay made himself virtually impossible to ignore, though Holmes and Hrdlička, easily his equals in certitude and stubbornness, tried as best they could.

Their task was not easy, given that they all operated in the same small world of Washington science, and Hay deliberately chose to bring his attack directly to the enemy camp. Hrdlička had argued that if Vero was legitimate, there ought to be ample evidence of a widespread Pleistocene human population in the Americas. Hay responded in *American Anthropologist* that there was, and he identified nineteen sites he believed testified "strongly to the fact that during early and middle Pleistocene times there existed in North America a population which was spread from the Atlantic Ocean to the Pacific." All the fallen classics made his list: the Calaveras skull was there, along with the Nampa Image and the Natchez pelvis. So too were sites that were little more than hearsay reports of artifacts in loess or other glacial deposits.[129]

As before, Hay acknowledged many of these sites had been criticized, especially by Holmes, and any one of them might be questioned. But just as one swallow does not make a summer, "when they come in flocks one has a right to conclude that at least the vernal season is on" (§7.4). This flock spoke of an antiquity on the American continent that stretched back possibly to the Pliocene and certainly to the Early Pleistocene. Naturally, "our good friends the anthropologists have set up over most of [these sites] the danger signal," but so far as Hay was concerned they "give too little weight to the cumulative effects of the[se] reported discoveries of Pleistocene man."[130]

There were few things Hay could have done to provoke Holmes's anger more than to cheerfully accept claims he had rejected, in some cases decades earlier and after bitter debate. But Hay hardly stopped with jabbing Holmes. There was also Hrdlička to poke in the eye.

7.6 FINDING VERO'S PLACE ON THE HUMAN FAMILY TREE

Hay believed that the mammalian fossil record attested to the fact that Asia and America were connected by a "wide land bridge" intermittently from the Pliocene onward, and thus there was "no reason why [humans] should not have accompanied on this journey the beasts with which he had associated."[131] Given the lack of evolutionary change in mammals over that time, there was also no reason to suppose a "race of men with about the grade of development of our North American Indians would have been out of harmony" with a fauna of Aftonian age coming out of Asia, where Hay assumed humans originated. Indeed, he viewed Europe as a human evolutionary backwater and the "Heidelberg man" a primitive anomaly, one that had nothing "to do with the present human races, except in having had, somewhere back in the Tertiary, an origin in common." Neanderthals were a species apart from *Homo sapiens*, which "attained a certain stage of bodily and mental development [but] then were exterminated by the superior races."[132] Hay's 1918 view of the place of Neanderthals on the human family tree was hardly an uncommon one.

Earlier that decade Marcellin Boule had conducted a detailed anatomical analysis of Neanderthal remains, focusing on the nearly complete La Chapelle-aux-Saints individual, but also the specimens from La Ferraissie, Krapina, Neanderthal, and Spy. He concluded Neanderthals were considerably more bestial than ourselves, a side branch in human evolution, long ago separated from the lineage leading to modern humans. At the same time, he discarded Dubois' *Pithecanthropus* as an ancestor, declaring it to be nothing more than a giant gibbon.[133]

The void in the human family tree left by their removal was almost simultaneously filled by the discovery in 1912 in England of the first Piltdown remains, a cranium and jaw that, on the basis of associated vertebrate fossils, were thought to be Late Pliocene to Early Pleistocene in age.[134] That made Piltdown comparable in age or older than any known human remains. For decades, England had been mired in the backwater of prehistory with early stone tools, including the much-debated Pliocene-age eoliths, but never fossil humans to rival those discovered on the continent. With Piltdown, *Eoanthropus dawsonii*, the English not only had the tool-maker, but with its large and modern-looking brain they saw what many anatomists long suspected: "In the evolution of man the development of the brain must have led the way." The Piltdown remains and eoliths were mutually reinforcing. If you had eoliths (dawn tools) there ought to be an *Eoanthropus* (dawn human) to make them; and the earliest humans were not likely to make artifacts readily distinguishable from naturally broken stones. Together, they made a compelling story.[135]

Better still, with its larger brain Piltdown neatly fit the preconception of what earlier human forms should look like than did either the "primitive" Neanderthals or the small-brained *Pithecanthropus* skulls. Moreover, it meant the *sapien* lineage (in England) could be traced independently of the decidedly more brutish and less human Neanderthals (from northern Europe) to an earlier period, thereby testifying to the early appearance of an essentially modern human form.[136] Piltdown helped force those other fossils off the evolutionary road to modern humans. As Boule put it:

> The Piltdown race seems to us the probable ancestor in the direct line of the recent species of man, *Homo sapiens*; while the Heidelberg race may be considered, until we have further knowledge, as a possible forerunner of *Homo neanderthalensis* [which, of course, led nowhere].[137]

All of which dovetailed nicely with Hay's view and made for a coherent evolutionary picture, for the vertebrate fossil record seemed to indicate long periods of anatomical and evolutionary stasis. To him nearly all extant species were survivals from the Pleistocene, with the difference between Pleistocene and Recent faunas (or fauna from different Pleistocene periods) being one of quantity rather than kind. Species might go extinct over the Pleistocene but not necessarily show significant evolutionary change. What, then, of humans? The answer all paleontologists and most anthropologists offered was there was no reason to suppose human evolution was more rapid or complex in the Pleistocene than evolution was in other mammals or vertebrates. (There was little understanding of the potentially amplifying role of the brain and culture accelerating anatomical and evolutionary change in humans.) That was not an unreasonable position, given the known fossil record, particularly with Piltdown a branch on the human family tree.

Hrdlička, however, vigorously disputed Piltdown's antiquity and argued that it was anatomically impossible for its essentially modern human skull to have held an apelike jaw. This "monstrous hybrid" did not fit theoretically: Piltdown was completely at variance with Hrdlička's understanding of biomechanics and the evolution of human craniofacial morphology. It implied that elements of the skeleton changed at different rates, and not in a graded, gradual evolution of the entire morphological pattern. Hrdlička expected to find as one moved back in time that earlier species of humans slowly dissolved in a mosaic of primitive features: Piltdown should have a primitive brain to match its jaw (§6.2). But then, the bones did not fit literally either, for the articular surfaces on the ascending ramus of the mandible were missing, making it impossible to check the articulation of mandible to skull.[138]

To accept Piltdown would have required Hrdlička to radically alter, if not reject, his understanding of evolution, and no fossil with as many problems as were presented by Piltdown was worth that sacrifice. Unlike Hay and most of his anthropological contemporaries, Hrdlička placed Neanderthals in the human evolutionary mainstream as a transitional form between modern humans on the one hand and more apelike forms such as *Pithecanthropus* on the other (§6.2).[139] There was no room in that scheme for Piltdown, let alone *any* modern humans in

Pleistocene times: not in Europe, and especially not in America, where all purportedly ancient skeletons fit readily within the anatomical range of modern Native Americans, and none had been found in secure Pleistocene deposits. Pleistocene-age humans (whether in Europe or America) ought to look different, and certainly not like modern Native Americans.

As it happens, European proponents of Piltdown (and opponents of Neanderthals as human ancestors), such as the English anatomist Arthur Keith, were just as keen to see anatomically modern humans *in* Pleistocene America. Keith willingly accepted the claims from Lansing, Long's Hill, Natchez, and even Trenton, supposing all of them dated to the prior interglacial, because these indirectly buttressed claims on behalf of Piltdown.[140] (That Keith cited Hrdlička in accepting that evidence presumably did not go over well in Washington.) The skeletal and geological evidence made it "plain [that] to account for modern man in Europe, in Asia, in America, long before the close of the Ice Age, we must assign his origin and evolution to a very remote period."[141]

7.7 VIOLATING THE SACRED CONFINES

In accepting the idea of geologically ancient but anatomically modern humans, Hay challenged publicly not only Hrdlička's position on human antiquity in America but also his views of human evolution in general. Hay kept up the pressure on both fronts. Close on the heels of his *American Anthropologist* paper, he reviewed several just-published papers on Vero, including Hrdlička's annual report in which he insisted the only "satisfactory explanation" of Vero was that the human remains resulted from intentional burial. Hay would have none of that:

> Naturally, this means satisfactory to [Hrdlička]; for six other men have furnished explanations on the same subject, each apparently satisfactory to its author, and all differing from that of Dr. Hrdlička. At least three of those six men are experts in the solution of geological problems, but not one of the six sustains Dr. Hrdlička in his theory of intentional burial. Meanwhile he hardly attempts to remove the difficulties which beset his assumption. His method may be defined as the easy one of solution by fiat.[142]

Hrdlička immediately asked MacCurdy to write a rebuttal to Hay's efforts to "force the matter down our throats." Holmes, he reported, was "quite annoyed" by Hay's "mischievous little article" and had already begun a response for *Science*.[143]

Holmes had been relatively quiet throughout the Vero discussion, contributing only a brief addendum on the site's pottery for Hrdlička's *Bulletin*. Now that Hay had raised the stakes and attacked everything he stood for, he had to respond, and not just about Vero. Notwithstanding Hay's claims on its behalf, Holmes dismissed Vero's purported antiquity as "fully offset by the interpretations of anthropologists of long experience." But a more sweeping response to Hay was necessary to counter his reopening "a large body of so-called evidence of geological antiquity which has long been discredited and relegated to the historic scrap heap where it should still remain." That Hay did so was indefensible, though perhaps understandable in light of the "peculiar and very strong fascination in the idea of hoary antiquity."[144] Taking the bait Hay dangled in the *American Anthropologist*, Holmes admitted it was he who had raised the "danger signal" (§7.5) in regard to those claims for great antiquity but he said he was fully justified in doing so:

> It is manifestly a serious duty of the archaeologist and historian of man to continue to challenge every reported discovery suggesting the great geological antiquity of the race in America and to expose the dangerous ventures of little experienced or biased students [Hay] in a field which they have not made fully their own.[145]

If those sites were not examined critically, as he had done over the years, "lamentable errors may become fixtures on the pages of history." As always, Holmes saw the dispute over a deep human antiquity in America in the starkest of terms, a struggle between polar forces:

> I do not wish for a moment to stand in the way of legitimate conclusions in this or any other field of research, but illegitimate determinations have been insinuating themselves into the sacred confines of science and history with such frequency and persistence that no apology is required for these words of caution.[146]

Harsh words, and they cut deeply. Within weeks of Holmes's publicly denouncing his work as "illegitimate," Hay reported to President Robert Woodward of the Carnegie Institution of Washington (his employer) that he would henceforth concentrate on putting "a new face on the history of Pleistocene vertebrates." The effort would "compel the geologists" to modify their views on the age of many deposits, but would also lead him into "the ticklish subject of anthropology." Hay was convinced his fresh look at the subject would firmly establish a deep human antiquity on this continent and make it impossible for "our friends Prof. W. H. Holmes and Doctor Hrdlička . . . to extricate themselves from the positions they have taken."[147] He would prove he was right about both Pleistocene animals and Pleistocene humans.

To accomplish that, Hay began delving even further into the attic of past claims, starting with Trenton, which he visited with Ernst Volk in the summer of 1918. He corresponded actively with Abbott, Volk, and George Frederick Wright, to each of whom he wrote that efforts such as the work of Leslie Spier and Wissler at Trenton (§6.15), but also new discoveries at places like Vero, had confirmed their views of the antiquity of the artifacts from the Trenton Gravels.[148] For Abbott's amusement, he joked he was "having some discussion with your old friend W. H. Holmes about the same matters."[149]

When Hrdlička's long-awaited BAE *Bulletin* on Vero appeared, Hay continued his offensive. He used the Trenton femur to shoot down Hrdlička's claim that all human remains were probably burials, for the Trenton discovery was a single fragment deep within a stratified deposit, hardly suggestive of a burial. Hay knew Hrdlička had an explanation for why the rest of this "burial" was missing:

it had become disassociated and fragmented owing to (as Hrdlička had put it) "movements, stresses, root action, and other agencies operating on or within the deposits enclosing the body."[150] Apparently at Trenton "the fragment of parietal was caught in its migrations 20 feet away." Perhaps, he added tartly, "we get a clue here to the reason why civilized people nail up their dead in good strong boxes."[151]

As for Vero, Hay proclaimed that if the presence of even a partially articulated skeleton indicated intentional burial as Hrdlička (and only Hrdlička) believed, then the nearly complete and articulated skeleton of a large alligator from Stratum 2 must have resulted from intentional burial by its fellow alligators. In Hay's view, the absurdity of that was matched only by the absurdity of Hrdlička's "declaration of independence" of anthropology from geology and paleontology. The way Hay saw it, that was only because "the geological test has not always resulted to his liking." And Hay insisted that measuring the antiquity of Vero against European anthropological standards of human evolution was far less reliable than using the fossil record of vertebrate paleontology.[152]

Barbour, still sore from his own disagreeable encounter with Hrdlička a decade earlier, cheerily wrote Hay to congratulate him for "touching up Dr. Hrdlička's disregard for geologic and paleontological evidence." Barbour thought it "the most seasonable article of the year." Others, such as Yale geologist Charles Schuchert, were no less pleased to see Hay confront Hrdlička, though they were quick to say they did not necessarily agree with Hay's conclusions about Vero, either.[153]

Archaeologist Frederick Sterns of Nebraska (a sometime colleague of Robert Gilder) had no such qualms and hopped on Hay's bandwagon, reviewing Vero for the *Scientific American Supplement*. Though writing as a "disinterested spectator," Sterns came down squarely for Vero's Pleistocene antiquity. He cited Hay with approval, laughed at his jokes, repeated many of his criticisms of Holmes and Hrdlička, and even added a few sneers of his own. Were Holmes's logic allowed to stand, Sterns declared, "no science of any kind is possible."[154] Hay sent Holmes a copy of Sterns's article, and Holmes snapped back:

Holmes to Hay: "Dr. Hay, I did not see this until today. You did not need to send it. Your opinion does not interest me and the researches of this young man can add nothing of value. Yours etc. WH Holmes"

Hay to Holmes: "Tut! Tut! Professor, go take something for your nerves. OPH"

Holmes to Hay: "You had better go to H[ell] WHH."

Their exchange took place on a small notepaper that Hay tore up when it came back to him the second time. Reconsidering, he carefully taped it together and placed it with his papers for posterity.[155]

Chamberlin, who believed the Vero discussion had gotten off to a good start because of the joint visit and the papers in his *Journal of Geology* (§7.5), saw the ugly turn toward "antagonism and propagandism" it had now taken—and was not pleased. He called for a cessation of this "lapse into polemics" and "criticisms from personal points of view" so that a "re-juvenated effort of the co-operative order" might be made. But Chamberlin also knew his audience:

If it is felt that an attack must be made, let it be arranged that the defense [be published simultaneously] with the attack in the spirit of ancient chivalrous combats and in the not less chivalrous spirit of the most approved form of modern polemics.[156]

If Wright happened to read Chamberlin's appeal, he could only have wondered why Chamberlin had not followed such fair-minded advice over the nearly three decades he had been publicly flaying Wright for a litany of perceived sins, geological and otherwise. On this occasion Chamberlin even invoked one of the themes he had often used in chastising Wright: the tendency of participants to push their particular views as though they were "determinate" and lacked dissent. Unless the "fact of such dissent is duly impressed on the reader, he [the reader] is placed under the disadvantage of partisan influence, and if he is keenly alive to his rights, he is likely to resent this."[157]

Chamberlin had not named names, but Hrdlička assumed Chamberlin's criticisms were not aimed at him. His criticisms of Vero were strictly "in a spirit of fairness, and merely for the purpose of setting right some facts where errors seemed to be of some importance."[158] Sellards, whose brief review of Hrdlička's BAE *Bulletin* tolled the several alternative hypotheses Hrdlička had studiously ignored and decried his misrepresentation of the situation at Vero, would have been astonished to hear such a claim. Those who were interested in human antiquity at Vero, Sellards warned, did themselves an injustice if they took Hrdlička's *Bulletin* as "a fair presentation of the subject."[159]

For his part, a feisty and unrepentant Hay shot back in *Anthropological Scraps* that "our great and honored geologist at Chicago" was a hypocrite. He believed Chamberlin, just as much as Vero's critics, had his mind made up at the outset and refused to change it even in the face of all contradictory evidence. Had Chamberlin been so biased in his work in glacial geology, "he would have accomplished little."[160]

Chamberlin's hope for reconciliation and his plan for reaching common ground came to naught. The problem, Hrdlička supposed and Chamberlin agreed, was that it was too late. Controversy could have been averted if the invitations to view the Vero site had been broadcast immediately after the human bones were found, and not after "compromising assertions were made in different periodicals which were then difficult to retract." To prevent that happening in the future, Hrdlička urged the establishment of a blue-ribbon panel under the auspices of the AAAS or National Research Council of "the foremost men in Geology, Anthropology, etc." to whom all such discoveries could be immediately referred, who would then investigate a locality before anything was removed or disturbed. Chamberlin thought that an excellent idea, save for the suggestion that he serve as its chair. He suggested Holmes. Hrdlička knew better: Holmes had for so long "opposed the hasty and unfounded claims of others in relation to man's great antiquity in America, that he would doubtless be accused of not being really impartial."[161]

Although Hrdlička offered the suggestion again on later occasions, nothing became of his blue-ribbon panel, whether for want of a proper chair, funding, or the difficulty of agreeing on its membership. Hrdlička had already made clear his opinion of the role of geology and geologists—they should be seen, not heard—and it is unlikely he could find one to his liking to serve on such a panel. But then, his opinion was not necessarily shared within the anthropological community, few members of whom assumed as he did that human physical remains were temporally diagnostic. Of course, there was also plenty of precedence starting with Trenton to indicate that site visits by panels, possibly even blue-ribbon ones, were not always able to achieve consensus on a site's antiquity. More often, they were exercises in incompatibility.[162]

Once again, the discussion of human antiquity reached an impasse marked by little forward progress, stubborn mistrust, and even a sense of apathy. In talking over possible topics for the 1918 American Anthropological Association meeting, Kroeber raised the possibility with Hrdlička of a symposium on human antiquity of the Americas, but then immediately withdrew the suggestion, realizing

> the sort of indefinite and aimless talk which would be brought out by announcing an open symposium on the antiquity of man in America. Talk which would begin nowhere in particular, end nowhere, and chiefly furnish opportunity for individuals to occupy the floor.[163]

Nelson, for his part, told Hay that since the "anthropologists . . . are weary of discussing the old finds," there was "nothing to do except lie low for the present" and look to the future: "Nevertheless, some day when Holmes retires, Dr. Hrdlička will probably find himself having a hot time, as I know of no one who would not enjoy scrapping with him. But then, is he sincere in his present attitude?"[164]

7.8 ERAS' ENDS

In the spring of 1919 Sellards resigned as Florida state geologist to take a position with the Bureau of Economic Geology in Austin, Texas. There, he rose through the ranks, ultimately to become state geologist.[165] His Florida departure came as his review of Hrdlička's BAE *Bulletin* appeared, and together these events effectively closed the chapter on Vero, at least for him. Eighteen years passed before Sellards again publicly spoke to the Vero evidence.[166] Not that he stopped thinking about it. Many of Sellards's actions over the course of his career can only be understood as efforts, directly or indirectly, to prove that he was right and Hrdlička wrong about Vero and human antiquity in the Americas (§1).[167]

About the time Sellards decamped for Texas, Wissler wrote Abbott to say Spier's long-awaited monograph on the Trenton work by the AMNH (§6.15) was about to appear.[168] Abbott was pleased, though he could not accept the news without a strong dollop of glum self-pity:

> Yours of yesterday received, and I will not wait for [Spier's] report to reply. It is just 48 years ago that I ventured to express the opinion as to the evidence of antiquity in my (then) neighborhood, and for this almost half a century,

I have had nothing, except at rare intervals, but abuse for my pain. Now at 76, I look backward and wonder I was so patient, but all's well that ends well and I am pleased to know that I have not labored in vain in my chosen pursuit of North American Archaeology. You spoke of a "first report." Is there to be another? If so, I trust to live long enough to see it. How very hard for some truths to be established. I certainly would not re-live my life for the satisfaction that has finally come to me.[169]

There was no additional report on Trenton, but Abbott would not have lived to see it. He died four months later and was buried at the Riverview cemetery beneath a large glacial boulder from the Delaware valley. His epitaph read much as he had planned (§6.15): "In this neighborhood Dr. Abbott discovered the existence of Paleolithic Man in America." Two months later, Volk was in an automobile accident and sustained severe injuries; two days later he died without recovering consciousness. In death as in life, Volk trailed behind Abbott.[170]

Wright wrote their joint obituary and defended their discoveries "in view of the fact that persistent attempts have been made to discredit them." The battle was not over, at least not for Wright. After two pages of vigorous arguments for a Pleistocene age of the Trenton paleoliths (though without any new evidence), and of equally vigorously criticisms of Holmes and Hrdlička (though without naming names), Wright returned to the business at hand and provided an elegy for the deceased:

> The work of Dr. Abbott and Mr. Volk illustrates the importance of having local observers interested in discoveries to be made about their own doors. They were both businessmen [*sic*] who turned aside to make and record observations which could be made only by those who were on the ground; and their observations have been carefully recorded and published, and their collections preserved where they are open to the observation of all scientific men.[171]

Wissler, who had far less of a stake in this fight, merely said Abbott's claims of three "superimposed cultures" in the soils of his estate had been systematically investigated by Volk and others—the "others" being Spier and colleagues—working under Wissler's direction.[172]

Wright's obituary of Abbott and Volk was a harbinger of his own. He died in April of 1921. Warren Upham, who wrote the principal obituary, spoke glowingly of Wright's contributions to glacial geology and archaeology, but with scarcely a word of the controversies in which Wright had been embroiled, nor of the fact that none of his claims on behalf of the unity of the glacial period, the Cincinnati Ice Dam, or an American Paleolithic of Pleistocene age had passed critical muster.[173] Wright's grave is marked by a large glacial erratic (a conglomerate with red jasper pebbles, transported from north of Lake Huron [§4.8]) that Wright had found on the Kentucky side of the Ohio River nearly forty years earlier, a key piece of evidence on which he had based his idea of an ice dam at Cincinnati (§4.8). He obviously felt it a fitting symbol of his life's work. His headstone is labeled "George

Frederick Wright. Geologist." No mention is made that he was also the Reverend George Frederick Wright, theologian.[174]

With Wright's passing, the last advocate of an American Paleolithic was gone from the scene. By then, Holmes had begun to pull back as well. His last major work in archaeology was the *Handbook of Aboriginal American Antiquities, Part 1: The Lithic Industries* (Part 2 was never written), which appeared in the fall of 1919.[175] There was little in it that was new, particularly in the chapter devoted to problems of chronology, the "problems" being mostly a recitation of the fatally flawed claims of human antiquity that had been championed over the years (Trenton, Newcomerstown, Little Falls, Lansing, and so forth). Still, that discussion gave Holmes the opportunity to take another victory lap for having been the one who had

> insisted on the closest scrutiny of all discoveries supposed to bear on the problems of chronology. His [Holmes's] restrictions, however, were regarded by some as ill-advised and as tending to embarrass or to endanger the acceptance of legitimate conclusions. The sequel seems, however, to justify the questioning attitude, since to-day, after 30 or 40 years of persistent research, a large part of the testimony advanced in support of geological antiquity is discredited and the remainder awaits a fate to be determined only by additional research.[176]

He proclaimed, as before, that he was open to the possibility that people had reached America "during any of the periods of mild climate which preceded and interrupted the Ice Age," but he still saw nothing in the archaeological, anatomical, or linguistic records to indicate an arrival "until after the final retreat of the glacial ice" from North America.[177]

Holmes's work was done. After decades at the BAE and USNM, he had become director of the National Gallery of Art, and by the late spring of 1920 he had had enough of holding multiple positions simultaneously. He resigned from his anthropological obligations to devote his attention to the National Gallery. Honored many times over,[178] his archaeological career seemingly past and his old antagonists dead, Holmes could afford to be slightly (but only slightly) more generous than he had been. In response to a query from Nelson as to how and when he had become involved in the human antiquity controversy, Holmes made an uncharacteristic concession:

> My first entry into the controversy over Paleolithic man resulted from a visit to the Salem Museum where Putnam had two cases filled with American paleoliths—so labeled. This was in the late 'eighties. . . . There was not a single specimen that showed definite signs of use. Investigations of shop and quarry sites which followed made it clear that these specimens were all mere rejectage of Indian arrowheads, spearpoints, knives, etc., and a little later—in the 'nineties—I felt justified in challenging the whole American Paleolithic "evidence." *I may have carried my opposition too far*, but so unyielding were the advocates of antiquity that decisive measures were demanded.[179]

There were still advocates out there, of course, and at least one of them was quite unyielding.

When Holmes's *Handbook* appeared, Hrdlička scrambled to have MacCurdy or some other "impartial and just person" review it for one of the major journals before "any of our 'friends' get a chance." Hay reviewed it anyway (in *Anthropological Scraps*), as he felt duty bound to turn his scientific confreres "from the errors of their ways."[180] Hay was doing what he could to keep Trenton, Vero, and the human antiquity issue alive and maintain a steady drumbeat of criticism of Holmes and Hrdlička.

His task should have been easy. Over the next several years more than a half-dozen finds of human skeletal remains of purportedly great antiquity were reported.[181] Yet most of these were skeletons from purportedly Pleistocene deposits (gravels, loess, and so on), with no associated fauna, extinct or otherwise. As all had been down that road already, few were willing to follow those claims. At the International Congress of the Americanists in the summer of 1922, Hrdlička dismissed these latest contenders for great antiquity with an easy wave, while Holmes trumpeted that

> There has not been collected either in North, Middle, or South America, a single archaeological object attributed to great antiquity that can stand the test of the scientific requirements here made plain; and the same is true of any skeletal finds, as has well been shown by Dr. Hrdlička, howsoever confidently announced as very ancient and by whatsoever authority. This is a challenge, though issued with the utmost confidence [and] quite open to counter-challenge.[182]

7.9 DANGEROUS TO THE CAUSE OF SCIENCE

Naturally, counterchallenges came. It was inevitable. But what neither Holmes nor Hrdlička expected was that the next one was from fieldwork funded by the BAE, conducted by a colleague at the USNM, and carried out with the help of the USGS. Alliances and loyalties were not as strong as in the days when Holmes was at the BAE and Chamberlin, McGee, and Salisbury were at the USGS. But in their unhappiness with those who ought to be allies but behaved like enemies, they had something in common with Hay, for that challenge came at a site just forty miles from Vero with a similar archaeological and paleontological record. Yet, its excavators considered it to be Late rather than Early Pleistocene in age. Hay had no choice but to criticize those he thought ought to be aiding the cause.

Like Vero, the Melbourne site was discovered bones first. In December of 1923, Amherst College vertebrate paleontologist Frederick Loomis was excavating an *Elephas (Mammuthus) columbi* skeleton when "a large rough flint implement, shaped like an arrowhead but larger" was found in the sand layer alongside the ribs and vertebrae of the mammoth and a nearby antler of *Odocoileus sellardsi*.[183] The artifact was made of black chert and was "roughly chipped, and with none of the fine workmanship characteristic of the Indians." Worried about what might have been missed, Loomis picked through the backdirt of a recently dug canal, and in doing so he found three bones he deemed worked by human hands: a bone handle (species unknown), a deeply grooved *Odocoileus* rib, and a partially drilled tibia.[184]

By the time excavations were complete a few weeks later, he had from that same stratum other extinct species, including *Mammut americanum, Equus complicatus, Chlamytherium septentrionalis,* and *Megalonyx jeffersoni.* The fauna was virtually identical to that at Vero, and on a brief visit to that site with Weills, Loomis was struck by how similar the two were in their "[geological] beds and manner of occurrence." Melbourne convinced him that humans were contemporaries with now-extinct Pleistocene fauna: the "close parallel with the finds at Vero [made] it doubly convincing."[185] In 1924 Loomis announced Melbourne's antiquity in the same journal (*American Journal of Science*) and with the same confidence that Sellards had eight years earlier. Loomis got a far faster and more malicious reaction.

Holmes was by then nearly eighty years old. Although retired from archaeology, he still kept a watchful eye on the field and his legacy in it.[186] When Loomis's Melbourne paper caught his attention, he was angry, perhaps in no small part because of who he learned was supporting the research. After his first stint of excavations at Melbourne, Loomis had invited USNM vertebrate paleontologist James Gidley to join him in work at the locality. Because the USNM had limited funds for fieldwork, Gidley requested support from the BAE.

The BAE's chief, Jesse Fewkes, was unconcerned about supporting work on an archaeological site of purportedly Pleistocene age. His only reluctance (it seemed to Gidley) was whether he was "going to get enough for himself out of it" to warrant the expense. He asked to have "first shot at the examination of any human artifacts that might be found." Gidley assured a nervous Loomis that "first shot" did not mean Fewkes was poised to attack the legitimacy of any artifacts or try to steal credit for finding proof of a deep human antiquity.[187] Fewkes was keenly aware of the BAE's longstanding, if unofficial, position in such matters and of whose now-retired toes he might be treading on. He and Holmes had overlapped in their tenure at the BAE, and a few years earlier Fewkes had even asked Holmes to clarify his views on the Paleolithic as they applied to America. Holmes responded with his usual lecture that rudely chipped stone artifacts were not old but merely manufacturing failures and thus were not implements (which, Holmes added, was probably true of a good many European Paleolithic artifacts as well, though this would have been a minority view in Europe).[188] Those who thought otherwise, Holmes asserted, were "blind to reason." Caution was necessary to avoid any errors. Knowing that as well as Holmes's strongly held views, Fewkes nonetheless was willing to allocate $1,000 of the BAE's budget for the 1925 excavation season at Melbourne.[189]

Holmes got wind of these plans in June and sent a preemptive strike to *Science*, with the urgent request that Cattell find space for it "and soon, as the Florida geologists are about to take the field again and may multiply their blunders."[190] Cattell found the space for Holmes's paper, but it did not appear until mid-September, well after any "blunders" might have been averted and after much of what Holmes had to say was rendered irrelevant by that summer's discoveries.

In "The Antiquity Phantom in American Archaeology," Holmes's stuck to his usual script. For nearly a half-century, he explained, "every rudely chipped stone has been in imminent danger of assignment by its form alone to great antiquity." He recounted his "bitter and long continued" battle and subsequent triumph. He proudly announced it was "difficult to find in any museum, American or foreign, a single American specimen" labeled as Paleolithic. Now Loomis had offered as "conclusive proof" of Pleistocene antiquity four meager artifacts. Yet, "all were obtained from superficial hammock deposits within three feet of the surface and are of types characteristic of the art of the Indian tribes of Florida." Worse, Loomis was oblivious to the "well-known fact that the Indians used chipped stone implements of every stage of rudeness . . . [and] the idea that either rudeness of refinement in the stone chipping art is exclusively characteristic of any period . . . is no longer entertained by archaeologists." It showed a "manifest lack of archaeological acumen."[191]

Not only was Loomis ignorant of archaeological methods, he and paleontologists generally (Holmes particularly had Hay in his sights) were "slow to recognize" that human remains from on or near the surface "are liable to intrusion by various means into older underlying deposits and that to very considerable depths." This was a theme Holmes had been sounding for more than three decades (§5.4, §5.10, §5.11), and paleontologists and geologists still were not allowing a "zone or horizon of doubt extending to the full depth of possible intrusion from above [and] reserving their final determinations until the evidences are so conclusive as to be beyond danger of error." Until they did so, Holmes warned he would

> feel it a duty to hold and enforce the view that the evidences of Pleistocene man recorded by Loomis at Melbourne, as well as those obtained by Sellards and others at Vero, *are not only inadequate but dangerous to the cause of science.* A similar attitude toward the illy considered announcements of followers of the phantom of antiquity should be rigidly maintained by all conservative students of the history of man in America.[192]

Even by the standards of this often harsh controversy, it was an ugly outburst, but perhaps it was understandable for one who perceived his contributions as being overlooked and his legacy tarnished, and who that same year suffered the loss of his wife and a lower leg to blood poisoning.[193]

Loomis was shocked but did not stand idle. Within a day of Holmes's "caustic attack" he crafted an aggressive rebuttal that said, in effect, that Holmes had no idea what he was talking about. If he had been "considerate enough" to approach Loomis beforehand with his questions and concerns, Loomis would have shared with him the evidence gathered that summer (though, of course, Holmes's paper was written before Loomis's summer work), which included a human skull and jaw "crushed as if trampled by an angry elephant." Had Holmes examined either the artifacts or the deposits at Melbourne, Loomis was sure he would not have said what he said about their depositional context and history. And had Holmes only seen the condition of fossil bones at Melbourne, he could have saved himself from making "absurd" statements about their reworking by Indians. It would be far better, Loomis lectured Holmes, for the question of human antiquity in America to be settled "by cooperation between geologists and anthropologists."[194]

Loomis sent Holmes a draft of his response, offering to hold off submitting it until Holmes had the opportunity to identify anything that ought not to appear in *Science* and prepare a reply. Although he extended that courtesy, Loomis conceded nothing:

> I have not taken your paper up paragraph by paragraph as I might have and shown how absurd most everyone is to one who has worked in the field in Florida, but if you will consult Dr. Fewkes or Mr. Gidley you will be in a much better position to have an opinion on these finds of which there are now many, and I am sure you will revise your opinion on many phases of this subject. You surely would if you examined the field.[195]

Holmes had no intention of following Loomis's advice, and merely replied that

> you have a perfect right to defend your work and your determinations, and I cannot have the least objection. If you are right, time will justify you and my apologies will then be due you.[196]

But then Holmes had second thoughts. Several correspondents urged him to reply, including Watson Davis at the *Science Service*, who was spoiling for a fight to report to his constituent newspapers around the country.[197] Holmes thought he might write a rebuttal that would "leave the whole scheme of the contemporaneity of the great mammals and man in Florida in a badly shattered condition," but he never did.[198] The *Science Service* had to settle for his noncommittal quote that "in fairness to the discoverers he [would] withhold comment until their report appear[ed]."[199] None ever appeared.

Indeed, the only public response by either Holmes or Hrdlička came at the December 1925 meeting of the Paleontological Society. The American Anthropological Association was holding its meeting a few blocks away, and when news of a talk by Gidley on Melbourne reached Hrdlička, he led the assembled anthropologists "en masse" down the street to hear the presentation. There followed a "spirited discussion" between Gidley and Hrdlička, with Hrdlička making the uncharacteristic (and to his listeners assuredly unbelievable) appeal for "closer affiliation between the ethnologists and geologists in studies of ancient man."[200]

7.10 WITH FRIENDS LIKE THESE

Those brief exchanges aside, Melbourne never sparked the volley of return fire Vero received from critics. That was not altogether surprising. Holmes was by then understandably weary of the fight, and there were also good reasons for Hrdlička not to bother. For one, Gidley and Loomis's Melbourne fieldwork included several expeditions to Vero that fully convinced them the geological setting and paleontological histories of the two localities were essentially identical. As a result, they linked the two sites in their presentations.[201] From Holmes and Hrdlička's perspective, there was no need to answer the Melbourne bell, as all their criticisms of Vero by extension applied to Melbourne. (Hrdlička did examine the Melbourne human remains when the bones were later transferred to him from the museum's vertebrate paleontology area. He had one of the "minor museum preparators" reconstruct the crania, which he characterized as "the usual type of the Indian crania found in the mounds in this part of Florida."[202])

Better still, criticism was already being leveled at Melbourne. The source of the criticism was surprising (and a twist on the "enemy of my enemy is my friend" theme), but it could not have been unexpected. In linking Melbourne to Vero and highlighting the strong similarity in their geology, stratigraphy, and vertebrate fauna, Gidley and Loomis naturally concluded that the human remains at both sites were the same age. But how old was that? During their 1925 fieldwork they realized they had made a mistake about the stratigraphic position of the human remains.[203] They had not come from Melbourne's equivalent of Vero's Stratum 2, but instead from the younger Stratum 3. Even though the human remains were associated with mammoth and mastodon, they were not Early Pleistocene but likely to be of "relatively recent date, either late Pleistocene or even post-Pleistocene." If one put the end of the Ice Age at 25,000 years ago, as Loomis did (an estimate not far off that used by Holmes as well), that indicated "man of the Neolithic type of culture" was in Florida 15,000 to 20,000 years ago.[204]

Their estimate of a late or postglacial antiquity for both Melbourne and Vero was affirmed by Wythe Cooke, a USGS geologist sent at the request of the Smithsonian Institution to examine the deposits at Melbourne and take a fresh look at Vero. On the basis of his reading of the position of the Atlantic shore at the time the fossil remains were deposited, he concurred that the human remains were old, though possibly somewhat older than Gidley and Loomis supposed. Marine shells not reworked by freshwater were found in the bone bed, indicating that the shore line was close by and that the time necessary for the shore (and sea level) to retreat to its present position, now four miles distant, demanded more than was available in the period since the closing stage of the Pleistocene.[205]

That still was not as old as Hay wanted to make the vertebrates at Vero and, by extension, the fauna at Melbourne. Cooke, Gidley, and Loomis were all aware of Hay's opinion, yet they explicitly rejected an Aftonian or Early Pleistocene age for the deposits. Cooke did so on the grounds that Hay was simply mistaken in restricting the fauna to the first interglacial. Gidley and Loomis took the same tack, arguing—as had been argued many times before—that the Pleistocene fauna lingered longer in Florida and other southern states than after the ice sheets had disappeared from the north.[206]

Hay was by now thoroughly exasperated. He expected opposition from Holmes and Hrdlička. He did not anticipate having to take the fight to his fellow vertebrate paleontologists and geologists. He had "been studying the vertebrates of the Pleistocene" for many years; he had detailed the species, genera, and even orders of the vertebrates as they had come and gone in relation to glacial periods; and he had carefully defined based on composition what the early and late Pleistocene faunas looked like. And yet, "having reached and announced these conclusions

I am impressed that some of my friends, paleontologists, geologists, and anthropologists, doubt the correctness of my views."[207] Once more he explained his evidence and the reasons why he thought certain genera were Early Pleistocene and not later Pleistocene in age. He defended using percentages to identify the age of a fauna and attacked the idea that the fauna lived longer in the south. If that were so, why had it changed so little over at least four glacial stages and three interglacial stages? And he challenged his critics to show evidence that the Melbourne and Vero faunas were Late Pleistocene in age. He was quite certain they had none.[208]

Instead, Hay suspected the age estimates for Melbourne and Vero that Cooke, Gidley, and Loomis proposed were based on their misreading of the paleontological and geological facts of the case and, worse, were a misguided "olive branch offered to the anthropologists." That was a concession Hay did not think worth making, for he scarcely believed "the late arrival of man in America is a fact so well demonstrated that geology and paleontology must yield unquestioning assent."[209]

Hay went after Cooke and the others again the following year (1927). He insisted, "It is not wholly a pleasant task to take up this subject again," but if he was not enjoying being embroiled in an internecine skirmish, Holmes watching from the sidelines certainly was. When Hay's paper bore down especially hard on what he saw as inconsistencies and errors in Cooke, Gidley and Loomis's interpretation of the age of Melbourne and Vero (as, for example, when Hay criticized the "contrast between the laxity of [their] argumentation, and the assertiveness of [their] conclusion"), Holmes emphatically wrote "*Good!*" in the margins of his reprint of Hay's paper. Holmes could see that the more questions and doubts Hay raised about Cooke, Gidley, and Loomis's interpretation, the less secure the larger claim for a Pleistocene antiquity of both Melbourne and Vero, and therefore the less effort required of him or Hrdlička.[210]

Hay obligingly kept up his drumbeat of criticism the following year, and finally Cooke and Gidley took the bait. (Loomis paid Hay's criticisms little mind because his funding dried up and he essentially ceded Melbourne to Gidley.) Each of them willingly conceded a bit of time in Hay's direction, though neither allowed Melbourne or Vero to be as old as the Aftonian. Both sites might contain taxa generally characteristic of the Aftonian, but when examined more closely species-level differences emerged (particularly in the proboscideans, horses, and camels), and there were also taxa not typical of the Aftonian (such as tapirs, edentates, capybara, and alligator) in those deposits.[211]

That was the last of it, for by 1928 the issues and evidence had changed dramatically, and with that the attention of anthropologists and paleontologists was focused elsewhere. Melbourne and Vero became afterthoughts, existing in the indefinite netherworld of sites about which, as had been Holmes's intent from the outset, any conclusions were held in abeyance (§7.4).

7.11 SPEAKING OF OLD EVIDENCE

Linguist Pliny Earle Goddard, who as editor had happily accepted Sellards's Vero paper for the *American Anthropologist*, was equally pleased when Loomis wrote on Melbourne for *Natural History* in the spring of 1926. Goddard relished "a cause to bring out his best; in a cause, his eyes lit up, the steel in him flashed, and he rejoiced in the cleanness of combat."[212] The antiquity of the Americas gave him a cause. In an accompanying commentary on Loomis's article, Goddard spoke of the reasons why a great human antiquity should be anticipated for the New World. There was the time required to domesticate and cultivate maize, to develop the "higher civilizations of Peru, Mexico, and our own Southwest," and especially to allow for the divergence of some "fifty [Native American] languages . . . so distinct from one another in their vocabularies, that no relationship can be traced."[213]

In his *Handbook of Aboriginal American Antiquities*, Holmes had dismissed the idea that the diversity of Native Americans—whether cultural, linguistic, or physical—bespoke a long period of divergence from a common ancestor. Not only were Native Americans not as diverse as they appeared (and on that point he had Hrdlička's testimony on physical evidence), but such diversity as existed could be readily explained not by in situ divergence but instead—in the case of language, for example—the migration of different groups speaking "radically different tongues."[214]

That was a reasonable argument, so far as it went: cultural, linguistic, or physical diversity could reflect either divergence over time from a single ancestral group, or multiple migrations of different groups. Diversity by itself could not demonstrate antiquity; that required independent geological evidence. Goddard was aware of that as well, and in his *Natural History* commentary he pointed to two recent discoveries that changed the situation. One was Loomis's evidence from Melbourne, which put humans in Florida "after, but immediately after, the end of the Ice Age." The other was evidence reported just a few months earlier, in November of 1925, by Harold Cook from the Lone Wolf Creek site in Texas, where "three worked flints [were found] under the fossilized skeleton of a bison of an extinct species." Together these sites (and for good measure Goddard threw in Williston's 12 Mile Creek) all testified, in his view, to a late or post–Ice Age presence in America that dated back some 25,000 years; this was long enough, he thought, to enable the descendants of the first Americans to "build civilizations, and for profound linguistic changes to take place."[215]

That amount of time was important to Goddard, as it was to virtually all linguists and social anthropologists of the time. This was partly because the linguistic and cultural evidence suggested as much, but also because a deeper prehistory provided crucial chronological evidence to help refute the hyperdiffusionist claims then in vogue that America's precontact civilizations could not have been indigenous or "home-grown," as there was insufficient time for them to have developed on their own, and therefore they must have had outside help, likely from the arrival of seafaring Egyptians carrying the ideas and tools to craft civilization.[216] Besides, pushing North American prehistory back had the added bonus, for Goddard at least, of irritating Holmes, toward whom he had a longstanding antipathy.[217] He made that clear in his ricochet of Holmes's angry *Science* outburst (§7.9): "Preconceptions and generalization . . . *are fatal to the open-mindedness necessary for progress in science*."[218]

Leaving aside the obvious pleasure Goddard took in settling a score, his championing of Lone Wolf Creek, and particularly his recycling of 12 Mile Creek, raised an obvious question. Chamberlin had handily disposed of 12 Mile Creek twenty years earlier. What exactly had changed in the intervening years that now made this site, or one apparently like it (Lone Wolf Creek), acceptable as evidence of a Pleistocene human antiquity? Goddard himself admitted there remained with Lone Wolf Creek the problem of the "exact division of the Pleistocene in which this bison became extinct and the stratum was laid down," though he was willing to let such questions be left for the future.[219] Nonetheless, he announced that as a result of the Lone Wolf Creek discovery the "hour has at last arrived for an extensive reorganization of our conception of the peopling of America."[220] Others were skeptical, in large part because of the nature of the Lone Wolf Creek discovery, but also because of who was promoting it.

CHAPTER EIGHT

In the Belly of the Beast, 1921–1928

READERS OPENING THE JULY 1926 issue of *Scientific American* were greeted by a photograph of a dour Aleš Hrdlička next to plaster busts of "Prominent Indians" looking equally somber. His article, "The Race and Antiquity of the American Indian," matched the mood: "There is no valid evidence that the Indian has long been in the New World," he announced. Of the claims of "arrow-points, fragments of pottery and even human bones . . . in association or in the same strata with the bones of the mastodon, fossilized buffalo, the glyptodont, and other extinct creatures," none withstood the "test of critique," despite the efforts of a generation of archaeologists and geologists (often "more enthusiastic than experienced").[1]

As for the Indians themselves, they came to this continent from northeast Asia, likely not in a single immigration wave but a "dribbling over," which could account for the variation in their languages and physical types.[2] Such variation notwithstanding, Hrdlička declared the American Indian one race that shared a variety of features, and of a type that clearly "did not reach America until after [humans] had attained a development superior to that of even the latest glacial (upper Neanderthal) man in Europe." That meant, he estimated, a "moderate, post-glacial and probably post-Aurignacian" arrival.[3] That being the case, what accounted for the wide range of opinion on human antiquity in America? Hrdlička blamed it "partly on the peculiarities of human nature and partly on differences in training, but in the main on the varying individual degrees of knowledge and experience."[4] It was the younger generation and "workers in collateral lines such as linguistics, geology and paleontology" who were most liable to err in such matters.

One of that younger generation of paleontologists was Harold Cook, who had never met Hrdlička but deeply disliked him and was spoiling for a fight. Earlier that year Cook announced to Henry Fairfield Osborn (a longtime mentor), "We have allowed preconceived notions and prejudices and ideas and notions of man's age and antiquity to dictate to us." He urged Osborn to "follow the evidence, no matter how many *Hrdličkas* [sic] may arise out of Washington or elsewhere, with their magnificent fund of inside information that primitive man could not have been in America."[5] Cook took "rather sharp issue" with Hrdlička's *Scientific American* article, and it gave him the opening for a counterattack, which he published in the same journal just four months later.[6]

Hrdlička's article was simply "splendid," Cook proclaimed, and no one was better qualified to write it than Hrdlička. Cook agreed it was "but natural that a man's opinion should be influenced by his training and that his point of view should be correspondingly affected." Yet, with expertise comes myopia, and "in fairness to the other branches of science" it was important to consider questions of origin and antiquity from the perspective of geology and paleontology, and not just anthropology.[7] Cook was happy to provide that perspective, for "it is the proper business of the true scientist to search out this evidence and to follow where it leads, not to try to prove or disprove theories—interesting and instructive as they may sometimes be as pets."[8] And any true scientist knew there was "no possible doubt" humans were in America in the Pleistocene, and Cook pointed to two sites in which he was directly involved—Snake Creek and Lone Wolf Creek. Evidence from these of artifacts in association with extinct fauna went a long way toward removing the "categorical denials of the possibility of glacial man in America" made by Hrdlička.[9]

Cook's roundhouse swing at Hrdlička delighted his fellow Nebraskans. Robert Gilder, still smarting from Hrdlička's attack on his Long's Hill claim (§6.14), took "a sort of savage satisfaction in seeing someone take a fall out of [Hrdlička], after the treatment he [Gilder] received at his hands." Cook's father-in-law, Erwin Barbour, was just as pleased, given "the way loess man was slammed" by Hrdlička. Hrdlička was not pleased: to him Cook's article was just "another head of the Hydra."[10]

Yet, even as he aimed to cut off this head, there were others. Several sites were discovered in the 1920s that promised a deep human antiquity. At each the chronological anchor was an association of artifacts with extinct Pleistocene fauna. This was not altogether a new kind of evidence, of course. It was the basis for claiming two dozen years earlier that 12 Mile Creek was Pleistocene in age, a claim dismissed by Thomas Chamberlin on the general principle that the precise ages of ostensibly Pleistocene faunas were poorly known (§6.9).

The vertebrate sequence was still not fully settled by the start of the 1920s. The extinct Pleistocene genera were reasonably well known, but as Vero and Melbourne showed there was spirited debate even among paleontologists as to when in the Pleistocene these disappeared (§7.10). The challenge of identifying extinct species within a nonextinct genus like *Bison*, as well as the timing of their extinction, was even greater, particularly at sites like 12 Mile Creek, which lacked independent chronological or geological testimony of Pleistocene antiquity. Regardless, by the 1920s a more widely accepted (even if imprecisely defined) agreement on the age of fossil bison had emerged, and with the demonstration late in the decade of spearpoints unequivocally associated with bison bones at a kill site in an arroyo outside of Folsom, New Mexico the Pleistocene barrier to human antiquity in North America was finally broken.

That end came about largely in spite of, and not because of, those promoting this evidence of great antiquity. Chief among those proponents were Cook and Jesse Figgins, with help of the peripatetic Oliver Hay. And because the Folsom claim did not rely on inherently "primitive" properties of artifacts or attributes of human skeletal morphology, neither Hrdlička nor William Henry Holmes could do much more than watch from the sidelines and smolder.

8.1 HAROLD COOK AND JESSE FIGGINS—WILLFUL REVOLUTIONARIES

Agate Ranch, Harold Cook's family homestead in far northwestern Nebraska, sits atop one of North America's richest Tertiary-age fossil localities.[11] Harold's father, James—who in the 1870s had brokered an agreement between the Sioux Chief Red Cloud and paleontologist O. C. Marsh that allowed Marsh to collect fossils on Sioux lands—was well aware of the scientific potential of his ranch and called it to the attention of scientists.[12] By the turn of the century its fossil deposits were attracting paleontological luminaries from institutions across the country, including Barbour, Richard Swann Lull (Yale), W. J. Sinclair (Princeton), William Holland (Carnegie Museum), Frederick Loomis (Amherst), J. W. Gidley (Smithsonian), and William K. Gregory, William D. Matthew, and Osborn (AMNH).[13] With them came rivalries, both personal and institutional. These were underlaid by a territorial conflict: Barbour was keen to keep his state's fossil riches in Nebraska and resented their plundering by Osborn and other well-financed eastern museums and universities (§6.11).

James gave Harold responsibility for choreographing the various field parties who were often looking to outmaneuver one another (and manipulate or pressure the family) in order to gain exclusive access to the richest of the fossil beds on the family ranch. Cook was not without loyalty to his state, but he did his best to straddle competing interests in what was often an awkward dance. He quickly learned that to insure that he and his family received credit for discovering localities and specimens, he had to assert control over the land and develop his own expertise in fieldwork and fossil identification.[14] He could see, however, that he was not on equal footing with the easterners, and this experience brought into sharp relief the strict hierarchy of scientific haves and have-nots, and where he and his family stood.

Cook developed a strong working relationship with Osborn and Barbour (whose daughter he married in 1910[15]), despite the fact that these two individuals and their institutions were rivals. (They had worked in the competing camps of Cope and Marsh and sparred over Long's Hill [§6.11].) Both valued the Cooks' role in bringing discoveries forward, and they saw the benefit of providing Harold, who showed a talent for fossil work, the opportunity to participate in excavations and occasional laboratory work. Over the years he gained valuable experience. There was formal study as well: at the University of Nebraska with Barbour (1907–1908), then at the AMNH and Columbia University (1909–1910) under Osborn, Matthew, and Gregory, where he studied zoology, paleontology, and comparative anatomy, analyzed fossil material, and wrote up results of the fieldwork he had done under the auspices of the AMNH and that resulted in his first publication, coauthored with Matthew, on fossils from western Nebraska Tertiary deposits.[16]

The AMNH was particularly solicitous and supportive of Cook's ambitions, providing him opportunities to analyze new specimens, name new species, and serve on their field staff on an almost equal footing as their chief collector,

Albert Thomson. It was an arrangement of mutual benefit. The AMNH came to enjoy a near-monopoly at Agate Ranch, while Cook gained a salary (but not enough that he could abandon ranch work), education and experience, and credibility as a paleontologist, though not a top-tier one.

Cook was, as so many before him, "ultimately subject to constraints that limited [his] power to act within the increasingly professionalized world of modern science."[17] His limited university training, his inability to break free of his ranch obligations, and the financial demands of supporting his family meant he never could. During his year in New York he occasionally rubbed shoulders with the Roosevelts (Teddy and daughter Alice) and the family of J. P. Morgan, the company that Osborn kept, but he could only look on from the edges of the room, wistfully observing how the scientific and social elite of the day "certainly did enjoy themselves."[18] Cook often declared, usually more strongly than circumstances called for, that he could care less about status. He insisted formal degrees meant little, though naturally he was pleased to make an exception when called to receive an honorary degree in 1952 from the South Dakota State School of Mines. Otherwise, "All any person has to have to get [a degree] is some time, money to live that long, have a temporarily retentive memory and a good phonographic reproducer."[19] Besides, he could boast of

> *FAR* more comprehensive and intensive and valuable training in the field than I have had within the narrow confines of college courses, and *AFTER* I had basic training in geology and zoology I instructed in minerals and rocks at the University of Nebraska, some weeks before I entered there as a freshman . . . It makes me smile when I occasionally hear of someone intimating that I have had insufficient basic training, because I have not taken the time to complete an orthodox college degree! Oh Lord—the work I have done under the men I have met and independently—had it been done in registered courses would have gotten me degrees as long as my arm! Because I have never cared anything about the handles . . . I have not taken the trouble to do this as registered college or University work. I wish I had, now—just because of the morale [*sic*] effect it has on scholastic hide-bound concepts, in many quarters.[20]

That large chip on his shoulder appeared early and proved lasting. As a teenager, he had dismissed as "a pin headed, egotistical, educated fool" the director of the Carnegie Museum of Natural History, who was patronizingly insistent that his museum deserved full control of the richest portion of the Agate Ranch fossil beds. A few years later, to head off a (literal) land office rush on the fossil-bearing hills beyond the edge of his family's holdings, the then-twenty-year-old year old Cook filed a homestead claim on the property. It meant abandoning his studies at the University of Nebraska to build a cabin to meet the terms of the Homestead Act, but it gave him leverage over the institutions that came calling.[21]

One of those was the newly established Colorado Museum of Natural History (CMNH). In the summer of 1916 its recently hired director, Jesse Figgins, having learned of the fossil wealth at Agate Ranch, wrote Cook wondering whether he might grant the museum permission to collect fossils.[22] Cook did, and he and Figgins soon established a strong working relationship. In late 1925 at age thirty-eight, Cook was appointed honorary curator of vertebrate paleontology. In 1928 the "honorary" was removed and he became an official curator, though still as a part-time employee.[23]

Cook's appointment paid just enough to reimburse his expenses.[24] Even so, the position meant status, though not as much as Cook envisioned or to which he aspired. He considered himself equivalent to the AMNH curators; they did not considered him likewise. Over the years Cook claimed to have received other offers (he never said from where) that would have brought him "into unusual prominence." He had to turn these down "on monetary grounds." Unless he could find a manager for his ranch, a part-time curatorship was the most he could afford.[25] The CMNH it was and would be.

From Figgins's point of view it was good that Cook stayed put. Figgins needed him—and for more than just access to Agate Ranch fossils. Even though he lacked extensive formal education in geology and vertebrate paleontology, Cook had far more expertise than Figgins. As Figgins himself admitted, he knew "nothing whatever, about archaeology and allied subjects, and but a smattering of paleontology and geology; certainly not enough to make my views of value or importance."[26] Still, when the time came he supposed that privileged his role in the human antiquity controversy: "Not being an 'ologist' of any sort that has a place in such a discussion, I cannot be accused of being prejudiced and 'influenced by training.'"[27]

Figgins's formal education ended at an early age. Born in Maryland, he had trained briefly for the ministry in a small religious college, but being more interested in ornithology he abandoned the effort and began collecting and studying birds and mammals on the Atlantic coastal plain. His efforts caught the attention of curators at the USNM, and he was hired to its temporary staff. That experience landed him a position as one of the naturalists under the Arctic explorer Robert Peary on his 1896 and 1897 Greenland expeditions. When Figgins's bird and mammal collections were turned over to the AMNH, he was asked to join its temporary staff to prepare exhibits, including habitat groups of native Eskimos. In 1905 Figgins gained a permanent position in charge of "the preparation of the higher animals and of plants for exhibition purposes, and the making of archaeological models and replicas" for the museum's department of preparation and installation.[28]

Although lacking a scientific pedigree or expertise in a subject area, Figgins had artistic talent (he painted the backdrops for the AMNH dioramas) and proved sufficiently skilled at exhibition and administration that he was hired in 1910 as the director of the CMNH. The museum had been founded the decade before and had been in its new building for only two years. As director, Figgins faced the formidable task of creating a museum largely from scratch, and he did so by hiring and ruling "with a firm hand" some of the "ablest preparators of mammals, birds and accessories in the country," including a strong staff of workers in fossil vertebrates. Or so Cook put it, making no mention that *he* was one of that strong and able staff.[29]

At the CMNH Figgins faced "endless troubles and difficulties," given the lack of curatorial and administrative infrastructure, the limited budget, the large tasks before him, and the fact that he was, as even fair-minded Matthew observed, "temperamental and high strung." From time to time in the mid-1920s in the midst of the museum's fieldwork at several putative Pleistocene sites, Figgins often spoke of being "on the ragged edge of a complete breakdown." Even when all was going well the during the first summer of fieldwork at Folsom in 1926, the site having already yielded museum-quality bison specimens and its first projectile point, Figgins fantasized about abandoning his obligations: "When I leave the house I will not know where I am going, nor will anyone. I won't know when I am coming back and there won't be any possible address that can reach me."[30]

Ultimately that happened, though whether by choice is unclear. In early 1928 Figgins boarded a ship in New York for a six-month South American expedition. The timing was odd. It was just months after the museum's pivotal Folsom discovery was celebrated across the scientific community, and as the CMNH was developing a major new excavation in Nebraska (the sort of project Figgins normally oversaw). Figgins's ostensible purpose in going to South America was to sketch, study, and collect birds and mammals from British Guiana and Brazil for museum display. But Cook privately spread the word that Figgins had been exiled by the museum's trustees after suffering a breakdown and committing acts that no institution "could countenance or tolerate, no matter whether caused by insanity or perversion."[31] It was hoped the South American sojourn would improve his disposition and resolve his problems.

Whether that was its true purpose—and one has to take into account that by the time Cook wrote that he was a hostile witness with a flair for the dramatic—the result was there was no victory lap for the CMNH at Folsom. Moreover, in Figgins's absence, but with his approval, Barnum Brown of the AMNH took charge of the excavations at Folsom the summer of 1928, effectively stealing the site and usurping the limelight from Cook. That ratcheted up already existing tension between Figgins and Cook, which only worsened on Figgins's return, when they exchanged icy letters about whether the arrangement with Brown had been fair and appropriate (not at all, Cook thought) and about what Brown and the AMNH were entitled to by way of an equitable division of Folsom artifacts and bones.[32]

Subsequent changes by the CMNH Board of Trustees separated the two so that Cook could bypass Figgins's direct oversight. That alleviated but did not eliminate the difficulties between them. In mid-1930 Cook angrily resigned his position. By his account it was an ugly divorce, riven with charges of duplicity, financial impropriety (on Figgins's part), and accusations of disloyalty and conspiracy.[33] Figgins was either more discreet or not nearly the malicious propagandist Cook accused him of being. Figgins lasted as director until early 1935, when he resigned over differences with the trustees—ironically, just a year after the museum's *Annual Report* opened with a flattering portrait of his work as director.

Cook and Figgins accomplished much during the years they were formally affiliated at the CMNH. They brought several sites to archaeological attention, became thoroughly entangled in the controversy over human antiquity in the Americas, and were ultimately responsible for the discovery that resolved it. But between the large chip on Cook's shoulder and his penchant for leaping to con-clusions, Figgins's volatile temperament and behavior that ricocheted from obse-quious to obsessive, and their shared deep insecurities and more-than-occasional bouts of paranoia, they often found themselves on the wrong side of controversy, even when their evidence was right. It was neither an easy nor a straightforward road to Folsom. For Cook in particular, the trouble began years earlier.

8.2 ANTHROPOID APES IN AMERICA?

During joint fieldwork in western Nebraska in 1908, Cook and Matthew re-ceived a tip from a ranch hand about fossils eroding out on a ridge in the Snake Creek drainage, some twenty miles south of where they were working. On visits to Snake Creek that summer they collected over a hundred genera of fossil mam-mals, birds, and reptiles from Miocene-Pliocene-age deposits.[34] That winter Cook worked with Matthew at the AMNH cleaning and describing their fossils for publication.

Their report is of particular interest for one passage that in retrospect takes on rich if unintended irony. In their observations of the teeth and jaws of *Pros-thennops*, an extinct peccary, they observed the "anterior molars and premolars of this genus of peccaries show a startling resemblance to the teeth of Anthropoidea, and might well be mistaken for them by anyone not familiar with the dentition of Miocene peccaries." They did not think anyone would make that mistake given the "improbability of finding Anthropoid teeth in a Lower Pliocene formation in this country." Although the Snake Creek deposits yielded an antelope associated in Europe with extinct Anthropoidea, "their occurrence at a corresponding hori-zon in this country was quite as little to be expected as the occurrence of Anthro-poidea would be, although of course not of equal scientific interest."[35]

Over the next decade AMNH field crews worked elsewhere, but Cook contin-ued to visit Snake Creek. In 1921, in what Matthew had designated the Pliocene-age Upper Snake Creek formation, Cook spotted a molar tooth "that very closely approaches the human type." In February of 1922, he sent the molar to Osborn at the AMNH.[36] Osborn instantly deemed the tooth "one hundred percent anthro-poid," just the sort of early American specimen he had been eagerly anticipating since his brush with Gilder's finds nearly twenty years earlier (§6.11). Osborn hur-ried down the hall to Matthew's laboratory, and the two combed drawings and casts to determine if it could be a higher primate. Their search yielded a small, previously overlooked tooth from Matthew and Cook's 1908 Snake Creek collec-tions that seemed to be "closely related generically . . . [if not] specifically" to this new specimen.

The second tooth fortified Osborn's impression that the animal was distantly related to humans, but the 1908 specimen was so badly worn he knew he needed to focus on Cook's latest find and get corroborative testimony from AMNH physical anthropologist William Gregory as to its anthropoid affinities. In the meantime, he told Cook, "we may cool down tomorrow, but it looks to me as if *the first anthropoid ape of America* had [*sic*] been found by the one man entitled to find it, namely, Harold J. Cook!"[37]

There was no cooling down. A few days later Gregory and Milo Hellman (on the faculty at New York University's Dental School) examined the specimen. Each was familiar with ape and human teeth and went further than Osborn dared hope, pronouncing the specimen as having a closer resemblance to *"Pithecanthropus* and with men rather than with apes."[38] Osborn rushed into print a type description for a new genus and species, *Hesperopithecus haroldcookii* ("an anthropoid of the Western World discovered by Mr. Harold Cook"). To achieve the greatest possible splash, slight variations of his announcement appeared simultaneously in the AMNH *Novitates* series, the *Proceedings of the National Academy of Sciences*, and then ten days later in *Science*.[39] The birth of *Hesperopithecus* was impossible to miss.

But what exactly was *Hesperopithecus*? The specimen bore only a remote resemblance to chimpanzees and had no close affinity to either *Dryopithecus* or *Sivapithecus*, the Eurasian anthropoid apes then known. Comparisons were made with closer forms, *Pithecanthropus* and even *Homo sapiens*, but ultimately Osborn concluded that the *Hesperopithecus* specimen

> cannot be said to resemble any known type of human molar very closely. The author agrees with Mr. Cook, with Doctor Hellman, and with Doctor Gregory, that it resembles the human type more closely than it does any known anthropoid ape type; consequently it would be misleading to speak of this *Hesperopithecus* at present as an anthropoid ape; it is a new and independent type of Primate, and we must seek more material before we can determine its relationships.[40]

Hesperopithecus arrived at an opportune moment. William Jennings Bryan, persistent presidential candidate and spokesman for Christian fundamentalism, had weeks earlier in a high-profile editorial in the *New York Times* rejected the theory of evolution as harmful and groundless. That an anthropoid ape was discovered in Bryan's home state gave Osborn a rich opportunity to score rhetorical points:

> It has been suggested humorously that the animal should be named *Bryopithecus* after the most distinguished Primate which the State of Nebraska has thus far produced. . . . The author [Osborn] had advised William Jennings Bryan to consult a certain passage in the Book of Job, "Speak to the earth and it shall teach thee," and it is a remarkable coincidence that the first earth to speak on this subject is the sandy earth of the Middle Pliocene Snake Creek deposits of western Nebraska.[41]

Unfortunately, as Osborn had to admit, what the earth was saying was "difficult and baffling."[42] All who examined the *Hesperopithecus* tooth were struck by how badly cracked, battered, and waterworn it was, which even made it difficult to determine whether it was a second or third molar and in turn raised questions about the reliability of the identification. Gregory and Hellman ably described the specimen in detail but cautiously detoured around its taxonomic identity:

[*Hesperopithecus*] combines characters seen in the molars of the chimpanzee, of *Pithecanthropus*, and of man, but, in view of the extremely worn and eroded state of the crown, it is hardly safe to affirm more than that *Hesperopithecus* was structurally related to all three.[43]

At least there seemed no doubt about its antiquity. Matthew, well versed in the Snake Creek stratigraphy, was certain the deposits in which Cook found the tooth were Lower Pliocene in age. But, finally recalling what he said of the *Prosthennops* teeth in 1909, he was unwilling to go on record about the specimen's taxonomic affinities, only its stratigraphic position.[44]

Gregory and Matthew were cautious. Others were skeptical *Hesperopithecus* was an anthropoid or had any affinities to *Pithecanthropus* or humans. In an odd convergence, its critics included Piltdown's staunchest defender, Arthur Smith Woodward of the British Museum (Natural History), who thought *Hesperopithecus* a primitive bear, and Piltdown's sharpest critic, Smithsonian zoologist Gerritt Miller, who declared *Hesperopithecus* a great ape. In the face of that criticism, and increasingly skeptical about the fossil, Gregory backed away from the possibility that *Hesperopithecus* was related to humans, even distantly. He knew that ascertaining its identity was going to require more fossil material.[45]

Osborn could see that too, and in the summer of 1922 he dispatched Thomson to Nebraska's "sacred ground" (Osborn's description) to find more of *Hesperopithecus*: "This animal will have a remarkable skull, and heaven grant that you may secure a bit of it." Yet, the landowner refused access, despite Thomson's offer to purchase the site and its surroundings.[46] The search for *Hesperopithecus* went on elsewhere in the Upper Snake Creek. And as additional fossil material came into the AMNH, Gregory began to suspect Cook's *Hesperopithecus* was just a heavily worn, badly eroded *Prosthennops* molar. Matthew and Cook's warning from two decades earlier had been prescient. Gregory successfully challenged Osborn's decision to have Charles Knight include *Hesperopithecus* in a new mural being painted for the AMNH Hall of Mammals.[47]

As resistance to *Hesperopithecus* was setting in, other bone fragments—among them one resembling a "trowel or paddle," another a "spatula" with a hole in it—were found by Thomson in the Upper Snake Creek formation. Cook and Osborn believed these were Pliocene-age artifacts. Others were doubtful. Barnum Brown urged Thomson to photograph any such specimens in situ and cautioned that many were similar to fossils he had seen shaped by erosion, natural breaks, etching by plants, and gnawing by rodents.[48] Cook granted these were broken bones found in a deposit rich in naturally broken bones, and there had obviously been "a great deal [of] natural wear, breaking and erosion" that could account for much of that breakage.[49] Still, even in the midst of "a whole lot of doubtful scrap" he was sure Thomson

still has a nucleus of material in which the evidence, viewed without bias, is so strong that it requires a straining stretch of a negatively inclined imagination to account for the conditions found in any way but the one which says "Made by Hand!"[50]

Nor was the great antiquity of the deposits in which they were found reason to doubt those were artifacts. As Cook saw it, "we have *NO PROOF* man may not have been well developed in the Lower Pliocene."[51]

Cook may well have thought so, but he was largely alone. Osborn quietly counted votes on the supposed artifacts—"*Con.* Brown, Matthew, Gregory. *Judicial.* H. F. O. [Osborn]. *Pro.* Thomson, Loomis of Amherst"—and realized he was surrounded "by an atmosphere of doubt which I may or may not be able to clear up." In the circumstances he advised Cook to proceed cautiously "because it will not do to take up a position and then later be obliged to back down and apologize."[52] Cook, thinking Osborn "want[s] the *glory*," ignored the advice.[53]

Cook had badly misread the situation. Osborn stayed true to his word and never published on the Snake Creek "artifacts" in any detail. At the 1926 AAAS meeting he mentioned in passing that these bone implements might be Middle Pliocene in age, perhaps as much as 4 million years old.[54] But Gregory followed Osborn at the lectern and voiced his skepticism about *Hesperopithecus* and the supposed artifacts. It was not at all contentious, as *Scientific American* editor Albert Ingalls reported to Cook. "Both men were friendly to each other. Gregory likes Osborn personally, as Osborn has always been his mentor ([but] he is not a blind worshipper, I notice)."[55] Osborn obviously was aware of Gregory's doubts, and by then had already divested himself of any investment in Snake Creek and was not bothered by his public disavowal.

But if neither Osborn nor Gregory were taking their differences too seriously, someone else in the audience was. Hrdlička had earlier spoken at the meeting on "Recent Findings in America Attributed to Geologically Ancient Man,"[56] and was not about to give Osborn a free pass. As Ingalls also reported:

> All during the presentation of [Osborn's] Nebraska paper at Philadelphia I watched Hrdlička sitting there, face getting red, eyes getting glassy. At the end, he rose and said of course he had never been able to accept any of the evidences, but he did not relish being put in a class which has shut its mind on the matter. He spoke with some mild temperature—not above 212 degrees, I should say. . . . Oh yes—he says the find should have been seen by "some authority." I wonder *who* he means?[57]

Cook, of course, was undeterred. *Hesperopithecus* and the Snake Creek bone tools figured proudly in his 1926 *Scientific American* rejoinder to Hrdlička.[58] Besides, by then he and Figgins had other candidates to bolster the case for a deep human antiquity in America. ,

8.3 ANOTHER HEAD OF THE HYDRA

In the spring of 1924 Nelson Vaughan, a fossil and artifact collector in Colorado City, Texas sent Figgins a photograph of a mammoth jaw he had excavated, hoping for a reward. Instead he was admonished by Figgins for the poor condition of the fossil caused by his haste in removing it from the ground. Still, Figgins realized Vaughan was in a region with good paleontological potential, and he urged the young man to keep his eyes open and provided tips on what to do if he found more fossils. Grateful for the opportunity and harboring hopes of a future as a salaried collector, Vaughan promised to do so.[59]

His efforts were soon rewarded with the discovery of a "huge skeleton" partially exposed along a stretch of Lone Wolf Creek on the edge of town. Having learned his lesson, Vaughan dug into the bank only far enough to satisfy himself this was "a real *find*."[60] Hearing the news, Figgins was keen to get the skull for display (though it is not clear he knew what the animal was) and there followed an awkward dance regarding how much Vaughan ought to be paid to recover the bones, each asking the other what they thought the effort was worth. Neither admitted to a figure, only to their circumstances: Vaughan had a wife and baby girl to provide for, Figgins a meager budget. When Vaughan admitted he hoped for a permanent position as a CMNH collector, Figgins exploited the opening, dangling the promise of one if Vaughan's excavation was successful. Vaughan took the job. Figgins instructed him on how to expose, shellac, and plaster the skeletal elements, and crate them for shipping to Denver. He also sent one of his occasional field hands, rancher H. D. Boyes of Yuma, Colorado to help with the excavation.[61]

Vaughan and Boyes excavated what appeared to be "not less than four or more complete skeletons" at Lone Wolf Creek in the summer of 1924 (Figure 8.1). Figgins came to regret his decision to have them work unsupervised. The bones were so densely packed that it was difficult to separate them in the field, and so they were taken out in blocks, though Vaughan and Boyes failed to apply enough shellac and made shipping crates too small to contain the bone-bearing slabs. Yet, instead of building larger crates Boyes trimmed the slabs, destroying many bones in the process. Rather than getting several complete bison skeletons, Figgins grumbled, the museum ended up with "but one and [it] had to be restored in many places."[62] Worst of all, Figgins learned only after the fact—and from a visitor who had happened to be at Lone Wolf Creek during the excavations—that two "arrowheads" were found with the bison.[63] Neither was spotted or photographed in place. Still, on the basis of what the visitor reported Figgins was convinced there was "no doubt of the arrowheads being associated directly with the fossils and no evidence that they might have reached that point [having been redeposited from] . . . the higher strata."[64]

Weeks later Boyes arrived at the CMNH with the artifacts and details of the excavation and sketched the site stratigraphy. Figgins told Cook the arrowheads were distinct from types routinely "picked up on the surface—they being long and slender, the edges being parallel without notches at the base and of fine workmanship." Cook readily grasped their potential significance:

I found your letter of [August] 27th about the bison, arrows, and part cranium! This is *splendid*! You probably recall Handel T. Martin found a similar thing [at 12 Mile Creek] some years ago, and no one would credit his find as being anything but accidental association. Some of these days, it will be forced down the necks, heels first, into some of the worthy mortals who known [*sic*] it all now,—as to when and where the race of Man came and went, that men *were* in America in late Pleistocene times, if not earlier!

Figure 8.1 H. D. Boyes (*left*) and Nelson Vaughan (*right*) at the Lone Wolf Creek site in the summer of 1924. The bison remains are exposed between them, with the dirt removed by their excavation forming a large talus cone below their feet. Courtesy of the Heart of West Texas Museum, Colorado City, TX.

How much earlier? From Boyes's description, but without having seen the site, Cook leapt to the conclusion that the bone-bearing deposits "would almost certainly be of Pleistocene age, if not older."[65]

Figgins announced the find in his *Annual Report* for 1924, and though he had only Boyes's description and sketch to go on, he nonetheless confidently put "one [artifact] immediately beneath a series of cervical vertebra and the other below a femur of the same animal." Their position made it "certain that they were of a date not later than the time at which the Bison remains were deposited there during the Pleistocene period."[66] That caught Oliver Hay's eye, who wrote Figgins for photographs of the bison skull and horn cores in order to narrow down the identification of the species and thus the age of the arrowheads. Suspecting Figgins was unaware of the controversy in which he was about to be ensnared, Hay sent a wish and a warning: "I hope your collectors took photographs and other means of establishing the positions of those arrow-heads. You know every find of that kind is attacked."[67]

Figgins, who did not know Hay well, put up a good front. Yes, he was aware this was a contentious area, but "the peculiar circumstances, nature of the matrix and overload [at the site] remove any and all doubt from my mind."[68] Perhaps. But Hay knew from long experience the issue was not settled just because Figgins thought so. And despite Figgins's bravado, he got Hay's point. Figgins quietly sought more details of the excavations from Vaughan, omitting any mention of what he had already learned from Boyes so as not to influence Vaughan's response. Vaughan corroborated what Boyes had said and filled in a few more details, namely that not only were the Lone Wolf Creek artifacts unlike any of the nearly 400 arrowheads in Vaughan's personal collection, the stone used in their manufacture did not outcrop in that area. It was the first inkling the artifacts had been carried in from elsewhere.[69]

Needing a better understanding of the stratigraphy and geology of Lone Wolf Creek than either Boyes or Vaughan could provide, Figgins dispatched Cook to the site. There Cook learned a third arrowhead had come out during the excavations. That specimen never made it to Denver, nor had Boyes mentioned it. When word of it reached Figgins, he was sure Boyes had stolen the artifact. Vaughan thought otherwise.[70] Regardless, building the case for Lone Wolf Creek as a Pleistocene archaeological site would have to be done post hoc. Cook surveyed the area, mapped the geology, and matched what he recalled of the matrix adhering to the artifacts with what he saw of the bison bone still in place. He was convinced they came from the same stratum and, from what Vaughan told him, were directly associated.

He said so in an article he submitted to *Science*, assuring its editor, James McKeen Cattell, that it would attract wide interest given its application to "the present unfortunate controversy as to man's origin." Cook also let Cattell know Barnum Brown had recently examined both the Lone Wolf Creek and 12 Mile Creek specimens and pronounced them to be of the "same type and workmanship." The age of 12 Mile Creek, of course, was "always questioned." Lone Wolf Creek might change that.[71]

Cook could not have been aware that a few weeks earlier Cattell had received Holmes's harsh rebuke of the work of Loomis at Melbourne and Sellards at Vero (§7.9).[72] It was only a coincidence that Cook opened his own *Science* paper by proclaiming Lone Wolf Creek was no Vero or Melbourne. Instead, it was "good, dependable definite evidence of human artifacts in the Pleistocene in America," for which "no reasonable doubt can exist that the artifacts and fossil animals found are contemporaneous, and that [both] . . . are in original, undisturbed Pleistocene deposits."[73]

Those deposits were in a channel cut into Triassic and Cretaceous bedrock, which then filled with Pleistocene sand and gravel. That fill was subsequently incised by the present drainage, "leaving undisturbed Pleistocene remnants along the sides." Evidence of its Pleistocene age came from the fossil fauna, which included "*Elephas* [*Mammuthus*], *Equus* and a camel [*Camelops*]," though none of these was associated with the artifacts, and the bison remains, which were from a "large species, considerably larger than the modern."[74] However, because no comparative study had been done, Cook could not offer a more specific taxonomic identification.

Granting a Pleistocene age to the formation, but keen to obtain a better estimate of the antiquity of the bison and artifacts, Cook observed the bones were overlain "by about five to seven feet of highly calcareous, cemented and undisturbed Pleistocene sands and gravels." The bones were presumably laid down during the earliest episode of Pleistocene valley filling. Cook could not resist another swipe at Vero: Lone Wolf Creek was "free from such cross-channeling as has worked over the materials at Vero." Sellards, still burning from Holmes's attack two months earlier, cried foul. The Vero deposits were not reworked, he snapped at Cook, and he was greatly disappointed that as state geologist of Texas he had not been informed of work being done in his domain.[75]

Given that one projectile point was evidently found under several articulated vertebrae and another beneath a femur, it seemed obvious the Lone Wolf Creek "bison had been shot and carried these flint points with him to the place where he finally died." Cook added that a third point was also recovered "in position with the body of this skeleton" before it was lost or stolen. He neglected to mention that all his information was thirdhand, the precise location of the artifacts was not recorded at the time, and there was no documentary or photographic support for his claims.[76] The points were, contrary to expectations Cook supposed one might have of artifacts of Pleistocene age, "very unexpectedly, of very fine workmanship, much more refined and beautifully worked than the arrow and spear points of the more recent types in that region." These were not paleoliths, Cook proclaimed, but nonetheless had been made "by a distinct people, of distinct culture."[77] Most important of all,

> there is no possibility of accidental inclusion of these artifacts with the bison, or of their being of later age. *They are certainly and positively contemporaneous with that fossil bison and the associated fauna of mammoths, camels and extinct horses*—of a type found elsewhere in beds of known Pleistocene age.[78]

So it also seemed to Pliny Goddard, cheering from the sidelines for a greater human antiquity to match the chronological demands of the linguistic evidence (§7.11): "We must assume that men were in America at the end of the Ice Age, or about 25,000 years ago."[79]

Hrdlička and Holmes would have none of that.[80] The "cocksureness of [Cook] makes one rather suspicious," Hrdlička snarled.[81] Of course, snarling would not resolve the matter. They needed help, and as the claim rested heavily on faunal evidence they turned to vertebrate paleontologist John Merriam.[82] By then Merriam was president of the Carnegie Institution of Washington but had done considerable work on Quaternary vertebrates (most especially at Rancho La Brea) during an earlier stint on the faculty at the University of California at Berkeley. He had assessed several discoveries of purportedly ancient human remains, including a 1924 find of six human skeletons found in deep trenching in Los Angeles. Merriam and vertebrate paleontologist Chester Stock[83] (Merriam's student and later successor) moved quickly. At the National Academy of Sciences (NAS) meeting a month later they reported that the human remains, though found nineteen to twenty-three feet below surface with no evidence of disturbance of the overlying deposits, closely resembled those of American Indians. They had none of the features seen in Neanderthals, nor were there any associated Pleistocene faunal remains.[84] Hrdlička, who had taken the opportunity to examine the human remains at the NAS before their talk, affirmed there was "not the slightest differences between these remains and the American Indians of the present day."[85]

That their investigation was completed quickly and definitively delighted Hrdlička, who wrote Merriam to say he "deserve[d] the earnest thanks of every anthropologist, for here was a case which was peculiarly liable to misinterpretation and a sensational deduction." Holmes agreed: "It is very gratifying to know . . . that you were present when at least one of the glacial men was found. I feel safe regarding conclusions when you are at hand." He then added, tongue fully in cheek, that "it will certainly be a matter of rejoicing to have it finally settled that the human race had its beginning in America."[86] Thus, when Cook's Lone Wolf Creek article appeared, Holmes immediately wrote to Merriam:

> You have doubtless seen an article by Harold J. Cook in the last issue of *Science*, in which the finding of arrowheads in the Pleistocene is constantly announced. The frequency of such risky announcements requires, if possible, that prompt attention be given in each case. . . . It occurs to me that you may find it possible to have Mr. Stock or some other well-qualified person visit the Texas site and report upon the evidence.[87]

A visit was not possible, but Merriam forwarded the inquiry on to Matthew, who (Merriam assured Holmes) was a paleontologist "equal to Cope and Marsh and thoroughly sound on all questions relating to the history [of] man." More important, he knew Cook well.[88]

Matthew had not seen the site but had examined the bones and artifacts in Denver. He was untroubled by the association of the horse and camel with the bison, or by Brown's opinion that the Lone Wolf Creek and 12 Mile Creek bison were the same species or that the "arrows show the same culture." But Matthew doubted Cook's assessment of the site's geology and age. The site needed to be visited by "two or three competent Pleistocene geologists," for it was "very easy to be mistaken in such a case as Harold describes," especially, as Matthew could attest first-hand, Cook's mistake-prone enthusiasm: "Harold has, as you know, a somewhat optimistic temperament, and I find it necessary to discount his geological conclusions more or less."

Matthew thought it impossible to resolve the antiquity of Lone Wolf Creek on stratigraphic evidence (he had little confidence in "peneplane geology") or preclude the possibility that the Pleistocene fauna in that region had survived later when they "would naturally be found associated with man."[89] Merriam agreed. Finding human remains with extinct animals was to be expected from time to time, but that may or may not indicate a Pleistocene age,[90] not that Cook accepted that premise.

Nor did Hay, but then he was pushing Cook hard in the opposite direction. He was not content with a Late Pleistocene antiquity. He wanted to know why Cook did not think Lone Wolf Creek was Middle or Early Pleistocene in age. It was the camels, Cook replied. Although they had gone extinct "about mid-Pleistocene times in the north," it seemed to him "they *might* have lived on longer, further south." Assigning a Late Pleistocene age to Lone Wolf Creek was the safer, more conservative estimate.[91]

Cook tried again the following year to press the case for Lone Wolf Creek, using Hrdlička's July 1926 *Scientific American* article as a starting point. Now he was getting personal.[92] In his opening salvo, Cook (who had little training in any field [§8.2]) declared that someone trained in anthropology (Hrdlička) could hardly "grasp the magnitude and breadth of the evidence" from fields such as geology and paleontology, even when that evidence was "too sound and definite to be ignored."

The site was as he had previously described it in *Science*, though now the bison remains were "under some eighteen feet" of undisturbed Pleistocene sand and gravel (as opposed to the five to seven feet previously reported), along with several other extinct animals, including a camel, two mammoth species, two primitive horse species, and a primitive deer. Given how well defined the stratigraphy was and the fragile condition of the articulated bison skeleton, Cook was confident any disturbance of the deposits would have been visible or, more likely, would have destroyed the bison skeleton and its association with the other fossils.[93] Hence, even if the precise age of the bison and its taxonomic designation were uncertain, it was assuredly Pleistocene in age because those other species had all been found "in other parts of America in beds of known glacial times." He admitted to one caveat, however:

> True, no one yet knows just how late some of these species may have survived in America. But on the other hand, they are typical Pleistocene species, and it is certain that the evidence as to time is far more convincing and more probably certain when we consider the anatomical evidence in *several* families, all known to have been evolving rapidly, than that of merely *one*: that is, man—as considered by the anthropologists who take that attitude.[94]

American Indians may not have changed "structurally and physically in many thousands of years," whereas over that same span other vertebrate species changed rapidly and markedly. It was therefore no surprise to Cook that the Pleistocene-age bison appeared distinct from modern *Bison bison* whereas Pleistocene human skeletal remains in America did not look "primitive" or resemble Neanderthals.

He put no stock in Hrdlička's core assumption that the evolutionary sequence of humans in Pleistocene Europe was relevant to America, nor did he side (at the moment) with Hay on the evolutionary stability of other vertebrates (§7.5).

Indeed, Cook thought the American evolutionary sequence began much earlier than Lone Wolf Creek. There was *Hesperopithecus*, "which antedates the classical *Pithecanthropus*," as well as the newly discovered proof of "humanoid" remains possibly hundreds of thousands of years old or older, notably the supposed artifacts from Snake Creek. Although these were dismissed by virtually all who saw the specimens, Cook was not abandoning them. The specimens just needed further study: "We do not want to fool ourselves. We intend to *know*."[95]

Hrdlička was not fooled, and he was quite certain Cook did not know. Within days of Cook's article Hrdlička once again wrote Merriam. Was it possible to have the Carnegie Institution send "some reliable man who would command the confidence of us all to look into the matter? It would be a substantial service to American Anthropology as well as to Paleontology."[96] Hrdlička followed up a week later, complaining that the Lone Wolf Creek discovery was not even made by Cook, nor was there "the slightest corroboration of the claims by any one in whom we all could have full confidence."[97]

Hay heard of Hrdlička's objections through the Smithsonian grapevine, and he passed them on to Figgins. Although far more sympathetic to the cause, Hay could see Hrdlička's point. He urged Figgins to publicly proclaim his confidence in Boyes and Vaughan's reliability and honesty. Figgins assured Hay he could address Hrdlička's objections, but he knew he needed to bolster the case.[98] He pushed Vaughan for more details about the Lone Wolf Creek artifacts, where the third point was found relative to the bison remains, what became of it, and why Boyes apparently refused to discuss the matter. Vaughan badly wanted to help, not least because Figgins represented the promise of gainful employment. But he had no more information to give. He defended Boyes, though he could not explain Boyes's reticence to speak to Figgins and admitted Boyes had been in too big a hurry and sloppy at times. As for the third point, Vaughan had not mentioned it, thinking there was "no use in saying anything about it as [no one] would believe it unless we have it to show." The last time he had seen it was in Boyes's room, which made Figgins certain Boyes had kept it. The best Figgins could gain from his extended interrogation was a signed affidavit from Vaughan attesting to what he had seen excavating two years earlier. It was hardly more than what he had had to begin with, and not enough to placate even Hay.[99]

The chance to fortify the Lone Wolf Creek case was past. But by then the CMNH had already begun excavating a new site near Folsom, New Mexico, one that also had bison bones and artifacts.

8.4 WHEN IT RAINS . . .

The Folsom site came to Figgins's attention in early 1926, though it was discovered nearly two decades earlier. On August 27, 1908, a heavy rainstorm deeply incised a portion of Wild Horse Arroyo, a tributary of the Dry Cimarron River just outside the town of Folsom. Sometime later George McJunkin, the foreman

on the local ranch, spotted bones freshly exposed at the base of the arroyo. There
is no evidence McJunkin saw any artifacts, but the bones piqued his curiosity,
and he told others about them, including Carl Schwachheim, the blacksmith in
nearby Raton.[100]

Schwachheim visited the locality in December 1922 with Fred Howarth, a
Raton banker, and collected some of the bison bone.[101] After failing to interest
the State of New Mexico in the Folsom site they visited the CMNH in January
of 1926, where they told Figgins and Cook of the locality.[102] They asked Howarth
to collect and send more of the bison teeth and bones to help eliminate the possibility that the animal was not a "common American bison." It was not. Figgins
and Cook instantly saw its similarities to the Lone Wolf Creek bison: if Folsom
was "not the same species, then it is a closely related form, apparently considerably
larger than our example." Howarth sent more of the bison bones, which confirmed their assessment of the animal's size.[103]

Figgins and Cook visited Folsom in March of 1926 with Howarth and Schwachheim. Bison bones were visible in the arroyo, and Figgins could see that an excavation had the potential to yield "mountable skeletons of high quality" for
display. Cook often grumbled, not unfairly, that finding mountable skeletons was
always Figgins's goal, usually crowding out more scientific aims.[104] Figgins subsequently insisted that was not all he sought when excavations began in the summer of 1926. He claimed to see "the possibility that additional evidence of man's
antiquity in America might be uncovered," and said he gave his crew

> explicit instruction that constant attention be paid to such discoveries—not
> with as much expectation of success, as in the belief that opportunities of
> that nature should not be neglected. It was therefore, something in the na-
> ture of an anticipated surprise when such a find was made.[105]

Or so he said with the clarity of hindsight. No contemporary evidence supports
that story. Figgins's crew was not instructed to look for, nor did they expect to
find, any archaeological remains. They were after a museum-quality bison skeleton.[106] Yet, others wondered about archaeological possibilities. On hearing Figgins's plans for Folsom, Brown suggested that if arrowheads were found with the
bones they ought to "remain in the matrix without the relationship being disturbed." Figgins ignored the suggestion.[107]

Schwachheim began work at Folsom in May of 1926, later joined by Figgins's
son Frank and, briefly, by Cook.[108] The work was slow, but by mid-June the crew
had removed the overburden and was in the bison bone layer. A month later, as
bones were being uncovered and prepared for shipment to the CMNH, an artifact was spotted in the loose dirt. Schwachheim nonetheless believed it had been
associated with one of the bison:

> It is a question which skeleton it was in, but from the position of them it
> must have been in the skeleton of the smaller one & just inside the cavity
> of the body near the back. It was found 8½ feet beneath the surface with

an oak tree growing directly over it 6 inches in diameter, showing it to have been there a great length of time.[109]

Learning of the find, Figgins insisted he had hoped for such a discovery all along, though that did not "lessen the surprise" ("shocking news" was how he described it Brown). With the lesson of Lone Wolf Creek still fresh, he straightaway had the artifact mailed to him in Denver to insure it was not "lost, strayed or stolen," and he now pressed his crew to be on the lookout for artifacts and especially any skeletal remains of "the maker of the arrowhead." If such a discovery was made, they were to "keep it absolutely quiet." He instructed Howarth, who was supervising the work on Figgins's behalf, to stress

> the importance of watching for human remains and then in no circumstances, remove them, but let me know at once. I would want to study and collect such material *in situ*. This find is of double importance in that it substantiates the other finds.[110]

One more projectile point was found later that season, but it too came from loose dirt.[111]

When the first point arrived at the museum, Figgins could see it was "nearly identical" to the two from Lone Wolf Creek. That made it the third such discovery of "arrowheads of the same cultural and geological periods found with Pleistocene Bison" (counting 12 Mile Creek). Appearing on the heels of Hrdlička's July *Scientific American* article, the cumulative weight of these discoveries seemed "likely to aid in putting a wrench in the cog wheels of Hrdlička's theories." I wonder, Figgins asked, "who will be the first to accuse us of finding *too many* arrowheads with *Bison*?"[112]

Although Figgins saw similarities among the three sites, differences were apparent as well. The Folsom bison was a "monster." He estimated it to be 15 to 20 percent larger than the Lone Wolf Creek bison. The Folsom arrowheads were "more pointed and superior in workmanship" than those at Lone Wolf Creek, a contrast Figgins supposed might be attributable to a "difference in material."[113] Folsom lacked other Pleistocene fauna, but Figgins was confident its bison and therefore its artifacts were Pleistocene in age. He deemed Folsom "merely confirmatory." Lone Wolf Creek was the site of "outstanding importance."[114]

Figgins kept quiet about Folsom, though he did tell Brown, who was naturally disappointed the artifact had not been found in place. ("That would settle all doubting Thomases.") But Brown appreciated its potential significance and began fishing for more information as well as the possibility of *Natural History* (which was published by the AMNH) featuring "a complete resume of all the bison finds where associated with artifacts."[115] For a variety of reasons, not least of all Figgins's belief that Brown had reneged on an earlier promise to publish Cook's paper on Lone Wolf Creek in *Natural History* (Cook was equally to blame in the matter), Figgins was in "no hurry" to make a decision as to where and how the Folsom story would appear.[116]

Publication became a higher priority two months later when, in the process of preparing one of the plaster-jacketed blocks of bison bone from Folsom, a sliver of a projectile point was spotted alongside a bison rib. One of the points found in the loose dirt that summer proved to be refit to that fragment. That they joined, and that the sliver had been found next to a rib, ought to remove any doubt that the artifact and the bison were associated. Or so it seemed to Figgins, who told Hay that although one point was dislodged during the excavation, "the second was found in place and is so preserved."[117] The reality, as was so often true, was not as straightforward.

8.5 BEARDING THE LION

That discovery in the first week of November of 1926 came just as Figgins heard from Hay of Hrdlička's objections to Lone Wolf Creek. Now Figgins had more ammunition. "By all means," he told Hay, "try to excite an expression from Dr. Hrdlička. We are fixed for a reply."[118] The reply ultimately came from both Figgins and Cook in articles they published in *Natural History* the next year.[119] Although Figgins insisted their articles should merely record the facts of Lone Wolf Creek and Folsom, he was so irritated by what Hay told him of Hrdlička's doubts that he at first could not follow his own direction. The opening paragraphs of his draft, as he crowed to Cook, had nothing to do with either site:

> I start off by quoting you [Cook's November 1926 *Scientific American* article] in saying what you did about a man's training influencing his opinions, and show the party skilled in the relationship of races and in skeletal anatomy is out of the discussion, because there are not sufficient skeletal remains for periods antedating recent times to enable him to make intelligent comparisons and determinations. That's durned good argument against the Hrdličkas [*sic*] don't you think? . . . Then, I go on to say, since there is a lack of skeletal evidence, time determinations are questions for the geologist, paleontologist and chemist, and since H[rdlička] is none of those things, perhaps the blatant yap will discover he is talking out of turn.[120]

Figgins, who admitted to not knowing enough archaeology, paleontology, or geology to have "opinions upon the subject," nonetheless felt he had sufficient authority to dismiss Hrdlička's.[121]

Once Hrdlička was put in his place, Figgins told Cook, they should turn to Lone Wolf Creek, particularly to why Cook thought the bison was Late Pleistocene in age ("that will spike Hay effectually"), and then move on to Folsom with mention of 12 Mile Creek along the way. Although Figgins thought the latter two sites ought to be discussed, he did not consider them comparable to Lone Wolf Creek:

> I would avoid, as does the devil holy water, all comment and introduction of any and all items, except those in strictest relation to the finds at Colorado [Lone Wolf Creek]. I have paved the way for this by saying that until the

studies now in progress are completed, the age of the Folsom finds cannot be determined. For the present, at least, I think we should view the Folsom artifacts in the light of contributory evidence.

"We are," he assured Cook, ". . . in a fair way to make history in the discussion."[122]

Figgins wrote Brown to tell him of the refit point from Folsom, then wondered whether he should "write it up or keep quiet."[123] He was not wondering at all, of course: he already had a manuscript under way. He was fishing for an invitation to publish in *Natural History*. Brown took the bait:

> By all means publish all the data *in extenso* . . . . Do this at once, and if you will let me have the data I will see that it is published here in *Natural History* as well as in some such paper as the *New York Times*.[124]

While Figgins was manipulating Brown, Hay was arranging "a showdown" in Washington. Because Figgins now had a plastered block with a point and bone in place (though an association made in the lab, not in the field), there seemed no reason to wait for Hrdlička's reaction. Hay wanted to bring the evidence to the Smithsonian to allow Hrdlička and other anthropologists, geologists, and paleontologists to see it for themselves. Hay foresaw a landmark outcome:

> If the specimen comes, the fossils which were collected with it ought to be represented, the camels, horses, etc. If Mr. Cook comes he could answer questions about the geology, etc. I hope that the arrowhead [in the block] has not been disturbed. That would give occasion to some people to question the find. I think that you have a case which may clinch the subject of Pleistocene man.[125]

Hay pressed the BAE's chief, Jesse Fewkes—who was not unsympathetic to claims of a Pleistocene human antiquity (§7.9)—to sponsor the visit.

Figgins was not averse to the idea, though he had a low opinion of archaeologists and anthropologists, declaring them "mostly jokes . . . too ignorantly prejudiced to give intelligence to the diseased meanderings of their warped imaginations."[126] Even so, he was not sure he wanted to face any. If there was to be a showdown in Washington, he told Hay, Cook should go to "be the goat."[127] Fortunately, the "goat" was enthusiastic. He credited his *Scientific American* article with having stirred things up and, evoking Ulysses S. Grant's famous Civil War declaration, proclaimed his readiness to "fight it out on that line if it takes all summer," as if a visit to the Smithsonian in 1927 was akin to the 1864 Wilderness campaign.[128]

In Washington Fewkes was puzzled. He was under the impression he had only been asked to pay the freight for a plaster block with artifact and bones.[129] Figgins corrected him: seeing the specimens by themselves was not enough, as "no great importance need be attached to the finds, other than the artifacts are of unusual form and workmanship." Cook had to be there to explain the sites' geological context and paleontological evidence. Better still, Figgins suggested that Fewkes send a representative to Denver to examine the bison as well the associated mammoth, horse, and camel remains from Lone Wolf Creek.[130] He recommended Hay.

Fewkes assuredly knew a visit from Hay was preaching to the choir, but it did not matter. Hay had just then realized the block with the artifact was from Folsom, not Lone Wolf Creek. Up to that moment he thought Figgins had a strong case, but "now I see difficulties ahead." The Folsom point might be indisputably associated with the bison, but Folsom had no indisputably extinct fauna to bolster its claim to a Pleistocene antiquity. Lone Wolf Creek, which had mammoth, horse, and camel as corroborative evidence, still rested entirely on the integrity and skill of the collectors who had so poorly excavated that site. They were back to where they were the year before. Hay anticipated that "the anthropologists will insist that it [Folsom] is of comparatively late origin."[131]

Again Hay asked Figgins what testimony he or Cook had that would satisfy the critics of the Lone Wolf Creek artifact association. Figgins was adamant that neither Boyes nor Vaughan had the character, wherewithal, motive, or even the intelligence to plant artifacts or lie about their association with the bones. Boyes was just an "ignorant rancher" who believed he was at Lone Wolf Creek "merely for the purpose of getting fossils." Of course, "getting fossils" and not looking for artifacts was precisely why Figgins had Boyes there. The fact that he had failed to mention finding artifacts—or evidently thought he could keep one—testified to that. As for planting artifacts, Figgins was "utterly unprepared to entertain even a suspicion that such occurred." Where would Boyes and Vaughan have found those artifacts? Why would they break one of the points (as happened during the excavations) if fraud was their intent?[132]

The more Figgins stewed, the angrier he became. He resented "with all the vehemence I can put on paper" the suggestion that there was crooked intent, and declared that if the "Hrdličkas attack . . . on the ground that a 'scientist' was not present—question the honesty or intelligence of those concerned—I shall attack." To Cook he promised a fight the instant Hrdlička criticized the finds in print:

> You see, I am a free lance and without responsibility in the matter of "scientific courtesy," so if a party tears a chunk of hide off my back, in a suave and courteous manner, there is nothing to prevent my removing three upper and two lower incisors, black one eye and gouge the other, after I have laid his hide across a barbed wire fence.[133]

"I am daring the whole miserable caboodle of them." Figgins toned down his rhetoric in writing to Hay, but the message was the same. "Yes, our opponents can say that the artifacts were "planted" by someone, but if they do, in print or in public, I shall make them regret it and hesitate to repeat such statements."[134]

Figgins's hostility toward Hrdlička was striking, if not puzzling. In late 1926 Hrdlička had not published any criticism of Lone Wolf Creek and assuredly was unaware of Folsom. All Figgins knew was what Hay reported from the Smithsonian, and Hay's information was largely secondhand, as he and Hrdlička hardly spoke. The criticisms certainly sounded Hrdlička-like, but it is difficult not to suspect Hay may have embellished what he heard (there is no evidence Hrdlička ever suggested the artifacts were planted). For his part Hrdlička would have been surprised to learn of Figgins's anger: there is little indication he knew Figgins or was aware he was involved at those sites. (He only knew Cook was.)

Hay, who was twenty years older than Figgins (Hay was then over eighty), far more experienced in dealing with Hrdlička, and overall a much cooler head, did not get caught up in Figgins's bluster. Instead, he merely suggested Figgins fortify the Lone Wolf Creek case. Hay deemed it a far stronger site than Folsom or, for that matter, 12 Mile Creek. His suggestions were motivated by the very concerns Hrdlička had and that Hay acknowledged (assuredly to Figgins's dismay) were legitimate: "The important thing is to connect those arrow-heads with that bison and that bison with the other [Pleistocene] fossils. *The anthropologists have a right to ask that.*"[135] The problem, simply, was that the occurrence of a point and bison bones, even locked together in a plaster-jacketed block, was necessary but not sufficient to prove geological antiquity. That required demonstrating the bison had not lasted past the Pleistocene—and that it was a new species.

8.6 WHAT'S IN A NAME?

Pleistocene bison taxonomy through the first two decades of the twentieth century remained much as paleontologist Frederic Lucas had summarized matters at the end of the nineteenth century (§6.9). By his tally, there was but one genus (*Bison*), which included a living species (*Bison bison*) and a half-dozen extinct ones, differentiated by the length, direction, and amount of curvature of their horn cores. The extinct taxa were *Bison antiquus, B. occidentalis, B. crassicornis, B. alleni*, and *B. latifrons*, plus one newly named on that occasion (the name did not stick). On the basis of the appearance of the extinct specimens and the conditions under which they were found, Lucas had suggested that "all species might have been coeval," although he supposed that was "highly improbable."[136]

In a 1914 synthesis of North American bison, Hay accepted Lucas' five extinct taxa and likewise introduced a sixth (though it too did not stick). Hay was then a few years from becoming embroiled in the human antiquity controversy, and said nothing of the antiquity of those species or the timing of their extinction (§7.1). Nor did he accord any particular significance to 12 Mile Creek's bison having been found with a "flint arrowhead," noting laconically "this fact has been treated by Williston."[137]

Bison taxonomy did not change appreciably over the next decade, and thus in the mid-1920s the Lone Wolf Creek and Folsom specimens could be assigned to *Bison bison* or one of the five previously identified extinct taxa. Unless, that is, they were sufficiently distinct to warrant a new species or even a new genus designation. As a scarred veteran of the disputes at Vero and Melbourne, Hay was now acutely aware of how much rode on narrowing the age of faunal remains associated with artifacts. By the 1920s he was insisting that "no extinct bison . . . lived after the Wisconsin glacial stage," and arguing that several disappeared much earlier. According to his continent-wide summary of Pleistocene vertebrate faunas (published in three large volumes in the 1920s), *B. antiquus* and *B. latifrons* had disappeared toward the end of the Iowan glacial period, which by then-current geological reckoning was the next to last glacial period (the last being the Wisconsin). The only species to survive to the end of the Wisconsin was *B. occidentalis*.[138]

Of course, the last appearance of extinct species provided only a minimum age. Determine how much older a particular fossil occurrence might be required looking at other geological evidence or extinct fauna that also occurred at a site in order to triangulate an age estimate, as Hay had done at Vero. Thus, when Hay made mention of 12 Mile Creek in 1924, this time around he stressed there were

> apparently equivalent deposits, about a mile away, [in which] Mr. Martin secured teeth of a fossil horse and some camel remains and a lower jaw of a wolf. . . . If the horse and camel remains were found in deposits of the same age as those inclosing the bison bones it would appear that the latter were as old as the Aftonian.[139]

Not coincidentally, that made 12 Mile Creek the same Early Pleistocene age as the human skeletal remains from Stratum 2 at Vero. (Kirk Bryan was exaggerating only slightly when he later remarked that "Hay thinks nearly all the finds are Aftonian."[140])

But were the Lone Wolf Creek and Folsom bison the same species, or the same as the one at 12 Mile Creek? Species identifications were based on skulls and horn cores, but "unhappily," Figgins told Hay, there were no crania from Lone Wolf Creek in good enough condition to make those determinations (thanks to Boyes and Vaughan's haphazard excavation).[141] It was for that reason that Figgins privately and Cook publically referred to the Lone Wolf Creek specimen as an "undescribed and unnamed" bison; it had to remain so pending the completion of comparative studies. Until then, the best clue to its age was its association with other known Pleistocene fauna, not its taxonomic identity.[142]

Brown made a cursory examination of the Lone Wolf Creek bison when he and Osborn were in Denver in July of 1925, and he was convinced it was *Bison occidentalis*, or possibly even the living wood bison, *B. bison athabaskae*. Figgins, anxious for it to be an entirely new extinct species to remove all doubt of it being Pleistocene in age, urged Brown "not [to] call it *occidentalis* as yet." Brown backed off, though he suggested he would not be surprised if the Lone Wolf Creek or Folsom bison (which he had yet to examine) proved to be a progenitor of *Bison bison athabaskae*.[143]

A careful, systematic study of these specimens was obviously in order, and Figgins proposed to Cook that he "go over *all* the *Bison* material with Hay." There were, Figgins supposed, advantages to collaborating with Hay on the study:

> You will both be able to help each other in many ways, through such a discussion. Since Hay is in close touch with the *Bison* group and as a means of saving yourself a great deal of trouble, due to our lack of comparative material, it is possible you might welcome him in the preparation of a joint paper, but such items are wholly for your personal preferences and decision.[144]

Figgins assured Cook this was "merely as a suggestion," not wanting his curator of paleontology to think his toes had been stepped on. Yet, without waiting for Cook's reply, Figgins wrote Hay to invite him to make the "fullest possible" investigation of the Lone Wolf Creek and Folsom bison, assuring him he would have Cook's complete cooperation. Figgins evidently had little intention of entrusting such a critical analysis to Cook alone. Because artifacts had been found with the bison, too much rode on correctly assessing their taxonomic identity and geological antiquity. Figgins made no effort to hide his own opinion, alerting Hay that he saw "three distinct races of Bison" at the two sites.[145]

Fortunately, Cook proved amenable to the plan that was already under way behind his back, and he was likewise certain the Lone Wolf Creek and Folsom bison were new species. Hay was agreeable as well, though he preferred (initially) that Cook do the groundwork preparing and describing the fossil remains and take the lead in publication.[146] In 1926 Hay had already experienced a decade of bruising fights over Pleistocene faunas in archaeological sites, and he was ready to ease into a supporting role. In the end, however, the old warhorse could not resist leading the charge.

But why did Figgins select the elderly Hay from among the available vertebrate paleontologists—especially, why not ask Brown, a long-standing friend who obviously had a keen interest? Unfortunately for Brown he was a contemporary of Figgins and in an unstated but real sense a competitor; worse, Figgins knew Brown relished the limelight. If given the chance Brown would steal as much credit as possible. Besides, Brown had already offered his view on the identity of the bison, and it was not to Figgins's liking. Figgins might not be able to sway Hay's taxonomic judgment, but at least he was assured Hay would "say nothing about them until we determine what we have."[147]

With all that settled, Hay and Cook embarked on their study in early 1927, and Cook and Figgins wrapped up their *Natural History* manuscripts. Just then another possible Pleistocene site appeared. This one seemed far less complicated, potentially a great deal older, and on its face more compelling than either Lone Wolf Creek or Folsom. And its antiquity did not depend on the nuances of bison taxonomy.

8.7 MAMMOTHS AND METATES

The site was dropped at Cook's doorstep. His *Scientific American* editor forwarded a letter he had received in response to Cook's article that told of large animal fossils and artifacts being found in a sand and gravel pit outside of Frederick, Oklahoma. Ingalls thought the writer, F. G. Priestley (a local physician), "was not the usual 'nut' that thinks a concretion is a Proterozoic homo or whatnot." Cook could decide for himself if the tip was worth pursing.[148] He wrote Priestly, who could not be certain of the scientific value of what he was finding but assured Cook there was "no snide" about it, these were bones, artifacts, and possibly even human skeletal remains that went "way farther back than any Indian or his works go."[149] That caught Cook's attention.

Within the week he and Figgins were en route to Frederick. They spent a day on-site with Priestly and quarry owner A. H. Holloman and were given most of the artifacts and fossil bones that had been found. Figgins listened to their testimony on where the finds were made, while Cook examined the stratigraphy in the quarry and the scatter of drag line trenches Holloman had dug across the site to gauge the extent of the sand and gravel deposit. In these he saw a cross-section of what appeared to be an ancient river channel in which the Frederick deposits were inset.

When they returned to Denver two weeks and 1,800 miles later (with detours through New Mexico and Texas to look at other localities), Figgins and Cook were "all excited" about Frederick. It "*overshadows* the Bison finds," Figgins announced to Brown. It was far older "and the evidence is too conclusive to leave a shadow of a doubt regarding its antiquity."[150] Of course, they had had only a brief visit, had not done any excavations or seen any artifacts in place, but nonetheless they judged Frederick so important that they began revising their *Natural History* manuscripts to include it. Figgins wanted to let the scientific world know of the find "as soon as possible." Between Snake Creek, Lone Wolf Creek, Folsom, and now Frederick, he and Cook could scarcely believe their run of luck. Figgins pressed Brown about when their papers might appear, adding a thinly veiled threat they would publish elsewhere if it was not soon enough. That was more than a little presumptuous, as they had yet to even submit their manuscripts.[151]

As to how much older Frederick might be, Cook was pushing deep into the Pleistocene. As he interpreted the geology, there were three main strata, the lowest of which was approximately fifteen to eighteen feet of sand and gravel. This stratum, the commercial deposit being mined by Holloman, Cook designated Bed A and supposed it had been laid down by an ancient river that had incised into Permian-age red beds. Bed A yielded artifacts and extinct fauna that included ground sloth, glyptodont, camel, horse, and three genera of proboscideans, including two types of mammoth and a *Trilophodon* (an ancestral mastodon). Atop that were nearly seven feet of silt and sand, Bed B, which represented "flood plain" wash. It likewise had fossils of extinct mammals and artifacts, though in this higher stratum the mammoth was a "more advanced form." Capping the Frederick section was Bed C, approximately five feet of aeolian silt.[152]

Bed C was not of paleontological or archaeological interest, but Beds A and B were, especially the former, which yielded a number of distinctive artifacts including "arrows" and "not less than five unquestionable metates." However, because the quarrymen did not appreciate the importance of those artifacts it seemed likely that many more were surely lost, either spread on local roads or buried in quarry refuse piles. Several specimens were discovered in Holloman's presence, and his testimony of their stratigraphic context satisfied Figgins and Cook.[153] Hay knew better:

> I suppose that [the artifacts] have been collected by the quarrymen. May not our opponents say that these objects may have fallen down from near the surface? Such an admixture is liable to occur. If you have a man there watching I hope that he can find something without doubt in place.[154]

As it happens, Figgins had a man there. After returning to Denver he dispatched Schwachheim to Frederick to watch the quarry for more fossils and artifacts (especially metates) to document them in situ.[155]

The extinct fauna from Beds A and B were one clue to the antiquity of the artifacts; another was the topography. The quarry was on a hill, yet its sand and gravel
deposits had obviously been laid down in a river valley. Cook attributed that topographic reversal to erosion. The red beds consisted of easily weathered clays, but
the partly cemented sand and gravel was more resistant. As the red beds eroded,
the more impervious deposits from the ancient valley were left as a high-standing
remnant. The removal of what Cook estimated to be several hundred of feet of
red beds took "quite a long period of time," an inference seemingly confirmed by
the finding of the *Trilophodon* and a primitive mammoth in Bed A and the "more
advanced" mammoth in Bed B.[156]

Cook had in mind just how many years were required to cause erosion of that
magnitude, but it was a number so startling he omitted it from his *Natural History* article, knowing the AMNH curators who had editorial control over the
magazine's scientific content were unlikely to let it pass. He ventured only that
Frederick was Early Pleistocene in age.[157] *Scientific American*'s Ingalls had fewer
qualms. In a manuscript Cook sent Ingalls he revealed what Early Pleistocene
meant in absolute years. Multiplying the nearly 300 feet of lost red beds by a erosion rate of one foot per 9,000 years resulted in the "appalling time of 2,520,000
years." That was too old even for Cook. On the assumption that erosion was more
rapid in the past, he halved his initial estimate and then halved it twice more to
"allow for other possible errors and to be conservative." The result was a whop-
ping (and arithmetically incorrect) 365,000 years.[158] The claim that people were
at Frederick that long ago cleared the editorial bar at *Scientific American*, but that
did not mean it would pass muster elsewhere, even among those who favored the
cause of a deep human antiquity.

8.8 BAITING THE TRAP

Figgins's and Cook's *Natural History* articles were submitted in late February of
1927 and published in the May–June issue, which appeared in August of that
year.[159] In later decades these papers were cited as marking the acceptance of
the Folsom site, and with it a Pleistocene human antiquity in North America. That
would have been news to Cook and Figgins, not to mention their contemporaries,
for these articles were written more than six months before a projectile point was
uncovered in situ at Folsom and its evidence accepted by independent observers.[160]

Moreover, Folsom was just one of several localities they discussed in *Natural History*. Folsom shared billing with Lone Wolf Creek and Frederick (12 Mile
Creek also made a cameo appearance in Figgins's article) and was not even the
headliner. Both Cook and Figgins considered Folsom the youngest and least important of their sites. They were betting on Frederick, the site for which they had
the least amount of evidence and information.[161]

Most of all the *Natural History* articles were written, so Figgins proclaimed, in
"a deliberate attempt to arouse Dr. H. [Hrdlička] and stir up all the venom there
is in him." Cook repeatedly insisted that their evidence of a deep human antiquity
was "too conclusive to admit of doubt or question."[162] Few readers of *Natural History* that summer of 1927 likely agreed—at least, not readers who were passingly

familiar with the long history of this controversy. It is apparent, in fact, that Figgins and Cook were largely unaware of the issues in play and what was required to make a compelling case for a Pleistocene archaeological site. It was not false modesty on Figgins's part to admit he had "not made a special study of this subject [and that] his opinions regarding the importance of the evidence would be valueless."

In fact, Figgins's opening diatribe against Hrdlička, editorially trimmed by Brown of its "personal [and] controversial matter,"[163] suggests Figgins's reading into the dispute did not extend much deeper than Hrdlička's 1926 *Scientific American* article. As Figgins portrayed it, the "technical opposition" to the idea of a deep human antiquity was fueled entirely by "an individual minority." He did not say who that was, but he did not have to after identifying that person as "learned in physical anthropology, comparative craniology and racial relationships," yet who failed to appreciate other lines of evidence and was unwilling to accept the conclusions of authorities in other fields. As a result, any opinions from that source were "distinctly misleading."[164] Figgins was unfamiliar with the other critics in this long controversy, all themselves authorities in their fields (for instance, he had never heard of Thomas Chamberlin[165]).

What Figgins did not lack was an inflated view of his own importance, one that strayed into narcissism. He had learned (presumably from Hay) that Hrdlička and the Washington anthropologists were "extremely uneasy and are having a lot of difficulty on formulating a plan of attack as they have not been able to discover a weak spot." Figgins obviously did not know Hrdlička well and vowed to lead the charge to "draw Hrdlička's fire, wrath and brimstone." He would brave the consequences:

> From what I gather, Hrdlička is waiting and preparing for an attack on what I have to say. . . . Naturally, Hrdlička would select me for attack and since I have tried my best to push as far under his hide as I am hopeful *Natural History* will permit, it is obvious he will dig me as deep as he is capable of and I am suffering for just that.

There is no evidence Hrdlička was aware of Figgins's forthcoming paper. Even so, Figgins was prepared to "take the Hrdlickas [*sic*] to a trimming, if they so much as peep."[166] Cook had a similarly robust ego and was just as naïve as Figgins, not realizing his assertion that the evidence "cannot be questioned" was little more than an invitation to do just that.[167]

After he landed his rhetorical punches, Figgins turned to the "arrowheads" from Lone Wolf Creek, Folsom, and 12 Mile Creek, drawing attention to their similarity and how different they were from arrowheads routinely found on the surface. These were distinctive in form, lacked notching, and were larger and better made. Figgins did not comment or even notice the fluting present on the Folsom and 12 Mile Creek points, or its absence on the Lone Wolf Creek specimens. The only difference he observed between them was the shape of their blades, those from Folsom having "decidedly more tapering at the point," and overall workmanship, the ones from Folsom being "quite superior."[168]

These points represented a "cultural stage quite distinct" from any previously seen, and because they were found with extinct bison Figgins assured his readers that geologists and paleontologists (by which he meant Cook and Hay) deemed them Pleistocene in age. Yet, having made that claim, neither he nor Cook provided compelling chronological evidence to back it up. They had yet to study the bison remains and could say little more than these were unusual and the animals possibly extinct, perhaps related to *B. occidentalis*, or entirely new species altogether. Figgins quoted Hay as supposing there might be as many as "three undescribed races" represented at Folsom and Lone Wolf Creek.[169]

Without knowing the taxonomic identity of the bison, Cook believed he could place them in the Pleistocene, at least at Lone Wolf Creek, where the only geological alternatives were the Cretaceous, Triassic, or post-Pleistocene. *When* in the Pleistocene was unspecified. Yet, there was something incongruous about even that estimate. The mammoth, horse, and camel bones from the "undisturbed calcareously cemented" stratum were scattered and fragmentary, "as is to be expected in stream deposits of sufficient force to transport such coarse material." However, the supposedly contemporaneous bison skeleton was nearly complete and articulated and thus in what appeared to be a primary context. That raised the question, which Cook ignored, of whether the washed-in yet ostensibly associated mammoth, horse, and camel bone fragments were contemporaneous with the intact bison and all of Pleistocene age. Cook could (and did) declare the artifacts to be immediately associated with the bison, but were those artifacts also in secondary context?[170]

No matter. There was Frederick, with its arrowheads and metates associated with mammoth and mastodon, horse, ground sloth, and glyptodont. As enthusiastic as Figgins and Cook were about this potentially much older site, there was one element of it that gave Cook brief pause. "Strangely enough," he wrote, "these implements show a degree of culture closely comparable with that of the nomadic modern Plains Indians."[171] But since neither he nor Figgins thought anything was amiss at Frederick and its association of artifacts and fossils, he rushed past that concern:

> It now appears to me quite probable that these very early people *were* closely comparable in most ways to our modern Indian, structurally as well as in culture; and if so, this would account for the errors in interpretation as to man's antiquity in America, fallen into by some talented scholars.[172]

AMNH archaeologist Nels Nelson saw an advance copy of *Natural History*. A very talented scholar by any measure, he was not buying what Cook and Figgins were selling. If everything they said was true, he told Figgins (as he had told Hay a decade earlier [§7.5]), "we shall have to revise our entire world view regarding the origin, the development, and the spread of human culture."[173] He was not ready to do that.

Few were, given the gaping holes in the Frederick evidence, the existence of which even Hay admitted. In a working sand-and-gravel quarry, stratigraphic control was difficult to maintain, admixture was likely, and the testimony of untrained quarry laborers was hardly reliable.[174] Schwachheim, who had been stationed at Frederick for nearly two months but had yet to see artifacts in the mammoth-bearing deposits, was increasingly doubtful that any had ever been found in the same level as the Pleistocene fossils.[175]

The scarcity of additional material from Frederick disappointed Figgins, who had planned fieldwork there that summer. In April he recalled Schwachheim and instructed him to prepare for another season at Folsom. It is revealing of Figgins's low opinion of Folsom that the 1927 excavations were an afterthought when Frederick did not pan out. In fact, that spring he saw other sites as higher priorities, including a mammoth find in Texas. He cautioned Schwachheim not to open too large an excavation block at Folsom in case he had to bolt for Texas. He thought a twenty-square-foot area ought to suffice to "give us all we need in the way of bison." Once again, Figgins failed to mention watching for artifacts (§8.4).[176]

8.9 FROM THE LION'S DEN . . .

Although the *Natural History* issue with Cook's and Figgins's articles was still months from appearing, Watson Davis of the *Science News-Letter* got wind of the Frederick story from a stringer in Oklahoma. He obtained an advance copy of *Natural History* (which caused an upset Figgins to briefly withdraw his and Cook's papers), then published a brief notice of Cook's and Figgins's sites in April.[177] The word was out, and with that, Hay renewed his efforts to have the evidence from the sites vetted in Washington.

He showed the fossil bones he already had to Smithsonian archaeologists Walter Hough and Neil Judd. Both conceded the fossils were probably of Pleistocene age; as archaeologists they were in no position to argue the point. But they were keen to see the artifacts reportedly associated with those bones. Hay understood why and guessed what their response might be:

> They will deny wholly that the artifacts are old unless they are of a rude style of make. "Otherwise [they will say] there would be no indication of progress." Our business will be to prove that the things were primarily associated with the animals and let the anthropologists anchor themselves to what they please.[178]

Certain that the anthropologists were "hitching their wagon to an extremely unsafe star" on the matter of "progress," Figgins told Hay he was traveling to Pittsburgh and was willing to visit the Smithsonian artifacts from their sites. "It might be good for the soul of the enemy to see them now."[179] Hay suggested a formal presentation, but Figgins balked:

> Outside of the fact that I am neither geologist, paleontologist nor anthropologist, I could not, officially or otherwise, submit any of the evidence to anthropologists. I would be glad to have them examine and study the artifacts, but this must be regarded as a courtesy and not a recognition of

anything that suggests that I concede it is their right and prerogative to pass upon such evidence.[180]

Figgins agreed to show the artifacts but drew the line at listening to any criticism made in regard to them.

A brave front, but by the time Figgins arrived at the Smithsonian in late May he fully expected "real cross words and scolding." To his surprise Hrdlička proved "extremely pleased and courteous." After examining the artifacts he merely expressed regret that none were found in place, and he offered a piece of advice: if artifacts were found in situ that summer they should be left there while telegrams were sent around the country inviting "two or three scientists from other institutions to make detailed studies of the whole occurrence." Figgins thought that perfectly reasonable. "Of course, I fully agree that would have been the course to pursue, had it not been for the fact that they were removed before their discovery."[181]

What Figgins failed to understand, however, were Hrdlička's motives. He was not going to be convinced by anything Figgins or Cook said about the site, its age, or any possible association of artifacts with extinct animals.[182] Claims for a great antiquity needed to be assessed on-site by experts. Neither Figgins nor Cook qualified. Few did, in Hrdlička's eyes, though it was not true, as Lucas joked, that Hrdlička would not be satisfied with the Folsom evidence unless he "fired the arrow himself." But he did want professionals called in when the arrow was unearthed.[183]

Although Figgins found his conversation with Hrdlička unexpectedly pleasant, he could see he had not changed Hrdlička's mind. He had no better success with Holmes. As they headed to Holmes's office, Hrdlička suggested they show him the artifacts but say nothing of where they were from or who Figgins was. It is questionable whether Hrdlička actually said this (it would be very uncharacteristic of him to play a practical joke, let alone on Holmes whom he revered). In any case, Figgins gleefully reported that

> Dr. Holmes took a brief look and declared [the projectile points] were "undoubtedly foreign—European—both in material and workmanship." He got no further, as Dr. Hrdlička broke in with an introduction of myself and the statement that they were the artifacts from Texas and New Mexico. I managed to retain a fairly straight face, and dignified demeanor. So much for the great Dr. Holmes. . . . Dr. Hrdlička was not happy.[184]

Judd joined them, and after seeing the artifacts he claimed he could duplicate them from "anywhere west of the Allegheny Mountains." Figgins called on Judd to prove it, then "enjoyed his failure to discover [in the collections] any example that was similar." He left his sessions with the "anthropologists, ethnologists, archaeologists and anarchists of the National Museum" certain he had won that round and convinced they would not attack Lone Wolf Creek or Folsom.[185]

But Figgins missed Holmes's point that proving contemporaneity required "arrowheads . . . imbedded in the bones of the latter in such manner as to prohibit question that they were shot into them while the animal was living." He attributed it to Holmes's inability to accept evidence "likely to interfere with his personal theories."[186] But Figgins was himself ignorant of the claims for antiquity that had failed for want of a secure association. In retrospect Holmes was both right and prescient.

Figgins departed, having changed the object of his obsession. It was no longer Hrdlička against whom he railed, but Holmes and Judd. And obsessed he was. Over the next six months Figgins told dozens of correspondents the story of his surprisingly enjoyable encounter with Hrdlička, Holmes's embarrassing blunder, and Judd's comeuppance. He particularly enjoyed tattling on Judd to other archaeologists and to Alexander Wetmore, who as director of the USNM was Judd's boss. But after Figgins repeated the story to Hay for the fourth time in as many months, Hay snapped, scolding Figgins that he was giving Hrdlička far too much credit. Having battled Hrdlička for more than a decade, Hay knew their adversary's true character. Figgins had been charmed by a snake.[187]

8.10 . . . TO THE BELLY OF THE BEAST

About the time Figgins was in Washington, Schwachheim began excavating at Folsom. Although he had not been instructed to do so, he knew to be on the lookout for artifacts.[188] The excavations reached the level of the bone bed in late June, but it was not until around noon on August twenty-ninth that he spotted a projectile point, partially buried but obviously in situ alongside bison ribs. Schwachheim stopped work and immediately wrote Figgins to alert the "doubters" to come see the evidence in place. When his letter reached Denver, Figgins sent telegrams around the country: "Another arrowhead found in position with bison remains at Folsom can you personally examine find [sic]." Hrdlička and Hay received copies, as did Nelson, Alfred Jenks (an anthropologist at the University of Minnesota), and Brown, who had passed through Denver a few weeks earlier and was still out west.[189] (Only two archaeologists received telegrams, either because of Figgins's low opinion of them or more likely because he hardly knew any.) While awaiting responses Figgins commanded Schwachheim to not let anyone remove the point for examination or even dig around it, "regardless of who it is or what reason they give." He was to keep his eyes on it "all the time it is uncovered."[190]

Hrdlička was in Washington, but as Wetmore explained to Figgins he was the only member of his department in town and it was "not practicable to send him." Hay himself was just back from Frederick and Lone Wolf Creek and naturally had a more cynical take on why Hrdlička did not go: he could not be bothered. It was just as well, Hay said. He had heard that Hrdlička claimed he had already "pulled chestnuts out of the fire for other people." The way Hay read it, Hrdlička feared he was going to "get his fingers burnt."[191]

The Smithsonian was represented nonetheless. Newly appointed BAE archaeologist Frank Roberts was in New Mexico at Alfred Kidder's inaugural Pecos Conference on southwestern archaeology.[192] On Wetmore's order Roberts left for Raton, where on September 4 he met up with Howarth, then traveled to Folsom, arriving on the heels of Figgins and Brown.[193] Figgins had not met Roberts during his trip to Washington (Roberts was still at Harvard finishing his PhD), but on

Figure 8.2   Carl Schwachheim (*left*) and Barnum Brown (*right*) at the Folsom site on September 4, 1927. Schwachheim is pointing to the in situ projectile point lodged between the ribs of the extinct bison. Courtesy of the Vertebrate Paleontology Archives, Division of Paleontology, American Museum of Natural History.

meeting him at Folsom he decided Roberts was open-minded and untainted by his new Washington colleagues.

That day on-site Brown examined the stratigraphy and geology, examined the in situ point and bison ribs (though he was careful to leave the specimens in the ground), and had the now iconic photograph taken of him and Schwachheim posing next to the specimen (Figure 8.2).[194] Roberts also carefully studied the association and two additional points Schwachheim had subsequently found (though not in situ). All agreed the point and the bison bones occurred in such a manner that their association could not be dismissed as adventitious, as so many prior claims to a deep human antiquity had been.[195] It was precisely the sort of evidence Holmes had told Figgins was necessary to prove the contemporaneity of an artifact and an extinct animal. Ironically, that moment came just as the *Natural History* issue with Cook's and Figgins's articles trumpeting Frederick as the oldest and most compelling of their sites was reaching subscribers and newsstands.

Back in Raton later that day, Roberts telegraphed Kidder urging him to visit Folsom while the point was still in place. That Roberts asked Kidder and none of the other participants at the Pecos conference is no surprise (§1.1). Kidder was, as Gordon Willey put it, the "outstanding American archaeologist of his time."[196] Roberts was well aware of the importance of having Kidder vet the Folsom association firsthand. That evening Roberts also spoke at length with Figgins about the projectile points, which he admitted were unlike any he had seen (which gave

Figgins another opportunity to twist the knife into Judd). And Roberts called Figgins's attention to a feature previously unnoticed: the "hollowed sides of the artifacts," which Figgins now saw were "*not* accidental."[197]

The next morning Figgins and Brown departed, but Roberts stayed behind. He returned to Folsom to take sediment samples and photographs, and to further examine the artifacts, observing that:

> They measure about 2 inches in length and about one in width and are very finely chipped. Down the center, on each side, from point to tang, is a groove formed by the knocking out of a single flake on each side. This resulted in a point similar in its nature to modern bayonets. The point is a little large for an arrowhead and a little small for a regular spear, it probably was for use in a javelin, perhaps with an atlatl or spear thrower.[198]

Roberts knew one might question the precise age of the site—that would be for geologists and paleontologists to determine—but despite his best efforts he "could find no way to explain the presence of the points by intrusion." They had entered the formation at the same moment as the bison.

Roberts went back to Folsom a third time on September 8, this time accompanied by Kidder and Cook.[199] Kidder completely agreed on the context and integrity of the find. As he put it to Figgins:

> As an archaeologist, I am of course not competent to pass either upon the paleontological or the geological evidences of antiquity, but I have paid great attention for many years to questions of deposition and association. On these points I am able to judge, and I was entirely convinced of the contemporaneous association of the artifact which you so wisely had left "in situ" and the bones of the bison.

Kidder was acutely aware of the decades of bitter controversy over human antiquity and thanked Figgins on behalf of anthropology "for having handled this material so carefully and so intelligently."[200] Of course, Figgins was just following advice Hrdlička gave him in Washington a few months earlier.

After Kidder and Roberts departed, Schwachheim carefully cut out and plastered a block of sediment surrounding the in situ point and bison ribs. On the last day of September he put the block, the other points, and the last plaster-jacketed bison bones on a train to Denver. At the CMNH the block was readied for display. (It has remained on more or less continuous exhibit there ever since.)

That same week Kidder was in Los Angeles lecturing on early man in America, a topic he had rarely spoken of previously, but Folsom had provided the opportunity to announce publicly what he had always hoped for privately (for reasons detailed in §9.12): the "first journey to the New World took place at least fifteen or twenty thousand years ago."[201] He then went on to the California Institute of Technology to brief Stock on what he had seen at Folsom. The two spoke at length of possible ways to develop investigations of Pleistocene-age archaeological sites. It was an opportune moment, for Merriam had just put Kidder in charge of archaeology within the Carnegie Institution's Division of Historical Research. Merriam and Stock, whose only prior experience in this realm had been to respond to alarms sounded by Holmes and Hrdlička (§8.3), were glad to hear Kidder's positive news of Folsom. Kidder in turn was pleased to tell Figgins that Merriam agreed to devote some of the Carnegie funds to investigations of this "most fascinating problem."[202]

Although Figgins surely appreciated the news, he may not have fully grasped the significance of Kidder's support. Figgins was not even sure who Kidder was, let alone did he understand the power of his testimony within the archaeological and scientific community. Figgins certainly could not have known Kidder had already been proposed for election to the NAS, or that just months earlier Hrdlička himself had written Merriam "with a deep sense of thankfulness" on learning Merriam had hired Kidder at the Carnegie Institution. Unlike Figgins and Cook, Kidder was someone Hrdlička trusted. In striking up a correspondence with Kidder, Figgins made sure to complain of the "chilly atmosphere" he had received in Washington that warm day months earlier, especially from Judd. He worried what would happen when Kidder and Roberts encountered Judd. Kidder was sympathetic but did not bother explain that he was not worried. Judd was a longtime colleague who had employed Roberts since 1925 and even orchestrated his hiring at the Smithsonian.[203] Most of all, Figgins seemingly had no idea that Kidder's acceptance of Folsom meant that the decades-long and bitter controversy over a Pleistocene human antiquity in America was essentially over.

8.11 SEEKING A NEW IDENTITY

There was still work to do, of course. The agreement that the in situ point was undeniably associated with the bison cleared the ground of one major obstacle to demonstrating a Pleistocene human presence. The co-occurrence of human remains and a marker of geological antiquity could not be rejected as adventitious. The "ever-present prairie dog," Kidder exclaimed, "didn't have anything to with that!"[204]

Figgins had anticipated and already disposed of a second potential obstacle. During his visit to Washington, Figgins repeatedly quizzed Judd as to whether "the form and workmanship of artifacts was a clew [sic] to their age." Judd insisted the answer was no, and Figgins "made him say it three times." He wanted to insure that Folsom was not rejected because the points looked modern, much as Hrdlička rejected any modern-looking human skeletal remains. But as was evident in Judd's answer, by 1927 it was no longer assumed that "older" in America had to have a "primitive" appearance, whatever that might mean.[205]

The remaining challenges were to show both that the Folsom bison was extinct and that it had vanished before the end of the Pleistocene. While at the site Brown reckoned the bison was possibly Middle to Late Pleistocene in age. Cook agreed, on the assumption the species was *Bison occidentalis*. Neither was yet ready to grant the specimen new-species status or assign it to a geologically earlier form.[206] Hay was, however. He had examined the bison from the 1926 excavations, and with additional information on the 1927 specimens he announced to Figgins that he was certain the Folsom bison was not *B. occidentalis* but instead "a very distinct and new species." This new species, he told Brown, was distinguished by its "long drooping horn-cores and by especially the great breadth of the face and the narrow muzzle."[207]

Brown was keen to have Hay assign a name to the new species, so keen he alarmed Hay, who could not understand the rush. (Brown had not explained he was scrambling to ready a presentation on Folsom for the American Anthropological Association (AAA) meetings three weeks later and wanted to announce the name at his talk.) Worried that Brown might scoop them on naming the new species, Hay and Cook quickly published a brief type description, "just enough to hold priority." *Bison taylori* (named after Frank Taylor, longtime president of the CMNH) debuted in early 1928, along with the Lone Wolf Creek bison, *Bison figginsi*, and a new species of mammoth from Frederick, *Elephas haroldcooki*.[208]

Hay likely intended these names to last long after Taylor, Figgins, and Cook had passed from the mortal scene, but their taxonomic status did not even survive the decade. By the time Hay and Cook completed their full analysis and description a year later, the Lone Wolf Creek bison specimen had become a new genus, *Simobison figginsi*. The Folsom specimens remained genus *Bison*, but now it was believed there were two species, *B. taylori* and *B. occidentalis*, at the site. The second was added against Brown's vigorous protest; he thought Hay was mistaking size differences and sexual dimorphism for species. Brown saw only one taxon, *B. taylori*. Hay offered a compromise: the smaller "type" specimen was identified as *Bison sp.* indeterminate. But that hardly ended the taxonomic debate. Figgins weighed in a few years later and rejected both of Hay and Cook's species and instead assigned the Folsom to a new genus (*Stelabison occidentalis taylori*) and a new species (*Bison oliverhayi*).[209]

That multiple bison species or even genera might be present at a single site partly reflected uncertainty about how to gauge the significance of the morphological variability in the specimens, but also something of taxonomic practices of the time. The period from 1928 to 1933 saw the "most concerted surge of taxonomic splitting" than any period before or since. During this brief span a total of two new genera, eight new species, and six new subspecies were named.[210] It was, as AMNH paleontologist George Gaylord Simpson observed a decade later, "chaotic, to say the least." That this was the period when the bison from Folsom and Lone Wolf Creek were described is no coincidence, for the eruption of taxonomic names is due largely to Cook, Figgins, and Hay.[211]

Yet, on one crucial point everyone, Brown included, agreed. Whatever the species of bison proved to be, or however many kinds there were, it was (they were) extinct. The remaining question was whether the bison went extinct during the Pleistocene, and if so *when* during the Pleistocene. Hay understood from Cook's geological interpretation that Folsom was Wisconsin in age. But did Cook mean the "culmination of the glacial stage," which implied climatic conditions that may have rendered the region uninhabitable, or did he mean the end of the Wisconsin? Climatic conditions during the latter were likely milder; this—and here Hay revealed his deeper concern—"would please the anthropologists for they could still further advance the time to suit themselves" (that is, bring Folsom into more

Table 8.1 The Pleistocene glacial and interglacial sequence as seen at the end of the third decade of the twentieth century

Stage name	Episode	Time of onset (years ago)
Recent (Holocene)	Interglacial	10,000
Wisconsin	Glacial	85,000
Sangamon	Interglacial	150,000–175,000
Illinoian	Glacial	200,000
Yarmouth	Interglacial	
Kansan	Glacial	500,000–750,000
Aftonian	Interglacial	
Nebraskan	Glacial	1,000,000

Note: Gray = Pleistocene glacial; white = interglacial. Stages from Leighton (1931) and Leverett (1929); absolute ages from Leverett (1930b).

recent times). But Hay saw an opportunity as well. If the site was occupied during times of milder climate, why could Folsom not date to a prior interglacial, one that occurred in the Middle or Early Pleistocene?[212]

It was the same line of questioning Hay had run past Cook regarding Lone Wolf Creek two years earlier (§8.3). If Folsom was from an earlier, post-Aftonian interglacial, there were by the 1920s only a couple of stages in which it could fall: the Yarmouth or the Sangamon (Table 8.1). Absolute ages for those periods were based "mainly on the erosion each [drift deposit] has suffered" and hence were approximate at best, but they implied an antiquity on the order of hundreds of thousands of years.[213] Hay pressed Cook to consider the possibility that Folsom dated to a previous interglacial.

Cook readily accepted that Frederick could be that old, but he was doubtful Folsom was. The Folsom bones were not mineralized, certainly not to same degree as those from Lone Wolf Creek and Frederick. Instead, they bore "the usual earmarks" of being much younger. The Folsom bison was less primitive (closer to *Bison bison*) than Lone Wolf Creek's, and it was found in sediments of a "muckhole, marsh or slough," which included aquatic invertebrates indicative of nonglacial conditions. Yet, having skirted close to the end of the Pleistocene, Cook quickly backed away:

> Do not think from all this that I believe [Folsom] to belong to Recent times! I believe there *is* evidence to be had which will throw important light on its age, beyond that which we now have, judging from conditions I have seen there; and I believe it will be proved that this deposit belongs to some interglacial stage of the Pleistocene.[214]

That satisfied Hay, though he suggested it was best not to be too specific about the age of the Folsom bison.[215] Still, by implying they dated to the Sangamon or Yarmouth they were fooling no one.

Cook, Figgins, and Hay were certain Folsom was much younger than Frederick or Lone Wolf Creek, yet (as Figgins was soon to announce) it was still hundreds of thousands of years old. In believing so, their estimates were an order of magnitude higher than anyone else supposed. In fact, as the *Science News-Letter* reported in the fall of 1927, Folsom was quickly becoming "accepted by a number of scientists as real evidence that America was inhabited . . . in the Pleistocene period," which they put at 15,000 to 20,000 years ago.[216] Although that was just a fraction of the age estimates bandied about by Cook, Figgins, and Hay, it still put Folsom safely within the confines of the Pleistocene, albeit the Late Pleistocene. Unless evidence emerged that the species of bison found at Folsom survived into the Recent period, it seemed likely to stay there.

8.12 HEDGING BETS

En route to New York in the fall of 1927 Brown stopped in Denver to see Figgins, who gave him the Folsom artifacts to take to the AMNH. Figgins had heard from Roberts and Kidder that Nels Nelson was the expert to consult on stone tools.[217] Figgins wanted Nelson's opinion but also hoped the artifacts might lure Nelson and AMNH vertebrate paleontologist Walter Granger out west. As he explained, only Brown, Kidder, and Roberts had visited Folsom. Others (Jenks and Hay) had been to one or the other of their sites, yet no one besides Figgins and Cook had been to all three. Knowing of Nelson and Granger's expertise and the respect they commanded, Figgins thought them well qualified to assess the archaeology, geology, and paleontology of the sites and arrive at a "definite and final conclusion" regarding their respective ages.[218] Despite Figgins's flattery and his assurance that they would need no more than a day at each site and thus be absent only "two weeks from Broadway," neither Nelson nor Granger took the bait.

Nelson was just back from the AMNH's two-year expedition to Mongolia and China and did not have the time. Besides, he had read the *Natural History* papers and had serious doubts about Lone Wolf Creek and Frederick (§8.8), and he knew Kidder had visited Folsom. Given those circumstances, he saw no need to go himself. Granger had a similar reaction: "Brown's opinion ought to be as good as mine anyhow." Both were likely also aware that drawing a final conclusion about a site from a day's visit was impossible. Disappointed but persistent, Figgins went over their heads to Osborn, hoping he would order them to visit the sites. Osborn was not convinced he needed to dispatch either of his curators; he ignored Figgins's request.[219]

Figgins then tried to entice Kidder and Roberts to visit Lone Wolf Creek and Frederick, using as bait the idea that those sites were older and even "more convincing" than Folsom.[220] The most he got was a noncommittal hope to visit.[221] With that, Figgins stopping trying, believing he had complied with the rules of the game:

> I briefly presented the evidence as accurately as I knew it. The localities are
> still there and wide open to the fullest study and investigation. . . . I have no

opinion to express and personally have but little interest in the ultimate decision. . . . It merely so happened that I followed up reports of fossils and sent men to the localities to dig fossils. Finding the artifacts was purely incidental, but recognizing their possible importance, I sent Mr. Cook to the locality to study the geology. Then I asked Dr. Hay to join Mr. Cook in a study of the fossils. Dr. Hay visited two of the localities. Barnum Brown went to Folsom with me and at my request. I invited Prof. Jenks to study the localities. . . . I invited Dr. Nelson to make a special study and wrote Prof. Osborn pressing [an] invitation to send Dr. Nelson and Walter Granger for joint studies and investigations. I invited Dr. Hrdlička.[222]

There was little more he thought he could do "towards demonstrating my desire to lay the whole thing open to the widest possible investigation." Even when contacted to make a statement about Folsom to the *Science News-Letter*, Figgins declined, suggesting the publication instead contact Brown, Roberts, or Kidder.[223]

Figgins chose deference as the best policy. What is not clear is whether he anticipated the risk of relinquishing the interpretation of the site and its evidence, or for that matter the risk of literally handing over the evidence (in this case the artifacts), to one as covetous and self-centered as Barnum Brown. (It is telling that Brown received the Folsom artifacts in October to deliver to Nelson, yet it was late December before Nelson got to see them.) There was obvious value to having others testify on behalf of Folsom, but it meant that other voices became the authoritative ones, not those of Figgins and Cook.

Figgins's newfound modesty and deference to the authority of others was a stark contrast to the Figgins who just a year earlier had boasted of laying Hrdlička's hide across a barbed wire fence, who had dismissed archaeologists and anthropologists as ignorant "jokes," and who had assumed that a case was proven because there was no doubt in *his* mind. Then, it was easy for Figgins to swagger when he had by his own admission "large ignorance of the subject."[224]

But that was before he spent time with Roberts at Folsom and had crash correspondence courses from Kidder—who finally convinced Figgins that Folsom points were not arrowheads but spear or dart points—and from Nelson on the history and issues involved in the debate over human antiquity in the Americas.[225] Nelson patiently explained that Holmes had had "good and sufficient reasons for dealing severely with 'evidence' and producers of evidence for Pleistocene man;" that it was appropriate to be skeptical of claims for a deep human antiquity given all the ones that had failed; that long-accepted evidence of mammals migrating from Asia to America certainly hinted that the Bering Strait was not always a barrier, but it was "still a question whether man was really in the region in Pleistocene times" and had crossed over then (the question of a Bering Strait came to the fore within the decade [§9.7]); and that the prehistory of America, though assuredly not identical to the Old World, nonetheless ought to show "some sort of consistency" between the hemispheres in that ancient (though not necessarily Paleolithic) artifacts ought to be found with extinct animals and more modern artifacts with a recent fauna.[226]

And in response to Figgins's harping on Judd having insisted he could dupli-cate Folsom points "from anywhere west of the Alleghenies," Nelson told him Judd might be a "young man of relatively limited experience" but in fact Judd had not gone far enough. Nelson had seen "plenty east of the Alleghenies" as well. In case Figgins was worried that had any bearing on the age of the points, Nelson assured him they were "doubtless very old types."[227] Figgins sputtered he had ex-amined "many thousands of arrowheads," yet none with that long broad flake re-moved. And in what might be charitably described as a white lie, Figgins insisted everyone who saw the Folsom artifacts "except the Washington folks" commented upon their "hollowed sides."[228]

Figgins soon realized fluted points were rare but not unique, and he and others searched collections for more specimens.[229] He also came to see that fluting made Folsom points distinct from those at Lone Wolf Creek, might be regarded as the "chief character" of the Folsom culture, and perhaps even provided a means of putting Folsom, Frederick, and Lone Wolf creek in chronological order.[230] As he later explained to Jenks, technology evolved from no fluting (Lone Wolf Creek) to less pronounced fluting (12 Mile Creek) to full fluting (Folsom). As for the Fred-erick point, it also lacked "concave chipping" (fluting), and given its cruder form and older associated fauna it was likely the "oldest of the lot."[231]

As Figgins became more cognizant of how little he knew and how much more complex the issues were, the remaining vestiges of his pugnacious swagger virtu-ally disappeared. He made mention a few times of replying "with bare arms" should criticism come, but by October of 1927 even these feeble boasts vanished from his correspondence.[232] He came to see that the word of individuals like Kid-der, Nelson, and Roberts in archaeology and Brown and Granger in vertebrate paleontology carried far more weight than his:

> Now, if the Washington people care to go counter to all of the opinions expressed—yourself [Hay], Mr. Cook, Barnum Brown, Prof. Jenks, Mr. Roberts and Dr. Kidder— . . . then I will be the first to congratulate their bravery, irrespective of how nearly it may approach what is sometimes re-ferred to as foolhardy. [But] I credit them with better judgment and more sincerity than such a course would indicate.[233]

Hay, who was keeping a close watch on the Washington anthropologists, reported back that they seemed to be "lying very low just now."[234]

8.13 WILL THE RISING TIDE LIFT ALL BOATS?

All was going well for Folsom, but neither Figgins nor Cook nor Hay was entirely satisfied. They continued to insist Folsom was the youngest and least important of their sites, and having just one Pleistocene site was not enough: they wanted the trifecta.[235] Through the remainder of 1927 and into the coming year they contin-ued efforts to shore up the evidence for their other ostensibly older sites.

Bolstering Frederick proved particularly challenging. Hay visited there in Au-gust of 1927 to see the site and interview Holloman, but he was frustrated in his efforts to pin Holloman down as to where and under what conditions artifacts had been found. Hay and Figgins enlisted Priestley's help to get written answers from Holloman about whether artifacts had been found cemented in place or in undisturbed sediment, or if they could have fallen in from above. Holloman provided answers but also sabotaged his credibility. He now insisted he had found a five-toed horse (impossible in deposits that old, Hay knew) and a stone on which there was an engraved flower (it proved to be a glyptodont carapace). Worse, Holloman said those things within earshot of a *Science News-Letter* reporter, who was preparing a story on Frederick.[236]

This was precisely the sort of claim that a shuddering Figgins saw could be "a more or less serviceable weapon in the hands of the enemy." Fortunately Watson Davis stopped by Hay's office and Hay was able to gain assurances (which he passed along to a relieved Figgins) that the *Science News-Letter* would say nothing about a five-toed horse. On the other hand, the story quoted Hay saying there could be no doubt the artifacts at Frederick were 500,000 years old, a statement that was received in the scientific community just as skeptically as news of a five-toed horse in those deposits would have been.

In the meantime, Cook was climbing out on even shakier limbs, arguing in *Scientific American* for a Pliocene antiquity of Snake Creek ("the oldest trace of humanity, by hundreds of thousands of years") and a Frederick occupation that reached back more than a quarter of a million years to the Early Pleistocene.[237] After reading his article, Matthew confessed he was "quite unable to take the Snake Creek 'artifacts' seriously," and wondered why Cook did not see what he saw: Folsom was the best evidence yet for "Pleistocene man in America [and] it looks mighty good to me." Cook merely shrugged.[238]

Evidence kept coming. Cook's latest *Scientific American* article elicited a letter from Cyrus Ray, who had been collecting bones and artifacts in gravel quarries around Abilene, Texas that he believed to be similar to those Cook described at Frederick. His letter convinced Cook these were sites that "*bear every evidence of being genuine and of the highest importance!!*" Cook urged Figgins to send him there. Figgins declined.[239] A few weeks later Ray wrote of more fossil bones and crudely worked artifacts in what Cook was certain (sight unseen) were Early Pleistocene gravels. Cook put more pressure on Figgins: "I really feel, again, that once more we have our fingers on another highly important discovery; and one that will again, if successfully carried out, bring the spotlight of scientific and popular attention of the finest sort to bear on the Colorado Museum." Figgins still refused.[240]

The problem from Figgins's perspective was twofold. Ray's sites were in geological settings that "contain little or nothing, of a nature suitable to our needs— mountable skeletons." The museum's scarce funds could not be spent chasing down scraps of bones, even on the off chance that these might be found with artifacts. That would require keeping workers constantly posted in the actively worked gravel pits, and "only the richest institutions can afford such luxuries."[241] And, second, if the CMNH could not have one of its observers on site, any discoveries would once again rest on the testimony of untrained observers in geologically complicated settings. That was more exposure than Figgins was willing to risk:

It is my opinion we should and must, avoid all expressions and all connections that do not apply to finds sponsored by this museum and that we have not personally investigated. . . . I feel we should not again record finds not made by representatives of the museum, unless the circumstances are such that they will stand the most rigid scrutiny.[242]

The lessons of Lone Wolf Creek and Frederick had been learned.

Yet, it was more complicated than that. Figgins was wary of collateral damage to all their sites, including Folsom. A slip at Frederick or elsewhere (say, at one of Ray's sites) "would be seized upon and would be bound to reflect upon us and damage the work and cause we have espoused."[243] To steer clear of trouble, Figgins repeatedly lectured Cook—but, it is telling, not Hay—that they needed to maintain silence and make no mistake "until the jury brings in its verdict."[244]

It was good advice, but neither Cook nor Hay nor even Figgins followed it. Hay, of course, had already pronounced Frederick 500,000 years old. Cook was writing his next article for *Scientific American*, opening with the statement that humans had lived in the Americas for hundreds of thousands of years and that the New World was "quite as old as the Old World, and it may have been inhabited by mankind quite as long." These were hardly uncontroversial claims within the scientific community, and they even tested the limits of Cook's very tolerant editor. Ingalls told him *Scientific American* "cannot risk [such statements] until there is more evidence" and insisted on adding a disclaimer:

In the first two paragraphs of his most interesting article, Mr. Cook, the author, makes claims concerning the proof of the antiquity of man in America—claims which the editor regards as requiring a still larger volume of substantiation than the available evidence affords. With Mr. Cook's friendly concurrence, the present statement, in which the editor disclaims all responsibility for their inclusion, is published.[245]

For his part, Figgins sent off two manuscripts that fall (greatly irritating Cook, who resented having been told by Figgins that he should not write anything[246]). The first, to *Outdoor Life*, put Folsom's age at "not less than 400,000 years." Even Hay, who rarely met an estimate of deep antiquity he did not like, was "not prepared to go so far." Neither was Cook, who, without the slightest hint of his own hypocrisy, "seriously question[ed] the advisability of such positive and definite statements being made at this time."[247] Cook had an editor who tried to save him from himself. At *Outdoor Life*, a popular magazine, Figgins did not. Fortunately an editor stepped in to protect Figgins on his other manuscript, an article sent to *Natural History* with news from the 1927 excavations at Folsom and the site visit. Included as well was the story of Figgins's encounter with the Washington anthropologists, and he named names. Brown acknowledged receiving the manuscript, but it never appeared in print. Brown perhaps realized it would be best if *Natural History* was not used to pick a fight with the Smithsonian.[248]

8.14 WHEREAS, FOLSOM

Word of the Folsom discovery flashed through the archaeological and scientific community in the fall of 1927, and it was welcome news. Folsom was hailed as a site "of the highest importance" (Wetmore) and even "the greatest event in American discoveries" (Osborn). More prosaically, there was this: "The Folsom finds were unequivocal evidence of man and extinct animals, and . . . everybody breathed a sigh of relief that finally what had been suspected was proven" (Emil Haury). Figgins basked in the glow of tributes such as the one Kidder offered: "The researches of yourself and Dr. Cook will go far towards opening a new era in the study of the question of Pleistocene man in the New World." Heady praise. Even Figgins realized Folsom "appears to have been almost unanimously accepted," quite different from the reception accorded Lone Wolf Creek or, for that matter, Frederick.[249]

Enthusiasm and interest in Folsom, coupled with the prospect of the long-running controversy over human antiquity having finally run its course, prompted Harvard's Roland Dixon to arrange a symposium on "Early Man in America" at the upcoming AAA annual meeting. Papers on that topic had been delivered over the decades at the AAA, but this was the first occasion in more than fifteen years that it was the focus of a special symposium. Dixon invited Nelson, Roberts, and a few others as well as Hrdlička, whose views he was especially keen to hear "in light of all the recent information."[250]

Hrdlička declined. He had already committed to attend the AAAS annual meeting that same week in Nashville where, as luck had it, Hay was scheduled to speak about Folsom, Frederick, and Lone Wolf Creek (which explains the subsequent report in *Science* that a "lively discussion" followed Hay's talk[251]). Nonetheless, Hrdlička considered the recent finds "of much interest," though far from "settling the problem of man's geological antiquity in America." As he had a decade earlier during the Vero debate (§7.7), he insisted that that would require a joint committee of representatives from the relevant sciences to whom would be

> reported or referred every serious discovery seeming to show ancient man, as soon as possible after the find was made and before the remains were removed from situ. . . . I believe that in some such way only can we arrive at conclusions that will command the confidence of every worker. Unless something of this nature is done, there will accumulate in the course of time many more or less ambiguous cases and claims, which will impede progress in American anthropology.[252]

Progress, however, was made at the AAA meeting without Hrdlička or his committee.

Brown, serving as a last-minute replacement for Roberts on the formal program, arrived armed with lantern slides of the site, the block with the bones and in situ point, and the mounted bison skeleton. He had as well the actual Folsom points he had brought back from Denver. (In later years Brown often carried the points to meetings.)[253] Nelson discussed the broader history of the search for early sites, whereas Jenks and Loomis focused on the artifacts and geology (respectively) of Folsom as well as Lone Wolf Creek and Frederick. Roberts and Kidder added extemporaneous remarks of what they had seen at Folsom. Warren Moorehead, who thirty-five years earlier had organized the explosive Madison showdown of the Great Paleolithic War (§5.13), culled the collections of Phillips Academy (a preparatory school that had developed an important museum and archaeological research program[254]) and displayed several points he had found that were similar to the Folsom type. Such specimens did indeed occur across North America, much as Judd and Nelson had said.[255]

Figgins heard conflicting reports of how the talks were received by the anthropological audience. Jenks assured him Brown's presentation was "clear, accurate, and convincing" and that "no one voiced question of the early man appearance of the data presented." Brown saw it slightly differently. The Lone Wolf Creek finds were accepted "without dissension," but Frederick was "questioned on very valid [geological] grounds." Figgins pinned that failure on Loomis, whose talk, Jenks tattled, was "typical second-class." Figgins thought it "venturesome" of Loomis to speak of sites he had not visited or spoken to Figgins about (Loomis's source was Cook, but that did not assuage Figgins's unhappiness with him).[256]

The issues with Lone Wolf Creek and Frederick notwithstanding, all agreed Folsom was enthusiastically received. Brown assured Figgins, "Your artifact finds at Folsom held the center of the stage at the Andover meeting, and . . . all the anthropologists and archaeologists present accepted the authenticity of the finds, saying they established a definite landmark in the history of prehistoric man in America."[257] (A decade later Roberts recalled "most of the anthropologists continued to doubt the validity of the discovery." But the contemporary record belies that claim, and his longtime colleague Henry Collins suspected his recollection was likely skewed by a chilly reception he had from Hrdlička in Washington afterward.[258]) In an unprecedented move the AAA endorsed Folsom with a resolution offered by Brown and Jenks and heartily approved at the annual business meeting:

> Be it resolved that, whereas, from the palaeontological [*sic*] and cultural data already at hand, gathered at the site and presented by competent experts at our annual meeting, December 28–30, 1927, the American Anthropological Association believes the Folsom, New Mexico, so-called "Bison quarry," holds most promising evidence in the disputed field of early man in America."[259]

That same day Figgins wrote to Gregory at the AMNH saying he was quite content to have the Folsom evidence "go before the final jury for ultimate opinion."[260] Unbeknownst to him, it just had.

The AAA symposium was a triumph, though more for some than others. It was Brown, Nelson, Kidder, Roberts, and others who spoke about Folsom. Neither Cook nor Figgins was invited to the meeting, let alone asked to participate; they were asked only to provide photographs, specimens, and information (§10.12). Lone Wolf Creek may have been accepted without dissension, but no resolution was offered in its favor, and certainly none was made on behalf of Frederick.

In preparing his *Annual Report* in December of 1927, Figgins led off with the discoveries made at Frederick. It was only afterward that he mentioned he had also sent a crew to Folsom that summer, admitting he did so to acquire more bison skeletal parts in order to complete a skeleton for mounted display (which the *Annual Report* proudly illustrated). That the crew happened to find several additional "arrows" and that "opinions are unanimous that the artifacts and extinct Bison were contemporaneous" seemed an incidental benefit.[261] It was a strange and subdued way to announce what Figgins had been assured was the opening of a "new era" in the study of human antiquity in the New World. But then, all was not well for Figgins as 1927 came to an end, and his difficulties directly affected his relationship with Cook and the museum and what followed at Folsom in 1928.

8.15 COMING APART AT THE (MU)SEAMS

The trouble began in Texas. In November of 1927 Cyrus Ray again contacted Cook (§8.13), this time raising the ante with news that parts of a human skeleton had been found in one of the gravel quarries yielding Pleistocene mammals. Cook pleaded with Figgins to send him to Texas, arguing this could be "the greatest chance that has ever come to an American institution, and I feel certain that we would make a serious mistake if we fail to capitalize it to the fullest extent in any legitimate and dignified manner."[262] This time Figgins did not refuse, and in late November he accompanied Cook on a 2,100-mile, two-week "flivver" trip with stops at Folsom, Lone Wolf Creek, Frederick, and Ray's Abilene sites.

Figgins said little afterward ("We found some new evidence of man in Texas, but further work is not necessary"[263]), but Cook's correspondents heard a great deal: about the sites, which Cook saw as yielding "real, valuable and first class authentic evidence here of [Pleistocene] man," and about Figgins. As Cook told it, trouble erupted on the Texas leg of the trip. The catalyst was a skull, "a thick-walled, low-browed type," that Figgins claimed to have found. Cook vehemently insisted that was untrue. (It was spotted in a museum collection; neither had actually "found" it.) Although both downplayed its possible significance, their haggling over credit for spotting it first indicates that they were taking no chances. Each had portions of the skull, and both sent their specimens to Gregory at the AMNH. His verdict came quickly: it looked no different from any Indian skull.[264] The find may not have been significant, but it made the long-simmering tension in Cook and Figgins's relationship boil over.

Given his many responsibilities and "high-strung" temperament (§8.1), Figgins had never been easygoing or collegial, particularly with Cook. Nor was Cook alone in having difficulties with Figgins. Alfred Bailey, who served under Figgins as ornithology curator from 1921 to 1926 (and then returned a decade later as the CMNH director), described Figgins as "the only man that ever went out of his way to make things uncomfortable for me."[265] Figgins had the potential, as Cook put it, to be "a trouble maker of the first order." Figgins may or may not have boasted he was "not happy unless he is 'getting' somebody or causing a fight between institutions or people," but his repeated belittling of Judd (§8.9–§8.10, §8.12) to Judd's friends and peers, such as Kidder and Nelson, or Judd's superior (Wetmore) strongly hints at that.

By late 1927, and for reasons now unclear, Figgins was under great stress and spiraling toward the "complete breakdown" he had almost promised in past years (§8.1). For the most part his blackening mood was invisible to correspondents, though there were signs. A comment by Wetmore about attendees of the American Ornithologists Union enjoying a steamer ride down the Potomac River elicited from Figgins the dark reply that he "would never get on a boat with that bunch for a trip on wider water than I can swim. I might not come back. Too many chances for an untimely and mysterious death by drowning."[266] Figgins was rarely given to humor and there is no indication he was kidding.

Evidence of Figgins's behavior that fall is known only through Cook's letters and therefore must be read with skepticism given Cook's penchant for exaggeration and his tendency to award himself a starring role in any drama. The root of the problem, as Cook saw it, was envy:

> [Figgins] is one of the sort who has to be the whole cheese, or he won't play; and he simply cannot *STAND* anyone else in his institution getting such recognition as I have had; and he has done every nasty contemptible little thing possible to hurt and handicap and undercut me, just as he has with every other person he has ever been associated with. . . . He is always perfectly *LOVELY* to my face, and behind my back telling others how he is going to "get" me.[267]

There may well have been true, though if Figgins was guilty of undermining Cook's reputation it must have taken place verbally, for there is no evidence of such in Figgins's external correspondence. But jealousy and hypocrisy were only part of the trouble. Cook hinted at far more scandalous matters:

> I have learned from authentic sources that he is a moral degenerate of a high order . . . that he has perverted museum funds to personal expeditions of a highly questionable character. Lately he has shown evidence of a mental imbalance that is serious, insidious, and dangerous in the extreme.[268]

Cook did not reveal the precise actions of which Figgins stood accused or his moral degeneracy, and he only alluded to a misuse of funds at Folsom by Figgins's son Frank as well as charges that Figgins's secretary and daughter Marien Lincoln was incompetent (about which Cook took credit for informing the CMNH board).[269] How much Cook told correspondents of Figgins seemed proportional to the odds of his comments reaching Figgins's ears. Ingalls, who did not know Figgins, got the unvarnished version; Howarth, Jenks, Loomis, Matthew, and Osborn heard more or less filtered accounts; Hay, who was corresponding regularly with Figgins, heard only hints of the trouble, at least initially.[270]

Cook himself was not altogether innocent of escalating tensions at the CMNH that fall, for he too was under significant strain. As Cook's daughter noted many years later, he was then "divesting himself" of his first wife (Barbour's daughter) in order to marry his second (his daughters' babysitter). She thought it "not unreasonable to suppose that, had his personal life been more serene at that time, his outlook on some of these matters might have been more balanced."[271]

To hear Cook tell it, Taylor and the CMNH board were deeply troubled by Figgins's behavior and the poisonous atmosphere between him and the staff. This prompted two actions: they administratively restructured Cook's position so he did not have to report directly to Figgins, and they temporarily exiled Figgins from the museum. Cook claimed Figgins left against his will. That cannot be corroborated, but the trip certainly materialized abruptly. Up to the last day of 1927 Figgins made no mention of any travel, but a week later he was alerting correspondents that he was about to leave for six months in South America. The trip came on so suddenly, according to Cook, because Figgins had been stirring up "every sort of serious trouble and difficulty in and out of the museum. He really was in pretty bad condition." Cook diagnosed the problem as "insanity or perversion . . . [of a] most insidious and dangerous variety." Taylor observed only that Figgins was "much run down in his nervous system & needed a change of scene and relief from supervisory duties." Figgins was hustled out of the country under the pretext of studying and acquiring birds and mammals for exhibits in the South American wing of the CMNH.[272]

Before his departure for South America Figgins sent a long letter to Taylor, detailing plans and work to be done by the staff and curators in his absence. The list included finishing the Folsom and Lone Wolf Creek bison skeletons for mounting and exchange, and seeing Hay and Cook's taxonomic descriptions into print. He ended on a melancholy note:

> Some have gone to South America who did not return. Should that occur in my case, I want you to know that I am deeply appreciative of your uniform kindness and consideration during the past seventeen years. Towards my many mistakes, you have been both kind and patient. I am grateful.[273]

Noticeably absent from his letter to Taylor was any mention of further excavation at Folsom. If fieldwork was to be conducted by the CMNH in 1928, Figgins recommended a site near Farmington, New Mexico, that offered the possibility of acquiring new species of fossil reptiles.

8.16 ONCE MORE, WITH FEELING

Figgins may not have seen the need for more work at Folsom, but Brown did. As soon as he heard Figgins was off to South America he asked about the Colorado Museum's intentions at Folsom in 1928. Figgins met with him in New York that January before sailing for South America and assured Brown the CMNH was agreeable to the AMNH taking over work at Folsom in the coming summer. That was news to Taylor and especially to Cook, who was aghast that Figgins had "practically turned the whole Folsom quarry over to Brown and the American

Museum." After a hasty meeting with Taylor, Cook let Brown know they indeed had a continued claim in Folsom and any work had to be done jointly. (Figgins subsequently wrote Taylor to say there was "an almost unanimous expression of opinion that that quarry should be worked out," though he had thought differently three months earlier.)[274]

Brown was weeks in replying, which made Cook anxious. Impatient for an answer, Cook wrote Osborn directly to make sure he understood the CMNH was not relinquishing its claim to Folsom regardless of what Figgins or Brown might say, and he pushed for an answer regarding the AMNH's intentions. Cook also let him know that the person (Cook apparently could not bring himself to use Figgins's name) "who had charge of former cooperative work, will not again be sent into the field for this institution . . . and every effort will be made . . . to send men who will really cooperate as they are expected to do, and maintain themselves in a proper way." Osborn, who likely had little idea of the trouble between Cook and Figgins, must have been perplexed.[275]

Cook's letter got results. Discussions were held at the AMNH and a week later Brown wrote to say funding had been approved for excavations at Folsom. More important, from Cook's vantage, Brown assured him "we are planning to cooperate with you . . . if the arrangement is agreeable." It was, with the proviso that Brown come through Denver in advance to "to be sure that everything possible is done to cooperate to the best advantage for all concerned."[276]

That proved to be impossible because Brown would not be able to reach Folsom until after the excavation season began, but that insured Cook had a prominent role setting out the summer's work.[277] Schwachheim was hired, and in late May Cook and the AMNH crew began excavating. Cook spent a week on-site, then he and Brown visited at intervals—Brown because he was busy with other projects, Cook because he never fully trusted Brown and wanted to keep a watchful eye on his and his museum's interests. (Over the years Cook grew increasingly vocal about Brown having usurped Folsom, his unhappiness exacerbated by later hagiographic portrayals of Brown's role at the site.)

The AMNH crew opened an area many times larger than that excavated by the CMNH in 1926–1927, the goal being to remove the "entire area underlain by bones" (they came close).[278] Throughout June and July bison remains were exposed, and along with them nine more Folsom points, several of which were left in situ. In an exchange that spring, Hrdlička recommended to Brown, as he had to Figgins the year before (§8.9), that telegrams be sent if anything of special interest appeared. It was hardly a necessary suggestion to a globe-trotting publicity hound appropriately (and literally) named after the proprietor of a three-ring circus. Barnum Brown knew "to run, not walk, to the nearest telegraph station." A score of telegrams were dispatched on July 23, 1928:[279]

> Three arrows now in place close to bones of four bison skeletons. Stop. Eleven arrows so far discovered in quarry all of same Folsom culture. Stop. Arrows will not be removed before August first please notify who is interested buy ticket to Raton New Mexico hire auto to quarry. Stop. Wire me Folsom New Mexico if coming.[280]

Those receiving the telegram included Fay Cooper Cole (University of Chicago), Cook, Hrdlička, Jenks, Moorehead, Earl Morris (National Park Service), Osborn, Roberts, Sellards, Wetmore, and Clark Wissler.

Over the next few weeks there were repeated gatherings of the scientific clan, as Cook put it. One of the more important visitors proved to be Judd, who assessed the situation at Folsom and deemed the finds of sufficient importance and enough in need of geological investigation that he urged Wetmore to request USGS geologist Kirk Bryan visit Folsom. Judd considered "no geologist better equipped than Bryan" for the task because Bryan knew "the danger of finding Pleistocene man in the Americas and he has the advantage of [an] archaeological point of view."[281] Wetmore sent the request, and Bryan, whose credentials even Cook admitted were impressive, arrived at Folsom a day later. He spent two days there, part of the time with Judd and Roberts. (Roberts had dropped everything as soon as he had received Brown's telegram and came without waiting for his supervisor's approval because he "didn't want to miss out on anything."[282])

The repeat site visit was a great success. All who saw Folsom deemed it "the most important find relating to man so far discovered in America, as the association is definite and beyond cavil."[283] After his visitors had come and gone, Brown returned to New York for the International Congress of the Americanists hosted by the AMNH. One session was devoted to "The antiquity of Man in America" and was largely a Brown-dominated presentation of Folsom. Following Brown's talk, Albrecht Penck, a German geomorphologist and Pleistocene geologist who had just been to Folsom, offered his opinion that the deposits were "ancient, but post-Pleistocene."[284]

That was not an unreasonable opinion from a geological perspective, or at least one based on a one-day site visit. On the paleontological side there was agreement that the Folsom bison was extinct and that, as Brown supposed, it dated to the "very late Pleistocene . . . notwithstanding Doctor Hay's determination of an earlier date" (Brown thought Hay's error was from mistaking a female *Bison taylori* for a *B. occidentalis*).[285] The key was to reconcile the geological and paleontological evidence. That was a challenge, as Bryan explained to Wetmore, because "we know so little about the Pleistocene faunas. Good localities are rare and those that have been exploited are not well tied to any geological chronology."[286]

To close that gap between the paleontological and geological records, Brown dispatched his crew late in the 1928 season to a large cave on a hillside south of the village of Folsom. Test excavations in the cave earlier that summer yielded what appeared to be a sloth jaw and a camel metatarsal, which Brown sent to the AMNH for vertebrate paleontologist George Gaylord Simpson to confirm. If these were, in fact, extinct Pleistocene mammals, Brown thought it their "only hope of dating the bison quarry," as the cave was in "a lava flow of the same period as the one that closes the lower end of the Folsom basin." The identifications were correct, but the cave deposits were "not prolific" and soon played out, leaving uncertain the association of the fauna with the lava flow.[287] Still, based on his estimate of the time it must have taken for the sediments of "highly restratified earth" overlying the bison bone at Folsom to accumulate, Brown was certain the site dated to the Late Pleistocene. To translate that estimate into years, he asked for the AMNH's "varve boxes and sectioning equipment" in order to determine the "time taken to decompose the twelve feet of shales overlying the Folsom [bison]." He never got the equipment, though it would not have made a difference. The bone bed was not overlain by shale that had weathered in place.[288] Even so, Brown put Folsom's age at 15,000 to 20,000 years ago on the basis of assumptions about sediment accumulation rates and the age then assigned to the Late Pleistocene period in glacial history.[289]

For his part, Bryan used remnant terraces and benches in the valley to reconstruct four main stages in the region's alluvial history—the last of which was represented by Wild Horse Arroyo. In each successive stage, rivers and streams cut below the level of older and broader valleys, leaving behind a stair-step of valley-floor remnants. The development of these "four successive stages," he estimated, "doubtless required all of Pleistocene time." The deposits in the most recent of the valleys were dominantly younger alluvium, little different from those of other New Mexico streams that contained "relics of the Pueblo Culture." But in Wild Horse Arroyo those deposits also contained a pocket of older alluvium, within which was the Folsom bison bone bed. If the younger alluvium in general was on the order of 1,000 years old, then the older alluvium at Folsom "on the ordinary criteria used by geologists . . . would be considered Early Recent or very Late Pleistocene."[290] Bryan, as he subsequently reported to Wetmore, agreed with Brown's central conclusions:

> 1. That the site was then as now a water course; 2. That the animals died at one time as a result of a "big kill" by the men who used the points; 3. That the antiquity is great but not earlier than late Pleistocene.[291]

Cook agreed that "no matter *what* the age of the deposit . . . we should do all in our power to get it *accurately* placed, if possible, and on evidence that the ultraconservatives cannot successfully attack."[292] Yet for Cook Folsom was much ado about very little. It might have "considerable antiquity" relative to the majority of other archaeological sites, but he was "confident as ever that it is far less old" than either Frederick or Lone Wolf Creek. With a striking lack of self-awareness, he mused that

> there will have to be a lot of talk and work before the Old Guard will ever believe anything for man's antiquity beyond a very short time. It is unfortunate that most Anthropologists who I have met seem to know extremely little about geology or paleontology, and to take little of any interest in any branch of natural history outside of their own line, no matter what fine men such men may be, or how well versed in their own subject, that condition necessitates a narrow perspective.[293]

This, of course, was from one who ignored the rejection of *Hesperopithecus* and concerns about Pliocene human artifacts at Snake Creek, or Early Pleistocene ones at Frederick. Even Cook could not help noticing, however, that a noose continued to tighten around each of those sites.

8.17 DEAD MEN WALKING

After his halfhearted defense of *Hesperopithecus* at the 1926 AAAS meeting (§8.2), Osborn made one last effort to keep it alive. In May of 1927 in *Science* he illustrated the development of humanity from the Pliocene to the present. At the base of the phylogenetic tree he inserted *Hesperopithecus* and its supposed fossil bone implements with their "possible evidence of [a] Middle Pliocene bone-tool age in America."[294] Within weeks of Osborn's *Science* article the *Hesperopithecus* case was sealed. Permission was finally granted to excavate where the type specimen was found (§8.2). Albert Thomson returned to Snake Creek, where Figgins, among others, hoped he would find the "skull to which those questionable teeth belong" and finally settle "one way or the other, the discussion regarding 'artifacts.'"[295]

Thomson recovered abundant *Prosthennops* fossils, but as William Gregory announced in *Science* in December, "Nearly every conspicuous character of the [*Hesperopithecus*] type can be matched in one or another of the *Prosthennops* teeth." Osborn, Matthew, and Gregory (at least initially) had been misled by the very badly worn and broken condition of the type *Hesperopithecus* molar (§8.2). This supposed anthropoid ape was an embarrassing taxonomic blunder. With Gregory's statement Osborn finally capitulated, though he tried to save face by seeking additional specimens with which to compare to what he now called the type specimen of *Prosthennops haroldcookii*. "I would like to clear up this matter myself rather than have others do it."[296] *Hesperopithecus* was dead, and no new discovery would change that fact, but that did nothing to dampen Cook's enthusiasm for Snake Creek's supposed artifacts. He had visited Thomson's excavations that summer and afterward urged Osborn to not jump ship:

> While a lot of these "artifacts" *COULD* have been made by erosion in water, it is that highly important nucleus of those that could *NOT* (save by such highly specialized conditions as to become absurd), that convinced me that there really *is* evidence here of primitive artifacts; and makes it highly probable that a whole lot of them really *ARE* what they seem to be.[297]

Osborn did not stay on board. He published nothing and privately said little more of the specimens. He died before he ever published a retraction.[298]

Nelson soon firmly closed the door on Snake Creek. He had examined all the purported artifacts Thomson recovered from the *Hesperopithecus* locality, and in March of 1928 he announced in *Science* that none of the nearly 3,000 specimens showed "positive evidence either of intentional design or of artificial workmanship." There were some eighteen specimens that somewhat resembled "accidental fragments" or "temporary" (expedient) artifacts from "cliff-dweller" sites, but that was a consequence of Thomson

culling over many thousands of fossil bone fragments. We have here, in other words, a close parallel to the selective procedure of which Europeans have made so much in the accumulation of eoliths. But, as in the case of eoliths, it is pertinent to remark that given the proper raw materials and the right natural conditions for their manipulation, nature produces many things more or less suggestive of human handiwork, and the collector by taking pains can easily gather an array of imitations which considered by themselves are sometimes deceptively impressive.[299]

On finishing Nelson's paper, Hrdlička responded with a hearty "amen" and "Thank God for this one solid and critical man."[300]

The Snake Creek "artifacts" were just waterworn bones, not that Cook was ever able to accept that conclusion or that *Hesperopithecus* was a peccary. *Hesperopithecus* "may be proved to be a peccary, a bear, a wart hog, fairy, or some other creature, more or less inhuman," Cook admitted, but he was adamant that those bone implements were humanly made. They were all the proof he needed of Pliocene or Early Pleistocene humans in North America, and as these were tools there must have been toolmakers (much as Piltdown provided support for eoliths and vice versa [§7.6]). *Hesperopithecus*, he believed, "was probably the direct ancestor of our Aftonian men from Frederick, Oklahoma!"[301]

Cook had virtually no allies in the fight. Evidently recognizing how few took him seriously on this matter, he mentioned Snake Creek only in letters to Barbour, Ingalls, Osborn, and Thomson. He never broached the topic in correspondence with Figgins, Hay, Jenks, Sellards, or any of the others to whom he regularly wrote on the topic of human antiquity.[302] And Cook stuck to his guns. Decades later he continued to insist that the Snake Creek specimens could not have formed naturally and that the "whole question as to Pliocene man or artifacts in America is far from being settled, either way."[303] He was alone in thinking so.

Frederick soon followed Snake Creek into archaeological obscurity, though here Cook had help putting up more of a rearguard action. After the *Science-News Letter* appeared in October of 1927 quoting Hay that Frederick was 500,000 years old, Charles Gould, director of the Oklahoma Geological Survey, decided to visit the site. He was accompanied by Leslie Spier, newly hired at the University of Oklahoma.[304] Hay was delighted. He had been trying for months to get a geologist "who so far has not been connected with the case" to independently confirm Cook's geological and chronological assessment (Cook had no objections), as well as an archaeologist to examine the specimens. He thought Spier "a good man," or at least supposed he might be on their side. The way Hay read it, Spier had shown that the artifacts in the Trenton gravels had a "definite arrangement" that indicated a Pleistocene age, Holmes's claims to the contrary. (Hay had misread Spier [§6.15].)[305]

Gould and Spier did not see any artifacts in the ground on their visit, and the ones they saw of Holloman's were unconvincing. They left with a large fragment of glyptodont carapace and the hope that Holloman was telling the truth about what he said he had been finding and where. Still, trust but verify. Gould wrote both Cook and Figgins to ask if they had seen artifacts in place at Frederick or could affirm Holloman's story that human bones and teeth had come from the pit.[306] The human remains, Cook explained, were actually a portion of mammoth tooth and a *Mylodon* caudal vertebra, misidentified as human by the local chiropractor. Neither he nor Figgins had seen anything in situ, but then the odds were against them since they had spent so little time on site. As for Holloman's honesty, Figgins had put it to a test:

> I went over the various items [Holloman had found] and repeatedly misconstrued what he had told me. Always he corrected me. . . . Next day I repeated like tactics to a degree that I am sure convinced him that I was exceedingly dull of comprehension.

One story Figgins repeatedly probed was the supposed finding of an arrowhead in Bed A, the layer that yielded the oldest of the fauna, including the jaw of a primitive *Elephas*. Holloman was emphatic that the artifact was so firmly cemented in this layer that it required a pack knife to free it.[307] On the strength of that testimony Figgins highlighted the specimen in his *Natural History* article.

Gould did not disagree with Figgins's appraisal of Holloman's honesty, but "we all realize that he has very little conception of the relative value of things which are important and those which are unimportant." And, of course, Holloman had plucked artifacts from the ground before scientific eyes saw them in place, and he tended to announce what he thought he had found. Those were hard habits to break, but Figgins and Gould tried. Gould assured Holloman he stood ready to drop everything and rush to Frederick if alerted to an artifact in situ, and Figgins pleaded with Holloman to "make no statements whatever, until future finds could be examined by competent judges."[308]

Cook and Figgins revisited Frederick on their December 1927 trip to Ray's sites (§8.15), and this time they saw fossil bone in place in the lower levels of the gravel bed. Cook now supposed it was only a matter of time before more artifacts appeared and would remove all doubt even from the minds of the "dyed-in-the-wool Doubting Thomases."[309] Cook had badly misjudged the Doubting Thomases.

Spier was willing to grant Cook and Hay's inference that the gravel bed at Frederick was Pleistocene in age, even that it dated to the Aftonian. And he agreed everyone considered Holloman truthful. Still, he warned in *Science* that the Frederick claim "must be received with caution." *Caution*. The same word Holmes had used in *Science* a decade earlier to attack Hay's Vero work. Whether he used it deliberately or not, in doing so Spier guaranteed he would soon be hearing from Hay. Spier raised several concerns: the artifacts had been found in the center of the quarry, and no "scientific man" had seen the overlying deposits of the original position of the surface relative to the artifacts or could eliminate the possibility that the artifacts had slid down from the surface. Some of the artifacts, particularly the metates, were possibly just waterworn slabs. Even if the stone "blades" were genuine, the site made no chronological sense. Metates were a "Neolithic art" and so too the blades. (Spier conceded they might be, at most, Upper Paleolithic Solutrean in technique.) Such artifacts were utterly incongruous with an Early Pleistocene fauna. In the Old World the oldest known artifacts, which were younger than the purported age of Frederick, were far more "roughly made." Old as Frederick was supposed to be, it implied a New World origin for humans, but if that was so, where were the anthropoid prototypes? *Hesperopithecus* was the only possible candidate, but as Spier pointedly observed it had just been given its last rites by Gregory.[310]

If Cook took the *Hesperopithecus* barb personally it went unmentioned, but he did declare he had been expecting just such an attack and was ready with new "unpublished and unannounced" evidence that would "quite effectively spike those guns."[311] Although he himself earlier admitted that artifacts may have come down from higher levels as Spier suggested, on his December visit Cook saw a way to eliminate any intrusive specimens. He pointed out that Bed C, the upper unit at Frederick, was permeated with hematite that stained everything in it a very dark red. Leaving aside the fact that Bed C itself was Pleistocene in age, if an artifact tumbled from Bed C into deeper, older units it "would certainly bear unmistakable evidence of its origin from this stain." Not only was red staining lacking on the Frederick metates, they were instead the much lighter-colored of the Beds A and B sediment.

The question of mixing settled to his satisfaction, Cook went after the larger issue. Why should one not see metates at a remote era in prehistory? After all, "need and a similar environment have caused different peoples at widely separated times and places to do similar things independently and reach similar results." It seemed to Cook no one could speak with authority on the "habits and customs" of people in Early Pleistocene time in America, let alone deny them their metates— especially not archaeologists, who studied only the comparatively recent cultures of the Americas and had not "in conjunction with geologists . . . worked out the sequence of Pleistocene events in this country, in relation to mankind." Of course, that is precisely what Spier was trying to do.[312]

Spier was not sufficiently troubled by Cook's comment to publish a reply. He instead sent a long letter lecturing Cook on what archaeologists knew (and Cook did not) about Native American cultures, human evolution (biological and cultural), and the absurdity of doing depositional and stratigraphic history by color: "One might as well argue that because the Fresno scraper left by Mr. Holloman in the exposed Permian bed is now a beautiful red that it is of Permian age." Spier circulated copies of his letter to others, including Hay and Hrdlička.[313] Hrdlička was pleased to see it, though perturbed all the same:

> We are evidently passing through another "epidemic" of claims by the paleontologists, and with some of these men the matter has become a regular obsession—I can find no better word with which to describe the state of affairs. The matter ought to be taken up energetically by our archaeologists, where archaeological evidence alone is concerned; [but] we almost seem to have no archaeologists.[314]

Hay said nothing, but he did not have to. His own rejoinder to Spier appeared the very next day in *Science*.

Spier had raised the possibility of surface artifacts falling into Early Pleistocene deposits, Cook had denied it on color, and Hay dismissed it as trivial. Who was to say specimens on the surface were not also Aftonian in age? Not that Hay thought there was much possibility that the ground had opened up and artifacts had tumbled down twenty to twenty-five feet, and were then cemented into the Early Pleistocene conglomerate.[315] To Hay that was the same tired script anthropologists had used for decades when geological results were not to their liking. Sure, Hay admitted, anything was possible and all manner of mechanisms (even whirlwinds!) might account for finding recent artifacts in Pleistocene deposits, but anthropologists bent "on making their own geology" needed to ask was that likely and had it occurred? To Hay it all bordered on the absurd:

> What is more probable than that, while men slept, a mighty wind arose and, gathering up a cache of Comanchean implements of the wigwam and the chase, hurled them with violence against the face of that quarry and drove them into the hard sand and conglomerate?

As for the incongruity between Neolithic artifacts and an Early Pleistocene fauna, which Spier supposed "seems not to have occurred to Dr. Hay," it had occurred to Hay. Where Spier was wrong was thinking Hay cared. He did not:

> There is an incongruity, but this is the creation of the anthropologists. They measure most things pertaining to human history in America by European standards. Because stone implements appear only late in Europe and are crude it is concluded that the art of working stone must have had a similar development in America.

Yet, that sequence must be wrong, given the evidence from Frederick, Lone Wolf Creek, and even Folsom.[316] Hay showed no sign of being aware his logic had just traveled in a circle.

Then there was "caution." Hay had heard enough of that word: "Our anthropological friends," he complained, always sound the warning to proceed with caution, but "do they exercise superior caution themselves?" Where was the caution when "they" (he meant Holmes) asserted without proof that the deposits at Trenton were mixed, or that the Vero skeletons were recent burials (Hrdlička)?

> Was even my friend Dr. Spier proceeding with caution when he suggested a depression where the Frederick artifacts were discovered, apparently without inquiring of the owner of the pit and of the workmen whether they had observed anything of the kind?

It was anthropologists who should be cautious about what was improbable or not, and to admit to not yet knowing "the climate of [the Early Pleistocene] nor much about the resources and arts of the people."[317]

Spier was unruffled, though this time he responded with a note to *Science*, reporting that on a recent visit to Frederick he had found "flint chips" on the surface. If Hay was correct, these were Aftonian in age and they ought to be in the lower Beds A and B as well—but they were not. Nor did the movement of artifacts from the surface to those lower strata require "a deep hollow" on top of the ridge, which, as Hay observed, "is incredible, and to which I [Spier] add, unnecessary." Spier saw no reason to change his opinion that the antiquity of Frederick was unproved.[318]

There the discussion of Frederick may have stalled, the evidence from paleontology and geology on one side and archaeology on the other, with several hundred thousand years dividing them. Each camp was convinced the other needed to concede. But then the archaeologists got unexpected help from the other side. Vertebrate paleontologist Alfred Romer, newly hired at the University of Chicago (he was Samuel Williston's replacement), reported in *Science* that July on the discovery of a camel skull deep in a Utah cave overlooking the floor of Pleistocene Lake Bonneville. The find was exceptionally well preserved, with dried muscle still attached to a portion of the skull. Yet, the cave was in lava that postdated the highest Bonneville stand, which was presumed to date to the Wisconsin. Therefore, this remarkably fresh-looking skull was likely Late Pleistocene in age.[319] If that was so, it had ruinous implications for Hay's cherished Pleistocene vertebrate sequence.

Hay had long insisted camels went extinct at the end of the Aftonian, and he had touted them as a particularly secure marker of the Early Pleistocene. But if Romer was correct, that meant this well-preserved camel skull was a half-million years old. That, Romer declared, was "utterly impossible." He added, pointedly, that it was more likely "our present views of the succession of Pleistocene vertebrate faunas [that is, Hay's] are much in need of revision." Romer was not content just calling into question Hay's decades of work compiling and synthesizing North American faunas. After the knife was plunged he gave it a vigorous twist. If camels and possibly other supposedly Aftonian forms were Late Pleistocene in age, then why should anthropologists in cases like Frederick not question the ages Hay assigned those sites?[320]

Years of battling Holmes and Hrdlička had inured Hay to such attacks, and he had learned to deflect. "My friend Dr. Romer," Hay replied, might be correct in believing this to be a fossil camel (Hay doubted it), and maybe Hay's life's work on Pleistocene vertebrates was wrong (Hay doubted that too). Perhaps Romer might even "restore the happy time when any species whatever of Pleistocene mammal might be expected to occur in any late Pleistocene deposit . . . especially if its occurrence appeared to have any bearing on human history." That, Hay knew, would be a relief to "many, if not to most, of our anthropologist colleagues." But Hay saw no reason why organic matter attached to a skull buried in a dry cave could not last a half-million years. Nor was he convinced the cave was as young as Romer supposed, ignoring the fact that Romer used Hay's estimate of the age of the Lake Bonneville high stand. Hay now put the Lake Bonneville high stand in the Aftonian. The evidence? Romer's camel. Hay had once again traveled in a logical circle.[321]

Yet, it was obvious that Romer's critique had struck a nerve. If the camel and other fauna associated with Bonneville and other Great Basin lakes were as young as Romer put them, that implied a high proportion—say 50 to 65 percent—of Pleistocene mammals survived past the Wisconsin glacial maximum and only then went extinct. Hay fairly snorted on the pages of *Science* at the absurdity of it all. If Romer had proof, he needed to show it:

The writer will continue to hope that the geologists, the paleontologists and the anthropologists who do not like his opinions on Pleistocene geology, paleontology and anthropology will speedily collaborate and impart to us their conclusions and their reasons therefor.[322]

Cook sent Hay "three hearty hurrahs and two added whoops of a right lusty order" for giving Romer his comeuppance. Cook himself had seen Romer a few weeks earlier and upbraided him "in no uncertain language." It all nicely reinforced Cook's self-image: "You and I and a few more of our ilk are not on the popular, ritual-as-accepted-by-almighty-order side of the Pleistocene." Cook was correct about the side he was on. He was less skilled at prophecy, however, assuring Hay that when the smoke cleared "it is going to be realized that you are *right* in your general contentions." Scarcely five years later Romer reexamined the Pleistocene vertebrate record, and left Hay's sequence in shambles (§9.5).[323]

By then the case for Frederick had likewise come apart. It did so despite Holloman's discovery in September of 1928 of an arrowhead in Bed A. This time the specimen was photographed in place by a friend of Gould's before being removed by Holloman (who, to everyone's dismay, still had not grasped the idea of leaving an artifact in place). But Holloman at least left that corner of the quarry untouched until Gould and Spier could visit the spot a few weeks later.[324] After examining the spot, a chagrined Spier admitted to Hay he now did not know what to make of Frederick. It appeared this artifact was in place in Bed A, though "at the same time it would be well to reserve final judgment until we are certain that the artifacts are not secondary inclusions," especially given that nagging incongruity between the artifacts and the fauna. Spier proposed "keeping after the [Frederick] case" to resolve the issue.[325]

But Hay was not going to wait on any further investigations. After a decade of baiting skeptics he finally had one on a hook, and this was news to be spread. He urged Gould to report on the site visit while he dashed off an article that included Spier's statement and explained its implications for Frederick and the larger human antiquity controversy. Gould's article affirmed that Frederick was likely 365,000 years old, just as Cook had estimated. Hay took a victory lap at Spier's expense, for in his view Spier's work at Trenton had shown the "futility" of trying to explain the presence of artifacts in Pleistocene deposits as the result of movement by natural agencies, the sort of explanation Spier had initially proposed for Frederick but was "now ready to withdraw or qualify most of these."[326]

That being the case, Hay asked, "what are the conclusions that we must reach?" It was a fair question if Hay had only Frederick in mind. But Hay leapt to a much broader conclusion, namely that Spier's Frederick concession opened the doors for Trenton, the various finds of human remains and artifacts in the loess of the Mississippi and Missouri valleys, the Natchez pelvis, McGee's Lake Lahontan biface, Vero, Melbourne, and, of course, Frederick (curiously, Lone Wolf Creek and Folsom were not included on the list).[327]

Hay hoped his and Gould's articles would be seen as vindication, yet it is not clear they were even widely seen. Unlike as in the previous year, when Frederick commanded attention in *Science*, these papers were published in the *Journal of the Washington Academy of Sciences*. Whether self-selection on Hay's part (he communicated his and Gould's papers at the Washington Academy meeting) or because *Science*'s editors declined to publish their papers is not known, but it may be telling that when Figgins submitted a brief note regarding Frederick to *Science* around that same time, it was rejected.[328] Regardless, the effect was to move what had been a national discussion out to the margins of the scientific community.

Debate over Frederick sputtered to an end, sped in part by a visit to Frederick by O. F. Evans of the University of Oklahoma's department of geology. Evans had "more experience than any of us regarding the streams of western Oklahoma." Evans showed that the supposed Aftonian fauna was not in primary context, and one might as well use the metates and arrowheads as indicators of the age of the Frederick deposits.[329] A half-dozen years later Figgins himself publicly retracted his previous statements about Frederick's antiquity:

> Since authors continue to refer to the Frederick City finds and cite publications in which they are referred to, I express a disavowal of longer attaching importance to them. This without the slightest reflection upon Mr. A. H. Holloman, who, I now believe, was the victim of a hoax.[330]

8.18 THE SOUND OF VICTORY, THE SILENCE OF DEFEAT

Even as the claims for Snake Creek and Frederick were collapsing, Cook was insistent that "at *no* time have we ever felt that [Folsom] had any such antiquity" as Frederick, though he could see Folsom had a "greater antiquity than our conservatives have been willing to admit for America."[331] As to what the conservatives themselves were willing to admit about Folsom, all was quiet until the spring of 1928, when Hrdlička spoke at the New York Academy of Medicine on "The Origin and Antiquity of Man in America."

There was little new in his presentation. He rehashed the long, futile search for ancient remains by eminent archaeologists; the evolutionary expectations that ought to be met but had not in any of these claims; his belief that there were "definite correlations between time and human culture, as well as physique;" and the scientific culture clash between geologists and paleontologists on one hand and archaeologists and anthropologists on the other. It was a divide he attributed to "potent human psychological peculiarities." Frederick Hodge, an old friend and BAE colleague now at the Museum of the American Indian in New York, was in the audience that evening and seconded the point: "I note that all of the geologists who are in favor of Pleistocene or even earlier man, always assume the highly unscientific attitude of endeavoring to prove the case without considering the evidence to the contrary. They aim to establish the existence of early man or bust their breeches in the effort."[332]

Hrdlička mentioned Folsom in his talk; it would have been too obvious had he not, but he easily danced by it as a find that would "necessitate prolonged and careful studies before any final judgment can be passed." He closed on the familiar refrain that there needed to be a "flying group or committee composed of the most experienced men in all the branches" who would be notified of any and all new discoveries and could then swoop in and examine them on the spot before the evidence was disturbed (§7.7, §8.14).[333]

As was customary, several commentators were invited to respond to Hrdlička's talk, among them Nelson and Brown. Nelson flattered: Hrdlička had given the topic "more serious study than any other man living." And Nelson agreed: Hrdlička's claim that all skeletal remains of supposed geological antiquity were "essentially modern in form" was true as well of the material culture, and that included the artifacts from Folsom. But then Nelson pivoted: fossils of anatomically modern humans in Europe were contemporary with Neanderthals, proof that skeletal forms with a "modern stamp" could have great antiquity.

The same principle did not hold for artifacts, culture being "an exceedingly plastic phenomenon." Thus, the discovery of "typically Neolithic" artifacts in America that supposedly reached back hundreds of thousands of years—as Cook, Figgins, and Hay supposed—could not possibly be correct. For if so it meant the Neolithic was much older in the New World than in the Old; that it lasted in America over vast time with no appreciable change; and that it "burst into full bloom without previous cultivation [a Paleolithic stage] or transplantation from without." It was better to assume human culture played worldwide on comparable time scales. Although that precluded a deep human antiquity in geological terms (§9.5), it still allowed the possibility of an arrival in North America "anywhere from 7,000 to 15,000 years" ago.[334]

Brown was not wasting time on a *possibility*. Holding the Folsom points high, he announced, "In my hand, I hold the answer to the antiquity of man in America. *It is simply a question of interpretation*." Not that Brown thought much interpretation was needed. These distinctive "fluted" points were undeniably associated with a newly defined extinct species of "considerable antiquity," in a geological setting that precluded "their having been made by historic men." (This may be the first use of the word "fluted" in publication.) Brown stepped through an interpretation of Folsom's geology, then estimated (citing Chamberlin and Salisbury) that "it would have taken twenty-four thousand years for this deposit to accumulate over the bison remains." That estimate, of course, was far short of the 400,000 years Figgins had published the month before in *Outdoor Life*, though to avoid embarrassing Figgins Brown made no mention of that. Deference only went so far, however. Brown reported that since their acknowledgment Folsom points were now recognized in museum collections from across North America. Figgins was wrong, Judd had been right (§8.9).[335]

Afterward, Hrdlička asked Brown for a copy of the photograph that showed a view of the Folsom site and particularly the "little ravine." He also suggested that if further work was done there Brown ought to clear the wall twenty feet on either side of where the bones and artifacts occurred to expose the stratigraphy, and then he ended by asking Brown to "give me your views as to how the artifacts may have come into association with the bisons [*sic*]." Brown replied with details of the site's geology and then assured Hrdlička that he "personally removed the five inches of stratified clay that immediately covered all of the fifth point . . . and this work was not done until I had made an exhaustive examination of the entire quarry. There is absolutely no possibility of any introduction of the points subsequent to the natural covering over of the bison skeleton." Hrdlička was not satisfied by that response. "There is only one additional point on which I should like your opinion and that is, the manner in which the arrow points or darts got into these places where they were found." With exasperation creeping into his tone, Brown asked what Hrdlička thought of the photograph he had sent that showed "one of the darts lying *in situ* between the two ribs." Brown was of the opinion that at least "three of these points were embedded in the animals when the carcasses were entombed." "Please write me," he added, "if I have not made the situation clear." Hrdlička backed off: "Thank you for your supplementary letter which is very satisfactory."[336]

But not entirely. Hrdlička had received the photographs Brown had sent, but not the one of the "little ravine" he had first requested. He sent another request, but Brown either never received it or failed to send that specific photograph. Several months later Hodge wrote, wondering whether Lone Wolf Creek, Frederick, or Folsom had been examined by any anthropologists besides Spier. Hodge thought every reported discovery needed to be followed up "from an archaeological point of view, as in the past," as "They all look so convincing *at first*!"[337] Hrdlička replied the finds had not been examined "except superficially" by anyone outside of those directly concerned (that was true), then added:

A curious incident happened to me with Barnum Brown. He showed, you may remember among others a picture that gave a general view at Folsom, and as this picture showed some suggestive conditions, I wanted a copy and so wrote him. In answer he sent me four other pictures quite useless, without mentioning the one I wanted; and when I wrote again specifying clearly what I was after, there was no answer. Such things naturally make one suspicious. There are many points about the Folsom find which need explanation; but they are not the points touched upon in the talks or reports on the find. Hay, Gidley, Barnum Brown, Harold Cook, Figgins, and Renaud—what a constellation of blossoms.[338]

Of course, within a couple of months Roberts, Judd, and Bryan were at Folsom making more than a superficial examination of the site (§8.16).

When Figgins returned from South America he could see Folsom "has gotten itself a reputation."[339] It was one not even Hrdlička could shake, though he took a swing at it. In October 1928 the *Science News-Letter* included a note on Brown's excavations that summer. Asked for comment, Hrdlička said Folsom had five weaknesses: Indians rarely killed game in large quantities; they made use of all parts including the bones; the artifacts' "strangely fine" workmanship could not be very ancient; the bison were likely mired at the spot trying to drink at the stream; and the "arrow" points might have been thrown into the stream much later by Indians making religious offerings.[340] Curiously, Hrdlička did not contest the geological antiquity of the bison, only their association with the artifacts. That battle was already lost. Kidder had seen to that.

A year after Folsom's debut at the AAA meeting it once again occupied center stage, this time at the December 1928 joint meeting of the AAAS and the GSA, held at the AMNH. Brown, as a host and member of the AAAS Special Committee on Entertainment, displayed one of the Folsom bison skeletons, along with a block with one of the points recovered in situ that past season. At a joint GSA-AAAS session, Brown spoke on "Folsom Culture and Its Age" and was followed by Bryan, who concluded on the physiographic evidence that the age of the sediments containing *B. taylori* and the Folsom artifacts "must be late Pleistocene or perhaps early recent." That agreed "reasonably well with the paleontologic argument of Mr. Brown."[341]

Hrdlička was at the meeting, though there is no indication he was present for Brown's talk. If he was, it is doubtful he said anything. For after probing for weaknesses in the Folsom case that spring, and following his few desultory comments in the *Science News-Letter* that fall, Hrdlička lapsed into complete silence about Folsom. He did so despite writing several articles over the next decade on human antiquity in the Americas. He avoided addressing Folsom by staying close to his customary script, stating there was no evidence of a pre-*sapiens* human skeleton in America. That was literally correct, if beside the point. Hrdlička did not explicitly speak to Folsom until 1942. On that occasion, a year before his death, he did not reject the site but merely offered the opinion that the Folsom bison may not have been extinct very long on an *absolute* time scale.[342] That too was true, and also irrelevant. By 1942 there was ample archaeological evidence to indicate humans were in North America in the Pleistocene (§9.6).

As for Holmes, Hay was certain he would attack Folsom.[343] But Hay had misjudged his longtime adversary, now retired from archaeology. To be sure, in the mid-1920s Holmes had reacted angrily to the work of Sellards at Vero and Loomis at Melbourne (§7.9), but by decade's end he had divested himself of any intellectual investment in the controversy. Fittingly, it was Sellards who broached the subject of human antiquity with Holmes in early 1930:

> At the time I was working on the Vero material in Florida I had one or two letters from you. . . . You felt very strongly that man could not have been in this country as early as the Pleistocene. It has now been nearly fourteen years since that work was done in Florida and it has been followed up, as you know, by Loomis and Gidley. In the meantime, a number of localities have come to light in this part of the country which may require careful and detailed consideration. I am wondering, therefore, what may be your feeling towards this problem at the present time?[344]

Holmes was gracious, if disingenuous, in reply:

> I remember taking part in the early discussion of the Pleistocene formations in Florida and my rash attempts to follow this discussion without actual personal knowledge of the geological formations. My discussions related only to my fear that the explorers were committing themselves to definite conclusions without sufficient knowledge of the dangers of misinterpretations. . . . I have now dropped the matter entirely, and am perfectly willing to accept the conclusions of the skilled men who are carrying on researches with the full

knowledge of the problems and the dangers. I wish them every possible success, and have no trace whatever left of the vigorous antagonism that arose from my early battles with the advocates of Paleolithic man and culture of Eastern United States.[345]

In the end, what is most significant is not what Holmes or Hrdlička said about Folsom but what they did not say. After decades of very publicly attacking any and all claims for a deep human antiquity in the Americas and tormenting those who made those claims, the silence of both Holmes and Hrdlička regarding Folsom spoke volumes. No mea culpa was offered, but then no one who knew them would have expected one.

CHAPTER NINE

Fast Forward, 1930–1941

IN NOVEMBER OF 1930 Oliver Hay passed away. He had lived just long enough to see his and Harold Cook's descriptive analysis of the Folsom, Frederick, and Lone Wolf Creek faunas appear in the *Proceedings of the Colorado Museum of Natural History* (§8.11). Cook's own affiliation with the CMNH had not even lasted that long, having ended several months earlier in an explosion of anger and accusations, the victim of his and Jesse Figgins's mutual mistrust and loathing (§8.15).[1]

With Hay's death, the passing that fall of Cook's teacher and longtime friend W. D. Matthew, and his relationship with Barnum Brown irrevocably damaged from their tussle over Folsom in 1928 (§8.16), Cook saw he was running out of allies. He pulled Smithsonian paleontologist J. W. Gidley closer, the two bonding over a lawsuit Gidley had against Figgins for medical costs incurred in a car accident in Denver (Figgins's wife was driving and Gidley's wife was badly injured).[2] He curried favor with Charles Gould and E. H. Sellards and he flattered his old mentor Henry Fairfield Osborn.[3] Cook was reaching out because he was worried that "prominent anthropologists and scientists" believed paleontologists had "not proved their case in relation to the succession and development of vertebrate life in America, since the opening of Pleistocene times." He was particularly anxious to eliminate any skepticism about the ability to separate Early from Late Pleistocene faunas. That distinction, of course, rested almost entirely on Hay's work, but with Hay gone Cook knew few would take his own word on the subject. He urged Osborn in the spring of 1931 to prepare an article on Pleistocene vertebrate history for *Science*, in order to reach a wide audience of "the highest standing." "You are in the strongest position," Cook cooed, "to write forcefully and with authority on the value and significance of such evidence, of any person living in America today."[4] Osborn declined.

Unfortunately for Cook, Alfred Romer had taken up Hay's challenge, made amid their debate over the age of the Utah camel (§8.17), to assess the entire Pleistocene vertebrate sequence. "As you may guess in advance," Romer told Cook, "my conclusions are radically different from those of Dr. Hay and I am extremely sorry this valiant old warrior is still not here to battle for his own point of view." The valiant old warrior's long-held belief that most Pleistocene species went extinct at the end of the Aftonian (§7.3-§7.5, §8.6) would be the principal casualty of Romer's "trenchant" assault.[5]

Romer's revision of Pleistocene vertebrate history was one of many significant developments in the years between Folsom and World War II, when research effectively ceased in most corners of science (save that related to the war effort). Either originating from outside of archaeology, as with Romer, or from within the discipline, these developments caused a sea change in the understanding of "early man" in America.[6] By 1941 many of the facts, ideas, and concepts that guided research and interpretation over the next half-century were put in place. And it happened because of what Folsom set in motion.

Folsom inspired dozens of investigators and several well-heeled institutions (most notably the Carnegie Institution of Washington [§8.10]) to search for early sites. Their efforts were highly successful, for Folsom offered a tangible lesson about search strategies. To find sites like it one needed to find large mammal bones, then carefully comb the localities for associated artifacts. That lesson was readily applied on the Great Plains in the 1930s as Dust Bowl drought and erosion exposed deeply buried Pleistocene-age deposits in arroyo channels and lake beds, and revealed bones of bison and mammoth. At nearly three dozen of those localities artifacts were associated with the bones, an apparent pattern of big-game hunting that strongly influenced subsequent interpretations.[7]

Folsom also meant, as Kirk Bryan put it, that the idea of a Pleistocene human antiquity in America "lost its heretofore hypothetical character."[8] With that, and the rapidly expanding corpus of data on those Pleistocene occupations, archaeologists and nonarchaeologists alike turned to broader questions related to the peopling process, such as where the newcomers came from; how often and by what route or routes they might have come; and what their strategies of "sustenance and maintenance" might have been.[9] The years that followed Folsom witnessed a remarkably rapid growth of knowledge about what Frank Roberts in 1940 called the "*Paleo-Indian*" occupation of North America. (The name stuck.)[10]

Archaeologists took the lead, as they had not before, in excavating these early sites. As it became apparent that not all points found with large mammal bones were the Folsom type, they sought to distinguish their diagnostic features and learn what they might of the relative age and geographic distribution of these materials, and what their sites, tool assemblages, and faunal remains might reveal of adaptations on a landscape where glacial ice still lingered to the north. Over this same period paleontologists redoubled their efforts to determine the timing of Late Pleistocene extinctions and perhaps its cause(s), and geologists parsed the geography of glaciation and evidence of Pleistocene climate for what these might reveal of the one or more routes taken by early peoples, and through breathtakingly long-distance correlations more precise ages for the sites. Aleš Hrdlička briefly landed back on center stage in 1937 as he and other physical anthropologists used recent discoveries of human remains to debate issues of race, evolutionary change, and population history. Even social anthropologists such as Alfred Kroeber and Franz Boas, who had previously scarcely noticed the school yard brawls of their scruffy archaeological brethren, now spoke knowingly of Folsom and took stock of the implications of America's newfound antiquity for larger issues of *their* concern—issues of race, language, and culture.

These matters were aired in national and international symposia devoted to early man in America, notably at the AAAS meeting in 1931 (the topic chosen to honor Boas, then president of the AAAS), the Fifth Pacific Science Congress in 1933, the 1935 meeting of the American Society of Naturalists, and the 1937 International Symposium on Early Man. (These were in addition to the meetings that took place in 1927 and 1928 in the immediate aftermath of Folsom [§8.14, §8.18].)[11] The issues were discussed at length in scores of articles and books and even became a subject worthy of a graduate degree to a new generation of researchers.

Human antiquity in America, a topic that just a few years earlier had been "virtually taboo,"[12] was now one that few could stop talking about. Not surprisingly, many now speaking and writing were newcomers with no residual animus from the long-running controversy. Moreover, many of the key players of past disputes had finally left the stage. Voices of participants that had been loud, often strident, and occasionally even "merciless"[13] through more than three decades of controversy were now forever silent, or otherwise much subdued even if still stubbornly unrepentant (Aleš Hrdlička).[14]

The aim of this book is and has been to understand the human antiquity controversy and its resolution at Folsom, rather than what emerged from it. Yet, it is valuable to summarize what was learned by the eve of World War II about the origins, antiquity, and adaptations of what were by then called Paleoindians; how it was learned and by whom; and what it was thought to mean in the broad sweep of the peopling of North America. Doing so highlights what had *not* been known and could not be answered or even asked through the previous decades of controversy, and shows just how rapidly knowledge could be, and was, created in the wake of resolving the fundamental antiquity issue. A detailed narrative of all the discoveries, research, discussions, and debates of those years is not required for such a summary, but a bit of historical detail is necessary to lay the groundwork for a more thematic discussion of the period. Accordingly, this chapter begins at the Clovis site on the High Plains of New Mexico, the first and unquestionably most important site discovered in the wake of Folsom.

9.1 LINING UP THE SHOT

Leading the Clovis investigation was Edgar B. Howard, a graduate of Yale's Sheffield Scientific School and a World War I veteran who had retired in 1928 (at the age of forty-one) from a highly successful career in the import-export business. In addition to a stately home in Bryn Mawr on Philadelphia's Main Line, he owned a ranch in the Texas Panhandle. Being interested in the archaeology, geology, and paleontology of the region, he had keenly followed the work at Folsom. On retiring he began graduate study at the University of Pennsylvania, where he earned a masters in anthropology in 1930 and a PhD in geology in 1935.[15]

In 1930 Howard began fieldwork in the Guadalupe Mountains of New Mexico. Although he was not looking for early remains, he found them in Burnet Cave, where his excavations uncovered Basketmaker (a pre-Pueblo culture) material along with, anomalously, bones of several extinct large mammals. He showed these to A. V. Kidder, who realized the bones must have come up when the cave's Basketmaker occupants disturbed underlying cave sediments. He advised Howard to dig "clear to the bottom" of the cave. Howard did the next summer, finding some four feet below the Basketmaker level the bones of extinct Pleistocene horse, camel, musk ox, a four-horned antelope, and, surprisingly, a projectile point that looked "very much like a Folsom point." At the 1931 Pecos Conference a few weeks later he showed the point to Kidder, Clark Wissler, and others, and all affirmed his identification. Kidder declared the find "the most important discovery in North America this year," providing as it did "the first definite stratification [of Folsom] below the Basketmaker level."[16]

Howard planned to meet Barnum Brown after the conference for a visit to the Folsom site. Now, with a point and Pleistocene fossils to show, Howard hoped he might persuade Brown to return with him to Burnet Cave to excavate more of the deposits and identify the faunal remains.[17] It did not take much convincing. A week later he and Brown were in Burnet Cave, and after seeing the site and recovering more of its fauna Brown agreed to join in fieldwork there the following summer.

The 1932 excavations in Burnet Cave proved less rewarding. More Pleistocene mammal bones were recovered, some charred and broken in suspicious ways, but no additional artifacts. By then Howard's relationship with Brown had soured. The publicity-hungry Brown (§8.6, §8.11, §8.16) had claimed in an interview in *Popular Science* that *he* had discovered the Burnet Cave projectile point, despite being in Montana when the find was made. Not content to just steal credit for the discovery, Brown also accessioned the Burnet Cave faunal specimens Howard sent for identification into the AMNH collections. This was particularly galling to Howard, who was a member of the Academy of Natural Sciences of Philadelphia (ANSP) and was trying to help revive its paleontological program and collections. (The ANSP's nineteenth-century glory days had long since passed.) He was also trying to bolster the archaeological visibility of the University Museum at the University of Pennsylvania. Brown's hijacking of the Burnet Cave collections was sabotaging those efforts.[18]

That killed only Howard's desire to work with Brown. It had no effect on his "driving mania" (as his student Loren Eiseley later disapprovingly called it[19]) to find early human remains in America. Even better, while at Burnet Cave Howard met A. W. Anderson, an artifact collector who reported seeing horse, bison, and mammoth bones and large spear points in old lake beds exposed in dune fields south of Clovis, New Mexico. After finishing at Burnet Cave in 1932, Howard traveled to Clovis to meet Anderson and two other collectors, George Roberts and Ridgely Whiteman.[20] As a teenager three years earlier, Whiteman had written Alexander Wetmore, trying to interest the Smithsonian Institution in working at Clovis. A paleontologist sent by Wetmore made a halfhearted inspection of the site but deemed it of little interest.[21] Howard, on the other hand, saw its potential and asked Roberts to watch the site. Howard left for Philadelphia a few days later carrying nearly a dozen Folsom points (including some from Whiteman's collections) and having seen many more on his visit. Although ostensibly collaborating with Brown, he was coy about this new find: "I have not given out any of this information yet as it seemed best to keep it quiet." Brown pushed for more information, but Howard stayed tight-lipped.[22]

That fall Howard heard from George Roberts that a road construction company had begun actively mining one of the lake beds near Clovis for gravel.[23] He asked Roberts to watch "the steam shovels at work" and alert him if any archaeological material turned up. In the meantime, Howard crafted an ambitious plan for the investigation of early man in America, one that would operate under the joint auspices of the ANSP and the University Museum. As he explained it to Charles Cadwalader, ANSP president, they would initially seek to cultivate and grow the "well recognized" public interest in the topic using the ANSP's collections—which included the venerable Natchez pelvis and associated *Megalonyx* bones (§2.2, §6.12)—to revamp its Pleistocene mammal displays and to make the ANSP and the University Museum research destinations. Fieldwork was a vital part of the plan. Under the auspices of these institutions Howard planned to continue his investigations in New Mexico.[24]

These efforts were seen as the first steps toward a larger and more ambitious goal: the creation of a national committee of archaeologists, physical anthropologists, geologists, and vertebrate paleontologists who would crisscross the country visiting "finds of importance" (Howard had carefully prepared a list of those) and bring the "whole question up to date in an impressive report reviewing all the evidence secured." Obvious candidates for such a committee would be Ernst Antevs, Kidder, and possibly even European scholars such as Sir Arthur Keith or the Abbé Henri Breuil. Just not Hrdlička, whom Howard thought too stubborn and biased. The credit-robbing Brown was not even mentioned. All this should begin soon, Howard urged, to get a jump on other institutions. If they were successful, "it does not require much imagination to see Phila[delphia] become a centre for the study of such subjects."[25]

They did not jump soon enough. Ten days later Howard had to relay "some very bad news." The idea of a national committee to investigate potential Late Pleistocene sites had been scooped with the formation of an American Subcommittee of the International Commission on Fossil Man, to be chaired by John Merriam of the Carnegie Institution and including Brown, Frank Roberts, Kidder, Bryan, Chester Stock, Nels Nelson, and others.[26] Many were on it, just not Howard. "I feel like a balloon," he groaned, "that has been stuck with a pin." He was particularly vexed that Brown had never said a word of those plans to him. Howard's only comfort was that it must have been a good idea if others thought so (Hrdlička, of course, had lobbied for years for such a committee [§7.7, §8.14]), but that was meager consolation indeed.[27]

Good news came soon enough, however. In mid-November Howard got word that a rich vein of large fossils was exposed in the Clovis gravel pit. He boarded the next train west.[28] When he reached the site a few days later he saw in a deep exposure a layer of blue sediment "literally full" of large mammal bones, along with stone flakes. He immediately telegraphed Horace Jayne (director of the University Museum) with the good news: "Extensive bone deposit at new site. Mostly bison also horse & mammoth. Some evidence of hearths along edges."[29]

Having lost the lead in forming a national committee, and wary of Brown or anyone else poaching the site, he wasted no time marking his territory. He tied up exclusive permission from the landowner to excavate the coming year, then boarded a return train and started writing. Immediately after reaching Philadelphia he sent a note to *Science Supplement* (it had a faster turnaround than *Science*, its parent publication), which appeared just two weeks after he had boarded the train to Clovis. Howard carefully crafted the notice to establish his priority ("the importance of this locality was recognized by Mr. Howard in the course of his explorations last summer"); his ongoing efforts at the site ("The [recent] discovery of man-made objects associated with fossil animal remains, made by road-builders working near Clovis, New Mexico, has led the Philadelphia Academy of Natural Sciences and the University of Pennsylvania Museum immediately to send to the site Edgar Howard"); and his institutions' intentions ("the University of Pennsylvania Museum and the Philadelphia Academy of Natural Sciences had been developing plans to study the archeological and paleontological problems of this site near Clovis next spring").[30]

Howard may have been relatively new to archaeology in 1932, but he was savvy in other ways. Once he had staked his claim he began a calculated courtship of Merriam at the Carnegie Institution, using Kidder (director of the Carnegie's Division of Historical Research [§8.10]) as his go-between. Howard wrote Kidder for advice on how to launch an interdisciplinary project at Clovis. The Clovis investigations, he professed, were "too big for him to handle alone." Kidder suspected that Howard was asking for more than just advice, but he liked him, and thought he had done good work at Burnet Cave. On a trip to Washington in early December Kidder stopped in Philadelphia to talk with Howard about Clovis. Impressed by what he heard and the artifacts he saw, Kidder spoke the next day to Merriam about the possibility of the Carnegie Institution cooperating in the venture. He sold the idea well. Merriam agreed to help, provided the project seemed "desirable."[31]

Judging that required digging deeper. Kidder returned to Philadelphia a week later, this time with Stock, a Carnegie Institution research associate and Merriam's longtime collaborator (and on whose judgment he often relied [§8.3]). Afterward, Kidder sent Merriam a long, candid letter, little of which was about Clovis. To be sure, he mentioned his and Stock's favorable opinion of the site's potential, but he focused more on Howard—his expertise and experience, making particular note of the fact that Howard was not the grasping type. Owing to his business success, he had "acquired a competency," retired to pursue scientific work, and drew no salary from the ANSP or the University Museum (in fact, he contributed to both). As for Cadwalader and Jayne, the former was a scion of an old Philadelphia family, a businessman, and a "booster and a go-getter with many of the somewhat jarring traits of such persons." Though mostly concerned with restoring the ANSP's luster, Cadwalader (an accomplished amateur ornithologist) had scientific interests at heart. Jayne's aims were similar to those of Cadwalader, though as an art historian by training he was "much more of an intellectual" and his institution "much more solidly entrenched in modern Philadelphia."[32]

Kidder thought these details important to pass along, as they potentially bore on the success of a collaboration (and perhaps on Merriam's decision whether to collaborate). He had seen joint projects go "on the rocks because of failure properly to envisage and provide against difficulties arising from division of material, apportionment of credit, etc." and from conflicts among parties involved. That was precisely the reason, Kidder told Merriam, why Brown and the AMNH were not in the picture: Howard saw Brown as

> over-anxious for publicity and much too grasping of material. Cadwalader, you will find, is pretty hostile to the American Museum. Apparently he [too] has been badly bitten.[33]

Not wanting to be tarred with the same brush as his fellow paleontologist, Stock had assured Cadwalader he was anxious to see the restoration of the ANSP's work in paleontology and was sure any division of the fossil spoils would satisfy all concerned. Cadwalader was so touchy on the subject that Kidder privately assured him Stock meant what he said, in addition to being "the best bet on the paleontological side."[34]

Surprisingly, it proved far easier to talk about money than about the disposition of fossils. Kidder expressed doubts the Carnegie could help much financially, and he would not commit to anything more than the possibility that Merriam might be willing to help raise funds. And the best way to convince Merriam that the effort was worthwhile, Kidder explained, was for the ANSP and University Museum to demonstrate their bona fides by getting the Clovis project off the ground. Cadwalader agreed: they would fund the first season of work. He would not give amounts, but Kidder doubted it would be much, privately telling Merriam that both Philadelphia institutions seemed "exceedingly hard-up . . . even desperately so."[35]

Cadwalader and Jayne had hidden their cards well. Their city and institutions suffered financially during the Great Depression (on his earlier visit to meet Howard, Kidder passed "scores of hopeless-looking men in the sun on the bridge, arms in the railing, watching the freight cars being shunted and the cars, many of them, loaded with Christmas trees"[36]). Yet, unbeknownst to Kidder, they had provided $1,500 to Howard for fieldwork the year before and would increase that to $2,000 the coming summer.[37] These were ample sums and more than Howard needed (he returned money at the end of each season).

The courtship of the Carnegie Institution was not about money; that was more an incidental benefit. Howard was a shrewd businessman who understood the concept of a loss leader, and he was using this opportunity to ingratiate himself with Kidder and Merriam. That had the less tangible but no less important goal of conferring credibility on the Clovis venture, enabling access to Carnegie research associates like Stock, and perhaps gain him membership on the American Subcommittee that had launched without him. Both sides knew it was too soon to enter a formal agreement (Kidder did not want to tie Merriam's hands), but they agreed on the next steps. Stock was en route back to California and would try to stop at Clovis to gain a better sense of its paleontological potential. Meanwhile, Howard, Cadwalader, and Jayne planned a trip to Washington the following week to make their pitch directly to Merriam.[38]

Howard picked his target well. Merriam had no interest in collections or inserting himself into the project and certainly did not want to diminish the enthusiasm of other institutions. Rather, he wanted the Carnegie and the American Subcommittee to collaborate in order to insure that the results of the Clovis work were as robust as possible. Given he had no ambitions beyond that, Merriam privately supposed this would insure that their holding Howard "to the highest [standard] of work will not be misunderstood."[39]

The meeting in Washington between Merriam, Howard, Cadwalader, and Jayne went well, though, as Kidder had warned, Cadwalader was "jarring" in his aggressive promotion of the ANSP. He opened with a territorial blast that *they* "had decided to do this work, that it was important to maintain paleontology at the Academy, and that they had already planned to secure the services of Stock and [Caltech's John] Buwalda before they learned of the American Committee." As a bemused Merriam listened, Cadwalader announced *he* expected "to control the operation and the publication, together with whatever relates to the problem," and indeed to "control the general study of early man in America."[40] Merriam hardly felt threatened, knowing he could walk away from the table at any time. Howard was relieved that Cadwalader and Jayne had to return to Philadelphia before Cadwalader could cause any real damage.

Howard stayed for an hour afterward. He assured Merriam that he would be glad to work on behalf of the American Subcommittee and that—Cadwalader's posturing notwithstanding—he (Howard) was certain a mutually agreeable plan could be worked out that would "not make too big a hole in the armor of pride and tradition that seems to surround all our institutions here in Philadelphia (?!!!)."[41] Merriam was sold, writing Kidder "I have agreed to do the utmost to aid with the Clovis work and have given Howard every assurance of assistance so far as we are able in this study." He expected that to take the form of covering Stock's expenses at Clovis the following summer.[42] Howard's scheme had worked beautifully.

Which is not to say Merriam was duped, or that his own motives were altruistic. The Carnegie Institution was supporting that summer's International Geological Congress (IGC) in Washington, which Merriam saw as an opportunity to advance "knowledge and the interpretation of early man in America." He had arranged to bring Arthur Smith Woodward, keeper of geology at the British Museum, to the IGC.[43] Merriam thought Woodward "the leading student of earliest man in England"[44] and wanted him to see some of America's early sites. He had already arranged for Woodward to join him on detour from the post-IGC meeting field trip to visit Folsom. After talking to Howard, Merriam realized that if all went well at Clovis a visit there might be preferable. He said nothing of the possibility to Howard, but it is telling that he also made no arrangements for anyone associated with the Folsom investigations to plan a meeting at that site.[45]

After their conference Merriam invited Howard to make suggestions regarding the American Subcommittee's research agenda. Howard's previous efforts to form a national committee turned out not to have been for naught, and he revised and updated his plan for Merriam.[46] As before, it seemed to Howard the most important task was to first unclutter the landscape by taking a hard look at the dozen or so discoveries that had been made in recent years. These sites needed to be reconsidered

not with any idea of proving that man was geologically ancient in America, but that it is probable that he existed, at least, at the end of the last glacial period. This is quite different than trying to bridge a gap of many more thousands of years and placing man in this country contemporaneous with Neanderthal man in Europe.[47]

It is difficult to imagine that in saying so Howard did not have the likes of Cook, Figgins, and Hay in mind, and indeed he singled out Lone Wolf Creek, Frederick, and Cyrus Ray's finds near Abilene as worthy of the committee's critical attention. These needed to be accepted or rejected as Late Pleistocene in age and not left in Aftonian limbo.[48] So too, new discoveries made since Folsom needed to be examined.

En route to the field in late April of 1933, Howard started the process. He visited Columbus, Ohio, where he was stunned by the hundreds of "Folsom type" points the Ohio Historical Society's Henry Shetrone had recorded across the state.[49] Then it was on to Lincoln to visit Erwin Barbour and Bertrand Schultz, who had been excavating bison- and artifact-bearing sites in western Nebraska, followed by a visit to Denver to meet with Figgins and Etienne B. Renaud (archaeologist at the University of Denver), who had sites in Colorado that produced Folsom and unfluted Folsom-like points that Renaud recently named "Yuma" points.[50] In Lincoln and Denver the conversation turned on what the relative age of these artifact types and sites might be. It was "beginning to look like a race to see who can arrive at definite results first," but Howard was confident he would win: "We have a pretty good start at Clovis."[51]

9.2 "SCATTERED AROUND LIKE A DOG BURIES BONES"

Howard reached Clovis in mid-May and with a small crew (including Whiteman) began scouting the exposed dry lake beds set within a dozen-mile stretch of Blackwater Draw, a sand-choked tributary of the Brazos River, as well as exploring the gravel pit that was located just off the Draw proper (the site that became the Clovis type locality). In the gravel pit and in many of those lake beds he saw the same "blue dirt" which he thought felt and looked like volcanic ash. (He variously referred to it as "blue clay" or "blue sand" because its texture varied.) It was rich with artifacts and large mammal bones, which he and his crew began to uncover.[52]

Stock arrived in late May on a day of howling winds and a "raving [*sic*] sandstorm." (Dust Bowl storms escalated sharply that summer.) Despite the miserable conditions, he and Howard examined several localities where bones and artifacts were in situ. Stock was especially intrigued by the blue sediment, recognizing it and the underlying sands as having been deposited during a wetter "pluvial period [under climatic] conditions quite different from those prevailing in the area today." Howard had found Folsom points on the floor of several of the lake basins and believed these were associated with bison and mammoth remains. Stock agreed. But because these were found loose on the surface they did not "furnish the right proof." That had to come from opening "an undisturbed section."[53]

To approach the problem "from three sides, archaeologically, paleontologically and geologically," Stock proposed that Howard excavate in the gravel pit and open an area of about twenty by twenty feet. That ought to be large enough to yield a representative sample of fossil bones (and in sufficient numbers to satisfy the institutions dividing the spoils), clarify the stratigraphic context of the remains and their mode of accumulation, and, it was hoped, reveal any artifacts in association with the extinct animals. Howard agreed with the plan, though he could not shake the feeling he was "looking for a needle in a haystack."[54] Fortunately for him, the haystack proved to have many needles.

During his visit Stock was guarded about what his own role might be. Afterward, however, he told Merriam he was prepared to devote his entire annual Carnegie research stipend (about $175) to send three of his graduate students to Clovis in July to help the excavations, and he requested an additional $225 to cover the full costs. Merriam made the extra funds available and planned his own visit to Clovis.[55]

Alerted that help was on the way, Howard decided to postpone the gravel pit excavations until Stock's students arrived. He and his crew instead spent June in the lake basins plaster-jacketing bones and searching for artifacts.[56] Merriam visited in late June and while there uncovered a point eroding out on the floor at Beck Forest Lake.[57] It was not the evidence Howard wanted, but Merriam could see the potential in the number of artifacts Howard had already recovered and the richness of the fossil record. He requested Howard provide a paper on Clovis for the IGC meeting.[58] He also asked Howard to go to Folsom in early August to meet the IGC postconference field trip participants to discuss his Clovis work.

Howard's paper, though not published for several years, reflects his understanding of the site as of early July of 1933. By then he had a sense of the stratigraphic sequence and how he might tie together the Clovis area sites. The key was the blue sediment, which proved to be a lacustrine diatomite. (The sandier portion represented a near-shore facies.) As exposed on the walls of the quarry pit, the diatomite was three to four feet thick and contained large quantities of bones, primarily bison and mammoth, which seemingly diminished in frequency with depth. The diatomite was underlain by various layers of pink and yellow or "speckled" sand and gravel, which contained horse, peccary, and turtle bones. Below that was a coarse commercial-grade gravel (the target of the quarrying). The sequence of diatomite over sand in the gravel pit appeared identical to that in the lake basins. There, erosion had left pillars of diatomite standing as erosional remnants, and in "every one of the lakes," Howard reported,

> teeth, tusks, or bones of mammoth and bones of bison have been found in and on top of the blue sand. . . . Most of the bones form a matted mass, such as might be produced around a water hole where animals, in a mad desire for water, trampled one another into the mud.[59]

The bison was yet unidentified, but Howard supposed the mammoth was *Elephas columbi* (Columbian mammoth) and the horse *Equus pacificus*. Obviously these were extinct. Howard was astonished they were so near the surface.

Artifacts were also found in the diatomite, mostly "typical Folsom points with grooves down the face and the characteristic fine flaking," but also "others without the grooves but with equally fine flaking." Unfortunately, none were found in situ in the gravel pit, only "a few flint chips and flakes were directly associated with bison bones."[60] Howard soon realized that to secure evidence of a Pleistocene antiquity at Clovis, he had to lower the bar of proof he and Stock originally sought.

That meant relying on evidence from the lake basins, where they were having more success finding artifacts associated with bones. In one of those, Anderson Basin No. 2, there was a large "island" of blue sand. When he showed the spot to the newly arrived Caltech students, their senior member, Francis Bode, spotted a point eroding from it just a few inches from a mammoth tusk Howard had shellacked days earlier. Howard was glad he had not seen the point himself, as "it helped to convince Bode that there was something to this whole thing." That same day one of Howard's crew "found a hearth at the bottom of the blue, bone-bearing sands, with a scraper, a handful of flint chips, and what is probably a point in place."[61]

The bones and artifacts in what Howard called "Erosion Island" were left in place, and he wrote Merriam and Stock to alert them that what they had been looking for had been found. But how could that be? It was not the point in situ with extinct faunal remains in the previously undisturbed quarry pit, but Howard declared it made no difference because "the formation [diatomite] can be correlated with that in the pit." Stock's men agreed and made plane table maps of the gravel pit and Anderson Basin to correlate the stratigraphy of the two areas.[62]

Merriam was delighted. And as Clovis had the advantage in being actively worked with in situ material, he quickly rerouted his post-IGC field trip from Folsom to Clovis, despite the complications that caused with the meeting being just two weeks away.[63] Merriam urged Howard to leave everything in situ until he and the other visitors arrived, in order that they might see "whether there is complete justification for acceptance of the idea of the association of artifacts with remains of the elephant and other extinct forms."[64] Once the distinguished visitors witnessed finds in place, he anticipated announcing their opinions, and "the clearer the data and the better the authority for statement, the more important will be the announcement when it comes."[65] The chance to make that announcement, Merriam admitted, was "one of the principal reasons for inviting Dr. Woodward."[66] The need for European approval of American sites (§3.6, §4.14) had not entirely vanished.

By late July Howard had "exhibits scattered around like a dog buries bones," all of which were to be fully uncovered when the visitors arrived, to give them "something substantial to think about." On the eastern edge of the gravel pit were bison bones and a horn core along with an in situ artifact. At Erosion Island there was a partially buried hearth with flakes and bones three feet down in the blue sand awaiting the visitors. Howard was anxious:

> I am personally convinced that the elephant and the bison were contemporaneous in this region, that the elephant existed into fairly recent times, and that man was living here at about the same time. If I can only prove this to Dr. Merriam and to Dr. Woodward, I believe that we shall have built one

Figure 9.1 The visit to the Clovis gravel pit, August 1, 1933. From left to right: unidentified female; unidentified male, possibly Francis Bode or H. D. Curry (Chester Stock's Caltech graduate students); Lady Smith Woodward (under umbrella); Victor Van Straelen (kneeling in the excavation); Arthur Smith Woodward (seated, under umbrella); John C. Merriam (standing); A. W. "Pete" Anderson (seated); Edgar B. Howard; unidentified male (possibly the driver for a car service). In this and another photo taken at about that same time, Howard stands modestly off to the side while the in situ material is exposed and discussed by the visiting dignitaries. Courtesy of Anthony T. Boldurian, the Clovis Archives, Latrobe, PA.

step in a theory that will enable us to fill in the gap between the "early recent" and the Basket maker.[67]

The visitors converged on Clovis on August 1, 1933, Merriam and Stock arriving from California, and Arthur and Lady Smith Woodward along with Victor Van Straelen, director of the Royal Museum of Natural History in Brussels, Belgium, from Washington. The next day was spent at the sites, examining the collections and exposing artifacts left in situ (Figure 9.1). At the gravel pit Woodward uncovered the half-buried artifact alongside the bison remains. The specimen proved to be a "knife-like scraper," and nearby was charcoal and burned bison bone. At Erosion Island Merriam excavated the hearth, exposing several "definitely flaked artifacts" mingled with charcoal and burned bison bone.[68]

On Merriam's advice Howard carefully recorded the testimony of the visitors as they gazed at the artifact Woodward exposed in the gravel pit:

Dr. Van Straelen stated that the implement found was in the blue sand. Dr. Smith Woodward agreed by saying there was no question about it. Lady Smith Woodward also examined it, saying there was no doubt that it was in the blue sand. Drs. Merriam, Woodward, Van Straelen and Stock agreed that it was contemporaneous with the bison bones.[69]

Howard's record of the testimonials, its solemnity almost comical, might seem unnecessary. After all, the association of artifacts and extinct bison had been well attested six years earlier at Folsom. Demonstrating that association at Clovis was not unimportant, but hardly a breakthrough on the scale of Folsom or one that warranted the approval of European savants. But Merriam was not thinking about bison when he advised Howard to take notes. The blue sand appeared to contain mammoth bones, and if artifacts were associated with the one (bison) they were by extension associated with the other. That was important to Merriam and why he had re-routed the trip to Clovis. He recognized the weakness of a case for a Pleistocene human antiquity that rested on an extinct bison species. The case was strengthened if the association was with mammoth, for there was no doubt mammoth were extinct. (The timing of their extinction, of course, was a separate issue.) Merriam chose his words very deliberately when he announced that the artifacts at Clovis were associated with "the elephant and a bison species which has been *considered* extinct."[70]

Late that night Howard telegraphed Jayne. Unlike the telegrams dispatched from Folsom in 1927 and 1928, the purpose was not to bring experts in for their opinion and assessment. The experts were already there. This telegram was to announce the outcome of their visit:

> Drs Merriam Stock Woodward and Van Straelen have examined excavations here today they agree evidence obtained indicates association of artifacts with extinct elephant and bison. Stop. Findings known to many persons here & might be garbled in press. Therefore might it not be advantageous for you to issue brief statement covering above report. Press here has respected confidence should like to give them chance to issue report simultaneously with your statement.[71]

Lest anyone question Howard's telegram or his authority to send it, Merriam telegraphed Jayne to affirm that he, Woodward, Van Straelen, and Stock had seen the telegram and approved its message. Howard did not mind being patronized. He was so relieved and excited that he had not been "turned down" that he could hardly sleep that night, as his mind "kept reviewing the entire details of our excursion."

Merriam thought it had all "worked out perfectly."[72] Other associations of artifacts with extinct mammals had been noted from time to time, but as he later explained in his *Annual Report* Clovis was of "exceptional interest by reason of the clear mingling of artifacts with remains of extinct forms in apparently undisturbed original deposits," but also "by reason of the wide-spread distribution and the interesting geological and paleontological relations represented."[73] The latter hinted at the site's potential to do more than merely associate humans and mammoth.

The following day the visitors boarded a westbound train to rejoin the main IGC excursion, while Howard and the crew returned to Erosion Island to finish the hearth excavation. In the process they found several more "blade-like" and "knife-like" scrapers, bison bone, and charcoal to cap the triumph.[74] A few days later, fieldwork was complete and Howard boarded a train for Philadelphia.

Both he and Merriam made productive use of the time on their respective trains. Merriam talked at length with Smith Woodward, Stock, and Van Straelen about what they had just seen and what was needed to advance research on North American early sites. Within two weeks a plan crystallized in Merriam's mind. There were, he wrote Kidder, several salient questions to answer, among them these: Artifacts were associated with mammoth and bison, but was the bison species undoubtedly extinct? What was the "stratigraphy and physiography" of the regions where these sites were found? What were the climate changes that had taken place since the Late Pleistocene, and how did these bear on the stratigraphy and character of the fauna? How should the artifacts be classified, and what was the sequence of the types and cultures in relation to the Basketmaker period? And—in many ways the most consequential question to those doing research in this area—what could the Carnegie do to "aid in one or more of these problems?"[75]

Merriam was inclined to contribute Carnegie funds and in-kind help, but as he confided to Kidder, "Some new arrangement will be necessary in order to get solution of each of the several types of problems." Notably, it was time to bring "expert judgment" to bear. Stock, of course, could handle paleontological questions, and Merriam thought geographer Isaiah Bowman (director of the American Geographical Society) and geologist Douglas Johnson (on the Columbia University faculty) were best suited to address the stratigraphy, physiography, and evidence of climate change at these sites. Merriam's choice of Stock made sense, the others less so. Years earlier Bowman had become a full-time administrator, and he was about to become president of Johns Hopkins University, with no time for research. At the outset of his career Johnson worked in arid regions, but by now he was a coastal geomorphologist, with expertise of little relevance to Clovis. Yet, what Bowman and Johnson both had that appealed to Merriam was status. They were members of the National Academy of Sciences.[76] Merriam, a fellow member, was a snob. And being a longtime administrator well past his own research career, he was also not necessarily aware of the younger stars in the field.

But what was to be done about Howard? Despite having told Cadwalader that he had confidence in Howard's work and scientific judgment,[77] Merriam thought that the broad effort he had in mind went beyond Howard's ability:

> I have nothing but praise for Howard's work and his intelligent following of the program, but in many ways his point of view is that of an amateur and I am learning that long experience is an asset of great value in practically every scientific field when one comes to final judgments.[78]

Had Howard heard those words they surely would have stung, but he also knew he needed to prove his merit to Merriam.[79]

Howard spent his return train trip writing. Six days after he reached Philadelphia a brief article appeared in *Science Supplement* reporting on the latest finds at Clovis. Clovis and Burnet Cave, Howard proclaimed, proved America was inhabited 15,000 years ago as "the last glacial sheet moved northward."[80] Ironically, he reached that conclusion about the Clovis site without actually having found any Clovis points in situ, let alone ones associated with mammoth remains; that was still several years away. The only typological distinction Howard saw in the artifacts recovered that first summer of work at Clovis was between the "typical Folsom points with grooves" and those "without the grooves but with equally fine flaking."[81] That is hardly surprising, as much of the attention was on the bison-rich diatomite, now known to be Folsom and post-Folsom in age. What is surprising is that Howard reported mammoth bones in the diatomite. It was several years before their stratigraphic position became clearer, though even then ambiguity remained (§9.8).[82]

9.3 STILL FIGHTING THE LAST WAR

It is perhaps no coincidence that just two weeks after Howard's Clovis report appeared that there was brief mention in *Science* of a forthcoming report by Jesse Figgins on the Dent site outside of Greeley, Colorado, where mammoth bones were found with two "spear-heads [of the] Folsom type," which showed the "artifacts belong to the same period as the mammoth."[83] If a spotlight was to be shined on sites with artifacts and elephants, Figgins evidently wanted his share.

The Dent site was found when large bones were spotted eroding out of a railroad cut along the Union Pacific line in the South Platte valley in the summer of 1932. The bones were brought to Father Conrad Bilgery at St. Regis College in Denver. Bilgery that fall began excavating the site with his students. They soon uncovered a point "in close proximity" to the mammoth bones. Figgins got word of the discovery, contacted Bilgery, and made arrangements for the CMNH to take over the excavations in the summer of 1933 (Figure 9.2). As was his custom, Figgins stayed in Denver awaiting news while his crew worked at the site (§8.3, §8.4). In July he was notified that a second artifact was spotted, and along with Bilgery and others he went to the site "where still and moving pictures" were made of the artifact in the ground.[84]

As Figgins interpreted the geology, the sand and gravel matrix in which the bones were found had been water laid. But judging by the size of its "bowlders" (much larger than those in the present river bed), he surmised they were deposited at a time when the South Platte "was a vastly greater and older stream." Unlike in previous years (§8.13), Figgins declined to estimate when that might have been,[85] though the fact that mammoth were present bespoke a Pleistocene antiquity. Figgins also saw the boulders as evidence of human hunting, believing they had been used to stone to death the dozen mostly young mammoths found at the site. The two points, larger and less refined versions of Folsom points (though not so distinct that Figgins thought they warranted a separate type name), presumably administered the coup de grâce. Bilgery disagreed: he thought the bones, boulders, and artifacts had washed in and the association was fortuitous. But he generously donated everything he had excavated to the CMNH. Figgins, as before (§8.4), was delighted by the "excellent prospect of having two mountable skeletons."[86]

Figure 9.2 Excavations at the Dent mammoth site, summer 1933. From left to right: Father Conrad Bilgery of St. Regis College; an unidentified St. Regis College student; Walter C. Mead (vice president of the CMNH); Charles H. Hanington (president of the CMNH); and Fred Howarth of Raton, NM, who had helped organize the Folsom excavations on Figgins's behalf. Copyright by Denver Museum of Nature and Science.

Figgins proclaimed Dent "among the most important discoveries that have a bearing upon the age of man in America," but in his subsequent report on the site, its importance was more narrowly defined:

> Critics opposed to the acceptance of the antiquity of man in America have advanced the explanation that the association of artifacts and the remains of extinct mammals is due to the intrusion of the former. There are, of course, possibilities of this nature, but until a satisfactory explanation of the selective processes is brought forward—why Folsom and Yuma types of artifacts are selected for intrusion, to the entire exclusion of the vastly more abundant modern points and blades, of equal and lesser proportions—the proposal is quite as untenable as it is illogical. In the light of the growing list of published discoveries of artifacts in association with extinct mammal remains . . . *only the uninformed and willfully perverse will longer deny the antiquity of man in America.*[87]

It is said generals are always preparing to fight the last war. Six years after Folsom, Figgins still had not realized the war was over and his side had won.

Had Howard done no more in 1933 than proclaim a Pleistocene antiquity for Clovis, there would have been little to distinguish his work from that of Figgins at Dent, or from what Cook and Figgins had done at Folsom six years earlier. At Folsom the primary goal, once the site's archaeological significance became known, was to document the association of artifacts with the bison skeletons and determine that these were indeed Pleistocene in age. But beyond Hay and Cook's study of the bison species and Bryan's situating the site in the region's geological history (§8.11, §8.16), as well as a few unpublished observations by Brown and Cook on the site sediments and snails, little more analysis was conducted.[88] Even less was done at Dent. Figgins took sole charge of its investigation and produced a seven-page article, nothing more.

But then, at Folsom and Dent (and, for that matter at Lone Wolf Creek, Frederick, and Snake Creek), the focus was on finding things, less so on finding things out. And the things of interest, given who sought them, were bones. It was no accident these sites were routinely referred to by the term "quarry" (as in the "Folsom bison quarry" [§8.16]). These were places paleontologists mined for vertebrate fossils and the occasional artifacts found with them.[89] Given that, it is perhaps understandable why it sufficed to have excavations done by minimally trained laborers under loose or long-distance and not-altogether expert supervision (such as Nelson Vaughan at Lone Wolf Creek or Carl Schwachheim at Folsom and Frederick), or as a weekend project by students (Bilgery and his crew at Dent). At the risk of overstating the case, all that was required of a crew at a fossil quarry was the ability to move earth, to expose fossil bones without damaging them, to leave artifacts in place when (if) found, and to know how to apply preservative or plaster to stabilize the bones for removal and shipment back to the museum. It was at the museum where the finer work of preparing material for display was done. (Of course, as Figgins learned to his chagrin at Lone Wolf Creek [§8.3], it was important to insure one's crew had those basic skills).

Beyond that there seemed little need to observe the stratigraphic details and depositional context, collect samples for analyses, or undertake specialized investigations to help understand, for example, the age or past environment of the site, or the manner in which the bones and artifacts had accumulated. To be sure, Figgins urged outside experts to visit Folsom, Lone Wolf Creek, and Frederick, but with the narrow aim of assessing the sites to arrive at a "definite and final conclusion" regarding their ages. His belief that this required no more than a day at each site (§8.12) highlights his naiveté and inexperience, but also the limits of his vision of what these sites offered beyond demonstrating a Pleistocene antiquity.

9.4 NOT JUST ANOTHER OLD SITE

Howard, in contrast, viewed documenting the Pleistocene antiquity of the Clovis site as "nothing more than a good start." After his first summer's fieldwork he drew up a detailed research design identifying the types of material from his excavations that needed to be analyzed—such as the vertebrate bones; charcoal from the hearths; diatoms from the lake beds; the pollen, gastropods, and mollusks from the sediments; and the sediments themselves—as well as a list of specialists to conduct those analyses. The archaeological study of the projectile points and other stone tools from Clovis and their comparison with artifacts from sites of similar age he assigned himself.[90] Howard had expressed the opinion to Merriam that the "only plan which will succeed must recognize the importance of cooperation in many different fields of research."[91] He followed his advice.

Over the next few years Howard orchestrated the publication of the various specialists' analyses. In addition, in 1934 he and Antevs spent several weeks at Clovis and other localities across the West seeking to correlate the Clovis stratigraphic sequence with climate changes on a regional and continental scale, and they also visited early sites and examined collections in an effort to understand the types and varieties of Folsom and "Folsom-like" forms that were emerging.[92] Nels Nelson, who was never easy to impress, was impressed by Howard's "all-round painstaking" effort.[93]

What was learned from the Clovis site over those several years was far more substantial and wide-ranging than what had come out of any early site previously excavated. Kenneth Lohman's (and later Ruth Patrick's) analysis of the diatoms, for example, showed that saline-tolerant forms increased upward through the blue diatomite, indicating a shallowing of the lake levels and increasing aridity. Henry Pilsbry identified among the gastropods several species that today inhabit higher latitudes and elevations, evincing far cooler climate condition than at present, a finding reinforced by the charcoal fragments identified by Ralph Chaney. That evidence, together with the fine-grained and undisturbed nature of the diatomite, pointed to a body of freshwater that existed during a period when rainfall was much greater than at present but steadily diminished over time.[94] The disproportionate frequency of bison remains toward the top of the diatomite suggested to Stock and Bode that in its later stages the lake became a rare source of water that lured animals on an increasingly dry landscape to this spot, but one where, evidenced by skeletal elements found in an upright position, they easily became mired. That made them, Howard supposed, a ready target for human hunters; this explained the presence of weaponry, the breakage patterns on the bones, and the hearths that contained burned bison bone.[95]

As to when those kills took place, the evidence for cooler and (early on) wetter conditions during the deposition of the diatomite was linked by Antevs to the high stand of pluvial Lake Estancia and with Late Wisconsin-age snow lines and glaciation in the southern Rocky Mountains. Those in turn he tied to the sequence of northern hemisphere glaciation. On that continent-spanning chain of reasoning, Antevs put the deposition of the artifacts in the diatomite at Clovis at "13,000 to 12,000 years before our time" (in retrospect a remarkably accurate estimate, though one that would not be confirmed until the advent of radiocarbon dating decades later). As Howard read the geological literature and learned from correspondence with several glacial geologists, including the venerable Frank Leverett (forty-five years after he had first discussed with Holmes and Putnam the challenges of determining the age of archaeological sites [§5.4]), that placed the hunters at the Clovis site squarely in the Late Pleistocene.[96]

These and other results appeared in a seven-part series of technical reports in the *Proceedings of the Academy of Natural Sciences* and in Howard's PhD dissertation, "Evidence of Early Man in North America," published in the University Museum's *Museum Journal*.[97] Howard's choice of publication venues insured both Cadwalader and Jayne got the credit they had wanted for their institutions' support (§9.1). His sly choice of a title to head all the reports in the *Proceedings* series—"The Occurrence of Flints and Extinct Animals in Pluvial Deposits Near Clovis, New Mexico"—insured that any historically informed reader would be reminded of the titles used by Joseph Prestwich and John Evans in their landmark 1860 papers (§2.1) that helped bring an end to the human antiquity controversy in Europe ("On the Occurrence of Flint-Implements, Associated With the Remains of Extinct Mammalia, in Undisturbed Beds of the Late Geological Period"). It was a surprisingly bold allusion to draw for someone as modest as Howard.

Merriam was kept apprised of the work and came to realize he had badly underestimated Howard. He made amends. In the spring of 1934 Howard was appointed as a research associate of the Carnegie Institution, and soon he was sprinkling his letters to Merriam with reference to "our Committee." It had taken more than a year, but Howard had worked his way onto the American Subcommittee (§9.1).[98]

Howard's Clovis work was pivotal in several respects. It was the first full-scale archaeological investigation of a Late Pleistocene site, yet it was also thoroughly interdisciplinary. It thus marked a radical change in the manner in which such sites were investigated and lured in a range of scientists previously uninvolved in such research. As such, it served as a model for later studies, such as Frank Roberts's extensive fieldwork at the Lindenmeier site (Colorado) that began in 1934.[99] It also provided a highly unflattering contrast with excavations and analyses that had been done previously.

Its success also rippled outward, for it convinced Merriam of the value of investing in research in this area. The Carnegie Institution began actively funding research on early sites; this in turn fueled the overall development of the study of such sites. That support was crucial at a time when the country was in the throes of the Great Depression. Although this was, ironically, a period flush with archaeological opportunities under the auspices of President Franklin Roosevelt's New Deal relief programs, those rarely involved Pleistocene-age sites. Their focus was not on particular time periods or research problems but rather on large sites, especially ones threatened by construction or inundation, readily visible on the surface, and accessible to a sizable work force (the principal goal of the relief work being to put as many unemployed laborers to work as possible). As a result virtually all sites investigated under New Deal auspices were from later periods. Occasionally "Folsom-like" points were found at these sites, but these were isolated occurrences, the points perhaps having been collected by the sites' inhabitants.[100]

Archaeological research projects aimed at Pleistocene sites were a luxury few institutions in the 1930s could afford. The Smithsonian was one, and it generously provided for Roberts's several seasons of excavations at Lindenmeier in mid-decade, but its funding was limited to its own projects.[101] Merriam had greater flexibility in selecting projects the Carnegie Institution could support and the financial wherewithal to do so. In 1935, for example, $1.15 million was budgeted for the Carnegie's seven research divisions, of which over $160,000 was earmarked for archaeology. (The lion's share went to Kidder's Maya program.) Also in that year's budget was $152,000 for discretionary disbursement (presumably by Merriam) in the form of "minor grants."

The amount of funding specifically for work on Late Pleistocene archaeological sites is difficult to ascertain, but at least $2,600 was provided to Antevs and Howard in 1935. (Each had received $1,300 the year before.) Comparable amounts were granted that year to M. R. Harrington for his work at Tule Springs and Smith Creek Cave in Nevada (he had previously received funding for his Gypsum Cave excavation), to Paul MacClintock for research on Yuma-Folsom sites in western Nebraska, and to Horace Richards for a study of the terraces around the Vero site.[102] All these individuals were funded over at least the next three years, as others—including John Cotter, Luther Cressman, Malcolm Rogers, and Paul Sears—were added to the Carnegie rolls. Until the Carnegie funding stream dried with Merriam's retirement in 1938 (his successor, Vannevar Bush, was uninterested in work outside the physical sciences), many tens of thousands of dollars were invested in research at Late Pleistocene archaeological sites.[103] By now fully confident in Howard's abilities and judgment, Merriam relied on him to identify projects worthy of Carnegie Institution support. (Kidder, as the Carnegie's senior archaeologist, previously had that role, but because of his other responsibilities he readily deferred to Howard.)[104] Merriam also provided in-kind support, making Carnegie Institution Research Associates Stock and Antevs available to provide expertise at archaeological sites. Antevs was also given a specific charge.

9.5 REFINING THE PLEISTOCENE

Even though there was general agreement that sites with artifacts and extinct fauna were Pleistocene in age, Merriam and Howard sought a means of ascertaining their ages independent of the fauna themselves.[105] Merriam turned the problem over to Antevs. In his native Sweden Antevs had helped develop varve chronologies in postglacial lakes, a method he imported to Canada and the northeastern United States in the early 1920s. Varves could be used to date ice retreat from a region with reasonable accuracy, and provided absolute rather than relative ages. By this method Antevs put the timing of "ice release" in New England at 11,500 years ago.[106]

Antevs first came to Merriam's attention in the 1920s when Carnegie Institution funds were awarded to Bowman, who, busy at his administrative post, transferred them to Antevs to investigate past climate and chronology in the Great Basin. In the absence of varves, Antevs used ancient lake levels and tree-ring records to develop a climate history, which he could link to global changes in climate and thus tie to a chronology, assuming a correlation existed between pluvial lakes and glacial expansion.[107]

Merriam wondered whether an archaeological site might be added to that chain of reasoning. Antevs supposed that was possible, but it would require tying the sites to distant, more securely dated glacial features:

> The only means, so far available, of dating finds of human remains and artifacts in the southern Great Plains area and in the Southwest is to correlate, directly or indirectly, the deposits in which the remains or artifacts occur with the continental glaciations in the northern half of the continent. So

far, three kinds of records for a remote moist and/or cool period have been studied in the Southwest, viz. mountain glaciers, pluvial lakes, and stream terraces.[108]

Antevs assumed there was at least an approximate correlation across the northern hemisphere in the timing of continental-scale ice advance and retreat, because the ice sheets were responding to the same underlying cause—changes in global climate.[109] He further assumed that geological features in areas south of the continental ice sheets, such as mountain glaciers, pluvial lakes, and stream terraces, were likewise responding to those same climatic changes, and therefore their ages could in turn be determined.[110]

The devil was in the details, of course, for the correlation between unglaciated and glaciated regions was not always precise nor necessarily in phase, given the varying rates at which different processes played out (glacial advance and retreat versus the filling and draining of a pluvial lake versus stream aggradation and incision). Likewise, there was the confounding influence of other aspects of climate. Precipitation, for instance, was influenced by topography, proximity to moisture sources, the changing position of air masses, and other "local" variables, and as a result might reflect a regional rather than a global climatic signal. (Thus, the timing of the "pluvial period," Antevs suspected, might vary even between adjoining states.) Antevs therefore reasoned that the "foremost basis" for correlating glacial, geological, and geomorphic processes across the hemisphere was via *summer* temperature, for this was the principal control over the "growth and dissipation of the ice sheets," evaporation (which could significantly influence pluvial lakes), and the "latitudinal distribution of plants and animals" (including the extinct forms that might appear in archaeological sites).[111]

With Carnegie support, Antevs spent the summer of 1935 mapping Wisconsin maximum glacial extent in the Rocky Mountains and comparing it to the modern positions, with the goal of using the latter and what was known of modern snowfall and summer temperatures to infer climatic conditions under which the Wisconsin ice reached its greatest extent. In effect,

> the difference in altitude between the late Wisconsin glaciation line and the modern glaciation line, and the decline of the latter with drop in temperature, furnish a means to calculate the approximate late Wisconsin temperature.[112]

The inferred temperature and position of the late Wisconsin snow line in turn provided a "fair idea of precipitation conditions." This result was then correlated with water levels in ancient pluvial lakes and stream terraces of the Great Basin and Great Plains, which Antevs began studying intensively in the summer of 1936 (and continued studying for many years to come).[113] Antevs, and soon Bryan as well, elevated this method of long-distance geological-climatic correlations—which sometimes required "a mental leap of more than a thousand miles"—to a high art, if not something of a scientific high-wire act.[114] Their efforts to determine the ages of sites by geological methods "fit admirably" into the work that Merriam and Howard envisioned at archaeological sites. Most especially, it meant they did not have to rely as much on the ages of associated faunal remains, which in their opinion were still unsettled.[115]

Few vertebrate paleontologists in the 1930s would have disagreed. As Romer admitted, the "assignments of age on the basis of faunal associations are of very dubious value." Stock— who had seen the evidence from Clovis firsthand and was convinced of the association between artifacts and bison, mammoth, and possibly camel—nonetheless said the site could be, but was not necessarily, Pleistocene in age, as these taxa may have survived "up into very late Quaternary time." This was a particular concern when dealing with a species of uncertain taxonomic identity, especially given the "uncertain state of our knowledge concerning the validity of a number of species recently described" (a none-too-subtle jab by Stock at Cook, Figgins, and Hay).[116] Stock wanted independent evidence to confirm the site's Pleistocene antiquity. (He ultimately accepted Antevs's geological reckoning of the age for Clovis.)

Still, paleontologists were not inclined to abandon their evidence or methods but instead sought to rectify the situation. Romer, as he had promised Cook (who likely took that promise more as a threat), systematically reexamined the geological occurrence of Pleistocene vertebrates. Romer focused on large-mammal genera, avoiding species, given the "dubious validity" of the many recently named bison species. (Like Stock, he could not resist a potshot at Cook, Figgins, and Hay.) He steered clear of small mammals because they were too poorly represented in most paleontological collections; their absence was an unreliable chronological indicator.[117]

Of the twenty-seven genera of large mammals Romer considered valid taxa, Hay had thought only ten survived into the Late Pleistocene. Of the other seventeen, Hay thought eleven had not made it out of the Early Pleistocene alive (including the camel, of course), whereas the other six survived the end of the Aftonian but were extinct by the end of the Middle Pleistocene. After his review of where those twenty-seven genera had been found and in what geological circumstances, Romer turned Hay's conclusions upside down. He argued that twenty-six of the twenty-seven genera (96.3 percent) in fact survived into the Late Pleistocene (Table 9.1).[118] Hay's claim that the major Pleistocene mammalian extinction event occurred at the end of the Aftonian had been off by several hundred thousand years. The vast discrepancy between the vertebrate paleontological and geological age estimates at sites such as Vero and Melbourne (§7.5, §7.9) instantly vanished.

The way Romer saw it, the discrepancy between Hay's results and his own was not solely or even mostly because of new discoveries of late-surviving genera, ones Hay had not had the opportunity to put in his study. Instead, he blamed it squarely on Hay's unshakeable faith that most extinctions occurred in Aftonian times. Romer ascribed that belief to Hay's having committed the "great fallacy" of ignoring geography and failing to realize that camels, horses, and imperial mammoths found in Aftonian deposits on the Great Plains were *southern* types that would not occur in Late Pleistocene deposits in northern areas. Hence, their absence in those areas was a result of habitat, not antiquity. Hay's faith was so deep-rooted that he dismissed any apparent Late Pleistocene occurrences as fossils

Table 9.1 Frequency (and percentage) of loss of Pleistocene large-mammal genera over time, according to Hay and Romer

	Taxa that went extinct before the Late Pleistocene (%)	Taxa that survived into the Late Pleistocene (%)	Sum (%)
Hay (various)	17 (63)	10 (37)	27 (100)
Romer (1933)	1 (3.7)	26 (96.3)	27 (100)

Note: From Romer (1933): Tables 2 and 3.

having been reworked from earlier deposits, or (in the case of horses) as accidental modern intrusions.[119]

The southern forms aside, the majority of those twenty-seven genera were known to occur in secure geological contexts in regions north of the terminal Wisconsin moraine, indicating that the animals were "still present after the retreat of the Wisconsin ice."[120] The pattern was so pronounced that Romer insisted that any genus that did not survive into the Late Pleistocene was the exception rather than the rule. Thus, across much of North America the Late Pleistocene forms included

> *Mastodon, Mammonteus* [mammoth], horses, fossil peccaries, camels, musk-oxen, bison (including very probably fossil species as well as that now living), *Felix atrox*, and the giant beaver. Other forms found here which are known to have existed elsewhere until late-Pleistocene times, if not later, include *Archidiskodon*, giant armadillo, glyptodont, sabre-tooth and the capybara.[121]

As far as most other vertebrate paleontologists were concerned (not Cook, of course), Romer got it right and Hay was wrong.[122] Hay would have turned somersaults in his grave.

What seemed most striking to Romer was not that so many taxa survived into the Late Pleistocene but rather what had become of them after that. For the evidence now suggested "a vast amount of extinction has taken place during the last score of millennia" (that is, the last 20,000 years). It was a conclusion Romer deemed "improbable" yet inescapable. Even so, it was difficult to know whether extinctions were synchronous across all those genera, at least not without more specific evidence on the date of the disappearance of each, particularly those taxa that seemed to have "persisted into practically Recent times."[123] That made it hazardous to argue a site had "great antiquity" solely on the presence of extinct taxa. Certainly if "man had entered North America as the last ice-sheet was shrinking" many of those extinct forms were present. The critical question was not how early humans had arrived, but how late the fauna had survived. Even with a "most spectacular" case like Folsom, Romer ruefully admitted, there was little one could say beyond acknowledging an association of artifacts with extinct forms. Otherwise, "positive statements are difficult."[124]

Nonetheless, Romer was willing to grant that the bison at Folsom was extinct (§8.11), though he remained skeptical of the value of this particular species (or any other newly named bison species) to serve as a reliable chronological marker except in the absence of this genus. That is, by the 1930s there was good evidence that bison did not arrive in North America until the Middle Pleistocene, making their absence from a paleontological assemblage an indicator of an Early Pleistocene fauna.[125] It was a useful fact, though one not relevant to archaeological questions (now that Hay, who had insisted sites like Vero were Early Pleistocene in age, was gone).

Yet, despite Romer, Stock, and other paleontologists' continued skepticism about the reliability of faunal ages being used to date archaeological sites, they parsed their words very carefully:[126]

> The association of man in America with certain fossil forms is unques-
> tioned. . . . Such contemporaneity, however, by no means indicates any re-
> mote geological antiquity for man on this continent, and there is at present
> almost no paleontological evidence suggesting his presence here at a time
> earlier than that of the withdrawal of the late Pleistocene ice sheet.[127]

The key phrase was "remote geological antiquity." By the 1930s, to speak of a site as having "great antiquity" or "geological antiquity" had come to mean it was Early or Middle Pleistocene and not Late Pleistocene in age (except to Hrdlička, for whom "great antiquity" was virtually synonymous with any premodern skel-etal forms, which by the Pleistocene time scale he used was not very old at all [§9.10]).[128] By then, however, only a "few extremists" (as Schultz and Eiseley put it) hazarded the idea that any North American archaeological site was that old (one eager individual claimed Clovis was two million years old; Howard wasted no time shooting down that notion).[129]

Romer did not reject the idea that humans were in North America in the Pleis-tocene. He insisted only that their antiquity could not be determined by Hay's timetable of vertebrate succession, which was a virtual guarantee of an Aftonian age (§8.6). Like most vertebrate paleontologists in the 1930s, Romer was willing to grant a Late Pleistocene arrival or, more specifically, a Late Glacial or early Postglacial one. The only problem, and it was a troublesome one, was just when those times were.

The core of the problem was how, when, and where one drew the boundary be-tween the Late Glacial and Postglacial, or more broadly between the Pleistocene and post-Pleistocene (Recent). In the 1930s there was only vague agreement about what those terms meant. The Pleistocene, or at least that portion of the Pleisto-cene (the Wisconsin) relevant to people, generally referred to the period when the continental ice sheets were at their maximum extent (in modern parlance, the Last Glacial Maximum). The Late Glacial (or for some, "post-Pleistocene") began as soon as the continental ice sheets started to recede.

The beginning of the Postglacial (or Recent) was much harder to pin down, for as geologist Richard Foster Flint observed, finding that boundary required choosing "some conspicuous, identifiable event during the last deglaciation" as the line of demarcation. But deglaciation was time-transgressive, given the south-to-north recession of the ice margin (shown early on by Antevs in varve sequences

Table 9.2 Estimates of the age for the onset of Postglacial time (and Late Glacial time, where available)

Source	Onset of Postglacial (years ago)	Postglacial markers
Antevs (1936:332)	8,500	Temperature like that of today (Late Glacial 36,000–8,500 years ago; Recent began at 4,000 years ago)
Daly (1934:34)	9,000	None specified (Late Glacial 30,000–25,000 to 14,000–9,000 years ago)
Johnston (1933)	9,000	Disappearance of ice sheets from north-central Canada (Late Glacial circa 40,000–9,000 years ago)
Leverett (1930b)	10,000	Temperature like that of today (Late Wisconsin 25,000– 10,000 years ago)
Leighton (Howard 1935b:103)	15,000	Temperature like that of today; most of continent free of ice
Romer (1933)	20,000	Deposits formed after retreat of the ice from the region south of the Great Lakes
Kay (Howard 1935b: 103)	25,000	Des Moines lobe began to retreat

Note: From Howard (1935b:103) and various additional sources, as noted.

as he moved north in the Connecticut valley[130]). Thus, estimates of Postglacial time were dependent on when a particular spot became ice free (or when modern climate conditions were reached in a particular region). This, by necessity, was locally determined and thus the results varied predictably (Table 9.2).[131]

For example, George Kay of the University of Iowa set his Postglacial clock by the onset of the retreat of the Des Moines lobe of the Laurentide; Romer, from his University of Chicago office overlooking Lake Michigan, declared the Late Glacial ended once Laurentide ice began to depart the Great Lakes; William Johnston of the Geological Survey of Canada drew the Postglacial line only after the ice sheet departed north-central Canada. Estimates for the onset of the Postglacial therefore ranged from as recently as 8,500 years ago to as much as 25,000 years ago, revealing the arbitrariness of the boundary, if not a certain absurdity to the process. The Clovis site, 12,000 to 13,000 years old by Antevs's reckoning, was Postglacial on Kay's Iowa scale but firmly within the Late Glacial using Johnston's Canadian timetable.

To see if consensus might be had, Howard polled twenty geologists across the country, asking them to specify their estimates for the relevant periods. Virtually all answered his letter, but few were inclined to give specifics and instead deferred to the authority of others, notably Antevs, Kay, Morris Leighton (of the Illinois Geological Survey), and Leverett.[132] There was no agreement among those experts, and the problem of marking the end of the Pleistocene lingered for decades (and to a degree still does).[133]

In effect, whether an archaeological site and its associated extinct faunal remains dated to the Late Pleistocene depended not only on whose Pleistocene it was, but where and when it was permissible to apply the "Pleistocene" label. Archaeologists Earl Bell and William Van Royen took the extreme position that "only if man entered America *before* the Wisconsin reached its greatest extension . . . we might, in truth, speak of Pleistocene man."[134] Anything short of that was post-Pleistocene, Late Glacial, or Postglacial but nonetheless had fully "respectable antiquity."[135] Theirs, however, was a minority opinion.

9.6 CONVERGING ON A CHRONOLOGY

Without a consensus on the age of the Pleistocene, and lacking firm extinction chronologies for the animals found in archaeological sites (a problem compounded when the genus survived), and without sites directly associated with glacial geological features or that could be correlated to such features long distance, how was it that claims for a Pleistocene human presence were in the 1930s nonetheless accepted? Obviously no single criterion or chronological marker sufficed, and so the answer for Howard, Roberts, and others was in triangulating lines of evidence (positive and negative) from as many sources as possible, particularly as these were set against the Basketmaker occupation, the oldest archaeological presence previously recognized and thus a handy chronological yardstick.[136] Among the lines of evidence were these:

- Folsom and Yuma projectile points were of a very distinctive type and technology and differed significantly from those associated with more recent Basketmaker occupations.[137]
- This distinction was also true of the remainder of the tool kit, as shown by Roberts at Lindenmeier, a residential site and one of the few in the 1930s to yield a range of artifacts such as end and side scrapers, gravers, flake knives, choppers, adzes, large blades and "rubbing stones" stained with red ocher. Although Nelson deemed it a "Neolithic" assemblage, Roberts showed that it differed in type and technology from later Basketmaker assemblages, being primarily a chipped rather than ground or polished stone technology (and altogether lacking pottery).[138]
- In sites with multiple components (such as Burnet Cave and Gypsum Cave), the Pleistocene material was found stratified well below Basketmaker material. Folsom and other fluted points were occasionally found in sites of later age, but these "casual occurrences" were deemed the result of erosion onto the same surface or points that had been picked up by later Indians.[139]
- Folsom and like points were associated with extinct genera (mammoth, ground sloth) and extinct species (bison) of large mammals, but they were never "in association with *Bison bison*." Likewise, although it was uncer-

tain when those taxa went extinct, their remains did not occur in later sites, nor were these animals part of the "historic American fauna."[140]

- Evidence from geology (the Clovis blue clay), vertebrate paleontology (the musk ox at Burnet Cave, the ground sloth at Gypsum Cave), invertebrate paleontology (snails at Clovis and Lindenmeier), and ecology (the charcoal and diatom record at Clovis) demonstrated that these occupations took place at a time when the climate was markedly different from that of the present, notably cooler and less arid, suggestive of glacial (and premodern) climate conditions.[141]

- And above all, whatever the absolute age of Clovis, Folsom, or any other such finds, there appeared to be a substantial gap in time separating them from other, later cultural periods. Basketmaker, the oldest of these, reached back only a couple thousand years.[142]

As Roberts summarized matters, when "archaeological manifestations occur in deposits that can be correlated with geologic phenomena identified as late Pleistocene and also are associated with extinct species of animals . . . the dating of the assemblage as late Pleistocene appears justified."[143]

Circumstantial geological evidence from far afield (literally) seemed to bolster that conclusion, for if the peopling of North America took place "through northeastern Asia through Alaska and the glaciated regions of northwestern North America," Antevs believed knowing the geological and climatic history of that region would "help to fix the probable age of arrival of the first men on this continent."[144]

9.7 THE PEOPLING PROCESS

It had long been accepted that ancestral Americans had traveled from the Old World to the New via the Bering Sea region (§2.2). Even saying so had become "trite and commonplace to American anthropologists" by the 1930s.[145] There was nevertheless uncertainty about whether the traverse was over land, ice, water, or some combination of the three. By the early twentieth century there was talk of a land bridge connecting Siberia to Alaska. Such talk was inevitable, given the similarity of New World and Old World animals, particularly those that could not have floated or flown from one land mass to the other. But when a land bridge was in place, and whether it spanned the north Pacific or north Atlantic, remained unknown. At a 1911 symposium on Native American origins, the U.S. National Museum's James Gidley ruled out an Atlantic land bridge, given the far greater similarity between Asian and American terrestrial faunas (a conclusion affirmed by his colleague Austin Clark, who showed from differences in marine invertebrates in the Pacific and Arctic oceans that there must have been a recent barrier across the Bering Strait). Based on vertebrate fossils, Gidley posited that the land connection was in place on at least two occasions. Initially at the beginning of the Pleistocene, to account for taxa such as mammoth, horse, bison and muskox appearing in the early fossil records of Siberia and Alaska, and then "as late as the close of the last glacial" given the similarity of mammals across the Arctic. Presumably, ancestral Native Americans could have crossed on either occasion, though in 1911 Gidley thought the latter more likely.[146]

Yet, at that same meeting William H. Dall, the venerable Arctic natural historian (who had conducted some of the earliest bathymetric surveys of the Bering Sea in the 1860s), categorically rejected the idea of a land bridge. The fundamental problem for Dall was the lack of a mechanism. He and others assumed that to create a bridge one had to elevate the Bering Sea floor. Only, there was no evidence of uplift in the terraces and land masses of Siberia and Alaska "when we might suppose primitive man to have invaded America." (He believed uplift had occurred much earlier in the geological record.) To his mind the animals on opposite sides of the Bering Sea were more indicative of long periods of separation, not connection. The most Dall was willing to grant was that

> the [human] migration from Asia took place when the culture of the invaders was sufficiently advanced for them to be able to cross the strait in canoes; or, like the present Eskimo, they may have during glaciation followed the marine mammals, the walrus and the seal, along the edges of immovable floe ice closing the strait perhaps for some centuries.[147]

There the quandary stood for more than a dozen years. If there was a land bridge, as circumstantial paleontological evidence indicated, how had it formed and when?

Antevs found the answer when he realized it was not necessary to raise the land to create a land bridge if there was a mechanism to lower the water. In 1928 he calculated from the extent and average thickness of glacial ice that global ice volume at the Wisconsin maximum was 32.8 million km^3 (36.9 million km^3 if southern hemisphere glaciation was synchronous with the northern hemisphere). If that much water was frozen on land and not returned to the oceans, he estimated global sea level would have fallen eighty-three to ninety-three meters below their present position (the lower estimate included a southern hemisphere contribution).[148] At the 1928 AAAS meeting he drew attention to the implications of a sea level fall of that magnitude for the Bering Sea region:

> A land connection of great extent may be assumed to have existed between Asia and America at almost any time during the Pleistocene, since the sea-level at that time was low enough to expose the bottom of the entire Bering Sea.[149]

It was instantly obvious to those listening to Antevs that this was "a most important paper," for it revealed people could have walked from Asia to America over a span that lasted many thousands of years on either side of the Wisconsin maximum. They did not have to wait for the close of the glacial period or the development of watercraft or been restricted to winter crossings.[150]

The relationship between sea level change and the emergence of a land bridge was elaborated on by Canada's Johnston and Philip Smith of the USGS.[151] Both were vague about timing, but because Johnston supposed the Bering Strait was open for all of his Postglacial time (Table 9.2), and that the land bridge formed during the full glacial (before 40,000 years ago, by his estimation), that meant Asia and America were joined for more than 30,000 years. Others estimated different ages, but there was increasing agreement that a land bridge "must have been made and unmade eustatically, several times" over the Pleistocene.[152]

The most prominent resistance to the idea of a land bridge came from Hrdlička in 1937 in the midst of debating Albert Jenks over "Minnesota Man" (§9.10). Jenks had innocently mentioned the land bridge to emphasize how easily descendants of *Sinanthropus pekinensis* (*Homo erectus*) could have traveled to North America. It was a throwaway line, but Hrdlička made him pay for it, insisting there had not been a land bridge in the Wisconsin period, citing Dall's statement of more than twenty-five years earlier and misrepresenting a more recent one by Smith.[153] Hrdlička's sharp reaction was in no small measure due to the fact that the Bering Sea was his first line of defense against a Pleistocene peopling of the Americas. As Harvard physical anthropologist Ernst Hooton drolly observed (with more truth than he may have realized),

> that formidable and indomitable veteran, Dr. Aleš Hrdlička, has stood like Horatius at the land bridge between Asia and North America, mowing down with deadly precision all would-be geologically ancient invaders of the New World. In fact, the story of alleged fossil man in America is virtually the tale of how well Hrdlička kept the bridge.[154]

Hrdlička aside, the land bridge solved one problem relating to how people could have crossed from Siberia, but it presented another one: if the crossing was over a land bridge it meant they arrived in Alaska when North America's Cordilleran and Laurentide ice sheets were at or near their maximum extent.[155] If at their maximum the two ice sheets joined one another they would have blocked the newly arrived migrants' passage south from Alaska. Unless, as Warren Upham supposed in 1895, there was an ice-free corridor between the ice sheets along the eastern flank of the Rocky Mountains, but he was so unsure of its presence and timing that he labeled it the "debatable tract" on his glacial map (see Figure 1.2).[156]

Three decades after Upham, the details of whether the ice sheets merged or when there was a corridor between them was still debatable, mostly for want of knowledge of the position of the respective ice margins at their maximum, especially in the northern reaches of Canada and Alaska, the point of departure for groups moving south. The challenge of gathering the necessary data was compounded by the difficulty of determining which moraines in those areas were from the Wisconsin maximum and which were older.[157] Even so, the idea that an ice-free corridor was present in the Late Pleistocene gained support. Johnston supposed that a gap at least 200 to 300 miles wide between the westernmost Laurentide and easternmost Cordilleran margins appeared soon after the ice began to retreat from its maximum position. He hedged slightly. If access to that corridor was by way of the Mackenzie valley, which he suspected stayed buried under ice longer, an unobstructed passage from Alaska to the northern Great Plains may not have opened until 30,000 to 25,000 years ago.[158]

Johnston realized there were other possible routes south. Groups could have followed the Yukon valley, a passage he thought less likely, as it led into the mountains of northern British Columbia and south between the Rockies and the Coast Range, a course "by no means easy to traverse even at the present day." It also was not open as early or as long as the corridor down the eastern slope of the Rockies, for Cordilleran ice lingered later in time in central British Columbia. Johnston supposed it may have been passable more than 10,000 years ago, but he declined to speculate further.[159]

Down the Pacific Coast was still another option for would-be Pleistocene migrants headed south, though Johnston doubted that was a viable route given the evidence that Cordilleran ice reached the sea and must have obstructed much of the coastline and required travel by boat. Further afield, he supposed migrants could have trekked around the northern edge of the Laurentide ice sheet, then turned south through the Hudson Bay region or traveled up the St. Lawrence seaway. These seemed unlikely alternatives, as neither appeared to have been open until well into Postglacial time.[160]

Antevs favored the ice-free corridor route as well, though he suspected its tundra environment would have made for a "hard and precarious" transit. Yet, it better explained "why most of the significant finds of earliest man in the United States have been made on the Great Plains." Antevs put the opening of the corridor around 20,000 years ago, well into the Late Glacial (Table 9.2), and because he supposed the land bridge was present by 20,000 years ago and could be crossed for 5,000 years after that, it meant one could walk uninterrupted from "central Asia to central North America" from 20,000 to 15,000 years ago. That hunters were at Clovis 13,000 to 12,000 years ago seemed to him no coincidence.[161]

The idea of migrating along an interior ice-free corridor as opposed to traveling down the coast appealed to most archaeologists of the 1930s, providing as it did both a route and a reason for travel. Roberts and others theorized that Siberian hunters followed game across the land bridge, then "drifted down along the corridor not long after the beginning of the glacial retreat."[162]

Howard was not content to merely theorize. In the spring of 1934 he drew up an ambitious plan to explore what he deemed the "perfectly logical route of migration" that might have been taken from Lake Baikal in Siberia through the Bering Sea region and down onto the Great Plains. It would have been impossible to conduct an archaeological survey along that entire route—Howard was ambitious, not irrational—but he supposed it would be "perfectly feasible to pick out critical points along it in an attempt to discover links in the chain of evidence."[163] At the University Museum he began working with an unnamed Russian who had recently been to Siberia and who would help him pursue evidence at that end of the "big arc" from Baikal to Folsom. And he wrote the National Museum of Canada's Diamond Jenness, a grizzled Arctic veteran, for advice about the feasibility of surveying Pleistocene terraces in the Mackenzie valley.

Jenness was blunt. There were hundreds of terraces to search over a vast area, he explained, requiring long slogs across muskeg (and "no one travels on muskeg who can avoid it") with "not a few mosquitos to help you carry your pack through country that will break both your back and your ankles." As for retracing the course of the Mackenzie valley 10,000 or 20,000 years ago, "that is the work of a lifetime, if it can be done in that time." Besides, even if people had been in Alaska then, they could have traveled south by a half-dozen possible routes within that broad corridor. An archaeological survey of the Mackenzie, Jenness growled, would accomplish "little or nothing." Howard needed to find more modest goals.[164]

Howard was discouraged but not defeated.[165] In the summer of 1935 he set sail for Russia. He aimed to find out if, in archaeological collections from Siberia, there were "analogies" of Folsom points or hints of sites where ancestors of the first Americans had once resided.[166] His time was spent entirely in Leningrad speaking with scholars and examining collections. What he learned—and it became a familiar refrain—was that there had not been enough work in Siberia "to be sure what is there archaeologically." He saw bones of bison reminiscent of "some of ours" but no evidence of Folsom points. If future work in Siberia failed to turn up any, it might prove that Folsom was "of purely American origin."[167]

That would be significant if Folsom was the earliest form of projectile point in America, but through the 1930s that seemed increasingly less likely.

9.8 RECOGNIZING VARIATION AND CHANGE

By then lanceolate projectile points had been recorded across North America. By virtue of their similar shape and size and in many instances the presence of fluting, these were labeled Folsom points, but as the discoveries snowballed variations in form and technology became more apparent. With that, questions emerged over whether these were part of the same archaeological culture as Folsom and of similar age.[168] In the 1930s archaeologists began to grapple with what fine-scale artifact differences—variation in degree rather than in kind[169]—meant in the Pleistocene past, and whether it reflected change over time, across space, in technology, function, or some complex combination of all. The effort to disentangle different types of early lanceolate points proved an enduring problem.[170]

The initial challenge was to distinguish Folsom points from the many hundreds of Yuma points (§9.1) from the Great Plains. Compounding the difficulty of sorting the two, leaving aside strictly typological concerns, was the absence of stratigraphic data to resolve their historical relationship. Most Yuma points were recovered on the surface of Dust Bowl "blow-outs." (Cook claimed to have Yuma points in situ at the Slim Arrow site, Colorado,[171] but his report was viewed with considerable skepticism, especially by Figgins, who by then was ill-disposed to accept Cook's word on anything [§8.15].)[172] In some cases Yuma points were said to be associated with extinct fauna, primarily bison but also mammoth. These claims too were suspect, as the association was often little more than that points and bones had been found on the floor of the same blow-out along with, as Figgins wrote with an almost audible sneer, "cattle bones, modern artifacts, bits of chinaware, iron wire, etc."[173]

Renaud attempted the first typology of Folsom and Yuma points, a classification with three main elements: point shape, base geometry, and flaking pattern of both the blade and the base.[174] It was a muddled scheme, with attributes of those elements often ill-defined and the resulting types not always mutually exclusive. As a result, a particular specimen could and often did fall into one or more of his types, with Yuma points routinely spread across much of the typology. Although this was a classification intended to remove "confusion and misunderstanding," it badly missed the mark.[175]

Even so, the effort highlighted for Renaud just how distinctive Folsom points were—as seen, notably, in the "typical and constant" presence of the "longitudinal groove" that occurred on Folsom points but not on Yuma points. That attribute, he supposed, was to facilitate hafting, but he also surmised it had the unintended consequence of weakening the point. Folsom points were shorter, on average, than Yuma points, indicating to him a greater susceptibility to breakage.[176] Therein for Renaud was a clue to their relative ages, for he perceived in his Yuma points a range in the length of basal thinning flakes. Assuming that those lengths had evolved over time (shorter thinning flakes were older, longer ones more recent), and that the increase represented improvement in knapping skill, then from Yuma points with short thinning flakes to fully fluted Folsom points were the "first timid steps, later bolder trials, [in the] march towards the possibly excessive and weakening fluting of the best Folsom artifacts." Admittedly, it was a "purely theoretical" scheme, with "no concrete means to establish such a claim" in the absence of geological or stratigraphic evidence. Nor could Renaud demonstrate that Yuma points with shorter basal thinning flakes were older than Yuma points with longer ones. Nonetheless, he was confident he was right.[177]

Others were skeptical. Cook was certain Folsom and Yuma points were contemporaneous. So were Schultz and Eiseley certain, on the basis of their Scottsbluff excavations. (A year later they changed their mind and considered the possibility that Yuma points were earlier.)[178] Figgins would have none of it: not only were Folsom and Yuma not the same age, he was peeved that Renaud gave pride of antiquity to Yuma. In his view the manufacturing techniques of Folsom and Yuma points were completely different, and Folsom was a distinct culture with its own "long line of evolutionary progression from cruder types" and not merely the "final stage of an earlier [Yuma] type."[179] Moreover,

> since there is no material similarity in form and detail between Folsom and Yuma artifacts, nor evidence of intergradation of their widely divergent characters, the ground upon which a relationship is proposed appears to be extremely insecure. . . . Their relative position is purely conjectural.[180]

Indeed, if "progressive improvement" was measured by the "quality of workmanship," then the more finely crafted Yuma points were superior and must follow, not precede, Folsom points in time.[181]

Figgins did not help his classificatory cause in his own definition of Folsom points. Although he described the type as having "wide spalls removed from the sides, beginning at the base and sometimes extending quite to the tip, producing a hollowed or 'fluted' effect," he thought the most diagnostic characteristics were the "relative proportions of width to length and the reduced width backwards of a point forward of the midsection," as well as the presence of edge retouch and deeply concave bases. Fluting might or might not be present.[182]

Yet, despite the variability within the type he was willing to accept, Figgins insisted, contra Renaud, that "not one of the . . . described characters [in Folsom points] are present in Yuma artifacts." That was because he saw one attribute—their distinctive basal morphology—as a necessary, sufficient, and universal feature of Yuma points. Folsom points simply "do not have squared bases, rarely paralleling edges, and never taper forward from a maximum basal width." But given the wide range of morphological variability in Yuma points, Figgins's insistence that all Yuma points were parallel sided, widest at their base, or had square bases was not compelling or useful as a definition.[183]

In the midst of this debate Roberts began fieldwork at Lindenmeier (§9.4), a site that produced hundreds of Folsom points in various stages of manufacture and use. These helped him reconstruct the fluting process and aspects of projectile-point use life.[184] Over the next several years he grappled with the definition of a Folsom point, a task simultaneously made easier and more difficult by the large sample from Lindenmeier. Before he started there, Roberts had little difficulty identifying the attributes of a Folsom point,[185] but the mounting number of specimens from the site were complicating the matter. Roberts knew that not all points met an idée fixe, least of all the specimen found in situ at Folsom in 1927. Although he labeled this the type specimen and considered it the standard against which other points were deemed Folsom or not, he could see it was not "typical" in any meaningful sense (few type points are).[186]

At Lindenmeier, in fact, he initially identified two classes of Folsom points. The first had a slightly rounded tip and was somewhat "stubby," with the broadest part of the blade "between the tip and a line across the center of the face" just beyond the end of the haft area. The Folsom type specimen fell in this group. The second, also present at Folsom but "rarely mentioned in discussions," was not stubby but rather "long and slender in outline" with "a tapering rather than a rounding tip."[187] At this juncture Roberts believed the difference in blade shape was an intentional by-product of manufacture and typologically significant. He did not realize the difference was a result of resharpening. He must have ascertained that soon thereafter, for that distinction shortly disappeared without comment from his publications.

Roberts considered blade shape important in differentiating subtypes within Folsom, but he saw fluting as the "essential feature" of the type, though doing so was not without complications. Two points from Folsom were fluted on only one side, and one from Lindenmeier was Folsom in outline and "style of chipping" and had been found in the Folsom-age deposits, yet it was unfluted, likely because it was so "extremely thin" as not to allow the removal of the channel flakes.[188]

Typologically awkward as unfluted Folsom points were, Roberts was at least certain they were not Yuma points (a few of which had also turned up at Lindenmeier), despite the "considerable confusion as to what constitutes [a Yuma] point." Keen to resolve the relative age of Yuma and make the designation "more specific than its present catch-all connotation," Roberts was pleased when stratigraphic hints emerged at Lindenmeier indicating that Folsom points were earlier than Yuma points.[189] Their relative stratigraphic position was confirmed in the 1936 excavations: if the two forms overlapped, it "was at best only a late contemporaneity . . . with a later survival of the Yuma."[190] Howard visited Roberts that summer, saw the evidence himself, and agreed, though he suspected that would not be the end of the discussion. He knew the "Nebraska crowd" (Barbour, Eiseley, Schultz) still believed Folsom and Yuma were contemporary, and they would be "sure to get up on their hind legs" when they heard the news.[191]

By that time, Howard had also realized the problem was greater than just distinguishing Folsom and Yuma points. Another variety of fluted point had begun to turn up in substantial numbers across the United States and even into Canada. Specimens of this type were larger, heavier, thicker, and not as finely trimmed as "true" Folsom points, and they had "grooving more apt to be irregular and to end more abruptly."[192] These specimens, archaeologist Henry Shetrone of the Ohio Historical Society estimated in 1936, numbered in the thousands. He had recorded nearly 150 in Ohio alone (§9.1).[193]

Howard realized he had seen such points before at the Clovis site. Figgins had too, at Dent. Although these points shared with Folsom that "specialized groove," in the wake of the much larger sample of points with which Howard was now familiar he recognized that to call them Folsom points rendered that term meaningless. But old habits are hard to shake, so these points were labeled "Folsom-like," "Folsomoid," "Generalized Folsom," and "Degenerate Folsom,"[194] names that guaranteed continued typological confusion. In 1935 Howard supposed there was "some sort of [chronological] relationship" among these types, though he was reluctant to say what that relationship was, for at that moment there was no geological evidence bearing on the question.[195] The following summer there would be.

In 1936 Howard sent teams back to Clovis under the supervision of John Cotter, a graduate student at the University of Pennsylvania and a Denver native who had worked with Renaud and Figgins and excavated with Roberts at Lindenmeier and Howard at Clovis. Cotter and the crew opened a large excavation in the gravel pit (the "Mammoth Pit," they called it, enjoying the pun). In that block they found partial remains of what turned out to be two mammoths. The bones were on the stratigraphic contact at the base of the blue clay (diatomite) and extended into the underlying speckled sands (§9.2).[196] In excavating down to the mammoth bones, Cotter spotted a stratigraphic trend in the vertebrate remains that Howard had not:

> With striking, though not quite absolute uniformity, a relatively dense debris of bison bones, occasionally articulated . . . occupies the upper two-thirds of the bluish clay layer. . . . Only two instances of bison bones lying in speckled sand were observed. . . . The lower third of the bluish clay material was found to contain, characteristically, the upper portions of mammoth bones, which lay an average of one-third of their bulk in the underlying speckled sand. Occasionally a mammoth bone lay entirely in the speckled sand immediately beneath the contact.[197]

Cotter thought there was nothing to indicate a "marked difference" in the absolute ages of the vertebrates, only that there was a "distinct tendency toward a relative sequence" of their occurrence.[198]

That relative sequence seemed to carry over to the artifacts as well. Found in situ and in direct association with the mammoth bones were several "Folsom-like" points, a half-dozen flakes (some retouched for use as scrapers), and two bone artifacts (later called "bone rods" or "bone foreshafts"). Cotter was certain the contemporaneity of the Folsom-like points and the mammoth "cannot be doubted." It was a significant discovery, one Cotter pointed out to site visitors (such as Antevs) examining the artifacts and bones in place.[199] Cotter also realized that no "true" Folsom points were associated with mammoth remains. That summer the only such specimens were found on spoil piles of diatomite on the edges of the gravel pit, making it impossible to determine their stratigraphic position or the vertebrates with which they might be associated. That was clarified in 1937 when Cotter found a direct association between Folsom points and bison remains.[200] Cotter saw an "implied relationship" between Folsom and Folsom-like points.[201] Possibly that relationship was historical, but he could not help but notice the two types had been aimed at animals of very different body sizes, the smaller and thinner true Folsom points at bison, the longer and thicker Folsom-like points at mammoth. Perhaps the difference was weaponry caliber.[202]

It was more than a decade before another crew returned to Clovis and confirmed Cotter's suspicions about the stratigraphic trend in vertebrate remains and artifacts.[203] But in the interim it appeared that geography might help separate the several point types. Folsom points were found primarily on the Great Plains. Assuming their presence there was not a by-product of better exposure from Dust Bowl erosion, this suggested that the region might be the "focal point of the complex."[204] Because of confusion over the Yuma type, its distribution was not as clear, but it appeared to be more widespread than Folsom, although primarily a phenomena of the Great Plains as well.[205]

In contrast, Generalized Folsom or Folsom-like points had a much wider range. They were especially abundant in the lowlands of central and southern Indiana and Ohio, in southern Kentucky and Tennessee, and along portions of the Coastal Plain from the Carolinas to New Jersey.[206] Unfortunately, the vast majority were surface finds with nothing to indicate their antiquity, save their absence from sites of more recent age. Lest anyone be tempted to infer their age from their geographic extent, Roberts issued a preemptive caution: "The fact that the eastern examples bear a striking resemblance to those in the West does not make them of equal antiquity." After all, they could represent the "survival of a highly specialized implement in later horizons."[207] Warning or no, Roberts could hardly keep from speculating about their antiquity. Invoking the age-area hypothesis (the greater the geographic extent, the older the type), "the eastern examples would indicate more antiquity than the western."[208]

Howard likewise thought Folsom-like points were older, though he arrived at that conclusion via a more circuitous route than Roberts took, having made a stop at one of Renaud's ideas along the way:

It seems to be a fair assumption that the true Folsom proved inefficient, on account of the weakness that developed as the result of the grooves. This we believe we can show by the fact that so many more broken points have been found than whole ones. . . . The true Folsom may, therefore, have been a development of the idea of thinning the point for hafting purposes, or for making the point lighter, which development, in the general New Mexico-Colorado-Nebraska region, proceeded from a cruder type, and which died out before spreading very far because it did not prove to be efficient, while at the same time the cruder, more sturdy forms, which retained the original idea of the groove, spread far across the country.[209]

Given this variation, Shetrone urged these points be grouped under the generic label "Fluted Flint Blade Culture Complex" (he thought it a far better term for a pancontinental phenomenon than the "name of an obscure village"), with subtypes named after their discovery site as in Folsom variant, Lindenmeier variant, and so forth. Roberts and others thought that an eminently sensible idea. Unfortunately, the idea was turned over to a committee that did not.[210]

9.9 A PHILADELPHIA STORY

The venue of that committee meeting was an international conference that Howard organized to help achieve his longtime goal of making Philadelphia a center for research into human antiquity.[211] In 1937 the ANSP would celebrate its 125th anniversary, and Howard planned to mark the occasion by convening a stellar cast of archaeologists, geologists, and paleontologists for an "International Symposium on 'Early Man.'" Coincidentally, it was almost fifty years to the month since Frederick Ward Putnam had called a special "Paleolithic meeting" at the Boston Society of Natural History and put paleoliths from America and abroad on display (§4.3), though of course only those from one side of the Atlantic proved genuine. American archaeology had, as Nelson wrote, "come a long way in [that] half century."[212]

The Philadelphia conference was an extraordinary gathering. Old World attendees presented the latest work on *Dryopithecus* from the Siwalik Hills, australopithecines in the Transvaal, *Homo erectus* in Java, and the newly found early human remains from Mt. Carmel, and they discussed the Paleolithic, Mesolithic, and Neolithic records of Europe and Eurasia. Even the reclusive Eugène Dubois sent a paper on his long-hidden *Pithecanthropus* fossil. The several dozen North Americas participants cut across disciplinary lines and included virtually everyone who had taken part in Late Pleistocene archaeological, geological, or paleontological research over the last several decades, including Barbour, then past eighty.[213] Merriam lent financial support and his name as chair to the venture.[214] The only ones missing from the event? The two individuals most responsible for sparking the decade of intensive research on a Pleistocene human presence in the Americas, Cook and Figgins. (The significance of their absence is addressed in §10.12.)[215]

It was a remarkably rich program, yet of all the topics presented over the several days of the conference only two triggered significant discussion that at times flared into intense debate.[216] In one, the catalyst was point typology; in the other it was Hrdlička, back from a decade of self-imposed silence.

An afternoon roundtable discussion on the typology and distribution of Folsom and Yuma points began with a presentation by Marie Wormington (the CMNH's new curator of archaeology) and Betty Holmes on the results of their study of some 500 points of both types, which attempted to sort them by flaking technique and form (these attributes seemed to correlate).[217] Their presentation released considerable pent-up typological angst, for the discussion that followed pushed the roundtable well beyond its allotted time and forced Howard to adjourn the session so participants could attend a reception and dinner. The discussion resumed the next morning, with the proposal that the best route to resolving the Folsom-Yuma problem would be to create type definitions "that would be acceptable to those at the meeting." Given that most in attendance had disagreed about type definitions the previous half-dozen years, this was at the very least unrealistically optimistic. A committee was charged with the task (more optimism) of deriving those definitions, and came back with ones that were by intent "generic and . . . not related to chronology or distribution." Howard reported their results:

> On this basis the Folsom point was defined as "a leaf-shaped blade. It has a varying base; neither barbed nor stemmed. It is fluted on one or both sides, wholly or partially. It is pressure-flaked from both sides." And the Yuma point was defined as "triangular. It runs from triangular, through parallel sides, to leaf-shaped. Its base is either straight or convex or concave. It is frequently stemmed, but when stemmed, has parallel sides—the sides of the stem are parallel. It is never fluted. It is pressure-flaked from both sides, the flakes being parallel."[218]

The committee was chaired by Harold Gladwin and included Cotter, Wormington, and Renaud, but it was the latter whose fingerprints are most obvious on the attributes identified and the wide range of attributes within each type. That Renaud had disproportionate influence is not surprising. Gladwin was new to the game, and Cotter and Wormington had been Renaud's students only a few years earlier. Despite their suspicions that not all fluted points were alike, and Cotter's recent evidence from the mammoth pit showing as much (a reconstruction of the excavation was on prominent display at the conference), it was difficult for them to push Renaud too far from a position he had stubbornly held for almost a decade.

The types defined by the committee were at best a temporary patch and at worst, in Roberts's view, an unfortunate setback. The term "Folsom" continued to be applied to all fluted points.[219] Still, after much discussion the definitions were approved by the assembly, which evidently agreed with Wormington that they were "acceptable as far as they go," even if they did not go far enough.[220]

A second afternoon roundtable was supposed to address the North American glacial chronology and whether ice advances in the New World and the Old were correlated. But that question could barely "hold the interest" of the participants, not after what Hrdlička said that morning about Minnesota Man.[221] Hrdlička had spent much of the previous ten years working in Alaska and writing about a range of topics related to human anatomy and evolution, but rarely about human antiquity in the Americas. With his gaze elsewhere, he had to be coaxed into attending the conference by Howard, who assured him that despite impatience with his "ultra-conservative viewpoint," his opinion was held in high regard. It would be a "sad commentary on the lack of cooperation" in the field, Howard told him, if he did not attend. Hrdlička, rarely given to humor, joked he could see that "unless I hurry [to accept] I run the danger of being accused of anything from a 'lack of cooperation' to murder."[222] But there was also incentive for Hrdlička to attend. In the previous decade a half-dozen purportedly Pleistocene-age human skeletons, Minnesota Man included, had been found.[223] And that put the game once again on his home court.

9.10 WHAT HAVE THE BONES TO SAY?

Hrdlička's views had not changed during his time away. He still rejected the idea that *Homo sapiens* was "somehow, cryptically, coeval from far back with the pre-Neanderthaler" (§6.2, §7.6). Neanderthals were not a separate species but instead a "phase" of human evolutionary history that graded morphologically and chronologically into modern humans. By the 1930s he was virtually alone in holding that view, as was obvious in the reaction to the lecture he gave in the late 1920s at the Royal Anthropological Institute of Great Britain on receiving from them the Huxley Medal. Despite having been urged by his hosts to not bring up Neanderthals and their place in human evolution, he did just that, and he was publicly chastised afterward by Grafton Elliot Smith, who sniped in *Nature* that

> Dr. Aleš Hrdlička has questioned the validity of the specific distinction of Neanderthal man, an issue which most anatomists imagined to have been definitely settled. . . . The only justification for re-opening the problem of the status of Neanderthal man would be afforded by new evidence or new views [but] I do not think Dr. Hrdlička has given any valid reasons for rejecting the view that *Homo neanderthalensis* is a species distinct from *Homo sapiens*.[224]

In Hrdlička's view Neanderthals were on the Pleistocene landscape from the third interglacial through the last glacial period, from 130,000 to 70,000 years ago. Modern humans did not appear until the Late Glacial, some 70,000 to 25,000 years ago.[225] Even in that later period, he surmised, these early modern humans ought to be anatomically distinct:

> The physical differences observable between Neanderthal and later man are essentially those of two categories, namely: (1) Reduction in musculature, that of the jaws as well as that of the body-with consequent changes in the teeth, jaws, face, and vault of the skull; and (2) Changes in the supraorbital torus, of the order known well to morphology as progressive infantilism. . . .

Further reduction of teeth, jaws, and the facial bones has taken place since [Postglacial] Magdalenian times and is now going on in more highly civilized man, of whatever racial derivation.[226]

By extension, that held true in America as well: a Late Glacial or Postglacial skeleton should show its age. On this very broad principle Hrdlička and others agreed.[227]Agreement vanished, however, when principle was put into practice. It was a threefold parting of the ways, manifest most visibly in the difference between Hrdlička and Hooton, the two major physical anthropologists of the day.

By Hrdlička's reckoning a skeleton that dated to the third interglacial or final glacial period had to be a Neanderthal. In contrast, for anthropologists of the 1930s who derived *Homo sapiens* from a "non-Neanderthaloid stock" from the Early Pleistocene (as most did[228]), it was

> not unreasonable, but on the contrary most probable, that migrants into the New World during the third interglacial period [Sangamon] would be comparable in evolutionary development to the Galley Hill type, the Cro-Magnon type, the Grimaldi type, or other late fossils types of essentially modern anatomical conformation.[229]

Hooton and others saw anatomically modern humans where (and when) Hrdlička saw only Neanderthals.

Compounding that difference in the interpretation of the types and timing of human evolutionary history was a different understanding of the process of anatomical and evolutionary change. Hooton, with one eye on Hrdlička and the other on Piltdown, argued there was an "essentially asymmetrical character" to evolution, which led to "morphological inequalities" in development. Some features might appear "modern" long before others—for example, Piltdown's very modern cranium yet ape-like jaw.[230] Hooton distinguished his view from Hrdlička's "implicit faith in an [*sic*] unilinear and symmetrical evolution, through the cumulative changes of minor variations [in a] slow but harmonious progression." In effect, both viewed morphological change as an evolutionary clock, but for Hrdlička the parts changed in sync (§6.2), whereas for Hooton some anatomical changes far outpaced others.

Finally, there were the questions of which anatomical features were useful for distinguishing ostensibly ancient skeletons from modern ones and whether those differences were significant and outside the range of variation in modern Native American populations, among whom there could be seemingly "primitive" features (even ones reminiscent of Neanderthals).[231] Answering these questions inevitably meant taking on Hrdlička in a contest of which he was a master. Nonetheless, several tried.

A human skull and postcranial elements were discovered in 1935 in the Cimarron River valley twenty miles downstream of Folsom. Figgins identified the individual as a new species, *Homo novusmundus*. Anticipating (correctly) that there would be resistance to his naming a species from a single specimen, he argued he was merely following taxonomic protocols of mammalogy and paleontology.

Were these nonhuman bones, there would be "no question" of the "propriety of recognizing them as of specific importance." Insofar as Figgins practiced taxonomy in those disciplines, that was the case. Two years earlier he had named one new genus (that from a single molar), three new species, and four new subspecies of bison (§8.11).[232] Left unsaid was that none of those taxa were accepted.[233]

Not surprisingly, the same fate befell *H. novusmundus*. At Roberts's request Figgins sent the specimen to the Smithsonian to be examined by Hrdlička, who then forwarded it to Hooton for study. From there it went to Harry Shapiro at the AMNH. Each had a different opinion of its affinities. (Hrdlička predictably sniffed, "Many and perhaps all of the features of the Figgins skull could [be] duplicated in the oblong-headed Pueblo and the Algonkin crania.") None thought it a new species.[234] Even if they accepted it as *Homo sapiens*, there was no secure evidence of its geological antiquity, notwithstanding Figgins's claim that it resembled the Pleistocene crania of Brünn and Predmost in Moravia.[235]

Of the other finds made that decade, Minnesota Man seemed to have the greatest chance of making it into the Pleistocene.[236] The remains were discovered in 1931 by a road repair crew who, after removing concrete pavement and grading off 2.5 feet of clay, spotted a human skull, whelk shell fragments, and an antler "dagger." They suspended grading to recover as much of the skeleton as they could, which they assembled by the roadside in anatomical position to "make a man out of it."[237] The find was reported to Jenks, who—from the descriptions and geological maps—suspected it was glacial lake clay covering the skeleton. A check of Highway Department records revealed nearly seven feet of overlying sediment had been removed to make the roadbed in 1929, putting the original depth of the skeleton at nearly ten feet below surface.

The following summer Jenks and a crew excavated from the shoulder of the road to its midline and recovered more skeletal fragments, along with portions of a turtle carapace, a loon metatarsal, a wolf's incisor, and a muskrat calcaneus, which Jenks interpreted as remains of a medicine bundle. Geologist George Thiel of the University of Minnesota confirmed the deposit was glacial lake clay, but according to Leverett it was outside the limits of Glacial Lake Agassiz and thus from an earlier glacial lake, which Jenks named Glacial Lake Pelican.[238] In January of 1933, Jenks announced Minnesota Man was actually a young woman (as was the custom of the time, he and others continued to refer to her as Minnesota Man[239]), with "distinct Mongoloid affinities" as well as features "unexpected in primitive American types." She had been recovered in lake clays that he estimated to be at least 20,000 years old.[240]

That assumed her remains were in situ, and Antevs for one was skeptical. Looking at Thiel's photographs of the varves and a sample of the sediment convinced him the deposit was disturbed, raising the possibility that "the person was buried by a landslide long after the formation of the silt."[241] Bryan visited the site with Jenks in 1935 and assured him (and later announced in *Science*) that Antevs must be mistaken. Bryan saw "no geologic phenomena in sight or processes in action that afford an indication of or reasonable probability of landsliding or similar disturbances which would lead to the intrusive deposition of the skeleton."[242]

Jenks's monograph on the discovery appeared the following year, now with a far more dramatic story line. The young woman had drowned in the icy waters of Glacial Lake Pelican, having fallen out of a boat or through the ice. The accident happened between 25,000 and 18,000 years ago, a conclusion that made prominent mention of Bryan's affirmation of the skeleton being in situ. Antevs's opinion went unmentioned.[243]

At that antiquity Jenks did expect the skeleton to appear modern, and perhaps not coincidentally his analysis supported that. He judged her teeth particularly large and unusual, reminiscent of the *Dryopithecus* pattern, with a face that displayed a prognathism not unlike that in European Paleolithic crania. Her "primitive" features were further affirmed when Jenks calculated her score using a quasi-quantitative system Hooton had developed a decade earlier to scale anatomical features from anthropoid apes to humans (this on his assumption that different features evolved at different rates). A human skeleton or modern group was scored on forty-one cranial and facial morphological attributes such as brow ridges, slope of foramen magnum, prognathism, size of mandibles, and projection of canine teeth. Possible scores went from 1 ("ultra-anthropoid") to 6 ("ultra-human").[244] The average and standard deviation for each fossil or group was calculated, the result said to indicate a skull's (or group's) antiquity and position on an evolutionary scale, which ranged from "ultra-anthropoids" (male and female gorillas and male orangutans), "typical anthropoids" (female orangutans, chimpanzees, and the *Pithecanthropus* and Heidelberg fossils), "subhuman" (Broken Hill, Piltdown,[245] and Neanderthals), "inferior human" (Eskimo, Australian, Combe Capelle, Negro, and Mongoloid), and the pinnacle of "typically human" (the Mediterranean, Nordic, and Alpine races).[246] (W. W. Howells, one of Hooton's most distinguished students, whose own work on skeletal populations undercut Hooton's, believed Hooton was not a racist, just misguided by his typological biases, craniometric methods, and poor understanding of statistics and sampling.[247] Howells was being generous).

Jenks calculated Minnesota Man as a 4.08 on the Hooton scale, which put her in the "inferior human" class, above the Eskimo (4.0) but below the Australian (4.10). Although she was Mongoloid, she had non-Mongoloid characteristics that hinted of a slight admixture with "early White and Negroid strains."[248] Hooton, not surprisingly, thought Jenks's study of Minnesota Man was "exemplary."[249]

Hrdlička did not, and lost no time saying so. He privately expressed his skepticism to Jenks in 1932, prompted in part by an "unsolicited letter from a western geologist" who cast doubt on the geological circumstances of the find.[250] If only Jenks had not made any claims about the antiquity of the find, the diagnosis could have been straightforward and hardly caused a "ripple of excitement." For Jenks's benefit, Hrdlička showed Jenks how his conclusion *ought* to have read:

A fairly ordinary skeleton of an Indian adolescent girl, approaching closely the Siouan type, with large, though not very exceptionally so, teeth and consequent rather pronounced alveolar protrusion, and with a few small anomalies. Also, in all probability an old burial, with intriguing but on the whole insufficient indications as to great age.[251]

Hrdlička doubted the skeleton had any geological antiquity, certainly not that could be demonstrated in light of the "serious and irreparable uncertainty as to the original status of the ground above and immediately about the skeleton."[252] The anatomical evidence likewise belied a claim of great antiquity because the posture of the body, Hrdlička declared, was not that of a drowned person. (Hooton, who rarely missed the opportunity to swing at one of Hrdlička's slow pitches, gleefully asked, "What is the ordinary posture of a drowned person?"[253]) Most important, Minnesota Man's anatomical features "must be adjudged as inseparable" from a large sample of female Sioux (the relevant comparison, Hrdlička noted, as the Sioux lived in the region historically).[254] Even her large teeth were known to be "especially frequent" among the Sioux. Jenks had not realized that few skulls "fail to show some exceptional, anomalous, or primitive features," and he lacked a sufficiently large comparative sample that could clarify whether a particular trait was characteristic, historically meaningful, or merely idiosyncratic.[255]

If they were Sioux, the remains could not be 20,000 years old. No group stayed in one territory that long, not without undergoing anatomical change. The completeness of the skeleton and the fact that two other burials were nearby (with metal artifacts of obvious historic vintage), supported Hrdlička's alternative interpretation that this was a young Sioux woman buried in glacial clays in relatively recent times.[256]

Hrdlička's critique of Minnesota Man appeared in January of 1937, and at the "Early Man" conference in Philadelphia two months later he used it as the centerpiece of a sweeping attack on all the latest skeletal finds of purported geological antiquity.[257] The moment Hrdlička finished speaking, Henry Retzek, who had hosted Bryan's 1935 fieldwork in Minnesota, leapt to his feet to announce he planned to speak on Minnesota Man at the "North American Chronology" roundtable that afternoon. He did, triggering "a lively discussion" over the age of the find between Hrdlička and Antevs on one side and Bryan and MacClintock and on the other.[258] (Jenks was absent; he published his response a year later.[259])

After ninety minutes of that "lively discussion" there was no consensus on whether Minnesota Man was modern or Pleistocene in age, but there was agreement by Antevs, Bryan, Kay, Leighton, and MacClintock that they would participate in a joint field conference at the site.[260] In June Antevs, Bryan, and MacClintock visited with Jenks and Retzek, but they had to satisfy themselves studying a section of road cut fifty feet from the site. In August the highway department closed half the road and diverted traffic, allowing Kay and Leighton, with Jenks and Thiel, to reopen the site.[261]

Antevs proclaimed that "all agreed that the varved glacial silt in the road cut about the site was disturbed, a condition heretofore denied." On the basis of the fauna in the medicine bundle, indicative of modern climate conditions, and the whelk shell that must have arrived through trade from the Gulf of Mexico 1,100 miles distant (trade that he learned was common in Hopewell and Mississippian times), Antevs concluded the young woman had been buried or died in a gully as recently as A. D. 1,000–1,500.[262] Bryan and MacClintock conceded there was some slight evidence of clay and sand movement, but it was so "trivial" it only rendered the varves useless for precise chronological purposes.[263] Otherwise, neither they nor Kay and Leighton saw evidence of mass wasting of sediment sufficient to have buried the skeleton beneath ten feet of overburden, nor did they see the gully Antevs offered as an alternative by which the skeleton had been buried. They came down firmly in favor of Jenks's interpretation.[264]

The question of whether Minnesota Man was Pleistocene in age was hopelessly deadlocked, and as had happened so many times in previous decades resolution was stymied by the fact that no one had seen the skeleton in situ.[265] Proponents were confident the "trusted employees of the State Highway Department" had years of experience observing stratigraphy. Yet, there was no detouring around Antevs's point that they saw the site only after it had been "heaved by frost, pressed by the weight of the tractor, and smeared and crumpled by the blade of the grader."[266] And no matter what the road crew saw in 1931 when they removed 2.5 feet of sediment, there was no hint—nor anyone to ask—what the other seven feet of overburden looked like in 1929 when the roadbed was first dug.

Jenks continued to proclaim a Pleistocene age for the young woman who was Minnesota Man[267] and used her anatomy as a measure of her antiquity. When two more skeletons from Minnesota were subsequently discovered, Browns Valley Man and Sauk Valley Man, he arranged them chronologically according to their anatomy (with help from Hooton's scale): Sauk Valley Man was in the middle because he was "more primitive than the Browns Valley Man, though not as primitive as the Minnesota Man." Because Browns Valley Man was associated with Folsom and Yuma points Jenks put its antiquity at between 8,000 and 12,000 years, with Sauk Valley Man older, but still younger than Minnesota Man, at 20,000 years.[268]

Hrdlička did not respond. But when he read Jenks's discussion of the Browns valley and Sauk valley remains (as he likely did—few escaped his notice), one can only wonder if he saw himself in the mirror. For Jenks was doing what Hrdlička had done many times before, using anatomy and morphology as a clock. When Luther Cressman subsequently approached Hrdlička to describe a potentially ancient skeleton from Catlow Cave (Oregon) and offered to provide Antevs's notes and interpretation on the geology of the site, Hrdlička curtly dismissed the offer: "If the bones do not speak, nothing else will."[269]

9.11 PROFILING

The lack of skeletal remains assignable to a Pleistocene antiquity meant there was "no possibility of suggesting relationships based on physical characteristics" among ancient or with modern Native American populations. Despite that, there was plenty of speculation throughout the 1930s about the type, variety, and population history of groups in the Americas, much of it based on one's assessment of the significance of the anatomical variation seen in more recent archaeological and modern populations, and whether that revealed anything of ancestral groups.[270]

Hooton laid down the gauntlet in the early 1930s with his analysis of 129 crania from Pecos Pueblo that he classified by their morphology in an attempt to "trace the affinities of these types outside of the American continents." His process was suspect because he visually sorted the crania into groups, then applied rudimentary statistics to show the groups were real (prompting the laconic observation from Howells that "given the process of selection, it is not surprising that the statistics appeared to support the distinction among the types").[271]

"Immigrant types," Hooton called the resulting groups based on his bedrock belief that race was a matter of biology, not culture; that such features were heritable; and that therefore similar features marked common ancestry.[272] Among the prehistoric inhabitants of Pecos he distinguished seven distinct craniofacial types (an eighth, the "Residual" type, was used for skulls that did not fit into the other seven): "Basket-Maker," "Pseudo-Negroid," "Pseudo-Australoid," "Plains Indian," "Long-faced European," "Pseudo-Alpine," and "Large Hybrid."[273] Although Hooton used labels highly suggestive of racial or geographic origin or both, they were allusions (hence the "pseudo" prefix) more so than implied direct links. His Long-faced European, for example, "was not in the least like any European type which [he] could adduce."[274]

What the labels signified for him were clues to deeper ancestry and admixture, anatomical echoes of racial origins and history, which became more apparent when he lumped the seven types into three broad cephalic groups (§6.4). There were the long-headed "dolichocephalic" types (Basket-Maker, Pseudo-Negroid, Pseudo-Australoid), which hinted at a distant mix of "Mediterranean, Negroid, and Archaic White" elements. The long-headed type was thought to have temporal priority and thus were the earliest Americans. "Palae-Americans," Hooton called them, and they were a "general mélange of strains fused together in the early invaders before their departure from the Asiatic continent."[275]

The long-headed group was "glossed over with Mongoloid traits from admixture" by later-arriving round-headed "brachycephalic" types (Long-faced European, Pseudo-Alpine, Large Hybrid). Anatomically these had the "familiar Mongoloid complex" along with "a greater capacity for civilization and cultural development," which meant they and not the dolichocephalics deserved credit for developing agriculture and New World civilization.[276] Unsure what to make of the intermediate "mesocephalic" Plains Indian type, Hooton dropped it in between as "a type resulting from the admixture of the long- and round-headed types.[277] Hooton published his Pecos skeletal study in a massive volume he knew Hrdlička would not approve. Knowing that, Hooton naturally dedicated the book to "Aleš Hrdlička, great student of the physical anthropology of the American Indian." It's hard not to admire such a brazen sense of humor.[278]

Hrdlička had, years before, declared it was impossible to attribute *any* American cranial variants to non-American racial groups, for two reasons. First, within the American Indian population cranial form ranged "from extreme dolichocephaly to hyper-brachycephaly; from low to high vault; and through all the other [cranial measures] as well as the intermediary shapes." Likewise, "there are further whole gamuts almost of variants in the breadth and length of the nose, the height, breadth, form and size of the orbits, [and] development of the teeth and jaws."[279] Any American cranium might resemble ones from the past and other parts of the world, but that was not evidence of a historical relationship. It only emphasized the importance and necessity of fully understanding the variety of Native American types, and appreciating that the differences were of degree and not kind (bearing in mind Hrdlička had no difficulty distinguishing a Sioux from a Shoshone by their skulls). And, secondly, for Hrdlička the American Indian past and present was part of "one large strain of humanity," which was derived solely from the "yellow-brown [race], whose old home was northern Asia."[280] No others participated in the mix.

Hooton was aware of Hrdlička's position, identifying it as one of four possible alternative explanations for the diversity he saw at Pecos, but he rejected it. It failed to explain, he argued, the non-Mongoloid features seen in the Pecos crania. For his part, Hrdlička did not accept that there were non-Mongoloid features in American crania, let alone any significant "racial differentiation" (which, of course, only reinforced his suspicions of any claims of great antiquity; if people had been in the New World since the Pleistocene, there ought to be differentiation).[281]

If Hooton swung too far to one extreme, Hrdlička went too far to the other, "oversimplifying the situation, in his somewhat impatient sweeping aside of all minor American differentiations as being nearly insignificant," as Kroeber put it.[282] This was not just about racial grouping. Embedded within their contrasting views were very different implications for the peopling process. Was it one of multiple migrations over a long span of time by different ancestral groups who had originated from across a vast geographic area of the Old World, then merged into a highly variable and admixed descendant population (Hooton)?[283] Or was it migration from a single geographic source of an "amorphous *Rassenkreis*" (a polytypic group) that, even if it involved "dribbling over from northeastern Asia . . . probably over a long stretch of time," was sufficiently homogeneous in its heterogeneity that however long the dribbling the result was always the same (Hrdlička)?[284]

The positions were irreconcilable, though one could—as Howells, McCown, and others did—attempt to navigate the terrain between the two. (As physical anthropologists, Howells and McCown had the wherewithal to do this; one feels greater sympathy for the burden under which Roberts, an archaeologist, labored in writing about the supposed early human skeletal remains, for he learned his physical anthropology from Hooton but Hrdlička was in an office just down the hall.[285]) The route Howells picked between those polar positions emphasized the wide variety of types within the Mongoloid family. Too often, he argued, the type was exemplified by the most anatomically specialized members of that group (the Chinese and Mongols), when it seemed to him there was a substantial "unspecialized" component observable in the peoples of central and northern Siberia and Tibet, who "exhibit the same vagueness of type and variation in form as do the Indians and share with them the same unspecialized and somewhat primitive configuration." But to demonstrate that point, McCown added, there needed to be more human skeletal evidence from Asia to ascertain the "time of origin and the precise character of the Asiatic Mongoloid peoples."[286]

In so moving toward Hrdlička, Howells simultaneously backed away from Hooton, arguing that the presumed non-Mongoloid skeletal features among prehistoric and contemporary American Indians were just the same unspecialized traits seen in Asia. In Howell's view, then, Native Americans were descendants of different migratory strains (Hooton's position, watered down), but all were ultimately northern, central, or eastern Asian in origin (Hrdlička's view).[287]

Kroeber took matters a step further. Recognizing that Howell's position had the advantage of going "less far afield geographically and racially" in search of the origin of American Indian diversity, he wondered if the process might have played out closer to home. Why not consider a "fairly heavy ingredient of internal differentiation" (that is, genetic drift, though Kroeber did not use the term) on the North American continent after arrival? There seemed no reason to preclude this possibility, as there had been at least 10,000 years for that internal differentiation to take place (and some diversity surely came with the group migrating in). Kroeber's position made smaller, or at least less "far-flung," assumptions about origins, and if and "only as a last resort" it proved necessary, one still could invoke "hypothetical pseudo-Australoids, Caucasoids, and the like."[288] Kroeber did not think that need would arise.

By decade's end, then, skeletal diversity and time could be reconciled, though doing so depended on the meaning and significance one attributed to skeletal diversity, how much variety one was willing to grant had arrived with in-migrating groups, how quickly it was assumed anatomical change occurred, and how much time one was willing to accept for prehistory on this continent.[289] It was a case of balancing the equation: if diversity was high, more time or greater preexisting variation, or both, were needed. If diversity was low, the time or diversity, or both, of in-migrating groups could be reduced, and so on. In the 1930s there was no consensus on the values to assign those variables in the equation.

9.12 FINDING THE TIME

Howard had seen this coming and was concerned that "we must not narrow down too much the margin of time for the development of our native cultures or else it will be difficult to account for such diversity of language, culture, and somatological characters as existed."[290] The "somatological" characters aside, what of linguistic and cultural development? Edward Sapir, Pliny Goddard, and others had insisted (§7.4) they needed more time than archaeologists had been willing to provide to account for the diversity of Native American languages and cultures (assuming they were descendants of a single group). Yet, as Wissler pointed out, there had never been a "satisfactory way . . . found to express this interval in years."[291] Though now in possession of 10,000 years and appreciating the "surprising swiftness"[292] by which language and culture could change, some linguists and social anthropologists were still not satisfied.

Native American languages, Franz Boas argued, were "so different among themselves that it seems doubtful whether the period of 10,000 years is sufficient for their differentiation." Multiple waves of migration might account for that diversity, but Boas was unhappy with that "arbitrary solution." He found it difficult to accept 10,000 years as a "final judgment," and demanded archaeologists continue the search for earlier sites. If none were found it would be necessary to either demonstrate there had been multiple waves of migration or, at far greater conceptual cost, "we [will] have to revise our opinions in regard to the stability of types and of fundamental grammatical forms."[293] He was not prepared to pay that cost.

Still, 10,000 years was sufficient to resolve another problem regarding cultural change. By the late 1920s the concept of diffusion, an important element of American historical anthropology under Boas,[294] had been taken to farcical extremes by the British hyperdiffusionist school. They assumed that human groups were by nature "uninventive" and that civilization was not likely to have developed independently but must have arisen in one region from where it was carried around the globe. Elliot Smith picked ancient Egypt as the source because here "the agricultural mode of life . . . furnished the favorable conditions of settled existence, conditions which brought with them the need for such things as represent the material foundations of civilization." To the suggestion that agriculture and civilization happened independently, Smith triumphantly asked why, if that was so, people nonetheless waited "all those hundreds of thousands, perhaps millions, of years before any of them took such a so-called obvious and inevitable step?"[295]

That chronological coincidence was evidence of a causal relationship for the diffusionists, but a hurdle for those like Kidder who were "100 percent American regarding the origin of the higher cultures of the Western Hemisphere." That is, it appeared as if complex society in the Americas sprang out of nowhere and no earlier than about 2,000 years ago. How could a "100 percent American" explain the origins of "respectable and highly ramified civilization" in the New World in those temporal circumstances?[296] Hyperdiffusionists had no such problem, attributing the seemingly sudden New World appearance of similar forms of agriculture, pottery, religion, pyramids, mummification, and other features to direct or indirect influence of seafaring Egyptians.[297] Could it be, Wissler asked tongue in cheek, that Native Americans were merely "disreputable plagiarist[s]?"[298]

Before 1927, it was difficult to say. So much archaeological attention had focused on "such spectacular remains as the cliff houses and mounds, the Maya temples and Peruvian cemeteries," that little was known of what came before. (Kidder blamed the museums that directed "practically all archaeological fieldwork" and were keen to have "interesting and striking finds.") To be sure, one could list many cultural elements present in the New World and absent in the Old (and vice versa).[299] But without evidence for indigenous origins of New World civilization, let alone for an initial peopling in Pleistocene times, such arguments were easily dismissed by hyperdiffusionists. Old World traits that were no longer present could have been lost in the centuries after their arrival.

It was because of the hyperdiffusionists that Kidder had been so quick to go to Folsom when Roberts called him in early September of 1927 (§8.10). Folsom said nothing about the origins of New World civilization per se, but it provided something just as valuable to Kidder: "ample chronological elbow room, even by Old World standards." And that, he added, ". . . is a great relief to me personally, as I think it must be to many other believers in the essentially autochthonous growth of the higher American cultures."[300] The relief was only temporary, however, for it still did not explain the origins of New World civilization, it only removed the chronological challenge of doing so. But Kidder was pleased enough to tackle that problem without the added burden of contending with the hyperdiffusionists.

Kidder also realized there was another large task looming after Folsom. What was going on, he asked, "in North America during the millennia that intervened between the time of . . . Folsom nomads and, say, the Basket Makers" or between "the Folsom epoch and the period of the well-developed farming, pottery-making cultures of [the Andean] region . . . ?"[301] Whatever the precise age of these sites, "a decisive and perhaps long gap separates them from the earliest remains which can be put into continuous archaeological sequence."[302] Attempts had been made to link the oldest of the North American sites, human remains, and tool assemblages with subsequent Indian tribes by using the distribution of "linguistic stocks, head form, chipping techniques, and physical environment," but—as Roberts observed—the resulting scenarios were so highly speculative that they were little more than "the play of fancy."[303] There was too great a chasm between the Late Pleistocene and the Late Prehistoric.

Folsom presented a challenge to American archaeology writ large, for though a great deal had been learned throughout the 1930s, by 1941 that only served to emphasize how little was known of the many thousands of years of prehistory that followed. That gap was unfortunate, Kroeber wrote, "but it must be admitted and should be emphasized." And something needed to be done to fill in the millennia between the known (the Late Prehistoric) and the remote unknown (the Late Pleistocene).[304] Meeting that challenge occupied much of the time and effort of American archaeologists in the decades to follow, involved a variety of new methods and approaches, and triggered the rise of Culture History, the approach that dominated American archaeology through the middle of the twentieth century.[305]

9.13 FAST FORWARD

In May of 1941, Bryan published in *Science* an address he had delivered twice at the AAAS meeting the previous December (once to Section E and once to Section H). The topic, "The [Geologic] Antiquity of Man in North America," was well received by both his audiences, though Hrdlička, not surprisingly, dismissed it as "bias & ignorance to the limits."[306] In it, Bryan trumpeted the advances made in understanding human antiquity in North America since the Folsom discovery in 1927, the most telltale measure of that being the fact that "we are no longer asking the question: What is the antiquity of the American Indian, but What is the antiquity of the Paleo-Indian?"

It was a clever turn of phrase, though Bryan knew it hardly began to capture all that had happened. For in those dozen or so years the antiquity dial had indeed been turned back, but far more had been learned than the fact there was a new starting date for the earliest peoples. This was so in part because of the remarkably rapid growth in the empirical record. Whereas in 1927 there was only one confirmed early site, by 1941 dozens had been excavated and hundreds of surface localities and occurrences were recorded.[307] One simple but illuminating measure of the increase in the body of evidence: it now required small monographs by Sellards and Wormington, or a lengthy article by Roberts, to adequately describe, discuss, or even list (Sellards) the archaeological and relevant nonarchaeological evidence from this period.[308]

Writing in 1940, Roberts knew studies of this period were in their infancy, yet he thought, as many did, that enough progress had been made to "provide a basis for a few broad conclusions."[309] Those were answers to questions about the origin, antiquity, and adaptations of early peoples that could not have been asked or answered a dozen years earlier. Some were more inference than fact, but they became the foundation for understanding North American Paleoindians through the end of the twentieth century.

Foremost among them was that North American prehistory began in the Pleistocene, but barely so. It was not a prehistory on an Old World time scale. To speak of an "early" New World occupation, Roberts explained, is to speak "of its relation to the modern Indians and not in that of its connotations in Old World studies."[310] The ghost of an American Paleolithic and the notion that prehistoric sequences on opposite sides of the Atlantic ought to be in sync were finally banished.[311] Even so, it was common in the 1930s for the flint-knapping skill of Folsom and like points to be likened to those of Upper Paleolithic Europe (notably Solutrean) or from the Danish and Egyptian Neolithic.[312] But no meaning beyond that was intended, given the "probable Asiatic derivation of the [American] tool-making industries."[313] Roberts was adamant. Any attempt to "correlate these artifacts with those of Europe or to insist that materials here should conform to the pattern there are based on erroneous conceptions." Still, this was not "a case of archaeological isolationism," he assured his readers, all of whom at the moment were acutely aware of the heated debate over whether the United States should enter the war in Europe, and would have understood his allusion.[314] No Charles Lindbergh, he.

Just what an "early" occupation of North America meant on an absolute time scale was uncertain, but there was virtual unanimity that people had been in North America 10,000 years and surely longer, though perhaps not much more than about 25,000 years.[315] In the mid-1930s Howard polled various parties for their opinions about the length of time "since man's arrival in the New World." Their estimates averaged roughly 14,000 years.[316] Hard numbers only came with Willard Libby's invention of radiocarbon dating after World War II. It was entirely fitting that Folsom was the first Paleoindian site Libby tried to date with his new technique, though Libby's effort failed because Harold Cook, never a stickler for details, provided a charcoal sample he had not collected, that was not actually from the Folsom site, and that yielded an age much younger than expected, which caused considerable confusion.[317]

Regardless of their precise antiquity, it was generally agreed that Paleoindians were on the landscape at a time when "lingering portions of the great ice sheet" still covered large parts of the northern half of the continent. It was obviously a world very different in climate and environment from that of the present, and even from that experienced by later Basketmaker groups. That, in turn, raised questions about glacial effects "on climatic conditions . . . on the routes of travel, the movements of peoples, and [on] the animals that were their main source of sustenance and maintenance."[318]

If, as seemed likely, migration was timed to the rhythms of glaciation, then for much of the Late Pleistocene travel to North America south of the ice sheets must have been a two-step process, for "passage to Alaska [via the Bering land bridge] would have been easy during the last glaciation," but "escape from Alaska would have been impossible until the recession of the ice."[319] With the slight differences in the environments of Siberia and Alaska, Bryan supposed there could have been repeated movement back and forth across the land bridge. Extending that reasoning, he argued that traffic headed south from Alaska must have been one-way only, for by the time migrants reached unglaciated North America they would have so thoroughly "changed their habits" in adapting to the different resources further south that they had effectively "built an environmental wall behind them." Besides, with the empty expanses of two great continents still in front of them there was "no inducement to backtrack," only to move forward.[320]

By 1941 the consensus was that the "escape" route south from Alaska was through an interior ice-free corridor rather than along the coast, the latter dismissed a priori as being more difficult because of the narrow coastal strip, the heavily glaciated terrain, and the presumed necessity of "a specialized fishing culture and the use of boats for movement." Most put the interior corridor through the Mackenzie valley and along the eastern slope of the Rocky Mountains. That route was thought to have been free of ice after the glacial maximum, though how soon after was still debated (§9.7). Some supposed it took no time at all to open, others that it was not available until 10,000 years later.[321]

That there was more than one interior route, coupled with differences emerging by decade's end in the archaeological record, led Roberts to speculate there might have been multiple migrations into North America (akin to Hrdlička's migratory dribbles [§9.11]). With the land bridge in place for thousands of years, different groups must have arrived in Alaska. Perhaps one departed after the initial opening of the Mackenzie valley, but then "minor oscillations" in the ice sheets subsequently closed that route (or made it less favorable). If the next group did not leave Alaska until later, they may have traveled along the newly opened Yukon valley route, which took them south between the Rockies and Coast Range. The earlier and later groups may have exited their corridors on opposite sides of the Rockies; if so, Roberts thought that could explain why Clovis and Folsom fluted points were common in Paleoindian sites on the Great Plains and to the east whereas a variety of nonfluted lanceolates (Lake Mohave, Pinto Basin, Cochise[322]) dominated the earliest sites in the Great Basin and far west. They were from different migratory "streams."

Roberts knew the idea of multiple migrations was an accommodative argument—a "broad interpretation of the record as it appears today" is how he put it—rather than a theoretical statement about the processes of colonization. Still, he liked it because it might also explain "the occasional appearance of an older physical type in recent groups," though there was hardly unanimity among physical anthropologists that any explanation was needed (§9.11), and Roberts himself appreciated that any statement about the physical characteristics of the first Americans was based on "flimsy evidence."[323]

Given that the migrants successfully traversed the Arctic environments of the Bering land bridge and ice-free corridor, it was reasoned that they must have been hunters, plant foods being sparse in those regions in the brief summers and absent altogether in winter. No sites were available to assess that inference, but certainly the archaeological record of the Great Plains sites seemed to mark Paleoindians as a "typical" hunting people "depending entirely upon game—mainly bison, but occasionally the mammoth and a stray camel, deer, and antelope."[324] Roberts granted the possibility that they supplemented their diet with seeds and "greens" but was quick to add that they did not cultivate plant foods. Cultivation required intimate knowledge of native plants and time to develop, and hence did not appear until the Basketmaker period.[325]

By virtue of their dependence on game, Paleoindians must have been highly mobile nomads who followed herds and had little incentive to migrate to a region where animals were absent. That they did not settle long in one place helped explain the scarcity of evidence for the shelters they may have used, and perhaps accounted for the puzzling scarcity of skeletal remains from this period.[326]

Paleoindian prey included extinct animals, but just how long all of those animals had been extinct was still an open question at decade's end.[327] And now a much larger question loomed. Having brought the majority of extinct animals into the Late Pleistocene and possibly the early Recent period, and with the apparent lack of evidence for marked Postglacial climate changes, the question was what role had human hunters played in their extinction (§8.17, §9.5)? In 1933 Romer thought the possibility that hunters were responsible for the extinction of the twenty-six genera that survived into the Late Pleistocene (Table 9.1) was "far from probable," given how few taxa had been found in archaeological sites (§9.5). If humans had an impact, he thought it indirect at best: perhaps their arrival threw "out of balance a fauna in delicate adjustment."[328]

Kroeber did not disagree, but he saw more nuance to the issue. There was no reason to believe that 10,000 years ago there were "sudden and decisive geological or climatic changes which simultaneously wiped out a considerable number of species." He also doubted all animals went extinct simultaneously. It was "inherently much more reasonable" that different genera disappeared at different times from the full glacial into the Recent. Extinctions that occurred early were likely due to climate change, as people were then few in number, their "arts and weapons [were] underdeveloped," and animals could live alongside them "without serious molestation." But when tribes were later "better equipped and organized," prey was more vulnerable to extinction as "large game animals quickly succumb entirely to man once he is in a position to hunt them."[329] The basic questions regarding the role or roles of human hunting, climate change, and mammalian extinctions were now in place. Answers, however, were nowhere in sight in 1941.

By decade's end, many thousands of the "unique and exasperating" (Nelson's opinion) Paleoindian projectile points had been described and their types debated, and in September 1941 much of the terminological fog lifted at a conference in Santa Fe (organized, yet again, by Howard). "Folsom-like" points were formally codified as "Clovis," Folsom points became a much more restricted class, and the catch-all Yuma category—which included so wide a range of unfluted points as to be essentially useless—was discarded altogether and replaced by less ambiguous site-based labels, just as Shetrone had earlier suggested (§9.8).[330] There was by then a general consensus that fluted points were the earliest types in North America, with Clovis in place before Folsom (§9.8). With that, a few began to wonder about the possibility of still-earlier occupations in North America. After all, if fluted points were of strictly American origin, as Howard supposed (§9.7), then *someone* must have come before. At decade's end, matters were still "decidedly blurred" with regard to who that may have been.[331]

The search for a pre-Clovis presence in North America started in earnest soon after the War.

CHAPTER TEN

Controversy and Its Resolution

THE GREAT ANTHROPOLOGIST ALFRED KROEBER was trying not to be condescending but having a hard time of it. "Incredible as it may now seem [the year was 1952] by 1915–1925 so little time perspective had been achieved in archaeology that [Clark] Wissler and I, in trying to reconstruct the native American past, could then actually infer more from the distributions and typology of ethnographic data than from the archaeologists' determinations." This struck Kroeber as curious because "even explorers as experienced as [William Henry] Holmes and [Jesse] Fewkes saw their archaeological pasts as completely flat." As to why that was so, Kroeber speculated it was a lack of willingness to venture into issues of chronology, possibly the result of not knowing how to attack chronological problems "by properly archaeological" methods, or perhaps just "American unhistorical mindedness" in which the past was seen "not as a receding stereoscopic continuum but as a uniform nonpresent."[1]

All three elements may have been in play, but as the foregoing narrative has shown, the causes of archaeology's "flat past" were far more complicated than Kroeber supposed. That flattening was so pervasive that Kroeber failed to realize he was as much a part of the problem as part of the solution.

Looking forward from the 1870s, it seemed as though the question of whether humans were in Pleistocene North America would be easily answered, and with less resistance than had marked the establishment of human antiquity in Europe. After all, that effort had laid the evidentiary and conceptual groundwork necessary for identifying a Pleistocene human presence using Louis Agassiz's glaciers and George Cuvier's extinct Pleistocene mammals (§2.1). Known too were the features by which genuine artifacts could be distinguished from naturally flaked stones, the forms and attributes of Paleolithic artifacts, and how these differed in material, workmanship, and type from those of Neolithic Europe.[2] Glimpsed as well were some of the distinctive anatomical characteristics of premodern humans—Neanderthals most especially—who lived in that distant past.

Equally important, by the 1870s the powerful and long-standing theological objections to a deep human antiquity had been effectively scuttled so far as the great majority of the scientific community was concerned, though within the broader populace theologians—and one would soon count George Frederick Wright among their number—and other writers continued to object to the idea. There was also the theoretical framework of evolution, both biological and cultural, for understanding and, indeed, anticipating a Pleistocene human past.[3] So too there was the expectation, at least initially, that geological and human history were the same in both the Old World and the New (§2.2). And most people readily accepted the possibility—based on the linguistic, cultural, and physical diversity of Native Americans—that their ancestors may have arrived a very long time ago (§2.1). The bar to a deep human antiquity on this continent appeared to be set rather low. But appearances were deceiving.

The search for and the ensuing controversy over a Pleistocene human presence in North America began in earnest on the banks of the Delaware River in Trenton, New Jersey in the 1870s. It did not end until the late 1920s in a small headwaters tributary of the Dry Cimarron River outside Folsom, New Mexico. There was not a straight line connecting those two points. In the five decades in between, some two dozen sites were heralded as proof of a Pleistocene human presence. Not all were alike in their evidence or the inferences drawn from them, which as the narrative has shown changed significantly over the decades. Yet, in every case but one (Folsom), the outcome was the same: each contender failed to pass critical muster on archaeological, anatomical, or geological grounds—or all three. There were many and varied reasons why they failed, aside from the fundamental, if not flagrantly presentist, observation that virtually none of those sites was Pleistocene in age. The 12 Mile Creek (§6.9) and Lone Wolf Creek (§8.3) sites were exceptions to that rule, but their antiquity became certain only after the Folsom discovery, which highlights just how ambiguous matters were up to the very end of the controversy (§10.15).

The repeated failure to demonstrate a Pleistocene human antiquity did not preclude the possibility that one existed, nor did it prevent it from being proved, testimony to the adage that the growth of knowledge can progress and ultimately be cumulative without being continuous or linear.[4] But that failure did create an atmosphere of profound skepticism that no archaeologist, Nels Nelson insisted, "can be expected to relinquish at once."[5] It led to mistrust between and among practitioners and disciplines (§7.5). And it made many in the archaeological community of the time reluctant, even fearful, to engage in the search for early sites and face the attacks that would inevitably follow.[6]

Though the narrative has provided the fine-grained historical details of this controversy, which in the end was all about whether there was depth beyond Kroeber's "flat past," a narrative takes us only so far in understanding what this controversy was about. Having explored and illustrated what, where, when, and how the dispute played out, I now step back from the narrative to explore the overarching themes and underlying dimensions that help explain why the controversy developed, defied solution for so long and so bitterly, and yet ultimately was resolved. Hence, this concluding chapter is organized thematically rather than chronologically, though I reference the narrative throughout so that this chapter can simultaneously serve as a review. I begin with the core empirical and conceptual issues that drove this dispute, then move outward to the broader nonepistemic influences and context. In this controversy the epistemic elements graded into and were ultimately inseparable from the larger social stage on which they played, a continuum that cannot be separated into tidy internal and external elements.[7]

10.1 THE MEDIUM IS NOT THE MESSAGE

Central to resolving the human antiquity controversy was gaining control over time, which proved an extremely difficult task, one still not fully settled (though settled enough) by the time of the Folsom discovery (§8.11). The central question was whether humans had been in North America since the Pleistocene, yet at the outset there was no agreement as to what a Pleistocene antiquity meant in relative terms or on an absolute time scale, how it might be measured, and especially whether the answer might be had from archaeological evidence alone.

Early on it was assumed artifact form could tell time in the "rude" or "primitive" artifacts that, as Charles Abbott declared at Trenton, differed in form and raw material from those used by known Native Americans and found in contexts stratigraphically and spatially separate from Indian village sites. Most significant, these bore an uncanny resemblance to artifacts of Paleolithic Europe (§3.1, §3.6), and if there was likeness in form, as argued by Henry Haynes and Thomas Wilson, whose European experience gave them credibility (however fleeting), then by analogy they ought to be comparable in evolutionary grade (Paleolithic) and geological age (Pleistocene) (§3.6, §4.14).

Abbott crafted a highly imaginative scenario by which European Paleolithic peoples made their way to North America, thence to the Arctic where they survived as Eskimos (§3.6). The European archaeological record was more than mere metaphor. In one respect Abbott was right: at a deep enough time, artifacts are indeed less sophisticated technologically. But the ones he was finding were not on that scale, which of course he did not know. He could only infer as much from what he knew of European Paleolithic artifacts and what he suspected (from what Henry Lewis told him) of the age of the Trenton Gravel (§3.5–§3.6). Others—famous Europeans, even—agreed and honored his discovery: America's "Boucher de Perthes," Abbott was called, and he was proud to accept the title. Like many American-born scientists (and easterners) of the time, he was anxious to put his work in the context of European scholarship.[8]

The American Paleolithic was at its core a visual argument, which is why American paleoliths were frequently displayed and illustrated alongside European specimens; why Wilson had no doubt that "comparison is as good a rule of evidence in archaeology as in law," thus allowing surface specimens to be identified as Paleolithic; and why George Frederick Wright was confident that once people knew what to look for and where (valleys like the Delaware), they would find paleoliths (§4.3). As Wright predicted, Trenton was followed by discoveries at sites such as Little Falls (Minnesota), Madisonville (Ohio), Claymont (Delaware), Medora (Indiana), Loveland (Ohio), New Comerstown (Ohio), and Washington, DC (§3.6, §4.1–§4.3, §4.5, §5.4). That so many sites were found in Trenton's wake was gratifying to American Paleolithic proponents and confirmed their long-held assumption that America *ought* to have tools alike in form, evolutionary grade, and antiquity as those of Pleistocene Europe. Nothing succeeds like success.

But Abbott and others had omitted a crucial step: artifact form could tell time, but only after time had been told for it. Their antiquity had to be confirmed independently. With glacial geology in its infancy and often causing more confusion than clarity, in the rush to the American Paleolithic that step was skipped. Abbott, of course, hardly thought it mattered (§3.6).

Holmes thought otherwise. From the Piney Branch quarry he launched a relentless attack on the conceptual underpinnings of the American Paleolithic (§5.10). He readily admitted that a rude appearance could be old, as the earliest forms of Paleolithic stone tools were unquestionably primitive. Evolutionary "progress" run backward saw to that. But with a nod to Ernst Haeckel's biogenetic law applied to the "ontogeny" and "phylogeny" of stone tool production, Holmes reasoned that rude forms might equally mark the early stages of stone tool production (§5.2). Instead of being ancient, a paleolith might be a recently made yet unfinished specimen, or one broken and jettisoned partway through the manufacturing process. He showed that two apparently distinct and immutable artifact types might be related by virtue of being different stages in the same chain of manufacture, and that the same forms could be finished tools of one kind or unfinished early stages in the manufacture of another.[9] This was a novel approach to material culture and made coherent what had been disjointed observations of "turtle-backs," hatchets, spears, and scrapers (§3.1). Holmes understood, as his contemporaries using a typological approach did not, that artifact forms were a static embodiment of an underlying dynamic of manufacture and use, and thus why turtlebacks might not appear in all sites. Holmes was in the main correct about the paleoliths of Abbott and others. These were (in modern terms) cores and early-stage bifaces, and not necessarily Pleistocene in age. The bulk of the Trenton archaeological sequence, in fact, is much more recent (see the Appendix to this volume).[10]

At a time when "object lesson" was a phrase taken literally, and order and control were prized in science (and in the increasing chaos of late nineteenth-century industrial America[11]), Holmes was a magician at making silent objects speak and reveal the complex yet coherent organizational structure invisible beneath their surface. Among those "who despised mystery and lauded simple, common-sense explanations, Holmes's breakthrough established him fully."[12] It earned him the Loubat Prize as well as praise that his work "revolutionized" American archaeology (§5.14). Even Franz Boas grudgingly admitted Holmes's "natural gifts lead him to a thorough appreciation of [the meaning of] visual objects."[13]

Viewing the process of stone tool technology as opposed to merely its static products, Holmes argued there was equifinality to artifact morphology. That meant a choice of analogy, and the choice had to be made carefully. Were "primitive" specimens like those of Paleolithic Europe or those from stone tool production sites like Piney Branch? In principle the possibility of equifinality was applicable to primitive stone artifacts of Europe, leading Holmes to suspect some of those venerable Paleolithic specimens were unfinished rather than ancient, and prompting Otis Mason to joke that if Holmes was right Boucher de Perthes might turn out to have been the "Dr. Abbott of France" (§5.11).

Yet, there was one key difference between Europe and America that Holmes could not erase. In Europe the Stone Age was long past. The odds favored a primitive specimen in England or France being Pleistocene in age, because no one had been making stone tools there for a very long time. Those odds did not apply in

America, where the past was present and in recent memory native peoples still made "Stone Age" artifacts (§2.3). The relative rudeness of an artifact might say something about its evolutionary grade (though exactly what was unclear), but by itself that said nothing about its age. "When grade, or stage of culture, is the concept," Mason observed, "the word 'prehistoric' does not refer to time at all."[14] In America, antiquity was not an inherent attribute of stone artifacts. Holmes insisted that must be determined independently by geological context (§5.2, §5.10).

Holmes was not plowing entirely new ground. Charles Rau and others, including Frederic Ward Putnam, Abbott's Peabody Museum champion, had earlier glimpsed the problem of assuming that form and age were congruent (§4.15). Holmes, however, saw more clearly the complications of artifact form itself. American Paleolithic proponents were nonetheless confident they could distinguish true paleoliths from Indian artifacts (§3.3, §5.3, §5.6). That belief proved to be one of the first casualties of the Great Paleolithic War. Proponents and critics failed to agree, even when looking at the same specimens, whether they were finished or unfinished, or paleoliths as opposed to early-stage manufacturing rejects (§5.2– §5.3, §5.6, §5.10–§5.11). The debate was an exercise in incommensurability, which played into Holmes's hands. He was certain he was right and American Paleolithic proponents were wrong about the form and antiquity of those specimens, but he was equally sure he did not need to prove the specimens were recent, only that they *could* be. That was sufficient to check American Paleolithic claims, and check them he did: by the mid-1890s the strong flow of discoveries of paleoliths slowed to a trickle.

Any lingering hope of recognizing paleoliths in America or of assigning them to a Pleistocene antiquity by morphology alone evaporated altogether in 1897 at the joint meeting of the AAAS and the BAAS. There, John Evans himself examined a box of Trenton specimens and rejected the lot. Whatever they were, they were not Paleolithic (§5.16). It was desperately important to many American Paleolithic proponents to have Evans's approval: Abbott had even fashioned one of his books after Evans's *The Ancient Stone Implements, Weapons and Ornaments of Great Britain* (§3.6). It was devastating when that approval was not forthcoming. Proponents put on a brave front but knew they had been beaten. New discoveries of American Paleolithic artifacts altogether ceased by century's end (§5.16). Nothing fails like failure.

As attention in the new century turned to human skeletal remains, the question of whether time could be read from morphology resurfaced. Hrdlička invoked the same premise Abbott had, that a primitive appearance (in Hrdlička's case, of human skeletal remains) indicated a Pleistocene age. Underlying that premise was Hrdlička's view that the interaction of heredity, adaptation, and environment over evolutionary time led to gradual but ultimately profound changes in skeletal morphology (§6.2). Accordingly, as one looked backward in time the features of the human skeleton ought to become increasingly primitive. But biological evolution ran in only one direction. With human skeletal anatomy there was no possibility (as Holmes declared there was with stone artifacts) that a primitive form might date to the Recent period—or the converse, that a modern form could be Pleistocene in age.

Thus, whereas Abbott sought to make artifacts in America old by virtue of their primitive form and resemblance to European paleoliths, Hrdlička aimed to make human skeletal remains young owing to their resemblance to recent Native Americans. It was the same principle, only applied in the opposite direction. Of course, whether one accepted Hrdlička's approach depended on one's view of how far back the Recent period extended and when anatomically modern humans first appeared. Hrdlička, unlike virtually all of his peers in America and abroad, rejected the idea that modern humans were geologically ancient, Piltdown notwithstanding (§6.2, §7.6, §9.10). He believed Neanderthals were the Pleistocene ancestors of modern humans, and that as recently as 15,000 years ago they "had not yet fully reached modern standards in physical development" (§7.4).[15] That being the case, any skeleton in the Americas older than that—as Lansing and Long's Hill were said by their proponents to be—*must* appear anatomically primitive (§6.6, §6.12).

Few agreed with him, believing that modern humans appeared much earlier in the Pleistocene or possibly the Pliocene (§6.8, §9.10).[16] Vertebrate paleontologist Oliver Hay was not alone in insisting Neanderthals were unrelated to modern humans, though he took that view to a breathtaking extreme: "If every other mammal made no perceptible physical advance in 500,000 or 1,000,000 years what right have we to assume that beings like those of the Neanderthal race have evolved into a condition of the highest races in one-tenth of that time?"[17] Hay granted a deep antiquity to anatomically modern humans and was equally generous on behalf of "Neolithic" artifacts, scarcely blinking at the prospect of pottery and metates dating back hundreds of thousands of years at Vero (§7.5) and Frederick (§8.17).

After the Piltdown discovery, many physical anthropologists agreed with Hay about how far back modern humans might go (§7.6, §9.10), but few archaeologists would travel that distance with those types of artifacts. Beyond the obvious problem that the tools associated with Piltdown were not Neolithic but instead "Eolithic" (and, indeed, not artifacts at all[18]), the suggestion that Early Pleistocene Americans were making Neolithic artifacts had troubling implications for the understanding of human cultural evolution, especially after the failure to prove that there was an American Paleolithic. Not that Hay cared. If deer, bear, wolf, and other species were "modern" deep in the Pleistocene, then modern humans were as well. Unless, that is, one believed humans did not abide by the rules and rates of evolutionary change that governed other species (§7.5–§7.6). Hay did not. For that matter, Hrdlička did not either, but only because he rejected the idea that any modern species reached deep into the Pleistocene, notwithstanding Piltdown and Galley Hill (§6.11, §7.6).

Hrdlička believed his method of "morphological dating" not only distinguished Pleistocene from recent skeletal remains but also trumped geological evidence. (Even Abbott grudgingly conceded geology could be relevant.) Thus, Hrdlička's approach was the same whether applied to the supposed anatomically primitive Burlington County and Riverview crania (§6.4) or Erwin Barbour's Long's Hill Neanderthal-like crania (§6.11), and even to those sites said to be Pleistocene in age by their geology rather than skeletal anatomy, such as Lansing (§6.8), Vero

(§7.4, §7.6), and Melbourne (§7.9). Hrdlička dismissed them all, convinced that their anatomical attributes were within the broad range of variability he perceived in historic and modern Native Americans and thus could not be Pleistocene in age no matter what the other evidence indicated. So certain was he on this point that he did not bother to explain why the Long's Hill individuals were *not* Neanderthal (§6.14). Many thought, as Kroeber did, that Hrdlička oversimplified the situation (§9.11), and even Hrdlička's allies in geology were nonplused by his dismissal of their evidence (§7.5, §10.8). Yet, none of those promoting the Pleistocene antiquity of these specimens could best Hrdlička's knowledge of human skeletal anatomy (at least until Ernst Hooton came along [§9.11]), or show that a particular specimen was outside the range of recent Native Americans, or match his formidable inability to admit error (though Hay came close).

In the end, virtually all participants in the controversy agreed, albeit reluctantly in some quarters (and not at all in Hrdlička's case), that a Pleistocene antiquity could not be demonstrated with artifacts or human skeletal remains alone, especially not if antiquity was based on similarities to European forms. But was it possible to *reject* a claim of Pleistocene antiquity in America using patterns known from Europe? Hay, Harold Cook, and Jesse Figgins did not think so (§8.8, §8.17). They rejected the incongruity of pottery with Early Pleistocene vertebrates at Vero (§7.5) and of metates with mammoths at Frederick (§8.7) as merely a "creation of the anthropologists." Just because "stone implements appear only late in Europe and are crude," Hay argued with homegrown boosterism, was no reason to believe that "the art of working stone must have had a similar development in America."[19]

By the time he made that declaration, however, enough was known about the broad trends of cultural evolution that archaeologists had no trouble swatting Frederick away. They might not know when metates first appeared in the archaeological record, but there was no reason to believe it was in the Early Pleistocene. They were not willing, as Nels Nelson put it, to "revise our entire world view" on the basis of Frederick and what it implied of the origin, development, and spread of culture (§8.8, §8.17). The New World archaeological sequence did not have to conform to the Old World sequence, but then it could not unduly violate certain of its parameters either. If it did, the chronological evidence had to be wrong.

That the chronological burden ultimately fell entirely on geologists and vertebrate paleontologists had far-reaching effects on the structure of the debate and its participants (§10.7–§10.8). For although it proved impossible to make the case for a Pleistocene antiquity solely on artifact or skeletal forms, it proved not much easier to do so with geological evidence. The challenge was twofold. First, one had to demonstrate that the artifacts or human skeletal remains were in primary context—that is, deposited at the same time as the formation in which they occurred or with the extinct vertebrate remains with which they were found. Then one had to ascertain the geological age of those formations or extinct vertebrates.

The principles involved were straightforward, or so Thomas Chamberlin declared: what was true of the "assignment of a pebble or a bowlder [to a glacial formation] . . . should hold good for human relics." Yet, Chamberlin was under few illusions about matters being easy when "human relics" were involved.[20]

10.2 CHALLENGING CONTEXT

It is a cardinal rule of archaeology, not to mention a perennial bane of its existence, that the only opportunity to determine with any confidence the depositional context or associations of artifacts or human skeletal remains is when they are still in situ.[21] The manner in which artifacts are "read" before they are removed from the ground determines how data are marshaled as evidence. Such concerns were not fully appreciated at the beginning of this controversy, for it was a time when there were few (if any) agreed-on strategies for fieldwork. After many painful lessons they were, by the end, well understood.

There were many reasons why context proved an especially knotty problem, beginning with the fact that most discoveries were not the result of controlled survey or excavation. Rather, they were largely haphazard, opportunistic, and incidental to other activities, as at Madisonville (§4.2), Vero (§7.2), Lone Wolf Creek (§8.3), and Folsom (§8.4). Abbott's decades of walking the banks of the Delaware valley and Ernst Volk's dutiful trudging behind the laborers digging Trenton's railroad and sewer lines were no exception to this rule (§3.15, §4.14, §6.1). Abbott inadvertently revealed a great deal about the nature of his fieldwork (and conceptual approach) when he titled one of his books *A Naturalist's Rambles about Home* (§4.2).

Furthermore, many of the discoveries were made by observers who could not recognize their potential significance, such as the Concannon brothers at Lansing (§6.6) and Nelson Vaughan at Lone Wolf Creek (§8.3). Even finds by individuals with archaeological knowledge and experience, such as Charles Metz at Madisonville (§4.2) and W. C. Mills at New Comerstown (§4.14), were taken from the ground before they could be examined in situ and their stratigraphic context assessed. Once specimens were removed from their original position, they could never be put back as they had been. Information vital to ascertaining whether they were in primary context, and from that assessing their age, was lost.

More complicating still, many if not most paleoliths were recovered from eroding valley walls or other active geomorphic settings that had complex stratigraphic sequences that challenged efforts to determine how the deposits formed, their relationship to distant glacial features, and their age (§10.3). It was also the case that since few specimens were recovered from excavation (controlled or otherwise), the stratigraphy was often not well exposed or readily examined. That prompted Holmes to follow up with his own excavations in two instances (he sent William Dinwiddie to Trenton [§5.4] and Gerard Fowke to Lansing [§6.8]), but no further remains were found in situ to help establish the context of the original discoveries.

Dispute was inevitable where individuals of varying and sometimes questionable skills worked largely alone through the "controversy-laden" process of recognizing stratigraphic units, teasing apart depositional and erosional histories, and trying to understand how and when artifacts or remains of human skeletons entered those deposits or came to be associated with faunal remains. Their descriptions and images (of which there were few) were poor proxies for those who were not there to see the evidence for themselves: far too much was riding on the outcome. Wright published a photograph showing where Mills found his paleolith at New Comerstown, but Holmes only sneered that a photograph "made of the tree after the bird has flown will not help us determining the bird" (to which Wright naturally replied it was strange to have Holmes "speak so slightingly of the value of a photograph . . . when he has himself given two fancy sketches, representing an impossible condition of things, to inform us how he thinks it might have been") (§5.11).[22]

In the absence of evidence witnessed in place, proponents routinely emphasized the experience and the integrity of the discoverer, who was often the sole individual to see the specimens in the ground. The only way around their testimony, Lucien Carr supposed, was to say "the boy lied"[23] (§4.15, §6.3), which as it happened was said on several occasions of Abbott, around whom there long lingered a whiff of untrustworthiness (§4.2, §5.6).

In fact, whether the "boy" was truthful about what he had seen and done was irrelevant to the critics. Wright thought Mill's character and experience made his testimony about the geological context of the New Comerstown specimen "as good as that of any expert," to which Holmes sneered that "observations on Mr. Mills' moral character, his education or business reputation [did not] diminish the danger of error" in properly identifying the deposit (§5.11).[24] The issue was not merely trust in the individuals to report honestly what they had seen, but trust that they could report the evidence accurately. The difference between proponents and critics, baldly put, was that critics were convinced honesty and intent were beside the point because most discoverers were inattentive amateurs (including Abbott §10.9) and few were sufficiently skilled in geology to differentiate undisturbed and disturbed deposits.

In his scorched-earth march through the American Paleolithic sites—accompanied by geologists such as Frank Leverett, WJ McGee, and Rollin Salisbury—Holmes enumerated the many ways (slope wash, mass wasting, tree throws, rodent burrowing, and the like) by which artifacts on the surface, and thus of recent vintage, might become incorporated in older, Pleistocene-age deposits (§5.10). He and others deemed stratigraphic mixing a problem so pervasive that they assumed any artifact found deep in an unconsolidated deposit was guilty of being an "adventitious inclusion"[25] until proved otherwise (§5.10). To that assumption Hrdlička added his own spin: the "universal" tendency of humans to bury their dead meant no human remains were ever likely to be the same age as the deposits in which they were found (§6.12, §7.4).

Naturally, these assumptions were vigorously disputed by proponents, who insisted it was meaningless to assume that artifacts or skeletal remains were in a secondary context without evidence that they had, in fact, been redeposited or buried (§4.15, §5.11, §7.7). Moreover, if mixing occurred, why were later-period artifacts not also in those ostensibly disturbed layers (§3.3, §5.10–§5.11)? Of course, because critics thought those artifacts and skeletal remains *were* recent in age, then they *did* occur in those layers.

Joint site visits, where proponents and critics examined the same ground—as at Trenton (§5.15), Lansing (§6.6), Long's Hill (§6.12), and Vero (§7.3–§7.4)—failed to alleviate critics' doubts regarding the ability of proponents to read stratigraphic evidence, or to convince proponents that critics could see past their preconceived biases. As a result, usually hovering just below the formal exchanges at these events was a tension that at times erupted into outright hostility between those who had done the work and those who swooped in for the kill. Because of the institutional and regional alignments of critics and proponents these confrontations in person and in print could also tap deep veins of resentment, whether between federal and state scientists, small museums and large ones, or even easterners and westerners. More often than not, and Folsom is the exception here (§8.10), site visits served to widen rather than narrow the gap between antagonists. This fact did not change after Folsom, as the events surrounding the Minnesota Man site showed (§9.10).

The result was considerable heat but little light on the question of primary versus secondary deposition. Of course, even if the context of the archaeological remains was satisfactorily resolved for all concerned, as it rarely was, one still had to ascertain the age of the deposits in which the remains were found or of the age of the fauna with which they were associated.

A final point regarding context: at all of the sites throughout the decades of controversy there was a fleeting moment when artifacts or human skeletal remains could be seen in the ground. The issue was not that specimens were never in situ but instead that opportunities to examine them in place were rare. That was not the case at Folsom (§8.10) as a result of Hrdlička's suggestion to Figgins that if further artifacts were found in place with the bison bones they should be left untouched and telegrams sent to independent observers to examine the find. Although Figgins thought the suggestion innocuous, it was not. Hrdlička did not trust his (or Cook's) reading of the find (§8.9). Hrdlička was hardly alone in this regard. For no matter one's viewpoint on this controversy, it was apparent by 1927 that the only means of resolving questions of context was for evidence to be viewed in the ground. Hence the importance of the telegrams sent from Folsom, but also the credibility of those who responded to those telegrams (§10.10–§10.11).

This is not to claim that had telegrams been sent out from 12 Mile Creek in 1896 or Lone Wolf Creek in 1924 it would have led to their acceptance as Pleistocene age sites. There was still the question of the antiquity of the bison associated with those artifacts (§6.9). All one might reliably draw from the historical record is that were it not for the telegrams broadcast by Figgins the outcome at Folsom might have been very different. Of course, had the antiquity of the Folsom bison been in question, it could also have led to a different result. But by 1927, the pieces necessary to determining the geological antiquity of the site had fallen into place, though that process had not been without difficulty.

10.3 ASCERTAINING ANTIQUITY

The challenge of recognizing the age of a site's deposits or its extinct fauna thoroughly entangled geologists and vertebrate paleontologists in the human antiquity controversy, despite the fact that few of them had strong opinions or much cared when humans reached America (§5.5). There were exceptions, most notably Wright and Hay. Wright's advocacy of a Pleistocene human antiquity was hamstrung by a theological unwillingness to accept deep time, with the consequence that he fought a constant rear-guard action against developments in glacial geology (§3.8, §4.9). Hay's championing a Pleistocene human antiquity was less a philosophical stand than it was a way of striking back at Holmes and Hrdlička for their cavalier dismissal of his arguments for Vero's antiquity on the basis of its vertebrate fauna (§7.7).

Even for geologists and paleontologists with little interest or investment in human antiquity the controversy was often unavoidable, as the effort to determine the age of those sites exposed conceptual gaps in their own field. Indeed, the USGS's John Wesley Powell, savvy politician that he was (§5.1), used the controversy to advantage, invoking the American Paleolithic debate to justify the Survey's mapping "the several stages of ice occupancy and of the formations that were produced thereby." Doing so, he intoned, was a "matter of intellectual importance and even of moral consequence."[26] What the moral consequences were of knowing the glacial sequence, Powell did not say, but the intellectual importance was obvious enough. Hilborne Cresson's Claymont paleoliths were reportedly found in situ in McGee's Columbia Formation (§4.5). If correct, it meant humans reached the New World in the penultimate glacial stage, which suggested a far deeper antiquity than anyone previously imagined.

The partitioning of the Pleistocene into separate stages was the work of many, though the effort was largely led by USGS geologists and most especially Chamberlin, longtime chief of the Survey's Glacial Division, and his USGS and University of Chicago colleagues Leverett and Salisbury (§4.6–§4.7). With a mandate to investigate the geology of the country, USGS geologists were able to map Pleistocene terminal moraines on a continental scale, unhindered by the borders that constrained state geologists. That made it easier to trace complex moraine systems from regions where they were spatially separated and readily distinguishable, such as the Midwest, into areas such as the Northeast and New England, where moraines of different ages often overlapped and were less readily delineated. However, the USGS geologists sometimes encountered resistance—from New England geologists, such as Yale's James Dana and Dartmouth's Charles Hitchcock (§4.7, §5.12), who staunchly resisted the idea of multiple glacial advances, as well as from state geologists over work they considered rightly theirs. That resistance resulted in occasional delays and statewide detours to avoid stepping on local toes, as well as episodes of hostility toward the Survey, most spectacularly during the Great Paleolithic War of 1892–1893 (§4.7, §5.8–§5.9).

Still, by the early 1880s Agassiz's single glacial stage had been divided in two. A decade later USGS geologists had pushed the number of stages to four, and by the early twentieth century to six (the number settled back and hovered around four or five by Folsom times [§4.7, §5.12, Table 6.1, Table 8.1]). By the first decade of the twentieth century there was broad acceptance of the idea that glacial ice had advanced and retreated multiple times, save for Wright's very public resistance to the notion, even while serving as a temporary field assistant for the USGS Glacial Division (§3.8, §4.9, §4.11, §5.12).

Wright's opposition was driven in no small measure by the implications of multiple glaciation for human antiquity. If, for example, the Claymont paleolith came from deposits dating to a penultimate glacial stage, that implied an antiquity on a scale Wright vigorously opposed. He had to either reject the find or reject the idea that there had been multiple glacial stages. To avoid being put in that same awkward position with future discoveries, he chose the latter strategy. Even by the 1880s it was a reactionary path (§4.6, §4.8) that put him on a collision course with Chamberlin. Wright accepted the inevitable: doctrine cannot be questioned without loss of faith. And because Wright's views of human antiquity were intimately linked with his ideas on glacial geology, and as Wright was the de facto geologist for many of the supposed American Paleolithic sites, it guaranteed the dispute would cross disciplinary lines.

Leaving interdisciplinary conflict aside for now (a thread picked up in §10.8), there were still areas of ambiguity glacial geologists needed to clarify. These included differentiating the physical deposits left by various advances, and whether features such as Wright's "fringe" represented separate stages—as Chamberlin and Salisbury argued—or were extramorainal debris from the single glacial event, as Wright supposed (§5.5, §5.12). Then too, as Chamberlin put it, there was the "misplaced confidence [of] even glacialists of ability and experience" in distinguishing a "true glacio-fluvial [outwash] plain" from material reworked from those deposits in postglacial time, possibly thousands of years after the ice retreated (§6.9). That difference might be trivial in geologic terms, he observed, but "of much moment in human history."[27] Most challenging of all was identifying glacial-age deposits from south of the southernmost terminal moraine that were not physically connected or readily linked to primary glacial formations. As all presumed Pleistocene archaeological sites were located in such regions, determining their age was no easy task.

The Lansing case well illustrates how complicated and contentious geological matters and their archaeological consequences could become. Differences emerged over whether the sediment bearing the human remains was "true" loess from the Iowan stage and so about 30,000 years old, as Newton Winchell insisted, or was a loesslike deposit reworked in Recent times from an older Pleistocene loess, as Chamberlin saw it (§6.8). That led to a vigorous debate over how to recognize loess, how much it varied, whether there was such a thing as true loess and what that looked like, how it could be identified in the field, and whether all loess deposits were the same (glacial) age (§6.9–§6.10).

That disagreement quickly became entangled in the larger dispute over loess origins. In the 1880s Chamberlin and Salisbury had rejected the idea that loess was deposited by wind, favoring instead the "aqueous" theory. But geomorphic anomalies plagued that idea (not least that "special fondness" of loess, as Chamberlin complained, for draping summit heights). This, coupled with evidence from Bohumil Shimek's snails, led Chamberlin and Salisbury a few years before Lansing to favor wind as the depositional mechanism (§6.7). Winchell and Warren Upham refused to go along, insisting loess was deposited by glacial floodwaters. Wright agreed, not least because if there had been vast floods perhaps one of them was Noah's. Wright was always on the lookout for geological testimony of Genesis, much to Chamberlin's disgust (§6.7).

The chronological implications for the Lansing remains were obvious. If the loess was deposited by the wind, the remains could be—but were not necessarily—Pleistocene in age (§6.8). As Samuel Calvin pointedly reminded Upham, the billowing clouds of dust they had seen on their visit to Lansing was loess forming "before our very eyes" (§6.6). But if the loess was deposited by water, then the remains had to be Pleistocene in age, though that also required glacial meltwater lakes that extended far beyond the glacial front and, more troubling for good uniformitarian geologists—as Chamberlin was and Wright was not—requiring crustal deformation and catastrophic epeirogenic tilting on an immense geographic scale (§6.7, §6.10).

There is little doubt that this multifaceted intradisciplinary debate over loess would have played out regardless of whether an archaeological site was involved. Still, it is fair to claim that Lansing triggered it and helped clarify the ambiguity surrounding loess and its origins, so much so that few of these issues arose at Long's Hill just a few years later (§6.13). Likewise, American Paleolithic sites such as Claymont and Trenton revealed the need to clarify the age of the gravels in which artifacts or skeletal remains were found and where these deposits occurred within the glacial sequence.

In fact, it was largely disputes over the age of these archaeological sites that prompted Chamberlin to introduce the "Studies for Students" section in his newly founded *Journal of Geology*. There he, Salisbury, and others lectured students and their colleagues on the challenges, complexities, and criteria for identifying glacial deposits from different advances, differentiating primary from secondary glacial deposits (the "bête noire of the incautious glacialist," as Chamberlin [§6.9] put it), and teasing apart the complicated geological processes that could deceive even "the very elect" among glacial geologists (§5.12, §6.9). Chamberlin had no doubt who was among the "very elect" and who was not (§10.10).

With the increasing understanding of glacial deposits came the realization that determining the age of artifacts or skeletal remains in them was at best chronologically imprecise, even leaving aside the problem of stratigraphic mixing (§10.2). Gravel or loess accumulated too slowly to pinpoint the age of the enclosed artifacts or human bones. As sites were found that contained the bones of extinct fauna, attention shifted from seeking to determine the age of archaeological remains from their surrounding deposits—their *context*—to the age of the vertebrates with which they were found—their *association*. Because fossil remains had the "merit of definite fixity in time and localization," they held out the promise of greater chronological resolution. The challenge, as Chamberlin observed in 1903, was that the Pleistocene faunal sequences had "yet to be developed in the main" (§6.9).[28]

That those sequences were not better resolved was due to the fact that studies of the North American Pleistocene faunas had a late start. At the end of the nineteenth century relatively little was known of the temporal range of extinct taxa, how far back into the Pleistocene each extended, the stages of the Pleistocene in which they were present, how recently they went extinct, and whether extinctions were synchronous across all taxa (§6.9, §7.1). Their occurrence in an archaeological site was not temporally diagnostic, especially given the suspicion that some of the Pleistocene fauna survived into the Recent.

Gaining chronological control proved particularly challenging for taxa that inhabited unglaciated North America or whose fossil occurrences otherwise could not be readily linked to deposits of known glacial age, as well as for apparently extinct species of still-living genera. In the case of the bison, which loomed large in the controversy, it was necessary to first demonstrate that the presumably extinct and extant species differed before the timing of their extinction could be settled (§6.9, §8.6, §8.11, §9.5). For many years it was difficult to avoid Chamberlin's "Janus-faced" dilemma of choosing whether bison in a site such as 12 Mile Creek (§6.9) meant "the antiquity of man or the recency of extinction of the animals."[29]

Again, archaeological questions provided the spark to resolve issues outside that discipline. The effort to ascertain the ages of apparently extinct taxa began in earnest at Vero, where the conflict between the anatomical, geological, and paleontological evidence spotlighted how little was known of the chronology of the Pleistocene vertebrates. That gap in knowledge, not to mention the sneers of Holmes and Hrdlička, launched Hay, then seventy-two years old, on a multiyear project to compile and map the distribution of Pleistocene vertebrate fossils, particularly as they related to glacial formations, in order to pin down their geological age and develop a vertebrate sequence (§7.4, §7.7).

That effort laid the groundwork for Hay and Cook's subsequent identifications of the Lone Wolf Creek and Folsom bison as an extinct Pleistocene species (§8.6, §8.11), and though their taxonomic assignment proved relatively uncontroversial, their effort to put its extinction in the Early Pleistocene—stoutly resisted by Barnum Brown, among others—was not (§8.11). Hay's compilations of the fossil record later helped Alfred Romer reject the ages Hay had assigned bison and other Pleistocene vertebrates, as well as dismantle the sequence of extinction he had created from those data (§9.5).

Would this geological and paleontological research have been undertaken had it not been driven by the effort to determine the age of archaeological sites? Almost certainly, but as it had not happened earlier or independently, the timing of these developments can be safely attributed to an archaeological catalyst.

10.4 NUMBERS GOING NOWHERE

In the decades before Folsom the central chronological question was posed in relative terms: did a site belong to the Pleistocene or the Recent? Few concerned themselves with numerical ages because methods to determine an absolute chronology were highly speculative and unreliable. It was, as Upham observed, "comparatively easy to determine the ratios or relative lengths of the successive geologic eras, but it is confessedly very difficult to decide beyond doubt even the approximate length in years."[30]

Nonetheless, there were occasional efforts to put years to events, notably the time elapsed since the Pleistocene ended. This was a span of obvious archaeological interest, one potentially amenable to calculation using geological features known to immediately postdate glacial retreat and that had changed since then. Estimating the rate by which onetime glacial kettle lakes filled with sediment and the rate of retreat of the St. Anthony Falls and Niagara Falls, Winchell and Wright by the 1880s put the end of the Pleistocene at 8,000 to 10,000 years ago, comparable to the amount of time G. K. Gilbert and Israel Russell estimated had elapsed since Pleistocene pluvial lakes Bonneville and Lahontan drained (§3.8).[31]

Chamberlin offered no absolute measure but rather a scale of units in which the Postglacial was one unit and previous episodes were multiples of that unit (Table 6.1).[32] If Chamberlin's "one" equaled 10,000 years, as one might suppose, then it was possible to infer numerical ages for archaeological materials that predated and postdated the last glacial stage. By it, Abbott's paleoliths were 10,000 years old (§3.8, §4.3, §4.5), with 10,000 to 30,000 years allotted to the Lansing and Long's Hill skeletal remains (§6.6, §6.11) and an age of 30,000 to as much as 150,000 years for the Claymont paleoliths (§4.5).

By the mid-1920s estimates of postglacial time came to rely on varve counting, which had its own liabilities (such as missing varves) but seemed more reliable (§8.11). Ernst Antevs's New England varve count put the end of the Pleistocene at roughly 11,500 years ago (§9.5). Because "postglacial" was often seen in local ice-free terms (Table 9.2), other estimates varied, but most were within a few thousand years of Antevs's mark. Such estimates, along with Brown and Kirk Bryan's determination that Folsom dated to the Late Pleistocene, provided the warrant for A. V. Kidder's fall 1927 announcement that the "first journey to the New World"[33] took place 15,000 to 20,000 years ago (§8.10, §8.11, §8.16).

As it happens, Kidder's estimate overlapped with one made by Hrdlička, who even before the Folsom discovery was willing to grant "ten or at most fifteen thousand years" for the onset of migration into the Americas. (The oft-made claim that Hrdlička allowed no more than a few thousand years for North American prehistory is untrue.[34]) That was far short of the 400,000 years Figgins gave Folsom, or the 365,000 or 500,000 years that Cook and Hay (respectively) deemed the age of Frederick. But then, they doubted these sites dated to the Late Pleistocene (§8.7, §8.11, §8.13). Still, neither their numbers nor those of Brown, Bryan, Hrdlička, or Kidder were anything more than ballpark estimates, and none could be "ground-truthed." Before Willard Libby's invention of radiocarbon dating in the late 1940s, absolute ages for sites were elusive targets. The precise age of the Folsom site was not known for decades.[35]

Not that it mattered. Absolute ages were neither necessary nor sufficient for the acceptance of a site as dating to the Pleistocene. That was as true in the decades before Folsom as it was in the decades after: the recognition of Clovis as a Pleistocene-age site is evidence of that (§9.2). All that was needed was secure geological evidence of a Pleistocene human antiquity on a relative time scale. That goal would have perhaps been easier to achieve had that relative time scale not been seen in absolutist terms.

10.5 FLATTENING THE PAST

In the years after 1890, as proponents and critics of a deep human antiquity in America pulled in opposite directions, one outcome of their tug-of-war was essentially to dichotomize the terms of the dispute. Geological antiquity was either Pleistocene or Recent; archaeological assemblages were Paleolithic or Neolithic; skeletal remains appeared primitive or modern (with the possibility that a modern form might be Pleistocene in age). Temporal shading was lost as cultural and biological change was viewed on an epochal scale, one that sought differences in kind not variation in degree.[36] Little attention was accorded to gradations, subtle or otherwise, between or within those units of time and change.

But it was more than just inattention, as Lee Lyman, Michael O'Brien, and others have argued. Archaeologists in America (one has to be specific on this point of geography) had not, by then, developed the analytical tools for measuring fine-scale temporal change in archaeological assemblages.[37] If there was a deep human past in North America, it had to look fundamentally different from the recent present. And, of course, it appeared there was not.

The persistent failure to find secure archaeological or geological evidence of a Pleistocene human presence inevitably led to the conclusion that North American prehistory was shallow rather than deep, and by extension that cultural and biological change over that recent span of time had been minimal, as Holmes and Hrdlička insisted. Yet, there was a none-too-subtle blurring of cause and effect on their part. Holmes and Hrdlička each had a priori theoretical reasons for rejecting an American Paleolithic (or premodern humans). To a large degree their conclusions were predetermined by their approach. That the conclusions were subsequently affirmed by geological testimony was almost unnecessary (for Hrdlička, geological affirmation *was* unnecessary [§7.7, §10.8]). Holmes and Hrdlička's actions, Kidder thought, were unmistakable proof of just "how strongly fashions of thought can influence judgment."[38]

That Holmes and Hrdlička thought about human antiquity in strikingly similar ways is not due to their having traveled the same intellectual paths (they did not), but it does explain why Holmes was keen to hire Hrdlička at the Smithsonian (§6.5). He recognized that Hrdlička's views dovetailed perfectly with his own and with the broader intellectual framework within which he and others steeped in Powell's BAE worked. It was a thoroughgoing uniformitarian framework: one worked from the known into the unknown, entering the archaeological past by way of the ethnographic present (§5.1). Within that approach there was no reason to expect any distinct Paleolithic (Pleistocene) races or cultures. Powell was insistent on this point. To accept "cataclysmic theories postulating intrusive or extinct races" who were non-Indian and dated from Pleistocene times yet seemingly left no succeeding generations would, he argued, introduce a massive chasm in the prehistoric record.[39] This not only violated uniformitarianism in its appeal to peoples not known in the present and its introduction of discontinuity into the prehistoric sequence, it disregarded the "known evolutionary laws" of progress and precluded the application of ethnographic data to the explanation of the archaeological record (§5.1).[40]

And apply ethnography they did, for BAE archaeologists assumed, as Bruce Trigger observed, that life in prehistoric times was not qualitatively distinct from that recorded in the centuries since European contact (substitute "anatomy" for "life" and one has Hrdlička here as well).[41] To be sure, their theoretical perspective implied that the passage of time ought to be reflected in material progress, yet try as he might the changes Holmes documented in material culture were as often at odds with the expectation of progress as they were consonant with it. The only choice was to regard these differences as a result of processes other than change through time. He was convinced the prehistoric record represented only a brief span. Any apparent differences between it and the historic or ethnographic records (changes in artifact styles, for example) were deemed insignificant and were attributed to the minor changes or movements expected over the period that separated the earliest and latest Native Americans.[42] The "mirroring" of past and present, and the lack of evidence of a deep human antiquity or major cultural changes in the shallow span of prehistory, resulted in Kroeber's "flat past" in which the archaeological record collapsed in on itself.[43] So it was that thoroughgoing cultural evolutionists, the intellectual offspring of Lewis Henry Morgan, had little interest in—and indeed an active resistance to—the idea that there had been major cultural change and evolution over the course of North American prehistory (§5.1).[44]

It is thus historically incorrect to claim it was the rejection of cultural evolution by Boas and his students that "forced the American archaeologists into a niche with a very limited [temporal] horizon."[45] Without question, archaeologists lacked a well-documented, long-term cultural sequence in the New World and had no sense of major cultural change or even a concept of microchange in culture. Yet, to suggest that this made them vulnerable to Boasian "historical particularists" is at best anachronistic. Boas's historical particularism began to reach a head only at the end of the first decade of the twentieth century, and it was only in the mid-1930s, Valerie Pinsky showed, that archaeology became widely incorporated into the anthropology graduate curriculum and directly influenced by Boasian perspectives.[46]

Moreover, BAE archaeologists accommodated a flat past within their cultural evolutionary approach, and there is no evidence Holmes ever followed Boas's intellectual lead, or that Boas's theoretical work had any influence, subtle or otherwise, on Holmes's approach to archaeology. The disregard was mutual. Boas early on (1902) recognized the "great need for well-trained students in archaeology" and considered offering such instruction at Columbia University, but he did not think Holmes up to the task.[47] In 1904, when Boas published a history of anthropology, he made no mention of Holmes or of *any* BAE personnel. But that was perhaps not a surprise to his peers (as opposed to his students). As William H. Dall put it, Boas "feels that wisdom will die with him."[48]

Finally, and without putting too fine a point on it, even if Holmes and Boas could have bridged the deep conceptual divide separating them, they intensely disliked one another for reasons legitimate and petty, further lessening the odds that Boas had influence on Holmes (or vice versa). Boas had little influence even on his few allies within the BAE, notably WJ McGee. McGee's theoretical screeds—regarding the evolutionary superiority of the Anglo-Saxon language, for example—appalled Boas, who chose to ignore them in order to maintain McGee as an ally within the BAE. (McGee's alliances, of course, were always inherently political, as evident in his several reversals on the human antiquity question [§5.7, §6.4].)[49]

The perception of the past as flat effectively subordinated archaeology to ethnology within the BAE program.[50] For unlike their nineteenth-century European counterparts who looked across space and saw deep into time, BAE archaeologists looked into time but saw only variation across space. It was neither accident nor coincidence that when Holmes summarized and synthesized the details of North

American prehistory he did so using maps, not chronological charts.[51] Variability in archaeological materials was perceived to be more significant on a geographic dimension than on a temporal one, just as it was among contemporary ethnographic groups. Archaeologists at the time were well aware of how European and American archaeology differed. As Frederick Sterns observed in 1915:

> The former studied relations of time and the latter relations of space. One considered cultural sequences and the other culture areas. To the European archeologist, peoples of one physical type and one kind of culture were succeeded by another people with different physical and cultural characteristics, and these in their turn were replaced by still other tribes or races. In America, on the other hand, the tendency has been to see all cultures as more or less contemporaneous except so far as the distinction has been made between the [Euroamerican] historic and the [Native American] prehistoric.[52]

For American archaeologists, geography was not merely a dimension but a cause. The physical, cultural, and linguistic diversity of modern Native Americans, long seen as evidence of great antiquity (§2.1), was attributed by Holmes to the passage of recent arrivals through the "new and constantly changing conditions" of the western hemisphere. These conditions "greatly accelerated differentiation" of groups, making moot the value of diversity as a measure of antiquity.[53] BAE archaeologists advocated what Julian Steward later named the "direct historical approach."[54] With it, the first and most important step was the "description and interpretation of the antiquities of the continent." The archaeologist, Holmes wrote, "observes the tribes of today, their cultural characteristics and environments, and acquaints himself with what is known of them historically."[55]

And because, according to Powell, "North American Indians were not largely migratory, but that the same peoples occupied the same districts of the country from generation to generation, and that such permanency of habitat extends far into the past,"[56] it made sense to assume a continuity from present to past. Holmes was sufficiently attuned to variation in material culture that he was aware one could not push the ethnographic record too far into the past, as the traces of particular peoples tended to fade out. Nonetheless, he affirmed the link between archaeology and ethnology. As he expressed it, the ethnologist saves the archaeologist from groping in the dark.[57]

Given these circumstances, it is hardly surprising that Kroeber used Holmes as a rhetorical foil to make his point about the flat past pervading archaeology. Both Holmes and Fewkes (Kroeber's other foil) were highly visible, well respected, and, of course, steeped in the BAE program that led to that flat past. Yet, Kroeber neglected to mention it was not just "explorers as experienced as Holmes and Fewkes" who at the turn of the century saw the archaeological past as completely flat. At that time most other American archaeologists and anthropologists did as well,[58] including Kroeber himself.

In his essay for a volume honoring Putnam, who fervently believed in a Pleistocene-age American Paleolithic, Kroeber observed that "neither archaeology nor ethnology has yet been able to discover either the presence or the absence of any important cultural features in one period that are not respectively present or absent in the other." Further, archaeology "at no point gives any evidence of significant changes of culture." The differences between past and present were only "in detail, involving nothing more than a passing change of fashion in manufacture or in manipulation of the same process."[59]

Likewise, Boas thought it "probable that the remains found in most of the archaeological sites of America were left by a people similar in culture to the present Indians." Despite the fact that Boas (as Douglas Cole argues) "made salient the idea that tribes were not stable units lacking in any historical development, but cultures in constant flux, each influenced by its nearer and more distant neighbors in space and time," he could not fully shed the idea of long-term stability. Hampered as he was by "a shallow temporal view of ethnic relationships," he tended to project historical entities back into remote prehistory.[60] It is little wonder that Boas supposed the "ethnological study of the Indians must be considered as a powerful means of elucidating the significance of archaeological remains."[61]

Boas was not the cause of the flat past, but he was in a real sense its beneficiary. Given the limited time perspective into which archaeology had fallen as a consequence of the broader failure to demonstrate a deep human antiquity, Boas deemed ethnology "a more likely approach to the history of man in the Americas than might otherwise have been the case."[62]

10.6 "SAVAGING" THE PRESENT

Boasian ethnology may have benefited from the idea of a flat past, but that was not true for Native Americans. They had long been perceived as "savages" at the lowest rung of the cultural evolutionary ladder.[63] That perception made them useful to nineteenth-century anthropologists, who, having gotten humans into the Pleistocene, needed to get them back out (§2.3). As George Stocking argued, history could not fill that breach in cultural time, nor could it be allowed to remain an empty void.[64] But just as Cuvier used modern animals to reconstruct extinct ones, so E. B. Tylor and others suggested that "by comparing the various stages of civilization among races known to history [and then] with the aid of archaeological inference from the remains of prehistoric tribes, it seems possible to judge in a rough way of an early general condition of man."[65]

John Lubbock thus saw Tasmanians and Native Americans as being to the archaeologist "what the opossum and the sloth are to the geologist" (§2.3). His uncharitable choice of lethargic animals as metaphors for indigenous peoples is telling. (Why not a more majestic species like the elephant? Cuvier used those too.) However, given the times and Lubbock's Victorian racism, it was hardly surprising. It symbolized the perception, shared by many, that Native Americans were slow to progress up the evolutionary ladder. (Not all societies were assumed to progress at the same rate, and some seemingly stagnated.)[66] Although that made Native Americans' ostensibly "fossilized" condition illustrative of ancient Europeans, it also meant their own history had stalled. After all, in the Old World "savagery" meant deep time and a measure of just how far Europeans had progressed since they had lived in that primitive state, as well as a stage of culture with living representatives. In the New World, "savagery" was one and not the other, a stage outside of time and history, and evidence of how little change had occurred. Indians were seen as culturally static, if not biologically inferior and indeed savage.[67]

Certainly that is how they were perceived at the beginning of the controversy. Abbott condemned Indians as a "usurping people" who drove off the earlier-arriving Paleolithic peoples (§3.1). That he expressed that view the same year as the Battle of Little Big Horn (1876) is surely coincidental, though it well expressed a prevalent societal attitude. The BAE response that there was not a separate race of non-Indian Paleolithic peoples (which, more coincidence, came the year [1890] of the Wounded Knee Massacre) reflects broader changes in societal views toward native peoples, though the BAE's championing of a flat past did nothing to change perceptions of their "primitive" status.

In searching for but not finding evidence of deep time or significant cultural change, BAE archaeologists "reconfirmed commonly held notions about the essentially unprogressive nature of American Indians."[68] Holmes was explicit on this point: "We infer from abundant evidence *a slow development of the* [Native American] *race in the past*, beginning with the earliest occupation of the continent and continuing up to the Columbian period." Their future hardly looked better, given the last "few hundred years of vain struggle ending with the pathetic present. . . . All that will remain to the world of the fated race will be a few decaying monuments, the minor relics preserved in museums."[69] Holmes captured their static past, if not their ill-fated future, in life-size plaster figures of Native Americans forever frozen in the act of quarrying stone and fashioning artifacts at Piney Branch. (Originally constructed for the World's Columbian Exposition in Chicago in 1893, the figures later did stints at the USNM and even the "sacred spot" [Holmes's opinion] of Piney Branch.)[70]

Thus, Steven Conn observes, Indians were placed outside the flow of history—literally outside. Here in America there was not, as there was in Europe despite the deep time of the Paleolithic, a seamless transition from prehistory to history. There was, instead, an unbridgeable divide: Euroamericans had "no filial interest in the man of Natchez" (§2.4).[71] As Bostonian F. E. Parker acidly put it in 1880, there was no point learning about the artifacts of a civilization "inferior to our own instead of superior."[72]

Instead, Native Americans were seen through the lens of nature.[73] With a few exceptions, Harvard's Peabody Museum most notably, collections of indigenous American archaeological and ethnographic materials were housed in natural history museums, "juxtaposed with . . . stuffed birds and mineral samples," rather than in museums of history and fine arts where "civilized" antiquities of the Old World classical period were put on display.[74] When Putnam announced Abbott's discoveries of what were thought to be the oldest artifacts in North America, he did so at the BSNH (§3.7, §4.3). When fellow Harvard archaeologist George Reisner brought back artifacts from his excavations in the royal tombs of Giza, they went on display at the Boston Museum of Fine Arts. (Reisner likewise had far greater success raising funds from Boston's wealthy elite for his excavations.)[75]

By the second decade of the twentieth century, overt expressions of "savagery" and cultural deprivation had greatly diminished, owing in large part to Boas's dismemberment of cultural evolution, his increasing dominance of anthropology's theoretical discourse (§10.5), his staunch advocacy of cultural relativism, and his antipathy toward the display of anthropological objects and life groups (of which Holmes was inordinately fond) in museums in general and natural history museums in particular.[76] Boas felt so strongly on this point that he insisted that if anthropology had to be in a natural history museum, "there ought to be no necessity for the visitor to come into contact with the natural-history exhibits while passing through the anthropological halls."[77] Yet Boas, who recognized that Native Americans had a history *and* a future, nonetheless perceived them as existing in the liminal space of the timeless "ethnographic present." Perhaps, as Conn supposed, "it asks too much of Boas to have transcended the process by which Native Americans had been divorced from the study of history."[78] (Physical anthropology, with its continued adherence to craniometry and racial typing, was largely immune to that change, as Earnest Hooton's scaling of "primitiveness" by anatomical form attests [§9.10].)[79]

Archaeology was initially not well established within the Boasian program (§10.5). By mutual, if informal, agreement, graduate training in archaeology was left to Putnam at Harvard (§10.7), where it had strong museum ties and a weak commitment to any significant theoretical position.[80] Still, in the second decade of the twentieth century its trajectory began to diverge from the notion of a timeless Native American past, one where archaeology and ethnology were seamlessly joined.

Ironically, that trend began under the watchful eye of Abbott sitting on his front porch, as Leslie Spier and Clark Wissler excavated in the summer of 1914 into Trenton's yellow sands in an attempt to understand the changes in the stratigraphic distribution of the argillite artifacts, and as they tried to avoid getting entangled in the Paleolithic dispute or any topic related to the possible antiquity of Trenton's artifacts (§6.15). They, along with Kroeber, Nelson, and Kidder, came to realize that even in the absence of deep time (Folsom was still more than a decade away) it was possible to measure cultural change on a nonepochal time scale by applying the techniques of stratigraphy and seriation to examine the waxing and waning of artifact types and styles and not merely their presence or absence.[81] It could now be shown archaeologically that past cultures were different to some degree, though change was still measured in degree and not in kind, from those documented by ethnographers.[82]

The notions of a uniform nonpresent and of the "Indian considered a creature of yesterday" (as Kidder put it[83]) were finally broken with the demonstration of a Pleistocene human antiquity at Folsom. The temporal and cultural gap between the Late Prehistoric and the Late Pleistocene was far too great to be overcome by working backward from the present, nor could it be closed, as it had been in Europe decades earlier (§2.3) with groups known ethnographically. By then Native Americans were neither stone age vestiges (or analogues) nor changeless savages, for Boas's critique of cultural evolution had run its course and routed the BAE's approach. There was no question that significant cultural change must have occurred, and thus the pressing question after Folsom was what prehistory looked like in the intervening millennia (§9.12).

In the wake of Folsom, Kidder, among others, urged the search for evidence of how ancestral Native Americans had transitioned (independently from any Old World help, he added) from Paleoindian hunters to the civilization builders of Middle and South America. Though he spoke of the "growth of the higher American cultures" and of the "degenerative transition" in flint-knapping from Folsom to that of more recent cultures, such terms were used mostly out of disciplinary habit, for by then they carried none of the explicit cultural evolutionary baggage they had in prior decades when used to arrange America's native peoples on a scale of progress.[84] By then there were also archaeologists ready to undertake that search, and that too was a significant change from the preceding decades.

10.7 HRDLIČKA'S LAMENT

In the 1870s and 1880s, the participants in the human antiquity controversy mostly identified themselves as archeologists (for example, Abbott, Frances Babbitt, Cresson, Haynes, Holmes, Henry Mercer, Metz, Mills, Warren Moorehead, Stephen Peet, Putnam, Volk, and Wilson). Geologists such as Chamberlin, Leverett, McGee, Salisbury, and Wright became involved but initially were in the minority. After the collapse of the American Paleolithic in 1897 (§5.16), that disciplinary imbalance was reversed. Archaeologists largely retired from the scene, and the active search for possible Pleistocene-age sites was led almost entirely by geologists and vertebrate paleontologists (Table 10.1a). From 1898 to 1927, few if any archaeologists initiated or even nominally participated in the search for early sites, a pattern that is statistically significant (Table 10.1b).

Hrdlička might not have been aware of the magnitude of that shift or when it occurred, but he certainly noticed its effects. In the spring of 1928, at the height of the debate over the association of metates and mammoths at Frederick (§8.7), he bemoaned what he considered an epidemic of claims for a Pleistocene human antiquity made by geologists and vertebrate paleontologists. Weary of leading the fight by himself (Holmes had by then retired [§7.9]), Hrdlička wished the effort might be "taken up energetically by our archaeologists." But as he complained to Spier, "We almost seem to have no archaeologists."[85]

Spier must have been taken aback by Hrdlička's comment. After all, he had just published a detailed critique of Frederick (§8.17). And the year before he had attended the first Pecos Conference, along with at least two dozen others actively engaged in archaeology. There was no dearth of archaeologists. Still, Hrdlička had a point. Spier was one of just a handful of individuals working in North American archaeology outside of the American Southwest, and one of the very few who had ventured into the controversy over human antiquity (§6.15, §8.17). Among the others on that short list, George MacCurdy participated in the Vero site visit (§7.3), Nelson and Sterns contributed brief commentaries on Vero, and Nelson also critiqued the supposed bone tools from Snake Creek (§8.17). But their roles were minor and had little effect on the course of the discussion. Indeed, the only detectable consequence of Sterns's Vero essay was that it provoked an angry Holmes to tell Hay to go to hell (§7.7).

Table 10.1 Investigations into putative Pleistocene-age sites initiated by archaeologists as opposed to nonarchaeologists in the decades before and after 1897

(a) Tally of the sites by time period and investigator		
Period	Investigation initiated by archaeologists	Investigation initiated by nonarchaeologists
Up to 1897	Trenton, NJ (1876—Abbott) Little Falls, MN (1883—Babbitt) Madisonville, OH (1885—Metz/ Putnam) Loveland, OH (1887—Metz/Putnam) Claymont, DE (1889—Cresson) Medora, IN (1889—Cresson) Washington, DC (1889—Wilson) New Comerstown, OH (1890—Mills) Trenton, NJ (1897—Mercer)	Nampa, ID (1890—Wright) 12 Mile Creek, KS (1896—Williston)
1898 to 1927	Trenton, NJ (1914—Wissler/Spier)	Lansing, KS (1902—Upham) Long's Hill, NE (1906—Barbour) La Brea, CA (1914—Merriam) Vero, FL (1916—Sellards) Snake Creek, NB (1922—Cook) Melbourne, FL (1924—Loomis) Lone Wolf Creek, TX (1925—Figgins) Folsom, NM (1926—Figgins) Frederick, OK (1927—Figgins)

Note: The several rounds of Trenton fieldwork were by different individuals for different purposes and hence represent effectively independent investigations.

(b) Statistical analysis of the tallies in (10.1a)			
Period	Investigation initiated by archaeologists	Investigation initiated by nonarchaeologists	Sum
Up to 1897	9 (1.48)	2 (–1.76)	11
1898 to 1927	1 (–2.06)	9 (1.48)	10

Note: $G = 12.13$, $p = .000$; Freeman-Tukey deviates are given in parentheses: all cells are statistically significant at $p = .05$ level (± 0.978).

Source: Author.

As for why so few archaeologists were actively engaged in the question of human antiquity in 1928, or for that matter in the several preceding decades, Hrdlička and Holmes were partly to blame. They had "swung so far to the right, they became so ultraconservative, [and] their attacks upon any find suggesting even respectable age were so merciless, that further purposeful search for early remains was most harmfully discouraged."[86] Or so Kidder judged. And though Kidder is surely a reliable informant (others of the time echoed his comments[87]), there was more to the dearth of interest than fear of Holmes and Hrdlička. Other archaeological subjects attracted more attention.

That is evident in the seventeen PhD dissertations written in archaeology in the years leading up to 1928. More than half focused on Latin American prehistory, primarily the archaeology of the Maya and other Mesoamerican complex societies, and many of the remainder were on the prehistory of the American Southwest.[88] Nearly all were on sites and complexes of relatively recent age. (The exception was MacCurdy's Yale University dissertation on the "Eolithic problem" in European archaeology.[89]) The first PhD directly related to Pleistocene human occupations in North America was Edgar Howard's 1935 University of Pennsylvania dissertation based on his work at Burnet Cave and Clovis (§9.4).

The focus on the more recent millennia of prehistory included the dissertations of all the Harvard University students who trained under Putnam. Thus, despite a lifelong interest in and eager search for evidence of a deep human antiquity, he groomed no successors to carry on his work. But then, as Curtis Hinsley and others have observed, Putnam's strengths were as an entrepreneurial founder of anthropology programs (at the Field Museum of Natural History, the AMNH, and the University of California) and as a science administrator through his decades-long service as secretary to the AAAS. His intellectual accomplishments and influence fall short of his institutional legacy (§10.10).[90]

Holmes and Hrdlička likewise had no intellectual offspring. Angry and frustrated as they became at discharging their self-appointed duty to guard against a deep human antiquity (in 1918, Holmes was already avowing it was "not from choice" that he was again criticizing such claims[91]), they were fated to fight alone. They were at an institution that trained no students and had few ties to the university system.[92] The best they could do was accept the trained students of others who might not share their intellectual perspective, such as Frank Roberts (§8.10). And the opportunities for bringing in those individuals were rare enough because, as Holmes ruefully joked, in the museums and bureaus of the federal government "few die and none resign."[93]

However, the apparent lack of interest among archaeologists over the first three decades of the twentieth century in the search for early sites may be more symptom than cause. For the turn of the century not only signaled the demise of the American Paleolithic, it also marked the realization that archaeological data was by itself insufficient to demonstrate a Pleistocene human antiquity (§10.1). The question of a Pleistocene human presence in North America was seen as a geological or paleontological problem, not an archaeological one, for only those disciplines could measure time (§10.3). It was no fluke that many of these sites were referred to as "bone quarries" (§8.14, §8.16, §9.2).

Table 10.2 Disciplinary involvement by time periods

Period	Investigation initiated by archaeologists	Investigation initiated by nonarchaeologists	Sum
Up to 1897	9 (1.13)	2 (–1.40)	11
1898 to 1927	1 (–2.40)	9 (1.82)	10
1928 to 1941	21(0.41)	14(–0.35)	35
Sum	31	25	56

Note: As in Table 10.1b, the pattern is not random ($G = 12.945$, $p = .000$). Freeman-Tukey deviates are given in parentheses; gray cells show statistical significance at $p = .05$ level (± 1.132).

Source: Author.

Once it was realized that matters were out of their hands, archaeologists drifted away from the search for early sites. This had consequences besides causing Hrdlička's lament. For one, because the search was driven by geologists and paleontologists, it kept the focus on matters of chronology and sequence (that was true as well in Europe before 1860[94]), or on the recovery of museum-quality specimens of extinct vertebrates (§8.4, §8.13). There was no extracting Pleistocene humans from an object-based natural history tradition in those circumstances, further isolating this aspect of archaeology from the Boas-led trend away from that tradition (§10.6). Concerns for the larger processes of human migration, adaptation, and material culture were rarely addressed, save in a few instances. There was Boas's Jesup North Pacific Expedition (1897–1902), a project aimed at settling questions of Native American origins, though with just a minor archaeological component, and a 1911 symposium on "The Problems of the Unity or Plurality and the Probable Place of Origin of the American Aborigines."[95]

Efforts to address the peopling process began in earnest only in the post-Folsom years, when a Pleistocene human antiquity in America, as Bryan put it, "lost its heretofore hypothetical character."[96] One could address antiquity without knowing anything of the process by which North America was peopled, but the reverse was more challenging because understanding migration routes and adaptations depended on knowing when during the shifting climate and environment of the Pleistocene the peopling occurred. Once attention no longer needed to focus on *when*, it shifted to questions of *how* and even *why* (§9.7–§9.8, §9.13). That shift was led by a new, younger generation of archaeologists—like Howard and Roberts (§9.1, §9.4)—who actively engaged in the search for Pleistocene sites.[97] With that, the overall participation and influence of geologists and vertebrate paleontologists began to wane (Table 10.2). Some, such as Barbour, Cook, and Sellards, continued to pursue early sites, but as before their focus was primarily on issues of time, not culture (§10.7).

But that disciplinary demographic shift happened only after Folsom. In the several preceding decades, having abandoned by virtue of inactivity the search for a Pleistocene human presence and thereby conceded the right to chart the course of investigation, archaeologists and anthropologists could do little more than react to discoveries and claims made by geologists and vertebrate paleontologists. For Holmes and Hrdlička, that meant decades of playing the spoiler role. They were not generating knowledge, only attacking it. And it meant that a key element of archaeology's identify—past time—was not under their control. Given these circumstances, it is hardly surprising that interdisciplinary border wars repeatedly broke out.

10.8 WHEN DISCIPLINES COLLIDE

In the spring of 1890, as Holmes was putting the finishing touches on his first Piney Branch publication, Powell offered a rose-colored view of research into human antiquity. Although he saw the solution as perhaps part geological and part anthropological, of one thing he was certain: the discussion across disciplinary boundaries was sure to be "done with little acrimony." This was so because the "researches of the last half of this century have taught scientific men lessons of wisdom and of tolerance for divergent opinions, and wider knowledge abates the asperities of controversy."[98] Powell was either very optimistic or very naïve. Regardless, he was very wrong.

Powell was writing on the cusp of a larger trend in American science and society, notably the diversification and increasing specialization of occupations, with individuals and groups (not just scientific ones) codifying and demarcating their identities through professional societies, specialized journals, national societies, meetings, and training—which in the sciences meant graduate academic training (§10.7). It was a rapidly evolving "credentialed society" that led to the unprecedented growth of a middle class that included, among the emerging professions, that of the salaried scientist.[99]

The onset of specialization within the sciences began in the 1880s and had hit full stride by the turn of the century. That process is readily illustrated in the segmentation of the AAAS, founded expressly to unify scientists beyond their regional and intellectual interests. The AAAS had served in the latter half of the nineteenth century as the common meeting ground for the scientific community. Over much of that time a few sizes fit all. From its 1848 founding to 1856, there were no discipline-specific sections. A new constitution in 1856 organized the AAAS into two Sections (A and B), "the manner of [their] division to be determined by the Standing Committee of the Association." In 1874 those divisions were made clearer but were still limited: Section A consisted of mathematics, astronomy, physics, chemistry, and mineralogy, and Section B of geology, zoology, botany, and anthropology.[100] By 1882 there was enough critical mass across a diversity of interests to allow the creation of nine discipline-specific sections. Some continued to be paired with others for lack of sufficient practitioners to warrant their own section (geology and geography were together in Section E, and anthropology was in Section H with psychology as a subsection). In 1902 the AAAS underwent the next round of sectional mitosis, and then the process recurred four more times in rapid succession in just the first two decades of the twentieth century, resulting in nineteen sections by 1920.[101]

As this was happening, national specialized scientific societies formed in large numbers. These included the GSA (founded in 1889), the AAA (1902), and the Paleontological Society (1908).[102] During the nineteenth century as a whole, sixty-one such organizations were created, but more than half of those (thirty-five) were established in just the last two decades of the century, and another forty-seven were founded in the first two decades of the twentieth century. With that explosive rise was a proportionally steep decline in the formation of the more traditional local, state, and regional natural history and scientific societies.[103] They were not missed. The view of Henry Rowland, Johns Hopkins University physicist, was not atypical, though perhaps more callous than most: "There are very many local societies dignified by high sounding names, each having its local celebrity, to whom the privilege of describing some crab with an extra claw, which he found in his morning ramble, is inestimable."[104]

Joining these new specialized societies were individuals with PhDs earned in America's expanding universities, which adopted the German model of research and graduate education in the late nineteenth century. Graduate school became the means of professional self-consciousness, legitimization, and inculcation, and increasingly after 1900 the primary route to research opportunities and scientific careers.[105] In 1898 the rising number of American-earned PhDs caught the attention of the National Research Council, which began an annual PhD count. From then until 1928 (the year of Hrdlička's lament, a relevant end date for purposes here), more than 9,600 PhDs were awarded across twenty scientific disciplines. Most were in chemistry ($n = 3057$), followed by zoology, physics, botany, and psychology. Anthropology lagged far behind in its total and had essentially flat growth in that span; only vertebrate paleontology's growth was lower. The rising academic tide had not fully lifted all disciplinary boats.[106]

The creation of formally credentialed discipline-specific identities inevitably led to territorial marking, a fragmentation of the American scientific community, and ever-sharper lines being drawn between the emerging class of professionals within the specialized sciences and the amateur practitioners who had previously been welcome members of the community. The AAAS, for decades a de facto "parliament of science" where common concerns were addressed and which served a concept of science rather than a specific discipline, increasingly became a venue for heated battles across disciplinary lines.[107]

Disciplinary specialization was not the cause of the human antiquity controversy, but it put it in a context that made it much more difficult to reconcile disciplinary differences or cope with the amateurs among them (§10.9). This emerging new world of science required detailed, highly specialized knowledge that was less likely to be understood by those outside the fold, let alone among the larger public. That lack of mutual comprehension was often compounded by the disciplinary myopia that specialization bred.[108] Thus, though Hrdlička insisted indisputable stratigraphic and geological evidence was necessary to make the case for a Pleistocene antiquity of human skeletal remains, that evidence was trumped whenever the skeletal remains failed to show "marked serial somatological distinctions" (§6.12). These were matters, of course, for which he considered himself the sole person qualified to judge. That was true no matter what geologists Upham or Barbour might say about the age of the loess at Lansing and Long's Hill (§6.8, §6.11), or what paleontologists Hay and Frederick Loomis might claim regarding the antiquity of the vertebrate fauna at Vero and Melbourne (§7.4, §7.9).

Hrdlička was hardly alone in resolutely insisting his evidence and interpretation were superior to those of others. Most had a similar conceit, accentuated by the fact that most of the individuals involved in this controversy were over forty years old at the time of their involvement and hence had well-formed and strongly held discipline-specific opinions. Their parochial stubbornness made it far more difficult to reach consensus on the significance, reliability, and merits of evidence or claims from different disciplines that were in direct conflict.[109] At the 1893 AAAS meeting, members of Sections E (geology) and H (anthropology) failed to find common ground in their daylong session on the American Paleolithic (§5.13), which led to "border warfare, in which neither geologists nor archaeologists are quite sure as to what belongs to them"[110] Each was nonetheless certain they were right. The AAAS was on many other occasions the venue for such exercises in incommensurability (§5.5, §5.16, §6.8 §7.3), but it was also where Brown and Bryan in 1928 gave the paleontological and geological evidence that affirmed a Late Pleistocene antiquity for Folsom while Hrdlička said nothing at all (§8.18).

Perhaps the most extreme case of irreconcilable disciplinary differences occurred at Vero, not least because, at Chamberlin's suggestion, its assessment had involved a larger number of participants across a wider disciplinary range, including geologists (E. H. Sellards, Rollin Chamberlin, T. Wayland Vaughan and Edgar Wherry), paleontologists (Hay and Robert Schufeldt), a paleobotanist (Edward Berry), and archaeologists and anthropologists (Holmes, Hrdlička, and Mac-Curdy, with a cameo by Nelson) (§7.3). It proved impossible for them to reach unanimity on any of the central questions of the Vero site. Their differences did not entirely conform to disciplinary lines, but they certainly tended in that direction (§7.4–§7.5).

Chamberlin urged the warring parties to reconcile their views, but no one took him up on it (§7.5, §7.7), for by this time their disciplinary conflicts were compounded by a deep well of suspicion and mistrust across disciplinary lines. Hay asked rhetorically whether the Pleistocene vertebrates might have survived later in the south, as "our friends the anthropologists appear to be willing to believe," then answered his own question: "The writer holds that the view is wholly wrong" (§7.5). He considered the question of human antiquity beyond the reach of anthropologists (§7.3, §7.4, §8.17). On the other side, Hrdlička deemed it "scarcely safe for the geologist or the paleontologist to assume that the problem of human antiquity is his problem" (§7.4, §7.7). Holmes was even less generous, publicly denouncing Hay's Vero claims as "dangerous ventures of little experienced or biased students in a field which they have not made fully their own"(§7.7).

Of course, that did not prove Hay was wrong, at least not about the paleontological evidence. And so as the investigations of early sites were increasingly conducted by geologists and paleontologists, Holmes and Hrdlička brought in allies from those disciplines to aid their cause. Where the claims for antiquity rested on the age of the geological deposits, they enlisted Chamberlin, Salisbury, and Leverett to assess the merits of the cases (§5.4, §5.6, §5.10, §5.15, §6.6, §6.8); when the evidence shifted to extinct vertebrates, they deputized paleontologists John Merriam, W. D. Matthew, and Chester Stock to help in the investigation.

Proponents of a deep human antiquity were not without allies of their own. When Sellards and Hay made the tactical decision to take the fight directly to the anthropologists' turf, they found linguist Pliny Goddard, editor of the *American Anthropologist*, a willing accomplice. Goddard needed a deeper prehistory for the diversification of Native American languages than Holmes or Hrdlička was willing to provide, but that Sellards and Hay would. He readily published Sellards's defense of Vero and Hay's wide-ranging tally of North American Pleistocene archaeological sites in the *American Anthropologist* (§7.4). That doing so irritated Holmes was an added benefit. (Goddard disliked Holmes, and the feeling was mutual.) Goddard later did the same for Loomis at Melbourne (§7.11) and cheered on Cook's claims for a Pleistocene antiquity at Lone Wolf Creek (§8.3).

Yet, after several decades of fierce interdisciplinary fighting, there was no such flare-up at Folsom. There are several reasons why that was so. For one, although Folsom was excavated by paleontologists, there was no ambiguity regarding its association of artifacts and bison bones (§10.2), as attested in 1927 and again in 1928 by witnesses such as Brown, Kidder, Roberts, and Neil Judd (§8.10, §8.16). That instantly nullified Holmes and Hrdlička's inevitable criticism that the remains were not "an original inclusion" in the deposit (§10.2).[111] Second, the close involvement of Brown provided a built-in intradisciplinary check on Hay's assessment of the taxonomic identity and antiquity of the Folsom bison, which, ironically, prevented an interdisciplinary conflict. In their several rounds privately debating Folsom bison taxonomy, Brown did not convince Hay of his point of view, but he made Hay understand that his claims would not go unchallenged (§8.11). It is no coincidence that Hay and Cook were effusive about the possibly Early Pleistocene age of the fauna at Frederick and Lone Wolf Creek where no independent vertebrate paleontologists were involved, but uncharacteristically silent (for Hay) regarding the antiquity of the Folsom bison.[112] Finally, in this case there *was* interdisciplinary agreement. Bryan's geological studies independently affirmed a Late Pleistocene age for the bison kill (§8.16).

But who were Brown and Bryan to judge, or Kidder, Roberts, and Judd to bear witness, and why did their say so carry so much weight in 1927? Asked another way, why had the word of Abbott, Putnam, and Wright not carried the day in 1897?

10.9 LAST DAYS OF THE TYRO

Obviously, significant differences in the empirical record and the closing of yawning conceptual gaps in the intervening decades are a large part of the answer. But there is more to it than that. At the outset of the controversy anthropology was broadly defined, open, and inclusive, centered in museums and local scientific and natural history societies. It was largely the purview of amateurs, there being few of what we might recognize today as professionals, and the line between amateur and professional was poorly drawn and highly porous.[113] There was no group recognized as competent arbiters of substantive and theoretical disputes. By the first decades of the twentieth century, that had changed (§10.10). A highly focused, closed, and exclusive professional discipline had emerged, one rooted in academic departments in universities (§10.7–§10.8).[114] With it came more sharply defined boundaries of scientific status and rank.

Anthropology was hardly unusual in this respect: this was an evolutionary process that was occurring at roughly the same time across virtually all of American science. Along with that transformation a sorting process took place, one that was at times deliberate and cold-blooded. The first phase of the process was aimed at discriminating (in the most prejudicial sense of that term) between amateur and professional, followed by the further establishment of a hierarchy within the ranks of the professionals. In this section I address the former; the latter is explored in the section that follows (§10.10).

The professionalization of a scientific discipline is a process that occurs over time. It is not an event, and one can rarely spot a (let alone *the*) historical tipping point that marks a phase change from an amateur endeavor to a professional one. Before the turn of the century the notion of a professional as customarily defined today—in which the divide is drawn along the lines of academic training, formal credentialing, full-time employment in the field, and membership in a specialized society—was largely irrelevant.[115] To be sure the founding of a national organization such as the AAA or the appearance of a journal catering to the needs of a specialized audience signal moments when individuals with shared interests formally organized themselves into a collective around a set of scientific questions or problems deemed worthy of investigation.

Yet, such tangible events more often indicate the *end* of the professionalization process, or at most the beginning of the end. There were inevitably kinks to be ironed out regarding, say, the criteria and exclusivity of membership (a cause of considerable friction at the AAA founding in 1902), a journal's standards, or the training and credentials necessary to be considered a professional.[116] The latter need not—and in anthropology could not until the twentieth century— have included university training (§10.7), which is not to suggest there were no professionals in anthropology before then (Boas's chauvinistic 1904 "The History of Anthropology" notwithstanding [§10.5]).[117]

The beginning of the professionalization process is even more difficult to spot, but as it happens there was one moment when that future could have been glimpsed with startling clarity. It happened at the AAAS meeting in August of 1883 when two vice presidents, anthropologist Otis Mason and physicist Henry Rowland, touched on the same topic in their respective section addresses. There was surely no collusion between them, but that they hit similar notes was not entirely coincidence. Each spoke to one of the central questions of the day in the scientific community: what did it mean to be an anthropologist or a physicist in late nineteenth century America? They were of very different minds on the answer.

Looking out across the membership of Section B (physics), the prickly Rowland announced, no doubt to the great discomfort of more than a few in his audience,

"It is a fact of nature, which no democracy can change, that men are not equal, that some have brains, and some have hands; and no idle talk about equality can ever subvert the order of the universe." Rowland had a practical goal in insulting his audience. He wanted individuals and institutions to recognize their obligation to insure that those with hands assisted those with brains.[118] But there was also no ducking his underlying message. Science, and here he was speaking to the larger scientific community, might be a democracy in the sense that all are entitled to participate, but it was also a meritocracy (at times a ruthlessly selective one), and not all participated with equal standing. This was the best kind of democracy, one where status was based "not on birth or wealth or social class but on intellectual originality and sheer hard work."[119] As David Hull put it a century later, it is precisely because selection in science is so intense that science is so elitist.[120]

Rather than urging his audience to take democratic pride in reducing everything to a common level in the unflattering sense of the term "common," Rowland urged the members to take a hard look at themselves and those around them. Not all were "men of genius," not all studies of the "facts and laws" of physics were equally important, and not every physicist achieved equal results. It was incumbent on the physics community to "form a true estimate of the worth and standing of persons and things," and then let the rare geniuses among them pursue that which is "great and good and noble" and let others follow and lend their hands.[121]

It was a kinder, gentler vice presidential address by Mason in Section H. Anthropology was a discipline where "everybody can study something," and he welcomed them all: "Mothers, school-teachers, those in charge of the insane, of the criminal, and of the defective classes, lawyers, mechanics, musicians, philanthropists [or] legislators." Anyone could contribute, for there was "no priesthood and . . . no sacred language." At a time when knowledge centered on objects, and objects like paleoliths were thought to have inherent meaning, all that seemed necessary to archaeological investigation was a set of eyes and a willingness to walk, observe, and collect. Finding things was equivalent to finding things out, and that allowed and encouraged amateur contributions. But only to a point.

Open as he was to the contributions of the masses, Mason was no dewy-eyed believer in their wisdom. *Someone* had to organize and make sense of what they had collected, a task best left to those whose "peculiar occupation" gave them the skills to do so: he thought of Holmes, with his artistic talent for "bringing order out of confusion," and of General Pitt-Rivers, who with his military background had ably worked out the history of the implements of war. "The day of tyros is gone," Mason announced, the multitude of collectors will "have to go to school to Professor Putnam, or Dr. Rau, or Dr. Thomas, before they will have the faintest conception of the significance of their treasures."[122]

Mason was sending the tyros to school; Rowland was telling many of his brethren their work was ordinary at best, mediocre at worst. The gap between their messages was partly one of personality, but it also reveals where their disciplines were in the professionalization process. In 1883 physics was already far too specialized and laboratory intensive to allow clerical workers, schoolteachers, farmers, or tradesmen to participate, only those among the tiny fraction of the American population who possessed college degrees and intellectual talent.[123] Anthropology in 1883 was only just beginning to define itself and set out the boundaries of its membership. As a result, it was soon to encounter a far more difficult problem than merely putting tyros in their place.

Namely, what of collectors and practitioners like Abbott (who would never tolerate being called a tyro), Babbitt, Haynes, Metz, Volk, and others who participated in the American Paleolithic debate? What of geologists such as Edward Claypole, Lewis, Upham, and Wright? The credentials separating them from Holmes, Putnam, Rau, and Thomas, or from Chamberlin, McGee, and Salisbury, were hardly significant. All were archaeologists (or glacial geologists) by virtue of experience, for few on either side of the dispute had formal education or training in archaeology or glacial geology.[124] It was an exaggeration, though not too far from the mark, for Wright to claim he had done "more fieldwork, extending over a wider and more representative area of glacial geology, than either [McGee or Salisbury], and almost more than both of them together." He dismissed them and Holmes as novices.[125]

Holmes, Chamberlin, Putnam, Salisbury, and others were "professionals" in one modern sense of the term: each was employed full-time in their discipline (so too was Abbott, but he did not last long in his position [§5.14]).[126] Of course, because the BAE, Peabody Museum, and University Museum were then the only institutions capable of supporting full-time archaeological work, and there were likewise a finite number of positions in the USGS and the state geological surveys, full-time employment was hardly a meaningful criterion for separating professionals from amateurs.

In fact, the majority of practicing archaeologists and glacial geologists in late nineteenth-century America lacked a position, a secure institutional base, and worked "at irregular intervals and in odd moments,"[127] and if they received compensation it was rarely adequate. (Wright was an exception, having academic credentials and employment, but his Oberlin position was in theology [§5.8].) Yet, at personal effort and expense, these part-time participants conducted research, presented results at meetings, and published in journals, though it is telling that they rarely published in Chamberlin's *Journal of Geology*, with its stringent review policy (§5.8, §6.11). Perhaps most important, they aspired to be more than mere collectors or amateurs. They wanted and even demanded to be taken seriously, as their role in this controversy attests.[128] Were they any less "professional" than those few fortunate enough to be employed? Were they any less authorities?[129]

These questions can be especially difficult to answer during times of controversy when the questions themselves become part of the polemic. A scientific controversy, David Oldroyd observes, is a contest, and to have one's views prevail is to win. In archaeology or glacial geology that often must happen without benefit of a definitive empirical test, for there are few "crucial experiments" (in Karl Popper's sense of the term[130]) to be had. The data can be equivocal and the conflicting interpretations irresolvable (though not, as it happened, in this particular controversy [§10.15]). Winning often is a zero-sum game of being right by proving others misguided or wrong, with the judgment of "what *counts* as truth at any given point" resting largely on the judgment of the contemporary scientific community. Of course, those making the judgment are also parties to the dispute, and for that reason scientists seek "to become *authorities*, so that their views will prevail."[131]

On what had been a level field, Powell and his BAE and USGS staffs sought to impose a standard of what it meant to be a professional and an authority in archaeology and glacial geology. It was one based on their shared background and vision. Nearly all were born and bred in the Midwest and were largely self-taught, having earned their knowledge on the Plains and in the mountains of the West (§5.1). They came east to Washington with a "populist science" chip on their shoulders and a healthy disdain for rigid and formalized European science, and they were convinced of the superiority of their approach.[132] They saw their work as revolutionizing archaeology and glacial geology: Holmes preened that he was at the forefront of a "new era . . . dawning in American archaeologic [*sic*] science."[133]

But in late nineteenth-century America, with the boundaries of membership in those disciplines drawn as loosely as they were, how was one to demarcate relative status and establish authority beyond one's own circle? On what grounds were Abbott, Haynes, Wright, and other proponents amateurs, and Holmes and Chamberlin and other critics authorities? In the absence of explicit, impersonal criteria for discriminating and defining relative status, Hinsley observes, "The Washington corps attempted to answer [these questions] by slander and storm."[134] The critics deliberately and at times venomously maligned their opponents. Holmes's drumbeat refrain was that the American Paleolithic was "hopelessly embarrassed" by the "blunders and misconceptions" of "amateurs," "relic hunters," and "enthusiastic novices with fertile brains and ready pens" (that would be Abbott), who had that "half wisdom half experience gives" (Haynes), possessing little scientific understanding of stone tool making, let alone of geological age and context (§5.10–§5.11). The American Paleolithic, Holmes insisted, was so "unsatisfactory and in such a state of utter chaos that the investigation must practically begin anew" (§5.11). Parody, caricature, and distortion, Hull observes, are routine in science, and so they were here.[135] Even the very act of discovery could not be left to amateurs unfamiliar with the new archaeology or the new geology. Items had to be found in situ and the context and geology of the find spot had to be properly understood (§10.2). Finding things was no longer tantamount to finding things out.

Without denying the polemical aims of the critics, it is indeed the case that those they were criticizing were perceived as amateurish, even by their allies. Wright agreed with Haynes that "distrust of Abbott's truthfulness and the untenable positions of Mr. Wilson have greatly embarrassed the subject of Paleolithic man in America."[136] Of course, both Haynes and Wright failed to see how they themselves could be targets, believing in their own expert authority.[137]

By the 1893 AAAS meeting the American Paleolithic, which just a few years earlier had been universally accepted, was now deemed unproven, even by sympathetic onlookers (§5.13). Holmes had successfully tarred his opponents. It was obvious that those considered authorities on the authenticity of paleoliths had been wrong, or at least had not been right. That had real-world consequences.

Abbott had parlayed his Trenton work into election as a AAAS fellow in 1883, the vice presidency of Section H in 1888, and in 1889 a coveted curatorship at the University of Pennsylvania (§4.14). But Holmes's Trenton critique took its toll, and Abbott's governing board feared for the scientific reputation of their institution. Given Abbott's poor job performance (he had little idea of what the position entailed), his board saw no reason to keep him: he was dismissed in 1893 (§5.14). Over the next two dozen years he cursed Holmes as a liar and saw every new discovery as vindication that he was right (§6.15, §7.2, §7.18). Beneath the bluster, however, Abbott was no fool. After 1893 he shied away from sending his articles to scientific journals, convinced that they would not pay him any mind and any claim of his made there might be referred to "some specialist."[138]

Haynes sought to recoup the respect he had lost to Holmes's attacks by insisting only a jury of "acknowledged pre-historic archaeologists of the *world* is competent to pronounce judgment upon this question," not mere geologists like Holmes whose archaeological studies were "limited to our native Indian tribes and their remains."[139] But his appeal to his European colleagues fell flat. Boyd Dawkins, who had accompanied Haynes on his trip to Trenton in 1880 and on the occasion praised Abbott's discovery (§3.7), ultimately deemed the "American evidence as to [the] Paleolithic . . . unsatisfactory." He did so for the very reason Holmes gave: "All [the specimens] that I have seen may readily be classified with Red Indian rude types, and therefore need not necessarily be paleoliths, even if they occur in undisturbed Pleistocene strata."[140] Haynes, whose credentials were based on his European experience and alliances, found his status undercut by those allies. He soon drifted from the scene.

Abbott and Wright tried appealing to an authority earned by experience. Abbott announced he knew "a smattering of gravel-ology" and insisted he could easily differentiate stratified and unstratified gravel (§5.6, §5.11). Wright agreed, adding that though Abbott was not a professional geologist he had spent long hours in the field, suggesting that conferred the knowledge to correctly interpret geological deposits (not to mention greater opportunities to find specimens). That was all utterly irrelevant to Holmes and Chamberlin. Amateurs could have experience, but experience was no guarantee of expertise, let alone the specialized knowledge of the "New Geology" (§5.6–§5.7, §5.11). That Wright fashioned himself a glacial geologist was proof of that. In their attack on his *Man and the Glacial Period*, Chamberlin, Salisbury, and McGee seized on errors in the work to expose in brutal terms Wright's ignorance of the fundamentals of glacial geology (§4.9, §5.6).

Because those critics were well-funded government scientists, defenders of Wright tapped into the thick vein of resentment toward the USGS within the larger geological community, and the atmosphere became hotly charged with claims of federal heavy-handedness and intimidation of self-proclaimed amateur geologists (§5.8–§5.9). Yet, the fact that these were government scientists was, in a sense, secondary. The intimidation was not driven by the institutions that employed Chamberlin, Holmes, McGee, and Salisbury (were that the case, one could as readily blame the University of Chicago), but by their individual efforts in a climate of professional porousness to distinguish professional from amateur and bring about consensus by the exclusion of nonconformists. That said, the marked funding advantage they enjoyed by virtue of their institutional base (§5.1, §5.7) assuredly made their power seem far more menacing and potent—and their opponents feel that much more vulnerable.

If we see the reaction to Wright as an effort to separate amateur from professional, it is evident his sins were far more grievous than mere ignorance of glacial geology, disagreeing with an official USGS line on Pleistocene glacial history, or even having failed to fairly consider alternative interpretations (§5.6). Just four months after hiring Wright, Chamberlin warned him that it was neither his right nor his privilege to present himself to the public as a spokesman for science. While Wright was in his USGS employ Chamberlin could leash him (though Wright slipped free on several occasions [§4.9]). But once out from under Chamberlin's control Wright casually ignored his increasingly menacing threats. He failed to realize how deeply hostile Chamberlin was to the prospect of Wright speaking directly to the public, or that Chamberlin deemed it his self-appointed duty to protect the public from what he considered inferior science (§4.9, §4.12, §5.6). Chamberlin had seethed privately in 1889 when Wright published *The Ice Age in North America and Its Bearing upon the Antiquity of Man* (§4.13). He did not contain himself three years later when *Man and the Glacial Period* appeared. Seeing the same errors from the first volume convinced him that Wright's "misapprehension and incompetency" was "conscious evasion."[141]

The attack on Wright's *Man and the Glacial Period* (§5.7–§5.9) was at its core driven by Wright's having overstepped the bounds of his amateur status.[142] "No one," Chamberlin thundered, "is entitled to speak on behalf of science who does not really command it. No one can be trusted to lead the public who does not himself know the way accurately" (§5.6). That Wright had the audacity to assume that role was in Chamberlin's eyes intellectual dishonesty and, given his view of the sacredness of science, profoundly immoral (§4.6). When James Dana admonished Chamberlin for seeking to suppress Wright's scientific voice, Chamberlin erupted. Wright's case, he snarled, lay

> entirely outside of considerations of legitimate freedom of opinion and publication. The impression that anyone is trying to suppress him seems to me quite aside from the mark. The effort is to suppress an abuse that seems to many of us, who are most intimately familiar with the facts, so pernicious as to call for abatement. I am fully persuaded that Professor Wright is deliberately sacrificing science to personal glorification, that he knowingly conceals or obscures facts of vital import, and, by adroit phraseology and an assumption of candor, creates an impression at variance with the plain truth.[143]

Wright, of course, saw it differently: "Having spread their own misinterpretations before the geological public, they [Chamberlin et al.] wished to prevent anybody else from discussing the subject, or getting his interpretations before the public."[144] But then Wright thought Chamberlin "pretty thoroughly discredited as an authority," viewed himself as Chamberlin's equal, and was certain "the forces which are coming up to my rescues are numerous and strong."[145] Such views reveal the depths of Wright's vanity and naiveté, if not his power of self-deception (which he took to his grave, or at least his gravestone [§7.8]).

Wright had badly misjudged the impact of the critics' assault and the weakness of his own position. Although it is likely true that he could take some credit for McGee having been driven from the USGS because of his role in attacking Wright (§5.13), he could hardly have failed to notice that James Geikie turned to

Chamberlin when he needed chapters on the North American Pleistocene for *The Great Ice Age and Its Relation to the Antiquity of Man* (§5.14). Nor could Wright have missed the fact that the groundswell of support he had anticipated in defense of his *Man and the Glacial Period* failed to materialize. Many, of course, were appalled by the assault, and though they rose in his defense it was rarely on substantive issues (§5.7), tacit admission that they saw Chamberlin and others as right and Wright as wrong. (Only Wright's staunch friend Warren Upham defended the unity hypothesis, but from behind a veil of anonymity out of fear of Chamberlin's retaliation [§5.12].) Most instead praised Wright's Christian suffering and framed his defense on moral grounds. Amateur geologists like Wright, Claypole and others argued, were the backbone of science, with much to contribute and the right to be heard, even by "official" science (§5.7–§5.8). Winchell, in fact, insisted that in publishing his book Wright had merely "exercised the privilege which every American citizen has."[146]

That was undeniably true, in America. But American science, as Rowland said, was no democracy.

10.10 ALL SCIENTISTS ARE EQUAL, BUT SOME ARE MORE EQUAL THAN OTHERS

In the decade after Mason welcomed all comers into anthropology, the previously hazy boundary separating amateur from professional came into sharper focus, and it was obvious on which side of the line Abbott, Claypole, Haynes, and Wright were standing. Likewise, the journals in which they published foundered; within the decade most were gone. Chamberlin's *Journal of Geology* survives to this day.[147] With the end of one century and the start of the next there was a subtle but unmistakable shift in the rhetoric of the controversy. Polemics were no longer aimed at amateurs, who by then had been firmly put in their place. Instead they sought to separate professionals from the professional elite.

Inequality in a scientific community is most visible during controversy, when the stakes are highest (§1.1). From the beginning of this dispute it was evident that not all participants' opinions carried equal weight or were on equal ground, even those that were on the same side. The influence of an individual was tied to that person's standing within the field, which in turn was based on their demonstrated ability to deliver reliable information and ideas. Authority depended on its point of origin, the consequence of which was an unspoken but unmistakable hierarchy of scientific status. Ultimately, advances were made because of, not in spite of, that inequality.[148] Understanding the relative status of individuals thus illuminates how, where, and with whom power resided; provides insight into why the controversy became so embittered; and, ironically, why it ended as quickly as it did when it did.

To ascertain the relative status of individuals in this controversy, there are sources both formal and informal.[149] In regard to the first, the American scientific community had by this time several formal marks of distinction earned by participants (Table 10.3).[150] In order of increasing selectivity and prestige, these

Table 10.3 Formal measures of status within the scientific community earned by participants involved in the human antiquity controversy or its resolution

Name	AAAS fellow (year elected)[1]	AMS star (volume)[2]	NAS member (year elected)
Abbott, Charles C.	1883[H]		
Babbitt, Frances	1887		
Baldwin, Charles C.	1891		
Barbour, Erwin	1898		
Blackman, Elmer			
Brinton, Daniel G.	1885[H]		
Brown, Barnum	1931		
Bryan, Kirk	1925	5	
Calvin, Samuel	1889[E]	1	
Carr, Lucien	1877		
Chamberlin, Rollin	1909	3	1940
Chamberlin, Thomas C.	1877[PE]	1	1903
Claypole, Edward W.	1882[E]		
Cook, Harold J.	1911		
Cope, Edward D.	1875		1872
Cresson, Hilborne T.			
Dall, William H.	1874	1	1897
Figgins, Jesse D.	1932		
Fowke, Gerard			
Gidley, James W.			
Gilder, Robert F.			
Goddard, Pliny E.	1912	3	
Gregory, William K.	1925[H]	3	1927
Hay, Oliver P.	1901	2	
Haynes, Henry W.	1884		
Holmes, William H.	1883[HH]	1	1905
Hrdlička, Aleš	1897[H]	1	1921
Ingalls, Albert G.			
Jenks, Albert E.	1902[H]		
Kidder, Alfred V.	1929[H]	3	1936
Knapp, George			
Kummel, Henry B.		1	

(continued)

Table 10.3 *continued*

Name	AAAS fellow (year elected)[1]	AMS star (volume)[2]	NAS member (year elected)
Lesley, J. Peter	1874		1863
Leverett, Frank	1891[E]	1	1929
Lewis, Henry C.	1880		
Loomis, Frederick B.	1913	4	
MacCurdy, George	1900[H]	4	
Mason, Otis	1877[H]	1	
Matthew, William D.	1906	3	
McGee, William John	1882[H]	1	
McGuire, Joseph	1891		
Mercer, Henry C.	1893		
Merriam, John C.	1905	2	1918
Mills, William C.	1902	3	
Moorehead, Warren K.	1890		
Nelson, Nels	1925	3	
Osborn, Henry F.	1883	1	1900
Owen, Luella	1911		
Peet, Stephen D.	1881		
Powell, John W.	1875[PH]		1880
Putnam, Frederic W.	1874[PH]	1	1885
Rau, Charles			
Roberts, Frank H.H.	1931	5	
Romer, Alfred	1924[F]	6	1928
Russell, Frank	1897		
Salisbury, Rollin D.	1890[E]	1	
Sellards, Elias H.		5	
Shaler, Nathaniel S.		1	
Shimek, Bohumil	1904[E]		
Shuler, Ellis	1931		
Spier, Leslie	1931	5	1946
Stock, Chester	1921	6	1948
Todd, James E.	1886[E]	1	
Upham, Warren	1880	1	

(*continued*)

Table 10.3   *continued*

Name	AAAS fellow (year elected)[1]	AMS star (volume)[2]	NAS member (year elected)
Vaughan, T. Wayland	1906	1	1921
Volk, Ernst			
Ward, Henry B.	1899	3	
White, Israel C.	1882[E]	1	
Whitney, Josiah D.[3]			1863[d]
Williston, Samuel W.	1902	1	1915
Wilson, Thomas	1888[H]		
Winchell, Newton	1874[E]	1	
Wissler, Clark	1906[H]	2	1929
Wright, George F.	1882		
Youmans, William	1889		
n (% of total roster)	**62 (82.7%)**	**38 (50.7%)**	**20 (26.7%)**

1. Letters in superscript indicate election to AAAS office, as follows: P = AAAS president; E, F, or H = Section vice president (E = geology, F = zoology, H = anthropology). Holmes was elected Section H vice president twice (in 1892 and 1909).

2. *AMS* 1 was published in 1906; *AMS* 2 in 1910; *AMS* 3 in 1921; *AMS* 4 in 1927; *AMS* 5 in 1933; *AMS* 6 in 1938.

3. Whitney was a charter member of the NAS but resigned in a snit in 1874. He dropped his membership in the AAAS that same year. It cannot be determined if he was a fellow of the AAAS.

Source: Author.

were election as a fellow of the AAAS, receipt of a star in the *American Men of Science* (*AMS*) volumes, and election to the National Academy of Sciences (NAS). Although these honors were tangible indicators of achieved status they are problematic in some respects. A few words of explanation of each is in order, especially of the long-abandoned *AMS* star system.

Election as AAAS fellow was a mark of distinction restricted to AAAS members who were "professionally engaged in science, or have by their labors aided in advancing science" (as the 1874 AAAS Constitution put it). During the years of this controversy there was no cap on the number of fellows who could be elected in a given year, nor were there particularly stringent criteria for election. Although the category was created to distance amateurs from professionals, the large number of individuals elected indicates the measure "did not distinguish very well between mediocre and very good."[151] The majority (*n* = 62/75, 82.7 percent) of the participants in the human antiquity controversy were AAAS fellows, including ones whose contributions were seen as limited (for example, Lucian Carr and Luella Owen). It was likely for reasons other than lack of accomplishment that individuals such as Gidley, Rau, and Sellards were not AAAS fellows; most obviously, they may not have joined the AAAS and thus not been eligible.

In contrast, any individual could be put up for election to the NAS, but very few were elected. The NAS had no formal quotas in place from 1871 to 1941 (they were in use before and after those years), but by virtue of its rigorous selectivity the NAS had a de facto quota system. Just 560 individuals across *all* sciences were elected from 1871 to 1941; only a fraction of those were anthropologists, geologists, or vertebrate paleontologists.[152] This made NAS membership the most exclusive honor one could achieve within the American scientific community, but it also meant not all elite scientists could be or were elected.

Nearly two dozen individuals who played significant roles in this controversy were NAS members. Those elected during the controversy included Chamberlin, Holmes, Hrdlička, Merriam, Osborn, Powell, Putnam, and Williston; elected later were Rollin Chamberlin, Kidder, Leverett, Spier, Stock, and Wissler. It is difficult not to notice that NAS membership tilted heavily toward USGS and Smithsonian scientists and those at the University of Chicago. Given their antagonism toward claims of a Pleistocene human antiquity (Williston excluded), it is evident that those seeking to make that case faced highly regarded opposition.

As measures of status, election as an AAAS fellow was too loose to be meaningful and membership in the NAS too restrictive to be comprehensive. But this statement can be hedged. Individuals elected as vice president of his or her respective Section or as president of the AAAS (Table 10.3) rose above the din of fellows, and NAS election identified elite members of the scientific community, just not all of them.

Falling somewhere between these measures of status was an *AMS* star. That honor was created by James McKeen Cattell, the entrepreneurial editor (of *Science, Popular Science Monthly,* and *American Naturalist*), and a Columbia University psychologist. By virtue of his contacts across the scientific community and because of a professional interest in the nature of scientific ability, Cattell began to compile at the turn of the century the first edition of the *AMS*. (Seven editions appeared in his lifetime.) Intending it as a scientists' version of *Who's Who in America*, he envisioned the *AMS* as contributing "to the organization of science in America" and helping scientists become acquainted with one another's work.[153]

There was another aim as well. In an age enthralled with using numerical measures to sort, segment, and impose order on disorder,[154] Cattell sought to identify within each of twelve scientific disciplines those whose "work is supposed to be the most important"—and thus the "stars" in their respective fields (and thus a convenient sample for his own studies).[155] To spot the stars Cattell asked "ten leading students" from each discipline to rank the individuals in their field in "order of merit."[156] American scientists turned from sorting culture and nature to sorting themselves.[157] Only 1,000 stars could be awarded, allocated among the twelve disciplines by the number of individuals in that field as a percentage of the total number of scientists ($n = 4,131$) in the first edition of *AMS*. The number of stars in anthropology was thus capped at twenty, and geologists and vertebrate paleontologists at one hundred.

The *AMS* star system was in place from the first (1906) through seventh (1944) editions of *AMS*, and with each the selection process repeated.[158] Yet the caps did not change despite the rapid rise in the total number of scientists in successive editions and in the number of members per discipline. Forging achievements that made one star-worthy was half the battle. One also often had to wait for the death of another star to free a slot to receive one's own. In the circumstances, "exclusion was not necessarily a dishonor."[159] Ultimately, more than three dozen ($n = 39$) participants in this dispute were starred. Rank orders are known for the anthropologists in *AMS* 1 through *AMS* 3, and for geologists and vertebrate paleontologists in *AMS* 1 (Table 10.4).[160]

The stars in anthropology in *AMS* 1 included several who were heavily involved in the human antiquity controversy, notably Holmes, McGee, and Putnam (Table 10.4a). Although Hrdlička was still new to the scene at the time of the *AMS* 1 voting, he was already recognized as a star but was not ranked among the top ten. Both Abbott and Haynes were listed in the *AMS*, but neither earned a star nor was even on the ballot to be considered for star status. Their reputations had already fallen too far. Hrdlička's star rose in *AMS* 2 (voting conducted in 1909), as did Putnam's, whereas McGee's plummeted, the result of his banishment from the BAE and anthropology after the failure of his power grab following Powell's death (§6.4).[161] By *AMS* 3 (voting in 1920) McGee and Putnam had died, Holmes's ranking had slipped, but Hrdlička's continued to rise. Among the new stars in that edition were several who were involved in the controversy, including Wissler, Nelson, Goddard, Kidder, and W. C. Mills, in that rank order. Later editions saw stars awarded to Hooton, MacCurdy, Spier, and Roberts. In keeping with the trend toward formal academic credentials, virtually all of the later stars had PhDs, a degree that by definition was exclusionary and elitist.[162]

Table 10.4 *American Men of Science* rank ordering of individuals who had a role in the human antiquity controversy, including stars and nonstars

(a) Ranking of anthropologists and archaeologists in *AMS* 1–*AMS* 3		
Individual	*AMS* 1 rank	*AMS* 2 rank	*AMS* 3 rank
Mason	2	Deceased	Deceased
Holmes	3	3	6
McGee	4	11	Deceased
Putnam	5	2	Deceased
Hrdlička	17	14	10
Wissler	No rank	15	3
Nelson	No rank	No rank	15
Goddard	No rank	25 (no star)	16
Kidder	No rank	No rank	19
Mills	No rank	No rank	20
McGuire	36 (no star)	Deceased	Deceased
Jenks	37 (no star)	32 (no star)	31 (no star)
MacCurdy[1]	No rank	34 (no star)	29 (no star)
Hooton[2]	No rank	No rank	40 (no star)[b]

(continued)

Table 10.4   *continued*

(b) Ranking of geologists and vertebrate paleontologists in *AMS* 1			
Starred in *AMS* 1		**Not starred in *AMS* 1**	
Individual	Rank	Individual	Rank
Chamberlin	1	Barbour	115
Shaler	8	Merriam[3]	134
Dall	18		
Osborn	20		
Salisbury	21		
Calvin	33		
McGee	35		
Williston	43		
Winchell	48		
Leverett	55		
Kummel	81		
Upham	88		
Vaughan	98		
Todd[4]	106		

Note: Ranking data are available only for the editions noted (data from JMC/LC; see also Meltzer 2002: Table 4).

1. MacCurdy received a star in *AMS* 4.

2. Hooton received a star in *AMS* 4.

3. Merriam received a star in *AMS* 2.

4. Despite a score above 100, Todd received a star in *AMS* 1; it is unclear why, unless an individual of higher rank passed away in the three years between the voting and the publication of the volume.

In geology and vertebrate paleontology, Chamberlin was ranked first among the nearly 250 individuals considered for star status (Table 10.4b). Virtually all of the geologists involved in the human antiquity controversy were starred. Broadly speaking, the higher-ranked geological stars were tilted toward those affiliated with the USGS, the University of Chicago, or both: Chamberlin, Rollin Chamberlin [starred in *AMS* 3], William H. Dall, Leverett, Salisbury, and Williston. Those outside that circle, such as Upham and Winchell, earned stars in *AMS* 1, but neither was high in the rankings. Upham did not make the top hundred on either Chamberlin or McGee's ballots. Winchell fared somewhat better (he broke the top fifty, though barely, on Chamberlin's list). Their low rankings were likely in part a consequence of timing, as the voting took place amid the Lansing dispute (§6.7–§6.10).

Of those who were considered for star status but did not make the cut, the most conspicuous was Wright. For the third edition of the *AMS*, Cattell solicited names to add to the star list. Upham (starred in *AMS* 1) nominated Wright, whom he insisted was long overdue for star status.[163] He need not have bothered. Chamberlin and McGee had poisoned that well nearly two decades earlier in their *AMS* 1 judging. McGee ranked Wright 245th out of 246 individuals. Chamberlin did not even deign to lowball Wright, explaining to Cattell:

> In one particular case I have not ranked the party because I think his work is positively bad and worse than valueless, and because it does not seem to me to fall within the canons of really scientific work.[164]

Another snubbed nominee was Erwin Barbour, whose long-standing work on the geology and vertebrate paleontology of Nebraska was obviously respected but evidently not considered star quality: he ranked 115th, missing the star cutoff by fifteen places. That Barbour did not gain a star cannot be read as prejudice against his provincial status: his midcontinent neighbor, Iowa State Geologist Samuel Calvin, was starred in *AMS* 1. Nor was it obviously the result of his vigorous advocacy of Long's Hill (§6.11), as the first round of *AMS* voting took place before the site was found. Moreover, his co-author on the skeletal material, Henry Ward, was starred in *AMS* 3. And Henry F. Osborn, who was also involved with that site, was likewise starred, but then he had a long-standing reputation and position that no gaffe—and his advocacy of *Hesperopithecus* (§8.2) was a major gaffe—could significantly tarnish.

Considering all three formal status measures, nearly two-thirds of the participants in this controversy who were AAAS fellows also earned *AMS* stars, and slightly more than half of the *AMS* stars were in turn elected to the NAS (Table 10.3). In effect, the climb in status was easier from AAAS fellow to *AMS* star than it was from *AMS* star to NAS election.

It is important to add that all three of these honors were permanent and reflect status gained, not lost. Yet, status can be lost, for although there is a degree of inertia to a scientific reputation (§10.12) it is also the case that, much like a position on the stock market, scientific status is contingent, mutable, and sometimes subject to volatile collapse. Over the course of this controversy, reputations declined when earlier work was reevaluated and found wanting (as happened to American Paleolithic advocates [§10.9]) and when later work was deemed of poorer quality or thought to lag behind advances in the field, such as Hay's increasingly marginalized claims about Pleistocene vertebrate history and antiquity (§8.17) and Upham and Winchell's persistent advocacy of an ever-more outmoded notion of loess formation (§6.10). Occasionally, status plunged for nonresearch related causes, as was the case with McGee's precipitous fall from scientific grace.[165]

Status changes are therefore more visible in less formal measures, most especially the unfiltered correspondence among members of the community, which can include candid assessments of the work and worth of individuals and thus a "real-time" measure of scientific stock as it rose or fell. Here one can see whose opinions mattered and to whom. For example, Kidder and others routinely identified Nels Nelson as "far and away the person best qualified" to assess the artifacts from Folsom (§8.12).[166] Likewise, correspondence can reveal whose opinions were considered of little value and by whom, such as Williston's (exaggerated) dismissal of Hrdlička as having "about as much knowledge of geology as my youngest boy"

(§6.13), or Matthew's revelation that he routinely discounted Cook's claims owing to his "optimistic temperament" (§8.3).

Signals of relative status can be detected in publications as well, notably in the manner in which individuals referred to one another's work. These can be a positive "tipping of the hat" in recognition, or quite the opposite.[167] These were sometimes explicit, such as Holmes's praise of Chamberlin's geological mastery—which, of course, was also obliquely self-serving praise (evidence he had chosen his allies wisely)—or Chamberlin's ferocious denouncement of Wright in the *Dial* (§5.6, §5.10, §6.8). On occasion the hints were more subtle, as, for example, the conspiracy of silence from critics in response to Winchell's defense of Lansing (§6.10) or Barbour's advocacy of Long's Hill (§6.14). Sometimes they were counterintuitive, notably Cook and Figgins's attacks on Hrdlička in *Scientific American* and *Natural History*. They knew whose authority they had to topple (§8.5, §8.8).

That the private commentaries were at times personal and gossipy, or that the public ones often had a sharp edge, is owed to the fact that most of the participants knew one another. In this small scientific community paths often intersected in fieldwork, training, or employment in complex anastomosing patterns. At the AAAS participants met annually and sometimes engaged in fierce rows (§5.5, §5.13, §5.16). They knew one another's tendencies, strengths, and weaknesses. Williston grimly amused himself by predicting to Barbour (and a decade later to Sellards) how Holmes and Hrdlička would react (§6.13, §7.3). This forced intimacy, coupled with the fact that many advocates and critics were opinionated, vocal, strong-willed, and immodest created a hothouse atmosphere as well as the often dark contempt that familiarity breeds.

To summarize these formal and informal indicators of status, it is useful to follow the analytical lead of Martin Rudwick and sort participants into one of three groups: amateur, accomplished, and elite. These fall along what he calls a "gradient of attributed competence," the relative status accorded by peers to an individual on the basis of their tacit judgment of that individual's ability to deliver reliable evidence, information, and ideas. The groups are discernible more by modal tendencies than sharply defined boundaries,[168] and though one could attempt more rather than fewer categories, that would likely only provide the illusion of historical precision. These groups approximate but do not coincide with election as an AAAS fellow, receiving an *AMS* star, or election to the NAS. However, those formal honors are useful for assigning individuals to each, just as the informal clues are valuable for watching movement between them.

Status provides one means of understanding the role of an individual in this controversy; degree of involvement provides another. One can sort the latter depending on whether, as with Holmes and Hrdlička or Cook and Figgins, the human antiquity controversy was a *major* theme of their research and writing; or if, as with Lewis, Claypole, or Goddard, their participation was relatively *minor*; or for those such as Dawkins, Evans, and Kidder, who made only *cameo* appearances. It is important to note that status and degree of involvement are neither correlated nor equally weighted in terms of their influence on the controversy.

Summarizing a complex community in this manner runs the risk of caricaturing participants and their roles, for there is inevitably a loss of information when continuous variables are treated in categorical terms, and one can always quibble about where specific individuals are assigned. Nor would I presume that my sorting would match the participants' view of their scientific landscape. But then, it should not: it is doubtful the participants could step outside themselves to give an objective view of these matters, let alone see their entire landscape and what went on behind the scenes on it. In many respects the view is better from the vantage that history provides a century later, and without any axes to grind.

At the base of the gradient of attributed competence are amateurs, whose opinions were rarely considered and whose roles were sharply limited. They might have a recognized competence, but one that was "local" in both a geographic and topical sense, such as familiarity with the archaeology or geology of an area, abilities in fieldwork, or basic competence in a related subject. The claims and opinions of this group were not taken at face value. They had either proved to be unreliable (Abbott, Babbitt, Haynes), had few opinions to offer, or implicitly understood it was not their place to offer them (Metz, Moorehead, Volk). There were also those in this category, Figgins and Cook most especially, who fashioned themselves giant killers and took on Hrdlička and Holmes (though the latter scarcely noticed their presence [§8.5]). Cook and Figgins, however, were victims of the Dunning-Kruger Effect,[169] lacking as they were the competence to judge their own incompetence, and thus they could not see how badly wrong their opinions were, even to the point that they believed their views trumped the experts in relevant disciplines (§8.8, §8.17). Ignorance, as Darwin said, more frequently begets confidence than does knowledge.[170]

For the sake of completeness there is a subset in this category that includes the often-invisible individuals who made the discoveries or were hired to carry out excavations, such as H. T. Martin and Thomas Overton at 12 Mile Creek (§6.9); Gilder at Long's Hill (§6.11); Frank Ayers and Isaac Weills, who found the Vero site (§7.2); and Vaughan, Fred Howarth, and F. G. Priestly, who brought Lone Wolf Creek, Folsom, and Frederick (respectively) to the attention of the CMNH (§8.3–§8.4, §8.7). Their views were discounted altogether, as Schwachheim learned when he tried to convince Figgins that Folsom points were not arrowheads. Figgins, who hardly knew better himself, accepted that was true only after he heard it from Kidder (§8.8).

The number of amateurs was much greater before the collapse of the American Paleolithic (§10.7), but they continued to play a small role into the new century. That role, often resisted if not resented, was to make the discoveries and then get out of the way to allow others to assess them.

The next group upslope, the accomplished members of the community, were those whose competence was broadly recognized, particularly if in a specialized field, but they were not granted the right or privilege of passing judgment on critical matters of archaeological or geological interpretation. Individuals in this category included Leverett, routinely consulted by Chamberlin, Holmes, and Salisbury for his detailed field knowledge of the glacial history of the Midwest

(§5.4, §5.12, §6.6, §6.8); Shimek, relied on for his knowledge of snails, which was sufficiently respected by Chamberlin that it helped tip his views on loess origins (§6.7, §6.10, §6.13); and Goddard, recognized for his expertise in Native American languages and their implications for human antiquity (§7.4 §7.11). Evidence and claims by individuals such as Upham, Barbour, Sellards, Loomis, or Gidley, all accomplished in their respective fields, were heard and taken seriously. They could not be dismissed a priori.

But they could be dismissed. Winchell thought himself Chamberlin's equal but was put firmly in his place when the imperial Chamberlin, who dismissed Winchell's work as "pitiably weak and incompetent," did what royalty (scientific royalty included) do to those beneath them. Following their disagreement over Lansing, he ever after pointedly ignored Winchell. Chamberlin took from Winchell what Robert Merton identified as the most widespread and basic form of scholarly recognition, that which comes with having one's work "used and explicitly acknowledged."[171] There was no mistaking the message Chamberlin's silence conveyed (§6.10, §6.14). Although the expertise in this group was recognized and valued at some level, their claims, opinions, or both could be ignored without loss.

That was the right and responsibility of the elite, those atop the gradient who viewed themselves and were regarded by others as the most competent arbiters of fundamental questions of theory and method, and of evidence and substantive results. It was their exchanges and arguments that influenced the course of controversy.[172] There was not a long list of individuals in this category. By the turn of the century Chamberlin, Holmes, Osborn, Putnam (though with a caveat, noted in a subsequent paragraph), Salisbury, and Williston were on it, and soon Hrdlička as well. In the coming years others joined that elite inner circle, among them Matthew, Merriam, Nelson, Spier, and, of course, Kidder.

Here there was more often respect across disciplinary boundaries. Highly accomplished scientists readily recognized one another (the reverse of the Dunning-Kruger Effect). Thus, Osborn made a 3,000-mile round trip expressly to see the Long's Hill skeletal material so that he might have the chance to announce the discovery of this "primitive" Nebraskan. Yet, before making that announcement he asked Hrdlička his opinion of the remains, and on hearing it he abandoned his own (§6.11). Likewise, Williston had a sterling reputation in paleontology but readily deferred to Chamberlin on critical geological questions and counseled others to do so as well (§6.6, §6.11). As noted, Holmes and Hrdlička at various times deputized Merriam, Matthew, and Stock, paleontologists who had the competence and credibility to respond to Hay's claims of great antiquity for the Vero, Melbourne, and Lone Wolf Creek faunas (§10.8).

But rank and respect had to be won on relevant grounds. Putnam was, by all formal measures, in the elite category. Yet, even his kindly biographer Alfred Tozzer struggled to identify significant contributions Putnam made to archaeology, ultimately settling on his having introduced improved field methods and stating that his "field notes are models of recording archaeological data." Damning with faint praise that, and Tozzer quickly moved on to praise Putnam's success as a serial founder of anthropology programs (§10.8).[173] Because Putnam had done so little significant work on archaeological theory, method, or analytical approach that was relevant to the human antiquity dispute, he was no match for Holmes. That much became painfully obvious when Putnam, seeing that Holmes was "annihilating Paleolithic man" (§5.5), had no defense to offer.[174] Instead, and despite being the only proponent with the status to stand up to Holmes, Putnam virtually disappeared during the heaviest fighting of the Great Paleolithic War, notwithstanding Wright's desperate pleas for help (§5.13). On his scorched-earth march through the Paleolithic, Holmes could and did ignore Putnam.

Chamberlin, who had a low opinion of most archeologists (save Holmes) and hardly deigned to recognize them, nonetheless singled out Putnam for "making a spectacle of himself" by believing there were human remains in glacial deposits (§6.9). He thought someone of Putnam's standing ought to know better. Chamberlin was no less forgiving of fellow elite geologists. He sent a blistering letter to Dana, a charter member of the NAS, for seeking a compromise between Chamberlin and Wright's view of glacial history.[175] Privately, Chamberlin blamed Dana's "reactionary" work for causing America's reputation in glacial geology to fall behind Europe's (§5.14). Chamberlin was not one for compromise, but then commitment to one's ideas, self-interest, and even bias are quite normal in science, and knowledge grows because of such characteristics, not in spite of them.[176]

Putnam aside, on questions regarding artifacts, association, context, and antiquity the opinions of the elite mattered, and mattered very much. It perhaps goes without saying that *they* thought so. Hrdlička long advocated the formation of "blue-ribbon panels" of recognized experts such as himself to look into claims of a deep human antiquity (§7.7, §8.8). Those who were not elite accepted that fact as well. Sellards knew exactly who to invite to Vero to assess its evidence (§7.3). In every paper Cook and Figgins wrote in the midst of this controversy, they took aim at Hrdlička and Holmes (§8.5). They recognized that however much they disliked the idea, and they disliked it intensely, the anthropological and geological elite had to be convinced of the veracity of their claims. It was only "right and proper," Cook admitted, that Holmes and Hrdlička "should not take without question such basic evidence as may seem necessary to establish a given fact beyond reasonable question."[177] Many waited and wondered, as paleontologist Frederic Lucas did, whether and when "brother Hrdlička [would] express himself" in regard to Folsom.[178]

The influence of the elite, even if they were not fully engaged, weighed heavily at moments both large and small in this controversy. Thus, for instance, in the summer of 1897 Henry Mercer, keen to make his mark as Abbott's replacement at the University of Pennsylvania Museum (§5.14), arranged for field parties to conduct new excavations at Trenton in hopes of resolving its chronology and stratigraphy. His plan for a joint publication in *Science* in advance of that summer's AAAS meeting was foiled when Holmes, Chamberlin, and Salisbury got wind of it and pounced on the geologists working with Mercer. Two of them had connections to the University of Chicago and feared crossing Chamberlin and especially the ill-tempered Salisbury. They caved to pressure both real (from Salisbury) and imagined (Chamberlin said nothing), and abandoned Mercer. George Knapp apologetically confessed to Mercer that he had relinquished his rights in favor of his material interests. "Low ethical standards?" he asked. "Perhaps so," he admitted (§5.15). Even kindly Putnam sabotaged Mercer, insisting he announce the results at the AAAS meeting, where Putnam would get more credit for his role as "instigator and director" of the work and gain credit with his donors (§5.15). When Mercer balked, Putnam undercut the young man by asking Wright to withdraw from his planned participation. Wright complied.

Mercer's pleas to Putnam, Holmes, and Salisbury got cold responses. He tried a last-ditch effort to publish his own paper in *Science*, but McGee thwarted that with a back-channel request to Cattell, *Science's* editor (§5.15). Mercer had been a rising member of the archaeological community, but that very year he abandoned prehistoric archaeology for Colonial American tools, an "abrupt change of direction" that Conn (and Mercer himself) attributed to "a new enthusiasm" for historic artifacts.[179] It is perhaps no coincidence he suddenly found that enthusiasm the moment his plans to make a name for himself in prehistoric archaeology were crushed.

That episode is perhaps the starkest example of the elites' ability to impose their will using harsh intimidation and behind-the-scenes mischief. It also reveals the underlying asymmetry of status and power. The substantial time and effort Mercer had invested was ultimately for naught, as the elite had influence far out of proportion to their involvement. It was a lesson Abbott learned years before Mercer came to ruin, and that Volk, Upham, Barbour, Sellards, Loomis, Figgins, Cook, and others learned in the years after. Influence in this controversy was driven by status, not time or effort. Dawkins sanctified the American Paleolithic on a one-day visit in 1880—Abbott believed he needed "no greater authority" (§3.7)—and it took Evans only a few minutes in 1897 to demolish it (§5.16). Kidder did not have to excavate the Folsom site himself: a day's visit was all it took for him to change the course of the controversy.

10.11 "BE SURE TO MENTION KIDDER"

Roberts was a young, relatively unknown member of the archaeological community at the time of his visit to Folsom in September of 1927. Yet, as a newly minted Harvard PhD and member of the BAE, he was well qualified: Hrdlička considered Roberts "very sensible." (That was written even *after* Folsom.)[180] Roberts was just that, for wheras Brown was confident he could identify the bison species at Folsom and thus its age, Roberts was reluctant to be solely responsible for the site's archaeological assessment and called Kidder to examine the in situ point (§8.10). That it was Kidder he called, and not any of the several dozen other archaeologists in attendance at the Pecos Conference who could as easily have made the trip, signifies that Roberts well understood who was best qualified to judge the meaning of Folsom. It was a opportune call for Kidder as well, who had his own reasons for wanting what Folsom might provide (§9.12).

Kidder was widely respected, even revered, within the archaeological community in 1927 (§8.10). That he examined Folsom, then within weeks publicly announced it was evidence of a Pleistocene human presence in the Americas, carried extraordinary weight—and not just among archaeologists but across the scientific community. In fact, recognition of the importance of his approval reached well beyond the sciences. Cook's *Scientific American* editor Albert Ingalls urged Cook to "be sure to mention Kidder and all the outfit that visited you" in an article Cook was then preparing. Cook did.[181]

There is no small irony in the fact that the legitimacy of Folsom as a Pleistocene-age site was attested to because Figgins took Hrdlička's advice and sent telegrams in August of 1927 (§8.9), as did Brown, who received the same advice and followed suit the next year (§8.16). By virtue of the fact that Folsom was a kill site containing the fossil remains of multiple bison (about thirty animals, by Brown's 1928 estimate), it was possible in 1927 and 1928 for successive groups of visitors, elite scientists among them, to witness a Folsom point in situ, repeatedly reinforcing the fact of contemporaneity.[182] It did not matter, as Lucas joked, that Hrdlička had not "fired the arrow himself" (§8.9), but it did matter to Hrdlička who was there to see it unearthed. Hrdlička could not have failed to notice, though it would have given him little cheer, that confirmation of the Pleistocene antiquity of the artifacts at Folsom involved an archaeologist from the BAE (Roberts) and a USGS geologist (Bryan). Times had changed.

Controversy in science, or at least nontrivial controversy, is not resolved by gaining consensus among the participants—a good thing, because consensus may not emerge on particularly contentious issues such as, for example, the position of Neanderthals on the human family tree (§6.11, §6.14, §7.6). Or consensus may emerge and be wrong, as evidenced by the once universal agreement about an American Paleolithic (§4.14). Rather, controversy is over when those most competent to judge, the elite within the community, deem it so.[183] Status matters, and it cuts both ways: it is what enabled Holmes and Hrdlička to repeatedly impose their will and influence the course of the debate over the decades leading up to that moment at Folsom, but it is also what made it possible for the controversy to be over so quickly once Kidder pronounced on the site. The rapid acceptance of Folsom attests to that, as does the deafening silence from Hrdlička and Holmes that followed (§8.18). With Folsom there was essentially none of the controversy that had dogged previous claims.

The combined impact of the testimonials at the AAAS meeting in December of 1927 from Kidder and Roberts on the site context, Nelson on the artifacts, and Brown on the bison bones caused a sea change in views of human antiquity in North America (§8.14). Once Folsom had their approval, they in turn brought in others who had the standing and wherewithal to amplify the results. Brown easily convinced Osborn and AMNH donor Childs Frick to support the 1928 season of excavations at Folsom (§8.17). Roberts's enthusiastic reports prompted the Smithsonian's Alexander Wetmore to throw the weight of his institution into the investigation of human antiquity in America, which took the tangible form of funding Bryan's 1928 fieldwork at Folsom and then later support for Roberts's excavations at Lindenmeier (§9.4, §9.8). On the West Coast soon after visiting Folsom, Kidder stopped in at the California Institute of Technology to meet with Stock (§8.10), with whom he "discussed questions of Pleistocene and Recent, and mapped out a tentative program for closer cooperation between their work and ours." "Ours" in this case was the Carnegie Institution. Merriam, who for many years had been a stalking horse for Holmes and Hrdlička (§8.3), now saw the research potential in this arena (§8.10), which ultimately led to the funding of Howard at Clovis and crucial financial support over the next decade, even during the worst years of the Depression, for virtually all of the fieldwork done during those years at Paleoindian sites (§9.4).

With the exception of Roberts, those who spoke to the Folsom evidence at the 1927 AAA meeting were established figures, and with the exception of Kidder all of them had been following events at the site. Still, none of them had spent more than a few days at Folsom, and none could be said to have had actual experience in the human antiquity realm. Brown had done only limited work with Pleistocene faunas, but none on an archaeological fauna (§8.6, §8.11). Roberts was not the Paleoindian archaeologist he would soon become (§9.8). Nelson was perhaps the best prepared to assess Folsom and its significance, having written on human antiquity in previous years and having had a hurried chance that December to examine the Folsom artifacts (once Brown reluctantly surrendered them [§8.12]).

In contrast, those who had far more intimate knowledge of the Folsom site, notably Cook and Figgins, did not speak at the AAA meeting, nor were they asked to participate. The closest they came was providing information to Brown and others and lending photographs and specimens (§8.14). This deliberate slight, if it was one, raises the question of why they were not called on to testify on behalf of their sites or to explain the larger implications of what they had found.

10.12 VICTIMS OF THE MATTHEW EFFECT

As a young man, the Smithsonian's Joseph Henry learned a piece of scientific wisdom from the eminent physicist Michael Faraday: early in a career one often has to share credit with those who have not earned it, but later one often received more credit than is due.[184] That precept was independently discovered more than a century later by Robert Merton, who codified it as the "Matthew Effect" after the Biblical passage "For whosoever hath, to him shall be given, and he shall have abundance, but whosoever hath not, from him shall be taken away even that he hath."[185]

As Merton applied the idea, it meant the "accruing of greater increments of recognition for particular scientific contributions to scientists of considerable repute and the withholding of such recognition from scientists who have not yet made their mark."[186] His formulation was based on co-authorship as well as cases of independent discovery in which the contributions of lesser known and (usually) younger authors were of equal or even greater importance, yet the better-known scientist received the lion's share of the credit, "minimizing or withholding . . . recognition for those who have not yet made their mark."[187] Merton's underlying assumption is that unequal credit is doled out for equal contributions.[188] It seems reasonable to presume the Matthew Effect is amplified in circumstances in which the work itself is not equal. That appears to have been the case for Cook and Figgins, as I have elsewhere discussed[189] and summarize briefly here.

In the aftermath of the Folsom find, there was an explosion of national and even international symposia devoted to human antiquity in America (eight in total from 1927 to 1937[190]), quite unlike the situation a decade earlier when Kroeber vetoed his own idea of having such a session at the AAA meeting (§7.17). Beyond affirming the spark created by Folsom, this spike in attention is also revealing in who participated in these sessions and who did not. Of the dozens of opportunities to speak in these symposia and meetings, not one was granted Cook or Figgins. Indeed, there were only two occasions when either was even asked to attend (but not speak at) the meeting.[191] Instead, in the human antiquity spotlight were Brown, Bryan, Howard, Merriam, Nelson, Roberts, Sellards, Stock, and Marie Wormington, among others (§8.14, §8.16, §8.18, §9.9). Even Hrdlička was invited to participate.[192] The interpretation of what Cook and Figgins found at Folsom was in the hands of others whose roles and influence were far out of proportion to their participation in uncovering the critical evidence.[193]

Cook and Figgins were not completely invisible: at the 1937 International Symposium at the ANSP (§9.9), Bryan credited them with the "*discovery* of Folsom points in association with the bones of extinct bison," but in that very same sentence declared that the find had been "*confirmed* by the masterly [1928] excavation of the site by Barnum Brown."[194] Given Cook's anger at Brown's perceived usurping of Folsom in 1928 (§8.16), Bryan's comment must have stung.

Worse, Bryan's comment, however unintended, put Cook and Figgins squarely in their place. It was apparent in the archaeological and geological communities that Cook and Figgins might be competent to make the Folsom find but not to give it meaning. Their credibility had by then been badly damaged by, among other things, their championing of Frederick and Snake Creek as legitimate archaeological sites, despite withering critiques by Spier and Nelson (§8.17), and their claim that these sites (along with Lone Wolf Creek) showed the New World was occupied upward of 500,000 years ago (§8.13, §10.4), which would make the human occupation in this hemisphere as old as or older than the Old World's. Cook declared that "mankind has lived on this continent for hundreds of thousands of years; and that even in early Pleistocene times had reached a stage of cultural development and civilization that was in some respects, at least, quite as advanced as any existing on earth at that time."[195] Figgins had an even more radical take, claiming "that America may necessarily have been the cradle of the human race and disproves the theory that man was not retrogressively crude in the matter of culture."[196] These were startling statements, balked at even by Cook's forgiving *Scientific American* editor (§8.13) and altogether rejected in the archaeological community. Accepting such claims would have demanded radically altering the hard-won understanding of "the origin, the development, and the spread of human culture," as Nelson put it (§8.8, §8.17). That would not occur, not on the basis of sites as inherently problematic as Frederick and Snake Creek.[197]

Equally damaging to Cook and Figgins's credibility was their judgment that Folsom was the youngest, weakest, and least significant of their localities.[198] They were betting on Frederick, whereas the archaeological and geological communities put their money squarely on Folsom (§8.13, §8.17). It is little wonder that Spier told Cook he had "a very likely case at Folsom," and that Spier thought so not as a result of what Cook said but of what Brown and Roberts presented at the AAA meeting in 1927 (§8.14).[199] Likewise, W. D. Matthew was not going to take

Cook's word on the site (he knew Cook too well [§8.3]); he wanted "two or three conservative expert opinions on the stratigraphic relations [at Folsom]. . . . And I would like to hear Nelson's opinion on the artifacts."[200]

Cook and Figgins were aware that few were buying what they were selling at Frederick, Lone Wolf Creek, and Snake Creek. Characteristically, Cook did not see that as a substantive difference of scientific opinion, but instead chalked it up to their being on the losing side of a popularity contest. It was the price they paid, he supposed, for not being on the "ritual-as-accepted-by-almighty-order side of this Pleistocene controversy" (§8.17). Figgins was more candid, admitting without false modesty that he knew "nothing whatever, about archaeology and allied subjects, and but a smattering of paleontology and geology" (§8.1). He thought it best for the museum that he and Cook "avoid all expression of opinion" concerning their finds, as their "opinions are valueless."[201]

Figgins was right, but in a sense so too was Cook. They *were* on the "wrong side," but not in the way Cook saw it. Rather, the divide was along status lines. In the crucial deciding moments of a scientific controversy, those whose opinions matter are individuals who have demonstrated their competence, who rank "first among equals," and who are deemed qualified by their peers to judge the meaning of evidence and its bearing on the continued course of controversy or its resolution (§10.10).[202] Brown, Bryan, Nelson, and especially Kidder were in that category. Cook and Figgins were not, though they (or at least Cook) might assert that right.

Cook and Figgins each had training (albeit informal), extensive experience, and professional positions. What they found at Folsom was profoundly significant. But the importance of what they found could not save their reputations after what they *said* about what they found. Following his visit to Folsom in September of 1927, Kidder took the opportunity to read Cook and Figgins's *Natural History* articles, yet his praise of their work then and later focused entirely on their having "found the materials [and] having brought them so promptly and liberally to the notice of other students."[203] No mention was made of their interpretation of the site.

Hence, neither Cook nor Figgins was invited to offer his views on human antiquity at any of those national and international symposia and, fairly or not, neither ever took to any of those stages to accept praise for what they had done to bring the controversy to an end. Both became victims of the Matthew Effect, for the Folsom discovery and the subsequent resolution of the human antiquity controversy were very separate events. Too much was riding on the latter to leave it to Cook and Figgins.

10.13 PREHISTORY REPEATS ITSELF

In this regard the histories of the establishment of a Pleistocene human antiquity in Europe and North America converge, for in both instances a clear dividing line existed between those who made the discoveries and those called on to judge the significance of the finds. There are other parallels as well.[204] In both instances the key discoveries at Brixham Cave and Folsom were an unintended consequence of excavations conducted for a nonarchaeological purpose, the acquisition of fossil remains (§2.1, §8.4). Both discoveries came at the end of decades of failure to demonstrate a Pleistocene human antiquity and thus emerged in a climate of skepticism, though skepticism was deeper and differently rooted in Europe, as there was no a priori presumption that human remains would or should be found in Pleistocene deposits. After 1859 that presumption no longer held, and indeed early on it was supposed that human antiquity in North America ought to go back as far as it did in Europe (§2.2). In America, skepticism toward a Pleistocene human antiquity resulted from a coupling of failure to prove as much, along with interdisciplinary suspicion and mistrust (§10.8).

Both key discoveries lent themselves to generalizations about how to find more sites like them. On both continents the search for early sites in prior years had been haphazard and led primarily by paleontologists and geologists. After those discoveries the search was reconstituted as a more systematic and dominantly archaeological enterprise.[205] In America those investigations were increasingly led by a new generation of PhD archaeologists who worked with teams of graduate students and specialists from other disciplines (§9.1–§9.2, §9.4). Interdisciplinary competition finally gave way to interdisciplinary cooperation.

The emergence of a class of formally trained professional archaeologists working in this research domain effectively closed the door on amateurs having anything but an ancillary role in the investigation of early sites.[206] It was no longer sufficient to hire the unemployed handyman or artifact collector (for example, Nelson Vaughan [§8.3]) or the nearby village blacksmith (Carl Schwachheim [§8.4]) to conduct unsupervised excavations (§9.3). Not only were field methods more demanding, as all concerned had by now learned the importance of context and association (§10.2), but also the realization that the past was not readily interpretable or understood by common sense was implicit knowledge. Otis Mason would have scarcely recognized archaeology's emerging priesthood with its sacred language (§10.9).

In both the Old World and the New the establishment of human antiquity came about after it was demonstrated that artifacts had been found in close association with the remains of extinct fauna. This was the only secure means available in those pre–radiocarbon days for telling time. However, that demonstration played out in different ways. Brixham Cave produced artifacts alongside the bones of Pleistocene fauna, but despite these having been found beneath a thick and apparently impervious calcareous layer there was no detouring around the fact that as a cave its stratigraphic evidence was suspect, hence the pilgrimages to the alluvial deposits of the Somme valley to examine Boucher de Perthes's sites and collections (§2.1). When John Evans and Joseph Prestwich visited there in the spring of 1859, they arrived in time to witness and photograph a hand ax in situ at St. Acheul.[207]

Yet, they did not identify that as the decisive moment that convinced them the artifacts from the Somme gravels were Pleistocene in age. They instead accepted the say-so of Boucher de Perthes and the quarry workmen that the artifacts occurred "in the position assigned to them." Prestwich trusted the workmen's "honesty and veracity," their long experience excavating in the gravels, and the knowledge that gave them to accurately interpret the stratigraphic context of their finds.[208] Beyond needing to document that association in a geological setting other than Brixham Cave, there was not in Europe in the 1850s the debilitating mistrust regarding archaeological context that permeated discussions in North America in the decades up to Folsom (§10.2).

Why, then, was trust so readily given to the anonymous workmen at St. Acheul? Or, to put it another way, why was trust in such short supply in North America in later decades? The answer cannot be because there was greater skepticism about a Pleistocene antiquity in America than in Europe. There were plenty of failed claims in Europe to warrant equal skepticism, if not a compelling a priori reason to doubt a Pleistocene antiquity existed. Nor was it because in Europe discoveries were made by professionals rather than untrained observers. Although that was true of Brixham Cave, it was not true of the finds made by the Somme valley workers. Rather, I suspect the answer can be found in the different frames of reference within which the search for a deep human antiquity played out in Europe and America.

In Europe in the 1850s the determination of whether archaeological materials were Pleistocene in age was largely a binary one; by the late nineteenth century in North America it was not. That is—at the risk of oversimplifying the case—in Europe the Pleistocene was identified by the presence of Cuvier's extinct animals, which inhabited the penultimate premodern world shrouded in Agassiz's glacier. The presence of distinctively primitive artifacts with the bones of those animals, once it was accepted that the deposit was undisturbed and the association was not obviously adventitious, was sufficient to prove a Pleistocene antiquity for humans (§2.1).

In North America the matter was much more complicated. By the 1880s there were thought to have been two glacial epochs, and soon four or perhaps as many as six. The Pleistocene fauna was known to range across most of those epochs, but it was not known how far back in the past they went or how close to the present they came. The occurrence of an extinct taxon was not equivalent to the Pleistocene, as it was in 1850s Europe almost by definition. One needed to determine which glacial (or interglacial) epoch the species belonged to and when it went extinct, which meant eliminating or at least reducing the possibility that it lingered into the Recent (which was particularly challenging with bison) and resolving just when the Pleistocene came to an end (§6.9, §8.11). The form of artifacts or human skeletal remains were of little help (§10.1). It was obvious only in retrospect that Folsom points looked nothing like what Pleistocene-age artifacts were expected to look like, but they were nonetheless Pleistocene in age (§8.18, §9.6). In effect, a Pleistocene human presence was a much more difficult target to hit in late nineteenth-century America than it had been in midcentury Europe.

But when those targets were hit, the results were significant. In Europe the establishment of human antiquity raised profound questions about human origins that were of compelling interest to specialists and the public alike. In North America the implications were less sweeping but still had profound implications for North American archaeology and prehistory, its relationship to the larger schema of human evolutionary history, and, indeed, a new view of the past and a new way of viewing the past (§9.12–§9.13).[209] It also meant that the American

Indian faded into an even more distant past, increasingly removed from American history (§10.6, §10.14).[210]

10.14 LIVING IN AN OLD NEW WORLD

The establishment of a Pleistocene human antiquity in North America raised many questions about the early prehistory of the continent: Who were the ancestral Americans and where had they come from? By what route or routes had they reached this continent and to what degree was their travel constrained or influenced by glacial ice? What was the climate and environment at the time of their arrival and spread across the continent? How rapidly had they dispersed, and what was the nature of their adaptation while doing so? And, of course, what was the course of North American prehistory after the initial arrival of people? These were questions that could have been asked, and on rare occasion were (§10.7), before the Folsom discovery. But until Folsom provided a secure chronological point of reference, they could not be addressed in any meaningful way, for only then did it become apparent *when* in time one needed to look for the answers (§9.7, §10.7).

The subsequent discovery and archaeological investigation of dozens of Late Pleistocene sites, coupled with geological improvements in age dating, a better understanding of extinct faunal sequences, and a more nuanced view of the glacial landscape and the "rhythm of climatic fluctuations,"[211] resulted in great strides through the 1930s in understanding when, how, and under what conditions people adapted and spread across Late Pleistocene North America (§9.4–§9.7, §9.13). From 1862, when Joseph Henry first published word of the discovery in Europe of a deep human past (§2.2), it took more than seven decades to demonstrate that humans were in America in Pleistocene times. Yet it required scarcely a dozen years after Folsom for the broad outlines and many of the key details of Paleoindian archaeology and culture to be established, an understanding that made possible the syntheses that Roberts, Sellards, and Wormington produced on the eve of World War II (§9.13). Inevitably, those works had to be revised in the decades after the war, but they stood the test of time well—far better and longer, it is not surprising, than did the syntheses of the American Paleolithic produced in 1889 by Abbott, McGee, Putnam, and Wright (§4.14). But then, those were written about a phenomenon that proved not to exist.

Important as Folsom was to the recognition and understanding of the peopling of the Americas, it also had implications that rippled outward in ways substantive and conceptual. Kidder had long struggled with the "most fundamentally important of all anthropological problems": whether the development of agriculture and complex society in the Americas resulted from diffusion or independent invention.[212] His efforts were hamstrung by the lack of the temporal on-ramp in the Americas necessary—or so the hyperdiffusionists insisted—for such developments to have occurred on their own (§9.12). He blamed that situation on archaeological efforts that had traditionally focused on identifying ancient sites with modern tribes, going forward rather than back, with the result that prehistory "was left without foundations." Folsom changed all that, to Kidder's great relief (§9.12). He was no "polygenist" when it came to American prehistory, and it was now possible for him and others to explore questions of origins without having to "look for answers overseas."[213]

Nor was it just questions of the origins of agriculture or of civilization for which archaeologists sought answers. Folsom presented the far greater challenge of filling in the details of North American prehistory over the thousands or possibly tens of thousands of years that suddenly appeared between Folsom and the Basketmakers (§9.12). Accomplishing that goal required a new conceptual framework and methodological tools, for the antiquity of Folsom rendered irrelevant the "general fashion" among archaeologists to connect all prehistoric remains with the historically known peoples occupying the regions in which such remains were found.[214] In effect, the long-standing core epistemological assumption about the direct relevance of the ethnohistoric present to the archaeological past was wrong. The use of ethnographic analogy, insofar as it traditionally assumed historical relatedness or at least proximity, became problematic at ever greater temporal remove and had to be approached in a different manner. Archaeologists of mid-twentieth-century America, Lyman and O'Brien observed, could not "escape the problems imposed by the remote past on keeping an ancestor-descendant line intact; rather, one [had] to create a type of analogy independent of that relationship."[215]

The once dominant BAE program was rooted in uniformitarianism (§5.1), both in the more substantive sense that the present and the past were fundamentally alike and in the methodological sense that the processes operative in the present could be extrapolated and applied in the past.[216] After Folsom, those two meanings had to be disentangled, for the first was no longer valid. That separation required archaeologists to approach the most fundamental of operations, archaeological classification, in new ways. The traditional arrangement of archaeological material was in ethnological and linguistic terms. Even if mediated by way of the direct historical approach (§10.5) through the concept of a culture area, that was no longer adequate. "In most instances," Will McKern wrote in 1939, "we cannot immediately bridge the barrier between pre-literate and historic or proto-historic cultural groups, and in many instances we can not reasonably hope ever to be able to do so." Not only were the culture areas of ethnologists (and BAE archaeologists [§10.5–§10.6]) doubtful, given that Native Americans rarely confined themselves to a particular region, it was evident too that "no cultural areas [are] devised to account for an *unlimited temporal factor*."[217] The consequences of Folsom dropping the bottom out of American prehistory had come home to roost.

With the ethnohistoric record applicable only to the latest segment of prehistory, but still needing to bring people from the Pleistocene out to the present, there was a flurry of activity in the post-Folsom decade aimed at classifying artifacts in strictly archaeological terms so as to detect diachronic change. That required thinking of time on a nonepochal scale and measuring artifact type frequencies and variants of traits to render finer-scale temporal resolution (§10.6). Lyman and O'Brien argue that developing an understanding of artifact variation and change over time, coupled with seriation and stratigraphy, were the keys to the twentieth-century chronological revolution in American archaeology.[218]

Seriation and stratigraphy were introduced in the decade before Folsom, but they gained far greater purpose and use after Folsom as archaeologists sought to close the temporal gap that discovery opened. The result was Culture History, the approach that dominated American archaeology in the middle of the twentieth century to a degree matched only by the dominance of the BAE program in the late nineteenth century. Culture historians sought chronologies of events using similarities in material culture across space and through time, and from that building out to local and regional prehistoric sequences. It is no coincidence the first efforts at large-scale time-space culture history sequences appeared only after Folsom.[219] They would have been irrelevant before then. American archaeology had regained control over past time.

Culture history was prehistory as chronicle. It was also prehistory as historical particularism. After Folsom, American archaeologists at last became Boasians, though they got there on their own accord and without Boas's help or influence, which helps explain why they got there decades later (§10.5).

Of particular interest and value to culture historians, given their principal analytical goal was identifying the temporal and spatial dimensions of archaeological material, were artifact classifications that yielded time-sensitive types (primarily style-laden artifacts). Abbott and Wilson's earlier assumption (§10.1) was finally realized: artifact form could tell time. It was by now obvious, however, that change in artifact style itself had no inevitable directionality; if it had, the Paleoindian projectile point sequence would have been far easier to resolve (§9.8).[220] It was also the case that to anchor a relative artifact sequence on an absolute chronological scale required independent geological testimony. Once that was done at Folsom and repeated at Clovis (§9.4), fluted points—with their distinctive form and technology—had value as time markers. Using those artifacts to identify a Pleistocene human presence was now "as good a rule of evidence in archaeology as in law" (§4.3). Unlike Abbott's paleoliths, there was no mistaking Folsom fluted points for quarry refuse, not after Roberts's work at Lindenmeier showed what early-stage fluted points looked like (§9.8).

With time and types also came the possibility of a deeper understanding of linguistic and anatomical diversity and of the rate and degree of change over time. Calibration of these relationships had been impossible in the absence of a secure measure of time depth. Folsom was welcome, as it reopened the possibility of an "original unity of speech" for Native Americans, though Boas still wanted more time than Folsom allowed. Even 10,000 years was not enough (§9.12).

Issues were more complicated in physical anthropology, where Hooton saw historical significance in minor anatomical variants. Like other self-styled "heriditarians," he believed these features were "welded together by mixture [and] not wonderfully adapted types made out of common clay by a creative environment." He had no use for "fanatical environmentalists" who refused to accept that a skull could "withstand the moulding [*sic*] and modifying influence of physical environment and . . . endure through generations as a racial earmark."[221] His position was more than faintly reminiscent of the American school of polygenesis,[222] but instead of separate creations his types marked separate migrations.

One of those "fanatical environmentalists" was Boas, who, in addition to having a low opinion of Hooton, had for decades steadfastly challenged claims of anatomical immutability and the significance of biological race. This was part of his larger effort to dampen the influence of racial thinking in physical anthropology, an effort that involved a bruising battle with Hrdlička and Merriam, among others, over the composition of the National Research Council Committee on Anthropology. (Boas sought unsuccessfully to keep eugenicists such as Madison Grant and Charles Davenport, Hrdlička's fellow travelers [§6.2], off the committee.[223])

Boas's approach to anatomical differentiation and type, Stocking observes, was akin to the concepts of "drift" and "population," with an emphasis on the effects of environment on biology.[224] For Boas the "marked local varieties" of Native Americans had clear ancestry in Asia, yet, as with languages, he thought 10,000 years was insufficient to account for that variety. Either the types differentiated at a faster rate, or there had been a number of distinct types coming one after another to the continent. Boas was skeptical of both possibilities, and it gave him further cause to insist archaeologists keep looking for older sites.[225]

Kroeber did not think that was necessary. He thought Folsom provided enough time for genetic drift to run its course and thus reduce the need for far-flung migrations of Hooton's hypothetical races of "pseudo-Australoids, Caucasoids, and the like," yet also allowed for the population differentiation that Hrdlička had otherwise impatiently swept aside (§9.11). The net result was to dampen the significance of race in the interpretation of American prehistory.

The proximate reason why the Folsom discovery had such wide-ranging influence and helped revolutionize American archaeology at its conceptual core and in its approach to prehistory is that it provided, for the first time in the history of the field, a secure measure of deep time. The ultimate reason it had that impact is that at Folsom the human antiquity controversy ended in *resolution*.

10.15 CONTROVERSY AND ITS RESOLUTION

It all started with a straightforward empirical question: how long had people been in North America? Initially, there seemed reason to expect, and evidence to prove, that humans had been here since the Pleistocene. But consensus on this point was fated to controversy in the fall of 1889, when the BAE, wrapping up its Mound Survey, turned its attention to the American Paleolithic (§5.1). Keen to understand human antiquity in its own theoretical terms and impose its vision of a new archaeology, the BAE was predisposed to reject the idea of and the evidence for a separate Paleolithic race (though not necessarily a Pleistocene human presence). So controversy began.

That the controversy became so highly visible and so heavily invested in by the major figures of late nineteenth- and early twentieth-century archaeology, anthropology, geology, and paleontology testifies to the fact that knowing the answer to the question mattered. It was clear to all that the outcome would reveal North America's allotment of past time, on which depended the understanding of its prehistory (and cultural changes over time), the relationship of its past to its present native peoples, the theoretical and methodological links between archaeology and anthropology (cultural and physical), and the synchrony of the archaeological and geological records of the Old and New Worlds—and with the latter, deeper insight into how North America's peoples fit into the grand scheme of human evolutionary history.

That the controversy lasted for decades is in no small measure a result of the fact that the solution required closing both factual and conceptual gaps. The human antiquity controversy was in fact a long string of individual disputes over specific types of evidence, sites, and claims. Although all the sites leading up to Folsom failed to withstand critical scrutiny on either archaeological or geological grounds, proponents stubbornly held to the idea that there was a Pleistocene human presence in the Americas and continued to seek evidence of it. And although critics propelled the discussion forward by the serial rejection of specific sites, they implicitly understood, however skeptical they were on this point, that failure of evidence was not proof that humans had not been here in the Pleistocene. The possibility of a deep human antiquity in North America never went away, regardless of the long failure to demonstrate it.

Each failed site was an independent event, but they were nonetheless linked in the aggregate in terms of the creation of archaeological knowledge. The empirical sifting process that was this controversy not only cleared the ground of many unsupported claims, it also had the important effect of continually exposing conceptual gaps in knowledge, particularly those related to context, chronology, and dating. As those gaps were closed in an adaptive response to what was, in effect, the intense scientific "selective pressure" that comes with controversy, much was learned, no matter how harsh the rhetoric at times became. The advances made in glacial geology, vertebrate history, and archaeology were largely cumulative.[226] In the end, a conceptual framework for recognizing a Pleistocene human presence in North America was built up by knocking down empirically unsupportable cases.

The fact that controversy embittered participants and had many toxic moments was undoubtedly due in part to the difficulty of reaching a solution, as well as to the skepticism and polarization of positions that followed. But there were other elements fueling it, epistemic and nonepistemic alike (*sensu* McMullin[227]). Among these were disparities in funding between institutions, which meant power for some and deep envy and resentment for others (§5.1, §5.3, §6.3, §6.13, §10.9); differences in the theoretical and methodological precepts brought to the investigation (§5.1–§5.3); tensions derived from the institutional and regional platforms of proponents and critics (§10.2–§10.3); and broader changes taking place in American science toward a concept of science that was increasingly nationalistic, more narrowly specialized, and above all deliberately and self-consciously professional and nonegalitarian (§10.9–§10.12).

These changes exacerbated intradisciplinary and interdisciplinary conflict (§10.8), creating deep mistrust across the controversial divide. It was an especially troublesome consequence in a discipline such as archaeology, in which factual evidence is finite as well as irreproducible once out of the ground, and on which so much depends on knowing and understanding the context and association of finds the moment they are made—something that is difficult to convey in either words or pictures (§1.1, §10.2). That difficulty was in turn compounded by the fact that many of the major figures on both sides of the controversy intensely disliked and deeply distrusted one another (§10.9–§10.10) and brought to the dispute their own conflicting agendas.

In spite of all this, the resolution to the controversy was achieved suddenly and with virtually no resistance at Folsom, even from Holmes and Hrdlička, whose uncharacteristic near-silence in response to Folsom spoke of acceptance in its own grudging fashion (§8.18).[228] In this regard it is a useful historical exercise, as James Griffin once suggested, to consider the list of sites Holmes and Hrdlička rejected before 1927. No locality they "checked out as postglacial or Recent in age," he observed, has "proven otherwise." Today the list of Pleistocene sites in North America does not include Claymont, Madisonville, or the New Comerstown paleoliths, the Holly Oak pendant, the Nampa Image, Lansing, Long's Hill, Vero, Melbourne, or Frederick, and for good reasons: none of these sites were the age they were claimed to be, and a few cases were not even legitimate finds. (The Appendix summarizes what is known today of some of the sites that figured in the controversy.)[229]

To be sure, there were exceptions to Griffin's observation, notably 12 Mile Creek (§6.9) and Lone Wolf Creek (§8.3), recognized now as genuine Paleoindian sites. Yet, these exceptions do not mean the critics (or Griffin) were wrong. Rather, just the opposite. They show, as noted at the outset (§1.1), that having the right site matters. However, what matters even more is having the knowledge necessary to recognize it as such. So it was that with the conceptual foundation in place by the late summer of 1927, all it took to bring the decades-long human antiquity controversy to an end was the convergence of a kill site with projectile points in situ between the ribs of an extinct bison, well-placed telegrams, and—at the risk of trivializing to make the point—Kidder's close proximity in Pecos.

That these elements converged at Folsom was a historical accident, but by the late summer of 1927 it was an accident just waiting to happen.

APPENDIX:
WHATEVER BECAME OF . . . ?

MANY OF THE LOCALITIES that played a role in the controversy over human antiquity, prominent or otherwise, have long since slipped into obscurity. These include virtually all the supposed American Paleolithic sites (for example, Claymont [§4.5], Lake Lahonton [§4.3], Little Falls [§4.1], Loveland [§4.2], Madisonville [§4.2], Medora [§4.5], and New Comerstown [§4.15]), and those finds that were little more than pranks (Frederick [§8.7] and Nampa [§4.15]). Vanished from the scene as well are some of the features identified in the course of the glacial dispute, notably George Frederick Wright's Cincinnati Ice Dam (§4.9) as well as his fringe (§4.8), which is now recognized as till from a pre-Wisconsin glacial advance. Thomas Chamberlin's Driftless Area (Figure 1.2, §4.11) and Kettle Moraine (Figure 4.12, §4.7) fared better, the latter now accepted as marking the Wisconsin-age terminal moraine. Also still around more than a century later, though under a different name, is the Debatable Tract identified by Warren Upham in 1895 (Figure 1.2, §9.7), now mapped as the ice-free corridor.

Other sites that were part of the pre-1927 controversy saw later fieldwork, analyses, or both, which in a number of cases clarified their character and antiquity. Following are brief summaries of what is now known of those sites, along with a couple of localities that were not considered important at the time of the controversy but in retrospect became so, for both good reason (Kimmswick) and bad (Holly Oak).

12 MILE CREEK: Interest in 12 Mile Creek was rekindled in the 1980s and led to renewed investigations of the site and its remains (Hill 2006; Rogers and Martin 1984). Analyzing the fauna recovered in the original excavations, Hill ascertained that thirteen bison were killed at this locality by hunters, perhaps during the late winter or early spring (Hill 2002). The bison bones show evidence of having been butchered, though the carcasses must have been buried soon afterward, given the lack of evidence of carnivore gnawing or weathering. The site dates to the late Pleistocene, with several radiocarbon ages on bison bone yielding an average age of $10,500 \pm 65$ [14]C years before present (B.P.) (Hill 2006:144). The projectile point found in the 1890s vanished—perhaps at a party at Williston's house—a fact that was long kept hidden for fear it would cast doubt on the site's authenticity (Hill 2006:149). When viewed in a contemporary illustration (Williston 1905a:338) and a surviving photograph (Rogers and Martin 1984: Figure 2d), the specimen could be either a Clovis point or a Folsom point. Rogers and Martin (1984:761) suggest the former, but on the basis of the age and associated fauna, assignment to the Folsom period seems more likely.

CALAVERAS: From the outset in the 1860s there was great doubt about the antiquity—and, indeed, the genuineness—of this "discovery" (§2.4). Those doubts were well founded. Radiocarbon dating of the specimen in the 1990s returned an age of 740 ± 210 ^{14}C years B.P. (Taylor et al. 1992:272–273), making it less than 1,000 years old. Calaveras was indeed a hoax, in which a relatively modern cranium was put into an ancient geological context.

FOLSOM: After the Colorado Museum of Natural History and the American Museum of Natural History finished their work at the site in the late 1920s, no further systematic excavations occurred in the bison bone bed until I returned to the site in 1997 (Meltzer 2006a). That reinvestigation was motivated by the fact that by then far more was known about other Folsom-age sites than about the type site, such as its age, the environment at the time of occupation, the size and nature of the bison herd, the processes of the kill and butchering of the animals, the source and character of the stone tool assemblages, and the like. Three years of intensive fieldwork at the site (1997–1999), along with detailed analysis of the faunal remains and artifacts recovered by the two museums in the 1920s, sought to answer these questions and resulted in a large volume on the site (Meltzer 2006a). Suffice it to say Folsom is indeed late Pleistocene in age—as Barnum Brown and Kirk Bryan originally supposed (§8.10, §8.16)—with numerous radiocarbon dates that now place its occupation at 10,490 ± 20 ^{14}C years B.P. As many as thirty-two bison, previously identified as *Bison taylori* (§8.11) and now as *Bison antiquus occidentalis* (McDonald 1981:94), were trapped in a knick-point arroyo and killed by a group of highly mobile Paleoindian hunters sometime in the late summer or early fall. The animals were thoroughly butchered, the meat readied for transport, and the hunters moved on—unaware that their brief activities at this spot would have such an enormous impact on American archaeology (Meltzer 2006a:307).

HOLLY OAK: Almost immediately after Frederick Ward Putnam's halfhearted presentation of Hilborne Cresson's alleged find of a whelk shell on which was engraved a mammoth (§4.15), the Holly Oak pendant disappeared from the archaeological scene. It did so likely owing to the suspicion that its striking resemblance to the La Madeleine mammoth engraving was no coincidence, a suspicion surely fueled by doubts about Cresson's trustworthiness among his peers (Meltzer and Sturtevant 1983). The pendant made a spectacular comeback nearly a century later, debuting in 1976 on the cover of *Science*, the accompanying article suggesting it was evidence of "an exciting new association of early man with the woolly mammoth in America" (Kraft and Thomas 1976:761). That claim triggered a systematic stylistic analysis of the mammoth image and historical investigations by me and William Sturtevant (Meltzer and Sturtevant 1983). We concluded there was a strong likelihood the engraving was done by Cresson himself (who had artistic training) on a whelk shell purloined from a Late Prehistoric site, likely a site from which he had been fired for stealing artifacts (Meltzer and Sturtevant 1983:343). Subsequent radiocarbon dating of the shell returned an age of 1530 ± 110 ^{14}C years B.P. (Griffin et al. 1988:579), confirming the forgery. That Cresson committed suicide in 1894 and left behind a note saying he feared the Secret Service was seeking to arrest him for counterfeiting (Meltzer and Sturtevant

1983:344) was a grim coda to this supposed evidence of a Pleistocene human antiquity.

KIMMSWICK: After Gerard Fowke's excavations at Kimmswick in the summer of 1902 failed to turn up human remains associated with mastodon bones, Kimmswick was dismissed as an archaeological site (§6.8). Yet it continued to attract attention as a vertebrate fossil locality, and sometime during excavations at the site from 1903 to 1907 by Charles Beehler, a Clovis fluted point was found. This was still decades before the Folsom and Clovis discoveries, and thus the significance of the specimen was unrecognized, though it—along with a knife-like biface—made its way into the collections at the Field Museum of Natural History (Graham 1979:42–43). In the late 1970s vertebrate paleontologist Russell Graham returned to Kimmswick and, during excavations, recovered two Clovis fluted points and other stone artifacts in association with mastodon and other extinct fauna indicating a late Pleistocene age for the site (Graham et al. 1981). Suitable material for radiocarbon dating is lacking, however.

KOCH MASTODON: As part of a mitigation program to salvage information before the closure of a dam on the Osage River (McMillan 2010:26), excavations were conducted in the 1960s and 1970s along the Pomme de Terre River at the locality where more than a century earlier Albert Koch reported having found projectile points along with his "Missouri Leviathan," one of the points ostensibly directly associated with a mastodon femur (McMillan 2010:39–40). Koch may have indeed recovered artifacts along with mastodon bones (§2.4), but the twentieth-century excavations revealed these were almost certainly not in primary context. Because of the complex spring-fed geology of the site—unnoticed or not understood by Koch—the artifacts Koch recovered had likely worked their way down into the deposits with the Pleistocene fossils (McMillan 1976:92). The projectile point found with the femur has long since disappeared, but from a nineteenth-century illustration it appears to be an Archaic-age corner-notched point, making it Early to Middle Holocene in age (Wood 1976:104–107). Radiocarbon dating of the peat in which the vertebrate fossils were found put their age at greater than 30,000 years B.P. (McMillan 1976:93). The association Koch claimed between humans and mastodons, which had been received with considerable skepticism by his contemporaries (§2.4), was indeed fortuitous (McMillan 2010:41).

LANSING: In the 1950s archaeologist Waldo Wedel reexamined the Lansing remains and concluded they were less like those of the historic tribes of the area and more akin to crania of older ceramic and preceramic groups (Wedel 1959:91–93). Bass subsequently (1973) reached a similar conclusion and arranged for radiocarbon dating of the specimens, ultimately obtaining four dates that ranged from 4610 ± 200 ^{14}C years B.P. to 6970 ± 200 ^{14}C years B.P. (Bass 1973: Table 1). A more precise age cannot be calculated given the lack of overlap among the ages even at two standard deviations (cf. Bass 1973:102). Regardless, and as Bass concluded, Hrdlička was right that Lansing was not Pleistocene in age; but Bass believed Hrdlička was wrong to have declared Lansing "practically identical with the typical male skeleton of a large majority of the present Indians of the Middle and Eastern states (Hrdlička 1903:328–329). Bass's own analysis suggested Lansing was "morphologically closer to the Early Archaic populations than to the later protohistoric or historic Plains populations" (Bass 1973:103).

LONE WOLF CREEK: Like 12 Mile Creek, Lone Wolf Creek was recognized as Pleistocene in age in the wake of the Folsom discovery, though no one considered it to be Middle Pleistocene in age, as Harold Cook, Jesse Figgins, and Oliver Hay had supposed [§8.13]). The Lone Wolf Creek projectile points are now identified as Plainview points, a Paleoindian type that falls in the centuries just before 10,000 ^{14}C years B.P. (Holliday 2000: Tables VIA–VIC). The bison species, originally named *Bison figginsi* (§8.11), is now—like the Folsom bison—identified as *Bison antiquus occidentalis* (McDonald 1981:94, Table 27). In the early 1990s Vance Holliday and I, using Cook's photographs from his 1925 visit to Lone Wolf Creek and aided by longtime Colorado City, Texas, resident (and Mayor) Jim Baum, relocated the area of the original find, with the aim of ascertaining if any of the site remained. Unfortunately, neither surveys of the area nor exploratory excavation trenches yielded bison bone or prehistoric artifacts. (Historic material was recovered, though not obviously dating to the 1920s excavations.) Either we searched the wrong location, or there is nothing left of the site. The latter seems likely, given that Lone Wolf Creek experiences occasional significant flood erosion and has been modified by the city.

LONG'S HILL: Like Lansing, the Long's Hill site was rejected by Hrdlička, but unlike Lansing—which saw follow-up analysis and discussion—Robert Gilder's discovery of human remains at Long's Hill was ever after ignored. Today it is virtually invisible in the archaeological and physical anthropological literature, and mentioned only in passing in histories of Plains archaeology (for example, Wedel 1981). The site nonetheless has a minor presence in the ill-informed corners of the Internet, where Erwin Barbour and Gilder's 1906–1907 articles are still cited as proof that there were once Neanderthals in Nebraska.

NATCHEZ: The Natchez pelvis resurfaced in the 1950s, its reemergence sparked by the announcement that a fluorine test had resolved the Piltdown fraud (Stewart 1951, Quimby 1956). It reminded Hrdlička's successor, T. Dale Stewart, that he had discovered in Hrdlička's files a copy of Thomas Wilson's (1895) fluorine test of the Natchez human pelvis and apparently associated sloth bone (§2.4, §6.12). Stewart called attention to Wilson's results and suggested this was "an important objective argument in favor of the antiquity of man in America" (Stewart 1951:392). In hopes of finding the site and recovering more skeletal remains, George Quimby visited the Natchez area but found no traces of the site, erosion having evidently removed the bone-bearing deposits sometime in the century since its discovery (Quimby 1956:78). In 1990 John Cotter arranged for radiocarbon dating of the Natchez bones: the human pelvis dated to 5580 ± 80 ^{14}C years B.P., and the sloth bone to 17,840 ± 125 ^{14}C years B.P. As Cotter put it, "If mylodon and man ever glimpsed each other in the New World it was not at Natchez" (Cotter 1991:39).

PINEY BRANCH: Despite its location just a few miles north of the White House, portions of the Piney Branch site remain intact more than a century after William Henry Holmes worked there. (Its sloping surface is strewn with quartzite cobbles and forms the backyards of a Washington neighborhood.) There has not been significant archaeological investigation at this locality since Holmes's investigations, and thus the age or ages of the quartzite quarrying are not known. On the basis of work elsewhere in the vicinity, and what has been learned of the prehistory of the region, it is suspected that the Piney Branch quarry activities fall within the Archaic Period, perhaps dating to about 4,000 years ago (Dent 1995:203). The site was not Pleistocene in age, as Holmes knew, but then it might not have been quite as recent in age as he suspected (§5.2).

RANCHO LA BREA: This is the only pre-1927 find of human skeletal remains that warranted inclusion in a recent compendium of ancient human remains in North America (Powell 2005)— ironic, given it was one of the very few finds of that era dismissed without controversy as *not* being Pleistocene in age (Merriam 1914; also Kroeber 1962). Even so, Merriam (1914) was correct to doubt the human remains were associated with the extinct fauna. Analysis and radiocarbon dating of the fossils revealed that the bones of the extinct mammals were in a distinct sedimentary matrix and showed less wear than the human remains, indicating different depositional contexts and histories. That there was stratigraphic mixing—as Merriam suspected—is also apparent in the radiocarbon ages. The human specimen dated to 9000 ± 80 ^{14}C years B.P.; the *Equus* bones John Merriam thought might be of a domesticated species in fact were Pleistocene horse dated to $15,700 \pm 530$ ^{14}C years B.P.; the Pit 10 bear bones (likely *Ursus arctos*) dated to 5270 ± 155 ^{14}C years B.P. (Marcus and Berger 1984:173). Powell (2005:200) examined the Rancho la Brea human and noted that this individual—a young female—was not (as Alfred Kroeber supposed) morphologically similar to historic Native Americans, but instead had a "dolichocranic form" more similar to other "ancient Americans." He attributed her distinctive phenotype to genetic drift rather than "race" or "a different migration" (Powell 2005:147, 200, 227).

SNAKE CREEK: After the rejection of *Hesperopithecus haroldcookii* (Gregory 1927) and the dismissal of the supposed artifacts from Snake Creek (Nelson 1928a), only Cook continued to insist there were traces of ancient humans in the area (for example, Cook 1952). Teams from the American Museum of Natural History nevertheless continued to investigate the drainage's rich paleontological deposits in later decades, which ultimately revealed that all of the fossil remains were from formations that ranged in age from the Early to the Late Miocene (Skinner et al. 1977:361). No further indication was ever found to suggest that traces of ancient humans were present in the Snake Creek beds. Given the geological age of these formations, that is a good thing.

TRENTON: There was extensive work in the Trenton area in the late 1930s under the auspices of the Works Progress Administration (Cross 1956), and again in the 1980s associated with highway and interstate construction (Hunter et al. 2009). On the basis of that work it was suggested that "some of the stone artifacts recovered by Charles Abbott, Ernst Volk and others 'from the Trenton gravels' *may perhaps*, under a modern chronological framework, be assignable to the Paleoindian or Early Archaic periods" (Hunter et al. 2009:5–10, emphasis mine). That noted, I have seen no diagnostic Paleoindian artifacts in Abbott's collections at the Peabody Museum, Harvard University, or the University Museum, University of Pennsylvania. Nonetheless, Paleoindian fluted points have been recovered in the area: Dorothy Cross reported finding four poorly made fluted points (1956:79–80, 169). The "paleoliths" Abbott thought were Pleistocene in age are temporally nondiagnostic bifaces and cores (as Holmes supposed); given their raw material and what is known of the archaeology of the region, these are likely Archaic in age. Abbott's overlying argillite culture (found in the yellow sand [§5.15]) constitutes the bulk of the Trenton archaeological record and "would mostly be assigned to the Middle Woodland period" (Hunter et al. 2009: 5–11).

VERO AND MELBOURNE: After Hrdlička's death Stewart reexamined the original remains from both Vero and Melbourne. He saw that the Melbourne cranium had been poorly reconstructed, and when he reassembled the recovered fragments he thought it "surprisingly similar to the Vero skull" in its morphology (Stewart 1946:19). Furthermore, the two crania, "instead of being like the majority of recent Florida Indians, are rather like their more long-headed neighbors to the north and therefore consistent with what little we know of the earlier American types" (Stewart 1946:22). It seemed to Stewart that there was "no reason from the viewpoint of morphology [Hrdlička's viewpoint] for denying [a Pleistocene] antiquity to the Melbourne and Vero remains" (Stewart 1946:22). In the early 1950s, Heizer and Cook (1952) and Sellards (1952) conducted fluorine and other chemical tests on the fossil bones from the two sites and found "reasonable consistency between the fossil human and animal bones" (Sellards 1952:93–94). Radiocarbon samples Sellards collected at Vero in 1952—discussed in chapter 1 of this book—were dated to 2500 ± 110 ^{14}C years B.P. and 1625 ± 200 ^{14}C years B.P. (Weigel 1962:9). Sellards went back to Vero again in December 1959 to collect more samples but died without knowing the age of the human remains. Their ages are still unknown. Vero recently reemerged in the archaeological literature following the discovery of a fossil animal bone on which is engraved the image of a proboscidean (Purdy et al. 2011). The bone is highly mineralized and could not be radiocarbon dated, nor is its precise find spot known. How it might relate to where Sellards worked a century earlier is also unknown. Although this latest find may be Pleistocene in age (a fact that remains unproved), one cannot draw from it the claim that it supports Sellards's "argument that the Vero site provided evidence of a Pleistocene human presence in North America" (Purdy et al. 2011:2912). Sellards was looking at different evidence.

NOTES*

Chapter One

1. Sellards to Hrdlička, July 17, 1916, AH/NAA. Virtually identical letters were sent the same day to all those mentioned.
2. Hrdlička to Sellards, July 20, 1916, AH/NAA.
3. Hrdlička to MacCurdy, October 13, 1916, AH/NAA. See also Hrdlička to Chamberlin, October 9, 1916, TCC/UC.
4. Chamberlin to Holmes, July 31, 1916, TCC/UC.
5. Sellards to Williston, November 4, 1916, BEG/UT.
6. Hrdlička to Holmes, November 9, 1916, AH/NAA.
7. Swanton 1944:39.
8. The views of all the participants were gathered by Thomas Chamberlin, who published them in the January 1917 *Journal of Geology*, in time for distribution before the holiday scientific meetings.
9. Holmes to Chamberlin, January 6, 1917, TCC/UC.
10. Holmes 1918:562, 1925:258. Wilmsen (1965:180) labels the latter passage one of the low points of American archaeology. In Holmes's defense, this outburst followed soon after the death of his wife.
11. Glen Evans believes it was not merely the virulence of the criticism that bothered Sellards, but the fact that it came from the leading scientists of the time. He wondered too if Sellards had not seen it coming, or supposed it would come "in the form of a gentlemanly peer review" (Evans 1986:13).
12. Sellards to Holmes, February 24, 1930, Holmes to Sellards, March 6, 1930, WHH/SIA.
13. The quote of "stones to heap on Hrdlička's grave" comes from Glen Evans (Evans, personal communication, March 12, 1991). Sellards's subsequent fieldwork included excavations at many Paleoindian localities in Texas and New Mexico (Sellards 1952).
14. Evans, personal communication, March 12, 1991, Austin, TX.
15. The samples Sellards collected in 1952 were radiocarbon dated. See the Appendix.
16. My attention is on North America (the United States, really), though for the sake of the prose I regularly drop the modifier. In a strict sense, the modifier would be redundant: it was North America, not South America, that provided the first evidence of human antiquity in the New World. No Anglo-centrism is intended: I am quite aware that Latin American archaeologists, like the Ameghino brothers, were grappling with these same issues. Their work deserves a full historical study.
17. Foster 1873b:79. Also Whittlesey 1869:271–272.
18. Abbott 1881b:126–127, emphasis in original.

---

*A list of acronyms used to identify manuscript sources can be found on pages 578–580.

19. Lewis 1880, 1884b; Shaler 1877; Wright 1881b, 1889b.
20. Wilson 1890a:679, 694; also Haynes 1881:135–137; Putnam 1888b:423–424; Wright 1890c:6–7.
21. For example, Abbott 1881b; Mason et al. 1889; McGee 1888b; Powell 1883a; Putnam et al. 1889; Wright 1889a.
22. Dawkins 1883; Keith 1915:274; Topinard 1893; Wallace 1887.
23. Cope 1885; Foster 1873b:79; Marsh 1878; Morse 1884.
24. Holmes 1890a:25.
25. Holmes 1892e:297. The revolution quote is in Powell to Holmes, June 8, 1894, in *Random Records* 5:138. Holmes in WHH/RR 1:35; Mason 1893:461.
26. For example, Holmes 1897b; Sapir 1916.
27. Holmes 1919:55–56. The same theme was sounded by Hrdlička (1926).
28. For example, Schneer 1978.
29. E. Haury, personal communication, September 10, 1981.
30. Hill 2006; Meltzer 1991.
31. McGee 1900:6; Thomas 1898:9. See also Haven 1864:37.
32. Moorehead 1910:1:250. See also Kidder 1936:145.
33. Hinsley 1981:29, 89–91; Trigger 2006:179–185.
34. Nelson to Figgins, August 16, 1927, JDF/DMNS.
35. Cook to Merriam, January 1, 1929, JCM/LC.
36. Rudwick 1985:431. Also Oldroyd 1990:344.
37. Meltzer 2005:436.
38. Fowke 1902:12.
39. Rudwick 1985:419. See also Grayson 1983, 1990; Merton 1988; Oldroyd 1990.
40. The development of a professional class within anthropology is discussed by Hinsley 1976, 1981; Parezo and Fowler 2007; Snead 2001.
41. Hinsley 1985:53–55. Fund-raising challenges plagued the AMNH (Snead 2001) and the University Museum, University of Pennsylvania (Conn 1998:92–94, 164).
42. Cravens 1978:89–105; Hinsley 1981:268–271; Meltzer 2002:232–234; Parezo and Fowler 2007:47.
43. It is because of their irreproducibility that McMullin suggests "historical sciences like paleontology or archaeology pose a particular challenge" in regard to finding resolution (McMullin 1987:66).
44. See, for example, Hinsley 1981:76; Manning 1967:117–121; Vetter 2008:274–275.
45. Wiebe 1967:121, 127. Also Kohlstedt et al. 1999:41–45.
46. Kidder 1936: 143–144.
47. Nelson to Hay, April 5, 1920, OPH/SIA.
48. Kroeber 1952:191. See, for example, Holmes 1919; Thomas 1898.
49. Kidder 1936:143–145.
50. Snead 2001:159. See also Lyman and O'Brien 2006:222–227; Meltzer 1983, 1985; Trigger 2006:180–181, 288–289.
51. Kidder 1936:145.
52. Meltzer 1983:40, 1985:256.
53. Meltzer 2005:450–451. See also Lyman and O'Brien 2006:183–184.
54. Boas 1933:360–363; Kidder 1936:143–145; Kroeber 1940a:461
55. Boas 1933:362–363.
56. Kroeber 1940a:461.
57. Dascal 1998:149; Oldroyd 1990:17, 367. Revolutions are not always the result of controversy; Cohen's (1985) massive compendium of scientific revolutions scarcely mentions any controversies.

58. Rudwick 2008:563; also Hull 1988.
59. McMullin 1987:53, 63.
60. McMullin 1987:51–52.
61. Gillespie 1976. Also DeBont 2003; Hull 1988.
62. Hull 1988; Olby 1989; Oldroyd 1990:6; Rudwick 1985:xxi.
63. Dascal 1998:149.
64. Dascal and Boantza 2011:2. Rudwick 2008:563; Secord 1986:12.
65. McMullin 1987:77–82. See also Oldroyd 1990:357.
66. Goodrum 2009:338. See also Conn 2003:165.
67. For example, see DeBont 2003; DeChadaverian 1996; Gillespie 1976; Olby 1989; Pettit 2006; Schneer 1978; Sloan 1976.
68. Oldroyd 1990:6; in order by publication—not necessarily stratigraphic—the books are *The Great Devonian Controversy* (Rudwick 1985); *Controversy in Victorian Geology* (Secord 1986); and *The Highlands Controversy* (Oldroyd 1990).
69. See, for example, Gifford and Rapp 1985:410–416; Hart 1976; Hinsley 1981:105–108, 280–281; Hinsley 1985:62–69; Judd 1967:14–15; Mark 1980:54, 149–154; Morison 1971; Mounier 1972; Noelke 1974:125–133; Numbers 1988; Rabbitt 1980:216–217; Schultz 1983:126–127; Spencer 1979; Trigger 1989:120–129, 2006:172–189; Willey and Sabloff 1980:47–50; Wilmsen 1965.
70. My previous publications on the topic are listed in the Bibliography ("C. Printed sources: Secondary"). I draw on some of that work here, in particular earlier chapters on Abbott (Meltzer 2003) and on the history of the Folsom site discovery, which I wrote about in the context of my archaeological investigations at the site (Meltzer 2006a). I am grateful to both the University of Alabama Press and the University of California Press (respectively) for permission to adapt that material for use here.
71. Hrdlička 1918:35, 60. On Hrdlička's sense of infallibility, see Judd 1967:70; Swanton 1944:36–39.
72. Figgins to Hay, December 28, 1926, HJC/AFB.
73. In this regard the establishment of human antiquity in America was very different from that of a half-century earlier in Europe, where antiquarian studies, geology, and vertebrate paleontology all converged to demonstrate a Pleistocene human antiquity. From that coalescence prehistoric archaeology emerged and laid claim, as Goodrum observes, "to research domains that previously resided partially within geology and partially within archaeology" (Goodrum 2009:342). The process by which that niche was created did not lead to any significant disputes between the contributing disciplines. See Goodrum 2002, 2009; Grayson 1983, 1990; Gruber 1965; Sackett 2000; Van Riper 1993.
74. Meltzer 2006a, 2009.
75. Meltzer 1991.
76. Meltzer 2005:434–435. Also Meltzer 1994.
77. Deloria, in Thomas 2000:xv.
78. Holmes 1984:132. See also Stocking 1968:9–12, 108–109.
79. Trigger (2006:39) labels this position "moderate relativism."
80. Hull 1988:25; also Merton 1968, 1988; Oldroyd 1990:8, 348; Rudwick 1985:10, 420; Rudwick 2008:554.
81. Rudwick 1985:6–7, 14. See also Secord 1986:12–13. There is no small measure of irony in historians invoking Geertz, for the borrowing has now gone full circle: when Geertz (1973:10) introduced the term to anthropology—it was not his to begin with but came from the philosopher and historian of philosophy Gilbert Ryle—he likened it to reading a historic manuscript that had faded over time.
82. Rudwick 1985:12–13.

83. Hull 1988:33.
84. Gould 1987:89; Oldroyd 1990:8. For my part, I visited and even excavated at some of the sites in the controversy; examined the artifacts, bones, or other materials from them; or both. As an archaeologist, seeing the sites and collections firsthand made it easier to appreciate the interpretive challenges these posed—especially given the knowledge of the time—and helped me better understand why they sparked often-unresolved debate.
85. Hull 1979; Oldroyd 1990:8; Stocking 1987:xv.
86. Although I assiduously tried to avoid projecting onto the historical record my own experiences in contemporary controversies (no easy task because we are grappling with many of the same questions and issues as a century ago), I did tap those experiences to help shed light on the behavior of the participants in the Great Paleolithic War. For these provided valuable clues of what to watch for in the behind-the-scenes maneuvering, offered lessons in reading the literature of controversy for what it reveals and what it attempts to disguise, and gave me a deeper appreciation for the process by which a purportedly ancient site gains acceptance (or not) and a participant in a controversy gains scientific credibility and status (or not). I cannot help but add that, difficult as it is to unpack and understand a controversy decades after it is over and when all parties have long since passed away, it is still far easier than it is to be *embroiled* in disputes in which one has a stake in the outcome and that involve contemporaries, friends, and colleagues.
87. Stocking 1968:8–9; also Mayr 1982; Trigger 1989.
88. Oldroyd 1990:8.
89. Hull 1979:13–14; Meltzer 1989:18.
90. Rudwick 1985:7–8; Secord 1986:12.
91. Williston to Barbour, January 3, 1908, EHB/NSM.
92. Chamberlin to Calvin, October 14, 1902, and Chamberlin to Upham, November 12, 1902, TCC/UC; Abbott 1907:9, Abbott 1908:10.
93. Hull 1988:31. Also Oldroyd 1990:16.
94. Hildebrand 1957:7. See also Gould 1987:85.
95. As Hull (1988) shows, turn up the heat high enough and the barrier evaporates.
96. Medawar 1982:79–80, 132; Rudwick 1985:434.
97. Holmes 1987:226, 229.
98. Medawar 1982:133.
99. Holmes 1987; Rudwick 1985:7. Rudwick further (1985:430) argues that historical evidence should not be dichotomized as published/unpublished or public/private, and that rather there is a continuum of relative privacy, from strictly private diaries to letters, circulated manuscripts, and published papers.
100. Knight 1987:8. Material intended for publication, such as early drafts of papers, has potential value too, as insights often come in the process of writing and revising papers—a process that often defines the analysis and conclusions themselves, as Holmes 1987:225–226 shows.
101. The Bibliography lists the manuscript sources, and the identifying acronyms used throughout.
102. Gould (1987:75) identifies the telephone as the greatest single enemy of historical scholarship; that mantle has now been passed to email. This controversy began as typewriters were put on the market (mid-1870s) and carbon paper became widely available, and having previously faced the challenge of translating near-illegible

handwriting on onion skin in letter-press blotted books, I was glad for it. On the advent of the typewriter and telephone usage, see Schlereth 1991.

103. Gruber 1966:21.
104. The publications and correspondence compiled in the course of this study are surely not all that were written then or even that survive now more than a century later. Even so, over the years of my research it became increasingly rare to encounter a previously unseen publication, or learn of an archival collection. Taking a page from sampling theory, that would suggest there are relatively few archival fruits still hanging in the tree.
105. As soon summarized by Lubbock 1865 and Lyell 1863.
106. Gibbs 1862.
107. It is from Oldroyd (1990:345–347) that I draw the term, and the idea for tracking hot spots through the frequency of publications. See also Rudwick 1985:426.
108. Alternatives might include using the frequency of correspondence among the participants, but because letters are not universally preserved their frequency would not be representative. Using the incidence of symposia and meetings dedicated to the topic is another possibility, but because of their irregularity and spacing over time they are not a very fine-grained barometer of activity.
109. Rudwick 1985:12.

Chapter Two

1. Leech (1941) paints an engaging portrait of Washington, DC, in the Civil War. Henry's woes were many. The War meant a loss of income from the Institution's substantial investment in Tennessee and Virginia state bonds, a sharp reduction in his operating budget, demands on the Smithsonian's large lecture halls by inventors wanting to exhibit their Union-saving inventions, and threats from the War Department to take over Smithsonian rooms as temporary troop quarters and its nearby fields as a firing range (Henry 1862:13, 38, 47–48). That was just 1861.
2. Henry 1862:13.
3. Henry 1861:38–39; Henry 1878:1–2.
4. Squier and Davis 1848; Haven 1856; see Meltzer 1998. Although Henry recognized archaeology's broad appeal, he also wanted to make it more of a science so as not to endanger the credibility of his Institution (Hinsley 1981; Meltzer 1998). Accordingly, he maintained tight editorial control of all Smithsonian publications on the topic (Hinsley 1981:34–37).
5. Squier and Davis 1848:305.
6. Grayson 1983.
7. Lubbock 1865:2.
8. Morlot 1861:284. See Hinsley 1981:41–42.
9. Haven 1856:153.
10. See reviews by Haven 1856:142–143 and Lubbock 1863:336.
11. Nott and Gliddon were motivated by venal racism and support for slavery, which they justified as the natural role for an "inferior" species. To buttress that argument, they embedded it in the general claim that all races were separately created species and could not have descended from Old World groups because they had been in America all along. As Nott said, "We may put forth the facts of Natural History as much as we please, but the *crowd* will not be convinced until it is shown

that the chronologies of nations are irreconcilable with the common readings of Moses" (Nott to Morton, June 1, 1847, SGM/PHS).

12. Gibbs penned those Instructions amid his daily linguistic and natural history work and nightly guard duty at the Capitol (Gibbs 1862:392). Like many others living in Washington, DC, at the time, Gibbs was pressed into guard duty in the months immediately following the fall of Fort Sumter and before the Capitol was firmly secured by Union troops. On Gibbs, see Hinsley 1981:51–55.

13. Goldstein 1994. In return, correspondents could request comparative specimens as well as Henry's *Annual Reports* and the articles they included on the latest work in science, and they received encouragement for their efforts, connections to others of like interest, and no small measure of self-esteem, all of which was especially valued by the more than half of the correspondents who resided in small towns (Goldstein 1994: Table 1). For many in rural areas, this was the closest they came to participating in something larger than themselves or their local community.

14. Gibbs 1862:392.

15. Gibbs 1862:395.

16. Morlot 1861:320.

17. Henry to T. A. Cheny, November 11, 1861, quoted in Hinsley 1981:40.

18. Henry 1862:34–35.

19. Henry 1871:1–2.

20. Henry 1878:1.

21. Meltzer 1998.

22. See, for example, Haven 1856.

23. Jefferson 1788:108. For a discussion, see Bedini 1990:376–378.

24. Haven 1856:22.

25. Rowly-Conwy 2006.

26. By the early 1800s, geology and the physical sciences had disentangled themselves from the literal Biblical chronology. Clearly the world was older than the 6,000 years allotted by the Biblical chronologies, though the "chronology of MOSES," as John Playfair (1802:126–127) put it, still applied "to the human race."

27. With one possible theological complication: Genesis provides two versions of Creation. The first (verses 1 to 2.3) places the creation of plants and animals *before* people; in the other version, taking place in Eden (verse 2.4 to 2.25), plants and animals are created *after* Adam. On theological and (later) fossil grounds, most accepted the former ordering.

28. Mayor 2005. In regard to native oral tradition and fossil forms, see also McMillan 2010.

29. Grayson 1983:44; Rudwick 1976:64. Earlier attempts to prove extinctions on the basis of fossil invertebrates were unsuccessful, because no matter how unlike a living shell a fossil shell might appear, that was no evidence the fossil species was extinct. It might yet live in the deep and unexplored ocean. Cuvier neatly sidestepped this problem, basing his proof on an animal so large that if any were still alive they surely would have been spotted by then (Grayson 1983:46).

30. Cuvier 1796, in Rudwick 1997:96.

31. Cuvier 1796, in Rudwick 1997:24.

32. Cuvier 1806, in Rudwick 1997:97.

33. Cuvier 1796, in Rudwick 1997:24; Cuvier 1806, in Rudwick 1997:96. Also Grayson 1990; Rudwick 2008:11–20.

34. Buckland 1823:146; Rudwick 2008:82–87, 193–194.

35. Buckland 1823:51. See also Grayson 1983:48; Rudwick 2008:79.
36. Rudwick 2008:345.
37. Rudwick 2008:180–189, 502–508.
38. Rudwick 2008:512–514.
39. Agassiz quoted in Wilson 1972:496.
40. Agassiz to Buckland, circa 1838, in Agassiz 1887:289. Rudwick 2008:532–533. Agassiz's ideas about glaciers were not entirely original, but he is rightly credited with elucidating and broadcasting them.
41. Grayson 1983:83, 202; Grayson 1990.
42. Lyell 1853:182. See discussion in Grayson 1983:81–82.
43. Parkinson 1833:467. See Gruber 1965:383.
44. Grayson 1983:55ff; Meltzer 2005:443–444.
45. Gruber 1965:385; Sackett 2000:42–44; Van Riper 1993:80–82.
46. Lyell 1830–1833, volume II:232, emphasis in original. Also Sackett 2000:44.
47. Evans 1872:466; Van Riper 1993:83–85.
48. Gruber 1965:385. Also Grayson 1983:181–182; Grayson 1990:9–10; Van Riper 1993:87–88.
49. Grayson 1983:185; Grayson 1990:10; Sackett 2000:46; Van Riper 1993:94–96.
50. Anonymous 1858:461.
51. Sackett 2000:46; Van Riper 1993:100–101.
52. On views of Boucher de Perthes, see Grayson 1983:185–186; Grayson 1990; Gruber 1965; Rudwick 1976; Sackett 2000.
53. Evans 1860:281. See also Prestwich 1859:53.
54. Lubbock 1865:269.
55. Darwin to Hooker, June 22, 1859 (in Burkhardt and Smith 1991:308); Darwin to Lyell, March 17, 1863 (in Burkhardt et al. 1999).
56. Boucher de Perthes 1860:96–97.
57. Falconer to Prestwich, November 1, 1858, in Prestwich 1899:119. Gruber 1965; Van Riper 1993:101–102.
58. Evans 1943:101–102; Prestwich 1899:123–124; Van Riper 1993:104–106. See the discussion of Prestwich and Evans's visit by Gamble and Kruszynski 2009, which includes their original photographs of the section and the in situ specimen.
59. Prestwich 1860; Evans, in Prestwich 1860:310, 312.
60. Evans 1943:103.
61. Grayson 1983; Meltzer 2005; Sackett 2000:47.
62. Grayson 1990:10; Van Riper 1993:111–113.
63. Lyell 1860:94.
64. Grayson 1983; Van Riper 1993:115–117.
65. Dawkins 1863.
66. Grayson 1983:208.
67. For example, Babbage 1860; Ramsay 1859; Worsaae 1859; Wright 1859.
68. Gruber 1965:374.
69. Lyell 1863. See also Christy 1865; Cochet 1857–1860, 1861; Dawkins 1862, 1863; Evans 1863, 1864; Lartet 1860; Prestwich 1861a, 1861b.
70. Estimates of the absolute age of humanity were wide ranging. Most reviewed estimates of the age of the Pleistocene, then concluded, as Lubbock (1865:312) did, speaking obliquely of the "enormous time" elapsed since the first appearance of humans. See also van Riper 1993:151–157.
71. Tylor 1871:198.

72. Gruber 1965; Stocking 1987:147.
73. Darwin 1859:488.
74. Grayson 1983:211–212.
75. Darwin 1871:3.
76. Wallace 1887:667.
77. Goldstein 1994:588–591, Goldstein 2008:528.
78. Morlot 1863:307–308.
79. Gibbs 1862:392.
80. Gibbs 1862:395.
81. Harlan 1825; Grayson 1984:7.
82. Dana 1863:567.
83. Agassiz 1840; Conrad 1839:240, 242.
84. Hitchcock 1841:253, emphasis in original.
85. Hitchcock 1841:258.
86. See, for example, Dana 1856; Hitchcock 1857; Rogers 1844, 1850. Merrill (1924:625–636) provides a useful, though somewhat biased, overview.
87. Dana 1856:26–28, 1863:546.
88. Hitchcock 1841:247.
89. Dana 1863:546.
90. Whittlesey 1869:271–272.
91. Dana 1863:132, 505–506.
92. Foster 1871:9. Whittlesey 1869:272. Foster's statement echoes the contemporary, widely held social evolutionary belief in a psychic unity of humankind (Stocking 1987).
93. Foster 1873b:79–81.
94. Gibbs 1862:395.
95. Stocking 1987:172.
96. Dawkins 1863:219; also Christy 1865; Dawkins 1874, 1880; Evans 1863; Geikie 1881; Lartet and Christy 1875.
97. Lubbock 1865:280.
98. Stocking 1987:172.
99. Stocking 1968:106; cf. Sackett 2000.
100. Morlot 1861:285; Nilsson 1868:lv. The metaphor of the thread and labyrinth would be used for decades. See Thomas 1894:21.
101. Lubbock 1865:336. For a discussion of Lubbock's "starkly ignoble view of savagery," see Stocking 1987:153–156.
102. Stocking 1987:173.
103. Grayson 1983:217.
104. Tylor 1871:198. Stocking 1987; Van Riper 1993:201–202.
105. Haven 1864:37; also Rau 1873:398; Wilson 1863a:300–301.
106. Squier and Davis 1848:213–214; Wilson 1863a:298–300.
107. See, for example, Foster 1873b:369; Lubbock 1863:335; Newberry 1870:367; Whittlesey 1869:281. None of the later authors cited Lubbock, whose estimate of the age of the mounds was one of the first offered after 1860.
108. Wilson 1863a:300.
109. Henry 1866:46, 50; also Foster 1873b:317–318.
110. Rau 1873:401; Wilson 1876:69.
111. Rau 1873:398; Wilson 1863a:302.
112. Henry 1866:50; Wilson 1863a:302.

113. Koch 1839:199.
114. McMillan 2010:39–40. The later nineteenth- and early twentieth-century exploration at Kimmswick is summarized in McMillan 2010:44–45. That Kimmswick ultimately proved to be an archaeological site and of Clovis age was realized and demonstrated only in the late twentieth century (Graham et al. 1981).
115. Quoted in McMillan 2010:40. Also Koch 1857:63.
116. Goddard 1841; Harlan 1843:69–70; Sillimans in Koch 1839:198.
117. Koch 1857; Wislizenus 1858.
118. Lubbock 1865:236. Excavations at Koch's Pomme de Terre more than a century later showed that the association of artifacts and mastodon remains was, indeed, fortuitous (McMillan 1976). See the Appendix.
119. Andrews 1875; Dana 1875b; Rau 1873:397.
120. Dickeson 1846:106–107.
121. Dickeson 1846:107.
122. Lyell 1849:196–198.
123. Grayson 1983:78–84. Rudwick 2008:310–312.
124. Lyell 1849:197–198.
125. Lyell 1863:200–203.
126. Summarized in Nott and Gliddon, 1854:337–338, 352–353; see also Hrdlička 1907:15.
127. Holmes 1860:Preface, iii, vii–viii. The taxonomic descriptions were provided by Joseph Leidy and published in Holmes's volume: he illustrated the pottery fragment alongside the mastodon molars (Holmes 1860: Plate 19). Leidy himself was skeptical of the association, observing that a number of the extinct forms, including the horse and mastodon, may have existed "at a period so recent [that] it is probable the red man witnessed their declining existence" (in Holmes 1860:v, vii). Hrdlička (1907:21) mentioned subsequent work by Leidy at the site, but it went unreported.
128. Whittlesey 1869:282–283.
129. Berthoud 1873; Wilson 1863a:296–297.
130. Whittlesey 1869:286.
131. Foster 1873b:58.
132. Evidence from these sites is summarized in Foster 1873a; Foster 1873b; Whitney 1867, 1879; Wilson 1863b:109–116.
133. Foster 1873b:71; Wilson 1876:56–57.
134. Wyman to Whitney, July 4, 1868, quoted in Dexter 1986:366.
135. Gibbs to Henry, June 6, 1867, SEC/SIA.
136. Foster 1871:8–9.
137. Harte 1866.
138. Foster 1873a, 1873b; Lubbock 1865.
139. Whittlesey 1869:287. See also Foster 1873b; Lubbock 1865; Lyell 1863; Newberry 1870; Whittlesey 1869; Wilson 1863b, Wilson 1876.
140. Newberry 1870:367.
141. Lubbock 1865:236. When first published in January of 1863, Lubbock's "North American Archaeology" dismissed a deep human antiquity in the Americas. Two years later, when he reprinted the essay in *Prehistoric Times*, Lubbock was more willing to grant that antiquity. The sites had not changed, only his opinion of them. His conversion was partly inspired by Lyell's *The Geological Evidences of the Antiquity of Man*, which Lubbock read and from which he borrowed liberally.
142. Lyell 1863:204–205.

143. Wilson 1876:57.
144. Henry 1869:33.
145. Rau 1873:400.
146. Hinsley 1981:46–47.
147. Henry 1878:1.
148. Short 1880:22, 130.
149. The quote is from Short 1880:23. Putnam penciled those notes on the same page on his copy of the book (now in the author's possession). The volume contains many notes by Putnam and his assistant Lucien Carr. Fortunately, the notes are easily identified, as Putnam and Carr each had distinctive handwriting and often initialed and occasionally dated their annotations.
150. A decade later Abbott told Henry Haynes that Short visited Trenton before he died and "acknowledged that I was right. I have had a letter from him, and he promised to make corrections in his second edition. But this I have not seen. It needed but a visit to Cambridge, as he wrote, to convince him" (Abbott to Haynes, March 1, 1889, HWH/MHS).

Chapter Three

1. In the nineteenth century, "drift" was a collective term for Ice Age deposits generally (Hamlin 1982:567).
2. Hinsley 1985:68.
3. Abbott's first paper in archaeology appeared in 1863 (he was then twenty years old) and reported on the discovery of a cache of about150 jasper "lance-heads" (large bifaces) from near Trenton (Abbott 1863). Four specimens from the cache are in the collections of the Peabody Museum, Harvard University; eleven specimens are curated at the University Museum, University of Pennsylvania.
4. Aiello 1967:210–211. Abbott received his MD from the University of Pennsylvania in 1865. His course of study there involved anatomy classes from Joseph Leidy, who encouraged Abbott's interests in natural science (Aiello 1967:210). Abbott's own view of his career path emerges in Anonymous (1887). The author was anonymous, but by its content and style I infer it was Abbott himself, or at least someone well coached by Abbott.
5. Meltzer 2003:49; also Anonymous 1887:549, Hinsley 1985:62. Abbott's father was a successful Philadelphia hardware merchant and then later a banker. In 1867 Abbott married Julia Boggs Olden, the daughter of a wealthy, well-connected Princeton family; they had a son and two daughters (Hunter et al. 2009:4–41).
6. Bynum et al. 1981:286.
7. See, for example, Lubbock 1865, Mortillet 1883, Pitt-Rivers 1875. The larger currents are detailed in Trigger 1989.
8. Meltzer 2003:81–82.
9. Meltzer 2003:54; Hinsley 1985:49–50; Hinsley 1999:142–143. Browman and Williams (2013:27, 57–59) describe the events of the "Salem secession." See also Mark 1980.
10. Abbott moved onto the homestead in 1874 (Anonymous 1887) and lived there until November 1914, when a disastrous fire on Friday the thirteenth destroyed Three Beeches (§6.15). Today the property is across the street from a shopping mall and immediately north of Interstate 295 but remains undeveloped and tree-covered, and the stone foundation of Three Beeches is still visible. See also Hunter et al. 2009.

11. Abbott 1872:144. Abbott wrote an extended version of this paper published under the same title in the Smithsonian Institution's *Annual Report* for 1875 (Abbott 1876e), but it did not appear until January 1877. Thus, this paper records Abbott's thinking in the early and not the mid-1870s. In fact, by the time the volume was finally published, Abbott was unhappy with its appearance, in some respects, but felt himself "helpless, in the matter, from impecuniarity" (Abbott Diary, January 8, 1877, CCA/PU).
12. Abbott 1872:145–146.
13. Abbott 1876e:248. Aiello (1967:211–212) claims that Abbott "began to notice" these cruder forms as early as 1867 and realized then that they "did not appear to fit into the general category of Lenni Lenape origin." However, there is no evidence to back this claim, and Abbott made no such statements before 1872.
14. Abbott 1873:209.
15. Meltzer 2003:50. See Abbott 1872:153, 156–157.
16. For example, Lubbock to Abbott, February 22, 1875, in Joyce et al. 1989:67. On Abbott's use of Nilsson and Lubbock, see Abbott 1872:146, 153–157, 199–200, 220–221. The volumes in question were Nilsson 1868 and Lubbock 1865 (the second edition of which appeared in 1869; Lubbock was Nilsson's translator). Through the 1870s, Abbott corresponded with Lubbock about artifact classification and function, in addition to sending him artifacts (for instance, Abbott Diary, April 13, 1875, February 10, 1876, CCA/PU; also Abbott 1872: 199, 205, 221). John Evans's (1872) *Ancient Stone Implements of Great Britain* subsequently became Abbott's primary source for artifact identification. See Meltzer 2003:50.
17. Abbott 1872:146; Meltzer 2003:50.
18. Abbott 1872:146.
19. Abbott 1876e:248.
20. Abbott 1873:209.
21. Abbott 1876e:248.
22. Comstock 1875; his specimens were likened to the one illustrated in Abbott 1876c:330.
23. Comstock 1875:260.
24. Abbott 1876d:431, emphasis in original.
25. Abbott 1876a:52.
26. Abbott 1876a:52.
27. Meltzer 2003:50.
28. Abbott 1876a:51; Abbott 1876b:66.
29. Abbott 1881b:490; also Abbott 1876c:332; Abbott 1877a:32.
30. Abbott 1877a:37. Meltzer 2003:51.
31. Wyman 1872:27.
32. Abbott 1876e:252.
33. Abbott 1876e:380. Meltzer 2003:51
34. Abbott 1873:205–206, emphasis in original.
35. Abbott 1873:209.
36. Abbott 1876b:72; Abbott 1876c:330–332.
37. Abbott 1876b:71.
38. Abbott 1876b:64, 66.
39. Abbott 1876c:330. Abbott estimated the age of the Indian remains in years by inferring rates of artifact change using an arbitrary measure of his own devising, as well as soil buildup atop sites. The crudest form of Indian arrowhead in his collection was

about 1,300 years old, according to the accumulation of forest litter. But because that particular class of arrowhead was already at the third of four "degrees of excellence" in the evolution of arrowhead making (a scale of Abbott's own invention), then a first-degree Indian arrowhead had to be at least 2,600 years older, hence around 4,000 years old. Thus, "from thirty-five to forty centuries ago [3,500–4,000 years], at least . . . . . the Indian appeared in what is now New Jersey" (Abbott 1876b:72).

40. Abbott 1876c:335. Abbott cited Croll's *Climate and Time in Their Geological Relations* (1875), and Geikie's *The Great Ice Age and Its Relation to the Antiquity of Man* (1874).

41. Abbott 1876b:72; Abbott 1876c:330, 333.

42. Meltzer 2003:53.

43. Abbott 1876c:334.

44. For example, Lartet and Christy 1875: 25.

45. Abbott 1876c:334.

46. Grote 1876:432. Grote was similarly critical of Abbott's claiming the Indians were usurpers, especially because Abbott's claim was based on "evidence from tradition" that Grote believed was "hardly to be trusted" (Grote 1876:432).

47. Grote 1877b:221. Grote also believed humans were in the far north before glacial times, and that the advent of the ice isolated these groups and "made something like an Esquimaux" of them, who then "accompanied the great glacier on its advance and retirement," surviving today as Eskimos (Grote 1877a:589; Grote 1877b:22). See also Mason 1877:310, and Abbott's response to Mason (Abbott 1877b).

48. Portions of this section previously appeared in Meltzer 2003.

49. Putnam's appointment followed the 1874 death of Jeffries Wyman. Asa Gray was acting curator until early 1875, when Putnam was appointed curator (Hinsley 1999:141–142). Putnam held that title until 1908, when he was named "Honorary Curator–Honorary Director" until his death in 1915.

50. Hinsley 1992:125–126; Hinsley 1999:142–143. Also Browman and Williams 2013:52–54.

51. Hinsley 1985:50–51; Conn 2003:168.

52. Peabody 1868:26.

53. Abbott Diary, February 7, 1876, CCA/PU.

54. Meltzer 2003:55.

55. Abbott Diary, September 1, 1876, CCA/PU.

56. Abbott Diary, September 3, 1876, CCA/PU.

57. Meltzer 2003:55.

58. Abbott Diary, September 7, 1876, CCA/PU.

59. Abbott Diary, September 6–11, 1876, CCA/PU.

60. Abbott Diary, September 8, 1876, CCA/PU.

61. Abbott Diary, September 13, 1876, CCA/PU.

62. Meltzer 2003:56. By comparison, in that same period the peripatetic Edward Palmer received nearly $2,000 for collecting in Utah and Mexico The sums paid Abbott, Palmer, and others are itemized in the Treasurer's reports in the tenth through twelfth *Annual Reports* of the Peabody Museum. In later *Annual Reports* individual payments were no longer itemized.

63. Meltzer 2003:56.

64. For example, Abbott to Wright, March 13, 1882, GFW/OCA.

65. Abbott Diary, July 13, 1883, CCA/PU. Hinsley (1992:126–131; 1999:144–147) discusses relationships Putnam had with other collectors, each of which was somewhat

different, but all of which were awkward to one degree or another. Abbott's was by the far the most difficult and volatile.

66. Putnam was finally appointed Peabody Professor in 1887. The reasons for the delay are explained in Hinsley 1992:133–134. See also Hinsley 1985:61; Hinsley 1999:149; Mark 1980:31.

67. Abbott Diary, December 14, 1876, CCA/PU. There were multiple reasons for Putnam's cautious and deliberate approach, not least his fear of embarrassing himself and his institution in front of donors, as well as his own considerable insecurities (Hinsley 1999:143).

68. Meltzer 2003:57.

69. Abbott 1877a, emphasis mine. Abbott later claimed credit for using the term "supposed," saying he realized "how important it was in this matter to make haste slowly" (Abbott 1881a:127). As that realization seems not to have carried much weight before this time, it seems more plausible to attribute the word choice to Putnam.

70. Abbott 1877a:31–32 Abbott 1878:226–228.

71. Abbott 1877a:32; see also Wadsworth 1881:146.

72. Abbott 1878:228–229.

73. Meltzer and Dunnell 1992:xiii.

74. Holmes 1879:250; also Holmes 1883:31–32.

75. Abbott 1877a:30; Abbott 1878:243.

76. Abbott 1877a:36, 39.

77. Abbott Diary, September 10, 1876, November 14, 1876, CCA/PU.

78. Abbott Diary, November 22, 1876, CCA/PU. For biographical information on Shaler, see Livingstone 1987 and Brace 2005.

79. Abbott Diary, February 14, 1877, emphasis in original, CCA/PU.

80. Shaler 1877:45–47.

81. Abbott 1877a:33.

82. Abbott Diary, April 21, 1877, CCA/PU.

83. Abbott 1877a:32–33.

84. Abbott 1877a:35.

85. Abbott 1877a:37. This was an argument Abbott would repeat (for example, Abbott 1881b:277).

86. Cook 1877:21–22.

87. Abbott 1877a:41.

88. Abbott 1877a:40–41; Abbott 1878:242.

89. Abbott 1878:241.

90. Pengelly 1877:323.

91. Marsh 1878:254 (delivered August 1877). Cope's find was reported in Mason 1878:125–126. Perhaps not coincidentally, it was a young Marsh who convinced his uncle, George Peabody, to fund Harvard's Peabody Museum of Archaeology and Ethnology.

92. Putnam 1877:11–12.

93. Wilson 1878:331–332. Although Wilson could not help but wonder whether the peculiarity of the drift implements was due partly to the "material of which they were fashioned" (Wilson 1878:332).

94. McGee 1879:208–209.

95. McGee 1879:210.

96. McGee 1879:210.

97. McGee 1879:226.
98. Abbott Diary, September 20, 1878, October 7, 1878, CCA/PU.
99. Abbott Diary, June 2, 1879, CCA/PU.
100. Abbott 1881a:130. Lewis was a gentleman scientist volunteering on the Geological Survey. He had graduated from the University of Pennsylvania a few years earlier with an MA in natural sciences and become affiliated with the ANSP. His work at the ANSP and on the Survey was in keeping with his distinguished family's tradition of public service (Upham 1888b:372).
101. Dana first established his "post-Tertiary" system in 1855, though the labels changed over the years. In the early 1870s he followed Lyell's lead and adopted the terms "Quaternary," "Pleistocene," and "Recent"—the basic content did not change (Dana 1894:941).
102. Dana 1875a:541–553, 590–591. Also Dana 1873. Other estimates, which Dana cited as well, put the Niagara retreat at one inch a year, which would mean an elapsed time of 380,000 years. He appeared to favor the younger (31,000) dates in 1875, however, and would certainly do so later.
103. Lewis 1880:298; Lewis 1881a:706–707; Lewis 1881b:539.
104. Lewis 1880:308.
105. Lewis 1881a:708–709; also Lewis 1880:298–299, 306–307.
106. Lewis 1881a:707.
107. Lewis 1881a:708–709; Lewis 1880:305.
108. Lewis 1880:304–305; Lewis 1881b:544–545.
109. Lewis 1881a:708.
110. Geikie 1874.
111. Croll 1864. Although Geikie relied on Croll's theory, he was subtle about it, not wanting to appear less the inductive researcher that Victorian geology demanded. Hamlin (1982) provides a thoughtful analysis of Geikie's very subtle use of (and citation to) Croll's theory. See also Fleming 2006.
112. Croll's theory was elaborated in his volume, *Climate and Time* (Croll 1875). For discussions of his theory, see Fleming 2006; Hamlin 1982; Macdougall 2004.
113. Geikie 1877:xiii. That comment of Geikie's was transcribed almost verbatim by Croll in the Preface to the third edition of *Climate and Time* (1885); he obviously appreciated the endorsement. Lyell himself came around as well, though with somewhat less enthusiasm; see Fleming 2006.
114. For a detailed chronicle of the literature in North American glacial studies, including the topic of multiple glaciation, from the perspective of the Chicago (Chamberlin) school, see Thwaites 1927. For a biting response, see Keyes 1928. See also Totten and White 1985:131–132 and White 1973.
115. Orton 1873; Winchell 1873.
116. McGee 1879:226, 229.
117. McGee 1878:339–341; also McGee 1879; Winchell 1873, 1875.
118. McGee 1879:227–229.
119. Chamberlin 1878:223.
120. Chamberlin 1878:228, emphasis in original.
121. Chamberlin 1878:231, 233–234.
122. Dana 1873, 1875a. Dana had several objections to Croll's theory. Among them, it demanded glacial conditions were present in only one hemisphere at a time and not both, and Dana doubted that was true. Moreover, geologists disagreed as to whether glaciation would occur where winters were colder (Croll's view), or in

the opposite hemisphere where summers were cooler, and the inability to resolve that point greatly weakened the theory. Finally, Dana dismissed Croll because he doubted that changes in eccentricity were sufficient to trigger "so great a result" as a glacial epoch (Dana 1875a:698).

123. Lewis 1880:305, 307–308.

124. Lewis 1880:308; Lewis 1881a:709.

125. When Abbott's paper (1876e) was published in the Smithsonian Institution *Annual Report*, Putnam urged Abbott to make it into a book. At the time, Abbott, demurred claiming "I've about got tired of Indian relics anyhow" (Abbott Diary, February 3, 1877, CCA/PU). He changed his mind, later even confessing that "it turns out, he [Putnam] was right after all" (Abbott Diary July 22, 1881, CCA/PU).

126. Abbott 1881b:vi.

127. On the opening lines, cf. Abbott 1881b:v and Evans 1872:v. Even the reviewer of Abbott's book for the *Nation* called attention to the "very noticeable" similarity between the two (Anonymous 1882).

128. Abbott 1881b:3.

129. Abbott 1881b:485–490, 511–512; Mortillet 1879.

130. Evans to Haynes, March 15, 1879, HWH/MHS.

131. Abbott 1881b:513–514.

132. Abbott 1881b:513–515.

133. Dawkins 1880:233, 242.

134. Abbott 1881b:515–516.

135. Abbott 1881b:517.

136. Abbott 1881b:517. See also Meltzer 1983:8–9.

137. Abbott 1881b:475–479.

138. Abbott 1881b:481.

139. Morison 1971:196–200.

140. Dawkins 1880; Haynes to Wright, October 20 and 26, 1880, GFW/OCA.

141. Meltzer 2003:58.

142. Abbott Diary, November 19, 1880, CCA/PU, emphasis in original; Meltzer 2003:58.

143. Dawkins to Haynes, November 20, 1880, HWH/MHS.

144. Dawkins to Haynes, February 24, 1881, HWH/MHS. Abbott subsequently changed his mind about Dawkins as well, writing a year after the visit that "I do *not* consider Prof. Dawkins one whit better authority than a half dozen others who have been here" (Abbott Diary, November 17, 1881, emphasis Abbott, inserted alongside the entry for November 19, 1880, CCA/PU). What prompted that change is unclear.

145. When Dawkins commented on Trenton publicly two years later, he wrote, "The implements are of the same type, and occur under exactly the same conditions as the river-drift implements of Europe." The conditions might have been the same—as in Europe, they were found in a "terrace of river gravel"—but he did not offer any testimony that the implements had been found in situ (Dawkins 1883:347).

146. Carr 1881:146.

147. Haynes 1881:132; Carr 1881:145; Putnam 1881:148.

148. Wright in the December 16, 1880 issue of the *Independent*, a clipping of which was carefully tipped in to Abbott's Diary in December of 1880, CCA/PU.

149. Haynes 1881:133–134.

150. Haynes 1881:137.

151. Putnam 1881:149.

152. Haynes 1881:133, 137.
153. Wright 1881a:143.
154. Wright 1881a:144.
155. Lewis to Wright, May 20, 1881, GFW/OCA.
156. Lewis 1881b:545–546.
157. Lewis 1881b:546.
158. Dawkins 1883:339. This perhaps explains why Dawkins reported that he visited Trenton with Lewis and Haynes, and neglected to mention Wright.
159. Lewis 1881b:547–549.
160. Wright 1881a:145.
161. Wright 1916:51, 54, 77.
162. Wright 1916:96–100.
163. Morison 1971:32; Numbers 1988.
164. Wright 1916:40.
165. Numbers 1988:625.
166. Russett 1976:29; Wright 1916:115–116, 132.
167. Wright 1882b:23–24, emphasis in original. This essay was originally published in the *New Englander* in October 1871 and was reprinted almost verbatim in this 1882 volume of essays.
168. Wright 1882b:25, emphasis in original.
169. Numbers 1988:625. See Wright 1916:116, 137–138. It was Wright who urged Gray to bring out his *Darwiniana* in 1876 and helped midwife the book (Numbers 1988:624–625).
170. Wright 1916:126.
171. Wright 1916:132. See also Morison 1971:196–200. The most significant of Wright's early forays into glacial geology of New England is Wright 1879.
172. Numbers 1988:625.
173. Wright 1916:128, 132.
174. Wright 1882b.
175. Wright 1882b:150.
176. Wright 1916:137–139.
177. Wright 1882b:155–157.
178. On this point Wright (1882b:101) quoted the good authority of Dana 1869:603.
179. Huxley 1863:159.
180. Wright 1882b:347, 350.
181. Wright 1882b:100–101. Asa Gray quoted in Numbers 1988:625.
182. Wright 1882b:365,372, 378, 357.
183. Wright 1882b:283.
184. Wright 1882b:311, 318. Late in life Wright's creed included the statement that conditions associated with the Glacial Epoch did not discredit the stories of the flood in Genesis (1916:426).
185. Wright 1882b:346.
186. Wright's arguments against Croll are detailed in Wright 1882b:326–346.
187. Wright 1882b:311, emphasis in original. Wright's comments on the Kettle Moraine are on p. 306.
188. Wright 1882a:7; Wright 1882b:331–336.
189. Wright 1881b:121–123; also Wright 1881a:145; Wright 1882a:6–7; Wright 1882b: 341–345.
190. Wright 1882b:346.

191. Lewis 1881b:550.
192. Abbott 1883a:359.
193. Powell 1883a. See Peet 1879a:218.
194. Abbott Diary, November 25, 1876, CCA/PU. Also Meltzer 2003:59–60.
195. Abbott Diary, June 14 and 18, 1878, CCA/PU.
196. Abbott to Wright, March 3, 1882, GFW/OCA.
197. Abbott Diary, December 4, 1883, CCA/PU.
198. Meltzer 2003:60; Abbott Diary, July 13, 1883, CCA/PU. See also Abbott to Wright, March 3, 1882, GFW/OCA. Also Hinsley 1985.
199. For example, Abbott Diary, June 14 and 18, 1878, July 16, 1883, CCA/PU.
200. Abbott to Putnam, November 20, 1881, PMP/HU, quoted in Hinsley 1985:65.
201. Meltzer 2003:56; Abbott to Wright, March 3, 1882, GFW/OCA.
202. Abbott Diary, July 13 and 16, 1883, CCA/PU.
203. Meltzer 2003:60.
204. Abbott Diary, November 27, 1883, CCA/PU. See also Hinsley 1985:65; Meltzer 2003:60.
205. Meltzer 2003:60.
206. Abbott Diary, December 8, 1883, CCA/PU.
207. Abbott Diary, April 22, 1885, CCA/PU.
208. Abbott 1881a:128. Abbott referred to finds by, among others, Berlin (1878) of paleoliths in drift deposits near Reading, Pennsylvania; Hoffman (1879) who found turtlebacks made of quartzite in the banks of the eastern branch of the Potomac (now Anacostia) River; and Jones (1873), who did not label or discuss the one specimen identified by Abbott (Jones 1873:Plate 16, Fig. 10) as a paleolith.

Chapter Four

1. Wright 1916:146–147.
2. Morison 1971:136–139; Numbers 1988:630.
3. Wright 1916:147–148. Wright was also supported by an unlimited pass on the New York, Pennsylvania, and Ohio Railroad, whose president (Jarvis Adams) was Wright's cousin.
4. Wright 1883a:271; Wright 1887a:176, 182–183. See also Haynes to Wright, April 15, 1883, GFW/OCA. As early as 1882, for example, in the Preface to his *Studies in Science and Religion* (Wright 1882b), Wright included two side-by-side figures: one a paleolith from Abbeville, the other a Trenton paleolith. Wright wanted everyone to know what they looked like.
5. For example, Mason et al. 1889; Putnam et al. 1881, 1888, 1889.
6. Hoffman 1879:113–114; Berlin 1878; Haynes 1882; Lewis 1883.
7. Mason 1885:6.
8. For a discussion of the role of women in science during this period, see Rossiter 1982. There were a few fields that were considered appropriate for or that otherwise welcomed women during this time. Archaeology was not one of them, for reasons discussed by Chester (2002).
9. Among the very few contemporary sources of biographical information on Babbitt is Brower 1902. It is perhaps telling that Babbitt published under the name Frances in 1880 in the *American Antiquarian*, but as Franc thereafter. The historical record is silent on why the change was made, but it may not be too far from the mark to observe that Franc, on the printed page, reveals less about the gender of an author

than Frances. In archaeology in those days, Babbitt may have believed that the views of a Franc might (however unconsciously) gain a wider acceptance than those of a Frances. Of course, Franc's contemporaries, ever the Victorian gentlemen, routinely referred to her as "Miss Babbitt." That she could not avoid. For a more detailed examination of Babbitt and her place in late nineteenth-century archaeology, see Chester 2002.

10. Winchell 1878:56, 58.

11. Winchell 1878:64.

12. Babbitt 1880:21–23.

13. Nathan Richardson, from *Little Falls Transcript*, July 23, 1880, quoted in Brower 1902:42.

14. The anonymous letter from "R" is the *American Antiquarian* 1881.

15. Babbitt to Haynes, September 11, 1882, HWH/MHS.

16. Babbitt 1883:369–370; Babbitt 1884a:388–390. Haynes, in Babbitt 1884b:705–706.

17. Although the meeting was held in nearby Minneapolis, *Science* reported Babbitt's paper was read by "Mr. Upton" [*sic*, Upham]; the *American Antiquarian* indicates Babbitt was present. Both, I think, are true. Upham did read the paper, for he himself said so in a letter to Chamberlin written just after the meeting (September 1, 1883, TCC/UC). However, Babbitt was likely present, judging by the proximity of the meeting to Little Falls and the comments in the *American Antiquarian*.

18. Upham, like George Frederick Wright, had begun his geological career in New England; he was a student of Hitchcock at Dartmouth, then a few years after graduation became an assistant on the New Hampshire Geological Survey. He moved west to join the Minnesota Geological Survey in 1879.

19. Babbitt 1884b:603–604, 702.

20. Babbitt 1884a:389; Babbitt 1884b:601–603.

21. Upham, quoted in Babbitt 1883:370 and in Babbitt 1884b:706–707; see Babbitt's footnote to Upham on page 707. Upham gave the same statement in letters to Wright (August 28, 1883, in GFW/OCA) and Chamberlin (September 1, 1883, in TCC/UC).

22. Babbitt 1884b:702.

23. Peet 1883:353. Also Babbitt 1883:370; Babbitt 1884b:706.

24. *Science* 1883a:370; *Science* 1883b:384. The fullest version of Babbitt's AAAS talk appeared in *American Naturalist* (Babbitt 1884b). The reasons to suspect Putnam had a strong hand in the broadcasting of Babbitt's paper are circumstantial. As AAAS permanent secretary, he was responsible for the publication of the papers, thought Babbitt's one of the most important of the meeting (according to Warren Upham, in a letter to Wright, August 28, 1883, GFW/OCA), and was close to *Science*'s editor, Samuel Scudder. I do not know if Putnam was on *Science*'s editorial board; Scudder kept this secret (Kohlstedt 1980:35–36).

25. *Science* 1883b:384.

26. For example, Peet 1878:46.

27. Abbott 1883b.

28. Mason 1885:6.

29. Babbitt 1885:594–595.

30. *Science* 1884b:344; Peet 1884:419.

31. *Science* 1884b:345. Tylor 1884:217–218.

32. Abbott Diary, September 12, 1884, CCA/PU. Tylor to Holmes, August 25, 1896, WHH/SIA.

33. Meltzer 2003:61.
34. Cope to Putnam, May 20, July 7, and July 11, 1884, FWP/HU.
35. Abbott Diary, September 30, 1884, CCA/PU.
36. Abbott Diary, November 21, 1884, CCA/PU.
37. Abbott Diary, May 7, 1885, CCA/PU. Whatever the cause, relations with Leidy were never completely mended, for Abbott later learned that Leidy strongly opposed his appointment as curator (December 28, 1889).
38. Abbott 1884; Abbott 1886b.
39. Putnam 1886a:408; Putnam 1886b:487, 491.
40. *Science* 1886.
41. Putnam 1886b:492–493.
42. See the discussion in Hinsley 1985.
43. Abbott 1886a. According to his diary, Abbott wrote the sketch in March of 1886; it appeared in September of that year.
44. The position was endowed as part of Peabody's original 1866 gift (Peabody 1868:27), though serious consideration of Putnam's candidacy did not begin until 1885, and then for nearly two years his path was blocked by Alexander Agassiz, the son of Louis Agassiz, likewise a geologist. The younger Agassiz inherited the directorship of Harvard's Museum of Comparative Zoology and shared his father's low opinion of Putnam. See the discussion in Hinsley 1985:49–50, 61.
45. Putnam 1888a.
46. Wright 1887b:255–256. Metz subsequently found two other apparent paleoliths at Loveland, as reported by Holmes (1893d:149–150).
47. Anonymous 1887.
48. Abbott Diary, August 20–21, 1887, CCA/PU.
49. Thomas to Fowke, October 31, 1887, BAE/NAA. See also Abbott 1889b:308.
50. Five of the years Wilson spent in Europe were as U.S. consul in Belgium, where he reported to the State Department on the Treaty of Ghent, and in France in Nantes and Nice. Just before visiting Trenton, he had spoken at the AAAS meetings on European archaeology. On Wilson's paper at the AAAS, see *Science* 1887:89; for biographical information on Wilson, see Mason 1902:288. This was the same Wilson who, four years earlier, had written Powell while still in France, hoping that when he returned to Washington they might have the chance to get together to compare notes and talk "of man before the flood" (Wilson to Powell, January 22, 1883, BAE/NAA).
51. Wilson 1890a:679.
52. Wilson 1890a:679. Wilson's paper to the Anthropological Society of Washington was titled "Paleolithic Man in the District of Columbia" and was delivered to that group on November 2, 1887. So far as I have been able to determine, no copy of it survives, but it seems reasonable to suppose that it was similar to Wilson's subsequent publications on the subject (for example, Wilson 1889, 1890c).
53. Henry 1878; Langley 1888.
54. Wilson to Wright, March 26, 1888, GFW/OCA.
55. Wilson to Haynes, December 6, 1888, HWH/MHS.
56. Wilson 1890a:679, 694.
57. There are no surviving copies of McGee's talk, "Conditions of the Accumulation of the Trenton Gravel," delivered on November 12, 1887. However, extensive portions of it were quoted by Abbott (1889b:305–307), and it is that source that was consulted and cited here.
58. McGee 1879:226.

59. Russell 1885:247, 267–270. McGee's report of the find was not given until a November 16, 1886 meeting of the Anthropological Society of Washington (see McGee 1887, 1889). Given McGee's radical reversal on the human antiquity question in 1889–1890, his 1889 paper may not accurately reflect his original view—or that reported by Russell. From the illustration of the specimen (Russell 1885:Figure 33), it was a large but nondiagnostic biface. The specimen disappeared long ago.

60. Putnam 1888b:422.

61. Putnam 1888b:424.

62. Abbott 1888:425

63. Wright 1884b, 1885, 1888:435.

64. Abbott to Wright, February 16, 1888, GFW/OCA. For Wright's earlier views, see chapter 3.

65. Upham 1888a:437, 446; Winchell 1887.

66. Putnam 1888b:422, 424.

67. The evidence from the gravels is summarized in Whitney 1879. See Putnam 1888b:448; Brinton 1888:298; for another, very negative reaction, see Peet 1879b and Peet 1888c:254–255.

68. Putnam 1888b:449, emphasis mine.

69. McGee to Youmans, May 26, 1888, WJM/LC.

70. McGee to Wright, March 8, March 20, and June 14, 1888, GFW/OCA; McGee to Youmans, May 26, 1888, WJM/LC.

71. Abbott 1889a; McGee 1888b:31, 36; McGee to Putnam, December 5, 1888, WJM/LC.

72. McGee 1888b:21–25.

73. McGee 1888b:25–27, especially Figure 1; also, Peet 1888a:47–48.

74. Abbott 1883c:101–102, emphasis in original. McGee did not mention that particular evidence in his published discussion, but in a handwritten note on the reprint of his 1888b article McGee stated that human bones had been found in the Trenton Gravel in the same deposits as the bones of mastodon, bison, and reindeer (reprint in possession of the author).

75. Putnam 1885, 1888a.

76. Putnam to Wright, June 26, 1888, GFW/OCA. Also Abbott to Putnam, June 15, 1888, PMP/HU. Putnam might have wanted it kept secret, but Cresson wrote Wright to tell him of the find, and he raised the possibility that it may have come from the Philadelphia Red Gravel but stated he would leave that to Wright to determine; Cresson to Wright, June 29, 1888, GFW/OCA.

77. Abbott to Putnam, March 22, 1888 and June 15, 1888, PMP/HU.

78. Peet 1887:50–51. See also the comments in Peet 1878:46; Peet 1879a:212; Peet 1888a:48; Peet 1888b:124.

79. Abbott 1889b:295.

80. Abbott 1889b:297.

81. Abbott 1889b:297–298. That spring Peet had claimed that paleolithic implements ought to have some "vitreous gloss" due to weathering which, though it might occur on specimens of more recent age, must occur on paleolithic implements. Those from Trenton and Little Falls, Peet observed, "are destitute of this vitreous gloss" (Peet 1888b:125).

82. Abbott 1889b:311–314.

83. Abbott 1889b:314.

84. Abbott to Wright, February 16, 1888, GFW/OCA; Abbott 1889b:309. He would expand on this issue in a paper the next year (Abbott 1889c). Putnam was also hopeful that a link would be found between the Paleolithic and the Eskimo, so much so that Cresson described him as "always straining a point toward supporting Abbott's Eskimo theories" (Cresson to Wright, March 2, 1889, GFW/OCA).

85. Abbott 1889b:315.

86. Abbott 1889b:304.

87. Abbott to Putnam, June 20, 1888, PMP/HU.

88. Abbott to Wright, June 18, 1889, GFW/OCA.

89. Cresson 1889a:149–150; Cresson to Wright, June 29, 1888, GFW/OCA; McGee 1888a:379. Cresson reported on these finds at the August 1888 meeting of the AAAS; however, his paper was not published in the *Proceedings* volume for that year. Instead, what is presumably the same or a very similar paper was given at a meeting of the BSNH, on December 19, 1888 and was published there. It is this version that is cited here.

90. Biographical information on Cresson is in Meltzer and Sturtevant 1983:339–341 and references therein.

91. On the details of the two discoveries, see Cresson to Wright, June 29, 1888, April 6, and July 22, 1889 (which has stratigraphic diagrams), GFW/OCA, and also Cresson 1889b:151. On Wright's visit to Claymont, see Cresson to Wright, October 18 and November 7, 1888, GFW/OCA. While in the area, Wright also paid another visit to Trenton; it was becoming an almost annual event. Abbott to Wright, November 7, 1888, GFW/OCA. Cresson's papers were listed only by title in the AAAS *Proceedings* for 1888.

92. On the photographs, see Cresson to Wright, November 7, 15, 1888, GFW/OCA. Wright 1889b:154.

93. McGee 1888a:379, 382, 461–463; McGee 1888b:21.

94. Abbott 1889b:296, 304–305; Putnam 1889a:158; Abbott to Putnam, June 15, 1888, PMP/HU.

95. McGee wrote to Cresson asking to be alerted to any new finds and offering to "aid in the exhumation." McGee to Cresson, March 8, 1889, WJM/LC.

96. Wright 1889b:155–156; McGee to Wright, October 17, 1888 and Cresson to Wright, November 15, 1888, GFW/OCA.

97. Wright to McGee, January 14, 1889, WJM/LC.

98. McGee to Wright, January 22, 1889, WJM/LC.

99. Wright 1889b.

100. Schultz 1976:17. As president of the University of Wisconsin twenty years later, Chamberlin spearheaded curricular reforms to deemphasize the classics, radically expand scientific offerings and research, and introduce graduate education and professional training (Schultz 1976:68). His presidency also involved several unpleasant events that led to his jumping at the offer to join the geology faculty at the University of Chicago (Fisher 1963:2–3).

101. Chamberlin 1934:312.

102. Chamberlin 1934:316–317; Schultz 1976:384.

103. Chamberlin, "Tentative Sketch of a Plan for the Development of Original Research in the Univ. of Chicago," in Schultz 1976:148. For his early and passing foray into religious and scientific accommodation, Schultz 1976:15–16. Also Chamberlin 1934:313.

104. Over time, that window would narrow, for reasons explored in Numbers 1988.

105. Wright 1916:100–102.
106. Chamberlin, quoted in Leith 1929:289.
107. The great diversity of the interest and goals of the various state surveys can be seen the compilation of their individual reports in Merrill 1920.
108. This material is drawn from the very thoughtful discussions in Schultz 1976, especially pages 146–150, and from Winnick 1970.
109. From Chamberlin's unpublished (circa 1896) essay, "Secular Theology," quoted in Schultz 1976:147.
110. Chamberlin 1888c:22.
111. Chamberlin 1888c:3–4.
112. Schultz 1976:48–50. On the "kitchen stove" quote, see, for example, Chamberlin to Fairchild, August 23, 1926, TCC/UC.
113. Powell to Chamberlin, June 17, 1881, quoted in Chamberlin 1934:316–317.
114. Chamberlin 1883a:271–272.
115. Chamberlin 1883b:295.
116. Chamberlin 1883b:295.
117. Chamberlin 1883b:302.
118. Chamberlin 1883b:310–313, 401.
119. Chamberlin 1883b:302, 314.
120. Alden 1929:299. See also *American Journal of Science* 1884 and *Science* 1884a.
121. Chamberlin 1883b:346.
122. Powell was more interested in topographic and geological mapping, paleontology, hydrology, and glacial geology, for example, than in the mining and minerals that had focused the survey's efforts under King (Rabbitt 1980:58, 92, 112).
123. Branner 1891:221.
124. Branner 1891:229.
125. Branner 1891:222. Increasing the strain was the gross inequity of funding. Generally, the USGS annual budget far exceeded the total budgets most state geologists received over periods of a decade or more. For summary data on state survey budgets, see Merrill 1920: Appendix 1. For contemporary comments on the relationship between the federal and state surveys, see Branner 1891. For more recent work, see Manning 1967:66–67, 117–121.
126. Chamberlin 1883b:341.
127. Chamberlin 1882:94. This paper was written after Chamberlin's first year with the USGS, but because of differences in the timing of publication it appeared before his paper in the USGS *Annual Report*.
128. Chamberlin 1882:97.
129. Lewis 1884a:43. Wright (1889a:iii) explained he was invited by J. Peter Lesley, director of the Pennsylvania Geological Survey, "to survey, in company with the late Professor H. Carvill Lewis, the boundary of the glaciated area across Pennsylvania." Later Wright implied theirs was a joint project (for example, Wright 1916:143–144). However, Lewis alone was given the commission for the survey, and it was he who suggested Wright as a volunteer assistant (Morison 1971:203). In the two years of fieldwork and write-up, Wright was directly involved just six weeks in the summer of 1881 (Lewis 1884a:vi, li).
130. Chamberlin 1882:94,97; Chamberlin 1883b:341.
131. Chamberlin 1883b:347.
132. Lewis 1881b:545–546.
133. Lewis 1884b:28, 40–45.

134. Lewis 1884a:38, 42–43.
135. Lewis 1884a:179, 201.
136. Wright 1889b:204. That the space between the moraine and the fringe showed no sign of glacial erosion was the basis of this inference (Lewis 1884a:201).
137. Lewis 1884a:201–202; Wright 1884d:760.
138. Lesley 1884:xli–xliv.
139. Lewis 1884b:285.
140. Wright 1882b:305–306.
141. Wright 1916:147–148. Wright reports that in the rural towns and areas in which they traveled he and his assistant were variously mistaken as circus advance agents, lightning rod salesmen and, on one occasion when they did not return on time with their livery rig, horse thieves.
142. Wright 1883a:271; Wright 1883b:44–46.
143. Wright 1883b:48–49; Wright 1884a:241–242; Wright 1889a:327. Wright came across the boulders in the summer of 1882 and collected several. When he died nearly forty years later, the largest of the boulders (perhaps the one photographed in Wright 1889a:Fig. 90) was used to mark his family plot in the Oberlin cemetery; the smaller ones marked the graves of individual family members (§7.8). For Wright those boulders, transported by glacial ice, symbolized "the story not only of their own travels, but of other most interesting events connected with the cause which transported them" (Wright 1889a:328).
144. Wright 1883a:270–271; Wright 1883b:49; Wright 1884c:203. As best I can determine from the modern geological literature, the ice sheet that crossed the Ohio River (from Ripley, OH to Carrollton, KY) was Illinoian in age, and it apparently did dam the waters temporarily, causing deposition of some lacustrine clays near Cincinnati (Goldthwait 1991:3).
145. Decades later he reported that in Cleveland "large audiences gathered" to hear his report on it to the Western Reserve Historical Society (Wright 1916:150).
146. White 1884:213.
147. Wright 1916:153.
148. Lesley 1884:ix. These were Lesley's written remarks. Wright's version of what Lesley said at the AAAS meeting is more colorful: "Providence has provided it [the high terraces] and Wright's dam will explain everything" (Wright 1916:154). See also Lesley 1883:436.
149. Chamberlin 1884b:212.
150. Chamberlin 1884a:210.
151. Chamberlin 1884b:212; Wright 1884d:760.
152. Upham 1884:222.
153. Upham to Wright, August 28, 1883, GFW/OCA.
154. Powell 1885b:xxvi; Manning 1967:130–131; Rabbitt 1980:116–117. Powell had that administrative flexibility because the USGS appropriation came as a lump sum, which he then distributed among the various divisions as he saw fit.
155. Chamberlin to Powell, July 12, 1884, quoted in Morison 1971:256–257.
156. Chamberlin to Wright, April 16, 1884 and Haynes to Wright, April 24, 1884, GFW/OCA. Haynes wrote his letter of recommendation to Powell. Powell had spent a year as a student at Oberlin (1858–1859), and though Wright was four years Powell's junior, he was then entering his fourth year, whereas Powell was entering his second (Darrah 1951:39). Morison reports the two knew each other (Morison 1971:254). Although a temporary employee, Wright was identified as an assistant

geologist by the USGS chief disbursing clerk (McChesney to Chamberlin, December 22, 1892, TCC/UC).

157. Wright 1882b:vi.

158. Chamberlin to Wright, April 14, 1885, GFW/OCA.

159. Many years later, Chamberlin would suggest that at the time he had a low opinion of Wright's competence, claiming that word going around at the time of Lewis and Wright's survey in Pennsylvania (1882) was that they had followed the Pennsylvania moraine into the middle of the state, where "they lost it" (Chamberlin to Fairchild, August 23, 1926, TCC/UC). Chamberlin hired Wright anyway. This would suggest those rumors were not circulating in the early 1880s but instead surfaced in the heat of the controversy of the late 1880s and early 1890s, as post hoc claims.

160. Chamberlin to Wright, April 16, 1884, GFW/OCA.

161. Wright 1884c:567.

162. Chamberlin to Wright, August 1, 1884, GFW/OCA.

163. Chamberlin to Wright, August 1, 1884, GFW/OCA.

164. Chamberlin to Wright, August 19, 1884, GFW/OCA.

165. Chamberlin to Wright, August 25, 1884, GFW/OCA; Morison 1971:258.

166. Chamberlin told Wright he had seen notice of the talk in the April 10, 1885 *Science*; a check of that issue, however, failed to turn up any such notice. Nor was it published in the previous week. (It could not have been published in the following week, as that would postdate Chamberlin's letter.) Chamberlin must have seen the notice elsewhere. Chamberlin to Wright, April 14, 1885, GFW/OCA.

167. Rabbitt 1980:85. Wright's hiring was made official on July 23, 1884 (Morison 1971:256–257).

168. Manning 1967:132–133; Miller 1970:144–146.

169. Rabbitt 1980:87–90, 102; Darrah 1951:289–293; Manning 1967:123–124;

170. Powell 1885a; Rabbitt 1980:105–109.

171. Rabbitt 1980:114–115.

172. Manning 1967:126–128. It would be interesting to learn, given that the dispute over the Cincinnati Ice Dam was likely reasonably well known in the city, whether Wright's differences with the USGS helped fuel the story in the Cincinnati *Commercial Gazette*.

173. Chamberlin to Wright, April 20, 1885, GFW/OCA. Cited in Morison 1971:260.

174. Chamberlin 1888a:78.

175. Chamberlin to Wright, October 28, 1885, GFW/OCA.

176. Agassiz 1885:255. Agassiz's letter was promptly reprinted in the *Nation* and in *Science*. Also, Rabbitt 1980:116; Manning 1967:135.

177. Rabbitt 1980:112, 117.

178. Darrah 1951:293–294; Manning 1967:134–135. Agassiz's antipathy toward paleontology might have been motivated by a deep dislike of Yale's O. C. Marsh, who was head of one of the USGS's Paleontology Divisions (Miller 1970:147).

179. Powell 1885c; Rabbitt 1980:120.

180. Manning 1967:136–137; Rabbitt 1980:122–123.

181. Manning 1967:138–141; Miller 1970:146–148.

182. Miller 1970:148.

183. Rabbitt 1980:124.

184. Manning 1967:147–150. Herbert did get in one good blow, however: in December 1886, he inserted into the Appropriations Committee discussions of the Sundry

Civil Expenses Bill a rider stipulating that "the estimates for all moneys wanted by the Geological Survey shall be itemized," and so it was passed by Congress and remained in force until 1950 (Rabbitt 1980:126).

185. Morison 1971:262.

186. Chamberlin 1888a:83.

187. Chamberlin and Salisbury 1885; Chamberlin 1888b.

188. Chamberlin and Salisbury 1885:212–214; Chamberlin 1888b:159–160.

189. Chamberlin and Salisbury 1885:286–289.

190. Chamberlin 1888b:159–160.

191. Chamberlin 1888b:210.

192. Chamberlin 1887:211.

193. Chamberlin (1891) was Chamberlin's first effort to develop a theory of the cause of glaciation; it was delivered at the GSA meetings but appeared only in abstract form. Much of Chamberlin's work on the origins of the earth was done in collaboration with University of Chicago astronomer F. R. Moulton (Moulton 1929:374).

194. Alden 1929:300; Totten and White 1985:133. See, for example, Leverett 1899, 1902; Leverett and Taylor 1915.

195. Chamberlin to Wright, October 28, 1885, GFW/OCA

196. Chamberlin to Wright, July 5, 1887, GFW/OCA.

197. Chamberlin to Wright, February 15, 1888, GFW/OCA.

198. Chamberlin transmitted Wright's volume for publication in May of 1889. Chamberlin 1890a:11.

199. Wright 1890a:73.

200. Wright 1890a:82–83, 97.

201. Wright 1890a:87, 89–90.

202. Wright 1890a:101.

203. White 1887:380.

204. Chamberlin 1890a:14.

205. Chamberlin 1890a:14–15, 34.

206. Chamberlin 1890a:15–17; also Chamberlin 1888b.

207. Chamberlin 1890a:17–18.

208. Chamberlin 1890a:19, 28–29.

209. Chamberlin 1890a:33, 36–38.

210. Chamberlin 1890a:37, emphasis in original.

211. Wright to Todd, January 23, 1890, GFW/OCA; also Morison 1971:263.

212. Wright to Powell, December 24, 1888, TCC/UC; Pilling (for Powell) to Wright, December, 27, 1888, TCC/UC.

213. Wright to Chamberlin, January 9, 1889, TCC/UC.

214. Chamberlin to Wright, January 24, 1889, TCC/UC, emphasis in original. Although the Wright archives contain what appears to be every letter Wright ever received, this letter from Chamberlin was not among them. I suspect that is no accident of preservation.

215. Chamberlin to Wright, January 24, 1889, TCC/UC.

216. Wright to Haynes, October 25, 1892, HWH/MHS.

217. Chamberlin to Powell, January 24, 1889, TCC/UC; Pilling [for Powell] to Chamberlin, January 29, 1889, TCC/UC.

218. Wright to Chamberlin, January 30, 1889, TCC/UC. Putnam to Wright, January 8, 1889, GFW/OCA.

219. Among those who preceded Wright as Lowell lecturers were Louis Agassiz and Asa Gray; the list of those who followed includes William James (1906), Franz Boas (1910), and Bertrand Russell (1914).

220. Morison 1971; Wright 1889a:144.

221. Wright 1887c; McGee to Wright, January 13, 1887, GFW/OCA. USGS geologists Todd, I. C. Russell, and Gilbert all had similar words of praise for Wright's Muir study (see Morison 1971:220). Wright made his observations of Muir over fourteen days. Because of inclement weather, the other two weeks were spent huddled in the tent (Wright 1889a:37).

222. Morison 1971:211–219; Wright 1916:156–159; Chamberlin to Wright, June 24, 1886, GFW/OCA.

223. Wright 1889a:vii.

224. Wright offered his opinion of the book's importance in Wright to Haynes, January 10, 1889, HWH/MHS. See also Morison 1971:221–224; Wright 1916:160–161. Wright's success may be measured also in the fact that he was asked back on two subsequent occasions to serve as a Lowell lecturer: in 1891–1892 (the basis for his *Man and the Glacial Period*) and in 1896, the experience from which came his book *Scientific Aspects of Christian Evidences* (1898).

225. See, for example, Geikie 1877:463; also Orton 1873; Winchell 1873. Geikie's volume first appeared in 1874. Geikie's second edition (1877) was available when Wright wrote his derivative volume. If Geikie was grateful for the flattery, he did not show it: when he came to prepare his third edition, he asked his friend Thomas Chamberlin, not Wright, to contribute the chapters on the "Glacial Phenomena of North America" (Chamberlin 1894a).

226. Wright to Chamberlin, January 30, 1889, TCC/UC.

227. Wright 1889a:330–336.

228. Wright 1889a:416–440. Wright also depended on a long critique of Croll that Dana had published in the *American Journal of Science* just a few years before (Woeikof 1886).

229. Wright 1889a:448.

230. Wright 1889a:452–475.

231. Wright 1889a:475. McGee had sent Wright an advance copy of his paper for the *Seventh Annual Report of the USGS*, in which he detailed the geological history of the Chesapeake Bay, concluding that even this region well south of the glacial terminus provided evidence of multiple glaciation. Wright was certainly in possession of the latest arguments and evidence in support of multiple glaciations (McGee to Wright, October 17, 1889, GFW/OCA).

232. Wright 1889a:479–482, 498–499.

233. Wright 1889a:477–478.

234. Wright 1889a:121, 144.

235. Wright 1889a:133–135.

236. Wright 1889a:179.

237. Wright 1889a:195.

238. Wright to McGee, January 14, 1889, WJM/LC; McGee to Wright, January 22, 1889, GFW/OCA.

239. Wright 1889a:179, also 195.

240. Wright 1889a:500.

241. Wright 1889a:viii.

242. Wright 1889a:549, 569–570 (on the relative age of Little Falls), and 553 (on the age of Claymont relative to the other sites).
243. Morison 1971:225–228.
244. Abbott to Wright, June 18, 1889, GFW/OCA; *American Journal of Science* 1889; American Geologist 1889—the suggestion that Winchell was the reviewer is in Winchell to Wright, July 12, 1889, GFW/OCA.
245. Davis 1889:118–119.
246. Chamberlin to Leverett, August 13, 1889, TCC/UC.
247. Abbott 1889c.
248. Mason et al. 1889.
249. Putnam to Wright, January 8, 1889, GFW/OCA.
250. Winchell to Wright, December 31, 1888, GFW/OCA. Also Winchell 1887.
251. Wright 1889a:511. Also Haynes to Wright, January 5, 1889, GFW/OCA.
252. Babbitt 1890:334. The "distinguished archaeological experts" Babbitt reported as having examined the Little Falls quartzes were Rau, Abbott, Haynes, Putnam, and Mason. (She reported them in that order, which presumably represents the order in which she sent the specimens out.)
253. Wright 1889a:513; Wilson 1889:237.
254. Mills to Wright, March 14 and 26, 1890, GFW/OCA.
255. Mills 1890; Wright 1890c. Also Mills to Wright, April 23 and June 5, 1890, GFW/OCA.
256. Wright 1890b:448. Also Upham to Wright, November 8 and 20, 1889, GFW/OCA.
257. Wright 1890b:447–449.
258. Putnam 1890; for details of the Holly Oak fraud, see Meltzer and Sturtevant 1983 and Meltzer 1990.
259. Meltzer and Sturtevant 1983.
260. For example, Putnam 1885, 1889d.
261. Cresson to Putnam, March 28, 1890, PMP/HU.
262. Meltzer and Sturtevant 1983.
263. Putnam 1888b:421; Wallace 1887. See also Topinard 1893; Wilson 1893.
264. Abbott to Putnam, June 21, 1889, FWP/HU; Meltzer 2003:64.
265. Madeira 1964:20. Abbott Diary, October 23 to December 2, 1889, CCA/PU.
266. Putnam to Pepper, November 8, 1889, UM/UP.
267. Putnam to Pepper, October 28, 1889, quoted in Hinsley 1985:65.
268. Putnam 1889c:71–72.
269. Abbott to Putnam, October 17, 1889, PMP/HU.
270. Chamberlin 1890b. Over the next decade there were several versions of this paper, and its essential theme was expressed in many of his other works (see Schultz 1976:116ff).
271. Chamberlin 1890b:93.
272. Chamberlin 1890b:94.
273. Wright 1890a:101.
274. Chamberlin 1890b:93.
275. This was actually the second meeting of the GSA, though the first annual meeting. The GSA was formed in December of 1888, and both Chamberlin and Wright were original fellows.
276. Chamberlin 1890c:472–474.
277. White in Chamberlin 1890c:477–478.

278. Chamberlin 1890c:478.
279. White, McGee in Chamberlin 1890c:479–480.
280. McGee to Wright, June 21, 1884, GFW/OCA.
281. Cresson to Wright, August 20, 1889, GFW/OCA; for the lead in to the visit, see Cresson to Wright, July 29, 1889, GFW/OCA.
282. Cresson to Putnam and Abbott, and to Putnam, both August 25, 1889, PMP/HU; Cresson to Wright, August 26, 1889, GFW/OCA.
283. Abbott Diary, August 24–31, 1889, CCA/PU; Cresson to Putnam, October 10, 1889, PMP/HU; McGee 1890.
284. McGee to Wright, October 17, 1889, GFW/OCA. McGee also asked Wright to return the advance copy of McGee's paper for the *Seventh Annual Report of the USGS*. It is not unreasonable to suppose that McGee, anticipating severing relations, wanted to first retrieve his belongings.
285. Proudfit 1889; Putnam 1889b:267, emphasis mine.
286. Holmes 1890a.
287. Carr to Henshaw, December 2, 1889, BAE/NAA.
288. Abbott to Henshaw, February 20, 1890, BAE/NAA.

Chapter Five

1. Holmes 1890a:15.
2. Meltzer 2003:67.
3. Henshaw to Abbott May 21, 1890, CCA/UPM.
4. Meltzer 2003:67.
5. Abbott 1890:9.
6. Holmes, *Undated sketches*, WHH/SIA.
7. Manning 1967:216; Mark 1980:145.
8. Dupree 1957; Lacey 1979.
9. Darrah 1951:319–321; Stegner 1954:251. Powell served under Ulysses S. Grant at Shiloh and by the Civil War's end was known to many of the country's future leaders (Goetzmann 1966:533). Carl Schurz, who recommended the appropriation that created the BAE, was a powerful alumnus of Grant's wartime staff.
10. For detailed histories of the four competing surveys and how the USGS emerged from them, see the discussions in Bartlett 1962; Goetzmann 1966; Rabbitt 1979, 1980.
11. Rhees 1901:818; also discussions in Hinsley 1981:147; Meltzer 1985; Noelke 1974: 53–55.
12. The BAE came under congressional scrutiny on occasion throughout its history, as for example in 1890 when the House formally requested an inquiry into "by what authority" the Bureau was founded (Rhees 1901:1539–1542), and in 1892 when a congressman complained about appropriating money "to hold inquests over the bones of Indians who died before Columbus landed on the shores of America" (Rhees 1901:1616–1623). On Powell's ability to handle such threats, see Stegner 1954:249.
13. Rhees 1901:857–863. Powell's annual budget through 1900 is based on data in Rhees (1901). For budgets from 1900–1910, see Noelke (1974:Appendix 1).
14. Putnam 1891:106.
15. Noelke 1974:56–57. Archaeology was not part of Powell's initial plan for the BAE and was likely forced on him through back-door maneuvering by Spencer Baird.

Powell nonetheless quickly saw the advantage of the situation and created the widely hailed Mound Survey (Meltzer 1985:250–251).

16. Hinsley 1981:151; Powell 1881a:xxxiii, 1883b:181–182. See also Hinsley 1981: 125–140.
17. Swanton 1944:34.
18. Powell 1883b:183; Hinsley 1981:127–131; Stocking 1987.
19. Powell 1883b:175–176; also Hinsley 1981:137.
20. Hinsley 1976:39–41; Hinsley 1981:125, 152; Rabbitt 1980:59.
21. Hinsley 1976:40; Mark 1980:144; Meltzer 1985:251; Pyne 1980:114.
22. Putnam 1886:29; also Hinsley 1976:41–42; Hinsley 1981:232.
23. Meltzer 1985:251. Powell's expectations of loyalty is from Walcott to Chamberlin, August 7, 1892, in Rabbitt 1980:213. On the animosity between Powell's hand-picked staff and the foreign-trained PhD holdovers from King's days, see Pyne 1980: 126–130.
24. Stegner 1954:155.
25. Thomas 1898:9, 22; Kidder 1936:145.
26. Powell 1881b:74; Powell 1886:lxiii.
27. Wilson to Goode, November 14, 1890, USNM/SIA.
28. Thomas 1894:21. See also Henshaw 1890:102.
29. Thomas 1898:23; Trigger 1989:124.
30. Powell 1890a:503. The "good queen" was Isabella of Spain, who—so the story goes—sold some of her jewels to help finance Columbus's voyages.
31. Powell 1890a:500.
32. Meltzer 1985:253–254.
33. Powell 1894:xxvi.
34. For detailed biographical information on Holmes, see Mark 1980; Meltzer 2010; Meltzer and Dunnell 1992; Swanton 1937.
35. Holmes was the youngest of three boys; his older brothers were evidently content to stay and work on the farm (Meltzer and Dunnell 1992:x–xi).
36. WHH/RR 1:30, 33; Holmes 1878:383.
37. See, for example, Holmes 1883, 1884, 1886a, 1986b, 1986c, 1888.
38. Meltzer and Dunnell 1992:xi–xiv, xxv. Holmes's stint at the Field Museum was brief and unhappy, as I discuss elsewhere (Meltzer 2010).
39. Holmes 1888:196; Holmes 1890c:139. For a discussion of Holmes's particular brand of evolutionism, see Hinsley 1981:103–104; Meltzer and Dunnell 1992:xxvii–xxxi; Thoresen 1977.
40. Holmes 1892a:248–249. Meltzer and Dunnell 1992:xxvii–xxxi.
41. Holmes 1886c:458 Holmes 1892a:246–247. Also, Hinsley 1981:104.
42. Holmes 1894 makes the most explicit use of the concept. In the late nineteenth century the idea that "ontogeny recapitulates phylogeny" was a prominent element of biological thought (Gould 1977).
43. Mark 1980:157; Swanton 1937:236.
44. Meltzer and Dunnell 1992; Holmes 1897a:61, 151.
45. Wilson 1890b:979–980.
46. Dunnell 1986:156–158.
47. This is not to say others were unaware of the processes by which artifacts were manufactured; they certainly were (for example, Putnam 1883:6–7). The difference is that others (at least early on) failed to see the connection between what modern natives were doing and the American Paleolithic.

48. Haven 1864:14.

49. Meltzer and Dunnell 1992.

50. See Proudfit 1889:245; Putnam 1889b; Wilson 1889. Powell later (1895:3–4) took credit for directing Holmes to Piney Branch, claiming he saw the significance of the quarry debris for resolving the Paleolithic issue. The contemporary records do not support that assertion. Holmes (1890a:4) identifies several people who previously noticed the workshop debris; Powell was not among them.

51. Holmes 1879:250; Holmes 1883:31–32.

52. Holmes 1890a:1. Late in life, Holmes reported his first foray into the controversy was in the late 1880s after he had "already differed with Dr. [Thomas] Wilson on the interpretation of [Paleoliths], since among all treasured in the Museum there was not a single specimen that showed definite specialization or signs of use" (Holmes to Nelson, December 2, 1921, BAE/NAA).

53. Holmes 1890a:10, 20.

54. Holmes 1890a:12–13, 15–19. In later papers, Holmes (1891a, 1893a) elaborated on the relationship between the technology used at the quarry and that used at non-quarry sites.

55. Holmes 1890a:14. Holmes did allow that on rare occasions a finished leaf-shaped blade might be inadvertently lost or dropped at the site. Otherwise, all was unfinished.

56. Holmes 1890a:3.

57. Holmes 1890a:13–14, emphasis mine.

58. Holmes 1890a:19–21, 24, emphasis mine.

59. Holmes 1890a:17, 21, 25.

60. Holmes 1890a:18, emphasis mine.

61. Holmes 1890a:23.

62. Holmes 1890b:391.

63. Wilson 1890b:979–980 (Wilson's report includes the quotes from Mason and Putnam); Abbott Diary, August 22, 1890, CCA/PU. Holmes's ideas were fast gaining exposure and acceptance outside North America as well. As part of a museum exchange, the University Museum at Oxford received several of Wilson's "paleoliths" and then some Piney Branch rejects along with Holmes's paper. Henry Balfour (the Oxford curator) wrote to say he was not convinced Wilson's specimens were true paleoliths, as their resemblance to European paleoliths was "comparatively slight." But he was convinced by Holmes's conclusion that they were quarry rejects. Balfour was especially struck by the "embryological aspect" of Holmes's argument (Balfour to Holmes, September 9, 1890, WHH/SIA).

64. Abbott 1890:9.

65. Abbott 1890:8–9.

66. Abbott 1890:10.

67. Meltzer 2003:69.

68. Abbott Diary, October 21, 1893, CCA/PU. On the earlier exchanges, see Putnam to Abbott, January 3 and 8, 1891, PMP/HU; Volk to Putnam, January 28, 1891, FWP/HU. Also Putnam to Abbott, April 13, 1891, FWP/HU; also Abbott to Putnam, April 4, 1891, FWP/HU.

69. Abbott 1890:11; see also Meltzer 2003:69.

70. Abbott to Council of the University Archaeological Association, May 4, 1891, CCA/UPM.

71. Mercer was described as a collector, traveler (extensively through Europe), and a "man of means." He had dabbled in archaeology before, publishing *The Lenape Stone*

(Mercer 1885), which seems to have fired his interest in archaeology. He joined the university's Archaeological Association in 1890 and was soon thereafter on the museum's governing board (Dyke 1989:45).

72. Abbott to Council of the University Archaeological Association, May 4, 1891, p. 3, CCA/UPM. Also Meltzer 2003:69–71.

73. Abbott 1892c:1, 19; Mercer to Abbott September 11, 1891, CCA/UPM.

74. Mercer's report is unpublished, but it exists in the form of a roughly twenty-five-page letter to Abbott, copied over in Abbott's hand and dated October 27, 1891, HCM/BCHS. The pagination follows this version, and the first page of the report is 75; in recopying the report Abbott had also appended it to his own draft report of the summer's work. During the recopying, Abbott added annotations to Mercer's text. Also Meltzer 2003:70.

75. Meltzer 2003:73.

76. Abbott 1892c:28. Abbott's redacted Mercer to Abbott, October 21, 1891, pp. 86–87, HCM/BCHS.

77. Wilson to Mercer, January 2, 1892, HCM/BCHS. Abbott to Mercer, December 1, 1891, HCM/BCHS, emphasis and relative type size as in the original. And just to make sure Mercer got the point, two days later Abbott snorted, "I am busy with Indian matters now and look on the other [paleolithic] question as *settled!*" (Abbott to Mercer, December 3, 1891, HCM/BCHS, emphasis in original.)

78. Abbott 1892c:4, 7.

79. Meltzer 2003:71.

80. Paschall to Abbott, April 5, 10, 24, 1890, UM/UP. Mercer himself was looking for argillite quarries near Trenton that winter (Mercer to Abbott, December 9, 1891, CCA/UPM), and a few years later he seized on the evidence from the Point Pleasant quarries to make the same point as Paschall and cast doubt on the veracity of the Trenton paleoliths (Mercer 1897:34ff; also Abbott 1897, 1908:43ff).

81. Paschall to Abbott, July 5, 1891, CCA/UPM. Abbott expressed his view in a note written on the bottom of Paschall's letter. The comments came amid a discussion of the Lenape Stone, which Abbott was convinced was a fraud (Abbott 1907:51).

82. Holmes 1891a, 1891b. Longer versions of these papers appeared as Holmes 1891c and 1893a.

83. Leverett to Chamberlin, December 23, 1889, TCC/UC.

84. Leverett 1890:585.

85. Leverett 1891b:361.

86. Holmes 1893d:149, 151.

87. Holmes 1893d:148–149, 151, 153.

88. Holmes 1893d:154–155.

89. Holmes 1893b:21; Holmes 1893f. Salisbury was a student of Chamberlin's, first at Beloit College, then at the University of Wisconsin, and a colleague of his in the USGS Glacial Division and finally at the University of Chicago.

90. Salisbury 1892a:102,104–106. Also Salisbury 1892b. Salisbury suspected this older glacial event had not been recognized because it did not produce distinctive moraines but had instead an attenuated margin (1892b:181). For biographical information on Salisbury, see Chamberlin 1931. Meltzer 2003:73–74.

91. Salisbury 1893c:106,113–114. On Wright's view of the gravel terrace, see Wright 1892a:244–245.

92. Holmes to Salisbury, July 2 and 12, 1892, RDS/UC. Holmes 1893b:20. Dinwiddie did not find "a trace of human handiwork," though as Holmes noted, "The first

paleoliths of Abbott were found on the exposed slope of the same gravels only a few yards away." Holmes's comments were written on the back of a large-format drawing of the stratigraphy of the trench examined by Dinwiddie. The drawing is entitled "Trenton Gravel, New Jersey, 1892," WHH/SIA.

93. McGee to Salisbury, July 26, 1892, WJM/LC; Holmes to Salisbury, July 26, 1892, RDS/UC.

94. Holmes 1893b:34.

95. Abbott Diary, July 23–24 and 29, 1892, CCA/PU; Meltzer 2003:74–75.

96. Culin to Holmes, August 1, 1892, WHH/SIA.

97. Abbott Diary, August 3, 1892, CCA/PU; Meltzer 2003:74–75.

98. Abbott to Putnam, August 28, 1892, and Putnam to Abbott, November 14, 1892, FWP/HU.

99. Putnam to Holmes, July 5, 1892, WHH/SIA.

100. Holmes 1892b, 1892c; Mercer 1892.

101. Fairchild 1893:10; the substance of George Frederick and Albert Wright's paper and the objections to them are in Upham 1892a:219, 221; see also Salisbury 1893e. For a later view, see Leverett 1934:7–8. Wright left the meeting early and was not around to hear McGee's version of the number and antiquity of the several glacial events delivered in Section E a few days later, but Upham filled him in on the details as soon as he got home (Upham to Wright, August 31, 1892, GFW/OCA).

102. Upham to Wright, August 31, 1892, and Claypole to Wright, December 4, 1892, GFW/OCA.

103. Holmes's letter of appointment to Chicago came through during the week of the AAAS meetings (see Goodspeed to Holmes, August 31, 1892, WHH/RR 7:19). Holmes's appointment in Chicago is discussed in Meltzer 2010. On the founding of Chicago's Department of Geology and the *Journal of Geology*, see Penrose 1929: 322–324.

104. Chamberlin to Salisbury, September 21, 1892, RDS/UC; Andrew Whitson was one of Salisbury's student assistants that summer; he later did graduate work at Chicago, and afterward took a faculty position at the University of Wisconsin. About this time, Salisbury detailed his views of the Trenton gravels in several places, including Salisbury 1892a:96 and Salisbury 1893e:106–114.

105. Chamberlin to Branner, January 12, 1893, JCB/SU. The papers that appeared in the first two *Journal of Geology* issues included Holmes (1893b, 1893d), Leverett (1893a), and Salisbury (1893b, 1893c).

106. Morison 1971:245–247.

107. Wright 1892a:242.

108. Wright 1892a:261–262, 299–301. The 25,000-year figure he derived from Prestwich; Wright himself dated the end of the ice age to between 7,000 and 10,000 years before the present, using as a basis the retreat of Little Falls and Niagara Falls (Wright 1892a:339, 346, 363–364). A table in Wright (1892a:324–325) that apparently was prepared by Upham divides the Glacial Period into two glacial events and an interglacial, but these are not to be construed as separate glacial epochs.

109. Baldwin to Wright, September 16, 1892, GFW/OCA. Upham voiced the same complaint but was equally gentle about it; the ingenuous Upham even suggested that for his next book Wright should publish on the substance of his recent Lowell Lectures, not realizing that he just had (Upham to Wright October 29, 1892, GFW/OCA).

110. Wright 1892a:212–218.

111. Wright 1892a:106–107, 109–110, 117–118.
112. Wright 1892a:117.
113. Wright 1892a:110.
114. Chamberlin to Pilling, June 6, 1890 and Chamberlin to Gilbert June 30,1892, TCC/UC; Noble to Wright, July 8, 1892, GFW/OCA. During his time in the employ of the USGS, Wright was paid $5.00 a day, plus expenses; the sum of Wright's wages and expenses from his time with the USGS came to $1503.59 (McChesney to Chamberlin, December 22, 1892, TCC/UC). Thirty years later Wright devoted only two sentences in his autobiography to his four years with the USGS (Wright 1916:155).
115. Abbott Diary, September 17, 1892, CCA/PU; Wright 1892a:viii.
116. Brinton 1892a; Peet 1892; Winchell 1892; others are cited in subsequent notes.
117. Upham to Wright, September 23, 1892, GFW/OCA; McGee to Chamberlin, October 15, 1892, WJM/LC.
118. Salisbury 1892c. The opinion of Salisbury comes from his friend and namesake, Rollin Chamberlin (son of Thomas), who had known Salisbury all his life (Chamberlin 1931:133).
119. Salisbury 1892d, 1892e; Garrison to Salisbury, October 24, and November 16, 1892, RDS/UC; Garrison to Wright, February 27, 1893, GFW/OCA. How Salisbury was able to place a review in *Nature* is unclear. He was corresponding with James Geikie about this time and may have sent a review to Geikie, who forwarded it to *Nature*.
120. Wright to Haynes October 25, 1892, HWH/MHS; Upham to Wright, October 29, 1892, and Starr to Wright, November 9, 1892, GFW/OCA.
121. Winchell to Salisbury November 21, 1892, RDS/UC. See Wright 1893b for his response to Salisbury.
122. Brinton 1892a.
123. Brinton 1892b:260.
124. Brinton 1892b:261.
125. Abbott 1892a:271. Starr was sorry Abbott's response was "quite so personal." He could not blame him for feeling that way but thought he should not let his feelings appear "quite so plainly." Starr thought Wright handled his reply much better (Starr to Wright, November 21, 1892, GFW/OCA).
126. Wright 1892c:275–276. N. D. C Hodges, editor of *Science*, seeing the beginnings of a controversy unfolding, "broadcast" that particular number among some 300 European anthropologists (Hodges to Wright, November 3, 1892, GFW/OCA).
127. Wright 1892c:276.
128. Wright 1892c:276.
129. Putnam to Abbott, November 14, 1892, PMP/HU.
130. Haynes to Wright, November 16, 1892, GFW/OCA; Wright to Haynes, November 25, 1892, HWH/MHS. Similar concerns were expressed by Starr to Wright, November 21, 1892, GFW/OCA.
131. Abbott to Wright, March 27, 1893, GFW/OCA.
132. Putnam [via Mead] to Wright, November 3, 1892, and Volk to Wright, November 7 and 22, 1892, GFW/OCA; Wright to Haynes, November 25, 1892, HWH/MHS.
133. Chamberlin 1892. Although published in Chicago, the *Dial* was by no means a local magazine: it was distributed nationwide, including in Oberlin, OH.
134. Chamberlin 1892:303.

135. Chamberlin 1892:303–304; Holmes 1892d; Wright 1892d.

136. Chamberlin 1892:305.

137. Chamberlin to D. Appleton and Co., October 7, 1892, GFW/OCA. Chamberlin's inquiry to the publisher was pointless because Wright would not conceive of removing the title, even if he had had the lead time to do so, as he was still officially on the USGS books. See also Morison 1971:269.

138. Starr to Wright, November 1892 (likely after November 23, 1892), and Baldwin to Wright, November 1892, GFW/OCA.

139. Dana to Wright, November 7, 1892; Wright to Dana, November 26, 1892; Dana to Wright November 28, 1892; all in GFW/OCA.

140. Wright 1892d.

141. Chamberlin 1892, 1893a; Wright 1892d, 1893a.

142. Leverett to Wright, January 23, 1893, GFW/OCA.

143. Claypole 1893b:779.

144. McGee to Chamberlin, November 11, 1892, WJM/LC.

145. Claypole 1893b:780.

146. McGee 1892.

147. Wright 1892d.

148. Wright 1892e:361.

149. McGee to Hodges, January 13 and 14, 1893, WJM/LC. Also, McGee to Salisbury, November 30, 1892, WJM/LC. McGee's brief reply to Wright was sent to *Science*, but the editor declined to publish it.

150. Doughty 1892.

151. McGee 1893a:95.

152. McGee 1893a:95.

153. The review was rejected first by the *Nation*, then *Goldthwait's Geographical Magazine*, then Winchell at the *American Geologist*. At the *Nation* and the *American Geologist*, McGee's review arrived too late: both journals already had Salisbury's reviews in hand. Winchell forwarded McGee's review to the *Literary Northwest*, where it was published (McGee 1893b). See McGee to the *Nation*, December 13, 1892; McGee to J. Redway, December 20, 1892; McGee to and from N. Winchell, December 16 and 22, 1892; all in WJM/LC.

154. McGee to Chamberlin, March 28, 1893, WJM/LC.

155. Dall to Wright, March 3, 1893, GFW/OCA; see also Haynes to Wright, March 6, 1893 and Wilson to Wright, February 6 and February 8, 1893, GFW/OCA.

156. Holmes to Branner, February 20, 1893, JCB/SU.

157. The full exchange is in Holmes to Wright, March 7, 1893, GFW/OCA; Wright to Baldwin, March 11, 1893, Baldwin to Wright, April 3, 1893, Wright to Holmes, April 5, 1893, Holmes to Wright, undated draft, but around April 7, 1893, Baldwin to Wright, April 10, 1893, Wright to Holmes, April 12, 1893; all in WHH/SIA; Holmes to Wright, May 3, 1893, GFW/OCA. Holmes's comment about not getting the original specimen is in Holmes 1893d:161.

158. For example, Starr to Wright, April 7, 1893; Dana to Wright, January 29, 1893; Winchell to Wright, January 9, 1893; Abbott to Wright, February 7, 1893; all in GFW/OCA. Similar views were voiced by Baldwin, Claypole, E. D. Cope, Cresson, Haynes, C. L. Herrick, W. K. Moorehead, S. Peet, R. S. Tarr, J. A. Udden, Upham, E. Williams, and Youmans, among others.

159. Wright to Haynes, October 25, 1892, HWH/MHS. See, for example, Dall to Wright, March 3, 1893, GFW/OCA. Winchell to Salisbury, November 25, 1892,

RDS/UC. Winchell told Salisbury it was hard to avoid the impression that his antagonism toward Wright over their differences on "New Jersey problems, and the 'unity of the glacial period'" colored his review.

160. Moorehead to Wright, January 28, 1893; Dana to Wright, April 9, 1893; Wilson to Wright, February 6 and February 8, 1893; Upham to Wright, January 11, 1893; all in GFW/OCA. Wilson could scarcely defend himself, let alone Wright. Wilson had for the last two years avoided directly confronting Holmes over the Washington paleoliths, and spoke out publicly in response to Holmes's Piney Branch work only in December of 1894 and after Holmes had left for Chicago (Wilson 1896). See Wilson to Holmes, December 5, 1894, WHH/SIA. Also McGee to Holmes, December 6, 1894, BAE/NAA.

161. Youmans to Wright, January 11, 1893 and Baldwin to Wright, April 3, 1893, WHH/SIA. Also Wright to Haynes, October 25, 1892, HWH/MHS.

162. Dana to Wright, March 22, 1893, GFW/OCA. The paper Dana saw as triggering the firestorm was Wright 1892b.

163. Winchell to Wright, January 9, 1893, GFW/OCA; similar comments were voiced by Williams to Wright, December 12, 1892 and January 23, 1893; Claypole to Wright, January 8, 1893; Baldwin to Wright, April 3, 1893; all in WHH/SIA; and Herrick to Wright April 14, 1893; Tarr to Wright April 15, 1893; Udden to Wright, April 22, 1893; all in GFW/OCA.

164. Claypole to Wright, December 4, 1892 and Udden to Wright, April 22, 1893, GFW/OCA.

165. Salisbury 1893d:198–199.

166. The actions surrounding the 1892–1893 Sundry Civil Expenses bill and its effects on the Survey are in Manning 1967:204–214 and Rabbitt 1980:203–214. See also Morison 1971:306.

167. Powell 1893a. See the response by Miller 1893. Manning 1967:212; Rabbitt 1980: 213–215.

168. C. C. Baldwin 1893:22.

169. Baldwin to Wright, January 28, 1893, GFW/OCA. Baldwin prepared his comments over the next several weeks, then sent the manuscript to Wright for "full criticism and abridgement" (Baldwin to Wright, February 18, 1893, GFW/OCA). The review was published a month or so later. For the exchange between McGee and Baldwin, see McGee to Baldwin, April 15, 1893, WJM/LC; Baldwin to Wright, April 21, 1893, GFW/OCA; McGee to Baldwin, May 10, 1893, WJM/LC.

170. Haynes to Wright, April 30, 1893, GFW/OCA.

171. Bain 1916. An editorial in the *American Geologist* applauded the Fifty-Second Congress for cutting the USGS appropriation, which they saw as healthy pruning of an agency that had grown too large too quickly, had neither well-defined goals nor a clear warrant, and that had engaged in "scientific ostracism" of those who criticized it and its work. Clearly, the congressional action was good news for the local and independent institutions, who now had a more "unobstructed line of progress" (*American Geologist* 1892). That same fall, it had also editorialized against the Survey's topographic mapping (*American Geologist* 1893a, also Winchell to McGee December 19, 1892 and McGee to Winchell, December 22, 1892, WJM/LC). In regard to McGee's attack, the journal editorialized that it was "eminently unbecoming to scientific literature and derogatory of the dignity of science" (*American Geologist* 1893b).

172. Winchell to Wright, January 9, 1893, GFW/OCA.

173. Chamberlin to editors and proprietors of the *American Geologist*, January 16, 1893, RTH/SMU; Chamberlin to Branner, January 12 and 24, JCB/SU.
174. Burnham 1990.
175. Chamberlin to Branner, February 27, 1893 and Winchell to Branner, February 3, 1893, JCB/SU. On the *American Geologist* editors' view of their role, see Bain 1916:58. In 1916, the *American Geologist* was folded into *Economic Geology*.
176. Claypole 1893a; Hitchcock 1893; James 1893; Leverett 1893b; Putnam 1893; Shaler 1893; Upham 1893b; Winchell 1893a; Wright 1893; Wright 1893d; Winchell to Wright, January 9, 1893, GFW/OCA.
177. Morison 1971:287–288.
178. Chamberlin to Branner, February 27, 1893, JCB/SU.
179. Abbott Diary, January 13, 1893, CCA/PU; McGee to Claypole, January 7, 1893, WJM/LC; Holmes to Salisbury, February 14, 1893, RDS/UC; Winchell to Wright, January 9, 1893, GFW/OCA. For his reply to some of the papers in that special issue, see Salisbury 1893e.
180. Youmans 1892; Branner to Youmans, December 14, 1892, TCC/UC; Youmans to Branner, April 19, 1893, JCB/SU. Branner sent copies to Chamberlin and Holmes; Chamberlin thought it "indeed 'salty' and will perhaps cause a little smarting but it will be wholesome in checking the tendency to septicemia that seems to have set in that quarter" (see Chamberlin to Branner, December 22, 1892, JCB/SU; also Holmes to Branner, February 20, 1893, JCB/SU).
181. Youmans to Wright, January 11, 1893, GFW/OCA.
182. Claypole 1893b:767.
183. Claypole to Wright, December 4, 1892, GFW/OCA; Claypole 1893b:773–774.
184. Claypole 1893b:777.
185. Claypole 1893b:768, 770.
186. Claypole 1893b:779–780. See also McGee to Mining Review Publishing Company, September 11, 1894, BAE/NAA.
187. *American Geologist* 1893b:112; Claypole 1893b:767; Wright himself wrote of being "under pressure from my theological predilections to discredit the [Paleolithic] evidence" (Wright 1893f:267). The Inquisition allusion is used in Youmans 1893a and in Todd to Wright, December 13, 1892; Peet to Wright, December 19, 1892; William to Wright, December 29, 1892; and Upham to Wright, January 28 and February 2, 1893; all in GFW/OCA.
188. Haynes to Wright, April 30, 1893, GFW/OCA.
189. The chair was created in his honor. See Morison 1971:313–317.
190. Youmans 1893a:841.
191. Youmans 1893a:842.
192. Abbott to Wright, March 27, 1893; Baldwin to Wright, April 21, 1893; Dana to Wright, March 9, 1893; all in GFW/OCA; Winchell 1893b:343. Also, Haynes to Wright, April 30, 1893, GFW/OCA. In the fall of 1892 Winchell heard Powell was to be appointed president of the University of California, and he quietly campaigned for the Survey directorship; he did so again two years later when Powell resigned (Winchell to Branner, October 31, 1892 and May 12, 1894, JCB/SU).
193. Powell 1893b:325. Powell began preparing his response in late March (see Holmes to Salisbury, March 22, 1893, RDS/UC; McGee to Chamberlin, March 28, 1893, WJM/LC). By mid-April, Powell's response was in Claypole's hands (Baldwin to Wright, April 15 and 21, 1893, GFW/OCA).

194. Powell 1893b:318–319, 321; Meltzer 1983:21. In response to Powell's claim that some Paleolithic proponents had been knighted, Claypole wrote it would "delight the archaeological world to be favored with a list of the Sir Knights who have received the accolade as a reward of their great powers and magnificent achievements on the hard-fought field of American archaeology" (Claypole 1893c:699).

195. Powell 1893b:325–326.

196. Youmans 1893b:413. Also, Claypole 1893c; Winchell 1893c.

197. McGee to Powell, June 17, 1893, WJM/LC; Dana to Wright, March 6, 1895, GFW/OCA. Youmans to Wright, November 1, 1893, GFW/OCA, makes a similar comment. For the history of the changes, see Rabbitt 1980:215–216, 230.

198. Branner to White, December 15, 1902, WHH/RR, and Branner to Purdue, December 15, 1902, JCB/SU.

199. Darrah 1951:348; Rabbitt 1980:238. C. D. Walcott was appointed Powell's successor, to the dismay of many geologists, Winchell perhaps foremost among them. Wright helped circulate a petition of grievances against the Survey with the hopes of forestalling Walcott's confirmation (Winchell to Branner, May 21, 1894, JCB/SU; see Wright to Haynes, May 30, 1894, HWH/MHS).

200. Holmes 1893g:36.

201. Holmes 1892d:295.

202. Holmes 1892d:296.

203. Holmes 1892d:296.

204. Holmes 1892d:296–297.

205. Holmes 1892d:297.

206. Geologist Clarence Dutton, for example, thought Holmes's piece was one of the most level-headed he had seen on the matter (Dutton to Holmes, December 10, 1892, WHH/RR 6:51–52). Upham praised the piece, telling Wright it was "a very satisfactory article" (Upham to Wright, December 3, 1892, GFW/OCA). Abbott Diary, November 25, 1892, CCA/PU.

207. Abbott 1892b:344–345. As Hull (1988:362) said, "Presenting one's views in an irresponsible manner is one way to decrease their influence." Abbott certainly pushed that boundary with his poetry.

208. Meltzer 2003:77.

209. Abbott 1892c:2.

210. Holmes to Salisbury, December 31, 1892, RDS/UC.

211. Holmes 1893c.

212. Holmes 1893a, 1893b 1893d, 1893f.

213. Holmes 1893b:16; Holmes 1893f:240. See also Hinsley 1981:107.

214. Holmes considered Abbott's work so vile, ad hominem, and error ridden as to be unmentionable "save to be damned," and he thought his critique of Trenton the best work he had ever done. Holmes to Salisbury, December 8 and December 31, 1892, RDS/UC.

215. Holmes to Salisbury, March 22, 1893, RDS/UC. See the response by Haynes 1893d:291.

216. The quote is from Holmes 1893b:27; see also Holmes 1893b:24–25; Holmes 1893d:149, 161; Holmes 1893f:236–239. For illustrations, see Holmes 1893b:Figures 2–6; Holmes 1893d:Figures 1–2.

217. Holmes 1893b:37.

218. Holmes 1893b:21, 29, 34–35; Holmes 1893d:147–150, 160, 163; Holmes 1893f:240.

219. Holmes 1893b:21, 23; Holmes 1893d:149.
220. Holmes 1893b:34. The Quaker reference was ostensibly intended by Holmes to amuse Abbott, and Holmes was offended when an otherwise friendly correspondent took it as a sneer at Quakers generally (Wistar to Holmes, April 7, 1893; Holmes to Wistar, April 8, 1893,WHH/SIA). Meltzer 2003:80.
221. Holmes 1893b:24.
222. Abbott Diary, undated but circa January 1893.
223. Wright 1893c.
224. Haynes 1893a:66.
225. Abbott 1893a; Abbott 1893b.
226. For the earlier discussion of separating finished (Paleolithic) and unfinished (Neolithic) implements, see Abbott 1892c:4–5, 13.
227. Abbott 1893b; Abbott to Wright, February 3, 1893, GFW/OCA.
228. Haynes 1881:132.
229. Holmes 1893e:135.
230. Haynes 1893c:209.
231. Haynes to Wright, April 30, 1893, GFW/OCA. Haynes had written the piece for the *Antiquarian* at Peet's request, though he "hate[d] to publish in such a trashy medium" as that journal (Haynes to Wright, November 16, 1892, GFW/OCA).
232. Haynes 1893b:38–42.
233. Haynes 1893c:208–209. The Holmes quote comes from Holmes 1893b:30.
234. Upham to Wright, May 4, 1893, GFW/OCA.
235. Wright to Holmes, April 12, 1893, WHH/SIA; Holmes to Wright, May 3, 1893, GFW/OCA.
236. Madisonville was a weak case, and Wright knew it: two of the three paleoliths had been found in talus debris. He passed over it quickly in his May article in *Popular Science Monthly* (Wright 1893e). Leverett visited the site and was skeptical as well (Leverett to Wright, May 4, 1893, GFW/OCA).
237. Wright 1893e; Mills to Wright, February 3, 1893, GFW/OCA.
238. Holmes 1893d:158.
239. Wright 1893f:267–268.
240. Holmes 1893d:160–161.
241. Haynes to Wright, June 6, 1893, GFW/OCA.
242. Haynes 1893e:318–319.
243. The visit was reported in Abbott to Culin, August 29, 1891, CCA/UPM. For a more detailed discussion of the American evidence with a different emphasis, see Topinard 1893.
244. Boule 1893:37–38.
245. Mason 1893:461.
246. Holmes 1893b:30; Mercer 1893; Brinton to Holmes, January 11, 1893, WHH/SIA.
247. Upham to Wright, August 31, 1892, GFW/OCA; Leverett to Wright, January 23, 1893, GFW/OCA. Upham appreciated Wright's defense of a single glacial epoch, but as a believer in Croll's astronomical theory he accepted the possibility of "two, three, or several glacial epochs" and found the evidence amassed by Chamberlin, Salisbury, McGee, and Leverett of well-marked and distinct epochs to be "very convincing" (from Upham to Wright, September 23, 1892 and September 26, 1891, respectively, both in GFW/OCA).
248. Dana to Wright, September 11, 1891, GFW/OCA.

249. Upham to Wright, October 11, 1892, GFW/OCA. See, for example, Upham 1892b.
250. Wright 1892b:357, 367, 371.
251. Wright 1892b:353, 356–357.
252. Wright 1892b:360–361.
253. Wright 1892b:363–364, 366–367.
254. Wright 1892b:369–370.
255. Wright 1892b:373.
256. Leverett to Chamberlin, December 23, 1889, TCC/UC. Leverett conveyed his doubts to Wright that winter but followed this moderate "scolding" with praise for Wright having done so much pioneering work at his own expense (Leverett to Wright, November 27 and December 4, 1889, GFW/OCA).
257. Leverett 1891a.
258. Leverett to Wright, November 27, 1892, GFW/OCA.
259. Leverett to Wright, November 27, 1892, GFW/OCA.
260. Leverett to Wright, January 23, 1893, GFW/OCA.
261. Dana to Wright, November 7, 1892, GFW/OCA. In telling Wright this, Dana urged him not to speak of it, as it might appear as "an improper divulging of Journal secrets."
262. Dana to Wright, February 27, 1893, GFW/OCA. Chamberlin 1893b:174.
263. Chamberlin 1893b:184–185.
264. Chamberlin 1893b:180.
265. Leverett 1902:27–28, 222, 224, 228; also Salisbury 1902.
266. Chamberlin 1893b:196–197; also Leverett 1902:240–242, 450.
267. Chamberlin 1893b:186–187,190, 198.
268. Chamberlin 1893b:171, 200; Geikie to Salisbury, November 10, 1892, RDS/UC; Chamberlin to Gilbert, March 21, 1890, TCC/UC. In this paper Chamberlin constructed a classification of views of glacial history: he put Wright among those holding the most "primitive" (Chamberlin 1894a:172–174).
269. Dana 1893.
270. Chamberlin 1893c:849.
271. Leverett 1893c:215.
272. Chamberlin to Branner, January 24, 1893, JCB/SU; Salisbury 1893b. For Salisbury's comments on Wright and glacial issues, though not necessarily his "Unity" paper, see Salisbury 1893a, 1893c,1893e.
273. Haynes to Wright, June 6, 1893, GFW/OCA.
274. McGee to Chamberlin, June 1, 1893, WJM/LC.
275. Leverett to Wright, May 4 and May 22, 1893, GFW/OCA.
276. On Putnam's whereabouts, see, for example, Dana to Wright, November 28, 1892, GFW/OCA. Wright to Putnam, January 13, 1893 and Mead to Wright, January 23, 1893, GFW/OCA; Putnam 1893.
277. Haynes to Wright, June 6, 1893, GFW/OCA.
278. Moorehead to Wright, May 9, 1893, GFW/OCA; Moorehead to McGee, May 9, 1893, WJM/LC. Moorehead had extended other invitations, including one to Abbott, who was urged repeatedly to come and defend Trenton; Abbott refused, saying he did not want to be there if Putnam was there and he was too busy anyway (Moorehead to Abbott, May 9, 17, and June 1, 1893, CCA/UPM). Haynes was invited and intended to go, but bad allergies prevented him from making the meeting (Moorehead to Haynes, May 25, 1893, HWH/MHS; Haynes to Wright, June 6, 1893, GFW/OCA).

279. *American Geologist* 1893c. Wright conceded that the Pattenburg and High Bridge deposits were till, as Chamberlin showed in his paper, but then Wright incorporated them into his fringe. Chamberlin's reaction is in Chamberlin to Dana, February 1, 1894, TCC/UC.
280. Moorehead 1893:171; *American Geologist* 1893c:173; Peet 1893:311.
281. Moorehead 1893:171. Other accounts of the meeting are in *American Geologist* 1893c and 1893d; McGee 1893c. Also Wright to Haynes, September 28, 1893, HWH/MHS; Wright to Winchell, April 15, 1907, NW/MHS. The primary participants included Brinton, Chamberlin, Claypole, Hovey, McGee, Mercer, Putnam, Upham, Van Hise, Volk, and Wright.
282. Wright to Haynes, September 28, 1893, HWH/MHS.
283. *American Geologist* 1893d:188.
284. *American Geologist* 1893d; McGee 1893c. McGee gave a reprint of his article to his wife, with the inscription to "Anita Newcomb McGee, M.D., a souvenir of the battle from a survivor. WJM." The reprint is in the author's collection.
285. *American Geologist* 1893d:188.
286. *American Geologist* 1893d:187. The editorial was anonymous, but Upham takes credit for it in Upham to Wright, September 23, 1892, GFW/OCA. The letter of the concerned friend was Stevenson to McGee, October 5, 1893, WJM/LC.
287. Leverett to Wright December 14 and December 17, 1893, GFW/OCA.
288. Chamberlin to Dana, February 1, 1894, TCC/UC.
289. Wright 1894a, 1894b.
290. Abbott to Wright, December 27, 1893; Upham to Wright, March 8, 1894. See also Upham to Wright, December 16, 1893, where he reports that Branner is a convert to the unity view. All in GFW/OCA.
291. Chamberlin to Salisbury, September 9, 1893, RDS/UC.
292. Chamberlin to Upham, March 12, 1895, TCC/UC.
293. Wright 1895a, 1895b, 1895c, 1895d, 1896a, 1896b, 1896c, 1896d. Youmans continued to welcome papers by Wright and gladly published his paper [Wright 1894a] on the Ice Dam, which he felt administered to Chamberlin the "unmerciful pummeling [which] he richly deserves" (Youmans to Wright, November 8, 1893, GFW/OCA.)
294. Wright expressed concerns about his reputation to Dana, who assured him he need not worry. After all, he had the endorsement of the *American Journal of Science* by virtue of having published in it. This was not the only occasion in which Dana assumed the world of science orbited about his journal (Dana to Wright, April 9, 1893, GFW/OCA). For Wright's Greenland trip, see Morison 1971:321–329 and the review in Chamberlin 1896c.
295. Geikie 1894:x. Chamberlin contributed two chapters to Geikie's book—see Chamberlin 1894a. It was small recompense for his worry that American glacial geology had slipped in the eyes of the world (§5.15).
296. Chamberlin to Davis, March 6, 1896, TCC/UC. Just a week earlier, Chamberlin had sent Leverett down the Ohio River to check on some terraces related to a forthcoming paper by Wright (1896b); Chamberlin to Leverett, February 27, 1896, TCC/UC. See also Chamberlin 1896a, 1896b.
297. Chamberlin 1934:337. Thomas Chamberlin produced two lengthy articles on his Greenland investigations (Chamberlin 1894b, 1894c).
298. On these investigations, see Salisbury 1894, 1895, and 1896.
299. Winchell to Wright, May 19, 1894, GFW/OCA; Powell 1895.

300. Hinsley (1981:238–246) provides an excellent discussion of McGee's misadventures in Seriland.
301. As detailed in Meltzer 2010. See Chamberlin to Holmes, December 5, 1893, TCC/UC; Chamberlin to Holmes, January 6 and 23, 1894, WHH/SIA; Holmes to Chamberlin, February 5, 1894, WHH/RR 7:29; and Boas to Putnam, February 18, 1894; Boas to and from Skiff, February 19, 1894; Holmes to Boas, February 21, 1894; Boas to Putnam, February 18, 1894; Putnam to Boas, March 7 and May 14, 1894; all in FB/APS.
302. Powell to Holmes, June 8, 1894, in WHH/RR 5:138. Holmes's Chicago sojourn was brief and unhappy. Three years later he was back in Washington (Meltzer 2010).
303. Abbott Diary, February 24, 1903, CCA/PU. Also Hinsley 1985.
304. Brinton to Holmes, March 9, 1893 and Culin to Holmes, March 29, 1893, WH/SIA; Madeira 1964:20.
305. For a fuller discussion, see Meltzer 2003:83–84.
306. Abbott Diary, October 13 and November 1, 1893, CCA/PU.
307. Abbott Diary, December 11, 1893, CCA/PU.
308. Abbott Diary, January 9 and September 10, 1895, CCA/PU. Abbott to Robins, July 14, July 16, and July 25, 1896, "Sunday A.M." 1896, April 14, April 21, May 22, May 29, and June 28, 1897, CCA/UPM.
309. Mercer to Wright, July 8, 1894, GFW/OCA; Abbott to Mercer, August 20, 1896, HCM/BCHS.
310. McGee to Holmes, September 21, 1896, WHH/SIA.
311. Mercer to Wright, May 22, 1897, GFW/OCA; Putnam to Abbott, June 26, 1897, PMP/HU. The details of the visit were worked out in Volk to Wright, June 2, 1897; Putnam to Wright, June 10, 1897; Volk to Wright and Mercer to Wright, June 21, 1897; all in GFW/OCA. Putnam himself chose to stay away lest there seem any appearance of influence on his part (Putnam to Mercer, June 30, 1897, HCM/BCHS).
312. The excavation trenches and two of the in situ specimens are show in Volk 1911:Plates 57–59. See also Volk 1911:90–91.
313. Abbott to Robins, June 28, 1897, CCA/ANS. Abbott to Putnam, June 30, 1897, PMP/HU. Abbott Diary, June 25–28, 1897, CCA/PU.
314. McGee, as chair of Section H, took responsibility for organizing the Section sessions.
315. Putnam to Mercer, June 30 and July 1, 1897, HCM/BCHS.
316. Putnam to Wright, June 30, 1897, GFW/OCA.
317. Kummel to Mercer, June 29, 1897, HCM/BCHS; Kummel to Wright, July 4, 1897, GFW/OCA.
318. Kummel to Mercer, July 8, 1897, HCM/BCHS; Knapp to Mercer, July 13, 1897, HCM/BCHS. See also Smock to Salisbury, July 19, 1897, RDS/UC.
319. Chamberlin to Kummel, July 14, 1897, TCC/UC; Kummel to Salisbury, July 17, 1897, RDS/UC.
320. Mercer to Wright, July 16, 1897, GFW/OCA.
321. The Knapp quotation is from Knapp to Mercer, July 29, 1897, HCM/BCHS. See also Kummel to Mercer, July 18, 22, and 27, 1897, HCM/BCHS. Knapp to Mercer, July 24 and 27, 1897, HCM/BCHS. Chamberlin to Kummel, July 26, 1897, TCC/UC. A month later Kummel, having heard nothing from Salisbury in reply to one of his letters, was anxious to insure he had not fallen from his good graces (Kummel to Salisbury, August 30, 1897, RDS/UC).

322. Mercer to Cattell, July 28, 1897 and undated but late July 1897, and McGee to Cattell, July 26, 1897, JMC/ LC.

323. Mercer to Salisbury, August 4, 1897; Salisbury to Mercer, August 2, 6, 1897; Holmes to Mercer, August 4, 1897; Putnam to Mercer, August 4, 1897; all in HCM/BCHS. See also Smock to Mercer, August 5 and 7, 1897, HCM/BCHS, which attempts to explain to Mercer why Knapp and Kummel went out to the site and why neither had a right to usurp Salisbury's claim.

324. Chamberlin 1931:132; Fisher 1963:4.

325. Holmes to Mercer, July 20, 1897 and Wilson to Mercer, July 28, HCM/BCHS. See also Volk 1911:91.

326. Composite stratigraphic description from Kummel 1898, Salisbury 1897, Wright 1897.

327. Kummel and Knapp were both absent but sent papers read by Salisbury (A. McGee 1897:510). Putnam 1898:347–348; Knapp 1898:350; Kummel 1898:349–350; Salisbury 1898:353–355.

328. Wright 1898:361–363; Mercer 1898:376–377; Hollick 1898:380.

329. Holmes 1898:366, 370.

330. Wilson 1898:382. Wilson was not in attendance but had his paper read for him.

331. Claypole 1897b; also Claypole 1898.

332. Putnam had a stenographer transcribe the discussion comments that afternoon and they follow the publications of the papers in the *AAAS Proceedings* volume for that year. The discussion was also summarized in A. McGee 1897.

333. Haynes 1893a:67; Haynes 1893d.

334. Evans may not have been the most influential prehistorian of the time (de Mortillet is perhaps the better candidate), but Evans was routinely cited in the American Paleolithic controversy, likely for several reasons. He played to a more popular audience and his work had international visibility—his *Ancient Stone Implements* was published in England but also in the United States, and there was a French edition as well. His work was copiously illustrated and, given that the American Paleolithic was at its core a visual argument, that mattered. Evans was also known personally to and in correspondence with some of the key players, such as Haynes.

335. McGee to Holmes, September 2, 1897, BAE/NAA.

336. McGee to Holmes, September 2, 1897, BAE/NAA. For other reports of the meeting, see Chamberlain 1897:582; Claypole 1897a, 1897b; Herdman 1897:487.

337. Abbott to Wright, April 5, 1899, GFW/OCA; Putnam May 13, 1899, PMP/HU.

338. McGee to Holmes, September 2, 1897, BAE/NAA; McGee to Cattell, September 8, 1897, WJM/NAA. Also A. McGee 1897; McGee 1897:334–336.

339. Abbott to Robins, October 25, 1897, CCA/ANS; Putnam to Wright, September 7, 1897, GFW/OCA; Wright to Mercer, September 9, 1897 and Hollick to Mercer, September 15, 1897, HCM/BCHS; Volk 1911:94–95. The *Times* column appeared on September 27, 1897 and was forwarded by Smock to Salisbury to see if the latter wanted to write a response.

340. Fowke 1902:12, 23; McGee and Thomas 1905:50; Thomas 1898:7. McGee and Thomas did not reject the Paleolithic outright or attack it as severely as McGee once had. This may reflect the fact that by this time (1905) McGee had lost his position at the BAE during a bitter struggle to succeed Powell. This fact pitted him against Holmes (Hinsley 1981). His softening on the Paleolithic issue may have more to do with his estrangement from Holmes than any change of heart about the Paleolithic case.

341. Holmes 1897b:824–825; Upham 1901; Winchell 1913; Wright 1899; Wright 1912:218–229.
342. Volk 1911:111–118.

Chapter Six

1. The specimen had been found several days earlier by a workman in a sand-and-gravel pit on the outskirts of Trenton (Volk 1911:111–112), but Volk returned to the scene to photograph the workman in front of the find spot (Volk 1911: Plates 88–87). For a summary of the archaeology of the area, see Cross 1956:1–10; Hunter et al. 2009.
2. Abbott Diary, March, 17, 1899, CCA/PU.
3. Abbott to Putnam, April 22 and May 13, 1899, FWP/HU. See also Abbott to Wright, April 5 and May 24, 1899, GFW/OCA.
4. Lamb to Putnam, April 22, 1899, NCN/AMNH; Putnam to Abbott, April 25, 1899, FWP/HU.
5. Abbott to Putnam, April 26, 1899, FWP/HU, emphasis in the original. Abbott sputtered at Putnam and questioned his taxonomic skills, but he also fired off a letter to Wright the next day warning him the bone was not human, but likely musk ox. He was probably sufficiently embarrassed for having previously proclaimed to Wright it was human that he did not want to bring up the possibility it was merely a cow (Abbott to Wright, April 27, 1899, GFW/OCA). On the "lies" of Holmes and others and the importance of publishing Volk's work, Abbott to Putnam, April 22, May 12, 13, and 22 and June 14, 1899, FWP/HU.
6. That assessment is from Hrdlička (1907:12), who thought "knowledge of the osseous structures of early man in other parts of the world is still meager." Recent commentators agree (for example, Ward 2003:78).
7. Volk to Putnam, February 17, 1912, FWP/HU. Stewart 1949; cf. Spencer 1979.
8. Abbott to Putnam, April 26, 1899 and Volk to Putnam, April 26, 1899, FWP/HU. This was an almost comically recurring theme. In December of 1900 Abbott wrote Putnam to say "I will trouble you no more. You are very amusing, but too irritating to be played with." That self-imposed recess did not last a year (Abbott to Putnam, December 27, 1900 and Putnam to Abbott, October 12, 1901, FWP/HU).
9. Abbott to Putnam, June 14, 1899, FWP/HU.
10. Putnam to True, May 2, 1899, NCN/AMNH. True sent Putnam a musk ox scapula from the collections at the USNM, and Putnam had compared it with the Trenton scapula in the company of Franz Boas and Ales Hrdlička, both of whom concurred that this specimen was indeed from that animal. This may have been one of the earliest occasions on which Boas and Hrdlička met.
11. Abbott to Wright, May 24 and June 5, 1899, GFW/OCA. Volk to Putnam, June 20, 1899, PMP/HU. Volk would not mention those remains in his final report on his work (Volk 1911).
12. Russell 1899:153.
13. Putnam 1888c:35, 1899; see also Hrdlička 1902:54. Abbott complained of Russell's misunderstandings in Abbott to Wright, April 5, 1899, GFW/OCA; Abbott to Putnam May 12, 1899, PMP/HU. Also Abbott 1908:12–13. For a general discussion, see Meltzer 1983:26; Spencer 1979:221, 227–228.
14. The discovery of the femur is carefully detailed in Volk 1911:112–117. His photographs of the find spot and specimens are in Volk 1911: Plates 89–91 and 103–107.

The railroad cut is the same one illustrated in Wright 1889: Figure 66 and again in Wright 1892: Figure 127. The site cannot be precisely relocated (see Hunter et al. 2009:6–4).

15. Volk to Putnam, December 1, 1899, PMP/HU.

16. Putnam to Volk, December 4, 1899, PMP/HU.

17. Abbott Diary, December 4–5, 1899, CCA/PU. Abbott was nonetheless angry that Holmes had visited Volk and seen "the musk ox remains, whereas he was not supposed to speak of them. If Holmes came around again to Trenton and Abbott spotted him, he snarled, Holmes would "be carried off a corpse" (Abbott to Putnam, undated but circa December 1899, PMP/HU).

18. Volk 1911:118.

19. Putnam to Volk, December 9, 1899, PMP/HU.

20. Spencer 1979:35–44; Spencer 1997a. Also Brace 2005:224.

21. Schultz 1946:305–306; Spencer 1979:47–53, 76, 80, 86–88.

22. Schultz 1946:306; Spencer 1979: 93, 102–105, 110ff.

23. See Blanckaert 1997, 2009; Brace 2005; Spencer 1979:116–117; Spencer 1997b:642.

24. Gould 1981:73, 86.

25. Spencer 1979:113, 117.

26. Blakey 1987:10–12, 15–16.

27. Brace 2005:225. Blakey (1987:13) observes that Hrdlička was not so far from the eugenists that he could not put two of them—Davenport and Kellogg—on the editorial board of his newly founded *American Journal of Physical Anthropology*. See also Stocking 1968:287–291. However, when the second edition of Grant's *Passing of the Great Race* appeared, Hrdlička told Boas he considered it "pretentious" and "badly biased" and feared that it would "influence men in important positions . . . and may even be used as a leverage for the establishment of a separate committee on 'Race' in connection with the Council" (National Research Council). He asked Boas to review it for the *American Journal of Physical Anthropology*, perhaps as a passive-aggressive stab at one of his own board members (Hrdlička to Boas, May 2, 6, and 29, 1918; Boas to Hrdlička May 4, 1918, AH/NAA).

28. Hrdlička 1928:426.

29. Hrdlička 1928:429.

30. Hrdlička 1927. See also Goodrum 2009; Reybrouck 2002.

31. Hrdlička 1907; see Stewart 1949.

32. Meltzer 1983:28–29.

33. Spencer 1979:125–126.

34. Spencer (1979:161–166) reports that Hrdlička met Putnam in November of 1896 at the autopsy of Qisuk, one of the six Polar Eskimos Robert Peary brought back from Greenland. That could not have happened because Peary did not return with the group until September of 1897. Hrdlička did participate in Qisuk's autopsy. For the sad story of this Eskimo group's sojourn, see Thomas 2000:77–83.

35. Spencer 1979:126, 169–170, 180. Franz Boas was sympathetic to Hrdlička's precarious employment situation and quietly made inquiries of WJ McGee at the BAE as to whether a position might be found for Hrdlička at the Army Medical Museum in Washington. Nothing came of it (Spencer 1979:171).

36. Meltzer and Dunnell 1992:xxii; Spencer 1979:218–219.

37. Swanton 1944:39; see also Schultz 1946:312; Spencer 1979:49, 191.

38. For example, Hooton 1937:102.

39. Schultz 1946:315; Spencer 1979:73.

40. Hrdlička 1911:247; on the modifications of the femur, Putnam to Volk, December 24, 1899, PMP/HU.

41. Wright 1911:243. Wright apparently visited in December of 1899 or January 1900, following Putnam's talk on the specimen at the New Haven meeting (Wright to Winchell, April 12, 1907, NW/MHS).

42. Putnam to Volk, December 13, 1899, PMP/HU.

43. Wilson to Putnam, January 3, 1900, PMP/HU.

44. Putnam to Volk, December 24, 1899, PMP/HU.

45. Volk himself, naturally, was even more "anxious" to know where his employment prospects stood (Volk to Putnam, February 28, 1900, PMP/HU). By 1902, they had begun to evaporate altogether, and Putnam advised Volk to look for employment elsewhere, and he did (Putnam to Volk June 28, 1904, FWP/HU; Volk to Holmes, April 6 and June 29, 1905, BAE/NAA).

46. Putnam to Mason, March 7, 1900, USNM/SIA, emphasis mine.

47. Putnam to Mason, March 7, 1900, USNM/SIA. Spencer 1979.

48. Hrdlička to Abbott, December 12 and 17, 1900, CCA/PU. Abbott's replies—sent on December 15 and 18, 1900—have not been located, but extracts appear in Hrdlička 1902. Putnam to Abbott, October 12, 1901 and Volk to Putnam February 13, 1901, FWP/HU.

49. The four cranial fragments found by Abbott were a molar, portions of a temporal and frontal bone, and a segment of a mandible (Hrdlička 1902). They were likely not from the same individual.

50. Hrdlička 1902:37–40.

51. Hrdlička 1902:25, Tables 1 and 2.

52. Boas 1899:449. Boas subsequently demonstrated in a massive multiyear study of immigrants who came through Ellis Island that the index varied among adults of a single group and even within the life of an individual. These results called into question the assumption of an absolute stability of human types (for example, Boas 1912a, 1912b). It was a conclusion decidedly antagonistic to a fundamental dogma of physical anthropology, including Hrdlička's physical anthropology (Stocking 1968:287–292).

53. Hrdlička 1902:42.

54. Hrdlička 1902:51–56; Russell 1899:152–153.

55. Hrdlička 1902:53, 56–57.

56. Volk to Wright, July 15, 1901, GFW/OCA. Also Winchell to Putnam, December 3, 1902, FWP/HU.

57. Hrdlička 1902:57.

58. On the geological context of the skulls, Hrdlička 1902:30–31. Abbott was convinced both of the crania were representatives of his "argillite man" (Abbott to Putnam, March 26, 1902; FWP/HU).

59. Abbott to Putnam, December 27, 1900, FWP/HU; Putnam to Mason, March 7, 1900, USNM/SIA.

60. Holmes 1907:61; Salisbury 1902:687. So far as Abbott was concerned, Salisbury's comments on ancient human remains were a "tissue of gross inaccuracies" (Abbott to Wright, January 12, 1912, GFW/OCA; see also Abbott to Putnam, April 26, 1902, FWP/HU; Winchell to Abbott, April 18, 1907, CCA/PU). Wright believed, erroneously, that Holmes had capitulated on the matter (in Wright to Winchell, April 18, 1907, NW/MHS).

61. Putnam to Holmes, November 19, 1907 (WHH/RR 9:205).

62. McGee to Fairchild, April 27, 1903, BAE/NAA. Abbott met up with McGee at the Congress of the Americanists and recorded in his diary that after McGee heard Putnam's discussion of the Trenton skeletal finds, "he was very changed in his attitude" toward the Trenton evidence (Abbott Diary, October 23, 1902, CCA/PU). Abbott's diary is suspiciously self-serving, but in a newspaper interview McGee admitted as much (unidentified clipping, WHH/RR 9:40). Wright reported the same (in Wright to Winchell, April 18, 1907, NW/MHS).

63. A.N. McGee to WJ McGee, November 13, 1903, ANM/LC. On the fight waged by McGee to be appointed Powell's successor, see Hinsley 1981:248ff; Spencer 1979:245ff; also, Meltzer and Dunnell 1992:xx–xxi.

64. Holmes 1899:636–637, 642–643; on the turn-of-the-century arguments over Calaveras, see Becker 1891; Blake 1899; Hanks 1901; Putnam 1906; Russell 1899; Sinclair 1908. The primary reference on the discovery is Whitney 1879.

65. Holmes 1897b:825.

66. Meltzer and Dunnell 1992:xxii; see also Holmes to Langley, November 5, 1902, WHH/SIA; WHH/RR 8:111.

67. Holmes to Hrdlička, April 8, 1903 and Langley to Hrdlička, April 17, 1903, AH/NAA. Boas's view is in Boas to Langley, December 7, 1903, FB/APS. See also Meltzer 1983:27–29, 1994:13; Spencer 1979:218–219, 249–251.

68. Stewart 1949.

69. Putnam to Hrdlička, February 5, 1907, FWP/HU.

70. Hrdlička was not anxious to leave the AMNH, but his position was precarious. When his USNM appointment came through he quietly slipped away, telling Putnam he had "not the heart to tell at once all the men in the Museum" (Hrdlička to Putnam, April 23 and May 2, 1903, FWP/HU).

71. On this occasion, as in previous and subsequent ones, proponents were at pains to testify to the reliability of the discoverers and to stress there was no evidence they were anything but honest brokers (for example, Williston 1905b:343; Winchell 1902:189). Critics accepted the testimony of the Concannon brothers as to where and how the bones had been found (for example, Chamberlin 1902:763).

72. The tunnel was ten feet wide, seven feet high with an arched top, and extended seventy-two feet into the bluff and some twenty feet below the surface (Upham 1902:a). Long's articles were in the *Kansas City Star* in March and July 1902 (Chamberlin to Leverett, July 30, 1902, TCC/UC; see also Upham 1902b:137).

73. Williston 1902a:196. On the role of human hunting in the extinction of the horse and mammoth, see Upham 1902b:150. For biographical details on Williston, see Shor 1971.

74. Williston 1902a:196.

75. Chamberlin to Williston, July 25, 1902, TCC/UC.

76. Chamberlin to Williston, July 30, 1902; also Chamberlin to Leverett, July 30, 1902, both in TCC/UC.

77. Upham sent invitations to Chamberlin, Leverett, and Holmes, but none could attend. In declining, Chamberlin did not miss the opportunity to let Upham know that the remains were likely postglacial (Chamberlin to Upham, August 6, 1902, TCC/UC).

78. Upham 1902a:356; Upham 1902b:139. Also Upham to Wright, August 27, 1902, GFW/OCA.

79. Upham 1902a:356; Chamberlin 1896d. The estimate obviously varied according to how much time one allotted to the post-Wisconsin period (one time unit).

80. Upham 1902b:148.
81. Upham 1902c:567; Upham 1902d:149; Upham 1902f:274.
82. Spencer 1988.
83. Upham 1902b:140–144; Winchell 1902.
84. Haworth to Wright, August 19, 1902, GFW/OCA; Chamberlin to Haworth, August 25, 1902, TCC/UC. Long also suggested a visit to Wright: he did not doubt Upham and Winchell, but Wright made a "special study" of loess and was a better judge (Long to Wright, August 23 and September 4, 1902, GFW/OCA).
85. Upham to Holmes, August 23, 1902 and Upham to Chamberlin, August 23, 1902, WU/MHS.
86. Chamberlin to Haworth, August 25, 1902 and Chamberlin to Upham, August 25 and September 5, 1902, TCC/UC.
87. Chamberlin to Leverett, August 26, 1902, TCC/UC.
88. The site visit took place September 20, 1902; see Chamberlin to Haworth, September 13, 1902 and Chamberlin to McGee, September 13, 1902, TCC/UC; Long to Wright, September 23, 1902, GFW/OCA; see Hinsley (1981:249) on McGee's activities around the time of Powell's dying.
89. Upham to Wright, September 25, 1902, GFW/OCA; Chamberlin to Calvin, October 10, 1902, TCC/UC.
90. Calvin to Winchell, September 30, 1902 and Upham to Winchell, October 4, 1902, NW/MHS.
91. Concerns about redeposition and it not being true loess were expressed by Leverett to Wright, who summarized Leverett's concerns to Upham. Upham assured Wright he was aware of the problem, and both he and Winchell carefully looked over the ground with this problem in mind, but found no evidence of it (Leverett to Wright, September 17, 1902 and Upham to Wright, September 22, 1902, GFW/OCA). See also Winchell 1902:193–194.
92. Upham to Calvin, October 4, 1902 [as attached to Upham to Winchell, October 4, 1902, NW/MHS].
93. Calvin to Upham, October 6, 1902, NW/MHS; Chamberlin and Salisbury 1885.
94. Chamberlin to Calvin, October 14, 1902 and Chamberlin to Calvin, October 10, 1902, TCC/UC. The "Epistle from Ephesus" is a play on the New Testament homily of Paul the Apostle's letter to the Ephesians urging, in effect, for everyone to please get along.
95. Chamberlin to Upham, November 4, 1902, TCC/UC.
96. Chamberlin and Salisbury 1885:287–307.
97. Schultz 1976:224–226.
98. Salisbury 1893f:859.
99. Chamberlin 1897:798.
100. Schultz 1976:224–226.
101. Shimek 1899:103; also Shimek 1896, 1898; also Totten and White 1985.
102. Shimek 1899:109. Shimek took special delight in mocking the "closet–naturalist" who was "too dainty to soil his fingers with the toil and exposure of field-work" (Shimek 1899:104).
103. Chamberlin and Shimek corresponded sporadically through the 1890s. Chamberlin 1897.
104. Chamberlin 1897:800.
105. Shimek 1899:98–99.
106. Chamberlin 1897:802.

107. Wright 1889a:362–363. Five years earlier, Wright had declared the loess "doubt-less a water deposit" (Wright 1884d:767); three years later, in *Man and the Glacial Period*, Wright said little of the loess but nonetheless maintained its aqueous origin.
108. Wright 1900:72.
109. Shimek to Wright, October 30 and November 21, 1901, GFW/OCA.
110. Shimek to Wright, December 4, 1901, GFW/OCA.
111. Wright 1902:134–138.
112. Wright 1901:138.
113. Wright 1901:138–139.
114. Chamberlin 1901. Although published anonymously, the paper is attributed to Chamberlin by Schultz, who supposes Chamberlin saw this as confirming his worst suspicions of Wright (Schultz 1976:254ff).
115. Wright to Winchell, March 11, 1907, NW/MHS.
116. Fowke went to Kimmswick to follow up on a report of a human fibula possibly as-sociated with mastodon remains (Holmes 1902a). He found nothing "to show that man existed here as a contemporary of the mastodon" (Fowke 1928:485).
117. The full listing of site visits at Lansing is as follows: August 9 (Haworth, Long, Upham, Williston, Winchell); September 20 (Thomas and Rollin Chamberlin, Dorsey, Fowke, Holmes, Salisbury, and Calvin); October 4 (Hrdlička); October 18 (Winchell and Brower); and October 26 (Chamberlin, Holmes) (for example, Holmes 1902b; Winchell 1903b:146; Chamberlin to Leverett, September 30, 1902 and Chamberlin to Winchell, November 4, 1902, TCC/UC).
118. For example, Upham to Wright, October 16, and October 24, 1902, GFW/OCA. Upham provided Wright with loess references and also, as noted, a copy of the twelve-point response Upham had sent to Calvin.
119. Upham 1902a–1902f.
120. Chamberlin to Upham, November 4, 1902, TCC/UC. Chamberlin had already drafted and mailed his paper to Calvin and Holmes several weeks earlier: Cham-berlin to Calvin, and to Holmes, October 10, 1902, TCC/UC.
121. Chamberlin to Upham, November 12, 1902, TCC/UC.
122. Three years later Upham asked Chamberlin to support a petition to Dartmouth College to award him an honorary doctoral degree. Chamberlin demurred on the grounds that "as a member of one faculty [who] urged upon another faculty to give a degree, it might embarrass the case" (Chamberlin to Upham, April 16, 1906, TCC/UC). Upham got the degree.
123. Chamberlin 1902:754.
124. Chamberlin 1902:763–770. Salisbury's comment is in Chamberlin 1902:778.
125. Chamberlin 1902:775.
126. Chamberlin 1902:772–773.
127. Chamberlin 1902:775.
128. Chamberlin to Holmes, October 16, 1902, TCC/UC.
129. Chamberlin to Leverett, August 13, 1902, TCC/UC; Schultz 1976:259.
130. Chamberlin to Calvin, October 10, 1902, TCC/UC. Calvin and Salisbury's com-ments appear in Chamberlin 1902:777–779.
131. Chamberlin 1902:768–769, 773; Salisbury in Chamberlin 1902:779. Williston (1905a:345) describes the discovery of the unio.
132. Hrdlička 1903:323. The session that included the Lansing and Trenton discussions took place October 22, 1902. Hrdlička was in Mexico from October through De-cember of 1902 (Spencer 1979:198).

133. Upham to Hrdlička, November 10, 1902, WU/MHS. The same letter was sent to Dorsey that day.
134. Upham to Winchell, November 11, 1902, WU/MHS. Upham sent out at least six virtually identical letters over a four-day period, November 8–11, 1902.
135. Holmes 1902b:744.
136. Holmes 1902b:744.
137. Upham to Winchell, February 14, 1903, WU/MHS.
138. Chamberlin to Winchell, November 4, 1902, TCC/UC.
139. Chamberlin to Holmes, November 4, 1902, TCC/UC.
140. Winchell 1903b:139.
141. Winchell 1903b:143–145; Shimek 1903:374.
142. Winchell 1903b:145, emphasis in the original; Shimek 1903:359.
143. Winchell 1903b:146.
144. Winchell 1903b:150.
145. Winchell 1903b:151–152.
146. Chamberlin to Shimek, December 18, 1903, TCC/UC.
147. Winchell 1903b:140–141.
148. Chamberlin to Holmes, October 16, 1902, TCC/UC, emphasis mine.
149. Winchell 1903b:152.
150. Chamberlin 1903:67–68, emphasis in the original.
151. Chamberlin to Davis, November 26, 1902, TCC/UC.
152. Chamberlin 1903:67–70.
153. Chamberlin 1903:70–77.
154. Chamberlin 1903:78–82.
155. Chamberlin to Barbour, November 27, 1906, TCC/UC.
156. Chamberlin 1903:78, 84.
157. Wright to Fowke, January 31, 1903, WHH/NAA.
158. Upham to Wright, March 17, 1903, GFW/OCA. On the BAE succession, Chamberlin to Branner, December 22, 1902, TCC/UC.
159. Williston 1897:301.
160. Williston 1902a, Williston 1902b. The 12 Mile Creek find and excavations are discussed by Hill (2006) and Hawley (2009). The exact time line of the discovery is ambiguous: Martin originally suggested in a newspaper interview that the find had been made in 1895. A few years later it was said to have been made following flooding of the site in 1896 (Hawley 2009:118–121).
161. Williston 1902b. Williston's paper from the meetings did not appear until several years later (Williston 1905a); it overlaps what he reported in the *American Geologist*, but it did not include illustrations.
162. Upham to Williston, November 21, 1902 and Upham to Wright, November 22 and December 16, 1902, WU/MHS. When Wright issued a new edition of his *Ice Age* in 1911, Williston's papers on 12 Mile Creek did not make it into his revised bibliography. Upham's comment was not intended to link Lansing and 12 Mile Creek. As Hawley observes, the two sites were not discussed together save in the media, although Martin joked to a reporter that the Lansing Man had shot the 12 Mile Creek bison (Hawley 2009:123, 125).
163. Chamberlin 1903:83–84.
164. Lucas 1898. See also Stewart 1897, who likewise assigned the species to *Bison antiquus*, but as Lucas subsequently observed—and here Williston agreed—Stewart "unfortunately confuses" the taxonomy of the 12 Mile Creek bison (Lucas

1899a:758, Lucas 1899b registers Williston's agreement). McClung (1908) subsequently affirmed Lucas's take on the taxonomic status of the 12 Mile Creek bison.

165. Lucas 1899a:756, 758.

166. Lucas 1899a; Williston 1902b. On other contemporary reactions, see Hill 2006. On Lucas's taxonomic efforts, see McDonald 1981.

167. Hill 2006:145; Rogers and Martin 1984:757. See also Meltzer 1991:30.

168. See especially Upham 1903a:31–33.

169. Chamberlin to Upham, February 19, 1903, TCC/UC. In sending reprints to his critics, Upham admitted he doubted they would agree with him (and Winchell) "on the mode of deposition of the loess." He was right. Upham wrote to Calvin, Shimek, Udden, all February 16, 1903, WU/MHS.

170. Upham 1903a. Winchell finished his paper in late February, 1903 and then sent it for comment to a number of correspondents, including Michael Concannon, Herman L. Fairchild, Luella Owen, J. E. Todd, Upham, and Wright (Fairchild to Winchell, March 2, 1903, NW/MHS; Owen to Wright, March 11, 1903, GFW/OCA; Todd to Winchell, Feb 18, 03, NW/MHS; Winchell to Wright, February 28, 1903, NW/MHS; Wright to Winchell, March 23, 1903, NW/MHS).

171. Fairchild to Winchell March 2, 1903, NW/MHS.

172. Winchell 1903b:265.

173. Winchell 1903b:289–290.

174. Winchell 1903b:273–274; cf. Chamberlin 1902:763.

175. Winchell 1903b:277–281; cf. Chamberlin 1902:767; Shimek 1904a.

176. Winchell 1903b:285–288.

177. Winchell to Wright, April 16, 1903, GFW/OCA; Winchell 1903b:287, 289.

178. Calvin to Winchell, September 30, 1902, NW/MHS.

179. Williston 1903:466, 473. Williston explained his position vis-à-vis Chamberlin in Williston to Barbour, December 11, 1906, EHB/NSM.

180. See Wright 1903, and Wright to Winchell, March 28, 1903, NW/MHS; Todd to Winchell, March 7, 1903, published in Winchell 1903b:291–294. Todd had earlier expressed his skepticism about Winchell's position, so Winchell would have been well aware of Todd's likely response to his paper (Todd to Winchell, February 18, 1903, NW/MHS).

181. Chamberlin to Long, May 4, 1903, TCC/UC. Chamberlin made it a point to ask Long to measure the high water mark of the Missouri River, which had flooded that month, but in the end the information was unneeded and unused.

182. Chamberlin to Leverett, June 10, 1903, TCC/UC.

183. Hrdlička 1903:324.

184. Hrdlička 1903:329.

185. Chamberlin to Shimek, December 18, 1903, TCC/UC.

186. Upham 1903b, 1903c; Wright 1903.

187. Shimek 1903, 1904b, 1904c; Wright 1904, 1905. Chamberlin to Shimek, December 18, 1903, TCC/UC.

188. Abbott to Wright, April 20, 1904, GFW/OCA. See also Winchell to Wright, January 6, 1904, GFW/OCA. Also Upham to Wright, April 18, 1904, GFW/OCA.

189. Shimek 1903:369; Winchell to Wright January 1, 1904, GFW/OCA; Leverett 1904.

190. Wright 1904:216, 200, 222. Wright later explained that the absence of aquatic snails in loess was due to either the ephemeral nature of glacial meltwater lakes, their cold temperatures, or their heavy silt load, all of which could have precluded snails living in this setting. Alternatively, their absence might have resulted from

weathering and dissolution. As to why terrestrial snails did not suffer the same effects, he suggested they were found "near the marginal portion of the loess where the destructive processes are least" (Wright 1911:410–411).

191. Wright to Winchell, March 26, 1907, NW/MHS. Shimek 1903:368.
192. Wright 1905:239.
193. Owen 1904:227–228. Owen had grown up on the loess hills of St. Joseph, MO, forty-five miles upriver from Lansing. She lacked a college education, hardly unusual at the time, but through self-teaching and experience at home and abroad (she toured the world in 1900, including a trip across China's loess plateau and through the loess hills of Germany), as well as correspondence with Winchell and Wright, she became an ally on the loess question (Owen 1904, 1905). Owen's support of the aqueous theory may have played against her: when her name was put forward by Wright for consideration as a fellow of the GSA, there was "some criticism . . . made of her qualifications" (Hovey to Winchell, March 4, 1907, GFW/OCA). Wright decided it was best to recuse himself from the process (Wright to Winchell, March 11, 1907, GFW/OCA).
194. Shimek 1904c:380–381. This was but one of several harsh comments Shimek aimed at Owen: in replying to her question of where the river bars were that could have acted as sources of windblown sediment, Shimek suggested that "Miss Owen might look out, at St. Joseph, across the Missouri to the west, and she can see some of them" (Shimek 1904c:373). As a colleague of Shimek's wrote, "You have not yet lost your fondness of a fight" (Pilsbry to Shimek, February 5, 1906, BHS/SIA).
195. Shimek 1904a; Owen to Wright, January 14, February 16, and March 1, 1905, and Winchell to Wright January 6, 1904, GFW/OCA.
196. Wright 1905:236–237.
197. Owen 1905:292–294, 296–297.
198. Shimek 1904b.
199. See Busacca et al. 2004 for a recent overview.
200. Gilder 1906.
201. Gilder 1907a.
202. Gilder 1906.
203. All quoted in Gilder 1906.
204. Gilder 1907c:378.
205. Regal 2002:93.
206. Osborn to Barbour, October 27, 1906, EHB/NSM; Gilder 1907c:379.
207. Osborn to Hrdlička, October 31, 1906, AH/NAA; Hrdlička to Osborn, November 1, 1906, VP/AMNH; Wissler to Osborn, November 20, 1906, EHB/NSM.
208. Spencer 1988. When found in the late 1880s, Galley Hill proved controversial; some supposed it was an intrusive burial. By the end of the first decade of the twentieth century, and especially after Piltdown, Galley Hill became emblematic of a pre-Neanderthal early modern human (for example, Keith 1915:187).
209. Osborn to Barbour, November 15 and November 23, 1906, EHB/NSM.
210. Barbour to Osborn, October 29, 1906, EHB/NSM; Osborn conceded on the matter of priority (Osborn to Barbour, November 15, 1906, EHB/NSM). On priority, see also Barbour to Gilder, November 24, 1906, EHB/NSM. Barbour made no mention of Wissler's letter in his reply to Osborn (Barbour to Osborn, November 27, 1906, EHB/NSM). Shor (1974) discusses the feud between Cope and Marsh and provides some details of the roles of Barbour and Osborn.
211. Barbour and Ward 1906a:628; Barbour and Ward 1906b:323. By Barbour and Ward's measurements the cephalic index for Long's Hill came in at 87.9, which is

brachycephalic. But as Hrdlička later observed (1907:73), there were variations in the crania, and several were incomplete; this made for inaccurate measurements.

212. Barbour was on site November 8, 1906. Barbour to Williston, November 10, 1906, EHB/NSM.

213. Williston to Barbour, November 11, 1906, EHB/NSM.

214. Barbour worked at Long's Hill about five days in November of 1906 and returned on other occasions, including when Hrdlička came to examine the site (Barbour to Hrdlička, January 25, 1907, AH/NAA).

215. Barbour to Williston, and Chamberlin, November 22, 1906, EHB/NSM.

216. Chamberlin to Barbour, November 27, 1906, EHB/NSM.

217. Chamberlin to Barbour, November 27, 1906, EHB/NSM. In the circumstances, it is curious that Chamberlin thought publishing in a scientific periodical would help sort out the Long's Hill case, as publication in *American Geologist*, nominally a scientific publication, had not helped sort out Lansing. Of course, Chamberlin likely did not see *American Geologist* as a scientific periodical.

218. Barbour to Chamberlin, December 31, 1906, Barbour to Williston, December 6, 2005, and Williston to Barbour, December 11, 1906, all in EHB/NSM.

219. Osborn 1907:375.

220. Barbour 1907a:414; Barbour 1907c:346. The Spy discovery, made in 1886 in the Grotto of Spy d'Orneau, Belgium, consisted of two skeletons. These were readily placed among the Neanderthals by anthropologists of the time (see Keith 1915:131).

221. Compare Ward 1907:413 to Barbour 1907a:414–415.

222. Barbour to Gilder, November 24, 1906, EHB/NSM.

223. Barbour 1907a:503; Barbour 1907b; Barbour 1907c:334–335. See Shimek 1908:253 in response.

224. Williston to Barbour, December 11, 1906, EHB/NSM.

225. Barbour 1907a:503; also Barbour 1907b:112; Barbour 1907c:347.

226. Gilder to Barbour, December 27, 1906, EHB/NSM; also Barbour to Jeannette Gilder, November 12 and December 5, 1906 and Jeannette Gilder to Barbour, November 26, 1906, EHB/NSM.

227. Upham to Blackman, November 17, 1906; Blackman to Upham, November 22, 1906; Upham to Blackman, January 12, 1907; all in WU/MHS. Upham recanted the disparaging remarks he had made about Gilder and his family in a subsequent letter sent later that same day (Upham to Blackman, January 12, 1907, WU/MHS).

228. Blackman to Upham, January 8, 1907 and Upham to Blackman, January 12, 1907, WU/MHS. See also Blackman 1907.

229. Holmes asked Barbour if he would mind a study visit from Hrdlička. After checking with the University chancellor and insuring the University's priority was safe, Barbour approved the visit. Holmes to Barbour, January 5 and January 23, 1907; Barbour to Andrews, January 15, 1907; Andrews to Barbour, January 16, 1907; all EHB/NSM.

230. Hrdlička to Holmes, January 29, 1907, WHH/RR 9:196–197.

231. Hrdlička 1907:13.

232. Hrdlička 1907:57, 64. Also Hrdlička to Putnam, March 6, 1906, FWP/HU.

233. Wilson 1895:725; Wilson to Goode, June 21, 1895, USNM/SIA; Wilson to Putnam, February 1, 1900, FWP/HU.

234. Hrdlička 1907:19. For a discussion of different perceptions of the utility of the fluorine test and its history, see Goodrum and Olson 2009; Meltzer 1989.

235. Hrdlička 1907:12–13.

236. Hrdlička 1907:13–14.
237. Both Barbour and Gilder send Hrdlička detailed letters on the history and plan of work at the site, as well as samples (Barbour to Hrdlička, January 25, March 2, and March 11, 1907 and Gilder to Hrdlička February 13 and March 7, 1907, AH/NAA). Much of this information, particularly from Gilder's letters, was edited by Hrdlička and then put in as if verbatim in Hrdlička 1907.
238. Hrdlička 1907:88–91.
239. Hrdlička 1907:92–97. The fact that Barbour, like so many others, assumed a low forehead and heavy brow ridge was a sign of antiquity prompted Hrdlička to add an Appendix explaining that these features might be present in geologically ancient humans, but not all individuals with those features were geologically ancient.
240. And with Putnam's permission, Hrdlička to Putnam, March 6, 1906, FWP/HU.
241. Hrdlička 1907:98, emphasis mine.
242. Gilder to Hrdlička, December 15 and December 20, 1907, AH/NAA.
243. Ward to Hrdlička, December 10, 1907, AH/NAA. In reply, Hrdlička distanced himself from Blackman, whom he could not even remember meeting, and apologized for mislabeling Blackman's title but downplayed its possible impact.
244. Blackman to Upham, December 10, 1907, WU/MHS; Blackman to Hrdlička, December 14, 1907, AH/NAA.
245. Ward to Hrdlička, December 10, 1907, AH/NAA.
246. Gilder to Hrdlička, December 15 and December 20, 1907, AH/NAA.
247. Blackman to Upham, December 10, 1907, WU/MHS; Barbour to Williston, December 1907, EHB/NSM. Barbour gives a slightly different version of Hrdlička's comments in Barbour to Wright, December 23, 1908, EHB/NSM.
248. Williston to Barbour, January 3, 1908, EHB/NSM.
249. Barbour to Gilder, December 25, 1907, EHB/NSM; Chamberlin to Shimek, December 21, 1907, TCC/UC; Gilder to Shimek, July 1, October 22 and December 14, 1907, BHS/SIA.
250. Williston to Barbour, January 3, 1908, EHB/NSM.
251. Barbour to Gilder, December 25, 1907; Barbour to Williston, late December, 1907; Gilder to Barbour, December 30, 1907; all in EHB/NSM. A year later Barbour was still convinced that Shimek had poisoned Hrdlička against them (Barbour to Wright, December 23, 1908, EHB/NSM). On Chamberlin's views, see Chamberlin to Shimek, December 21, 1907, TCC/UC.
252. Shimek 1908:250–253. While he was once again on the topic of human remains in supposed loess, Shimek took the opportunity to take a few swipes at Winchell (Winchell 1907).
253. Barbour to Wright, December 23, 1908, EHB/NSM.
254. Blackman to Upham, July 19, 1907, WU/MHS. See also Gilder 1908:70.
255. Gilder 1907b, 1908, 1909; the last article contains the skeletal descriptions by Hrdlička. Gilder to Shimek, October 6 and November 10, 1908, BHS/SIA. Gilder's complaints about Hrdlička and Shimek are in Gilder to Barbour, January 6 and February 6, 1908, EHB/NSM.
256. Gilder to Shimek, November 10, 1908, BHS/SIA.
257. Hull 1988:287.
258. Barbour to Wright, December 23, 1908, EHB/NSM.
259. See Winchell 1907, Wright 1907. Wright to Winchell, March 11, 1907, NW/MHS. Winchell helped Barbour seek other records of Neanderthal-like specimens recovered in the Midwest (Winchell to Barbour January 29, 1907, EHB/NSM).

260. Winchell 1917:151. Hrdlička's copy is preserved in his reprint library, in the department of Anthropology, NMNH. It was uncut when I saw it in the 1980s.

261. When Piltdown was discovered in 1912, there was a radical shake-up of the human family tree, with Neanderthals pushed to a side branch leading nowhere and Piltdown seen as a Late Pliocene/Early Pleistocene ancestor of modern humans (Keith 1915). At that point Europeans disengaged anatomy from time. Hrdlička, who steadfastly held to Neanderthals as human ancestors (Hrdlička 1927), steadfastly insisted anatomy and time were related. See the discussion in §7.6.

262. Wright 1911:683.

263. Hrdlička 1907:92.

264. By coincidence, that same weekend Gilder wrote that Barbour gave his daughter in marriage to Harold J. Cook, who spent much of the 1920s firing at Hrdlička—as discussed in chapter 8 (Gilder to Barbour, October 13, 1910 and Barbour to Gilder, October 24 and November 17, 1910, EHB/NSM.)

265. Gilder 1911:157–158.

266. Gilder 1911:162, 168, emphasis in the original. Among the misstatements Hrdlička is accused of is how deeply frozen the ground was on the day of his visit in January of 1907. Gilder claims Hrdlička put the frozen depth at three or four vertical feet, when it was actually less than eight inches (Gilder 1911:159). Three years earlier Gilder himself had stated that on the day of Hrdlička's visit, "frost penetrated the ground for three feet" (Gilder 1908:64).

267. Barbour to Ward, January 11, 1912, EHB/NSM. In reply, Ward thought Gilder's article "good" but not as effective from a scientific standpoint as a reply from Barbour would be (Ward to Barbour, January 16, 1912, EHB/NSM).

268. Hrdlička 1912:vi–vii, 387–388 provides a listing of Ameghino's publications. Hrdlička hoped to have Shimek accompany him, but funding was unavailable (see Hrdlička to Shimek, December 6, 1909; Holmes to Shimek, January 27 and March 9, 1910; Shimek to Holmes, March 18, 1910; all in BS/SIA). USGS geologist Bailey Willis went instead (Holmes to Shimek, April 1, 1910, BHS/SIA).

269. Hrdlička to Holmes, May 15, 1911, AH/NAA. Holmes was, in fact, one of Hrdlička's collaborators on the South American volume, analyzing stone tools and ceramics that Hrdlička had collected in Argentina.

270. That was Abbott's take on Nebraska (Abbott Diary, December 23, 1906, CCA/PU). He had a similar response to Lansing (Abbott Diary, October 22, 1902, CCA/PU).

271. Abbott to Winchell, May 3, 1907, NW/MHS; Putnam to Abbott, July 10, 1902 and Putnam to Volk, June 28, 1904; FWP/HU; Volk to Holmes, June 29 and October 21, 1905, BAE/NAA.

272. Misunderstanding a request from Otis Mason for a brief paper to include in a forthcoming Smithsonian Institution *Annual Report*, Putnam offered Volk's report instead. Mason quickly quashed that idea: the Trenton Gravels were far too controversial for the Smithsonian's tastes (Mason to Putnam, October 23, 1907; Putnam to Mason, October 24, 1907; Mason to Putnam, November 8, 1907; all FWP/HU).

273. Abbott 1907:9, Abbott 1908:10. Holmes is mentioned by name and called a liar (or some variant thereof) in eighteen of the seventy letters (approximately 25 percent) that Abbott sent Putnam over a span of nine years in the first decade of the twentieth century. Abbott's explanation for Putnam's "suppression" of Volk's report is in Abbott to Winchell, April 22, 1907, NW/MHS.

274. Abbott to Winchell, May 3, 1907, NW/MHS.

275. Volk to Putnam, June 6, 1905, FWP/HU.

276. Mead to Wright, April 17, 1907, FWP/HU. The *Archæologia* series was bankrolled by Moses Taylor Pyne, a Princeton millionaire who arranged for Abbott's collections to go to Princeton University. Abbott otherwise described Princeton University as the deepest of "all woefully dreary, fanatical, un-scientific holes" (Abbott to Putnam, October 18, 1911, FWP/HU).

277. Abbott 1907:9, 44; Abbott 1908:49; Abbott 1909:11–12.

278. Abbott Diary, November 18, 1906, CCA/PU.

279. Abbott 1907:3–11, 22; also Abbott 1909:12–13. Volumes 2 and 3 of the *Archæologia* series also made disparaging remarks about Holmes, though not in such a toxic concentration, and swiped at Mercer (Abbott 1908:43ff) and Hrdlička (Abbott 1909:76ff). On holding himself back, see Abbott to Winchell, May 3, 1907, NW/MHS).

280. Wright to Abbott, April 12, 1907, CCA/PU. Abbott had sent copies to Hrdlička and Winchell. The former found the volume "enjoyable" (Hrdlicka to Abbott, April 16, 1907, CCA/PU); the latter was perplexed.

281. Holmes to Putnam, November 4, 1907, FWP/HU; Putnam to Holmes, November 19, 1907, WHH/RR.

282. Abbott to Wright, May 27, 1911, GFW/OCA.

283. Winchell to Abbott April 30, 1907, CCA/PU; Winchell to Abbott, July 7, 1909, NW/MHS. All later published as Winchell 1909.

284. Winchell 1909:250–251.

285. Winchell 1909:252.

286. Abbott to Winchell, July 12, 1909, NW/MHS. Winchell dedicated *The Weathering of Aboriginal Stone Artifacts* to Abbott and in the Preface likened Abbott to Boucher de Perthes (Winchell 1913:vii).

287. Abbott to Wright, May 27, 1911, GFW/OCA.

288. Abbott to Putnam, August 3, 1911, FWP/HU.

289. Putnam to Abbott, August 12, 1911, FWP/HU. Putnam's letter brought tears to Abbott and Volk both (Volk to Putnam, August 18, 1911, FWP/HU). Abbott reports on the resumption of his relations with Putnam in Abbott to Wright, October 11, 1911, GFW/OCA.

290. Putnam, in Volk 1911:vi; Putnam to Wright, August 6, 1911, GFW/OCA.

291. Volk 1911:110–111; Abbott to Putnam, August 18 and 28, 1911, FWP/HU.

292. Volk 1911:120, 126–127.

293. Volk 1911:128.

294. Putnam, in Volk 1911:v, emphasis mine.

295. See appendices by Woodman, Woodworth, and Wright, in Volk 1911; Salisbury 1902.

296. Abbott to Wright, October 11, 1911 and January 12, 1912, GFW/OCA. For his stray shots at Volk, see Abbott 1912:122–123, 161–162.

297. Putnam to Abbott, August 12 and September 24, 1911 and Abbott to Putnam, August 14, 1911, FWP/HU; Putnam to Wright, August 6, 1911, GFW/OCA.

298. Abbott to Putnam, August 18 and 28, 1911, FWP/HU; Wright to Winchell, March 16, 1907, NW/MHS.

299. Wright 1912; Wright to Putnam, August 21, 1911, FWP/HU.

300. Winchell 1913:95.

301. Winchell 1913:32–34, Plates 8 and 9.

302. Winchell 1913: x–xi, 37.

303. Winchell 1913:173–174.

304. Winchell 1913:vii.
305. Abbott to Putnam, February [no date] 1913 and July 12, 1913, FWP/HU. Holmes registered his opinion in his copy of Winchell's book, now in the Library of the Department of Anthropology, Smithsonian Institution. Winchell nonetheless reported to Abbott that "all who have expressed an opinion commend it" (Winchell to Abbott, November 19, 1913, NW/MHS).
306. Abbott to Wright, December 28, 1914, GFW/OCA.
307. Wright to Abbott, December 30, 1914, CCA/PU.
308. Winchell to Upham, March 25, 1914, MHS/MHS; Abbott to Putnam, May 4 and September 18, 1914, FWP/HU.
309. The fire occurred Friday November 13, 1914. Volk to Wright, December 21, 1914 and Abbott to Wright, December 28, 1914, GFW/OCA. See also Hunter et al. 2009:5–13.
310. Spier 1913:676, 679. See also Truncer 2003, who provides a detailed discussion and assessment of Spier's work at Trenton, plus its longer term consequences.
311. Wissler 1917; see also Browman and Givens 1996.
312. Spier 1916:182.
313. Spier 1916:186, 189.
314. Abbott to Putnam, June 3, 1914, FWP/HU. When Abbott got wind of Wissler's project from Volk, he initially was unhappy (Abbott to Putnam, April 15, 1913, FWP/HU). In the end he was pleased.
315. Wissler 1916b:197.
316. Wissler to Putnam, March 5, 1915, FWP/HU; Abbott to Wright, December 28, 1914, GFW/OCA.
317. Holmes to Kummel, November 15, 1915, and Kummel to Holmes, November 16, 1915, WHH/SIA. Kummel did not mention the fact that he had visited the AMNH excavations on at least one occasion; he was there in June of 1914 and was photographed alongside Abbott, Charles Reed (the AMNH geologist), Skinner, Spier, and others (Abbott to Putnam, June 3, 1914, FWP/HU). Holmes likely learned of the Trenton work when Wissler sent Hrdlička a draft copy of his paper on Trenton in November of 1915 (Wissler to Hrdlička, November 8, 1915, ANTH/AMNH).
318. Wissler to Holmes, January 26, 1916 and Holmes to Wissler, February 16, 1916, ANTH/AMNH.

Chapter Seven

1. Wissler 1916a:234.
2. Wissler 1916a:234–235.
3. Chamberlin 1903; Wissler 1916a:234.
4. Merriam 1914:201.
5. For example, Sellards 1916a.
6. Kroeber 1962 (originally written in 1914).
7. Merriam 1914:202–203.
8. Sellards 1916a:9.
9. Thomson 2008 provides a recent assessment. See also Manning 1967:205–212; Rabbitt 1980:164; Romer 1941; Shor 1974.
10. Hay 1908.
11. Lull 1931:31.

12. Holmes to Stanton, April 21, 1911; Chamberlin to Stanton, May 26, 1911; see also Branner to Stanton, March 30, 1911; Merriam to Stanton, April 17, 1911; all in GC/CIW.
13. Hay 1909:890.
14. Hay 1909:892–893. One of Hay's earliest works (Hay 1902) was the massive "Bibliography and Catalogue of the fossil Vertebrata of North America," which helped secure his reputation among vertebrate paleontologists.
15. Lull 1931:34. Osborn to Woodward, May 27, 1918, GC/CIW.
16. Hay 1912:13.
17. Hay 1912:15. These divisions would be somewhat revised by Hay in later years (for example, Hay 1925).
18. Hay 1912:15–16.
19. Hay 1914.
20. Hay 1914:169–170.
21. Lull 1931:32, 34.
22. For biographical information on Sellards, see Evans 1961, 1986; Krieger 1961.
23. Sellards 1916a:4–7; also Sellards 1916c:127–130.
24. Sellards 1916a:9–10; also Sellards 1916c:132–134.
25. Sellards 1916a:10.
26. Sellards 1916a:13–14.
27. Sellards 1916a:14–15; also Sellards 1916c:123.
28. Sellards 1916a:17–18.
29. Hay 1916:76.
30. Hay 1916:76; Sellards 1916c:143.
31. Sellards to Hay, October 2, 1916, OPH/SIA; see Sellards 1916c:147.
32. Sellards to Hrdlička, July 17, 1916 and Hrdlička to Sellards, July 20, 1916, AH/NAA; Sellards to Osborn, July 17, 1916, VP/AMNH.
33. Holmes to Chamberlin, July 27, 1916, TCC/UC.
34. Chamberlin to Holmes, July 31, 1916, TCC/UC.
35. For example, Sellards to Williston, August 10, 1916, EHS/UT; Matthew to Sellards, September 8, 1916, VP/AMNH; Holmes to Chamberlin September 30, 1916, Chamberlin to Holmes, October 5, 1916, Hrdlička to Chamberlin, October 9, 1916, all TCC/UC; Sellards to Hay, October 9, 1916, OPH/SIA; Hrdlička to MacCurdy, October 13, 1916 and Chamberlin to Hrdlička, October 14, 1916, AH/NAA.
36. For example, Sellards to Matthew, July 25, 1916; Sellards to Osborn, July 28, 1916; Matthew to Osborn, October 2, 1916; Osborn to Sellards, October 2, 1916; Sellards to Osborn, October 6, 1916, all in VP/AMNH.
37. Sellards 1916b:615–616; also Sellards 1916c:136–139, 160.
38. Hrdlička to Chamberlin, October 9, 1916, TCC/UC; MacCurdy to Hrdlička, October 16, 1916, USNM/SIA.
39. Sellards to Abbott, November 27, 1916 and Abbott Diary, November 29, 1916, CCA/PU.
40. Williston to Sellards, October 17, 1916, EHS/UT, emphasis in original.
41. Williston to Sellards, October 19, 1916, EHS/UT.
42. Holmes to Chamberlin, October 23, 1916, TCC/UC.
43. Most were at Vero over several days in late October, during which all experienced insects, rain, and a partial flood. Hrdlička 1918:23; Sellards to Williston, November 4, 1916, EHS/UT.

44. Chamberlin to Sellards, October 30, 1916, TCC/UC.
45. Vaughan to Chamberlin, November 3, 1916 and Holmes to Chamberlin, November 3 and 7, 1916, TCC/UC. At the time Holmes wrote, Hrdlička was still traveling in Florida, examining other sites for evidence he would use against Vero.
46. Sellards made slight revisions to the paper he had submitted earlier to the *Journal of Geology*, the one that triggered the site visit (Sellards to Chamberlin, November 4, 1916, TCC/UC). The rest wrote papers for the occasion. Chamberlin to Sellards, October 30, 1916, TCC/UC; Chamberlin to Hrdlička, October 31, 1916, AH/NAA; Vaughan to Chamberlin, November 3, 1916 and Sellards to Chamberlin, November 4, 1916, TCC/UC; Hrdlička to Chamberlin, November 8, 1916, MacCurdy to Hrdlička, November 10, 1916, and Hrdlička to MacCurdy, November 11, 1916, all in AH/NAA.
47. Chamberlin 1917:2–3.
48. Sellards to Williston, November 4, 1916, EHS/UT. Hrdlička exchanged letters with Sellards while their papers were being written, but these concerned noninterpretive matters such as, for example, whether Sellards would allow Hrdlička to submit some of the Vero bone for chemical analysis (Hrdlička to Sellards, November 25, 1916 and Sellards to Hrdlička, December 2, 1916, AH/NAA).
49. Hrdlička 1917:48–49; Sellards 1917a:22; see also Hay 1917a:54; Vaughan 1917:41.
50. Hrdlička 1917:44, 49; Hrdlička to Holmes, November 9, 1916 and Hrdlička to MacCurdy, November 21, 1916, AH/NAA. For Hrdlička's prior comments on mineralization, see Hrdlička 1907 and discussion in §6.12.
51. Hrdlička to Sellards, November 25, 1916 and Sellards to Hrdlička, December 2, 1916, AH/NAA. Wherry's results are in Hrdlička 1918:61–63.
52. Chamberlin 1917a:35–36, 39; Vaughan 1917:41–42.
53. Meltzer 2006a:206–208. Hay's penchant for splitting taxa and naming new species reached almost absurd heights in his later work on the Lone Wolf Creek and Folsom bison (§8.11).
54. Hay 1917a:54–55.
55. Hrdlička 1917:50; Holmes to Hrdlička, December 1, 1916, AH/NAA, reprinted in Hrdlička 1917:51.
56. MacCurdy 1917a:57–59. Hrdlička to MacCurdy, November 13 and 21, 1916, MacCurdy to Hrdlička, November 17, 1916, AH/NAA.
57. MacCurdy 1917a:60–62.
58. Chamberlin, editor, 1917.
59. Hrdlička to Chamberlin, December 29, 1916, AH/NAA.
60. Bassler 1917:24. Sellards attended the meeting at the urging of Osborn (Osborn to Sellards, July 21, 1916, EHS/UT). Matthew's opinion is in Matthew to Osborn, October 2, 1916 and Matthew to Sellards, October 18, 1916, HFO/AMNH.
61. Sapir 1916 (reprinted in Mandlebaum 1949:454).
62. Goddard to Sellards, January 30, 1917, EHS/UT.
63. Sellards 1917b:240–241; MacCurdy 1917b:258.
64. Sellards 1917b:243–247.
65. Sellards 1917b:249–250, emphasis in original. Sellards to Hay, January 15, 1917, OPH/SIA.
66. Sellards 1917a:20–21; Sellards 1917b:248, 251.
67. Sellards 1917b:251.
68. Sellards 1917b:250.
69. MacCurdy 1917b:255–256.

70. MacCurdy 1917b:260.
71. MacCurdy 1917b:259, 261.
72. For example, Sellards 1917c:76–77.
73. Sellards to Hay, March 10, 1917, OPH/SIA.
74. Sellards to Hay, April 2, 1917, OPH/SIA.
75. R. Chamberlin to T. C. Chamberlin, March 24, 1917, TCC/UC.
76. Chamberlin 1917b.
77. Chamberlin 1917b:673, 676, 678–679. Sellards agreed with Chamberlin on this point and admitted that Stratum 2 and Stratum 3 "are not greatly separated in time" (Sellards to Hay, April 2, 1917, OPH/SIA).
78. Chamberlin 1917b:683.
79. Hay 1917b:358–359. Behind the scenes, Sellards provided Hay with localities to add to his listing of early sites, Sellards to Hay, April 30, 1917, OPH/SIA.
80. Hrdlička to Cattell, May 17, 1917, JMC/LC. Sellards, the "young man" in question, was only six years younger than Hrdlička.
81. Hay 1917b, 1918a.
82. Hrdlička 1917:50; Hrdlička to Chamberlin, December 29, 1916, AH/NAA.
83. For the occasion, Hrdlička solicited updated information from Merriam; Hrdlička to Merriam, November 18, 1916 and Merriam to Hrdlička, November 25 and December 6, 1916, AH/NAA.
84. Hrdlička 1918:35, emphasis in the original.
85. Hrdlička unpublished, but circa 1916–1917.
86. Hrdlička 1918:35–36.
87. Hrdlička 1918:36.
88. Hrdlička 1918:42–43. Hay not-so-gently mocked Hrdlička's generalization about the "endowment of patience" possessed by Indians (Hay 1918c:460).
89. Hrdlička 1918:47–48.
90. Swanton 1944:36.
91. Hrdlička 1918:49.
92. Hrdlička 1918:55.
93. Hrdlička 1918:52, 59. Hrdlička used the phrase "ordinary Indian" repeatedly to emphasize how little reason there was to conceive of these as having any evolutionary or chronological significance. He was struck by some differences in the Stratum 3 remains, but he was certain these were Indian, though perhaps of "a higher or more modern development" (Hrdlička 1918:59).
94. Keith 1915.
95. Hrdlička 1918:37.
96. Hrdlička 1918:36.
97. Meltzer 1983:33.
98. Hrdlička 1918:52, 64.
99. Sellards 1917d:142.
100. Berry 1917; Hay 1917c; Sellards 1917c, 1917d; Shufeldt 1917.
101. Berry 1917:31–33.
102. Hay 1917c:66–67. Berry was similarly convinced that there was "no hiatus between beds 2 and 3, and that there is no great difference in age" (Berry 1917:19).
103. Sellards 1917c:81.
104. See also Chamberlin 1919:332. Hay was even more unhappy a decade later when paleontologist Frederic Loomis downgraded the Melbourne fauna—and by extension Vero's—to the later Pleistocene (§7.11).

105. Chamberlin 1919:331.
106. Holmes to Chamberlin, January 6, 1917, TCC/UC. This is a position that Hay urged Hrdlička to look into (Hay 1921b:20).
107. Holmes to Hay, December 14, 1917, author's collection. The strike-through was Holmes's.
108. Chamberlin 1919:322.
109. Chamberlin 1919:324–325, 331.
110. Chamberlin 1919:311.
111. Chamberlin 1919:331.
112. Holmes 1910:161; Hay 1918a:36.
113. Chamberlin 1919:332, 334.
114. Hay 1918a:13. Hay later (1919c) devoted a long paper to rebutting the claim that Pleistocene vertebrate fauna had survived later in the south.
115. Hay 1919a, 1919b, 1920, 1921a, 1921b. The quote is from Hay 1921a:17.
116. Chamberlin 1919:320.
117. Hrdlička 1918:60. On Hrdlička's supreme sense of infallibility, see Judd 1967:70.
118. Hay 1917a; Hay 1918a:10–14; Hay 1918b:371. Also Hay to Woodward, July 25, 1918, GC/CIW.
119. Chamberlin 1919:311–312, 325; Chamberlin 1917b:673; cf. Hay 1919c.
120. Hay 1918a:13–14; Hay 1918b:370–371; Hay 1919c.
121. Hay 1918a:32; Hay 1918b:371.
122. Snead 2001:112.
123. Nelson 1918a:395.
124. Nelson 1918a:395.
125. Nelson 1918b:102. That is what he said publicly. Privately he wrote to Sellards, "The possibilities of artifacts having found their way some distance down into this sand are many and until objects of human origin can be shown to lie underneath or in some other exceptionally natural juxtaposition to fossils the case for the antiquity of man is weak" (Nelson to Sellards, June 6, 1917, EHS/UT).
126. Nelson 1918a:395.
127. Hay 1918a:33–34.
128. Hay 1918a:32, 34–35.
129. The association of artifacts "with the remains of extinct animals is always a matter of much scientific interest," Holmes wrote after his investigations at Afton, "but it appears that in this case the association has little significance, the fossil bones belonging in the original geological formations of the region, while the human relics are of recent introduction into the spring" (Holmes 1902a:128).
130. Hay 1918a:15, 29.
131. Hay 1918a:32–33.
132. Hay 1918a:32.
133. Boule 1911–1913. For a discussion of Boule's impact on views of Neanderthals, see Hammond 1982; Reybrouck 2002; Sommer 2006.
134. Keith 1915:308–311.
135. Smith 1913:131. MacCurdy 1914. On the "codependency" of eoliths and Piltdown, see Grayson 1986 and Spencer 1990.
136. Spencer and Smith 1981:436; Spencer 1988, 1990.
137. Boule 1911–1913:245–246. See also Spencer and Smith 1981:442.
138. Meltzer 1989:5–6; Spencer and Smith 1981:442–443.

139. His most explicit discussion on this point is in Hrdlička 1927:270.
140. Keith 1915:274–285.
141. Keith 1915:278.
142. Hay 1918b:370.
143. Hrdlička to MacCurdy, April 20, 1918, AH/NAA.
144. Holmes 1918:562.
145. Holmes 1918:562.
146. Holmes 1918:562
147. Hay to Woodward, July 25, 1918, GC/CIW. Osborn had written Woodward a couple of months earlier and spoken to the importance of Hay's work on human antiquity; his letter was sent less than two weeks before Holmes's paper appeared in print, so it was not prompted by it—unless Hay or Osborn, or both, knew Holmes's paper was coming out (Osborn to Woodward, May 27, 1918, GC/CIW).
148. Hay to Abbott, August 13, 1918, CCA/PU; also Hay to Abbott, April 14, 1919 and Hay to Volk, May 19, 1919, CCA/PU; Hay to Wright October 27, 1919, GFW/OCA.
149. Hay to Abbott, August 13, 1918, CCA/PU.
150. Hrdlička 1918:47–48.
151. Hay 1918c:460.
152. Hay 1918c:460–461.
153. Barbour to Hay, January 8, 1919; Schuchert to Hay, December 16 and 21, 1918; Weills to Hay, January 16, 1919; all in OPH/SIA.
154. Sterns 1919:119. This was, in fact, the second of two reviews Sterns wrote for the *Scientific American Supplement*. His previous one (Sterns 1918) was also in favor of Vero and also took its cues from Hay.
155. Holmes-Hay-Holmes, February 1919, OPH/SIA.
156. Chamberlin 1919:323
157. Chamberlin 1919:323
158. Hrdlička to Chamberlin, October 20, 1919, AH/NAA.
159. Sellards 1919.
160. Hay 1919b:8.
161. Hrdlička to Chamberlin, October 20, 1919; Chamberlin to Hrdlička, October 22, 1919; Hrdlička to Chamberlin, October 25, 1919; all in AH/NAA.
162. Meltzer 1994:17–18.
163. Kroeber to Hrdlička, February 19, 1918, AH/NAA.
164. Nelson to Hay, April 5, 1920, OPH/SIA.
165. Sellards to Hay, April 15 and May 10, 1919, OPH/SIA.
166. Sellards 1937.
167. In 1958, just a few years before Sellards's death, the University of Kansas presented him with the Erasmus Haworth Distinguished Alumnus Award. That he was honored with an award bearing the name of one of Lansing's early skeptics is not without irony.
168. Spier 1918.
169. Abbott to Wissler, March 22, 1919, ANTH/AMNH.
170. Wright 1919. Abbott died July 27, 1919 and Volk September 17, 1919. See also Hunter et al. 2009:5–13–14, Plate 5.8.
171. Wright 1919:453.
172. Wissler 1920:70.

173. Upham 1922. The only allusion made by Upham to controversy was that Wright's proposed Cincinnati Ice Dam was met with an "attempted refutation" by Chamberlin (Upham 1922:18).

174. Upham (1922:22) made sure to note that Louis Agassiz was likewise buried under a glacial erratic. (Agassiz's grave is at Mount Auburn cemetery, Cambridge, MA.)

175. Holmes 1919; Holmes to Hewett, June 25, 1920, NMJ/NAA; Meltzer and Dunnell 1992. Nelson judged Holmes's volume to be largely a republication of "old scattered papers." The sort of treatise he was hoping for was evidently reserved for a second volume, which Nelson surmised "will never be written" (Nelson to Hay April 5, 1920, OPH/SIA). Nelson was correct.

176. Holmes 1919:94. For his general summaries of the errors made by proponents of a deep human antiquity, see Holmes 1919:75, 93–94.

177. Holmes would not be pinned down to any specific age for that arrival, merely noting that ice retreat began around 20,000 years ago in lower-latitude areas of North America (the Ohio and Delaware valleys) and around 10,000 years ago further north in the Great Lakes region (Holmes 1919:73, 94).

178. By then, Holmes had received two honorary degrees (an AB from Hopedale Normal College in 1889 and a doctor of science from George Washington University in 1918), a volume of essays in his honor, two Loubat prizes, and was elected (1905) to the National Academy of Sciences (Meltzer and Dunnell 1992:xxiv–xxv, xxxix).

179. Holmes to Nelson, December 2, 1921, BAE/NAA, emphasis added. This was Holmes in private; publicly, he showed no contrition (Holmes 1922).

180. Holmes 1919; Hrdlička to MacCurdy, October 14, 1919, AH/NAA. MacCurdy gave the book a favorable review in the *American Anthropologist* (MacCurdy 1920). Hay was less charitable (Hay 1920).

181. Listed in Hrdlička 1922. These included the Stanford skull (Willis 1922; also Heizer and McCown 1950), the Lagow sand pit specimen (Shuler 1923), and multiple human skeletal remains found at depths of nineteen to twenty-three feet below the surface during the digging of a trench in Los Angeles (Merriam 1924; Stock 1924). The Lagow specimen was the only one for which a claim of great antiquity was made—by Ellis Shuler, a Harvard-trained geologist and paleontologist at Southern Methodist University. Hrdlička did not deign to reply, though on the margin of his reprint of Shuler's paper he scrawled his opinion: "Biased from the start and inaccurate" (AHL/NMNH). Harold Cook, not surprisingly, had a very different reaction to Shuler's report, excitedly urging Erwin Barbour to read it (Cook to Barbour, March 24, 1923, EHB/NSM).

182. Holmes 1922:175; Hrdlička 1922.

183. Loomis 1924:503. *O. sellardsi* was identified and named by Hay in his detailed report on the Vero fauna (1917c). No specimens of this species have been recorded since, and whether it represents a distinct species has not been ascertained (according to the *Paleobiology Database*, available at http://paleobiodb.org/#/http://flatpebble.nceas.ucsb.edu/cgi–bin/bridge.pl).

184. Loomis 1924:506.

185. Loomis 1924:508.

186. That legacy was foremost on his mind, for Holmes was then in the midst of compiling his twenty-volume *Random Records of Lifetime in Art and Science*, in which his triumph over the American Paleolithic looms large.

187. Gidley to Loomis, January 30, 1925, FBL/AC.

188. Holmes to Fewkes, January 11, 1922, BAE/NAA. Fewkes joined the BAE in 1895 and served as its chief from 1918 to 1928 (Judd 1967:26–28).
189. Gidley to Loomis, February [no date], 1925, FBL/AC. On the dates of the fieldwork at Melbourne and Vero, see Gidley 1926b:24 and Gidley 1927:169–170.
190. Holmes to Cattell, June 22, 1925, WHH/SIA.
191. Holmes 1925:257
192. Holmes 1925:258, emphasis added.
193. Even so, at least one of his contemporaries applauded Holmes's piece (for example, Hodge to Holmes October 24, 1925, WHH/SIA). Wilmsen (1965) deemed it one of the low points in American archaeology.
194. Loomis 1925:258. Cooke (1928:416) made the "angry elephant" observation; Gidley and Loomis refrained from such a dramatic claim.
195. Loomis to Holmes, September 22, 1925, WHH/SIA.
196. Holmes to Loomis, September 24, 1925, WHH/SIA.
197. Davis to Holmes, September 17, 1925 and Holmes to Davis, September 29, 1925, WHH/SIA.
198. Holmes to Fowke, September 29, 1925, WHH/NAA.
199. *Daily Science News Bulletin* 1925.
200. Stewart 1946:1; Gidley 1926a.
201. Gidley 1926a, 1926b, 1926c; Gidley and Loomis 1926; Loomis 1926.
202. Stewart 1946:4; see Hrdlička 1937b.
203. Loomis to Cook, January 7, 1926, HJC/AFB.
204. Gidley 1926c:310; Loomis 1926:262; Loomis to Cook, January 7, 1926, HJC/AFB.
205. Cooke 1926:451–452.
206. Cooke 1926:451; Gidley and Loomis 1926:263–264.
207. Hay 1926:387–388.
208. Hay 1926:389–391.
209. Hay 1926:392.
210. Hay 1927a:277, 280. Holmes's comments appear in the margins of his reprint of Hay 1927a, the copy of which was in the BAE reprint library.
211. Hay 1928c, 1928d; Cooke 1928:420; Gidley 1929:499
212. Kroeber 1929.
213. Goddard 1926:257–258. Goddard was a curator at the AMNH, which published *Natural History*, and it is likely he had an editorial hand in the journal.
214. Holmes 1919:55; Hrdlička made the same argument in regard to physical evidence (for example, Hrdlička 1925:485).
215. Goddard 1926:258–259. Goddard sounded the same theme in a review of Kidder's *Introduction to the Study of Southwestern Archaeology*, criticizing Kidder's age estimates for prehistoric events of the Southwest as being woefully inadequate to account for linguistic variation and the "long upward struggle of the beginning of agriculture and the arts" (Goddard 1925:463).
216. For a contemporary view of the diffusionist controversy, see Smith et al. 1927. Harris 1968:379ff. provides a historical overview of the German and British hyperdiffusionists.
217. Goddard's feelings toward Holmes possibly dated back nearly a decade. Goddard, a linguist trained at the University of California (PhD 1904), had moved to New York in 1909, accepting a position at the AMNH and later an appointment at Columbia University, where he became a close colleague of Boas. Goddard was a

staunch defender of Boas during the Boas Censure in December 1919. Goddard was hit by collateral damage: within a year he was removed as editor of *American Anthropologist* (Stocking 1968:294).

218. Goddard 1926:259, emphasis added.
219. Goddard 1927:262.
220. Goddard 1927:262.

Chapter Eight

1. Hrdlička 1926:7.
2. Hrdlička 1926:9.
3. Hrdlička 1926:8–9.
4. Hrdlička 1926:8.
5. Cook to Osborn, January 8, 1926, HJC/AFB, emphasis in original.
6. Cook to Barbour, October 3, 1926, EHB/NSM; Cook 1926.
7. Cook 1926:334.
8. Cook 1926:335.
9. Cook 1926:336.
10. Cook to "Miss Marjorie" (the assistant to Erwin Barbour at the Nebraska State Museum), November 17, 1926, EHB/NSM. Also Cook to Figgins, November 17, 1926, HJC/AFB. Hrdlička's "Hydra" annotation is on his copy of Cook's article in AHL/NMNH.
11. In the 1960s a portion was set aside as the Agate Fossil Beds National Monument.
12. J. Cook 1925:227–228, 274–276.
13. J. Cook 1925:279; Cook 1968:200–203. For a history of the paleontological work at Agate Ranch, and the relationship between the Cooks and the scientists who worked there, see Vetter 2008.
14. Vetter 2008:299.
15. Cook married Barbour's daughter Eleanor in October of 1910 (Barbour to Gilder, October 24, 1910, EHB/NSM).
16. Matthew and Cook 1909; Cook 1968:197–203.
17. Vetter 2008:302.
18. Cook 1968:201, 204.
19. Cook to Barbour, July 10, 1926, EHB/NSM.
20. Cook to Ingalls, July 2, 1927, HJC/AFB.
21. Vetter 2008:288–291.
22. Figgins to Cook, August 8, 1916, HJC/AFB.
23. On December 4, 1928 Cook was promoted from honorary curator to curator, a title he retained until he left the museum in 1930. The details can be found in various *Annual Reports* of the CMNH for the years 1925 through 1930. Cook also had brief appointments in 1925–26: he was a lecturer at Chadron State Teachers College in Nebraska, and in 1929 taught at Western State College of Colorado.
24. Figgins to Cook, December 5, 1925, HJC/AFB.
25. Cook to Barbour, July 10, 1926, HJC/AFB.
26. Figgins to Hay, December 21, 1926 and February 23, 1927, OPH/SIA.
27. Figgins to Cook, November 16, 1926, HJC/AFB.
28. Biographical information on Figgins is sparse but can be found in Cook (1944), Wormington (1946), and Cook to Colbert, June 20, 1944, VP/AMNH. His position at the USNM was at a sufficiently low level that he is not listed as a staff member of

any title in its *Annual Reports*. His position at the AMNH did not result in a sub-stantial footprint either, though records of his activity do appear sporadically in the *Annual Reports*.

29. Cook 1944:28.

30. Figgins to Cook, February 20, 1925, HJC/AFB; Figgins to Frank Figgins, August 16, 1926, JDF/DMNS. See also Cook to Barbour, October 3, 1926, EHB/NSM and Matthew to Cook, January 25, 1928, HJC/AFB, on Figgins's "temperament." At some point Cook must have found his way into the museum's correspondence files, for with the Cook papers is a page with typed extracts from letters that Figgins wrote in 1925 and 1926 to Taylor and to his son in which Figgins complains about Cook, writes of the stress he is feeling, and displays ill temper. One can only surmise this was done by Cook to document earlier instances of Figgins coming unhinged (the page is attached to the otherwise innocuous letter of Figgins to Cook, August 3, 1926, HJC/AFB).

31. Cook to Matthew, January 20, 1928, HJC/AFB.

32. Cook to Figgins, December 17, 1928, HJC/AFB; Figgins to Cook, December 19, 1928, DIR/DMNS.

33. Cook complained about Figgins loudly and often starting in late 1927. The first inkling of trouble is in Cook to Ingalls, December 14, 1927, HJC/AFB. Other in-stances are later cited in the text. By the spring of 1929 Cook was writing long, angry, and harmful letters, often to prominent individuals and colleagues. The theme of Brown usurping Folsom runs through many of them—for example, Cook to Merriam, February 25, 1929, JCM/CIW; Cook to Jenks, February 26, 1929 and AFB Cook to Gould, March 5, 1929, HJC/AFB; Cook to Hay, March 12, 1929, OPH/SIA. One of the few letters in which Figgins speaks disparagingly about Cook is Figgins to Brown, March 12, 1934, VP/AMNH.

34. Cook 1968:197–198, 203; Skinner et al. 1977:277–278. Matthew and Cook collected from several formations of varying ages within the Snake Creek locality. In general, these formations range from Early to Late Miocene (Skinner et al. 1977:290–315, 361).

35. Matthew and Cook 1909:390.

36. Cook to Osborn, February 25, 1922, in Osborn 1922a:1. Skinner et al. (1977:355) state that Cook made the find in July 1917, and thus held on to it for five years before sending it to Osborn. Cook himself was vague about precisely when the discovery was made, saying only that "I have had here, for some little time, a molar tooth. . . ." (Cook to Osborn, February 25, 1922). All that is known is that the find was made before 1922, because Cook speaks of sending Loomis a paper on the discovery, to be read at the December 1921 meeting of the Paleontological Society (Cook 1922). The account that follows is adapted from and builds on that in Meltzer 2006a:27–28.

37. Osborn to Cook, March 14, 1922, in Osborn 1922:1, emphasis in original.

38. Cook 1922; Cook to Osborn, February 25, 1922; Osborn to Cook, March 15, 1922; Gregory and Hellman to Osborn, March 23, 1922; all in Osborn 1922a:1. Also Gregory and Hellman 1923a:14.

39. Osborn 1922a:2, Osborn 1922b, Osborn 1922c.

40. Osborn 1922a:4. Also Gregory and Hellman 1923a:14

41. Osborn 1922b.

42. Osborn 1925. Osborn brandished *Hesperopithecus* at Bryan over the next sev-eral years, though by the time of the Scopes trial—in which Osborn had an "off-camera" role—he backed off somewhat (Osborn 1925:800–801). By then he was

aware that invoking this specimen might backfire (see Wolf and Mellet 1985; Regal 2002:156–162).

43. Gregory and Hellman 1923a:14.

44. Matthew, in Gregory and Hellman 1923a:12–13; Dingus and Norell 2010:232. Skinner et al. (1977:310) equate the Upper Snake Creek beds of Cook with the Laucomer Member of the Snake Creek formation, Middle Miocene in age.

45. Gregory and Hellman 1923b:518, 526. Smith Woodward would not have wanted Piltdown to lose pride of place among the earliest humans; Miller was simply skeptical of any humans at that antiquity.

46. Osborn to Thomson, June 27, 1922, HFO/AMNH; see also Cook 1927b:115; Skinner et al. 1977:281–282. Barnum Brown attributed the resistance of the landowner (a Mr. Ashbrook) to Cook's having been "instrumental in getting a highway surveyed through Ashbrook's property," which tainted the AMNH's relations with Ashbrook (Dingus and Norell 2010:230).

47. Regal 2002:149.

48. Dingus and Norell 2010:231.

49. Cook to Osborn, January 8, 1926, HJC/AFB. Also Cook 1927b:115–116; Skinner et al. 1977:282.

50. Cook to Osborn, January 8, 1926, HJC/AFB.

51. Cook to Osborn, January 8, 1926, HJC/AFB, emphasis in original.

52. Osborn to Cook, January 27, 1926 and Osborn to Cook, May 16, 1927, HJC/AFB. In Cook's correspondence, he mentioned the Snake Creek artifacts only when writing to Osborn and to Barbour (for example, Cook to Barbour, December 5 and December 10, 1926, EHB/NSM).

53. Cook to Barbour, December 10, 1926, EHB/NSM.

54. Osborn 1927: Figure 1; Osborn to Cook, May 16, 1927, HJC/AFB.

55. Ingalls to Cook, May 10, 1927, HJC/AHC.

56. Kidder and Terry 1927.

57. Ingalls to Cook, May 10, 1927, HJC/AHC, emphasis in original.

58. Cook 1926; Cook to Barbour, October 3, 1926, EHB/NSM.

59. Figgins to Vaughan, March 25 and April 5, 1924, JDF/DMNS and Vaughan to Figgins, around March 30, 1924, JDF/DMNS. Vaughan rarely dated his letters, but their approximate date can be inferred. This discussion of Lone Wolf Creek is adapted and greatly expanded from Meltzer 2006a:28–30.

60. Vaughan to Figgins, around April 23, 1924, JDF/DMNS, emphasis in original; cf. Figgins 1927:229.

61. Figgins to Vaughan, April 5, April 28, and May 16, 1924 and Vaughan to Figgins, around May 1, 1924, JDF/DMNS.

62. Figgins to Cook, August 5, 1924, HJC/AFB; Figgins to Vaughan, March 31, 1925 and Figgins to Hay, March 11, 1925, JDF/DMNS. Figgins's estimate of the number of bison in the deposit was based on the presence of four right humeri from bison (Figgins to Vaughan, March 16, 1925, JDF/DMNS).

63. Figgins to Vaughan, March 16, 1925, JDF/DMNS. These were spear points rather than arrowheads, but as that was not known at the time, this usage is followed.

64. Figgins to Cook, August 5, 1924, HJC/AFB.

65. Figgins to Cook, August 27, 1924, HJC/AFB; Cook to Figgins, September 7, 1924, JDF/DMNS. Hill (2006:146) suggests that it was Barnum Brown who first connected the dots between 12 Mile Creek and Lone Wolf Creek, citing a letter from

Brown to Martin on September 11, 1925. Given that Cook and Figgins were discussing the matter a year earlier, that suggestion is incorrect.

66. Figgins 1925:17.
67. Hay to Figgins, March 5, 1925, JDF/DMNS.
68. Figgins to Hay, March 11, 1925, JDF/DMNS.
69. Figgins to Vaughan, March 16, 1925 and Vaughan to Figgins, March 20, 1925, JDF/DMNS.
70. The existence of the third projectile point is first mentioned in Figgins to Vaughan, May 9, 1925, JDF/DMNS. One positive outcome of Vaughan's revelation of a third point was that his candor on the matter impressed Figgins sufficiently that he promised Vaughan he would see about getting him a "permanent connection" with the museum (Figgins to Vaughan, May 9, 1925, JDF/DMNS). It never happened.
71. Cook to Cattell, July 25, 1925, HJC/AFB. (Note: Cook misdated this letter as June 25, 1925; the date on the original manuscript put its completion in July.) Cattell accepted the paper for *Science* a week later; Cattell to Cook, August 3, 1925, HJC/AFB. On Brown's observations of the Lone Wolf Creek and 12 Mile Creek specimens, see also Figgins to Cook, July 27, 1925, HJC/AFB. Cook had pushed Figgins in February to write up the site for *Science*, but Figgins was content to have Cook do so.
72. Holmes 1925:258.
73. Cook 1925:459.
74. Cook 1925:460.
75. Cook 1925:460. Sellards to Cook, November 25, 1925 and AFB Cook to Sellards December 12, 1925, HJC/AFB. Alfred Romer wrote to say that between Cook's article and Loomis and Gidley's Florida work, "we may be getting somewhere on Pleistocene man," and he agreed that Lone Wolf Creek was older (Romer to Cook, December 5, 1925, HJC/AFB).
76. Cook 1925:459–460.
77. Cook 1925:459.
78. Cook 1925:460, emphasis in original.
79. Goddard 1926:259. See discussion in chapter 7.
80. Meltzer 2006a:30.
81. Hrdlička to Merriam, via Gilbert, November 28, 1925, JCM/CIW.
82. Biographical information on Merriam is from Stock 1951.
83. Biographical information on Stock is from Simpson 1952.
84. Merriam 1924; Stock 1924.
85. Merriam to Stock, May 7, 1924, JCM/LC.
86. Hrdlička to Merriam, May 5, 1924, JCM/LC; Holmes to Merriam, March 29, 1924, USNM/SIA. Holmes also appended a note to Hrdlička's letter.
87. Holmes to Merriam, November 23, 1925, WHH/SIA.
88. Merriam to Holmes, November 27, 1925, WHH/SIA.
89. Matthew to Merriam, November 27, 1925, JCM/LC.
90. Merriam to Matthew, November 30, 1925, JCM/LC.
91. Hay to Cook, April 27, 1926, HJC/AFB; Cook to Hay, December 23, 1926, OPH/SIA. Cook blamed the eight-month delay in replying to Hay on not receiving Hay's letter. He must have received it at some point, for the letter is with the Cook papers.
92. Ingalls to Cook, January 12, 1927, HJC/AFB.
93. Cook 1926:335–336.

94. Cook 1926: 335–336. Hrdlička was doubtful on this point and put a question mark next to this passage in Cook's article.

95. Cook 1926:334–336.

96. Hrdlička to Merriam, November 10, 1926, AH/NAA.

97. Hrdlička to Merriam, November 15, 1926, AH/NAA.

98. Hay to Figgins, November 1, 1926, JDF/DMNS; Figgins to Hay, November 6, 1926, OPH/SIA.

99. Figgins to Vaughan, December 21, 1926; Vaughan to Figgins, January, 8, 1927; Figgins to Vaughan, January 11, 1927; Vaughan to Figgins, January 27, 1927; Figgins to Vaughan, February 25, 1927; Vaughan to Figgins, March 3, 1927; Figgins to Vaughan, March 7, 1927; all in JDF/DMNS. See also Cook to Hay, December 23, 1926 and Figgins to Hay, December 21, 1926, OPH/SIA.

100. There are different accounts of when and how the Folsom site was discovered (Meltzer 2006a:33–34), Cook's (1947) being the most egregious and inaccurate in its failure to properly credit McJunkin. In 1928 Brown made "rather careful inquiries . . . of Schumacher [Lud Shoemaker] and the Owens family," who lived on the ranch where the site was located, and they agreed McJunkin was the first one to see the bones (Brown to Figgins, May 28, 1937, VP/AMNH).

101. A photograph taken on the occasion of the December 10, 1922 visit appears in Meltzer 2006a:34. The date of the visit is from Schwachheim's diary (in Meltzer 2006a: Appendix 1).

102. Cook (1947) claimed that he and Howarth were prior acquaintances and that Howarth's reading of his article on Lone Wolf Creek prompted him to visit Denver. Both claims may be true, but there are enough factual errors in Cook's account to raise doubt about it.

103. Howarth to Figgins, February 4 and February 16, 1926 and Figgins to Howarth, February 6 and February 23, 1926, JDF/DMNS; Cook to Barbour, February 15, 1926, EHB/NSM.

104. Figgins to Brown, May 26, 1926, BB/AMNH; Figgins to Taylor, June 21, 1926, JDF/DMNS. Cook to Osborn, July 9, 1927, VP/AMNH.

105. Figgins 1927:232. See also Figgins to Hay, November 6, 1926, OPH/SIA.

106. Meltzer 2006a:35.

107. Brown to Figgins, June 2, 1926 and Figgins to Brown, June 18, 1926, BB/AMNH.

108. The specifics of the museum's excavation plan and procedure and how it unfolded are described in Meltzer 2006a:84–86.

109. Schwachheim Diary July 14, 1926, in Meltzer 2006a:317–318.

110. Figgins to Frank Figgins, July 20, 1926 and Figgins to Howarth, July 22, 1926, JDF/DMNS; Figgins to Brown, July 23, 1926, BB/AMNH.

111. Figgins to Brown, July 23, 1926, BB/AMNH; Meltzer 2006a:35–36.

112. Frank Figgins to Figgins, July 10 [*sic*], 1926 and Figgins to Taylor, July 20, 1926, DIR/DMNS. Also, Figgins to Brown, July 23, 1926, BB/AMNH. The date on Frank Figgins's letter of July 10 is incorrect; the first point was discovered on July 14, 1926. This conclusion is based on several pieces of evidence, including Schwachheim's diary and other letters by Figgins and Howarth over that same period.

113. Figgins to Brown, June 18, July 14, and July 23, 1926, BB/AMNH. Also Cook to Barbour, July 10, 1926, EHB/NSM.

114. Figgins to Cook, September 25, 1926, HJC/AFB.

115. Brown to Cook, August 26, 1926, HJC/AFB; Brown to Figgins, BB/AMNH.

116. Figgins to Cook, September 25, 1926, HJC/AFB. The complicated thread of that minor tempest can be traced through Brown to Figgins, September 23 and September 25, 1925; Figgins to Brown, September 26, 1925; Brown to Figgins, October 8, 1925; Cook to Brown, November 3, 1925; Brown to Figgins, June 2, 1926; Figgins to Brown, June 18, 1926; all in BB/AMNH. Although it pertained to an article by Cook, Figgins—for reasons that are unclear—assumed "it was doubtless to hit at me personally" (Figgins to Cook, December 6, 1926, HJC/AFB). There is no record of why or whether that was the case.

117. Figgins to Hay, November 6, 1926, OPH/SIA.

118. Figgins to Hay, November 8, 1926, OPH/SIA.

119. Although their articles were written about the same time and published in the same issue, there was limited coordination between Figgins and Cook, as Figgins explained to Brown: "My only suggestion to Cook was that he confine himself strictly to the geology and palaeontology of the immediate vicinity of the points at which the fossils and artifacts were found associated, and he did not show me his account when he had written it" (Figgins to Brown, April 9, 1927, BB/AMNH).

120. Figgins to Cook, November 16, 1926, HJC/AFB.

121. Figgins to Hay, December 21, 1926 and February 23, 1927, OPH/SIA.

122. Figgins to Cook, November 16 and 23, 1926, HJC/AFB. Figgins subsequently wrote Martin for photographs of the Twelve Mile Creek projectile point (Figgins to Martin, December 16, 1926, JDF/DMNS).

123. Figgins to Brown, November 16, 1926, BB/AMNH.

124. Brown to Figgins, November 30, 1926, BB/AMNH; Figgins to Cook, December 6, 1926, HJC/AFB, discusses Brown's reply.

125. Hay to Figgins, November 17, JDF/DMNS.

126. Figgins to Hay, December 28, 1926, HJC/AFB.

127. Figgins to Howarth November 27, 1926, JDF/DMNS.

128. Cook to Barbour, December 5, 1926, EHB/NSM. That Cook likened the task to the one that faced Ulysses S. Grant's troops at Spotsylvania in 1864 is in keeping with his penchant for rhetorical excess and his inflated view of what was at stake.

129. Fewkes to Figgins, November 27, 1926, JDF/DMNS.

130. Figgins to Fewkes, December 1, 1926, JDF/DMNS.

131. Hay to Figgins, December 17, 1926, JDF/DMNS.

132. Figgins to Hay, December 21, 1926 and Cook to Hay, December 23, 1926, OPH/SIA.

133. Figgins to Cook, December 28, 1926, JDF/DMNH.

134. Figgins to Hay, December 21, 1926, OPH/SIA.

135. Hay to Figgins, December 31, 1926, JDF/DMNS, emphasis mine.

136. Lucas 1899a:756.

137. Hay 1914:170.

138. Hay 1923:14–15, 258.

139. Hay 1924:195.

140. Bryan to Wetmore, August 16, 1928, SEC/SIA.

141. Hay to Figgins, March 5, 1925 and Figgins to Hay, March 11, 1925, JDF/DMNS.

142. Cook 1926:335.

143. Brown to Figgins, June 2, 1926; Figgins to Brown, June 18, 1926; Brown to Figgins, August 26, 1926; all in BB/AMNH.

144. Figgins to Cook, November 30, 1926, HJC/AFB.

145. Figgins to Cook, November 30, December 6, and December 21, 1926, HJC/AFB.

146. Figgins to Hay, November 30, 1926 and Cook to Hay, December 23, 1926, OPH/ SIA; Hay to Cook, December 31, 1926, HJC/AFB.

147. Hay to Figgins, January 18, 1927, JDF/DMNS.

148. Ingalls to Cook, January 12, 1927 and Ingalls to Slosson, March 24, 1927, HJC/AFB.

149. Priestly to Cook, January 19, 1927, HJC/AFB. Priestly asked in return that Cook assure the editor at *Scientific American* that his letter to them had been of value to Cook, in the hopes that they would take seriously any subsequent correspondence he sent (Priestly to Cook, January 31, 1927, HJC/AFB). Figgins to Brown, January 21, 1927, JDF/DMNS.

150. Figgins to Brown, February 8, 1927, BB/AMNH, emphasis added.

151. Figgins to Brown, February 8, 1927, BB/AMNH; Cook to Martin, February 8, 1927, HJC/AFB; Figgins to Hay, February 7, 1927, OPH/SIA.

152. Cook 1927a:246–247.

153. Cook to Hay, February 25, 1927, OPH/SIA; Cook 1927a:246–247; Figgins 1927:235.

154. Hay to Cook, March 11, 1927, JDF/DMNS.

155. Schwachheim was in Frederick from late February to late April, 1927 (Schwachheim Diary).

156. Cook 1927a:247.

157. Cook 1927a:247.

158. Cook 1927b:117. Cook's *Scientific American* article not only provided absolute age estimates for Frederick, it also ignored both Lone Wolf Creek and Folsom and brought in the evidence from Snake Creek (Cook to Ingalls April 10, 1927, HJC/ AHC). Figgins had attempted to dissuade Cook from publishing the paper, though mostly because he feared Cook was merely replicating his *Natural History* article (Figgins to Cook April 13, 1927, HJC/AHC). He need not have worried.

159. Proofs for the papers were sent out only in early July (Cook to Figgins, July 6, 1927, JDF/DMNS). The official date of publication was August 3, 1927 (Cook to Sellards, January 21, 1931, EHS/UT).

160. For example, Daniel 1975:275; Willey and Sabloff 1980:121; Wilmsen 1965:181; Wormington 1957:23–25. Even after Folsom was confirmed, their contemporaries tended not to cite their papers, or at least not without also citing Brown, Bryan, or Kidder (for example, Roberts 1939, 1940a).

161. Cook claims he could not work out the "exact relations" of the geology of Folsom area "for lack of time" (Cook 1927a:244), though he was there for weeks. Yet, his one day at Frederick was sufficient for the task. The number of words devoted to each site in Figgins (1927) and Cook (1927a) provide one measure of their sense of the importance of each site (the numbers are rounded):

Site	Figgins 1927	Cook 1927a
Lone Wolf Creek	760	770
12 Mile Creek	130	0
Folsom	810	620
Frederick	1,400	1,240
Total	3,100 words	2,630 words

162. Figgins to Hay, February 23, 1927, OPH/SIA; Figgins to Brown, April 9, 1927, VP/ AMNH; Cook 1927a:243, also 241–242, 247.

163. Brown to Figgins, April 4, 1927, VP/AMNH. Figgins admitted he could see Brown's point about needing to tone down his introduction, but he explained he had made it pointed so as to "excite resentment and draw all the fire and brimstone" to himself; Figgins to Brown, April 9, 1927, VP/AMNH. Cook's article had to be revised as well, though not because it was personal or controversial but because it was not lively enough for *Natural History* (Brown to Figgins, April 4, 1927, VP/AMNH).

164. Figgins 1927:229.

165. Figgins to Hay, October 22, 1927, OPH/SIA.

166. Figgins to Brown, February 16 and April 9, 1927, VP/AMNH.

167. Cook 1927a:241–242.

168. Figgins 1927:230–231, 234. Schwachheim pushed Figgins to see that the points were likely spear points rather than arrowheads (Schwachheim to Figgins, December 23, 1926, DIR/DMNS). Figgins ignored Schwachheim, coming around only that December when corrected by A. V. Kidder (Kidder to Figgins, December 15, 1927 and Figgins to Kidder, January 6, 1928, JDF/DMNS).

169. Cook 1927a:243–244; Figgins 1927:234.

170. Cook 1927a:241–242.

171. Cook 1927a:247; also Cook 1927b:117; Figgins 1927:237–239.

172. Cook 1927b:117, emphasis in original.

173. Nelson to Figgins, August 16, 1927, JDF/DMNS. For biographical information on Nelson, and particularly the stratigraphic work in which his reputation was earned, see Snead 2001; also Lyman and O'Brien 2006.

174. Hay to Cook, March 11, 1927, DIR/DMNS.

175. Figgins to Schwachheim, March 4, 1927 and Schwachheim to Figgins, March 10, 1927, JDF/DMNS.

176. Figgins to Schwachheim May 6, 1927, JDF/DMNS; see also Figgins to Cook, April 11, 1927, HJC/AFB; Figgins to Brown, April 28, 1927, BB/AMNH; Howarth to Cook, May 4 and June 8, 1927, HJC/AFB.

177. Davis 1927a. The appearance of the story in *Science News-Letter* caused a minor tempest. Davis originally contacted Figgins for more details. Figgins replied that his and Cook's manuscripts were being published by *Natural History*. By that Figgins meant they could not be shared, but Davis took it to mean he should contact the AMNH for copies. Davis did, and received copies. Davis then visited Hay at the Smithsonian to see the fossil specimens, and he prepared a story that he sent to Figgins (Davis to Matthew, March 5, 1927, VP/AMNH; Davis to Figgins, March 24, 1927, DIR/DMNS; Hay to Figgins, March 25, 1927, JDF/DMNS). Figgins was deeply unhappy to see the story and immediately sent multiple telegrams to Barnum Brown trying to ascertain if someone at the AMNH had supplied Davis with the manuscripts, ultimately withdrawing both his and Cook's manuscripts because "Davis' course has spoiled them," rendering later publication redundant, and doing so "as means of relieving *Natural History* of embarrassment" (Figgins to Brown, telegram of March 29, 1927, and letters of March 30 and April 9, 1927, BB/AMNH). Brown telegrammed back to ask Figgins to reverse his decision, explaining that *Natural History* saw "no reason why *Science Service* should not publish the news part of your article because it would not interfere with *Natural History* treatment (Brown to Figgins April 4, 1927, JDF/DMNS). Figgins, obviously, consented to publication, though in his account of the events to Cook he remained unhappy (Figgins to Cook, April 13 and April 14, 1927, HJC/AHC). At *Scientific American* Albert Ingalls was upset as well, because the *Science News-Letter* story made no

mention of his magazine's role in connecting Priestly with Cook, and thus they lost the "credit and publicity" attached to the story, which as he complained to Cook "has *real value* to any magazine" (Ingalls to Slosson, March 24, 1927 and Davis to Ingalls, March 25, 1927, HJC/AHS, emphasis in original).

178. Hay to Figgins, April 5, 1927, JDF/DMNS.

179. Figgins to Hay, April 11, 1927, OPH/SIA.

180. Hay to Figgins, April 25, 1927, JDF/DMNS; Figgins to Hay, April 29, 1927, OPH/SIA.

181. Figgins to Hay, July 1 and September 29, 1927, OPH/SIA; also Figgins to Brown, June 8, 1927, BB/AMNH; Lucas to Figgins, November 18, 1927, JDF/DMNS.

182. Meltzer 2006a:37.

183. Lucas to Figgins, November 18, 1927, JDF/DMNH.

184. Figgins to Brown, June 8, 1927, BB/AMNH.

185. Figgins to Brown, June 8, 1927, BB/AMNH.

186. Figgins to Brown, June 8, 1927, BB/AMNH.

187. Hay to Figgins, October 18, 1927, JDF/DMNS. Figgins subsequently backpedaled, explaining he was only referring to the fact that Hrdlička was the most courteous of the Washington anthropologists (Figgins to Hay October 22, 1927, OPH/SIA). Figgins repeated the tale and made similar disparaging remarks about Holmes, Judd, and Walter Hough (briefly present as well) in Figgins to Brown, June 8 and June 30, 1927, BB/AMNH; Figgins to Cook, July 22, 1927, JDF/DMNS; Figgins to Hay, July 1, September 7, September 29, October 13, and October 22, 1927, OPH/SIA; Figgins to Kidder, October 17, 1927; Figgins to Nelson, August 24 and September 27, 1927; Figgins to Wetmore, September 7 and September 12, 1927; the last three in JDF/DMNS.

188. Figgins to Schwachheim, December 16, 1926 JDF/DMNS; also Schwachheim to Figgins, December 23, 1926, June 4 and June 11, 1927, JDF/DMNS. Unusually heavy rains in mid-June further slowed Schwachheim's efforts. For details on the fieldwork at Folsom in 1927, see Meltzer 2006a.

189. For example, Figgins to Brown, August 30, 1927, BB/AMNH; Figgins to Hay, August 30, 1927, OPH/SIA. Figgins's telegram to Hay mentioned he had telegraphed Hrdlička as well. Hay returned to Washington from a two-week trip to see the Frederick site just in time to receive his telegram. Jenks was in Taos, NM at the time, but roads "in the worst wet conditions" prevented him from reaching Folsom "at *the* right moment" (Jenks to Figgins, September 15, 1927, JDF/DMNS, emphasis in original). Jenks did see Frederick and Lone Wolf Creek as a consolation but admitted it "will always . . . be a sore regret to me that we did not hazard the trip over to Folsom" (Jenks to Figgins, October 3, 1927, JDF/DMNS).

190. Figgins to Schwachheim, August 31, 1927, JDF/DMNS.

191. Hay to Figgins, August 31, 1927 and Wetmore to Figgins, September 2, 1927, JDF/DMNS. I previously understood Hrdlička was in Alaska in late August of 1927 (Meltzer 2006a:37), and on the basis of Wetmore's letter that is incorrect.

192. Woodbury 1993.

193. Dorsey to Roberts, September 1, 1927, BAE/NAA; Wetmore to Figgins, September 2, 1927, JDF/DMNS. Schwachheim to Figgins, September 4, 1927, JDF/DMNS and Schwachheim Diary, September 4, 1927 records arrived over those several days. Roberts's later (1935:5) recollections of the dates of his visit do not square with the contemporary archival material.

194. Figgins to Hay, September 15, 1927, OPH/SIA, describes in some detail what the photograph shows of the stratigraphy. His description is quoted in Meltzer 2006a:112.
195. Figgins to Hay, September 7, 1927, OPH/SIA; Figgins to Nelson, September 7, 1927 and Figgins to Wetmore, September 7, 1927, JDF/DMNS.
196. Willey 1967:293, 298. See also Givens 1992; Willey and Sabloff 1980.
197. Figgins to Wetmore and Figgins to Hay, both September 7, 1927, JDF/DMNS, emphasis in original. Figgins to Nelson September 27, 1927, JDF/DMNS. Schwachheim alerted Figgins to the fact the Folsom point was new to Kidder as well (Schwachheim to Figgins, September 11, 1927, JDF/DMNS).
198. Roberts to Fewkes, September 13, 1927, BAE/NAA.
199. Schwachheim Diary, September 6 and 8, 1927 in Meltzer 2006a:323.
200. Kidder to Figgins, October 13, 1927, JDF/DMNS.
201. Kidder 1927:5, also Kidder 1936.
202. Kidder to Merriam, October 7, 1927, JCM/CIW; Kidder to Figgins, October 13, 1927, JDF/DMNS. Merriam would prove to be good on his word (§9.1–§9.2).
203. Figgins to Kidder, October 17 and December 8, 1927 and Kidder to Figgins, November 19 and December 15, 1927, JDF/DMNS. On Roberts, see Judd 1966:1227. Hrdlička to Merriam, June 2, 1927, AH/NAA. On Kidder's nomination to the National Academy of Sciences, see Boas to Fewkes, November 28, 1927, AH/NAA and Fewkes to Boas, November 30, 1927, FB/APS. Kidder was elected to the NAS in 1936. See also Meltzer 2006a:42.
204. Cook to Matthew, September 9, 1927, HJC/AFB. The prairie dog also figures prominently in the *Science News-Letter* story on Folsom (Davis 1927c). It is presumably used here as a symbol for all burrowing animals, for it is in fact a relatively uncommon species in the Folsom region.
205. Figgins to Brown, June 8, 1927, BB/AMNH; Figgins to Hay, July 8, 1927, OPH/SIA.
206. Cook to Matthew, September 9, 1927, HJC/AFB; Figgins to Hay, September 7, 1927, OPH/SIA; Figgins to Jenks, September 19, 1927 and Figgins to Nelson, September 7, 1927, JDF/DMNS; Figgins to Osborn, September 12, 1927, VP/AMNH; Figgins to Wetmore, September 7 and September 12, 1927 and Wetmore to Figgins, September 8, 1927, JDF/DMNS.
207. Hay to Figgins, November 24, 1927, JDF/DMNS; Hay to Brown, December 5, 1927, BB/AMNH.
208. Brown to Hay, December 12, 1927, BB/AMNH; Hay to Figgins, December 16, 1927 and Figgins to Hay, December 20, 1927, JDF/DMNS. The type descriptions appear in Hay and Cook 1928.
209. See Hay and Cook 1930; Figgins 1933b:19–22. Within a few years the Folsom bison was synonymized with *Bison antiquus* subspecies *figginsi* (Skinner and Kaisen 1947). Today it is identified as *B. antiquus occidentalis* (McDonald 1981:94; Meltzer 2006a:209).
210. McDonald 1981:40.
211. Simpson to Eiseley, August 26, 1942, LCE/UP. For a detailed discussion, see Meltzer 2006a:206–209.
212. Hay to Cook, January 17, 1928, HJC/AFB.
213. Leverett 1930b:193–194. Also Leverett 1930a.
214. Cook to Hay, January 25, 1928, OPH/SIA, emphasis in original. Also Cook to Loomis, December 20, 1927, HJC/AFB.

215. Cook to Hay, January 25, 1928, OPH/SIA.
216. Davis 1927c.
217. Schwachheim to Figgins, September 11, 1927, JDF/DMNS; Schwachheim Diary, September 29–20, 1927. On sending the artifacts to New York for study, see Figgins to Cook, October 12, 1927, HJC/AFB; Figgins to Hay October 13, 1927 and Figgins to Schwachheim, October 11, 1927, JDF/DMNS. Hay to Figgins, October 24, 1927, JDF/DMNS reports that Brown was at the USNM in Washington the day before and showed the artifacts to various individuals, including Merriam. Clark Wissler also evidently tagged Nelson as the expert to consult; see Figgins to Hay, September 29, 1927, OPH/SIA.
218. Figgins to Granger, September 30, 1927, VP/AMNH; Figgins to Schwachheim, September 13, 1927, JDF/DMNS. Jenks had come close, visiting Frederick and Lone Wolf Creek on his trip west in August of 1927, and Figgins regretted that he had not been able to get to Folsom, because "no other anthropologist has visited the Oklahoma and Texas localities, and [Figgins] was in hopes [Jenks] would see all" (Figgins to Jenks, October 10, 1927, JDF/DMNS).
219. Figgins's efforts and the replies he received are in Figgins to Nelson, September 7 and 27, 1927, JDF/DMNS; Figgins to Granger, September 30, 1927 and Figgins to Osborn, September 12, 1927, VP/AMNH; Granger to Figgins, September 26, 1927 and Nelson to Figgins, September 13, 1927, JDF/DMNS. Figgins even tried manipulating Osborn by having Wetmore write him about the importance of Nelson and Granger visiting the sites (Figgins to Wetmore, September 12, 1927, JDF/DMNS). Figgins insisted Osborn's lack of reply did not bother him, but he mentioned it on several occasions (for example, Figgins to Cook, November 8, 1927, HJC/AFB and Figgins to Kidder, December 8, 1927, JDF/DMNS).
220. Figgins to Kidder October 17, 1927, JDF/DMNS.
221. Figgins to Kidder, October 17 and December 8, 1927 and Kidder to Figgins, December 15, 1927, JDF/DMNS; Figgins to Wetmore, November 14 and December 9, 1927 and Wetmore to Figgins, November 19, 1927, JDF/DMNS.
222. Figgins to Kidder, December 8, 1927, JDF/DMNS.
223. Figgins gave the green light to others to speak to the *Science News-Letter* editor in Figgins to Brown, October 15, 1927 and Figgins to Wetmore, October 22 and November 14, 1927, JDF/DMNS. The other quote is from Figgins to Kidder, December 8, 1927, JDF/DMNS.
224. Figgins to Nelson, August 24, 1927, JDF/DMNS.
225. Kidder to Figgins, November 19 and December 15, 1927, JDF/DMNS. Schwachheim had tried to convince Figgins on this point a year earlier, but Figgins did not believe him (§8.8).
226. Nelson to Figgins, August 16 and September 13, 1927, JDF/DMNS. In *Outdoor Life*, Figgins (1928b:82) stated, "We will never know the locality of the cradle of the human race," putting it in a context that suggested he was willing to grant the possibility that cradle was in America, a point he had made explicitly in correspondence (Figgins to Gould, October 14, 1927, JDF/DMNS).
227. Nelson to Figgins, September 13, 1927, JDF/DMNS. Nelson repeated his assurances in Nelson to Figgins, December 20, 1927, JDF/DMNS.
228. Figgins to Nelson, September 27, 1927, JDF/DMNS.
229. Figgins to Nelson, September 27, 1927; Figgins to Kidder October 17, 1927; Kidder to Figgins November 19 and December 15, 1927; Figgins to Kidder, December 8 and December 15, 1927; Nelson to Figgins, September 13, 1927; all in JDF/DMNS.

230. Figgins to Nelson, September 27, 1927, JDF/DMNS.
231. Figgins to Jenks, December 16, 1927, JDF/DMNS.
232. The "bare arms" quote is from Figgins to Hay, September 29, 1927, OPH/SIA. See also Figgins to Hay, October 13 and October 22, 1927, OPH/SIA. By 1928, defending Folsom's honor—if such a defense was needed—had become Brown's problem, and Figgins did little more than offer advice: "Go after them if they so much as cheep. They need it on general principles" (Figgins to Brown, December 17, 1928, BB/AMNH).
233. Figgins to Hay, September 29, 1927, OPH/SIA.
234. Hay to Figgins, October 18, 1927, JDF/DMNS.
235. Cook to Loomis, December 20, 1927, FBL/AC.
236. Hay to Priestly, September 14, 1927, OPH/SIA; Figgins to Holloman, September 30, 1927 and Hay to Figgins, September 16 and October 7, 1927, JDF/DMNS; Gould to Hay, October 10, 1927, OPH/SIA; Figgins to Gould, October 14, 1927, JDF/DMNS.
237. Cook 1927b:116.
238. Matthew to Cook, August 11, 1927 and Cook to Matthew, September 9, 1927, HJC/AFB.
239. Cook to Figgins, September 8, 1927, HJC/AFB, emphasis in original.
240. Ingalls to Cook, September 29, 1927 and Cook to Figgins, September 30, 1927, HJC/AFB.
241. Figgins to Howarth, October 10, 1927, JDF/DMNS. Figgins was a sufficiently adept administrator that he happened to mention Ray's discoveries in a letter to Childs Frick, who funded much of the paleontological work at the AMNH, and mused aloud his "regret that we cannot keep a man constantly engaged about the gravel pits" (Figgins to Frick, October 12, 1927, JDF/DMNS). Frick must have shown some slight interest, for Figgins gently pushed the matter in a subsequent letter (Figgins to Frick, October 25, 1927, JDF/DMNS), but ultimately funding was not forthcoming from Frick.
242. Figgins to Cook, October 12, 1927, HJC/AFB.
243. Figgins to Cook, October 12, 1927, HJC/AFB.
244. Figgins to Cook, October 12, 1927, JDF/DMNS.
245. Cook 1928a:38; Ingalls to Cook, March 28, 1928 and Cook to Ingalls, April 6, 1928, HJC/AFB.
246. Cook to Hay, January 29, 1928, OPH/SIA.
247. Figgins 1928b:82; Cook to Hay, January 29, 1928, OPH/SIA; Hay to Cook, February 13, 1928, HJC/AFB.
248. Figgins to Brown, October 15, 1927, JDF/DMNS; Figgins to Berger, November 17, 1927, VP/AMNH. Brown told Figgins that his paper would appear in the March-April issue of *Natural History*, but in the same letter he mentioned that the magazine was getting a new editor and that issue would be the first "under the new regime" (Brown to Figgins, December 8, 1927, JDF/DMNS). There is no record that Figgins asked Brown why his paper had not appeared.
249. Brown to Figgins, December 8, 1927; Kidder to Figgins, October 13, 1927; Wetmore to Figgins, November 3, 1927; Figgins to Wetmore, November 14, 1927; all in JDF/DMNS. Haury to Meltzer, September 10, 1981, author's personal collection. Expressions similar to Haury's are recorded in Mason 1966:196.
250. Dixon to Hrdlička, November 27, 1927, AH/NAA. Dixon was primarily an ethnologist but had kept an interest in archaeology. See Browman and Williams 2013:

211–215. That the meeting sessions were organized just a month before is testimony to how much has changed in the last century.

251. Hay 1927c; Cole 1928.

252. Hrdlička to Dixon, December 2, 1927, AH/NAA. Harry Shapiro of the AMNH took Hrdlička's spot on the program.

253. Brown to Figgins, December 8 and 19, 1927, JDF/DMNS.

254. Willey 1967:297.

255. Jenks to Figgins, January 4, 1928; JDF/DMNS; Roberts 1937b:154–155.

256. Brown to Figgins, January 10, 1928; Figgins to Brown, January 10, 1928; Figgins to Jenks, January 9 and January 16, 1928; Jenks to Figgins, January 4 and January 11, 1928; all in JDF/DMNS. Cook briefed Loomis in Cook to Loomis, December 12 and December 20, 1927, FBL/AC.

257. Brown to Figgins, January 10, 1928, JDF/DMNS.

258. Roberts 1935:5–6; Roberts 1937b:155. Collins also knew Hrdlička well and was witness to the awkward dynamic between the long-ensconced and inflexible Hrdlička and the newly hired Roberts—which is why he suspected Roberts was remembering more his own conversations at the Smithsonian with Hrdlička and not what happened at the AAA meeting (Meltzer 1983:36). James B. Griffin thought Roberts was "being a bit dramatic" (personal communication, December 30, 1980).

259. Hallowell 1928. The final resolution is identical (save for the addition of a "Be it re-solved") to the draft offered by Jenks, which is in Jenks to Figgins, January 4, 1928, JDF/DMNS.

260. Figgins to Gregory, December 30, 1927, JDF/DMNS.

261. Figgins 1928a.

262. Cook to Figgins, November 1, 1927, HJC/AFB.

263. Figgins to Jenks, December 12, 1927, JDF/DMNS. Also Figgins to Kidder, December 8, 1927, JDF/DMNS.

264. Cook to Ingalls, December 14, 1927, HJC/AHC; Cook to Loomis, December 20, 1927, FBL/AC; Figgins to Gregory, December 12, 1927 and Gregory to Figgins, December 19, 1927, JDF/DMNS.

265. Bailey to Cook, February 12, 1947, HJC/AHC . Also Cook to Barbour, October 3, 1926, EHB/NSM.

266. Wetmore to Figgins, November 19, 1927 and Figgins to Wetmore, December 9, 1927, JDF/DMNS.

267. Cook to Ingalls, December 14, 1927, HJC/AHC.

268. Cook to Ingalls, December 14, 1927, HJC/AHC.

269. Figgins himself was aware his son was being irresponsible with museum funds: for example, Figgins to Frank Figgins, July 22 and July 24, 1926; Figgins to Howarth, July 22 and July 24, 1926; Howarth to Figgins, July 22, 1926; all in JDF/DMNS. Cook provided some (though, again, not many) of the details a year or so later in Cook to Hay, March 12, 1929, OPH/SIA; Cook to Jenks, April 24, 1929 and AFB Cook to Osborn, April 24, 1929, HJC/AFB.

270. Cook to Howarth, April 7, 1928, HJC/AFB; Cook to Ingalls, December 14, 1927, HJC/AHC; Cook to Jenks, May 20, 1928; AFB Cook to Loomis, December 20, 1927;AFB Cook to Matthew, January 20, 1928;AFB Cook to Osborn, March 21, 1928; the last four HJC/AFB.

271. Dorothy Cook Meade to D. Meltzer, July 9, 1993. See also Figgins to Brower, October 13, 1927, JDF/DMNS.

272. Cook to Matthew, January 20, 1928, HJC/AFB; minutes of the CMNH Board of Trustees, February 10, 1928 and November 9, 1928, DIR/DMNS (the board passed a resolution "earnestly" hoping the trip would do Figgins good).

273. Figgins to Taylor, January 18, 1928, JDF/DMNS.

274. Brown to Figgins, January 10, 1928, JDF/DMNS; Brown to Taylor, February 6, 1928, DIR/DMNS; Cook to Brown, February 17, 1928, BB/AMNH; Cook to Ingalls, February 27, 1928, HJC/AHC; Cook to Howarth, April 7, 1928, HJC/AFB; Figgins to Taylor, March 9, 1928, JDF/DMNS.

275. Cook to Osborn, March 21, 1928, HJC/AFB.

276. Brown to Cook, March 29, 1928; Cook to Brown, March 30, 1928; Brown to Frick, August 8, 1928; all in VP/AMNH. Taylor to Figgins, February 15, 1928, DIR/DMNS; minutes of the CMNH Board of Trustees, February 10, 1928,DIR/DMNS.

277. Brown to Howarth, April 7, 1928, BB/AMNH; Brown to Cook, April 9, 1928, HJC/AHC; Cook to Brown, April 22, 1928 and Brown to Cook, April 27, 1928, BB/AMNH.

278. Brown to Cook, April 9, 1928, HJC/AHC. Details of the fieldwork are from Meltzer 2006a, chapter 4.

279. Hrdlička to Brown, March 19, 1928, AH/NAA; Barton 1941:311.

280. Brown to Cook, July 23, 1928, HJC/AHC.

281. Judd to Kidder, February 8, 1926, NMJ/NAA.

282. Cook to Ingalls, August 9, 1928, HJC/AHC; Judd to Wetmore, July 30, 1928 and Wetmore to Bryan, July 30, 1928, SEC/SIA; Cook to Hay, July 18, 1927, OPH/SIA; Roberts to Dorsey, August 9, 1928, BAE/NAA. Roberts was grateful his travel was reimbursed but would not have regretted it if it had not been.

283. Brown to Frick, August 8, 1928, VP/AMNH.

284. Boas 1928:363; Brown to Hay, September 10, 1928, VP/AMNH.

285. Brown to Frick, August 8, 1928, VP/AMNH. Brown made sure that Hay heard they had obtained more female cow skulls that were the same as what Hay was calling *B. occidentalis*: Brown to Hay, August 27, 1928, BB/AMNH. For differing contemporary views on the post–Pleistocene survival of bison, see Cook to Hay, January 25, 1928, OPH/SIA; Romer to Cook, May 26, 1931, HJC/AFB; Simpson to Brown, July 25, 1928, VP/AMNH. The problem of precisely when these bison went extinct lingered: see Antevs 1935b:302–303; Bryan 1941:507; Colbert 1937; Howard 1936a:401; Howard 1936b:317; Kidder 1936:144; Roberts 1936b:345; Roberts 1940a:104–105; Romer 1933:70. But see also §9.6.

286. Bryan to Wetmore, August 16, 1928, SEC/SIA.

287. Brown to Simpson, August 8 and October 1, 1928; Brown to Kaisen, September 4, 1928; Kaisen to Brown, September 5 and September 17, 1928; all in VP/AMNH. See also Meltzer 2006a:93.

288. Brown to Frick, August 8, 1928, VP/AMNH; Brown 1929:128. The "shingle shale" overlying the bone bed had washed in during Late Pleistocene times; see Meltzer 2006a: chapter 5 and pp. 229–234.

289. Brown 1928a:828; Brown 1929; Brown 1932:82.

290. Bryan 1929:129, presented more fully in Bryan 1937:142–143.

291. Bryan to Wetmore, August 3, 1928, SEC/SIA.

292. Cook to Hay, September 4, 1928, OPH/SIA.

293. Cook to Ingalls, August 9, 1928, HJC/AHC. Also Cook to Loomis, November 12, 1928, HJC/AFB; Cook to Ingalls, January 6, 1929, HJC/AHC; Cook to Wissler, March 25, 1929, HJC/AFB; Cook 1928b.

294. Osborn 1927:Figure 1.
295. Figgins to Brown, June 30, 1927, BB/AMNH.
296. Osborn to Cook, January 27, 1926, HJC/AFB; Gregory 1927:580–581. Gregory intended to publish his rejection of *Hesperopithecus* in *Natural History*, but its editors—likely with one eye on Osborn—declined to publish his paper, and so it appeared in *Science* (Regal 2002:149). Skinner reports on Osborn's efforts to make amends in Skinner et al. 1977:355. See also Dingus and Norell 2010:231.
297. Cook to Osborn, July 9, 1927, VP/AMNH, emphasis in original.
298. Cook tried to get Osborn to say something as late as the 1929 AAAS meeting, but Osborn declined, though he did lend Cook his lantern slides: Osborn to Cook, December 18, 1929 and Cook to Osborn, March 11, 1930, HJC/AFB.
299. Nelson 1928a:316–317. See also Skinner et al. 1977:282.
300. Hrdlička to Nelson, March 23, 1928, AH/NAA.
301. Cook to Ingalls, February 27, 1928, HJC/AHC.
302. For example, Cook to Barbour, December 5 and December 10, 1926, EHB/NSM; Cook to Loomis, November 12, 1928, HJC/AFB. Other letters, as cited above. By 1928, however, Osborn was unwilling to help, Barbour had become Cook's ex-father-in-law and was no longer replying to his letters, and neither Ingalls nor Thomson had the scientific credibility to come to the aid of the cause.
303. Cook 1952:17–18.
304. Davis 1927b; Gould to Hay, October 10, 1927, OPH/SIA.
305. Hay to Cook July 15 and August 13, 1927, HJC/AFB; Hay to Figgins, July 15, August 8, September 16, and October 7, 1927, JDF/DMNS; Figgins to Hay, September 29, 1927, OPH/SIA; Figgins to Cook, October 12, 1927, HJC/AFB.
306. Gould to Cook, October 11, 1927, HJC/AFB; Gould to Figgins, October 11, 1927, JDF/DMNS. It is telling that Gould devoted more space in his memoirs to that *Glyptodon* discovery than he did to the artifacts supposed to have come from the site (Gould 1959:225–227).
307. Figgins to Gould, October 14, 1927, JDF/DMNS. Also Cook to Gould, October 22, 1927, HJC/AFB; Figgins to Hay, October 13, October 22, and October 28, 1927 and Cook to Hay, December 6, 1927, OPH/SIA; Hay to Figgins, October 7, 1927, JDF/DMNS; Hay 1928a:442.
308. Figgins to Hay, October 22, 1927 and Gould to Hay, October 10, 1927, OPH/SIA. Gould to Figgins, October 18, 1927, JDF/DMNS reported it was only a six- to seven-hour drive to the site from Norman.
309. Cook to Hay, December 6, 1927, OPH/SIA. Also Cook to Loomis, December 20, 1927, HJC/AFB.
310. Spier 1928a; Gregory 1927.
311. Cook to Brown, February 17, 1928, VP/AMNH; Cook to Hay, February 20, 1928, OPH/SIA. In advance of his reply, Cook sent Spier his *Science, Scientific American,* and *Natural History* papers, perhaps in hopes of convincing him there was plenty of evidence already in hand of a deep human antiquity in the Americas.
312. Cook 1928b: 372; Cook to Gould, October 22, 1927, HJC/AFB. In private Cook was even less charitable to archaeologists and their ability to appreciate and understand geological and paleontological evidence—for example, Cook to Hay, January 25, 1928, OPH/SIA.
313. Spier to Cook, April 26, 1928, HJC/AFB.
314. Hrdlicka to Spier, May 7, 1928, AH/NAA.
315. Hay 1928a:442.

316. Hay 1928a:443.
317. Hay 1928a:443–444.
318. Spier 1928b.
319. Romer 1928:19. To be certain the camel was a fossil form, Romer compared its skull with that of modern camels. That it was not a fossil form was a possibility not to be overlooked, as camels were imported into the United States in the 1850s by then–Secretary of War Jefferson Davis, who aimed to have the Army use them for transport in the arid West. When the experiment failed, the camels were released to the wild.
320. Romer 1928:20. Romer was inspired to become a vertebrate paleontologist as an Amherst undergraduate after taking a course from Frederick Loomis, though a decade before Loomis became involved in the human antiquity controversy at Melbourne.
321. Hay 1928b:299–300. The camel skull was subsequently submitted for radiocarbon dating and returned an age of 11,075 ± 255 ^{14}C years B.P. (Nelson and Madsen 1979). Romer had been right and Hay wrong.
322. Hay 1928b:300.
323. Cook to Hay, October 3, 1928, OPH/SIA; Romer 1933. Schultz and Eiseley came to Hay's defense as well, though largely because they themselves were unhappy with Romer having also insisted that their extinct bison species lived into the Recent period (Schultz and Eiseley 1935:311–315).
324. Gould to Cook, January 3, 1929, DIR/DMNS; Gould to Hay, December 7, 1928 and January 4, 1929, OPH/SIA; Hay to Figgins, October 16 and October 25, 1928, JDF/DMNS; Spier to Hay, October 25, 1928, OPH/SIA.
325. Spier to Hay, October 25, 1928, OPH/SIA, and quoted in Hay 1929:94.
326. Gould 1929; Hay 1929:96.
327. Hay 1929:97–98.
328. Figgins to Hay, November 8 and December 17, 1928, OPH/SIA; Figgins to Gould, October 26, 1929, DIR/DMNS.
329. Spier to Hay, April 5, 1929, OPH/SIA. Evans's summary findings were given to Gould in October of 1929. Gould forwarded them to Cook and Figgins, who naturally disagreed with the findings, Cook scoffing that Evans's conclusions were "rather ridiculous" even if Gould thought Evans "may be right" (Gould to Cook, October 17, 1929 and November 4, 1929 and Cook to Gould, November 2, 1929, HJC/AFB; Gould to Figgins, October 17, 1929 and Figgins to Gould, October 26, 1929, DIR/DMNS). Evans published a formal report a year later in the *Journal of the Washington Academy of Sciences*, Hrdlička communicating the paper to the Academy (Evans 1930). By then Hay had passed away, and only Cook (1931b) replied. Afterward, only Sellards seemed to care, visiting Frederick and assessing its geology (Sellards 1932; Sellards to Cook, November 13, 1931, HJC/AFB).
330. Figgins 1935b:4. By the time Roberts summarized the matter five years later (Roberts 1940a:58), Frederick warranted only a dismissive footnote: "The discoveries at Frederick, Okla., are entirely omitted from this paper because of their obviously spurious nature." See also Cotter 1939:154–155.
331. Cook to Ingalls, January 6, 1929, HJC/AFB.
332. Hrdlička 1928. Hodge to Hrdlička, June 1, 1928, AH/NAA.
333. Hrdlička 1928:807, 809, 812.
334. Nelson 1928b:820, 823.
335. Brown 1928b:825, 828. It is not known whether the term "fluted" originated with Brown. Brown invited Hay to serve as a discussant, and though Hay could "hardly restrain myself from undertaking it," in the end he declined and promised to send

Brown talking points. It is not clear what, if any, elements of Brown's comments were based on material sent by Hay. See Brown to Hay, February 9, 1928 and Hay to Brown, February 9 and February 17, 1928, VP/AMNH.

336. Hrdlička to Brown, March 15, 1928; Brown to Hrdlička, March 16, 1928; Hrdlička to Brown, March 19, 1928; Brown to Hrdlička, March 21, 1928; Hrdlička to Brown, March 22, 1928; all in AH/NAA. A fuller reprinting of the exchange is in Meltzer 2006a:37–38.

337. Hodge to Hrdlička, June 1, 1928, AH/NAA, emphasis in original.

338. Hrdlička to Hodge, June 7, 1928, AH/NAA.

339. Figgins to Schwachheim, October 1, 1928, DIR/DMNS. Two months later, however, Figgins feared there might be "bloodshed" at Brown's presentation on Folsom at the AAAS-GSA meeting. He was wrong; Figgins to Brown, December 17, 1928, VP/AMNH.

340. Davis 1928:209.

341. Brown 1929; Bryan 1929:129. Brown invited Figgins to attend the meeting, though the invitation did not include the opportunity to speak. Figgins declined because of other obligations (Brown to Figgins, November 30, 1928 and Figgins to Brown, December 13, 1928, DIR/DMNS).

342. Hrdlička 1942:54.

343. Hay to Figgins, October 18, 1927, JDF/DMNS.

344. Sellards to Holmes, February 24, 1930, WHH/SIA.

345. Holmes to Sellards, March 6, 1930, WHH/SIA.

Chapter Nine

1. Cook complained about Figgins often and at length, including in Cook to Gidley, November 26, 1930; Cook to Gould, January 25, 1931; Cook to Howarth, November 17, 1930; Cook to Ingalls, December 16, 1930; Cook to Jenks, January 26, 1931; all in HJC/AFB. Figgins was far more circumspect about Cook, at least in writing. One of the few letters is Figgins to Brown, March 12, 1934, VP/AMNH.

2. Cook to Gidley November 26 and November 27, 1930 and Gidley to Cook, November 23, 1930, in HJC/AFB.

3. Cook to Sellards, January 24, 1931 and Cook to Gould, January 25, 1931, HJC/AFB.

4. Cook to Osborn, April 25, 1931, VP/AMNH.

5. Romer to Cook, May 26, 1931, HJC/AFB. In the spring of 1931 Romer prepared his article for a volume that was to be distributed in advance of the Fifth Pacific Science Congress, originally slated for British Columbia in June of 1932. The conference was subsequently moved to June 1933, and the volume (Jenness, ed. 1933) was presented as planned. But his results were known well before then, as he summarized them at the 1931 AAAS symposium on "Early Man in America" (Danforth 1931:118).

6. In the 1930s "early man" became the common phrase denoting Pleistocene peoples, as the term "Paleoindian" had not been invented and "Paleolithic" was obsolete. I recognize the gender bias of "early man" but use it in this chapter in the interests of historical accuracy.

7. Meltzer 2006b: Table 1; see also Howard 1936c:320; Roberts 1940a:55; Wormington 1939:5, 9.

8. Bryan 1941:507, 514.

9. Roberts 1940a:102.

10. Roberts 1940a:109. This appears to be the first use of the term "Paleo-indian," as indicated by Bryan 1941:507 (see also Holliday and Anderson 1993). Most authors subsequently omitted the hyphen.

11. Danforth 1931; Jenness [ed.] 1933; and MacCurdy [ed.] 1937. It appears Boas may have selected the theme for the symposium in his honor; he certainly played a role in organizing it, including inviting Hrdlička to participate (Boas to Hrdlička April 21, 1931 and Hrdlička to Boas, April 25, 1931, AH/NAA).

12. Roberts 1940a:52. Also Kidder 1936:145; Roberts 1939:534.

13. Kidder 1936:143–144.

14. There were a few other longtime participants still around, such as Erwin Barbour, who, although in his middle to late seventies during this period, continued to do research on early sites, working with his successor, C. B. Schultz; and Elias Sellards, by then in his late fifties, became reenergized and even more active than before, and continued to defend Vero's virtues (for example, Sellards 1937).

15. Published biographical details on Howard are available in Mason 1943, Roberts 1943.

16. Howard to Jayne, September 4, 1931 and Jayne to Howard, September 22, 1931, EBH/UPM. See also Howard 1930; Howard 1932a.

17. Howard to Jayne, September 5, 1931, EBH/UPM. On that September 1931 visit, Brown found a Folsom point on the AMNH backdirt pile (Specimen 18, described and illustrated in Meltzer 2006a: Table 8.2 and Figure 8.4b).

18. Howard to Brown, March 15, 1932, EBH/ANS. Howard describes his unhappiness with Brown in Howard to Cadwalader, March 3 and March 15, 1932, EBH/ANS. Howard complained of the matter to others, including Kidder, who reported it in Kidder to Merriam, December 19, 1932, DHR/CIW. In the end Howard worked on the Burnet Cave fauna with paleontologist C. B. Schultz of the Nebraska State Museum (Schultz and Howard 1935).

19. Eiseley worked for Howard in the Guadalupe Mountains region in 1934, and he spoke about his time with Howard in his autobiography, though he did not mention his name.

20. Howard 1935b:81–82.

21. Whiteman to Wetmore, February 5, February 25, and April 2, 1929 and Wetmore to Whiteman, February 15 and March 12, 1929, SEC/SIA. At Neil Judd's suggestion, Wetmore had Smithsonian vertebrate paleontologist Charles Gilmore visit the site; Judd suggested that if Gilmore deemed it of interest Frank Roberts might follow up with an archaeological investigation (Judd to Wetmore, March 8, 1929, SEC/SIA). Gilmore did, but after an hour's inspection in the company of Whiteman, Gilmore deemed the locality insufficiently rich to warrant archaeological work. Howard hired Whiteman to work at the site in 1933 (Boldurian and Cotter 1999:11).

22. Howard to Brown, June 22, 1932 and Brown to Howard, June 25, 1932, VP/AMNH. Howard shared more details with Kidder, as reported in Kidder to Merriam, December 19, 1932, DHR/CIW.

23. Howard 1935b:81–82.

24. Howard's plan is in Howard to Cadwalader, October 1932, EBH/ANS. The steam shovel quote is from Kidder to Merriam, December 19, 1932, DHR/CIW.

25. Howard to Cadwalader, October 1932 EBH/ANS.

26. *American Journal of Physical Anthropology* 1932. The full listing was John Merriam (chair), Frank Roberts (secretary), W. C. Alden, Barnum Brown, Kirk Bryan,

M. Harrington, Ernest Hooton, Aleš Hrdlička, A. V. Kidder, Nels Nelson, H. F. Osborn, and Chester Stock.

27. Howard to Cadwalader, October 1932, EBH/ANS.

28. Pete [Anderson] and George [Roberts] to Howard, November 12, 1932, EBH/UPM.

29. Howard to Jayne, November 16, 1932, EBH/UPM.

30. Howard 1932b.

31. Merriam, memorandum of conversation with Kidder, December 6, 1932 and Kidder to Merriam, December 15 and December 19, 1932, JCM/CIW. Kidder's home base was in Andover, MA, and he regularly took the train to and from Washington, DC. Kidder recorded the meeting in his diary: "At 3 went into session with Edgar Howard, who has done a tremendous job of correlating the scattered finds of Folsom points that have been made all over the country. He's also found a site near Clovis, New Mexico, that has a gravel layer of fossil bones associated with artifacts" (Kidder Diary, December 5, 1932, AVK/HU).

32. Kidder to Merriam, December 19, 1932, JCM/CIW.

33. Kidder to Merriam, December 19, 1932, JCM/CIW.

34. Kidder to Cadwalader, December 17, 1932, GC/ANS.

35. Kidder to Merriam, December 19, 1932, JCM/CIW.

36. Kidder Diary, December 5, 1932, AVK/HU.

37. Howard's budget accounting for the 1932 and 1933 seasons are in "Accounts 1932, E. B. Howard" and "Expenses to Clovis Apr. 25th to August 12th, 1933," EBH/UMP. In 1932 he spent almost $1,100 of his $1,500 allocation; in 1933 he spent just under $990 of the $2,000 available to him. The University Museum had an endowment, the Harrison Fund, earmarked for the support of fieldwork, and so despite a complete cutoff of city funds to the museum and cutbacks in support from the university, Jayne was still able to send fourteen expeditions into the field in 1931 (Winegrad 1993:98).

38. Kidder to Merriam, December 19, 1932, JCM/CIW.

39. Kidder to Merriam, December 19, 1932, DHR/CIW; Merriam to Kidder, December 20, 1932, JCM/CIW.

40. Merriam to Kidder, December 21, 1932, DHR/CIW.

41. Kidder to Merriam, December 27, 1932, JCM/CIW. The quote from Howard is from a letter he wrote to Kidder, and which Kidder in turn quoted to Merriam; the punctuation was Howard's.

42. Merriam to Kidder, December 21 and December 27, 1932, JCM/CIW.

43. Merriam to Stock, July 14, 1933, JCM/LC.

44. Merriam to Cadwalader, August 14, 1933, GC/ANS..

45. Kidder to Merriam, December 15, 1932, DHR/CIW; Merriam to Kidder, December 20, 1932 and July 10, 1933, JCM/CIW.

46. Howard's initial version was done in October 1932. A round of extensive revisions was made in November, 1932 (the much-marked-up draft is in EBH/UMP). The version that Merriam received two months later is likewise revised from the November iteration.

47. Howard to Merriam, February 10, 1933, JCM/LC.

48. Howard to Merriam, February 10 and April 21, 1933, JCM/LC.

49. Later published in Shetrone 1936.

50. Renaud 1932.

51. Figgins offered to send a couple of men to help Howard with the excavations and watch for any extra fossil material the CMNH might acquire in exchange. (Figgins, typically, wanted another complete skeleton for display, and he put a small Pleisto-

cene camel on the table as trade bait.) After telling Jayne and Cadwalader of these inquiries, Howard was quick to assure them he saw no harm in their visiting because "they recognize our priority there" (Howard to Jayne, May 1, 1933, EBH/UPM; also Howard to Cadwalader, May 1, 1933, EBH/ANS). The "armour of pride and tradition" in Philadelphia would remain intact. Cadwalader, showing a surprising lack of confidence in Howard's judgment, was not taking any chances. He quietly wrote Stock to find out, in essence, if Barbour and Figgins were honorable men. Stock assured him they were (Cadwalader to Stock, May 16, 1933; Stock to Cadwalader, May 24, 1933; Figgins to Cadwalader, May 28, 1933; all in GC/ANS).

52. Howard to Cadwalader, May 22, 1933, EBH/ANS. The excavation of the bones in the gravel pit proved more difficult than it had been the previous November; a half-year later the sediment had hardened "almost to concrete" and the bones had become fragile from the exposure.

53. Stock to Merriam, June 2, 1933, JCM/LC.

54. Howard to Cadwalader, June 4, 1933, EBH/ANS; Howard to Jayne, June 24, 1933, EBH/UMP; Stock to Merriam, June 2, 1933, JCM/LC. Stock wrote Cadwalader to assure him there would be enough fossil specimens to go around; Stock to Cadwalader, June 7, 1933, GC/ANS.

55. Stock to Merriam, June 2, 1933 and Merriam to Stock, June 5, 1933, JCM/LC.

56. Howard also made a quick trip down to the Guadalupe Mountains to appease Bill Burnet, who was engaging in "frantic efforts" to lure him back to the region with the promise of newly discovered caves. Howard to Cadwalader, June 4 and June 11, 1933, GC/ANS.

57. Howard 1935a:113. The specimen Merriam found is a parallel-flaked Late Paleoindian lanceolate illustrated in Howard 1935a: Plate 39, Number 3.

58. Howard to Jayne, June 24, 1933, EBH/UMP. By IGC rules, which Merriam was first unaware of, papers could not be presented in absentia. Neither Howard's papers nor Merriam's, though published in the Congress proceedings, were delivered at the IGC, to Merriam's displeasure (Gilbert to Kidder, July 13, 1933 and Merriam to Kidder, July 15, 1933, DHR/CIW).

59. Howard 1936a:1330.

60. Howard 1936a:1331–1332

61. Howard to Cadwalader, July 9, 1933, GC/ANS.

62. Howard to Cadwalader, July 9, 1933, GC/ANS; Howard to Jayne, July 15, 1933, EBH/UMP.

63. The trip's nominal leader, Princeton geologist Richard Field, was unhappy about this last-minute change and resisted it until Merriam's assistant (pushed by his anxious boss) telephoned and threatened that "if it proved absolutely necessary to stick to the Folsom plan, we would need to have a pretty careful explanation of the reason for such action." Kidder also put his thumb on the scale, telling Field that "Clovis is infinitely more important than Folsom, and there should be no slip-up as to the routing." Kidder to Gilbert, July 20, 1933, JCM/CIW. See also Howard to Cadwalader, July 9, 1933, GC/ANS; Merriam to Cadwalader, July 10, 1933, GC/ANS; Merriam to Howard, July 14, 1933, JCM/LC.

64. Merriam to Howard, July 14, 1933, JCM/LC. Also Stock to Merriam, July 20, 1933, JCM/LC.

65. Merriam to Howard, July 15, 1933, JCM/LC.

66. Merriam to Kidder, August 14, 1933, JCM/CIW. At the IGC meeting Woodward gave a "masterly presentation" (Kidder's view) defending the antiquity of Piltdown

and detailing the evidence that put it in the Lower Pleistocene because of the occurrence of eoliths and associated vertebrate species (Woodward 1933; Kidder to Merriam, July 25, 1933, JCM/CIW).

67. Howard to Jayne, July 28, 1933, EBH/UPM.
68. Howard 1935b:93–95.
69. Howard 1935b:93.
70. Merriam to Cadwalader, August 14, 1933, GC/ANS, emphasis mine. Stock and Bode 1936:238. See also Romer 1933:53.
71. Howard to Jayne, August 3, 1933, EBH/UMP, punctuation added. The telegram is dated August 3, though it was sent late on August 2 from Clovis and arrived just after midnight in Philadelphia.
72. Howard to Merriam, August 3, 1933, EBH/UMP; Merriam to Kidder, August 14, 1933, JCM/CIW.
73. Merriam and associates 1933:323.
74. Howard to Merriam, August 3, 1933, EBH/UMP.
75. Merriam to Kidder, August 14, 1933, JCM/CIW. A similar list appears in Merriam and associates 1933:323–324.
76. Bowman, Johnson, Kidder, Stock, and Merriam himself were all members or were soon to be elected into the NAS. Of course, in 1933 Merriam's own active research career and contributions were long behind him.
77. Merriam to Cadwalader, July 10, 1933, GC/ANS.
78. Merriam to Kidder, August 14, 1933, JCM/CIW. Kidder had a higher opinion of Howard's work; Kidder to Cadwalader, September 18, 1933, GC/ANS.
79. Howard to Jayne, July 28, 1933, EBH/UPM.
80. Howard 1933:8.
81. Howard 1936a:1331. Of the forty-five projectile points found that summer, four were Clovis, eleven were Folsom, and the remaining thirty were a variety of unfluted or basally thinned points, including Midland, Plainview, Agate Basin, Eden, Scottsbluff, and various parallel oblique forms. Only eleven points were found at the gravel pit, whereas Beck Forest Lake ($n = 22$) and Anderson Basin No. 2 ($n = 9$) yielded the majority of the remainder. Of the four Clovis points, three were surface finds at Beck Forest Lake and the other was found on the floor of the gravel pit. Data on the number of artifacts recovered during the 1933 field season are from Boldurian and Cotter 1999: Table 2b.
82. Howard 1935b:109. It should be noted, however, that the precise stratigraphic position of Clovis points and mammoth remains at Clovis is somewhat ambiguous—see, for example, Haynes 1995:384–385.
83. *Science* 1933:188.
84. Figgins 1933a:4.
85. Figgins 1933a:7.
86. Figgins 1934a:13.
87. Figgins 1933a:7–8, emphasis mine.
88. Meltzer 2006a.
89. For example, Barbour and Schultz 1932b; Brown 1928a; Cook 1927a; Figgins 1933a; Schultz and Eiseley 1935:313.
90. Howard, *"Memorandum Regarding Next Steps in Following up Clovis Work,"* GC/ANS.
91. Howard to Merriam, January 13, 1934, JCM/LC.
92. That summer they visited Burnet Cave, Folsom, Frederick, Lone Wolf Creek, and Cyrus Ray's Abilene area sites, as well as the more recently discovered localities of

Gypsum Cave (being excavated by M. R. Harrington), Barbour's Nebraska sites, the newly reported Nall site in far western Oklahoma, and Albert Jenks's Minnesota Man site near Pelican Rapids. Antevs 1934 Fieldbook, EA/UA; Howard and Antevs 1935:309–310.

93. Nelson 1936:238.

94. Howard 1935b; Patrick 1938.

95. Stock and Bode 1936: 228, 233, 238.

96. Antevs 1935a:310; Howard 1935a:301; Howard 1935b:87–91. Antevs was estimating the age of the diatomite and the Folsom and Late Paleoindian artifacts (now known to date between 12,000 and 13,000 calibrated radiocarbon years ago). There were objections to Antevs's methods, but even those objecting thought he was likely correct; these are discussed in Wormington 1939:17–18.

97. For example, Antevs 1935a; Clarke 1938; Cotter 1937a, 1938; Howard 1935a, 1935b; Patrick 1938; Richards 1936; Stock and Bode 1936. Howard completed his dissertation in June of 1935 (Howard to Merriam, June 25, 1935, JCM/LC); see the very positive review of Howard's dissertation in Nelson 1936.

98. Howard to Merriam, January 13, 1934; Merriam to Howard, March 29, 1934; Howard to Merriam, April 17, 1934; all in JCM/LC.

99. Roberts 1935:11–15; Roberts 1936b:340–342.

100. Seven fluted points were recovered from Parrish Village. Three of these were collected from the surface before work at the site and four were found during the excavations, though none were associated with any of the features at the site (Webb 1951:437). Two fluted points were found at the Carlston Annis site in Kentucky (Webb 1951:437).

101. For example, Roberts to Stirling, July 29 and August 10, 1935, BAE/NAA.

102. Aggregate financial data and disbursements for the Carnegie Institution are available in its *Year Book*. Specific amounts from 1935 for research on Paleoindians are in Merriam to Howard, May 23, 1935, JCM/LC. Hodge reports on funding from Carnegie for Gypsum Cave work (Harrington 1933:2).

103. Those who received support were identified by examining *Year Books* of the Carnegie Institution.

104. For example, Howard to Merriam, March 19, 1934 and April 12, 1935 and Merriam to Howard, May 23, 1935, JCM/LC; Merriam to Howard, March 2, 1936 and Howard to Merriam, March 4, 1936, GC/CIW. When the Carnegie Institution was founded decades earlier, Holmes had had a role identifying projects that warranted funding (Holmes to Walcott, January 9, 1903, JCM/CIW).

105. Merriam to Howard, March 2, 1936 and Howard to Merriam, March 4, 1936, GC/CIW.

106. Antevs 1936:332; Antevs 1922. Richard Foster Flint was particularly distrustful because of the possibility of the loss of varve layers to erosion (Flint 1941:31–32), but also because of his view of ice retreat and subsequent radiocarbon dating. However, work in the last few decades by the North American Glacial Varve Project has shown the essential correctness of Antevs scheme (http://geology.tufts.edu/varves). See Ridge 2004.

107. Antevs 1925; Haynes 1990:55–56.

108. Antevs to Merriam, May 6, 1935, GC/CIW.

109. Antevs 1936:331. Antevs resisted the idea of wiggle-matching varve sequences across the Atlantic (Ridge 2004:176). See also Bryan 1941:509; Daly 1934:84, 111.

110. Antevs 1936:331. Also Bryan 1941:509, 511; Bryan and Ray 1940. See the discussion in Haynes 1990.

111. Antevs 1936:333–334.

112. Antevs 1935c:317.
113. Antevs to Merriam, May 6, 1935 and February 25, 1936, GC/CIW. Also Antevs 1936:332–333; Antevs 1937b, 1938a.
114. Bryan 1941:511. See Haynes 1990.
115. Merriam to Howard, March 2, 1936 and Howard to Merriam, March 4, 1936, GC/CIW.
116. Romer 1933:70. Stock and Bode 1936:240. Stock was even reluctant to declare the Clovis site bison as *Bison taylori* (the same species as at Folsom), given the fragmentary nature of the remains from the Clovis site. See also Stock 1936:324, Stock 1941:150–151.
117. Romer 1933:50–51, 62–63, 74.
118. Romer 1933: Table 2, Table 3.
119. Romer 1933:58–60, 67, 75.
120. Romer 1933:68.
121. Romer 1933:70.
122. Colbert 1942:19–21; Schultz and Eiseley 1935; Stock 1936:330; Stock 1941:148.
123. Romer 1933:70, 77. Also Howard 1936c:316–317; Schultz and Eiseley 1935; Stock 1936:329.
124. Romer 1933:78–80.
125. Romer 1933:74.
126. Stock 1941:154. Also Colbert 1942:23; Schultz and Eiseley 1935:306–307.
127. Romer 1933:81.
128. Antevs 1935b:305; Schultz and Eiseley 1935:311–312. Also Hrdlička 1937b:104; Roberts 1940a:51; Romer 1933:81.
129. Schultz and Eiseley 1935:311–312. See Bryan 1938; Howard 1939; also Sellards 1940:398.
130. Antevs 1922; for a recent appraisal of the accuracy of Antevs's varve results, see Ridge 2004.
131. Antevs 1935b:309; Bell and Van Royen 1934:49; Howard 1935b:102.
132. Howard 1935b:101. See also Roberts 1940a:106.
133. Meltzer 2009:48–49.
134. Bell and Van Royen 1934:49, emphasis mine.
135. Schultz and Eiseley 1935:312.
136. Kidder 1936:144.
137. Howard 1935b:149; Howard 1936c:317.
138. Nelson 1937:320; Roberts 1935:32; Roberts 1936b:344; Roberts 1937b:159–160; Wormington 1939:18–20.
139. Harrington 1933:161–162, 169–171; Howard 1935b:149; Roberts 1935:6.
140. Roberts 1940a:57, 70; Schultz and Eiseley 1935:313. Also Harrington 1933:164–166, 186; Howard 1936b:401.
141. Harrington 1933:166–168; Howard 1936c:316–318; Roberts 1940a:57, 70–71; Wormington 1939:37.
142. Barbour and Schultz 1936:433; Howard 1935b:150; Kidder 1936:144; Kroeber 1940a:474; Roberts 1940a:57.
143. Roberts 1940a:105.
144. Antevs 1935b:305.
145. Hooton 1933:142. Also Hooton 1930:351–352; Howells 1940:3.
146. Gidley 1912:20, 22–23; Clark 1912:27. For a summary history of the Bering land bridge, see Hopkins 1967.

147. Dall 1912:17–18. Also Bartsch 1912:49.
148. Antevs 1928:81. Antevs later (1935b:306) put the decline in sea level at seventy-three meters during the glacial maximum. In 1885 Chamberlin and Salisbury estimated, from presumed ice thickness and areal extent, that global sea levels fell 1,000 to 1,200 feet in glacial times (Chamberlin and Salisbury 1885:294). They did not perceive its implications for a land bridge.
149. Antevs's 1928 AAAS presentation was titled "Quaternary Climatic Conditions and Their Relation to the Peopling of America" but was not published under that title. Danforth et al. (1929:124) reported on it.
150. Danforth et al. 1929:124.
151. Johnston 1933:28. Johnston supposed he was the first person to link continental glaciation, lower sea levels, and the emergence of a land bridge, which is odd given he knew Antevs and was very familiar with his work. Antevs did not seem to mind not getting the credit.
152. The quote is from Daly 1934:193–194. See also Boas 1933:359; Bryan 1941:505–506; Johnston 1933:32, 43–44; Smith 1937:88–91. Wormington (1939:61) suggested that the land bridge was not available "at the time of man's migration," but that was a minority opinion by the time she wrote it.
153. Hrdlička 1937b:186–187. In response to an inquiry from Hrdlička, Smith replied "There is no known evidence which proves a 'land bridge' connection between the American and Asiatic continents during the Wisconsin glacial time" (Smith to Hrdlička, January 8, 1937, in Hrdlička 1937b:187). In 1937 that was literally true: there was no *proof* of a land bridge. However, as Smith also made clear three months later, though the presence of a land bridge could not be demonstrated, changes in global sea levels during the Pleistocene were "sufficient to form extensive land connections between the two continents" (Smith 1937:90).
154. Hooton 1937:102.
155. Howells 1940:4; Johnston 1933:43–44; Smith 1937:88–91.
156. Upham 1895: Plate 2. See also Leverett 1899: Plate 1.
157. Johnston 1933:14–22.
158. Johnston 1933:22–24, 44.
159. Johnston 1933:42, 44.
160. Johnston 1933:42–44.
161. Antevs 1935b:306–307.
162. Roberts 1939:542; Wormington 1939:62. On skepticism regarding a coastal migration, see Bryan 1941:506; Roberts 1940a:104; Wormington 1939:60–61.
163. Howard to Merriam, February 25, 1934, JCM/LC.
164. Jenness to Howard, March 12, 1934 and Howard to Merriam, March 14, 1934, JCM/LC.
165. Howard to Merriam, March 17, 1934 and Merriam to Howard, March 19, 1934, JCM/LC.
166. Howard to Merriam, February 25, 1934 and June 25, 1935, JCM/LC.
167. Howard to Jayne, August 8, 1935, EBH/UMP.
168. Renaud 1932:5.
169. Dunnell 2003:16; Lyman and O'Brien 2006:226–241, 251.
170. For useful contemporary summaries, see Roberts 1940a; Wormington 1939. For details of the efforts related to defining the Folsom type, see Meltzer 2006a:338–344, from which this discussion is partly drawn.
171. Cook 1931a.

172. Figgins 1934b:2; Figgins 1935b:3–4.
173. Barbour and Schultz 1932a; Bell and Van Royen 1934; Cook 1931a; Schultz and Eiseley 1935. Cf. Figgins 1934b:2–3; Figgins 1935b:3–4; Renaud 1931:16; Roberts 1935:9; Wormington 1939:27.
174. Renaud 1931:7–10.
175. Renaud 1931:13.
176. Renaud 1931:13–14; Wormington 1939:9.
177. Renaud 1931:15.
178. Cook 1931a:103; Schultz and Eiseley 1935:309; Schultz and Eiseley 1936:522–523.
179. Figgins 1934b:2; Figgins 1935b:6. Shetrone (1936:251–252) arrived at a similar conclusion and saw Renaud's scheme of the evolution of fluting as "somewhat illogical."
180. Figgins 1934b:2.
181. Figgins 1934b:5–6; Figgins 1935b:7.
182. Figgins 1934b:3; Figgins 1935b:2–6.
183. Figgins 1934b:4; Figgins 1935b:5–6.
184. Roberts 1935:18–21; Roberts 1936a:19–20; Roberts 1936b:343; Wormington 1939:21–22.
185. Roberts 1934:18.
186. Roberts 1935:5; Roberts 1939:533.
187. Roberts 1935:16.
188. Roberts 1935:18.
189. Roberts 1936a:21.
190. Roberts 1936a:34–35; Roberts 1936b:21–22. By 1940 Roberts rejected the claim of a greater antiquity for Yuma as "only conjectural typological seriation" (Roberts 1940a:62).
191. Howard to Merriam October 6, 1936, JCM/LC.
192. Howard 1936b:404–405. Also Howard 1935a:119–121; Roberts 1935:8; Roberts 1937b:161; Shetrone 1936:244, 246.
193. Shetrone 1936:244, 246.
194. Howard 1935b:109; Roberts 1935:8; Roberts 1936a:21; Roberts 1939:543; Shetrone 1936:251. Figgins (1935b:5) seemingly opposed on principle the "introduction of an hyphenated designation."
195. Howard 1935b:121–122; Roberts 1935:8.
196. Cotter 1937a:6.
197. Cotter 1937a:10. Also Cotter 1938:115–116, which offers a slightly different take on the position of the bison bone. For a retrospective summary, see Boldurian and Cotter 1999:56–72.
198. Cotter 1937a:15. Howard thought Cotter's results that season had not "added as much to the general knowledge of man's antiquity in North America as they have confirmed what other works had already known or suspected" (Howard, in Cotter 1937a:1). It was a curious, if uncharitable, assessment.
199. Cotter 1937a:6–7, 15.
200. Cotter 1938:116–117.
201. Cotter 1937a:13, 15.
202. Cotter 1937b:30–31; Cotter 1938:117. See also Roberts 1939:544; Roberts 1940a:56; Wormington 1939:11.
203. Stratigraphic confirmation came after the World War II in Sellards's excavations in the Clovis gravel pit (Sellards 1952:31).

204. Cotter 1937b:31; see also Howard 1935b:119; Roberts 1936a:8; Roberts 1939:543; Wormington 1939:9–10.
205. Barbour and Schultz 1936:443; Cotter 1937b:29; Roberts 1940a:65; Wormington 1939:25–26.
206. Cotter 1937b:33. See also Howard 1935b:121–122; Howard 1936b:404; Roberts 1939:543; Wormington 1939:11.
207. Roberts 1939:544.
208. Roberts 1939:544.
209. Howard 1935b:122.
210. Shetrone 1936:243. Roberts 1937b:161; Roberts 1939:543; Wormington 1939:11.
211. Howard to Cadwalader, October 1932, EBH/ANS.
212. Nelson 1936:239.
213. Barbour and Schultz 1937.
214. Howard to Merriam, October 9, 1936, JCM/LC.
215. Participants included Ernst Antevs, Erwin Barbour, Kirk Bryan, Edwin Colbert, John Cotter, William K. Gregory, Ernst Hooton, Edgar Howard, Aleš Hrdlička, George Kay, Morris Leighton, Paul MacClintock, John C. Merriam, Nels Nelson, E. B. Renaud, Horace Richards, Frank Roberts, C. B. Schultz, Paul Sears, E. H. Sellards, Philip Smith, Herbert Spinden, and Marie Wormington, and even several who had had only peripheral roles, such as Neil Judd, George Grant MacCurdy, Alex Wetmore, and Clark Wissler. This list of participants was gleaned from Howard 1937 and MacCurdy, ed. 1937.
216. One minor tempest from this meeting was Spinden's (1937:114) assertion that "only four thousand years" were available for prehistory in the American West. That assertion was widely perceived as "extremely radical and opposed to the general opinion" (Mason 1966:195), and no one appears to have accepted it a decade after Folsom rendered such a claim obsolete.
217. Wormington 1939:23–25.
218. Howard 1937:444.
219. Roberts 1939:543.
220. Wormington 1939:23.
221. Howard 1937:442.
222. Howard to Hrdlička, November 20 and December 14, 1936 and Hrdlička to Howard, December 17, 1936, AH/NAA. Schultz 1946:309 summarizes Hrdlička's Alaska work.
223. For a full listing, see Roberts 1940a:99.
224. Smith 1928. See Spencer 1979:570.
225. Hrdlička 1927:255, 270–273.
226. Hrdlička 1927:272–273. Also Hrdlička 1937b:93.
227. Hooton 1937:103; Howells 1938:326; Hrdlička 1937b:93; McCown 1941:204, 211.
228. Hooton 1930:350–351. See also Bryan 1941:507; Hooton 1925:125; Hooton 1937:103–104; Keith 1915:308; Woodward 1933. Keith believed anatomically modern types were separated from Neanderthals as early as the Middle Pliocene more than 400,000 years ago (Keith 1915:502).
229. Hooton 1930:351.
230. Hooton 1925:126–127.
231. Hrdlička 1937b:93–94.
232. Figgins 1935a:2–3. On Figgins's efforts at bison taxonomy, see Figgins 1933b. See also Simpson to Eiseley, August 26, 1942, LCE/UP.

233. All of Figgins's taxonomic designations were rejected, the process of rejection starting with Skinner and Kaisen (1947). See the detailed discussion in Meltzer 2006a:206–209.

234. Figgins to Roberts, November 8, 1935; Hrdlička to Roberts, November 21, 1935; Hooton to Roberts, December 16, 1935; Woodbury to Roberts, December 16, 1935; Shapiro to Roberts, May 28, 1936; all in FHHR/NAA. Hooton and Woodbury thought the skull similar to Hooton's pseudo-Australoid; Hooton to Roberts, December 16, 1935 and Woodbury to Roberts, December 16, 1935, FHHR/NAA. Roberts 1937a.

235. Bryan 1941:507; Hooton 1937:104; McCown 1941:204, 207, 211; Roberts 1940a:100.

236. Jenks 1936:171.

237. Jenks 1936:4, 7, 10, 12.

238. Jenks 1932; Jenks 1936:5–6, 168–169.

239. Jenks 1936:40. Because the remains had been known since 1932 as Minnesota Man, Jenks thought it less confusing to stay with the incorrect gender label (as did others—for example, Wormington 1939:54).

240. Jenks 1933:5–6.

241. Antevs 1935b:305.

242. Bryan 1935:171.

243. Jenks 1936:33, 47, 177

244. Hooton 1925:127.

245. Not surprisingly, Piltdown had the greatest standard deviation, a consequences of its "inferior human" scores on its braincase, and its "anthropoid" scores on its mandible and teeth (Hooton 1925:132). Hooton did not suspect it was a human cranium and an ape jaw, he just noted it scored that way.

246. Hooton 1925: Table 3 shows the calculated scores, Figure 1 their ranking by group. Hooton admitted the attributes and scoring he used were arbitrary, but that was inevitable in any quantitative scoring of qualitative traits. Hooton claimed his scheme offered "mathematical accuracy" (means and standard deviations), and was more rigorous and avoided the "evolutionary stalemate" that resulted from arguments over whether thick lips and curly hair represented a more or less advanced state in humans (a jab at Franz Boas). Hooton dared readers unconvinced by his approach to develop their own method and scale: he was confident it would "not seriously affect the positions and rankings of the types and races" he had found (Hooton 1925:125–126, 129, 131).

247. Howells 1992:5–7. See also Brace 2005:234–235.

248. Jenks 1936:170–174.

249. Hooton 1937:104.

250. Hrdlička to Jenks, November 14, 1932, AH/NAA.

251. Hrdlička 1937a:184.

252. Hrdlička 1937a:185–187.

253. Hooton 1937:104. In a subsequent reply to Hrdlička, Jenks reported there was no particular position drowning victims assumed, according to his conversations with police and coroners. Neither the posture nor the articulation of the skeleton ruled out death by drowning (Jenks 1938:331).

254. Hrdlička 1937a:188.

255. Hrdlička 1937a:189, Hrdlička 1937b:95; McCown 1941:205. Also Howells 1938:322.

256. Hrdlička 1937a:196, 199; also Hrdlička 1937b:103. McCown (1941:204) accepted Hrdlička's point in general but wondered if Minnesota Man was "just another Sioux female." Bryan and MacClintock, neither of them a human anatomist, nonetheless contended that the extraordinarily large teeth were "hardly compatible with her alleged status as an ordinary Sioux maiden" (Bryan and MacClintock 1938:280).

257. Hrdlička 1937b:95, 101–104.

258. Howard 1937:441.

259. Jenks (Jenks to Hrdlička, May 7, 1937, AH/NAA) alerted Hrdlička that he planned a response, which he evidently originally intended for Hrdlička's *American Journal of Physical Anthropology*. For reasons unknown to me, it was published instead in the *American Anthropologist* (Jenks 1938). For the view of a dispassionate observer of this particular exchange, see Howells 1938:322, 323–326.

260. Kay and Leighton 1938:268–269. The visit to the site was funded by the Carnegie Institution. Antevs 1935 Fieldbook, EA/UA.

261. Because of scheduling conflicts, not all participants could be at the site at the same time.

262. Antevs 1937c:338; Antevs 1938b:295; Bryan and MacClintock 1938:291–292; Kay and Leighton 1938:278.

263. They were sarcastic in their concession: "We cannot withhold our admiration of Antevs's technique in discovering and demonstrating these small structures" (Bryan and MacClintock 1938:283).

264. Antevs 1937c:335, 338; Bryan and MacClintock 1938:283, 291; Kay and Leighton 1938:274–276.

265. Roberts 1940a:101; Wormington 1939:49–50.

266. Antevs 1937c:336; Bryan and MacClintock 1938:290; Kay and Leighton 1938:277.

267. Jenks 1938.

268. Jenks 1937; Bryan et al. 1938; Jenks and Wilford 1938:167–168. Because the geological context at all three sites was ambiguous, only Brown's valley with its associated Folsom and Yuma points was generally accepted as a Postglacial age occurrence (for example, McCown 1941:205; Roberts 1940a:101; Wormington 1939:27–28). Radiocarbon dating subsequently showed that Minnesota Man dates to 7840 ± 70 ^{14}C years B.P. and Brown's Valley Man has two ages of 8,700 ± 100 ^{14}C years B.P. and 8900 ± 80 ^{14}C years B.P., making both of the specimens Early Holocene in age (Myster and O'Connell 1997: Table 7.1).

269. Cressman to Merriam, October 11, 1937, JCM/CIW. When the Catlow Cave bones spoke to Hrdlička, they spoke in Shoshonean (Hrdlička to Cressman, October 22, 1937, JCM/CIW). Hooton, however, put the remains in his Basketmaker type (Hooton to Cressman, January 12, 1938, JCM/CIW). See also Cressman 1938; Cressman et al. 1940:7.

270. Howells 1940:9; Roberts 1940a:102.

271. Hooton 1933:156–157, 161; Howells 1992:7.

272. Hooton 1930:355; Hooton 1933:134–135.

273. Hooton 1933:157–160. The entire collection of skeletal remains from Pecos Pueblo was repatriated in 1999 following a thorough inventory and a variety of analyses (Morgan, ed. 2010). None tried to replicate Hooton's cranial types, in part because of the high incidence of artificial cranial modification (Morgan 2010:34; Weisensee and Jantz 2010:43, 45). Nonetheless, Weisensee and Jantz responded to those who have criticized Hooton's cranial types, arguing that their analysis affirms the heterogeneity of the Pecos population and its "long and complicated biological history"

(Weisensee and Jantz 2010:55). Although their point regarding variation within the Pecos population is fair—if not unsurprising, given that the site was long a major center for trade and exchange—it is not a valid defense of Hooton's cranial types, which remain suspect because of the data on which they were based, the analyses conducted, and the assumptions behind them.

274. Hooton 1933:160; also Hooton 1930:359–360. McCown (1941:209–210) could only wonder why, if these loaded terms meant so little, Hooton used them in the first place.

275. Hooton 1930:355, 357. Also Roberts 1940a:99; Roberts 1940b:335; Wormington 1939:64.

276. Hooton 1930:358, 362; Hooton 1933:161.

277. Hooton 1930:360.

278. Hooton 1930. In his copy, Hrdlička passed over without annotation the last chapter in which Hooton conjured his scenarios of racial types and origins.

279. Hrdlička 1937b:94.

280. Hrdlička 1926:8–9; Hrdlička 1937b:94–95.

281. Hooton 1930:362–363; Hooton 1933:152–154; Hrdlička 1928:815–816; Hrdlička 1937b:94–95. On the differences between Hooton and Hrdlička, see Howells 1940:7 and McCown 1941:209–210.

282. Kroeber 1940a:461. Also Boas 1933:360; Howells 1940:7; McCown 1941:209.

283. Hooton 1930:355–357. Also Kroeber 1940a:460.

284. Hrdlicka 1926:9. Also Howells 1940:7.

285. See especially the very carefully worded statements in Roberts 1940a:98–99.

286. Howells 1940:8; McCown 1941:210.

287. Howells 1940:8.

288. Kroeber 1940a:461.

289. Roberts 1940a:52. Also Boas 1933:362.

290. Howard 1936b:410; cf. Roberts 1940a:52.

291. Wissler 1933:216; see Goddard 1926, 1927; Nelson 1933:97–98; Sapir 1916.

292. Roberts 1940a:52.

293. Boas 1933:362–363.

294. The reaction to nineteenth-century cultural evolution, led by Boas, challenged the idea that similarities in cultural traits were necessarily a function of the uniformity of history operating along parallel evolutionary lines (Boas 1896:906–908). He argued it could as readily be the result of historical contact or diffusion of traits among groups. But though Boas used diffusion to correct for the prior overemphasis on independent invention, he did not think it the sole mechanism of cultural change, as did the hyperdiffusionists of the British school.

295. Smith 1928:20, 25, 32. See also Smith et al. 1927; discussion in Harris 1968: 380–382.

296. Kidder 1936:144–145; Kroeber 1940a:474.

297. Smith in Smith et al. 1927:20–25.

298. Wissler 1933:167.

299. Boas 1933:360–361, 363–364; Spinden, in Smith et al. 1927:67–90.

300. Kidder 1936:145.

301. Kidder 1936:144. Also Howard 1935b:148, 151; Howard 1936c:315; Kroeber 1940a:476.

302. Kroeber 1940a:474. Also Harrington 1933:171.

303. Roberts 1940a:108. Also Roberts 1940b:335.

304. Kroeber 1940a:474.

305. Meltzer 1983:40.

306. For his Section H audience Bryan omitted the term "geologic." Hrdlička registered his low opinion of the paper in a handwritten note he penned on his reprint of Bryan's paper AHL/NMNH.

307. Bryan 1941:507; Meltzer 2006b.

308. Roberts 1940a; Sellards 1940; Wormington 1939.

309. Roberts 1940a:102. Also Bryan 1941:507; Wormington 1939:5.

310. Roberts 1940a:55.

311. Almost banished: Renaud reported artifacts resembling Lower and Middle Paleolithic artifacts from the Black Fork's valley of Wyoming. The claim was not taken seriously (Wormington 1939:47–48).

312. For example, Harrington 1933:177, 186–188; Howard 1936b:410; Nelson 1937:320.

313. Roberts 1940a:107. See also Antevs 1935b:304; Harrington 1933; Howard 1935b:137; Howard 1936b:410; Kidder 1936:144; Nelson 1933:115–116, 130; Nelson 1937:320; Renaud 1931:13; Renaud 1932:15; Renaud 1934:11; Spinden 1933:224.

314. Roberts 1940a:105–106. The debate over American isolationism ended abruptly on December 7, 1941.

315. Boas 1933:362; Bryan 1941:511; Howard 1935a: 151; Kidder 1936:144; Kroeber 1940a:474; Roberts 1940a:107; Wormington 1939:63.

316. Howard to Merriam, May 17, 1935, JCM/LC.

317. Arnold and Libby 1950; Roberts 1951. See the discussion in Meltzer 2006a:115–116.

318. Roberts 1940a:102.

319. Howells 1940:4. Also Bryan 1941:506.

320. Bryan 1941:506.

321. Bryan 1941:506; Roberts 1940a:103–104.

322. Antevs 1937a; Campbell and Campbell 1935; Campbell et al. 1937. See contemporary summaries in Roberts 1940a:85–91; Wormington 1939:35–43, 45–46.

323. Roberts 1940a:99, 108–109.

324. Roberts 1939:541. Also Bryan 1941:506; Howard 1936b:410; Roberts 1940a:104.

325. Bryan 1941:507–508; Roberts 1939:541; Roberts 1940b:336; Wormington 1939:22.

326. McCown 1941:207; Roberts 1939:541–542; Wormington 1939:22. Perhaps because they were mobile, Paleoindians did not intentionally bury their dead (Howard 1936c:319–320).

327. Kroeber 1940a:474; Roberts 1940a:104–107; Sellards 1940:410–411.

328. Romer 1933:76–77; see also Colbert 1942:27.

329. Kroeber 1940a:475.

330. Howard 1943:227–229; Nelson 1937:320.

331. Howard 1936b:410; Roberts 1940a:105. The first significant call to search for a pre–Clovis presence was made by Krieger 1953:238–239.

Chapter Ten

1. Kroeber 1952:151. Kroeber attributed the "unhistorical or even antihistorical orientation . . . widely prevalent in our country" to the belief that "we are effective as a nation because our history is short" (Kroeber 1952:149). Kroeber was far more charitable in 1952 than he had been a decade earlier, when he said that "most archaeologists of those days [early in the century] had little sense of problem, solution, and new problem; and

no sense of time whatever. They loved camping, they loved digging, they loved finding something; and the longer it lasted the better they like it" (Kroeber 1940b:2).

2. Evans, in Prestwich 1860:310, 312; also Grayson 1990:9–10; Van Riper 1993:120–122.

3. Grayson 1983. See also Livingstone 2008.

4. Merton 1988:620.

5. Nelson 1918a:394; Wissler 1916a:235.

6. Kidder 1936:143–144.

7. My use of "epistemic" and "nonepistemic" follows McMullin 1987:59–60. The dichotomy of "internal" and "external" approaches in the history of science (Kuhn 1977:118–120; Oldroyd 1990:356) makes discrete that which is arguably continuous and tends to sort along disciplinary lines (Knight 1987:7; Meltzer 1989:17–19).

8. Meltzer 1983:9. Also Sinclair 1979.

9. Meltzer and Dunnell 1992:xxx–xxxiii.

10. Hunter et al. 2009.

11. Schlereth 1991:3; also Hinsley 1981:256; Jenkins 1994:260.

12. Hinsley 1981:107, 256; also Meltzer and Dunnell 1992:xxxii–xxxiv.

13. Boas to Bell, August 7, 1903, FB/APS. Although Boas appreciated that particular skill, he was quick to add that Holmes's "interest in that part of anthropology which deals with ideas alone is slight." Holmes was not without a theoretical disposition; it just happened to be one that was anathema to Boas.

14. Mason, unpublished and undated, quoted in Conn 2004:215.

15. Hrdlička 1918:37; Hrdlička 1927:255.

16. See, for example, Keith 1915:272–285. Keith accepted claims for Trenton, Lansing, Long's Hill, and even the Calaveras skull, yet Keith cited Hrdlička with approval. That must have grated on Hrdlička.

17. Hay 1927b:5.

18. For a discussion of the Eolithic controversy, see de Bont 2003; Grayson 1986. The inability to demonstrate an American Paleolithic precluded any controversy in America over eoliths.

19. Hay 1928a:443.

20. Chamberlin 1903:66; Chamberlin 1919:309–310.

21. Holmes 1922:173. Also Chamberlin 1903 and Hrdlička 1907:13, 1912:2–3.

22. Holmes 1893d:158.

23. Carr to Henshaw, December 2, 1889, BAE/NAA.

24. Holmes 1893d:158; Wright 1893f:267–268.

25. McGee 1891:73.

26. Powell 1893c:114–115.

27. Chamberlin 1903:76.

28. Chamberlin 1903:69, 83.

29. Chamberlin 1903:84.

30. Upham 1893a:210. This was not a difficulty limited to North American sites: European archaeologists grappled with it as well (Van Riper 1993:216, 227–228).

31. See also the summary in Upham 1893a:210.

32. In later papers Chamberlin provided numerical age estimates; see Chamberlin 1919:311.

33. Kidder 1927:5, also Kidder 1936.

34. Hrdlička 1925:491 (although published in 1925, the quote comes from the Smithsonian Institution's 1923 *Annual Report*). The notion that in the pre–Folsom years prehistory extended back only 2,000 to 3,000 years occurs in many publications

(for example, Stocking 1976:24; O'Brien 2003:64). That notion was debunked by Mason (Mason 1966:194; see also Roberts 1940a:107).

35. The Folsom bison kill dates to 10,490 ± 20 radiocarbon years before the present (Meltzer 2006a:151).

36. Meltzer 1983:39. See also Hinsley 1981:29, 281; Lyman and O'Brien 2006:223–224; O'Brien 2003:70; Trigger 1980:664–665; Trigger 2006:180.

37. Lyman and O'Brien 2006:224–227; Holmes 1890a:19–21, 24.

38. Kidder 1936:145.

39. Powell 1881b:74; Powell 1886:lxiii.

40. Powell 1890b. The insistence on continuity was a central element in cultural evolutionary uniformitarianism (Stocking 1968:106).

41. Trigger 2006:184.

42. For example, Holmes 1919; Thomas 1898. Meltzer and Dunnell 1992:xxxvii–xxxviii. See also Lyman and O'Brien 2006:221–223; Trigger 2006:180.

43. Kroeber 1952:151. See also Trigger 1980:665–666.

44. Hinsley 1981:238–245; see also Cravens 1978:95; Meltzer 1983:39. American archaeologists were hardly alone in believing that modern Native Americans, as well as aboriginal Australians and African peoples, were culturally unchanged over a long period (Stocking 1987).

45. Willey and Sabloff 1974:86–87. That notion was criticized by Meltzer (1983:38–39), Lyman and O'Brien (2006:223–224), Pinsky (1992:164, 175–176), and Rowe (1975). Willey accepted my criticism, writing "You are probably right in your interpretation of the 'flat past' and it is incorrect to blame 'Boasianism'" (Willey to Meltzer, personal correspondence, September 19, 1983). However, he and Sabloff later argued that I misunderstood: they were not saying that antievolutionism caused the lack of time depth, but "rather that the lack of time depth subsequently weakened the ability of archaeologists to counter this anti-evolutionism" (Willey and Sabloff 1993:95). Here is what is repeated in all editions of their *A History of American Archaeology*: "The distrust of evolutionary thinking and the marked historical particularism of American anthropology [that is, Boas] forced the American archaeologist into a niche with a very limited horizon" (Willey and Sabloff 1974:86–87; Willey and Sabloff 1980:80; Willey and Sabloff 1993:90). I stand by my point.

46. Pinsky 1992:176. Pinsky notes that "archaeology was not particularly well-established within the Boasian program" and indeed was marginal to it (Pinsky 1992:169, 175). On the rise of Boas's historical particularism, see Cole 1999:127–130, 263–265; Hinsley 1981:98–100.

47. Boas tried unsuccessfully to win support for the teaching of archaeology at Columbia, in Boas to Murray (president of Columbia University), November 15, 1902, FB/APS (reprinted in Stocking 1974:292). On Boas's view of the four fields, see Boas 1904:523. See also comments in Cole 1999:267; Pinsky 1992:169–173; Stocking 1968:282; Stocking 1976:9.

48. Boas 1904. Boas wrote this soon after Holmes was appointed Powell's successor at the BAE instead of Boas's own candidate, WJ McGee. Boas was still smoldering over that outcome. Dall's opinion is in Dall to Branner, October 26, 1914, WHD/SIA.

49. Cole 1999:236–241; Hinsley 1981:242–254; Meltzer 2010:191–196; Pinsky 1992:167–172; Stocking 1968:281–292.

50. Lyman and O'Brien 2006: 224; Stocking 1976:24.

51. Compare Holmes 1910 with Holmes 1914 or Holmes 1919. See also Trigger 1980:664–665, Trigger 2006:180–181.

52. Sterns 1915:121.
53. Holmes 1919:55–56; see also Hrdlička 1926.
54. Steward 1942.
55. Holmes 1914:413. Such an approach never gained purchase among European archaeologists. Although they viewed archaeology as providing an account of their own cultural development—for them, history extended seamlessly into prehistory—it was obvious their prehistory went deep into the Pleistocene, rendering an ethnographic baseline utterly irrelevant to their distant past (MacCurdy 1913:572).
56. Powell 1890b:646; the point is echoed in Thomas 1898:23. See also Trigger 2006:183–184.
57. Holmes 1913:567.
58. Berthold Laufer, who exhorted his colleagues in 1913 to develop chronology as "the nerve electrifying the dead body of history," was a lone voice in this archaeological wilderness (Laufer 1913:577).
59. Kroeber 1909:4–5, 15–16. See also Lyman and O'Brien 2006: 226–227.
60. Cole 1999:257.
61. Boas 1902:1. See also Boas 1904:521–522; Stocking 1976:24.
62. Stocking 1976:24. See also Trigger 2006:184–185.
63. See, for example, discussions in Bieder 1986; Conn 2004; Cravens 1978; Hinsley 1981; Jenkins 1994; Stocking 1968; Stocking 1987.
64. Stocking 1968:11, 105–106; Stocking 1987:172–173; Van Riper 1993:200–201.
65. Tylor 1871:19.
66. Lubbock 1865:336. See Stocking 1987:153–156; Van Riper 1993:201–202.
67. Conn 2004:143–144, 215; Hinsley 1981:22, 68; Meltzer 1983:39–40; Trigger 1980:664–665; Trigger 2006:177, 179, 184–185, 188–189. Cravens (1978:94) observes that "the evolutionary or neo-Lamarckian point of view reduced culture to nature and assumed certain fixed stages of human 'progress' through which all people must develop if they were to rise to the high plateau of white civilization."
68. Conn 2004:143–144; Hinsley 1981:68; Trigger 1980:664. Willey and Sabloff (1993:95) object to the idea that BAE archaeologists and anthropologists denigrated Native Americans, finding it "hard to believe that Powell would have—or just as importantly countenanced—an anti–Native American bias." They are surely correct: Powell was a child of pre–Civil War abolitionists and deeply sympathetic toward Native Americans. But that hardly stopped him or others in the BAE from viewing American Indians as living in the savage stage of human culture (Hinsley 1981:149–150).
69. Holmes 1910:161, emphasis mine. See also Hinsley 1981:284–285.
70. WHH/RR 6:56, 69; Meltzer and Dunnell 1992:xvii. See also Hinsley 1981:106–108; Jenkins 1994:246; Mark 1980:155.
71. Quoted in Conn 2004:215. See also Hinsley 1981:22; Hinsley 1985:59–62.
72. Quoted in Hinsley 1985:55.
73. Mark 1980:5; also Cravens 1978:98.
74. Conn 2004:143–144, 182–183, 214–215; also Conn 1998:79; Hinsley 1985:55–56; Hinsley 2003:4–5; Mark 1980:5; Trigger 2006:187; Van Riper 1993:218.
75. Hinsley 1985:55, 59–62.
76. Stocking 1968:280–283; Trigger 1980:667. There is, of course, a vast literature on Boas's influence on anthropological theory and anthropology's development as a discipline, which need not be detailed here.
77. Boas 1907:929. Boas's antipathy toward museums developed out of a vision of anthropology in which the study of language, thought, customs, and ideas was paramount.

Material objects, the focus of museum anthropology, were less significant, for in no case, he argued, "do specimens alone convey the full idea that a collection is intended to express" (Boas 1907:928). In contrast, archaeology at the turn of the century was strongly object oriented (Conn 1998:80–81, 110–111; Conn 2004:194–196; Hinsley 1992:141–142; Hinsley 2003:18; Jenkins 1994:260–268).

78. Conn 2004:196. Also Jenkins 1994:246.

79. See Blakey 1987:10–20; for Boas's role, and his uneasy relationship with Hrdlička regarding these issues, see Stocking 1968:188–189, 287–292.

80. Pinsky 1992:167, 172–173,175–176; also Hinsley 1992:141–142; Stocking 1968:278–279, 282.

81. Dunnell 2003:16; Lyman and O'Brien 2006:226–241; 251. Lyman and O'Brien argue—rightly, in my view—that the methods of stratigraphy and seriation by themselves were insufficient to bring about this change: artifacts had to be thought of in different ways than they had been. Were that not the case, American archaeologists would have been measuring time a great deal earlier.

82. Lyman and O'Brien 2006:233.

83. Kidder 1936:145.

84. Kidder 1936:144–145.

85. Hrdlička to Spier, May 7 1928, AH/NAA.

86. Kidder 1936:143–144.

87. For example, Roberts 1939:534; Roberts 1940a:52.

88. Bernstein 2002: Appendixes A and B list dissertations written over the first several decades of anthropology in North America. Nearly one hundred PhD degrees had been awarded in anthropology by 1928, the great majority in ethnology, and most of those with Boas at Columbia University.

89. MacCurdy 1905.

90. Tozzer 1933:129, 137. See also Hinsley 1992:131; Hinsley 1999:143, 152; Mark 1980:14, 51–52; Stocking 1968:278–279.

91. Holmes 1918:561.

92. Meltzer 2002:232, 235–236; Stocking 1968:278; Stocking 1976:9.

93. Holmes to G. Stanley Hall, April 25, 1905, quoted in Noelke 1974:313. Also Stocking 1974:284.

94. Van Riper 1993:205–206.

95. That was Boas's stated goal, and it drew Jesup's interest and financial support, but as Freed et al. (1988:9) observe, Boas seemed little concerned about Native American origins, believing the answer was already obvious. On the 1911 AAA symposium, see Fewkes et al. 1912.

96. Bryan 1941:507, 514.

97. The average age of initial involvement before 1927 for all active participants was 40.9 years ($n = 21$; this tally does not include those who merely commented from the sidelines); for archaeologists after 1927, the average age was 36.5 ($n = 12$). That age difference is not statistically significant.

98. Powell 1890b:640.

99. Schlereth 1991: 29, 35, 252; see also Wiebe 1967:112–113.

100. The AAAS constitutions of 1856 and 1874 are available on the AAAS Web site: archives.aaas.org/docs/.

101. Data from the *Proceedings of the American Association for the Advancement of Science*: the "meiotic" events occurred in 1856 (creating Sections A and B), 1882 (resorting Sections A and B and creating Sections C through I), 1902 (Section K),

1907 (Section L), 1915 (Section M), and 1920 (Sections N through Q). Anthropology became the sole possessor of Section H in 1920, when psychology was shifted to Section I. Despite the two fields sharing a section label for nearly four decades, an anthropologist was elected vice president of the Section in most of those years, suggesting that anthropologists had a greater critical mass (or stronger voting bloc). See also Kohlstedt et al. 1999:43–44.

102. Many of the newly created national scientific societies had insufficient numbers of members to support independent meetings until the first or second decades of the twentieth century, and they often held their meetings along with the AAAS, which served as the crossroads for gathering anthropologists, geologists, and vertebrate paleontologists.

103. Data were compiled by examining the list of the AAAS Affiliated Organizations, all of which are specialized associations on a national scale, then checking each organization's individual Web site to determine the year it was established. The affiliates list is at www.aaas.org/aboutaaas/affiliates/. Data for the decline in local and state scientific societies can be found in Goldstein 2008: Table 1.

104. Rowland 1884:123–124.

105. Wiebe 1967:121; also Reingold 1976:53, 64; Schlereth 1991:250.

106. These data are compiled primarily from *Science*, which in 1898 began publishing an annual tally of the number of PhDs; some data for later years are from reports by the National Research Council.

107. Debated at the AAAS were questions of the definition of science, the place of amateurs, status among scientists, the appropriateness of efforts at science popularization, and, not least, the role of the federal government in science (Kohlstedt et al. 1999:45–48, 51–52).

108. Conn 1998:17.

109. Chamberlin 1919:309–310, 331.

110. Peet 1893:311.

111. Holmes 1922:173; Hrdlička 1907:13, Hrdlička 1912:2–3.

112. Hay and Cook 1928, 1930.

113. An issue ably discussed in Hinsley 1976, 1981, and 1985.

114. Hinsley 1992:141–142; Meltzer 2002:232–234.

115. Hull et al. 1978:719; Goldstein 1994:591–592, 594; Reingold 1976:34–35, 37.

116. Stocking 1960; Stocking 1968:283–286. Boas insisted that only persons who "have contributed to the advancement of anthropology either by publication or by high-grade teaching" were professionals (Boas to McGee, January 25, 1902, in Cravens 1978:103). There was likewise tension in the AAAS as increasingly specialized sciences sought to raise the level of the AAAS membership, their efforts resisted by those—such as Putnam, the longtime secretary of the AAAS—who wanted to maintain the open-door policy of amateur participation within the organization (Kohlstedt et al. 1999:45–47, 51–52).

117. Boas 1904.

118. Rowland 1884:120. These comments were made in the context of a wide-ranging "plea for pure science" in which Rowland argued the necessity and value of unburdening great thinkers ("brains") from their mundane obligations and providing them with the financial support—not to mention the "hands" of research assistants—so that they might achieve significant results in fundamental research.

119. Rudwick 2008:554. See also Hull 1988:389, 513–514.

120. Hull 1988:361.

121. Rowland 1884:122–123.
122. Mason 1884:379. Also Mason 1883, quoted in Hinsley 1981:86; also Hinsley 1976:39–41.
123. Conn 2003:165–166; Goldstein 1994:589.
124. Hinsley 1976:50–51.
125. Wright to Haynes, February 1, 1893, HWH/MHS.
126. Hinsley 1976:40–41; Hinsley 1999:142; Meltzer 1985:249–251.
127. Jones 1873:vii.
128. Hinsley 1985:64.
129. Hinsley 1976:51.
130. Popper 1963:112.
131. Oldroyd 1990:344–345.
132. Powell paid no heed to academic degrees, and his overall philosophy and approach to science created tension even within the USGS between his self-taught staff and those, like Samuel Emmons and Arnold Hague, who had had formal graduate training in Germany and France. Darrah 1951:279, 323–324; Pyne 1980:114. Also Hinsley 1976:42; Meltzer and Dunnell 1992:xiv, xl.
133. Holmes 1893e:135.
134. Hinsley 1976:51.
135. Hull 1988:289–290.
136. Wright to Haynes, November 25, 1892, HWH/MHS.
137. Wright to Haynes, January 10, 1889, HWH/MHS. Long after Holmes had made his arguments, Wright continued to seek Haynes's expert judgment regarding paleoliths he had found; Wright to Haynes November 23, 1895, HWH/MHS.
138. Abbott to Putnam, October 18, 1911, FWP/HU. Abbott tended to publish in popular magazines and even newspapers.
139. Haynes 1893a:66–67, emphasis mine.
140. Dawkins to Haynes, June 2, 1894, HWH/MHS.
141. Chamberlin to Dana, February 1, 1894, TCC/UC.
142. See also Hinsley 1976:48–50.
143. Chamberlin to Dana, February 1, 1894, TCC/UC.
144. Wright to Haynes, February 1, 1893 and September 28, 1893, HWH/MHS.
145. Wright to Haynes, January 3, 1893, HWH/MHS.
146. Winchell 1893b:194.
147. The *American Geologist* and Peet's and Moorehead's archaeological journals barely survived the first decade of the new century. The *Journal of Geology*, founded in 1893, is still viable today.
148. Rudwick 1985:419, 425; also Grayson 1983:207–208; Merton 1988:619–620; Oldroyd 1990:345, 348.
149. Visher 1947:4; also Rossiter 1982:287; Rudwick 1985:419.
150. The data for this table were compiled from volumes of the *Proceedings of the American Association for the Advancement of Science*, which provided a directory of fellows and the year they were elected; the first seven editions of the *American Men of Science*; and the NAS directory (www.nasonline.org/member-directory/), which includes all living and deceased members.
151. Kohlstedt et al. 1999:44–45. The AAAS constitutions are available at archives.aaas.org/docs/.
152. Only eighteen anthropologists were elected over that seventy-year span. In 1899, with an eye on increasing and diversifying its membership, the NAS created six

standing committees to identify and screen nominations in different fields. Powell and Putnam served on the committee for anthropology; it is perhaps no coincidence that Boas was elected in 1900 and Holmes in 1905.

153. Cattell 1906:v. See also Meltzer 2002:221–222.

154. Chudacoff 1989:5; see also Wiebe 1967:77, 111–132, 156–159.

155. Cattell 1906:vi; also Cattell 1903:566. A star was, literally, an asterisk next to one's name in the *American Men of Science* volume.

156. Cattell 1906:vi. Meltzer 2002: Table 1 lists the panel of anthropology judges for *AMS* 1 and *AMS* 2, and Tables 2 and 3 in that same article provide their individual rankings. The original data are in the Cattell Papers, JMC/LC. Among the judges of the stars for the first edition of the *American Men of Science* (which appeared in 1906) were Chamberlin and McGee ranking the geologists and vertebrate paleontologists, and Boas, Holmes, Hrdlička, and McGee (again) ranking the anthropologists. The identity and rankings of anthropology judges for the later editions is not known (Meltzer 2002:234).

157. The mechanics of the *AMS* star system are detailed in Meltzer 2002, Rossiter 1982, and Visher 1947.

158. The awarding of stars ended with the last volume produced before Cattell's death. For a variety of reasons, not least of which were criticisms of the merit of the star system, Cattell's son, who inherited the task of compiling and editing the *AMS* volumes, discontinued the practice (Meltzer 2002).

159. Rossiter 1982:290. See also the discussion in Meltzer 2002:233–235, Visher 1947:5, 23, and Cattell and Cattell 1933:1261.

160. Detailed rankings are in Meltzer (2002).

161. Hinsley 1981:249–253; Meltzer 2002:231–232; Meltzer and Dunnell 1992:xx–xxi.

162. Meltzer 2002:234–236. Mills had obviously escaped any ill effects from having discovered the New Comerstown paleolith three decades earlier, but that is not surprising because that was a one-off event for him. (His primary work was on mound complexes in Ohio and was well regarded.)

163. Upham to Cattell, November 12, 1919, JMC/LC.

164. Chamberlin to Cattell, March 6, 1903, JMC/LC. One caveat: Chamberlin did not explicitly identify Wright as his subject, but given that he thought of Wright in those terms and in fact did not rank him, it is likely Wright he was speaking of.

165. Hinsley 1981:249–253; Meltzer 2002:231–232; Meltzer and Dunnell 1992:xx–xxi.

166. Figgins to Kidder, October 17, 1927 and Kidder to Figgins, November 19, 1927, JDF/DMNS; Figgins to Nelson, August 24, 1927, DIR/DMNS.

167. Hull 1988:286–287; Merton 1988:621–622.

168. Rudwick 1985:419–424. Rudwick likens such groupings to vegetation or climatic zones on a mountainside, their differences perceptible once fully within them, but the boundaries between them gradual and indistinct. Also Hull 1988:283.

169. Dunning 2011; also Kruger and Dunning 1999.

170. Darwin 1871:3.

171. Merton 1988:620. See also Hull 1988:283, 287.

172. Rudwick 1985:420; also Hull 1988:309–310; Rudwick 2008:556.

173. Tozzer 1933:129, 137.

174. In regard to human antiquity, Tozzer again found it difficult to identify any substantive contribution Putnam had made, save his administrative support for the years of investigation at Trenton, concluding, "However much one may differ with

Professor Putnam as to the subject in question, no one can fail to admire the persistence of his search for the truth" (Tozzer 1933:130). More faint praise.

175. Chamberlin to Dana, February 1, 1894, TCC/UC.

176. Hull 1988:22; Oldroyd 1990:16.

177. Cook to Hay, December 23, 1926, OPH/SIA.

178. Lucas to Cook, November 3, 1927, HJC/AFB.

179. Conn 1998:167. Mercer's archive after 1897 contains virtually no correspondence pertaining to prehistoric archaeology.

180. Hrdlička to Hewett, December 8, 1934, AH/NAA.

181. Ingalls to Cook, January 31, 1928, HJC/AHC; Cook 1928a:39.

182. The exact totals were not known in 1927–1928; see the discussion in Meltzer 2006a:212–213, 252–252.

183. Grayson 1983:207–208; Oldroyd 1990:345; Rudwick 1985:420.

184. Henry to Squier, November 24, 1847, SEC/SIA.

185. Merton 1968:58; also Oldroyd 1990:355.

186. Merton 1988:609.

187. Merton 1968:57–58, 62; Merton 1988:607, 609.

188. Merton 1988:607.

189. Meltzer 2006a:42–48.

190. The list includes the 1927 AAA meeting and three meetings in 1928 at the New York Academy of Medicine, the International Congress of Americanists, and the GSA; sessions were also held at the 1931 AAAS meetings, the 1933 Fifth Pacific Science Congress, a 1935 meeting of the American Society of Naturalists, and, finally, at the 1937 "International Symposium on Early Man" sponsored by the ANSP—some of which are discussed in the text (§8.16–§8.18, §9.9).

191. Gregory to Boas, April 27, 1931, FB/APS. Howard considered inviting Figgins to the 1935 symposium at the American Society of Naturalists, but for reasons unknown he never did (Howard to Merriam November 16, 1935, JCM/LC). Howard invited Cook to the 1937 meeting in Howard to Cook, December 31, 1936 and Cook to Howard, March 15, 1937, HJC/AFB.

192. Hrdlička 1928, 1937b; see also Dixon to Hrdlička, November 27, 1927; Boas to Hrdlička, April 21, 1931; Howard to Hrdlička, November 30, 1936; all in AH/NAA.

193. Bryan 1941; Howard 1935b; Merriam 1936; Nelson 1933; Roberts 1940a.

194. Bryan 1937:139–140, emphasis mine. A similar expression was made by Gladwin 1937:133.

195. Cook 1928b:38.

196. Figgins 1928b:19.

197. Spier 1928a; Spier 1928b; Roberts 1940a:58.

198. Cook 1927b:117; Cook to Hay, January 25, 1928, OPH/SIA; Cook to Ingalls, April 6 and August 9, 1928, HJC/AHC; Cook to Ingalls, January 6, 1929 and January 6, 1932 and Cook to Loomis, December 12 and December 20, 1927, FBL/AC; Cook to Wissler, March 25, 1929 and AFB Figgins to Cook, September 25, 1926, HJC/AFB.

199. Spier to Cook, March 22, 1928, HJC/AFB.

200. Matthew to Cook, August 11, 1927, HJC/AFB.

201. Figgins to Gregory, December 30, 1927, DIR/DMNS.

202. Rudwick 1985:419–421.

203. Kidder to Figgins, December 15, 1927, DIR/DMNS; the same sentiment is expressed in Kidder to Figgins, October 13, 1927, DIR/DMNS.

204. See also Meltzer 2005:455; Meltzer 2006a:45.

205. Van Riper 1993:206–210.

206. Van Riper 1993:215–216.

207. Gamble and Kruszynski (2009: Figures 2 and 5) provide the original photographs taken that April day in 1859, as well as photographs of the discovery specimen, which they found in the Prestwich collection at the Natural History Museum (Gamble and Kruszynski 2009:468 and Figures 3 and 4).

208. Prestwich 1860:298. Evans, in a letter to Prestwich that was inserted as an appendix, also expressed great faith in the "concurrent testimony of the workmen" (Evans, in Prestwich 1860:312).

209. Van Riper 1993:222.

210. Van Riper 1993:219.

211. Bryan 1941:514.

212. Kidder 1936:144.

213. Kidder 1936:151.

214. Kidder 1936:145; Kroeber 1940a:474–475.

215. Lyman and O'Brien 2006:183–184.

216. I use Gould's (1965) distinction drawn between substantive and methodological uniformitarianism, but the meanings of uniformitarianism are arguably not so clear-cut, hence my hedging with the word "more." For the multiple meanings that can be applied to uniformitarianism, see Oldroyd 2003:95–97; Rudwick 1971; and especially Rudwick 2008:258–260, 295, 303–305, 315–330, 356–361, 558–559.

217. McKern 1939:302–303, emphasis mine. See also Kidder 1936:145, cf. Lyman and O'Brien 2006:181.

218. Lyman and O'Brien 2006:224–227, 250; also Dunnell 2003:16.

219. Ford and Willey 1941.

220. Lyman and O'Brien 2006:181–182.

221. Hooton 1933:134, 162.

222. Meltzer 1998; Stanton 1960.

223. Stocking 1968:287–292. As Stocking explains, Boas's opposition to the National Research Council Committee composition helped ignite the firestorm that became the Boas censure in 1919.

224. Stocking 1968:189.

225. Boas 1933:362–363.

226. Hull 1988:377.

227. McMullin 1987:59–60.

228. It should be said at this juncture that this was not an instance of Planck's Principle, the idea that a new scientific truth emerges "because its opponents eventually die, and a new generation grows up that is familiar with it" (Hull et al. 1978:718; Hull 1988:379–382). Both Hrdlička and Holmes were alive in 1927, though Holmes had little investment in the field (§8.18). That they were both virtually silent in response to Folsom reveals they saw all too clearly how the tide was running.

229. Griffin 1977:4. Vero recently reappeared on the list of possible Pleistocene sites (Purdy et al. 2011), but that case is as yet unproven, and in any case is based on very different evidence than was brought forward a century ago.

BIBLIOGRAPHY

A. Manuscript sources

Archival material was found for many of the major players in the controversy and were quite rich in some cases, notably those of Erwin Barbour, Barnum Brown, Thomas Chamberlin, Harold Cook, Jesse Figgins, Aleš Hrdlička, William Henry Holmes, WJ McGee, Frederic Ward Putnam, Warren Upham, Newton Winchell, and George Frederick Wright. The Wright papers were by far the most extensive, containing thousands of letters, material sent to him by his clipping services, unpublished papers, sermons, and just about anything else one might want from his professional and personal life—except for a few letters sent to him by Chamberlin. These letters were so harsh that Wright, perhaps understandably, chose not to save them. I know of these only because Chamberlin kept copies. Others among the major players have less available material. There is little from E. H. Sellards's early years at Vero, save for the letters he sent out that are in other collections. The papers of Rollin Salisbury, Bohumil Shimek, and Thomas Wilson are somewhat more limited as well.

Archivists at the Smithsonian Institution had puzzled over why Oliver Hay Papers consisted almost solely of correspondents whose names began with C, F, or S. The answer, I discovered, is that Hay saved his correspondence related to the human antiquity controversy and with Cook, Figgins, and Sellards. Missing are what must have been extensive files relating to his long career in vertebrate paleontology. Perhaps he considered his role in the human antiquity debates more important and wanted to leave guideposts for future historians. In fact, at one point he tore up a small, sharply worded, and intemperate note from Holmes, then reconsidered, carefully pieced it back together, and kept it with his papers, along with annotations for posterity (§7.7).

The Holmes papers, housed at the Smithsonian Archives and the National Museum of American Art (which has his twenty-volume *Random Records of a Lifetime in Art and Science*) are in many ways equally selective. Holmes discarded many letters but carefully kept those that were favorable or complimentary, especially in compiling *Random Records*—essentially a scrapbook of his life. John Swanton applauded the organization and annotations of *Random Records*, saying it would "make the task of any biographer a joy" (Swanton 1935:237). Not a joy, actually: *Random Records* hides more than it reveals.

The most unfortunate gap in archival coverage pertains to Charles Abbott. He was a prolific correspondent, but virtually all of his papers burned in a tragic house fire in 1914 (§6.15). Snatches of his correspondence from earlier years, along with his diaries, survive at the Firestone Library at Princeton University and elsewhere. The Putnam and Peabody Museum papers at Harvard University, and the Winchell papers at the Minnesota Historical Society, contain many Abbott letters, mostly from the last years of his life (after he and Putnam mended fences).

A few of the players, like Edward Claypole and William Youmans, appear not to have left any papers behind, or at least none I could find. Unpublished works of these individuals will remain known only through correspondence that survives in the collections of others, until such time (if ever) their papers are found, deposited in an archive, and reported.

Most of the archival material is correspondence, some 3,500 letters altogether. Diaries are known for only a few of the participants, namely Abbott, Alfred Kidder, and Warren Moorehead. Kidder was an inveterate diarist, but unfortunately a decade of his diaries are missing, and that gap includes his visit to Folsom in September of 1927.

Archival material is identified in the endnotes using acronyms listed below. For convenience, there are two listings. One is of collections identified (and ordered alphabetically) by an individual's acronym (initials/archive). The other is of collections from departments within museums, research bureaus, and so on. These are ordered by institution, and where they provide key materials for specific individuals those individuals are so identified.

INDIVIDUALS

AH/NAA   Aleš Hrdlička papers, National Anthropological Archives, Washington, DC

ANM/LC   Anita Newcomb McGee papers, Library of Congress, Washington, DC

AVK/HU   A. V. Kidder papers, Harvard University, Cambridge, MA

BB/AMNH   Barnum Brown papers, American Museum of Natural History, New York, NY

BHS/SIA   Bohumil Shimek papers, Smithsonian Institution Archives, Washington, DC

CCA/ANS   Charles C. Abbott–Mrs. Robins letters, Academy of Natural Sciences, Philadelphia, PA

CCA/PU   Charles C. Abbott papers, Firestone Library, Princeton University, Princeton, NJ

CCA/UPM   Charles C. Abbott curatorial records, University of Pennsylvania Museum, Philadelphia, PA

EA/UA   Ernst Antevs papers, Department of Geosciences, University of Arizona, Tucson, AZ

EBH/ANS   Edgar B. Howard Clovis papers, Academy of Natural Sciences, Philadelphia, PA

EBH/UPM   Edgar B. Howard curatorial records, University of Pennsylvania Museum, Philadelphia, PA

EHB/NSM   Erwin H. Barbour papers, Nebraska State Museum, Lincoln, NE

EHS/UT   Elias H. Sellards files, University of Texas, Austin, TX

FB/APS   Franz Boas papers, American Philosophical Society, Philadelphia, PA

FBL/AC   Frederic B. Loomis papers, Amherst College, Amherst, MA

FHHR/NAA   Frank H. H. Roberts papers, National Anthropological Archives, Washington, DC

FWP/HU   Frederick Ward Putnam papers, Harvard University, Cambridge, MA

GFW/OCA   George Frederick Wright papers, Oberlin College Archives, Oberlin, OH

HCM/BCHS   Henry Mercer papers, Bucks County Historical Society, Doylestown, PA

HFO/AMNH   Henry Fairfield Osborn papers, American Museum of Natural History, New York, NY

HJC/AFB   Harold J. Cook papers, Agate Fossil Beds National Monument, Scottsbluff, NE

HJC/AHC   Harold J. Cook papers, American Heritage Center, University of Wyoming, Laramie, WY

HWH/MHS   Henry W. Haynes papers, Massachusetts Historical Society, Boston, MA

JCB/SU   John C. Branner papers, Stanford University, Palo Alto, CA

JCM/CIW   John C. Merriam papers, Carnegie Institution of Washington, Washington, DC

JCM/LC   John C. Merriam papers, Library of Congress, Washington, DC

JDF/DMNS   Jesse D. Figgins, Director, Denver Museum of Nature and Science, Denver, CO

JMC/LC   James McKeen Cattell papers, Library of Congress, Washington, DC

LCE/UP   Loren C. Eiseley papers, University of Pennsylvania, Philadelphia, PA

NMJ/NAA   Neil M. Judd papers, National Anthropological Archives, Washington, DC

NW/MHS   Newton Winchell papers, Minnesota Historical Society, St. Paul, MN

OPH/SIA   Oliver P. Hay papers, Smithsonian Institution Archives, Washington, DC

RDS/UC   Rollin D. Salisbury papers, University of Chicago, Chicago, IL

RTH/SMU   Robert T. Hill papers, Southern Methodist University, Dallas, TX

SGM/PHS   Samuel G. Morton papers, Pennsylvania Historical Society, Philadelphia, PA

TCC/UC   Thomas C. Chamberlin papers, University of Chicago, Chicago, IL

WHD/SIA   William H. Dall papers, Smithsonian Institution Archives, Washington, DC

WHH/NAA   William Henry Holmes letters, National Anthropological Archives, Washington, DC

WHH/RR   William Henry Holmes, *Random Records of a Lifetime in Art and Science*, National Museum of American Art, Smithsonian Institution, Washington, DC

WHH/SIA   William Henry Holmes papers, Smithsonian Institution Archives, Washington, DC

WJM/LC   WJ McGee papers, Library of Congress, Washington, DC

WJM/NAA   WJ McGee papers, National Anthropological Archives, Washington, DC

WU/MHS   Warren Upham papers. Minnesota Historical Society, St. Paul, MN

INSTITUTIONS

AHL/NMNH   Aleš Hrdlička library, National Museum of Natural History, Washington, DC

ANTH/AMNH   Department of Anthropology correspondence, American Museum of Natural History, New York, NY

BAE/NAA   Bureau of American Ethnology papers, National Anthropological Archives, Washington, DC

BEG/UT   Bureau of Economic Geology records, University of Texas, Austin, TX

DHR/CIW   Division of Historical Research files, Carnegie Institution of Washington, Washington, DC

DIR/DMNS   Papers of the Director (Figgins, Bailey), Denver Museum of Natural History, Denver, CO

ER/UP   Expedition records (Mercer, Howard), University of Pennsylvania Museum, Philadelphia, PA

GC/ANS   General correspondence, Academy of Natural Sciences, Philadelphia, PA

GC/CIW   General Correspondence files, Carnegie Institution of Washington, Washington, DC

MHS/MHS   Archaeology Division papers, and general correspondence (Upham, Winchell), Minnesota Historical Society, St. Paul, MN

PMP/HU   Peabody Museum papers, Harvard University, Cambridge, MA

SEC/SIA   Office of the Secretary of the Smithsonian Institution, Smithsonian Institution Archives, Washington, DC

UM/UP   University Museum papers, University of Pennsylvania, Philadelphia, PA

USNM/SIA   United States National Museum, administrative files, Smithsonian Institution Archives, Washington, DC

VP/AMNH   Department of Vertebrate Paleontology correspondence, American Museum of Natural History, New York, NY

B. Printed sources: Primary

Abbott, C. C. (1863) Description of a collection of jasper "Lance-Heads" found near Trenton, New Jersey; and remarks on the locality, with reference to Indian antiquities. *Proceedings of the Academy of Natural Sciences of Philadelphia* 15:278–279.

Abbott, C. C. (1872) The stone age in New Jersey. *American Naturalist* 6:144–160, 199–229.

Abbott, C. C. (1873) Occurrence of implements in the river drift of Trenton, New Jersey. *American Naturalist* 7:204–209.

Abbott, C. C. (1876a) Antiquity of man. *American Naturalist* 10:51–52.

Abbott, C. C. (1876b) Indications of the antiquity of the Indians of North America, derived from a study of their relics. *American Naturalist* 10:65–72.

Abbott, C. C. (1876c) Traces of an American autochton. *American Naturalist* 10:329–335.

Abbott, C. C. (1876d) Western worked flakes and the New Jersey rude implements. *American Naturalist* 10:431– 432.

Abbott, C. C. (1876e) The stone age in New Jersey. *Smithsonian Institution Annual Report for 1875*:246–380.

Abbott, C. C. (1877a) On the discovery of supposed Paleolithic implements from the glacial drift in the Valley of the Delaware River, near Trenton, New Jersey. *Peabody Museum Annual Report* 10:30–43.

Abbott, C. C. (1877b) The classification of stone implements. *American Naturalist* 11: 495–496.

Abbott, C. C. (1878) Second Report on the Paleolithic implements from the glacial drift in the Valley of the Delaware River, near Trenton, New Jersey. *Peabody Museum Annual Report* 11:225–257.

Abbott, C. C. (1881a) Historical sketch of their discovery. *Proceedings of the Boston Society of Natural History* 21:124–132.

Abbott, C. C. (1881b) *Primitive Industry*. Salem: George A. Bates.

Abbott, C. C. (1883a) Paleolithic man in Ohio. *Science* 1:359.

Abbott, C. C. (1883b) Evidences of glacial man. *Science* 2:437–438.

Abbott, C. C. (1883c) A recent find in the Trenton gravels. *Proceedings of the Boston Society of Natural History* 22:96–104.

Abbott, C. C. (1884) *A Naturalist's Rambles About Home*. New York: D. Appleton.

Abbott, C. C. (1886a) Sketch of Frederick Ward Putnam. *Popular Science Monthly* 29: 693–697.

Abbott, C.C. (1886b) *Upland and Meadow*. New York: Harper and Brothers.

Abbott, C. C. (1888) The antiquity of man in the valley of the Delaware. *Proceedings of the Boston Society of Natural History* 23:424–426

Abbott, C. C. (1889a) A correction. *Popular Science Monthly* 34:411.

Abbott, C. C. (1889b) Evidences of the antiquity of man in eastern North America. *Proceedings of the American Association for the Advancement of Science* 37:293–315.

Abbott, C. C. (1889c) The descendants of Paleolithic man in America. *Popular Science Monthly* 36:145–153.

Abbott, C. C. (1890) Report of the Curator of the Museum of American Archaeology, University of Pennsylvania. *Annual Report* 1:1–54.

Abbott, C. C. (1892a) Paleolithic man in America. *Science* 20:270–271.

Abbott, C. C. (1892b) Paleolithic man: A last word. *Science* 20:344–345

Abbott, C. C. (1892c) Recent archaeological explorations in the Valley of the Delaware. *University of Pennsylvania Series in Philology, Literature and Archaeology* 2(1):1–30.

Abbott, C. C. (1893a) Are there relics of man in the Trenton gravels? *Archaeologist* 1:81–82.

Abbott, C. C. (1893b) The so-called "cache implements." *Science* 21:122–123.

Abbott, C. C. (1897) Review of "*Researches upon the Antiquity of Man in the Delaware Valley and the Eastern United States*," by H. C. Mercer. *Critic* 796:350.

Abbott, C. C. (1907) *Archaeologia Nova Caesarea.* Trenton: MacCrellish and Quigley.

Abbott, C. C. (1908) *Archaeologia Nova Caesarea, no. 2.* Trenton: MacCrellish and Quigley.

Abbott, C. C. (1909) *Archaeologia Nova Caesarea, no. 3.* Trenton: MacCrellish and Quigley.

Abbott, C. C. (1912) *Ten Years' Digging in Lenape Land, 1901–1911.* Trenton: MacCrellish and Quigley.

Agassiz, A. (1885) The Coast Survey and "Political Scientists." *Science* 6:253–255.

Agassiz, E. (1887) *Louis Agassiz. His Life and Correspondence.* Boston: Houghton and Mifflin.

Agassiz, L. (1840) *Études sur les Glaciers.* Jent et Gassmann: Neuchatel.

American Antiquarian (1881) Preglacial man. *American Antiquarian* 3:145.

American Geologist (1889) Review of "*The Ice Age in North America and Its Bearing upon the Antiquity of Man.*" *American Geologist* 4:106–108.

American Geologist (1892) Editorial: The United States Geological Survey. *American Geologist* 10:179–181.

American Geologist (1893a) Editorial: The topographical work of the National Geological Survey. *American Geologist* 11:47–55.

American Geologist (1893b) Editorial: Some recent criticism. *American Geologist* 11:110–112.

American Geologist (1893c) Editorial: Pleistocene papers at the Madison meetings. *American Geologist* 12:165–181.

American Geologist (1893d) Editorial: Glacial man in America. *American Geologist* 12:187–188.

American Journal of Physical Anthropology (1932) Committee on early man. *American Journal of Physical Anthropology* 16:279.

American Journal of Science (1884) Review of "*Second Annual Report of the United States Geological Survey to the Secretary of the Interior, for the Year 1880–1881.*" *American Journal of Science* 27:66–69.

American Journal of Science (1889) Review of "*The Ice Age in North America and Its Bearing upon the Antiquity of Man.*" *American Journal of Science* 38:412–413.

Andrews, E. (1875) Dr. Koch and the Missouri mastodon. *American Journal of Science* 10:32–34.

Anonymous (1858) Report on the twenty-eighth meeting of the British Association for the Advancement of Science: Section C, geology. *Atheneaum* 1615:461.

Anonymous (1882) Review of Abbott's *"Primitive Industry."* *Nation* 34:37–39.

Anonymous (1887) Sketch of Charles C. Abbott. *Popular Science Monthly* 30:547–553.

Antevs, E. (1922) The recession of the last ice sheet in New England. *American Geographical Society Research Series* 11.

Antevs, E. (1925) On the Pleistocene history of the Great Basin. *Carnegie Institution of Washington Publication* 352.

Antevs, E. (1928) *The Last Glaciation: With Special Reference to the Ice Retreat in Northeastern North America.* New York: American Geographical Society.

Antevs, E. (1935a) The occurrence of flints and extinct animals in pluvial deposits near Clovis, New Mexico. Part 2: Age of Clovis lake beds. *Proceedings of the Philadelphia Academy of Natural Sciences* 87:304–312.

Antevs, E. (1935b) The spread of aboriginal man to North America. *Geographical Review* 25:302–309.

Antevs, E. (1935c) Climate of the Southwest during the Late Wisconsin glaciation. *Year Book—Carnegie Institution of Washington* 34:317–318.

Antevs, E. (1936) Dating records of early man in the southwest. *American Naturalist* 70:331–336.

Antevs, E. (1937a) Climate and early man in North America. In *Early Man*, ed. G. G. MacCurdy, pp. 125–132. Philadelphia: Lippincott.

Antevs, E. (1937b) Studies on the climate in relation to early man in the Southwest. *Year Book—Carnegie Institution of Washington* 36:335.

Antevs, E. (1937c) The age of the "Minnesota Man." *Year Book—Carnegie Institution of Washington* 36:335–338.

Antevs, E. (1938a) Studies on the climate in relation to early man in the Southwest. *Year Book—Carnegie Institution of Washington* 37:348.

Antevs, E. (1938b) Was "Minnesota Girl" buried in a gully? *Journal of Geology* 46:293–295.

Arnold, J. R. and W. F. Libby (1950) *Radiocarbon Dates.* Chicago: Institute for Nuclear Studies, University of Chicago.

Babbage, C. (1860) Observations on the discovery in various localities of the remains of human art mixed with bones of extinct races of animals. *Proceedings of the Royal Society* 10:59–72.

Babbitt, F. (1880) Ancient quartz workers and their quarries in Minnesota. *American Antiquarian* 3:18–23.

Babbitt, F. (1883) Vestiges of glacial man in central Minnesota. *Science* 2:369–370.

Babbitt, F. (1884a) Vestiges of glacial man in central Minnesota. *Proceedings of the American Association for the Advancement of Science* 32:385–390.

Babbitt, F. (1884b) Vestiges of glacial man in central Minnesota. *American Naturalist* 18:594–605; 697–708.

Babbitt, F. (1885) Exhibition and descriptions of some Paleolithic quartz implements from central Minnesota. *Proceedings of the American Association for the Advancement of Science* 33:593–599.

Babbitt, F. (1890) Points concerning the Little Falls Quartzes. *Proceedings of the American Association for the Advancement of Science* 38:333–339.

Baldwin, C. C. (1893) Review extraordinary of "Man and the glacial period" by a member of the United States Geological Survey, with annotations and remarks thereon. Privately printed.

Barbour, E. (1907a) Prehistoric man in Nebraska. *Putnam's Magazine* January:413–415, 502–503.

Barbour, E. (1907b) Evidence of man in the loess of Nebraska. *Science* 25:110–112.

Barbour, E. (1907c) Evidence of Loess man in Nebraska. *Nebraska Geological Survey* 2:329–347.

Barbour, E. and C. B. Schultz (1932a) The mounted skeleton of *Bison occidentalis*, and associated dart points. *Bulletin of the Nebraska State Museum* 32:263–270.

Barbour, E. and C. B. Schultz (1932b) The Scottsbluff bison quarry and its artifacts. *Bulletin of the Nebraska State Museum* 34:283–286.

Barbour, E. and C. B. Schultz (1936) Palaeontologic and geologic consideration of early man in Nebraska. *Bulletin of the Nebraska State Museum* 45:431–449.

Barbour, E. and C. B. Schultz (1937) Pleistocene and post-glacial mammals of Nebraska. In *Early Man*, ed. G. G. MacCurdy, pp. 185–192. Philadelphia: Lippincott.

Barbour, E. and H. Ward (1906a) Discovery of an early type man in Nebraska. *Science* 24:628–629.

Barbour, E. and H. Ward (1906b) Preliminary report on the primitive man of Nebraska. *Nebraska Geological Survey* 2:319–327.

Bartsch, P. (1912) The bearing of ocean currents on the problem. *American Anthropologist* 14:49–50.

Bassler, R. (1917) Proceedings of the Eighth Annual Meeting of the Paleontological Society. *Geological Society of America Bulletin* 28:189–234.

Becker, G. (1891) Antiquities from under Tuolumne Table Mountain in California. *Geological Society of America Bulletin* 2:189–200.

Bell, E. and W. van Royen (1934) An evaluation of recent Nebraska finds sometimes attributed to the Pleistocene. *Wisconsin Archaeologist* 13:49–70.

Berlin, A. F. (1878) Paleolithic implements. Notes on the discovery of Paleolithic implements found within the limits of the city of Reading, Pennsylvania. *American Antiquarian* 1:10–12.

Berry, E. W. (1917) The fossil plants from Vero, Florida. *Florida Geological Survey Annual Report* 9:19–33.

Berthoud, E. (1873) Antiquities on the Cache La Poudre River, Weld County, Colorado Territory, Colorado. *Smithsonian Institution Annual Report for 1871*:402–403.

Blackman, E. (1907) Prehistoric man in Nebraska. *Records of the Past* 6:76–79.

Blake, W. (1899) The Pliocene skull of California and the flint implements of Table Mountain. *Journal of Geology* 7:631–637.

Boas, F. (1896) The limitations of the comparative method. *Science* 4:901–908.

Boas, F. (1899) The cephalic index. *American Anthropologist* 1:448–461.

Boas, F. (1902) Some problems in North American archaeology. *American Journal of Archaeology* 6:1–6.

Boas, F. (1904) The history of anthropology. *Science* 20:513–524.

Boas, F. (1907) Some principles of museum administration. *Science* 25:921–933.

Boas, F. (1912a) Changes in the bodily form of descendants of immigrants. *American Anthropologist* 14:530–562.

Boas, F. (1912b) The instability of human types. In *Papers on Interracial Problems Communicated to the First Universal Races Congress*, edited by G. Spiller, pp. 99–103. Boston: Ginn and Co.

Boas, F. (1928) The twenty-third International Congress of Americanists. *Science* 68: 361–364.

Boas, F. (1933) Relationships between North-west America and North-east Asia. In *The American Aborigines: Their Origin and Antiquity*, edited by D. Jenness, pp. 357–370. Toronto: University of Toronto Press.

Boucher de Perthes, J. (1860) *De l'Homme Antediluvian et de ses Oeuvres*. Paris: Jung-Treuttel.

Boule, M. (1893) L'homme paleolithique dans l'Amerique du Nord. *L'Anthropologie* 4:36–39.

Boule, M. (1911–1913) L'homme fossile de La Chapelle-aux-Saints. *Annales Paléontol* 6:111–172 (1911); 7:21–192 (1912); 8:1–70 (1913).

Branner, J. C. (1891) The relations of the state and national geological surveys to each other and to the geologists of the country. *Proceedings of the American Association for the Advancement of Science* 39:219–237.

Brinton, D. (1888) Prehistoric chronology of America. *Proceedings of the American Association for the Advancement of Science* 36:283–301.

Brinton, D. (1892a) Review of *"Man and the Glacial Period." Science* 20:249.

Brinton, D. (1892b) On quarry rejects. *Science* 20:260–261.

Brower, J. V. (1902) *Kakabikansing: Memoirs of Explorations in the Basin of the Mississippi*, vol. 5. St. Paul: Minnesota Historical Society.

Brown, B. (1928a) Recent finds relating to prehistoric man in America. Discussion of "The Origin and Antiquity of Man in America" by A. Hrdlička. *Bulletin of the New York Academy of Medicine* 4(7):824–828.

Brown, B. (1928b) Letter. *Natural History* 28:556.

Brown, B. (1929) Folsom culture and its age. *Geological Society of America Bulletin* 40:128–129.

Brown, B. (1932) The buffalo drive. *Natural History* 32:75–82.

Bryan, F. (1938) A review of the geology of the Clovis finds reported by Howard and Cotter. *American Antiquity* 4:113–130.

Bryan, K. (1929) Discussion of "Folsom Culture and Its Age," by B. Brown. *Geological Society of America Bulletin* 40:128–129.

Bryan, K. (1935) Minnesota man—A discussion of the site. *Science* 82:170–171.

Bryan, K. (1937) Geology of the Folsom deposits in New Mexico and Colorado. In *Early Man*, ed. G. G. MacCurdy, pp. 139–152. Philadelphia: Lippincott.

Bryan, K. (1941) Geologic antiquity of man in America. *Science* 93:505–514.

Bryan, K. and L. Ray (1940) Geologic antiquity of the Lindenmeier site in Colorado. *Smithsonian Miscellaneous Collections* 99(2).

Bryan, K. and P. MacClintock (1938) What is implied by "disturbance" at the site of Minnesota Man? *Journal of Geology* 46:279–292.

Bryan, K., H. Retzek, and F. McCann (1938) Discovery of Sauk Valley man of Minnesota, with an account of the geology. *Bulletin of the Texas Archeological and Paleontological Society* 10:114–135.

Buckland, William (1823) *Reliquiæ Diluvianæ; or Observations on the Organic Remains Contained in Caves, Fissures, and Diluvial Gravel, and on Other Geological Phenomena, Attesting the Action of an Universal Deluge.* London: John Murray.

Campbell, E. and W. H. Campbell (1935) The Pinto Basin site. *Southwest Museum Papers*, no. 9.

Campbell, E., W. H. Campbell, E. Antevs, C. Amsden, J. Barbieri, and F. Bode (1937) The archeology of Pleistocene Lake Mohave: A symposium. *Southwest Museum Papers*, no. 11.

Carr, L. (1881) Statement relating to the finding of an implement in the gravel. *Proceedings of the Boston Society of Natural History* 21:145–146.

Cattell, J. M. (1903) American Society of Naturalists. *Homo scientificus americanus. Science* 17:561–570.

Cattell, J. M. (1906) *American Men of Science: A Biographical Directory.* New York: Science Press.

Cattell, J. M. and J. Cattell (1933) *American Men of Science: A Biographical Directory,* 5th ed. New York: Science Press.

Chamberlain, A. (1897) Anthropology at the Toronto meeting of the British Association. *Science* 6:575–583.

Chamberlin, R. (1917a) Interpretation of the formations containing human bones at Vero, Florida. *Journal of Geology* 25:25–39.

Chamberlin, R. (1917b) Further studies at Vero, Florida. *Journal of Geology* 25:667–683.

Chamberlin, T. C. (1878) The extent and significance of the Wisconsin kettle moraine. *Wisconsin Academy of Science Transactions* 4:201–234.

Chamberlin, T. C. (1882) The bearing of some recent determinations on the correlations of the eastern and western terminal moraines. *American Journal of Science* 124:93–97.

Chamberlin, T. C. (1883a) *Geology of Wisconsin.* Geological Survey of Wisconsin 1.

Chamberlin, T. C. (1883b) Preliminary paper on the terminal moraine of the second glacial epoch. *United States Geological Survey Annual Report* 3:291–402.

Chamberlin, T. C. (1884a) The character of the outer border of the drift. *Proceedings of the American Association for the Advancement of Science* 32:210.

Chamberlin, T. C. (1884b) The terminal moraines of the later epoch. *Proceedings of the American Association for the Advancement of Science* 32:211–212.

Chamberlin, T. C. (1887) An inventory of our glacial drift. *Proceedings of the American Association for the Advancement of Science* 35:195–211.

Chamberlin, T. C. (1888a) Report of Prof. T. C. Chamberlin. *United States Geological Survey Annual Report* 7:76–85.

Chamberlin, T. C. (1888b) The rock-scourings of the great ice invasions. *United States Geological Survey Annual Report* 7:155–248.

Chamberlin, T. C. (1888c) *The Ethical Functions of Scientific Study: An Address Delivered at the Annual Commencement of the University of Michigan, June 28, 1888.* Ann Arbor: University of Michigan.

Chamberlin, T. C. (1890a) Introduction. In *The Glacial Boundary in Western Pennsylvania, Ohio, Kentucky, Indiana and Illinois,* by G. F. Wright. *United States Geological Survey Bulletin* 58:13–38.

Chamberlin, T. C. (1890b) The method of multiple working hypotheses. *Science* 15:92–96.

Chamberlin, T. C. (1890c) Some additional evidences bearing on the interval between the glacial epochs. *Geological Society of America Bulletin* 1:469–480.

Chamberlin, T. C. (1891) The present standing of the several hypotheses of the cause of the glacial period. *American Geologist* 8:237.

Chamberlin, T. C. (1892) Geology and archaeology mistaught. *Dial* 13:303–306

Chamberlin, T. C. (1893a) Professor Wright and the Geological Survey. *Dial* 14:7–9.

Chamberlin, T. C. (1893b) The diversity of the glacial period. *American Journal of Science* 45:171–200.

Chamberlin, T. C. (1893c) Editorial. *Journal of Geology* 1:847–849.

Chamberlin, T. C. (1894a) Glacial phenomena of North America. In *The Great Ice Age and Its Relation to the Antiquity of Man,* 3rd ed., ed. J. Geikie, pp. 724–774. New York: D. Appleton.

Chamberlin, T. C. (1894b) Glacial studies in Greenland. *Journal of Geology* 2:649–666.

Chamberlin, T. C. (1894c) Glacial studies in Greenland, II. *Journal of Geology* 2:768–788.

Chamberlin, T. C. (1896a) Review of "*New Evidence of Glacial Man in Ohio*" by Professor G. Frederick Wright. *Journal of Geology* 4:107–112.

Chamberlin, T. C. (1896b) [Untitled response to Wright 1896b]. *Journal of Geology* 4:219–221.

Chamberlin, T. C. (1896c) Review of "*Greenland Ice Fields and Life in the North Atlantic*," by G. Wright and W. Upham. *Journal of Geology* 4:632–636.

Chamberlin, T. C. (1896d) Editorial. *Journal of Geology* 4:872–876.

Chamberlin, T. C. (1897) Supplementary hypothesis respecting the origin of the loess of the Mississippi Valley. *Journal of Geology* 5:795–802.

Chamberlin, T. C. (1901) Remarkable discoveries. *Science* 13:987–990.

Chamberlin, T. C. (1902) The geologic relations of the human relics of Lansing, Kansas. *Journal of Geology* 10:745–779.

Chamberlin, T. C. (1903) The criteria requisite for the reference of relics to a glacial age. *Journal of Geology* 11:64–85.

Chamberlin, T. C. (1917) Editorial note. Symposium on the age and relations of the fossil human remains found at Vero, Florida. *Journal of Geology* 25:1–3.

Chamberlin, T. C. (1919) Investigation versus propagandism. *Journal of Geology* 27:305–338.

Chamberlin, T. C. and R. D. Salisbury (1885) Preliminary report on the Driftless Area of the Upper Mississippi Valley. *United States Geological Survey Annual Report* 6:199–322.

Chamberlin, T. C. and R. D. Salisbury (1906) *Geology*. 3 vols. New York: H. Holt.

Chamberlin, T. C., ed. (1917) Symposium on the age and relations of the fossil human remains found at Vero, Florida. *Journal of Geology* 25:1–62.

Christy, H. (1865) On the prehistoric cave dwellers of southern France. *Transactions of the Ethnological Society*, London n.s. 3:362–372.

Clark, A. (1912) The distribution of animals and its bearing on the peopling of America. *American Anthropologist* 14:23–30.

Clarke, W. (1938) The occurrence of flints and extinct animals in pluvial deposits near Clovis, New Mexico. Part 7: Pleistocene mollusks from the Clovis gravel pit and vicinity. *Proceedings of the Philadelphia Academy of Natural Sciences* 90:119–121.

Claypole, E. (1893a) Preglacial man not improbable. *American Geologist* 11:191–194.

Claypole, E. (1893b) Prof. G. F. Wright and his critics. *Popular Science Monthly* 42:764–781.

Claypole, E. (1893c) Major Powell on "Are There Evidences of Man in the Glacial Gravels?" *Popular Science Monthly* 43:696–699.

Claypole, E. (1897a) The anthropological session at Toronto. *Science* 6:450.

Claypole, E. (1897b) American Association for the Advancement of Science. *American Geologist* 9:194–202.

Claypole, E. (1898) Paleolith and Neolith. *American Geologist* 21:333–344.

Cochet, L'abbé (1857–1860) Hachettes diluviennes du basin de la Somme. *Mémoire Societé Emulàtion Abbeville* 1857–1860:607–623.

Cochet, L'abbé (1861) Revue des decouvertes archeologiques dans le department de la Seine inferieure. *Revue Archeologique* V:16–22.

Colbert, E. H. (1937) The Pleistocene mammals of North America and their relation to Eurasian forms. In *Early Man*, ed. G. G. MacCurdy, pp. 173–184. Philadelphia: Lippincott.

Colbert, E. H. (1942) The association of man with extinct mammals in the western hemisphere. *Proceedings of the Eighth American Scientific Congress* 2:17–29.

Cole, F. C. (1928) Section H (anthropology). Report of the sessions and sections and societies at the second Nashville meeting. *Science* 67:125.

Comstock, T. B. (1875) Geological report. In *Report upon the Reconnaissance of Northwestern Wyoming, Including Yellowstone National Park, Made in the Summer of 1873*, ed. W. Jones, pp. 85–292. Washington, DC: Government Printing Office.

Conrad, T. A. (1839) Notes on American geology. *American Journal of Science* 35:237–251.

Cook, G. (1877) *Annual Report of the State Geologist for the Year 1877*. Trenton: Naar, Day & Narr.

Cook, H. J. (1922) Tooth of almost human type from the Lower Pliocene Snake Creek beds of Western Nebraska. *Geological Society of America Bulletin* 33:192

Cook, H. J. (1925) Definite evidence of human artifacts in the American Pleistocene. *Science* 62:459–460.

Cook, H. J. (1926) The antiquity of man in America. *Scientific American* 137:334–336.

Cook, H. J. (1927a) New geological and paleontological evidence bearing on the antiquity of mankind in America. *Natural History* 27:240–247.

Cook, H. J. (1927b) New trails of ancient men. *Scientific American* 138:114–117.

Cook, H. J. (1928a) Glacial age man in New Mexico. *Scientific American* 139:38–40.

Cook, H. J. (1928b) Further evidence concerning man's antiquity at Frederick, Oklahoma. *Science* 67:371–373.

Cook, H. J. (1931a) More evidence of the "Folsom Culture" race. *Scientific American* 144:102–103.

Cook, H. J. (1931b) The antiquity of man as indicated at Frederick, Oklahoma: A reply. *Journal of the Washington Academy of Sciences* 21:161–167.

Cook, H. J. (1944) Jesse D. Figgins. *News Bulletin of the Society of Vertebrate Paleontology* 12:27–28.

Cook, H. J. (1947) Some background data on the original "Folsom Quarry" site: And of those connected with its discovery and development. Unpublished manuscript on file, Agate Fossil Beds National Monument, Scottsbluff, Nebraska.

Cook, H. J. (1952) Early man in America. Unpublished manuscript on file, Agate Fossil Beds National Monument, Scottsbluff, Nebraska.

Cook, H. J. (1968) *Tales of the 04 Ranch: Recollections of Harold J. Cook, 1887–1909*. Lincoln: University of Nebraska Press.

Cook, J. H. (1925) *Fifty Years on the Old Frontier: As Cowboy, Hunter, Guide, Scout, and Ranchman*. New Haven: Yale University Press.

Cooke, C. W. (1926) Fossil man and Pleistocene vertebrates in Florida. *American Journal of Science* 12:441–452.

Cooke, C. W. (1928) The stratigraphy and age of the Pleistocene deposits in Florida from which human bones have been reported. *Journal of the Washington Academy of Sciences* 18:414–421.

Cope, E. D. (1885) The occurrence of man in the Upper Miocene of Nebraska. *Proceedings of the American Association for the Advancement of Science* 33:593.

Cotter, J. L. (1937a) The occurrence of flints and extinct animals in pluvial deposits near Clovis, New Mexico. Part 4: Report on the excavations at the Gravel Pit in 1936. *Proceedings of the Philadelphia Academy of Natural Sciences* 89:1–16.

Cotter, J. L. (1937b) The significance of Folsom and Yuma artifact occurrences in the light of typology and distribution. In *Twenty-fifth Anniversary Studies Philadelphia Anthropological Society*, ed. D. S. Davidson, pp. 27–35. Philadelphia: University of Pennsylvania Press.

Cotter, J. L. (1938) The occurrence of flints and extinct animals in pluvial deposits near Clovis, New Mexico. Part 6: Report on the field season of 1937. *Proceedings of the Philadelphia Academy of Natural Sciences* 90:113–117.

Cotter, J. L. (1939) A consideration of "Folsom and Yuma Culture Finds." *American Antiquity* 5:152–155.

Cressman, L. S. (1938) Early man and culture in the northern Great Basin in Oregon. *Year Book—Carnegie Institution of Washington* 37:341–344.

Cressman, L. S., H. Williams, and A. D. Krieger (1940) *Early Man in Oregon: Archaeological Studies in the Northern Great Basin.* University of Oregon Monographs. Studies in Anthropology, No. 3. Eugene: University of Oregon Press.

Cresson, H. J. (1889a) Early man in the Delaware Valley. *Proceedings of the Boston Society of Natural History* 24:141–150.

Cresson, H. J. (1889b) Remarks on a chipped implement, found in modified drift, on the east fork of the White River, Johnson County, Indiana. *Proceedings of the Boston Society of Natural History* 24:150–152.

Croll, J. (1864) On the physical cause of the change of climate during geological epochs. *Philosophical Magazine* 28: 121–37.

Croll, J. (1875) *Climate and Time in Their Geological Relations: A Theory of Secular Changes of the Earth's Climate.* New York: D. Appleton.

Croll, J. (1885) *Climate and Time in Their Geological Relations: A Theory of Secular Changes of the Earth's Climate*, 3rd ed. New York: D. Appleton.

Cuvier, G. (1796) Mémoire sur les éspèces d'elephans tant vivantes que fossils, lu à la séance publique de l'Institut National le 15 germinal, an 4. *Magasin Encyclopédique*, second year, 3:440–45.

Cuvier, G. (1806) Sur les éléphans vivans et fossils. *Annales du Muséum d'Histoire Naturelle* 8:1–58, 93–155, 249–69.

Daily Science News Bulletin (1925) Denies antiquity of man in Florida. *Daily Science News Bulletin*, September 17, 1925. Washington, DC: Science Service.

Dall, W. H. (1912) On the geological aspects of the possible human immigration between Asia and America. *American Anthropologist* 14:12–18.

Daly, R. A. (1934) *The Changing World of the Ice Age.* New Haven: Yale University Press.

Dana, J. D. (1856) Address of Professor James D. Dana, President of the American Association for the year 1854, on retiring from the duties of president. *Proceedings of the American Association for the Advancement of Science* 9:1–36.

Dana, J. D. (1863) *Manual of Geology.* Philadelphia: Theodore Bliss.

Dana, J. D. (1869) *Manual of Geology*, revised ed. Philadelphia: Theodore Bliss.

Dana, J. D. (1873) On the Glacial and Champlain eras in New England. *American Journal of Science* 5:198–211.

Dana, J. D. (1875a) *Manual of Geology*, 3rd ed. New York: Ivison, Blakeman, Taylor.

Dana, J. D. (1875b) On Dr. Koch's evidence with regard to the contemporaneity of man and the mastodon in Missouri. *American Journal of Science* 9:335–346.

Dana, J. D. (1893) On New England and the Upper Mississippi basin in the glacial period. *American Journal of Science* 46:327–330.

Dana, J. D. (1894) *Manual of Geology*, 4th ed. New York: American Book.

Danforth, C. (1931) Section H (anthropology) and related organizations. *Science* 74: 117–118.

Danforth, C., J. A. Mason, and R. Tanner (1929) Section H (anthropology) and related organizations. *Science* 69:122–124.

Darwin, C. (1859) *On the Origin of Species by Means of Natural Selection, or the Preservation of Favoured Races in the Struggle for Life.* London: J. Murray.

Darwin, C. (1871) *The Descent of Man, and Selection in Relation to Sex.* London: J. Murray.

Davis, W. (1927a) Was man in America during Ice Age? *Science News-Letter* 11(312):207.

Davis, W. (1927b) Early man in America. *Science News-Letter* 12(338):215–216.

Davis, W. (1927c) Ancient man in America. *Science News-Letter* 12(345):333.

Davis, W. (1928) Man's antiquity in America debated. *Science News-Letter* 14(391):209.

Davis, W. M. (1889) Review of *"The Ice Age in North America and Its Bearing upon the Antiquity of Man."* *Science* 14:118–119.

Dawkins, W. B. (1862) On a hyaena-den at Wookey Hole, near Wells. *Quarterly Journal of the Geological Society* 18:115–125.

Dawkins, W. B. (1863) Wookey Hole hyaena-den. *Somersetshire Archaeological and Natural History Society Proceedings* 11:197–219.

Dawkins, W. B. (1874) *Cave Hunting: Researches on the Evidence of Caves Respecting the Early Inhabitants of Europe.* London: Macmillan.

Dawkins, W. B. (1880) *Early Man in Britain and His Place in the Tertiary Period.* London: Macmillan.

Dawkins, W. B. (1883) Early man in America. *North American Review* 137:338–349.

Dickeson, M. W. (1846) Fossils from Natchez, Mississippi. *Proceedings of the Philadelphia Academy of Natural Sciences* 3:106–107.

Doughty, F. (1892) Evidences of man in the drift. Address at the American Numismatic and Archaeological Society, privately printed.

Evans, J. (1860) On the occurrence of flint implements in undisturbed beds of gravel, sands, and clay. *Archaeologia* 38:280–307.

Evans, J. (1863) Account of some further discoveries of flint implements in the drift on the Continent and in England. *Archaeologia* 39:57–84.

Evans, J. (1864) On some recent discoveries of flint implements in drift deposits in Hants and Wilts. *Quarterly Journal of the Geological Society* 20:188–194.

Evans, J. (1872) *The Ancient Stone Implements, Weapons and Ornaments of Great Britain.* London: Longmans.

Evans, O. (1930) The antiquity of man as shown at Frederick, Oklahoma: A criticism. *Journal of the Washington Academy of Sciences* 20:475–479.

Fairchild, H. L. (1893) Proceedings of the fourth summer meeting, held at Rochester, August 15 and 16, 1892. *Geological Society of America Bulletin* 4:1–12.

Fewkes, J. W., A. Hrdlička, W. Dall, J. W. Gidley, A. H. Clark, W. H. Holmes, A. C. Fletcher, W. Hough, S. Hagar, P. Bartsch, A. F. Chamberlain, and R. B. Dixon (1912) The problems of the unity or plurality and the probable place of origin of the American Aborigines. *American Anthropologist* 14:1–4.

Figgins, J. D. (1925) Report of the director. *Colorado Museum of Natural History Annual Report for 1924*:10–24.

Figgins, J. D. (1927) The antiquity of man in America. *Natural History* 27:229–239.

Figgins, J. D. (1928a) Report of the director. *Colorado Museum of Natural History Annual Report for 1927*:9–20.

Figgins, J. D. (1928b) How long man has hunted: The light fossils shed on the quest. *Outdoor Life* 61:18–19, 82.

Figgins, J. D. (1933a) A further contribution to the antiquity of man in America. *Proceedings of the Colorado Museum of Natural History* 12(2):4–8.

Figgins, J. D. (1933b) The bison of the western area of the Mississippi Basin. *Proceedings of the Colorado Museum of Natural History* 12(4).

Figgins, J. D. (1934a) Report of the director. *Colorado Museum of Natural History Annual Report for 1933*:9–19.

Figgins, J. D. (1934b) Folsom and Yuma artifacts. *Proceedings of the Colorado Museum of Natural History* 13(2):2–8.

Figgins, J. D. (1935a) New World man. *Proceedings of the Colorado Museum of Natural History* 14:1–9.

Figgins, J. D. (1935b) Folsom and Yuma artifacts, part 2. *Proceedings of the Colorado Museum of Natural History* 14(2).

Flint, R. F. (1941) Glacial geology. In *Geology 1888–1938: Fiftieth Anniversary Volume*, ed. R. Berkey, pp. 17–41. New York: Geological Society of America.

Ford, J. and G. Willey (1941) An interpretation of the prehistory of the eastern United States. *American Anthropologist* 43: 325–363.

Foster, J. (1871) Address of J. W. Foster, ex-president of the association. *Proceedings of the American Association for the Advancement of Science* 19:1–19.

Foster, J. (1873a) On the antiquity of man in America. *Transactions of the Chicago Academy of Science* 1:227–257.

Foster, J. (1873b) *Prehistoric Races of the United States of America.* Chicago: S. Griggs.

Fowke, G. (1902) *Archaeological History of Ohio.* Columbus: Ohio State Archaeological and Historical Society.

Fowke, G. (1928) Archaeological investigations—2. *Bureau of Ethnology Annual Report* 44:399–540.

Geikie, J. (1874) *The Great Ice Age and Its Relation to the Antiquity of Man.* New York: D. Appleton.

Geikie, J. (1877) *The Great Ice Age and Its Relation to the Antiquity of Man*, 2nd ed. New York: D. Appleton.

Geikie, J. (1881) *Prehistoric Europe, a Geological Sketch.* London: E. Stanford.

Geikie, J. (1894) *The Great Ice Age and Its Relation to the Antiquity of Man*, 3rd ed. New York: D. Appleton.

Gibbs, G. (1862) Instructions for archaeological investigations in the United States. *Smithsonian Institution Annual Report for 1861*:392–396.

Gidley, J. W. (1912) Paleontological evidence bearing on the problem of the origin of the American Aborigines. *American Anthropologist* 14:18–23.

Gidley, J. W. (1926a) Fossil man in Florida. *Geological Society of America Bulletin* 37:240.

Gidley, J. W. (1926b) Investigations of evidences of early man at Melbourne and Vero, Florida. *Smithsonian Miscellaneous Collections* 78(1):23–26.

Gidley, J. W. (1926c) Fossil man associated with the mammoth in Florida: New evidence of the antiquity of man in America. *Journal of the Washington Academy of Sciences* 16:310.

Gidley, J. W. (1927) Investigating evidence of early man in Florida. *Smithsonian Miscellaneous Collections* 78:168–174.

Gidley, J. W. (1929) Ancient man in Florida: Further investigations. *Geological Society of America Bulletin* 40:491–502.

Gidley, J. W. and F. B. Loomis (1926) Fossil man in Florida. *American Journal of Science* 12:254–264.

Gilder, R. (1906) Nebraska Man. *Omaha World-Herald*, Sunday, October 21, 1906.

Gilder, R. (1907a) A primitive human type in America—The finding of the "Nebraska Man." *Putnam's Monthly and the Critic* 1:407–409.

Gilder, R. (1907b) Archaeology of the Ponca district, east Nebraska. *American Anthropologist* 9:702–719.

Gilder, R. (1907c) Nebraska Loess. *American Antiquarian* 29:378–381.

Gilder, R. (1908) Recent excavations at Long's Hill, Nebraska. *American Anthropologist* 10:60–73.

Gilder, R. (1909) Excavation of Earth-Lodge ruins in eastern Nebraska. *American Anthropologist* 11:56–84.

Gilder, R. (1911) Scientific "inaccuracies" in reports against probability of geological antiquity of remains of Nebraska Loess man considered by its discoverer. *Records of the Past* 10:157–169.

Gladwin, H. (1937) The significance of early cultures in Texas and southeastern Arizona. In *Early Man*, ed. G. G. MacCurdy, pp. 133–138. Philadelphia: Lippincott.

Goddard, P. B. (1841) [Untitled remarks]. *Proceedings of the Philadelphia Academy of Natural Sciences* 1:115–116.

Goddard, P. E. (1925) Review of "*An Introduction to the Study of the Southwestern Archaeology,*" by A. V. Kidder. *American Anthropologist* 27:461–463.

Goddard, P. E. (1926) The antiquity of man in America. *Natural History* 26:257–259.

Goddard, P. E. (1927) Facts and theories concerning Pleistocene man in America. *American Anthropologist* 29:262–266.

Gould, C. N. (1929) On the recent finding of another flint arrow-head in the Pleistocene deposit at Frederick, Oklahoma. *Journal of the Washington Academy of Science* 19:66–68.

Gregory, W. K. (1927) *Hesperopithecus* apparently not an ape nor a man. *Science* 66:579–581.

Gregory, W. K. and M. Hellman (1923a) Notes on the type of *Hesperopithecus haroldcookii* Osborn. *American Museum Novitates* No. 53.

Gregory, W. K. and M. Hellman (1923b) Further notes on the molars of *Hesperopithecus* and of *Pithecanthropus*. *American Museum of Natural History Bulletin* 48:509–530.

Grote, A. (1876) Were the oldest American people Eskimos? *American Naturalist* 10:432–433.

Grote, A. (1877a) The early man of North America. *Popular Science Monthly* 10:582–594.

Grote, A. (1877b) On the peopling of America. *American Naturalist* 11:221–226.

Hallowell, A. I. (1928) Proceedings of the American Anthropological Association for the year ending December 1927. *American Anthropologist* 30:532–543.

Hanks, H. (1901) *The Deep Lying Auriferous Gravels and Table Mountains of California*. San Francisco: F. H. Abbott.

Harlan, R. (1825) *Fauna Americana: Being a Description of the Mammiferous Animals Inhabiting North America*. Philadelphia: Finley.

Harlan, R. (1843) Description of the bones of a new fossil animal of the order Endentata. *American Journal of Science* 44:69–80.

Harrington, M. R. (1933) Gypsum Cave, Nevada. *Southwest Museum Papers* 8.

Harte, B. (1866) To the Pliocene skull: A geological address. In *The Poetical Works of Bret Harte*. Boston: Houghton and Mifflin.

Haven, S. (1856) The archaeology of the United States. *Smithsonian Contributions to Knowledge* 8:1–168.

Haven, S. (1864) Report of the librarian. *Proceedings of the American Antiquarian Society*, April 1864:30–52.

Hay, O. (1902) Bibliography and catalogue of the fossil Vertebrata of North America. *United States Geological Survey Bulletin* 179.

Hay, O. (1908) *The Fossil Turtles of North America*. Washington, DC: Carnegie Institution of Washington.

Hay, O. (1909) The geological and geographical distribution of some Pleistocene mammals. *Science* 30:890–893.

Hay, O. (1912) The recognition of Pleistocene faunas. *Smithsonian Miscellaneous Collections* 59(20).

Hay, O. (1914) The extinct bisons of North America, with description of one new species, *Bison regius. Proceedings of the United States National Museum* 46:161–200.

Hay, O. (1916) Descriptions of some Floridian fossil vertebrates, belonging mostly to the Pleistocene. *Annual Report of the Florida State Geological Survey* 8:39–76.

Hay, O. (1917a) The Quaternary deposits at Vero, Florida, and the vertebrate remains contained therein. *Journal of Geology* 25:52–55.

Hay, O. (1917b) On the finding of supposed Pleistocene human remains at Vero, Florida. *Journal of the Washington Academy of Sciences* 7:358–360.

Hay, O. (1917c) Vertebrata mostly from stratum No. 3, at Vero, Florida, together with description of new species. *Florida Geological Survey Annual Report* 9:43–68.

Hay, O. (1918a) Further consideration of the occurrence of human remains in the Pleistocene deposits at Vero, Florida. *American Anthropologist* 20:1–36.

Hay, O. (1918b) A review of some papers on fossil man at Vero, Florida. *Science* 47:370–371.

Hay, O. (1918c) Doctor Ales Hrdlička and the Vero Man. *Science* 48:459–462.

Hay, O. (1919a) A study of skulls and probabilities. *Anthropological Scraps* 1:1–4.

Hay, O. (1919b) On Pleistocene man at Trenton, New Jersey. *Anthropological Scraps* 2:5–8.

Hay, O. (1919c) On the relative ages of certain Pleistocene deposits. *American Journal of Science* 47:361–375.

Hay, O. (1920) Bulletin 60, Bureau of American Ethnology. *Anthropological Scraps* 3:9–12.

Hay, O. (1921a) The newest discovery of "ancient" man in the United States. *Anthropological Scraps* 4:13–16.

Hay, O. (1921b) The people which sat in darkness saw great light. *Anthropological Scraps* 5:17–20.

Hay, O. (1923) *The Pleistocene of North America and Its Vertebrated Animals from the States East of the Mississippi River and from the Canadian Provinces East of Longitude 95°*. Washington, DC: Carnegie Institution of Washington.

Hay, O. (1924) *The Pleistocene of the Middle Region of North America and Its Vertebrated Animals*. Washington, DC: Carnegie Institution of Washington.

Hay, O. (1925) A revision of the Pleistocene period in North America, based especially on glacial geology and vertebrate paleontology. *Journal of the Washington Academy of Sciences* 15:126–133.

Hay, O. (1926) On the geological age of Pleistocene vertebrates found at Vero and Melbourne, Florida. *Journal of the Washington Academy of Sciences* 16:387–392.

Hay, O. (1927a) *The Pleistocene of the Western Region of North America and Its Vertebrated Animals*. Washington, DC: Carnegie Institution of Washington.

Hay, O. (1927b) A review of recent reports on investigations made in Florida on Pleistocene geology and paleontology. *Journal of the Washington Academy of Sciences* 17:277–283.

Hay, O. (1927c) An account of three recent important finds of relics of man associated with remains of Pleistocene Mammalia. Unpublished paper presented at the American Association for the Advancement of Science meeting, Nashville, TN.

Hay, O. (1928a) On the antiquity of relics of man at Frederick, Oklahoma. *Science* 67:442–444.

Hay, O. (1928b) An extinct camel from Utah. Science 68:299–300.

Hay, O. (1928c) Again on Pleistocene man at Vero, Florida. *Journal of the Washington Academy of Sciences* 18:233–241.

Hay, O. (1928d) Characteristic mammals of the early Pleistocene. *Journal of the Washington Academy of Sciences* 18:421–430.

Hay, O. (1929) On the recent discovery of flint arrow-head in early Pleistocene deposits at Frederick, Oklahoma. *Journal of the Washington Academy of Science* 19:93–98.

Hay, O. and H. J. Cook (1928) Preliminary descriptions of fossil mammals recently discovered in Oklahoma, Texas and New Mexico. *Proceedings of the Colorado Museum of Natural History* 8:33.

Hay, O. and H. J. Cook (1930) Fossil vertebrates collected near, or in association with, human artifacts at localities near Colorado, Texas; Frederick, Oklahoma; and Folsom, New Mexico. *Proceedings of the Colorado Museum of Natural History* 9:4–40.

Haynes, H. W. (1881) Their comparison with paleolithic implements from Europe. *Proceedings of the Boston Society of Natural History* 21:132–137.

Haynes, H. W. (1882) Some indications of an early race of men in New England. *Proceedings of the Boston Society of Natural History* 21:382– 390.

Haynes, H. W. (1893a) Palaeolithic man in North America. *Science* 21:66–67.

Haynes, H. W. (1893b) Palaeolithic man in North America. *American Antiquarian* 15:37–42.

Haynes, H. W. (1893c) The Palaeolithic man once more. *Science* 21:208–209.

Haynes, H. W. (1893d) The Palaeolithic man in Ohio. *Science* 21:291.

Haynes, H. W. (1893e) Early man in Minnesota. *Science* 21:318–319.

Henry, J. (1861) Report of the secretary. *Smithsonian Institution Annual Report for 1860*:13–54.

Henry, J. (1862) Report of the secretary. *Smithsonian Institution Annual Report for 1861*:13–48.

Henry, J. (1866) Report of the secretary. *Smithsonian Institution Annual Report for 1865*:13–74.

Henry, J. (1869) Report of the secretary. *Smithsonian Institution Annual Report for 1868*:12–53.

Henry, J. (1871) Report of the secretary. *Smithsonian Institution Annual Report for 1870*:13–45.

Henry, J. (1878) Circular in reference to American archaeology. *Smithsonian Miscellaneous Collections* 316:1–15.

Henshaw, H. (1890) The descendants of Paleolithic man in America. *American Anthropologist* 3:100–102.

Herdman, W. (1897) Anthropology at the British Association. *Nature* 56:486–488.

Hitchcock, C. (1893) A single glacial epoch in New England. *American Geologist* 11:194–195.

Hitchcock, E. (1841) First anniversary address before the Association of American Geologists, at their second annual meeting in Philadelphia. *American Journal of Science and Arts* 41:232–276.

Hitchcock, E. (1857) Illustrations of surface geology. *Smithsonian Contributions to Knowledge* 9(3):1–155.

Hoffman, W. J. (1879) The discovery of "turtle-back" celts in the District of Columbia. *American Naturalist* 13:109–115.

Hollick, A. (1898) Appendix. *Proceedings of the American Association for the Advancement of Science* 46:378–380.

Holmes, F. S. (1860) *Post-Pleiocene Fossils of South Carolina*. Charleston: Russell & Jones.

Holmes, W. H. (1878) Report on the ancient ruins of southwestern Colorado, examined during the summers of 1875 and 1876. *U.S. Geological and Geographical Survey of the Territories Annual Report for 1876* 10:383–408.

Holmes, W. H. (1879) Notes on an extensive deposit of obsidian in the Yellowstone National Park. *American Naturalist* 13:247–250.

Holmes, W. H. (1883) Art in shell of the ancient Americans. *Bureau of Ethnology Annual Report* 2:179–305.

Holmes, W. H. (1884) Ancient pottery of the Mississippi Valley, a study of the collection of the Davenport Academy of Sciences. *Davenport Academy of Sciences Proceedings* 4:123–196.

Holmes, W. H. (1886a) Pottery of the ancient pueblos. *Bureau of Ethnology Annual Report* 4:257–360.

Holmes, W. H. (1886b) Ancient pottery of the Mississippi Valley. *Bureau of Ethnology Annual Report* 4:361–436.

Holmes, W. H. (1886c) Origin and development of form in ceramic art. *Bureau of Ethnology Annual Report* 4:437–465.

Holmes, W. H. (1888) A study of the textile art in its relation to the development of form and ornament. *Bureau of Ethnology Annual Report* 6:189–252.

Holmes, W. H. (1890a) A quarry workshop of the flaked stone implement makers in the District of Columbia. *American Anthropologist* 3:1–26.

Holmes, W. H. (1890b) Aboriginal stone implements of the Potomac Valley. *Proceedings of the American Association for the Advancement of Science* 39:391.

Holmes, W. H. (1890c) On the evolution of ornament: An American lesson. *American Anthropologist* 3:137–146.

Holmes, W. H. (1891a) On the distribution of stone implements in the tide water province. *Proceedings of the American Association for the Advancement of Science* 40:366.

Holmes, W. H. (1891b) Aboriginal novaculite quarries in Arkansas. *Proceedings of the American Association for the Advancement of Science* 40:366–367.

Holmes, W. H. (1891c) Aboriginal novaculite quarries in Garland County, Arkansas. *American Anthropologist* 4:49–58.

Holmes, W. H. (1892a) Evolution of the aesthetic. *Proceedings of the American Association for the Advancement of Science* 41:239–255.

Holmes, W. H. (1892b) Aboriginal quarries of flakeable stone and their bearing upon the question of Paleolithic man. *Proceedings of the American Association for the Advancement of Science* 41:279–280.

Holmes, W. H. (1892c) On the so-called paleolithic implements of the Upper Mississippi. *Proceedings of the American Association for the Advancement of Science* 41:280–281.

Holmes, W. H. (1892d) Modern quarry refuse and the Palaeolithic theory. *Science* 20:295–297.

Holmes, W. H. (1893a) Distribution of stone implements in the Tidewater Country. *American Anthropologist* 6:1–14.

Holmes, W. H. (1893b) Are there traces of man in the Trenton gravels? *Journal of Geology* 1:15–37.

Holmes, W. H. (1893c) Gravel man and Palaeolithic culture: A preliminary word. *Science* 21:29–30.

Holmes, W. H. (1893d) Traces of glacial man in Ohio. *Journal of Geology* 1:147–163.

Holmes, W. H. (1893e) A question of evidence. *Science* 21:135–136.

Holmes, W. H. (1893f) Vestiges of early man in Minnesota. *American Geologist* 11:219–240.

Holmes, W. H. (1893g) Man and the glacial period. *American Antiquarian* 15:34–36.

Holmes, W. H. (1894) A natural history of flaked stone implements. In *Memoirs of the International Congress of Anthropology*, ed. C. Wake, pp. 120–139. Chicago: Schulte Publishing.

Holmes, W. H. (1897a) Stone implements of the Potomac-Chesapeake Tidewater province. *Bureau of Ethnology Annual Report* 15:3–152.

Holmes, W. H. (1897b) Primitive man in the Delaware Valley. *Science* 6:824–829.

Holmes, W. H. (1898) Primitive man in the Delaware Valley. *Proceedings of the American Association for the Advancement of Science* 46:364–370.

Holmes, W. H. (1899) Preliminary revision of evidence relating to auriferous gravel man in California. *American Anthropologist* 1:107–121, 614–645.

Holmes, W. H. (1902a) Flint implements and fossil remains from a sulphur spring at Afton, Indian Territory. *American Anthropologist* 4:108–129.

Holmes, W. H. (1902b) Fossil human remains found near Lansing, Kansas. *American Anthropologist* 4:743–752.

Holmes, W. H. (1907) Antiquity. In *Handbook of American Indians*. Bulletin 30, part 1, 59–62. Washington, DC: Bureau of American Ethnology.

Holmes, W. H. (1910) Some problems of the American race. *American Anthropologist* 12:149–182.

Holmes, W. H. (1913) Remarks by W. H. Holmes on "The Relation of Archaeology to Ethnology." *American Anthropologist* 15:566–567.

Holmes, W. H. (1914) Areas of American culture characterization tentatively outlined as an aid in the study of the antiquities. *American Anthropologist* 16:413–446.

Holmes, W. H. (1918) On the antiquity of man in America. *Science* 47:561–562

Holmes, W. H. (1919) *Handbook of Aboriginal American Antiquities, Part 1: The Lithic Industries*. Bulletin 60. Washington, DC: Bureau of American Ethnology.

Holmes, W. H. (1922) Pitfalls of the Paleolithic theory in America. *Proceedings of the 20th International Congress of Americanists* 2:171–175.

Holmes, W. H. (1925) The antiquity phantom in American archaeology. *Science* 62:256–258.

Hooton, E. (1925) The asymmetrical character of human evolution. *American Journal of Physical Anthropology* 8:125–141.

Hooton, E. (1930) *The Indians of Pecos Pueblo: A Study of Their Skeletal Remains*. New Haven: Yale University Press.

Hooton, E. (1933) Racial types in America and their relation to Old World types. In *The American Aborigines: Their Origin and Antiquity*, ed. D. Jenness, pp. 133–163. Toronto: University of Toronto Press.

Hooton, E. (1937) *Apes, Men, and Morons*. New York: G. P. Putnam's Sons.

Howard, E. B. (1930) Archaeological research in the Guadalupe Mountains. *Museum Journal*, University of Pennsylvania Museum 21:189–202.

Howard, E. B. (1932a) Caves along the slopes of the Guadalupe Mountains. *Bulletin of the Texas Archeological and Paleontological Society* 4:7–19.

Howard, E. B. (1932b) Arrowheads found with New Mexican fossils. *Science* 76:12–13.

Howard, E. B. (1933) Early man in America. *Science Supplement* 78:7–8.

Howard, E. B. (1935a) The occurrence of flints and extinct animals in pluvial deposits near Clovis, New Mexico. Part I: Introduction. *Proceedings of the Philadelphia Academy of Natural Sciences* 87:299–303.

Howard, E. B. (1935b) Evidence of early man in North American. *Museum Journal*, University of Pennsylvania Museum 24: 2–3.

Howard, E. B. (1936a) Early human remains in the Southwestern United States. *Report of the Sixteenth Session, International Geological Congress* 2:1325–1333.

Howard, E. B. (1936b) An outline of the problem of man's antiquity in North America. *American Anthropologist* 38:394–413.

Howard, E. B. (1936c) The association of a human culture with an extinct fauna in New Mexico. *American Naturalist* 70:314–323.

Howard, E. B. (1937) Minutes of the International Symposium on Early Man held at the Academy of Natural Science of Philadelphia. *Proceedings of the Academy of Natural Science of Philadelphia* 89:439–447.

Howard, E. B. (1939) The Clovis finds are not two million years old. *American Antiquity* 5:43–51.

Howard, E. B. (1943) The Finley site: Discovery of Yuma points, in situ, near Eden, Wyoming. *American Antiquity* 8:224–234.

Howard, E. B. and E. Antevs (1935) Studies on antiquity of man in America. *Year Book—Carnegie Institution of Washington* 33:309–311.

Howells, W. W. (1938) Crania from Wyoming resembling "Minnesota Man." *American Antiquity* 3:318–326.

Howells, W. W. (1940) The origins of American Indian race types. In *The Maya and Their Neighbors*, ed. C. Hay, S. Lothrop, R. Linton, H. Shapiro, and G. Valliant, pp. 3–9. New York: D. Appleton.

Hrdlička, A. (1902) The crania of Trenton, New Jersey, and their bearing upon the antiquity of man in that region. *American Museum of Natural History Bulletin* 16:23–62.

Hrdlička, A. (1903) The Lansing skeleton. *American Anthropologist* 5:323–330.

Hrdlička, A. (1907) *Skeletal Remains Suggesting or Attributed to Early Man in North America*. Bulletin 33. Washington, DC: Bureau of American Ethnology.

Hrdlička, A. (1911) Report to Prof. F. W. Putnam on the human femur and parietal. In *The Archaeology of the Delaware Valley*, by Ernst Volk. *Papers of the Peabody Museum of Archaeology and Ethnology* 5:244–247.

Hrdlička, A. (1912) *Early Man in South America*. Bulletin 52. In collaboration with W. Holmes, B. Willis, F. Wright, and C. Fenner. Washington, DC: Bureau of American Ethnology.

Hrdlička, A. (1917) Preliminary report on finds of supposedly ancient human remains at Vero, Florida. *Journal of Geology* 25:43–51.

Hrdlička, A. (1918) *Recent Discoveries Attributed to Early Man in America*. Bulletin 66. Washington, DC: Bureau of American Ethnology.

Hrdlička, A. (1922) Man's antiquity in America. *Proceedings of the Twentieth International Congress of Americanists* 2:57–61.

Hrdlička, A. (1925) The origin and antiquity of the American Indian. *Smithsonian Institution Annual Report for 1923*:481–494.

Hrdlička, A. (1926) The race and antiquity of the American Indian. *Scientific American* 135:7–9.

Hrdlička, A. (1927) The Neanderthal phase of man. The Huxley Memorial Lecture for 1927. *Journal of the Royal Anthropological Institute of Great Britain and Ireland* 57:249–274.

Hrdlička, A. (1928) The origin and antiquity of man in America. *Bulletin of the New York Academy of Medicine* 4(7):802–816.

Hrdlička, A. (1937a) The Minnesota "Man." *American Journal of Physical Anthropology* 22: 175–99

Hrdlička, A. (1937b) Early man in America: What have the bones to say? In *Early Man*, ed. G. G. MacCurdy, pp. 93–104. Philadelphia: Lippincott.

Hrdlička, A. (1942) The problem of man's antiquity in America. *Proceedings of the Eighth American Scientific Congress* 2:53–55.

Huxley, T. (1863) *Evidences as to Man's Place in Nature*. London: Williams and Norgate.

James, J. (1893) The Cincinnati Ice Dam. *American Geologist* 11:199–202.

Jefferson, T. (1788) *Notes on the State of Virginia*. Philadelphia: Prichard and Hall.

Jenks, A. E. (1932) Pleistocene man in Minnesota (a preliminary announcement). *Science* 75:607–608.

Jenks, A. E. (1933) Minnesota Pleistocene *Homo*: An interim communication. *Proceedings of the National Academy of Sciences* 19:1–6.

Jenks, A. E. (1936) Pleistocene man in Minnesota: A fossil *Homo sapiens*. Minneapolis: University of Minnesota Press.

Jenks, A. E. (1937) Minnesota's Brown's Valley man and associated burial artifacts. *Memoirs of the American Anthropological Association* 49.

Jenks, A. E. (1938) Minnesota Man: A reply to a review by Dr. Aleš Hrdlička. *American Anthropologist* 40:328–336.

Jenks, A. E. and L. Wilford (1938) The Sauk Valley skeleton. *Bulletin of the Texas Archeological and Paleontological Society* 10:136–168.

Jenness, D., ed. (1933) *The American Aborigines: Their Origin and Antiquity*. Toronto: University of Toronto Press.

Johnston, W. A. (1933) Quaternary geology of North America in relation to the migration of man. In *The American Aborigines: Their Origin and Antiquity*, ed. D. Jenness, pp. 11–45. Toronto: University of Toronto Press.

Jones, C. (1873) *Antiquities of the southern Indians particularly of the Georgia tribes*. New York: D. Appleton.

Kay, G. and M. Leighton (1938) Geological notes on the occurrence of "Minnesota Man." *Journal of Geology* 46:268–278.

Keith, A. (1915) *The Antiquity of Man*. London: Williams and Norgate.

Keyes, C. (1928) Theory of multiple glaciations. *Pan-American Geologist* 50:131–144.

Kidder, A. V. (1927) Early man in America. *Masterkey* 1(5):5–13.

Kidder, A. V. (1936) Speculations on New World prehistory. In *Essays in Anthropology*, ed. R. Lowie, pp. 143–151. Berkeley: University of California Press.

Kidder, A. V. and R. Terry (1927) Section H. Anthropology. The Philadelphia sessions of sections and societies. *Science* 65:111–112.

Knapp, G. (1898) On the implement bearing sand deposits at Trenton. *Proceedings of the American Association for the Advancement of Science* 46:350.

Koch, A. (1839) The mammoth (mastodon? Eds.). *American Journal of Science* 36:198–200.

Koch, A. (1857) Mastodon remains in the state of Missouri, together with evidence of the existence of man contemporaneous with the mastodon. *Transactions of the Academy of Science of St. Louis* 1:61–64.

Krieger, A. (1953) New World culture history: Anglo-America. In Anthropology Today: An Encyclopedic Inventory, prepared under the chairmanship of A.L. Kroeber, pp. 238–264. Chicago, University of Chicago Press.

Kroeber, A. (1909) The archaeology of California. In *Putnam Anniversary Volume*, ed. F. Boas, R. Dixon, A. Kroeber, F. Hodge, and H. Smith, pp. 1–42. New York: Stechert.

Kroeber, A. (1929) Pliny Earle Goddard. *American Anthropologist* 31:1–8.

Kroeber, A. (1940a) Conclusions: The present status of Americanistic problems. In *The Maya and Their Neighbors*, ed. C. Hay, S. Lothrop, R. Linton, H. Shapiro, and G. Valliant, pp. 460–476. New York: D. Appleton.

Kroeber, A. (1940b) The work of John R. Swanton. *Smithsonian Miscellaneous Collections* 100:1–9.

Kroeber, A. (1952) *The Nature of Culture*. Chicago: University of Chicago Press.

Kroeber, A. L. (1962) The Rancho La Brea skull. *American Antiquity* 27:416–417.

Kummel, H. (1898) The age of the artifact-bearing sand at Trenton. *Proceedings of the American Association for the Advancement of Science* 46:348–350.

Langley, S. P. (1888) Circular concerning the Department of Antiquities. *Smithsonian Institution, United States National Museum, Circular* 36.

Lartet, E. (1860) On the coexistence of man with certain extinct quadrupeds, proved by fossil bones, from various Pleistocene deposits, bearing incisions made by sharp instruments. *Quarterly Journal of the Geological Society* 16:471–479.

Lartet, E. and H. Christy (1875) *Reliquiae Aquitanicae: Being Contributions to the Archaeology and Paleontology of Perigord and the Adjoining Provinces of Southern France.* London: Williams and Norgate.

Laufer, B. (1913) Remarks by Berthold Laufer on "The Relation of Archaeology to Ethnology." *American Anthropologist* 15:573–577.

Leighton, M. (1931) The Peorian Loess and the classification of the glacial drift sheets of the Mississippi Valley. *Journal of Geology* 39:45–53.

Lesley, J. P. (1883) Wright's ice-dam at Cincinnati. *Science* 2:436.

Lesley, J. P. (1884) Letter of transmittal. In The terminal moraine of Pennsylvania, by H. C. Lewis. *Second Geology Survey of Pennsylvania* Report Z:v–xlix.

Leverett, F. (1890) Glacial studies bearing on the antiquity of man. *Proceedings of the Boston Society of Natural History* 24:585–586.

Leverett, F. (1891a) The Cincinnati Ice Dam. *Proceedings of the American Association for the Advancement of Science* 40:250–251.

Leverett, F. (1891b) Relation of a Loveland, Ohio, implement bearing terrace to the moraines of the ice-sheet. *Proceedings of the American Association for the Advancement of Science* 40:361–362.

Leverett, F. (1893a) The glacial succession in Ohio. *Journal of Geology* 1:129–146.

Leverett, F. (1893b) Supposed glacial man in southwestern Ohio. *American Geologist* 11:186–189.

Leverett, F. (1893c) Relation of the attenuated drift border to the outer moraine in Ohio. *American Geologist* 11:215–216.

Leverett, F. (1899) The Illinois glacial lobe. *Monographs of United States Geological Survey* 33.

Leverett, F. (1902) Glacial formations and drainage features of the Erie and Ohio Basins. *Monographs of United States Geological Survey* 41.

Leverett, F. (1904) The Loess and its distribution. *American Geologist* 33:56–57.

Leverett, F. (1929) Pleistocene glaciations of the northern hemisphere. *Science* 69:231–239.

Leverett, F. (1930a) Problems of the glacialist. *Science* 71:47–57.

Leverett, F. (1930b) Relative lengths of Pleistocene glacial and interglacial stages. *Science* 72:193–195.

Leverett, F. (1934) Glacial deposits outside the Wisconsin terminal moraine in Pennsylvania. *Pennsylvania Geological Survey, Fourth Series, Bulletin G 7.*

Leverett, F. and F. Taylor (1915) The Pleistocene of Indiana and Michigan and the history of the Great Lakes. *Monographs of United States Geological Survey* 53.

Lewis, H. C. (1880) The Trenton Gravel and its relation to the antiquity of man. *Proceedings of the Academy of Natural Sciences of Philadelphia* 32:296–309.

Lewis, H. C. (1881a) The antiquity of man in eastern America, geologically considered. *Proceedings of the American Association for the Advancement of Science* 29:706–710.

Lewis, H. C. (1881b) The antiquity and origin of the Trenton gravel. In *Primitive Industry*, by C. C. Abbott, pp. 521–551. Salem: George A. Bates.

Lewis, H. C. (1883) On a supposed human implement from the gravel at Philadelphia. *Proceedings of the Academy of Natural Sciences of Philadelphia* 35:40–43.

Lewis, H. C. (1884a) The terminal moraine of Pennsylvania. *Second Geology Survey of Pennsylvania* Report Z.

Lewis, H. C. (1884b) On supposed glaciation in Pennsylvania south of the terminal moraine. *American Journal of Science* 28:276–285.

Loomis, F. (1924) Artifacts associated with the remains of a Columbian elephant at Melbourne, Florida. *American Journal of Science* 8:503–508.

Loomis, F. (1925) The Florida man. *Science* 62:436.

Loomis, F. (1926) Early man in Florida. *Natural History* 26:260–262.

Lubbock, J. (1863) North American archaeology. *Natural History Review* 3:1–30.

Lubbock, J. (1865) *Prehistoric Times*. London: Williams and Norgate.

Lucas, F. (1898) The fossil bison of North America. *Science* 8:678.

Lucas, F. (1899a) The fossil bison of North America. *Proceedings of the United States National Museum* 21:755–771.

Lucas, F. (1899b) The characters of *Bison occidentalis*, the fossil bison of Kansas and Alaska. *Kansas University Quarterly* 8:17–18.

Lyell, C. (1830–1833) *Principles of Geology, Being an Attempt to Explain the Former Changes of the Earth's Surface, by Reference to Causes Now in Operation*, vols. 1–3. London: John Murray.

Lyell, C. (1849) *A Second Visit to the United States of America*. London: John Murray.

Lyell, C. (1853) *Principles of Geology: Or the Modern Changes of the Earth and Its Inhabitants Considered as Illustrative of Geology*, 9th rev. ed. London: John Murray.

Lyell, C. (1860) On the occurrence of works of human art in post-Pliocene deposits. *British Association for the Advancement of Science Report* 1859:93–95.

Lyell, C. (1863) *The Geological Evidences of the Antiquity of Man with Remarks on Theories of the Origin of Species by Variation*. London: John Murray.

MacCurdy, G. G. (1905) The Eolithic problem: Evidences of a rude industry ante-dating the Paleolithic. *American Anthropologist* 7:425–479.

MacCurdy, G. G. (1913) Remarks by George Grant MacCurdy, "On the Relation of Archaeology to Ethnology from the Quaternary Standpoint." *American Anthropologist* 15:567–573.

MacCurdy, G. G. (1914) The man of Piltdown. *American Anthropologist* 16:331–336.

MacCurdy, G. G. (1917a) Archaeological evidence of man's antiquity at Vero, Florida. *Journal of Geology* 25:56–62.

MacCurdy, G. G. (1917b) The problems of man's antiquity at Vero, Florida. *American Anthropologist* 19:252–261.

MacCurdy, G. G. (1920) Review of "Handbook of Aboriginal American Antiquities. Part I: The Lithic Industries," by W. H. Holmes. *American Anthropologist* 22:75–78.

MacCurdy, G. G., ed. (1937) *Early Man*. Philadelphia: Lippincott.

Marsh, O. C. (1878) Address of Professor O. C. Marsh. *Proceedings of the American Association for the Advancement of Science* 26:211–258.

Mason, J. A. (1943) Howard-Edgar Billings, 1887–1943. *American Antiquity* 9:230–234.

Mason, O. (1877) Anthropological news. *American Naturalist* 11:308–311.

Mason, O. (1878) Pliocene man. *American Naturalist* 12:125–126.

Mason, O. (1884) The scope and value of anthropological studies. *Proceedings of the American Association for the Advancement of Science* 32:367–383.

Mason, O. (1885) An account of the progress in anthropology in the year 1884. *Smithsonian Institution Annual Report for 1884*:677–717.

Mason, O. (1893) Notes and news. *American Anthropologist* 6:459–462.

Mason, O. (1902) In memoriam. Thomas Wilson. *American Anthropologist* 4:286–291.

Mason, O., WJ McGee, T. Wilson, S. V. Proudfit, W. H. Holmes, E. R. Reynolds and J. Mooney (1889) Symposium on "The aboriginal history of the Potomac Tidewater region." *American Anthropologist* 2:225–268.

Matthew, W. D. and H. J. Cook (1909) A Pliocene fauna from western Nebraska. *Bulletin of the American Museum of Natural History* 26:361–414.

McClung, C. E. (1908) Restoration of the skeleton of *Bison occidentalis*. *Kansas University Science Bulletin* 4:249–252.

McCown, T. D. (1941) The antiquity of man in the New World. *American Antiquity* 6:203–213.

McGee, A. N. (1897) Anthropology at American Association for the Advancement of Science. *Science* 6:508–513.

McGee, WJ (1878) On the relative positions of the forest bed and associated drift formations in northeastern Iowa. *American Journal of Science* 15:339–341.

McGee, WJ (1879) On the complete series of superficial formations in northeastern Iowa. *Proceedings of the American Association for the Advancement of Science* 27:198–231.

McGee, WJ (1887) On the finding of a spear head in the Quaternary beds of Nevada. *Scientific American Supplement* 23:9221–9222.

McGee, WJ (1888a) Three formations of the Middle Atlantic slope. *American Journal of Science* 35:120–143, 328–330, 367–388, 448–466.

McGee, WJ (1888b) Paleolithic man in America: His antiquity and his environment. *Popular Science Monthly* 34:20–36.

McGee, WJ (1889) An obsidian implement from Pleistocene deposits in Nevada. *American Anthropologist* 2:301–312.

McGee, WJ (1890) Some principles of evidence relating to the antiquity of man. (Abstract). *Proceedings of the American Association for the Advancement of Science* 38:333.

McGee, WJ (1891) Some principles of evidence relating to the antiquity of man. *American Antiquarian* 13:69–79.

McGee, WJ (1892) Letters to the editor. Man and the glacial period. *Science* 20:317.

McGee, WJ (1893a) Man and the glacial period. *American Anthropologist* 6:85–95.

McGee, WJ (1893b) A geologic palimpsest. *Literary Northwest* 2:274–276.

McGee, WJ (1893c) Anthropology at the Madison meeting. *American Anthropologist* 6:435–448.

McGee, WJ (1897) Anthropology at Detroit and Toronto. *American Anthropologist* 10:317–345.

McGee, WJ (1900) Cardinal principles of science. *Proceedings of the Washington Academy of Sciences* 11:1–12.

McGee, WJ and C. Thomas (1905) *Prehistoric North America*. Philadelphia: G. Barrie and Sons.

McKern, W. (1939) The Mid-western taxonomic method as an aid to archaeological study. *American Antiquity* 4:301–313.

Mercer, H. (1885) *The Lenape Stone: Or, the Indian, and the Mammoth*. New York: Putnam.

Mercer, H. (1892) Pebbles chipped by modern Indians as an aid to the study of the Trenton gravel implements. *Proceedings of the American Association for the Advancement of Science* 41:287–289.

Mercer, H. (1893) Trenton and Somme gravel specimens compared with ancient quarry refuse in America and Europe. *American Naturalist* 27:962–978.

Mercer, H. (1897) Researches upon the antiquity of man in the Delaware Valley and the eastern United States. *University of Pennsylvania Series in Philology, Literature and Archaeology* 6.

Mercer, H. (1898) A new investigation of man's antiquity at Trenton. *Proceedings of the American Association for the Advancement of Science* 46:370–378.

Merriam, J. C. (1914) Preliminary report on the discovery of human remains in an asphalt deposit at Rancho La Brea. *Science* 40:198–203.

Merriam, J. C. (1924) Present status of investigations concerning antiquity of man in California. *Science* 60:1–2.

Merriam, J. C. (1936) Present status of knowledge relating to antiquity of man in America. *Report of the Sixteenth Session, International Geological Congress* 2:1313–1323.

Merriam, J. C., J. Buwalda, N. Hinds, R. Kellogg, and C. Stock (1933) Paleontology. *Year Book—Carnegie Institution of Washington* 32:323–330.

Miller, S. A. (1893) Criticism of the U.S. Geological Survey. *Science* 21:67–68.

Mills, W. C. (1890) Account of discovery. In *Discovery of a Paleolithic Implement at New Comerstown, Ohio,* by W. C. Mills and G. F. Wright, pp. 3–4. Western Reserve Historical Society, Tract 75. Cleveland: Western Reserve Historical Society.

Moorehead, W. K. (1893) The meeting of the American Association for the Advancement of Science. *Archaeologist* 1:170–172.

Moorehead, W. K. (1910) *The Stone Age in North America,* 2 vols. New York: Houghton and Mifflin.

Morlot, A. (1861) General views on archaeology. *Smithsonian Institution Annual Report for 1860*:284–343.

Morlot, A. (1863) An introductory lecture to the study of high antiquity. *Smithsonian Institution Annual Report for 1862*:303–315.

Morse, E. (1884) Man in the Tertiary. *Proceedings of the American Association for the Advancement of Science* 33:3–15.

Mortillet, G. de (1879) Présentations. *Bulletins de la Société d'Anthropologie de Paris* 10:439–440.

Mortillet, G. de (1883) *Le Préhistorique: Antiquité de l'Homme.* Paris: C. Reinwald.

Nelson, N. (1918a) Review of *"Additional Studies in the Pleistocene at Vero, Florida."* *Science* 47:394–395.

Nelson, N. (1918b) Chronology in Florida. *Anthropological Papers of the American Museum of Natural History* 22(2).

Nelson, N. (1928a) Pseudo-artifacts from the Pliocene of Nebraska. *Science* 67:316–317.

Nelson, N. (1928b) Discussion of *"The Origin and Antiquity of Man in America"* by A. Hrdlička. *Bulletin of the New York Academy of Medicine* 4(7):820–823.

Nelson, N. (1933) The antiquity of man in America in the light of archaeology. In *The American Aborigines: Their Origin and Antiquity,* ed. D. Jenness, pp. 87–130. Toronto: University of Toronto Press.

Nelson, N. (1936) Review of *"Evidence of Early Man in North America"* by Edgar B. Howard. *American Antiquity* 1:237–239.

Nelson, N. (1937) Review of *"Additional Information on the Folsom Complex"* by Frank H. H. Roberts. *American Antiquity* 2:317–320.

Newberry, J. S. (1870) On the earliest traces of man found in North America. *Nature* 2:366–367.

Nilsson, S. (1868) *The Primitive Inhabitants of Scandinavia.* London: Longmans, Green.

Nott, J. C. and G. R. Gliddon (1854) *Types of Mankind.* Philadelphia: Lippincott.

Orton, E. (1873) Report on the third geological district: Geology of the Cincinnati group; Hamilton, Claremont, Clarke Cos. *Ohio Geological Survey Report* 1:365–480.

Osborn, H. F. (1907) Discovery of a supposed primitive race in Nebraska. *Century Magazine* 73:371–375.

Osborn, H. F. (1922a) *Hesperopithecus,* the first anthropoid primate found in America. *American Museum Novitates* 37.

Osborn, H. F. (1922b) *Hesperopithecus,* the first anthropoid primate found in America. *Proceedings of the National Academy of Sciences* 8:245–246.

Osborn, H. F. (1922c) *Hesperopithecus,* the first anthropoid primate found in America. *Science* 55:463–465.

Osborn, H. F. (1925) The earth speaks to Bryan. *Forum* 73:796–803.

Osborn, H .F. (1927) Recent discoveries relating to the origin and antiquity of man. *Science* 65:481–488.

Owen, L. (1904) The loess at St. Joseph. *American Geologist* 33:223–228.

Owen, L. (1905) Evidence on deposition of loess. *American Geologist* 35:291–300.

Parkinson, J. (1833) *Organic Remains of a Former World: An Examination of the Mineralized Remains of the Vegetables and Animals of the Antediluvian World; Generally Termed Extraneous Fossils.* London: Sherwood, Neely and Jones, etc.

Patrick, R. (1938) The occurrence of flints and extinct animals in pluvial deposits near Clovis, New Mexico. Part 5: Diatom evidence from the gravel pit. *Proceedings of the Philadelphia Academy of Natural Sciences* 90:15–24.

Peabody, G. (1868) Foundation. *First Annual Report of the Trustees of the Peabody Museum of American Archaeology and Ethnology,* pp. 25–28. Cambridge, MA: John Wilson.

Peet, S. (1878) The field we occupy. *American Antiquarian* 1:45–46.

Peet, S. (1879a) A comparison between the archaeology of Europe and America. *American Antiquarian* 1:211–224.

Peet, S. (1879b) Neolithic implements found in gravel beds in California. *American Antiquarian* 2:177.

Peet, S. (1883) Archaeology at the American Association. *American Antiquarian* 5:352–353.

Peet, S. (1884) Editorial. Anthropology in the British and American Associations. *American Antiquarian* 6:412–421.

Peet, S. (1887) The antiquity of man in America. *American Antiquarian* 9:49–53.

Peet, S. (1888a) Paleolithics and Mound-builders: Their age and date. *American Antiquarian* 10:46–50.

Peet, S. (1888b) Paleolithics and Neolithics. *American Antiquarian* 10:124–126.

Peet, S. (1888c) Paleolithics. *American Antiquarian* 10:254–255.

Peet, S. (1892) Review of *"Man and the Glacial Period,"* by G. Frederick Wright. *American Antiquarian* 14:359.

Peet, S. (1893) The advance of anthropology. *American Antiquarian* 15:311–314.

Pengelly, W. (1877) [Untitled]. *Nature* 407:323.

Pitt-Rivers, A. H. L. (1875) On the evolution of culture. *Proceedings of the Royal Institution of Great Britain* 7:496–520.

Playfair, J. (1802) *Illustrations of the Huttonian Theory of the Earth.* Edinburgh: William Creech.

Powell, J. W. (1881a) Report of the director. *Bureau of Ethnology Annual Report* 1:xi–xiii.

Powell, J. W. (1881b) On limitations to the use of some anthropologic data. *Bureau of Ethnology Annual Report* 1:73–86.

Powell, J. W. (1883a) The Trenton gravels. *Science* 1:525–526.

Powell, J. W. (1883b) Human evolution. *Transactions of the Anthropological Society of Washington* 2:176–208.

Powell, J. W. (1885a) The administration of the scientific work of the general government. *Science* 5:51–55.

Powell, J. W. (1885b) Report of the director. *United States Geological Survey Annual Report* 6:xv–xxix.

Powell, J. W. (1885c) Answers to charges affecting the Geological Survey. *Science* 6:424–425.

Powell, J. W. (1886) Report of the director. *Bureau of Ethnology Annual Report* 4:xxvii–lxiii.

Powell, J. W. (1890a) Prehistoric man in America. *Forum* 8:489–503.

Powell, J. W. (1890b) Problems of American archaeology. *Forum* 8:638–652.

Powell, J. W. (1893a) The work of the U.S. Geological Survey. *Science* 21:15–17.

Powell, J. W. (1893b) Are there evidences of man in the glacial gravels? *Popular Science Monthly* 43:316– 326.

Powell, J. W. (1893c) Report of the director. *United States Geological Survey Annual Report* 14:11–165.

Powell, J. W. (1894) Report of the director. *Bureau of Ethnology Annual Report* 12:xxi–xlviii.

Powell, J. W. (1895) Stone art in America. *American Anthropologist* 8:1–7.

Prestwich, G. (1899) *Life and Letters of Sir Joseph Prestwich, M.A., D.C.L., F.R.S., Formerly Professor of Geology in the University of Oxford.* Edinburgh: W. Blackwood and Sons.

Prestwich, J. (1859) On the occurrence of flint-implements, associated with the remains of extinct mammalia, in undisturbed beds of a late geological period [Abstract]. *Proceedings of the Royal Society of London* 10:50–59.

Prestwich, J. (1860) On the occurrence of flint-implements, associated with the remains of extinct Mammalia, in undisturbed beds of the late geological period. *Philosophical Transactions of the Royal Society of London* 150:277–318.

Prestwich, J. (1861a) On the occurrence of flint-implements associated with the remains of animals of extinct species in beds of a late geological period, in France at Amiens and Abbeville, and in England at Hoxne. *Philosophical Transactions of the Royal Society* 150:277–318.

Prestwich, J. (1861b) Notes on some further discoveries of flint implements in beds of post-Pliocene gravel and clay; with a few suggestions for search elsewhere. *Quarterly Journal of the Geological Society* 17:362–368.

Proudfit, S. V. (1889) Ancient village sites and aboriginal workshops in the District of Columbia. *American Anthropologist* 2:241–246.

Putnam, C. (1886) *Elephant Pipes and Inscribed Tablets in the Museum of the Academy of Natural Sciences, Davenport, Iowa.* Davenport, IA: Glass and Axtam.

Putnam, F. W. (1877) Report of the curator. *Peabody Museum Annual Report* 10:7–12.

Putnam, F. W. (1881) Concluding remarks. *Proceedings of the Boston Society of Natural History* 21:147–149.

Putnam, F. W. (1883) Remarks upon some chipped stone implements. *Bulletin of the Essex Institute* 15:3–8.

Putnam, F. W. (1885) Man and the mastodon. *Science* 6:375–376.

Putnam, F. W. (1886a) Report of the curator. *Peabody Museum Annual Report* 18:401–418.

Putnam, F. W. (1886b) Report of the curator. *Peabody Museum Annual Report* 19:477–501.

Putnam, F. W. (1888a) [Untitled remarks]. *Proceedings of the Boston Society of Natural History* 23:242.

Putnam, F. W. (1888b) [Untitled remarks]. *Proceedings of the Boston Society of Natural History* 23:421–424, 447–449.

Putnam, F. W. (1888c) Report of the curator. *Peabody Museum Annual Report* 22:31–60.

Putnam, F. W. (1889a) [Untitled remarks]. *Proceedings of the Boston Society of Natural History* 24:157–159.

Putnam, F. W. (1889b) Discussion of "The aborigines of the District of Columbia and the lower Potomac." *American Anthropologist* 2:266–268.

Putnam, F. W. (1889c) Report of the curator. *Peabody Museum Annual Report* 23:73–81.

Putnam, F. W. (1889d) The Peabody Museum of American Archaeology and Ethnology in Cambridge. *Proceedings of the American Antiquarian Society* 6:180–190.

Putnam, F. W. (1890) [Untitled remarks]. *Proceedings of the Boston Society of Natural History* 24:468–469.

Putnam, F. W. (1891) Report of the curator. *Peabody Museum Annual Report* 24:87–107

Putnam, F. W. (1893) What was the origin of post-glacial man? *American Geologist* 11:195.

Putnam, F. W. (1898) Early man of the Delaware Valley. *Proceedings of the American Association for the Advancement of Science* 46:344–348.

Putnam, F. W. (1906) Evidence of the work of man on objects from Quaternary Caves in California. *American Anthropologist* 8:229–235.

Putnam, F. W., C. C. Abbott, H. W. Haynes, G. F. Wright, L. Carr, and M. E. Wadsworth (1881) Symposium on "The Paleolithic implements of the Valley of the Delaware." *Proceedings of the Boston Society of Natural History* 21:124–149.

Putnam, F. W., C. C. Abbott, G. F. Wright, W. Upham, H. W. Haynes, and E.S. Morse (1888) Symposium on "Paleolithic man in eastern and central North America." *Proceedings of the Boston Society of Natural History* 23:419–449.

Putnam, F. W., H. T. Cresson, G. F. Wright, and C. C. Abbott (1889) Symposium on "Paleolithic man in eastern and central North America," part 3. *Proceedings of the Boston Society of Natural History* 24:141–165.

Ramsay, A. (1859) Works of art in the drift. *Athenaeum* 1665:83.

Rau, C. (1873) North American chipped stone implements. *Smithsonian Institution Annual Report for 1872*:395–408.

Renaud, E. B. (1931) Prehistoric flaked points from Colorado and neighboring districts. *Proceedings of the Colorado Museum of Natural History* 10(2).

Renaud, E. B. (1932) Yuma and Folsom artifacts. *Proceedings of the Colorado Museum of Natural History* 11(2).

Renaud, E. B. (1934) The first thousand Yuma-Folsom artifacts. University of Denver, Department of Anthropology.

Richards, H. (1936) Mollusks associated with early man in the southwest. *American Naturalist* 70:369–371.

Roberts, F. H. H. (1934) Scientist describes true Folsom points. *Literary Digest* 118(4):18.

Roberts, F. H. H. (1935) A Folsom Complex: Preliminary report on investigations at the Lindenmeier site in northern Colorado. *Smithsonian Miscellaneous Collections* 94(4):1–35.

Roberts, F. H. H. (1936a) Additional information on the Folsom Complex: Report on the second season's investigations at the Lindenmeier site in northern Colorado. *Smithsonian Miscellaneous Collections* 95(10):1–38.

Roberts, F. H. H. (1936b) Recent discoveries of the material culture of Folsom Man. *American Naturalist* 70:337–345.

Roberts, F. H. H. (1937a) New World man. *American Antiquity* 2:172–177.

Roberts, F. H. H. (1937b) The Folsom problem in American archaeology. In *Early Man*, ed. G. G. MacCurdy, pp. 153–162. Philadelphia: Lippincott.

Roberts, F. H. H. (1939) The Folsom problem in American archaeology. *Smithsonian Institution Annual Report for 1938*:531–546.

Roberts, F. H. H. (1940a) Developments in the problem of the North American Paleo-indian. *Smithsonian Miscellaneous Collections* 100:51–116.

Roberts, F. H. H. (1940b) Pre-pottery horizon of the Anasazi and Mexico. In *The Maya and Their Neighbors*, ed. C. Hay, S. Lothrop, R. Linton, H. Shapiro, and G. Valliant, pp. 331–340. New York: D. Appleton.

Roberts, F. H. H. (1943) Edgar Billings Howard. *American Anthropologist* 45:452–454.

Roberts, F. H. H. (1951) Radiocarbon dates and early man. *American Antiquity, Memoir* 7:20–22.

Rogers, H. D. (1844) Brief history of the recent labors of American geologists and a rapid survey of the present condition of geological research in the United States. *American Journal of Science* 47:137–160, 247–278.

Rogers, H. D. (1850) On the origin of the drift and of the lake and river terraces of the United States and Europe, with an examination of the laws of aqueous action connected with the inquiry. *Proceedings of the American Association for the Advancement of Science* 2:239–255.

Romer, A. S. (1928) A fossil camel recently living in Utah. *Science* 68:19–20.

Romer, A. S. (1933) Pleistocene vertebrates and their bearing on the problem of human antiquity in North America. In *The American Aborigines: Their Origin and Antiquity*, ed. D. Jenness, pp. 49–83. Toronto: University of Toronto Press.

Romer, A. S. (1941) Vertebrate paleontology. In *Geology, 1888–1938 Fiftieth Anniversary Volume*. New York: Geological Society of America.

Rowland, H. (1884) A plea for pure science. *Proceedings of the American Association for the Advancement of Science* 32:105–126.

Russell, F. (1899) Human remains from the Trenton gravels. *American Naturalist* 33:143–155.

Russell, I. C. (1885) Geological history of Lake Lahontan: A quaternary lake of north-western Nevada. *Monographs of United States Geological Survey* 11.

Salisbury, R. (1892a) A preliminary paper on drift or Pleistocene formations of New Jersey. *Annual Report of the State Geologist for the year 1891*:35–108.

Salisbury, R. (1892b) Certain extra-morainic drift phenomena of New Jersey. *Geological Society of America Bulletin* 3:173–182.

Salisbury, R. (1892c) Review of "*Man and the Glacial Period*." *Chicago Tribune*, October 22, 1892.

Salisbury, R. (1892d) "*Man and the Glacial Period*," by George Frederick Wright. *Nature* 1207:148.

Salisbury, R. (1892e) Wright's "*Man and the Glacial Period*." *Nation* 55:496–497.

Salisbury, R. (1893a) Man and the glacial period. *American Geologist* 11:13–20.

Salisbury, R. (1893b) Distinct glacial epochs, and the criteria for their recognition. *Journal of Geology* 1:61–84.

Salisbury, R. (1893c) Surface geology: report of progress. *Annual Report of the State Geologist for the year 1892*:33–166.

Salisbury, R. (1893d) Editorials. *Journal of Geology* 1:198–200.

Salisbury, R. (1893e) The older drift in the Delaware Valley. *American Geologist* 11:360–362.

Salisbury, R. (1894) Surface geology: Report of progress. *Annual Report of the State Geologist for the Year 1893*:33–328.

Salisbury, R. (1895) The Greenland expedition of 1895. *Journal of Geology* 3: 875–902.

Salisbury, R. (1896) The Philadelphia brick clays, et al. *Science* 3:480–481.

Salisbury, R. (1897) On the origin and age of the relic bearing sand at Trenton, N.J. *Science* 6:977–981.

Salisbury, R. (1898) On origin and age of the relic bearing sand at Trenton. *Proceedings of the American Association for the Advancement of Science* 46:350–355.

Salisbury, R. (1902) *The Glacial Geology of New Jersey*. Trenton: MacCrellish & Quigley.

Sapir, E. (1916) Time perspective in aboriginal American culture: A study in method. *Geological Survey Canada, Anthropological Series* 13.

Schultz, C. B. and E. B. Howard (1935) The fauna of Burnet Cave, Guadalupe Mountains, New Mexico. *Proceedings of the Academy of Natural Sciences of Philadelphia* 87:273–298.

Schultz, C. B. and L. C. Eiseley (1935) Paleontological evidence for the antiquity of the Scottsbluff bison quarry and its associated artifacts. *American Anthropologist* 37:306–319.

Schultz, C. B. and L. C. Eiseley (1936) An added note on the Scottsbluff quarry. *American Anthropologist* 38:521–524.

Science (1883a) Proceedings of Section H—anthropology. *Science* 2:358–370.

Science (1883b) Notes and news. *Science* 2:384–386.

Science (1884a) Review of *"Annual Reports of the United States Geological Survey to the Secretary of the Interior, ii, iii."* *Science* 4:62–71.

Science (1884b) Proceedings of the Section of anthropology. *Science* 4:343–346.

Science (1886) The Abbott collection at the Peabody Museum. *Science* 7:4–5.

Science (1887) Proceedings of the American Association. Section H. *Science* 10:88–89.

Science (1933) Scientific notes and news. *Science* 78:186–189.

Sellards, E. H. (1916a) On the discovery of fossil human remains in Florida in association with extinct vertebrates. *American Journal of Science* 42:1–18.

Sellards, E. H. (1916b) Human remains from the Pleistocene of Florida. *Science* 44:615–617.

Sellards, E. H. (1916c) Human remains and associated fossils from the Pleistocene of Florida. *Annual Report of the Florida State Geological Survey* 8:122–160.

Sellards, E. H. (1917a) On the association of human remains and extinct vertebrates at Vero, Florida. *Journal of Geology* 25:4–24.

Sellards, E. H. (1917b) Further notes on human remains from Vero, Florida. *American Anthropologist* 19:239–251.

Sellards, E. H. (1917c) Review of the evidence on which the human remains found at Vero, Florida, are referred to the Pleistocene. *Annual Report of the Florida State Geological Survey* 9:69–84.

Sellards, E. H. (1917d) Supplement to studies in the Pleistocene at Vero, Florida. *Annual Report of the Florida State Geological Survey* 9:141–143.

Sellards, E. H. (1919) Literature relating to human remains and artifacts at Vero, Florida. *American Journal of Science* 47:358–360.

Sellards, E. H. (1932) Geologic relations of deposits reported to contain artifacts at Frederick, Oklahoma. *Geological Society of America Bulletin* 43:783–796.

Sellards, E. H. (1937) The Vero finds in the light of present knowledge. In *Early Man*, ed. G. G. MacCurdy, pp. 193–210. Philadelphia: Lippincott.

Sellards, E. H. (1940) Early man in America, index to localities and selected bibliography. *Geological Society of America Bulletin* 51:373–431.

Sellards, E. H. (1952) *Early man in America.* Austin: University of Texas Press.

Shaler, N. (1877) On the age of the Delaware gravel beds containing chipped pebbles. *Peabody Museum Annual Report* 10:44–47.

Shaler, N. (1893) Antiquity of man in eastern North America. *American Geologist* 11:180–184.

Shetrone, H. (1936) The Folsom phenomena as seen from Ohio. *Ohio Archaeological and Historical Quarterly* 45:240–256.

Shimek, B. (1896) A theory of the loess. *Iowa Academy of Science, Proceedings* 3:82–86.

Shimek, B. (1898) Is the loess of aqueous origin? *Iowa Academy of Science, Proceedings* 5:32–45.

Shimek, B. (1899) The distribution of the loess fossils. *Iowa Academy of Science, Proceedings* 6:98–113.

Shimek, B. (1903) The loess and the Lansing man. *American Geologist* 32:353–369.

Shimek, B. (1904a) The Lansing deposit not loess. *Bulletin from the Laboratories of Natural History of the State University of Iowa* 5: 346–352.

Shimek, B. (1904b) Loess and the Iowan drift. *Bulletin from the Laboratories of Natural History of the State University of Iowa* 5:352–368.

Shimek, B. (1904c) Evidences(?) of water-deposition of loess. *Bulletin from the Laboratories of Natural History of the State University of Iowa,* 5:369–381.

Shimek, B. (1908) Nebraska "Loess" Man. *Geological Society of America Bulletin* 19:243–254.

Short, J. T. (1880) *The North Americans of Antiquity.* New York: Harper and Brothers.

Shufeldt, R. W. (1917) Fossil birds found at Vero, Florida. *Florida Geological Survey Annual Report* 9:35–41.

Shuler, E. W. (1923) Occurrence of human remains with Pleistocene fossils, Lagow Sand Pit, Dallas, Texas. *Science* 57:333–334.

Sinclair, W. (1908) Recent investigations bearing on the question of the occurrence of Neocene Man in the auriferous gravels of the Sierra Nevada. *University of California Publications in American Archaeology and Ethnology* 7:107–131.

Skinner, M. and O. Kaisen (1947) The fossil *Bison* of Alaska and preliminary revision of the genus. *American Museum of Natural History Bulletin* 89:123–256.

Smith, G. E. (1913) The Piltdown skull. *Nature* 92:131.

Smith, G. E. (1928) *In the Beginning: The Origin of Civilization.* New York: William Morrow.

Smith, G. E., B. Malinowski, H. J. Spinden, and A. Goldenweiser (1927) *Culture: The Diffusion Controversy.* New York: W. W. Norton.

Smith, P. (1937) Certain relations between Northwestern America and Northeastern Asia. In *Early Man,* ed. G. G. MacCurdy, pp. 85–92. Philadelphia: Lippincott.

Spier, L. (1913) Results of an archaeological survey of the state of New Jersey. *American Anthropologist* 15:675–679.

Spier, L. (1916) New data on the Trenton Argillite culture. *American Anthropologist* 18:181–189.

Spier, L. (1918) The Trenton argillite culture. *Anthropological Papers of the American Museum of Natural History* 22(4).

Spier, L. (1928a) Concerning man's antiquity at Frederick, Oklahoma. *Science* 67:161–162.

Spier, L. (1928b) A note on reputed artifacts from Frederick, Oklahoma. *Science* 68:184.

Spinden, H. (1933) Origin of civilizations in Central America and Mexico. In *The American Aborigines: Their Origin and Antiquity*, ed. D. Jenness, pp. 219–246. Toronto: University of Toronto Press.

Spinden, H. (1937) First peopling of America as a chronological problem. In *Early Man*, ed. G. G. MacCurdy, pp. 105–114. Philadelphia: Lippincott.

Squier, E. and E. Davis (1848) Ancient monuments of the Mississippi Valley. *Smithsonian Contributions to Knowledge* 1.

Sterns, F. H. (1915) A stratification of cultures in eastern Nebraska. *American Anthropologist* 17:121–127.

Sterns, F. H. (1918) The Pleistocene man of Vero, Florida: A summary of the evidence of man's antiquity in the New World. *Scientific American Supplement* 2214:354–355.

Sterns, F. H. (1919) The Pleistocene man of Vero, Florida: A review of the latest evidence and theories. *Scientific American Supplement* 2251:118–119.

Steward, J. H. (1942) The direct historical approach. *American Antiquity* 7:337–343.

Stewart, A. (1897) Notes on the osteology of *Bison antiquus* Leidy. *Kansas University Quarterly* 6:127–135.

Stock, C. (1924) A recent discovery of ancient human remains in Los Angeles, California. *Science* 60:2–5.

Stock, C. (1936) The succession of mammalian forms within the period in which human remains are known to occur in America. *American Naturalist* 70:324–330.

Stock, C. (1941) Prehistoric archaeology. In *Geology 1888–1938: Fiftieth Anniversary Volume*, ed. R. Berkey, pp. 137–158. New York: Geological Society of America.

Stock, C. and F. Bode (1936) The occurrence of flints and extinct animals in pluvial deposits near Clovis, New Mexico. Part 3: Geology and vertebrate paleontology of the late Quaternary near Clovis, New Mexico. *Proceedings of the Philadelphia Academy of Natural Sciences* 88:219–241.

Thomas, C. (1894) Report on mound explorations of the Bureau of Ethnology. *Bureau of Ethnology Annual Report* 12:3–730.

Thomas, C. (1898) *Introduction to the Study of North American Archaeology*. Cincinnati: Robert Clarke.

Thwaites, F. (1927) The development of the theory of multiple glaciation in North America. *Wisconsin Academy of Sciences, Arts and Letters, Transactions* 23:41–164.

Topinard, P. (1893) L'anthropologie aux États-Unis. *L'Anthropologie* 4:301–351.

Tylor, E. B. (1871) *Primitive Culture: Researches Into the Development of Mythology, Philosophy, Religion, Language, Art and Custom*. London: John Murray.

Tylor, E. B. (1884) Some American aspects of anthropology. *Science* 4:217–219.

Upham, W. (1884) The Minnesota Valley in the ice age. *Proceedings of the American Association for the Advancement of Science* 32:213–231.

Upham, W. (1888a) The recession of the ice-sheet in Minnesota in its relation to the gravel deposits overlying the quartz implements found by Miss Babbitt at Little Falls, Minnesota. *Proceedings of the Boston Society of Natural History* 23:436–447.

Upham, W. (1888b) Professor Henry Carvill Lewis and his work in glacial geology. *American Geologist* 2:371–379.

Upham, W. (1892a) Pleistocene papers at the Rochester Meeting. *American Geologist* 10:217–224.

Upham, W. (1892b) Review of "*Man and the Glacial Period.*" *Popular Science Monthly* 42:266–269.

Upham, W. (1893a) Estimates of geological time. *American Journal of Science* 45:209–220.

Upham, W. (1893b) Man and the glacial period. *American Geologist* 11:189–191.

Upham, W. (1896) The Glacial Lake Agassiz. *Monographs of United States Geological Survey* 25.

Upham, W. (1901) Derivation and antiquity of the American race. *American Antiquarian* 23:80–88.

Upham, W. (1902a) Man in Kansas during the Iowan stage of the Glacial period. *Science* 16:355–356.

Upham, W. (1902b) Man and the Ice Age at Lansing, Kansas and Little Falls, Minnesota. *American Geologist* 30:135–150.

Upham, W. (1902c) Glacial man in Kansas. *American Anthropologist* 4:566–568.

Upham, W. (1902d) Primitive man and his stone implements in the North American loess. *American Antiquarian* 24:413–420.

Upham, W. (1902e) Primitive man in the Ice Age. *Bibliotheca Sacra* 59:730–743.

Upham, W. (1902f) The fossil man of Lansing, Kansas. *Records of the Past* 1:272–275.

Upham, W. (1903a) Valley loess and the fossil man of Lansing, Kansas. *American Geologist* 31:25–34.

Upham, W. (1903b) How long ago was America peopled? *American Geologist* 31:312–315.

Upham, W. (1903c) Editorial comment. The antiquity of the fossil man of Lansing, Kansas. *American Geologist* 32:185–187.

Upham, W. (1922) Memorial of George Frederick Wright. *Geological Society of America Bulletin* 33:15–30.

Vaughan, T. W. (1917) On reported Pleistocene human remains at Vero, Florida. *Journal of Geology* 25:40–42.

Volk, E. (1911) *The Archaeology of the Delaware Valley. Papers of the Peabody Museum of Archaeology and Ethnology* 5.

Wadsworth, M. (1881) On the lithological character of the implements. *Proceedings of the Boston Society of Natural History* 21:146–147.

Wallace, A. R. (1887) The antiquity of man in North America. *Nineteenth Century* 22:667–679.

Ward, H. (1907) Peculiarities of the "Nebraska Man." *Putnam's Monthly and the Critic* 1:410–413.

Webb, W. S. (1951) The Parrish Village site, Site 45 Hopkins County, Kentucky. *University of Kentucky Reports in Anthropology* 7(6).

White, I. C. (1884) Relation of the glacial dam at Cincinnati to the terraces of the upper Ohio and its tributaries. *Proceedings of the American Association for the Advancement of Science* 32:212–213.

White, I. C. (1887) Rounded boulders at high altitudes along some Appalachian Rivers. *American Journal of Science* 34:374–381.

Whitney, J. D. (1867) Notice of a human skull recently taken from a shaft near Angles, Calaveras County, California. *American Journal of Science* 43:265–267.

Whitney, J. D. (1879) *The Auriferous Gravels of the Sierra Nevada of California.* Harvard Museum of Comparative Zoology, *Memoir* 6 (1–2).

Whittlesey, C. (1869) On the evidences of the antiquity of man in America. *Proceedings of the American Association for the Advancement of Science* 17:268–288.

Willis, B. (1922) Out of the long past: A story of the Stanford skull. *Stanford Cardinal,* October:8–11.

Williston, S. (1897) The Pleistocene of Kansas. *University Geological Survey of Kansas* 2: 299–308.

Williston, S. (1902a) A fossil man from Kansas. *Science* 16:195–196.

Williston, S. (1902b) An arrow-head found with the bones of *Bison occidentalis*, Lucas, in western Kansas. *American Geologist* 30:313–315.

Williston, S. (1903) The fossil man of Lansing, Kansas. *Popular Science Monthly* 67:463–473.

Williston, S. (1905a) On the occurrence of an arrow-head with the bones of an extinct bison. *Proceedings of the International Congress of Americanists*, 1902, Thirteenth Session:335–337.

Williston, S. (1905b) On the Lansing man. *American Geologist* 35:342–346.

Wilson, D. (1863a) Physical ethnology. *Smithsonian Institution Annual Report for 1862*:240–302.

Wilson, D. (1863b) *Prehistoric Man: Researches Into the Origin of Civilization in the Old and the New World.* Cambridge, UK: Macmillan.

Wilson, D. (1876) *Prehistoric Man: Researches Into the Origin of Civilization in the Old and the New World*, 3rd ed. London: Macmillan.

Wilson, D. (1878) Address of Professor Daniel Wilson. *Proceedings of the American Association for the Advancement of Science* 26:319–334.

Wilson, T. (1889) The Paleolithic period in the District of Columbia. *American Anthropologist* 2:235–241.

Wilson, T. (1890a) Results of an inquiry as to the existence of man in North America during the Paleolithic period of the stone age. *United States National Museum Annual Report* 1888:677–702.

Wilson, T. (1890b) Anthropology at the American Association for the Advancement of Science. *American Naturalist* 24:975–984.

Wilson, T. (1890c) The Paleolithic period in the District of Columbia. *Proceedings of the United States National Museum* 12:371–376.

Wilson, T. (1893) Primitive industry. *Smithsonian Institution Annual Report* 1892:521–534.

Wilson, T. (1895) On the presence of fluorine as a test for the fossilization of animal bones. *American Naturalist* 29:301–317, 439–456, 719–725.

Wilson, T. (1896) Piney Branch (D.C.) quarry workshop and its implements. *American Naturalist* 30:873– 885, 976–992.

Wilson, T. (1898) Investigations in the sand-pits of the Lalor Field, near Trenton, New Jersey. *Proceedings of the American Association for the Advancement of Science* 46:381–383.

Winchell, N. (1873) The surface geology. *Geological and Natural History Survey of Minnesota Annual Report* 1:61–62.

Winchell, N. (1875) The geology of Mower County. *Geological and Natural History Survey of Minnesota Annual Report* 3:20–36.

Winchell, N. (1878) Primitive man at Little Falls. *Geological and Natural History Survey of Minnesota Annual Report* 6:53–64.

Winchell, N. (1887) Winchell on quartz implements. *American Antiquarian* 9:46–47.

Winchell, N. (1892) Review of "*Man and the Glacial Period*." *American Geologist* 10: 387–389.

Winchell, N. (1893a) Professor Wright's book a service to science. *American Geologist* 11:194.

Winchell, N. (1893b) Editorial comment. Prof. Youmans and the United States Geological Survey. *American Geologist* 11:342–343.

Winchell, N. (1893c) Editorial comment. Major Powell in a current issue of *Popular Science Monthly*. *American Geologist* 12:115–118.

Winchell, N. (1902) The Lansing skeleton. *American Geologist* 30:189–194.

Winchell, N. (1903a) Was man in American in the glacial period? *Geological Society of America Bulletin* 14:133–152.

Winchell, N. (1903b) The Pleistocene geology of the Concannon farm near Lansing, Kansas. *American Geologist* 31:263–308.

Winchell, N. (1907) Pre-indian inhabitants of North America. *Records of the Past* 6:145–157, 163–181.

Winchell, N. (1909) Newton Winchell to C. C. Abbott. *Records of the Past* 8:249–252.

Winchell, N. (1913) *The Weathering of Aboriginal Stone Artifacts No. 1—A Consideration of the Paleoliths of Kansas.* St. Paul: Minnesota Historical Society, *Collections* 16.

Winchell, N. (1917) The antiquity of man in America compared with Europe. *Bulletin of the Minnesota Academy of Science* 5(3):121–151.

Wislizenus, A. (1858) Was man contemporaneous with the mastodon? *Transactions of the Academy of Science of St Louis* 1:168–171.

Wissler, C. (1916a) The present status of the antiquity of man in North America. *Scientific Monthly* 2:234–238.

Wissler, C. (1916b) The application of statistical methods to the data in the Trenton Argillite culture. *American Anthropologist* 18:190–197.

Wissler, C. (1917) The new archaeology. *American Museum Journal* 17:100–101.

Wissler, C. (1920) Charles Abbott. *American Anthropologist* 22:70–71.

Wissler, C. (1933) Ethnological diversity in America and its significance. In *The American Aborigines: Their Origin and Antiquity*, ed. D. Jenness, pp. 167–216. Toronto: University of Toronto Press.

Woeikof, A. (1886) Examination of Dr. Croll's hypotheses of geological climates. *American Journal of Science* 31:161–178.

Woodward, A. S. (1933) Early man and the associated faunas in the Old World. *Science* 78:89–92.

Wormington, H. M. (1939) *Ancient Man in North America.* Denver: Colorado Museum of Natural History.

Wormington, H. M. (1946) Jesse Dade Figgins, 1867–1944. *American Anthropologist* 48:75–77.

Wormington, H. M. (1957) *Ancient Man in North America*, 4th ed. Denver: Denver Museum of Natural History.

Worsaae, J. (1859) Letter. *Athenaeum* 1679:889–90.

Wright, A. A. (1893) Older drift in the Delaware Valley. *American Geologist* 11:184–186.

Wright, G. F. (1879) The kames and moraines of New England. *Proceedings of the Boston Society of Natural History* 20:210–220.

Wright, G.F. (1881a) On the age of the Trenton gravel. *Proceedings of the Boston Society of Natural History* 21:137–145.

Wright, G. F. (1881b) An attempt to calculate approximately the date of the glacial era in North America. *American Journal of Science* 21:120–123.

Wright, G. F. (1882a) Glacial phenomena of North America and their relation to the question of man's antiquity in the valley of the Delaware. *Bulletin Essex Institute 14:71–73.*

Wright, G. F. (1882b) *Studies in Science and Religion.* Andover: Warren F. Draper.

Wright, G. F. (1883a) Glacial phenomena in Ohio. *Science* 1:269–271.

Wright, G. F. (1883b) Recent investigations concerning the southern boundary of the glaciated area of Ohio. *American Journal of Science* 26:44–56.

Wright, G. F. (1884a) The terminal moraine in Ohio and Kentucky. In The terminal moraine of Pennsylvania, by H. C. Lewis. *Second Geology Survey of Pennsylvania* Report Z:203–243.

Wright, G. F. (1884b) The Niagara River and the glacial period. *American Journal of Science* 28:32–35.

Wright, G. F. (1884c) The theory of a glacial dam at Cincinnati and its verification. *American Naturalist* 18:563–567.

Wright, G. F. (1884d) The glaciated area of North America. *American Naturalist* 18:755–767.

Wright, G .F. (1885) The Niagara gorge as a chronometer. *Science* 5:399–401.

Wright, G. F. (1887a) The relation of the glacial period to archaeology in Ohio. *Ohio Archaeological and Historical Society* 1:171–183.

Wright, G. F. (1887b) Pre-glacial man in Ohio. *Ohio Archaeological and Historical Society* 1:254–256.

Wright, G. F. (1887c) The Muir glacier. *American Journal of Science* 33:1–18.

Wright, G. F. (1888) The age of the Ohio gravel-beds. *Proceedings of the Boston Society of Natural History* 23:427–436.

Wright, G. F. (1889a) *The Ice Age in North America and Its Bearing upon the Antiquity of Man.* New York: D. Appleton.

Wright, G. F. (1889b) The age of the Philadelphia Red gravel. *Proceedings of the Boston Society of Natural History* 24:152–157.

Wright, G. F. (1890a) The glacial boundary in western Pennsylvania, Ohio, Kentucky, Indiana and Illinois. *United States Geological Survey Bulletin* 58:39–100.

Wright, G. F. (1890b) The Nampa image. *Proceedings of the Boston Society of Natural History* 24:424–450.

Wright, G. F. (1890c) Report. In *Discovery of a Paleolithic Implement at New Comerstown, Ohio,* by W. C. Mills and G. F. Wright, pp. 5–14. Western Reserve Historical Society, *Tract* 75. Cleveland: Western Reserve Historical Society.

Wright, G. F. (1892a) *Man and the Glacial Period.* New York: D. Appleton.

Wright, G. F. (1892b) Unity of the glacial epoch. *American Journal of Science* 44:351–373.

Wright, G. F. (1892c) Man and the glacial period. *Science* 20:275–277.

Wright, G. F. (1892d) Man and the glacial period. *Dial* 13:380.

Wright, G. F. (1892e) Excitement over glacial theories. *Science* 20:360–361.

Wright, G. F. (1893a) "The Ice Age in North America." A closing word with the reviewer. *Dial* 14:40–41.

Wright, G. F. (1893b) Some of Prof. Salisbury's criticisms on "Man and the glacial period." *American Geologist* 11:121–126.

Wright, G. F. (1893c) Some detailed evidence of an Ice-Age man in eastern America. *Science* 21:65–66.

Wright, G. F. (1893d) Additional evidence bearing upon the glacial history of the Upper Ohio Valley. *American Geologist* 11:195–199.

Wright, G. F. (1893e) Evidences of glacial man in Ohio. *Popular Science Monthly* 43:29–39.

Wright, G. F. (1893f) Mr. Holmes's criticisms upon the evidence of glacial man. *Science* 21:267–268.

Wright, G. F. (1894a) The Cincinnati Ice Dam. *Popular Science Monthly* 45:184–198.

Wright, G. F. (1894b) Continuity of the glacial period. *American Journal of Science* 47:161–187.

Wright, G. F. (1895a) Account of the discovery of a chipped chert implement in undisturbed glacial gravel near Steubenville, Ohio. *American Geologist* 16:225–226.

Wright, G. F. (1895b) New evidence of glacial man in Ohio. *Popular Science Monthly* 45:157–165.

Wright, G. F. (1895c) New evidence of glacial man in Ohio. *American Naturalist* 29:951–953.

Wright, G. F. (1895d) Dr. Holst on the continuity of the glacial period. *American Geologist* 16:396–399.

Wright, G. F. (1896a) The age of the Philadelphia brick clay. *Science* 3:242–243.

Wright, G. F. (1896b) The age of the second terrace on the Ohio at Brilliant, near Steubenville. *Journal of Geology* 4:218–219.

Wright, G. F. (1896c) Fresh relics of glacial man at the Buffalo meeting of the A.A.A.S. *American Naturalist* 30:781–784.

Wright, G. F. (1896d) Fresh geological evidence of glacial man at Trenton, New Jersey. *American Geologist* 18:238.

Wright, G. F. (1897) Special explorations in the implement-bearing deposits on the Lalor Farm, Trenton, New Jersey. *Science* 6:637–645.

Wright, G. F. (1898) Special explorations in the implement-bearing deposits on the Lalor Farm, Trenton, New Jersey. *Proceedings of the American Association for the Advancement of Science* 46:355–364.

Wright, G. F. (1899) The truth about the Nampa figurine. *American Geologist* 23:267–272.

Wright, G. F. (1900) Remarks on the loess in North China. *Science* 12:71–73.

Wright, G. F. (1901) Geology and the Deluge: Remarkable geological discoveries in central Asia and southern Russia, showing that the Noachian Flood is a scientific possibility. *McClure's Magazine* 17:134–139.

Wright, G. F. (1902) Origin and distribution of the loess in northern China and central Asia. *Geological Society of America Bulletin* 13:127–138.

Wright, G. F. (1903) The age of the Lansing skeleton. *Records of the Past* 2:119–124.

Wright, G. F. (1904) Evidence of the agency of water in distributing the loess in the Mississippi Valley. *American Geologist* 33:205–222.

Wright, G. F. (1905) Professor Shimek's criticism of the aqueous origin of loess. *American Geologist* 35:236–240.

Wright, G. F. (1907) Recent geological changes as affecting theories of man's development. *American Anthropologist* 9:529–532.

Wright, G. F. (1911) *The Ice Age in North America and Its Bearing upon the Antiquity of Man*, 5th ed., rev. Oberlin: Bibliotheca Sacra.

Wright, G. F. (1912) *Origin and Antiquity of Man*. Oberlin: Bibliotheca Sacra.

Wright, G. F. (1916) *Story of My Life and Work*. Oberlin: Bibliotheca Sacra.

Wright, G. F. (1919) Charles Conrad Abbott and Ernest Volk. *Science* 50:451–453.

Wright, T. (1859) Flint implements in the drift. *Athenaeum* 1651:809.

Wyman, Jeffries (1872) Report of the curator. In *Fifth Annual Report of the Trustees of the Peabody Museum of American Archaeology and Ethnology*, pp. 5–30. Boston: A. A. Kingman.

Youmans, W. J. (1892) Sketch of George Frederick Wright. *Popular Science Monthly* 42:258–262.

Youmans, W. J. (1893a) The insolence of office. *Popular Science Monthly* 42:841–842.

Youmans, W. J. (1893b) The attack on Prof. Wright. *Popular Science Monthly* 43:412–413.

C. Printed sources: Secondary

Aiello, L. (1967) Charles Conrad Abbott, M.D., pick and shovel scientist. *New Jersey History* 85:208–216.

Alden, W. C. (1929) Thomas Chrowder Chamberlin's contributions to glacial geology. *Journal of Geology* 37:293–319.

Bain, H. F. (1916) N. H. Winchell and the *American Geologist*. *Economic Geology* 11:51–62.

Bartlett, R. A. (1962) *Great Surveys of the American West*. Norman: University of Oklahoma Press.

Barton, D. R. (1941) Father of the dinosaurs. *Natural History* 48:308–312.

Bass, W. (1973) Lansing man: A half century later. *American Journal of Physical Anthropology* 38:99–104.

Bedini, S. (1990) *Thomas Jefferson: Statesman of Science*. New York: Macillan.

Bernstein, J. (2002) First recipients of anthropological doctorates in the United States, 1891–1930. *American Anthropologist* 104:551–564.

Bieder, R. (1986) *Science Encounters the Indian, 1820–1880: The Early Years of American Ethnology*. Norman: University of Oklahoma Press.

Blakey, M. (1987) Skull doctors: Intrinsic social and political bias in the history of American physical anthropology, with special reference to the work of Aleš Hrdlička. *Critique of Anthropology* 7:7–35.

Blanckaert, C. (1997) L'anthropologie lamarckienne à la fin du XIXe siècle: Matérialisme scientifique et mésologie sociale. In *Jean-Baptiste Lamarck 1744–1829*, ed. G. Laurent, pp. 611–619. Paris: Comité des Travaux Historiques et Scientifiques.

Blanckaert, C. (2009) *De la Race à l'Évolution. Paul Broca et l'Anthropologie Française. 1850–1900*. Paris: L'Harmattan.

Boldurian, A. and J. Cotter (1999) *Clovis Revisited: New Perspectives on Paleoindian Adaptations from Blackwater Draw, New Mexico*. Philadelphia: University Museum.

Brace, C. L. (2005) *"Race" Is a Four Letter Word: The Genesis of the Concept*. New York: Oxford University Press.

Browman, D. and D. Givens (1996) Stratigraphic excavation: The first "new archaeology." *American Anthropologist* 98:80–95.

Browman, D. and S. Williams (2013) *Anthropology at Harvard: A biographical history*. Cambridge, MA: Peabody Museum Press.

Burkhardt, F., and S. Smith (1991) *The Correspondence of Charles Darwin*, vol. 7, 1858–1859. Cambridge, UK: Cambridge University Press.

Burkhardt, F., D. Porter, S. Dean, J. Topham, and S. Wilmot (1999) *The Correspondence of Charles Darwin*, vol. 11, 1863. Cambridge, UK: Cambridge University Press.

Burnham, John C. (1990) The evolution of peer review. *Journal of the American Medical Association* 263:1323–1329.

Busacca, A., H. Markewich, D. Muhs, N. Lancaster, and M. Sweeney (2004) Eolian sediments. In *The Quaternary Period in the United States*, ed. A. Gillespie, S. C. Porter, and B. Atwater, pp. 275–309. New York: Elsevier Science.

Bynum, W., E. Browne, and R. Porter, ed.s (1981) *Dictionary of the History of Science*. Princeton: Princeton University Press.

Chamberlin, R. T. (1931) Memorial of Rollin D. Salisbury. *Geological Society of America Bulletin* 42:126–138.

Chamberlin, R. T. (1934) Biographical memoir of Thomas Chrowder Chamberlin, 1843–1928. *Biographical Memoirs of the National Academy of Sciences* 15:307–407.

Chester, H. (2002) Frances Eliza Babbitt and the growth of professionalism of women in archaeology. In *New Perspectives on the Origins of Americanist Archaeology*, ed. D. Browman and S. Williams, pp. 164–184. Tuscaloosa: University of Alabama Press.

Chudacoff, H. P. (1989) *How Old Are You? Age Consciousness in American Culture*. Princeton: Princeton University Press.

Cohen, I. B. (1985) *Revolution in Science*. Cambridge, MA: Harvard University Press.

Cole, D. (1999) *Franz Boas: The Early Years, 1858–1906*. Seattle: University of Washington Press.

Conn, S. (1998) *Museums and American Intellectual Life, 1876–1926*. Chicago: University of Chicago Press.

Conn, S. (2003) Archaeology, Philadelphia, and understanding nineteenth century American culture. In *Archaeology and Archaeologists in Philadelphia,* ed. D. Fowler and D. Wilcox, pp. 165–180. Tuscaloosa: University of Alabama Press.

Conn, S. (2004) *History's Shadow: Native Americans and Historical Consciousness in the Nineteenth Century*. Chicago: University of Chicago Press.

Cotter, J. L. (1991) Update on Natchez man. *American Antiquity* 56:36–39.

Cravens, H. (1978) *The Triumph of Evolution: American Scientists and the Heredity-Environment Controversy 1900–1914*. Philadelphia: University of Pennsylvania Press.

Cross, D. (1956) *Archaeology of New Jersey,* vol. 2: *The Abbott Farm*. Trenton: New Jersey State Museum.

Daniel, G. (1975) *A Hundred and Fifty Years of Archaeology*. Cambridge, MA: Harvard University Press.

Darrah, W. (1951) *Powell of the Colorado*. Princeton: Princeton University Press.

Dascal, M. (1998) The study of controversies and the theory and history of science. *Science in Context* 11:147–154.

Dascal, M. and V. Boantza (2011) *Controversies within the Scientific Revolution*. Amsterdam: John Benjamins.

De Bont, R. (2003) The creation of prehistoric man: Aimé Rutot and the Eolith controversy, 1900–1920. *Isis* 94:604–630.

De Chandaverian, S. (1996) Laboratory science versus country-house experiments: The controversy between Julius Sachs and Charles Darwin. *British Journal for the History of Science* 29:17–41.

Dent, R. J. (1995) *Chesapeake Prehistory: Old Traditions, New Directions*. New York: Plenum.

Dexter, R. (1986) Historical aspects of the Calaveras skull controversy. *American Antiquity* 51:365–369.

Dingus, L. and M. Norell (2010) *Barnum Brown: The Man Who Discovered* Tyrannosaurus rex. Berkeley: University of California Press.

Dunnell, R. C. (1986) Methodological issues in Americanist artifact classification. *Advances in Archaeological Method and Theory* 9:149–207.

Dunnell, R. C. (2003) The first new archaeology and the development of chronological method. In *Picking the Lock of Time: Developing Chronology in American Archaeology,* ed. J. Truncer, pp. 9–21. Gainesville: University Press of Florida.

Dunning, D. (2011) The Dunning-Kruger Effect: On being ignorant of one's own ignorance. *Advances in Experimental Social Psychology* 44:247–296.

Dupree, A. H. (1957) *Science in the Federal Government*. Cambridge, MA: Harvard University Press.

Dyke, L. F. (1989) Henry Chapman Mercer: An annotated chronology. *Mercer Mosaic Journal of the Bucks County Historical Society* 6:35–65.

Evans, G. (1961) Elias Howard Sellards. *Bulletin of the American Association of Petroleum Geologists* 25:1904–1906.

Evans, G. (1986) E. H. Sellards' contributions to Paleoindian studies. In *Guidebook to the Archaeological Geology of Classic Paleoindian Sites on the Southern High Plains, Texas and New Mexico,* ed. V. T. Holliday, pp. 7–18. College Station: Texas A&M University Press.

Evans, J. (1943) *Time and Chance: The Story of Arthur Evans and His Forebears*. London: Longmans, Green.

Fisher, J. (1963) *The Seventy Years of the Department of Geology University of Chicago, 1892–1961.* Chicago: University of Chicago Press.

Fleming, J. R. (2006) James Croll in context: The encounter between climate dynamics and geology in the second half of the nineteenth century. *History of Meteorology* 3:43–53.

Freed, S., R. Freed, and L. Williamson (1988) Capitalist philanthropy and Russian revolutionaries: The Jesup North Pacific Expedition (1897–1902). *American Anthropologist* 90:7–24.

Gamble, C. and R. Kruszynski (2009) John Evans, Joseph Prestwich and the stone that shattered the time barrier. *Antiquity* 83:461–475.

Geertz, C. (1973) *The Interpretation of Cultures.* New York: Basic Books.

Gifford, J. and R. Rapp (1985) The early development of archaeological geology in North America. In *Geologists and Ideas: A History of North American Geology*, ed. E. Drake and W. Jordan, pp. 409–421. Boulder: Geological Society of America.

Gillespie, N. (1976) The Duke of Argyll, evolutionary anthropology, and the art of scientific controversy. *Isis* 68:40–54.

Givens, D. (1992) *Alfred Vincent Kidder and the Development of Americanist Archaeology.* Albuquerque: University of New Mexico Press.

Goetzmann, W. (1966) *Exploration and Empire: The Explorer and the Scientist in the Winning of the American West.* New York: W. W. Norton.

Goldstein, D. (1994) "Yours for Science": The Smithsonian Institution's correspondents and the shape of scientific community in nineteenth-century America. *Isis* 85:573–599.

Goldstein, D. (2008) Outposts of science: The knowledge trade and the expansion of scientific community in post–Civil War America. *Isis* 99:519–546.

Goldthwait, R. (1991) The Teays Valley problem: A historical perspective. In *Geology and Hydrogeology of the Teays-Mahomet Bedrock Valley System*, ed. by W. N. Melhorn and J. P. Kempton. Geological Society of America, *Special Paper* 258:3–8.

Goodrum, M. R. (2002) The meaning of Ceraunia: Archaeology, natural history and the interpretation of prehistoric stone artefacts in the eighteenth Century. *British Journal for the History of Science* 35:255–269.

Goodrum, M. R. (2009) The history of human origins research and its place in the history of science: Research problems and historiography. *History of Science* 47:337–357.

Goodrum, M. R. and C. Olson (2009) The quest for an absolute chronology in human prehistory: Anthropologists, chemists and the fluorine dating method in palaeoanthropology. *British Journal for the History of Science* 42:95–114.

Gould, C. (1959) *Covered Wagon Geologist.* Norman: University of Oklahoma Press.

Gould, S. J. (1965) Is uniformitarianism necessary? *American Journal of Science* 263:223–228.

Gould, S. J. (1977) *Ontogeny and Phylogeny.* Cambridge, MA: Belknap.

Gould, S. J. (1981) *Mismeasure of Man.* New York: W. W. Norton.

Gould, S. J. (1987) *An Urchin in the Storm.* New York: W. W. Norton.

Graham, R. W. (1979) The Kimmswick bone beds: A historical perspective. *Living Museum* 42: 40–44.

Graham, R. W., C. V. Haynes, D. L. Johnson, and M. Kay (1981) Kimmswick: A Clovis-mastodon association in eastern Missouri. *Science* 213: 1115–1117.

Grayson, D. K. (1983) *The Establishment of Human Antiquity.* New York: Academic.

Grayson, D. K. (1984) Nineteenth century explanations of Pleistocene extinctions: A review and analysis. In *Quaternary Extinctions: A Prehistoric Revolution*, ed. P. S. Martin and R. G. Klein, pp. 5–39. Tucson: University of Arizona Press.

Grayson, D. K. (1986) Eoliths, archaeological ambiguity, and the generation of "middle-range" research. In *American Archaeology Past and Future*, ed. D. Meltzer, D. Fowler, and J. Sabloff, pp. 77–133. Washington, DC: Smithsonian Institution Press.

Grayson, D. K. (1990) The provision of time depth for paleoanthropology. In *Establishment of a Geologic Framework for Paleoanthropology*, ed. L. Laporte. Geological Society of America, *Special Paper* 242:1–13.

Griffin, J. B. (1977) A commentary on early man studies in the northeast. In *Amerinds and Their Paleoenvironments in Northeastern North America*, ed. W. Newman and B. Salwen, pp. 3–15. New York Academy of Sciences, *Annals* 288.

Griffin, J. B., D. J. Meltzer, B. D. Smith, and W. C. Sturtevant (1988) A mammoth fraud in science. *American Antiquity* 53:578–582.

Gruber, J. W. (1965) Brixham Cave and the antiquity of man. In *Context and Meaning in Cultural Anthropology*, ed. M. Spiro, pp. 373–402. New York: Free Press.

Gruber, J. W. (1966) In search of experience. In *Pioneers of American Anthropology: The Uses of Biography*, ed. J. Helm, pp. 3–27. Seattle: University of Washington Press.

Hamlin, C. (1982) James Geikie, James Croll, and the eventful Ice Age. *Annals of Science* 39:565–583.

Hammond, M. (1982) The expulsion of the Neanderthals from human ancestry: Marcellin Boule and the social context of scientific research. *Social Studies of Science* 12:1–36.

Harris, M. (1968) *The Rise of Anthropological Theory*. New York: T. Y. Crowell.

Hart, K. (1976) Government geologists and the early man controversy: The problem of "official" science in America, 1879–1907. Unpublished PhD dissertation, Department of History, Kansas State University.

Hawley, M. F. (2009) The Gilded Age "Bone Wars" and the birth of Paleoindian archaeology: Williston, Martin, Overton, and the 12 Mile Creek site. *North American Archaeologist* 30:105–140.

Haynes, C. V. (1990) The Antevs–Bryan years and the legacy for Paleoindian geochronology. In *Establishment of a Geologic Framework for Paleoanthropology*, ed. L. Laporte, Geological Society of America, *Special Paper* 242:55–68.

Haynes, C. V. (1995) Geochronology of paleoenvironmental change, Clovis type site, Blackwater Draw, New Mexico. *Geoarchaeology* 10:317–388.

Heizer, R. and S. Cook (1952) Fluorine and other chemical tests of some North American human and fossil bones. *American Journal of Physical Anthropology* 10:289–304.

Heizer, R. F. and T. McCown (1950) The Stanford skull, a probable early man from Santa Clara County, California. *Reports of the University of California Archaeological Survey*, No. 6.

Hildebrand, J. (1957) *Science in the Making*. New York: Columbia University Press.

Hill, M. E. (2002) The Folsom-age 12 Mile Creek bison bonebed in western Kansas. *TER-QUA Symposium Series* 3:53–70.

Hill, M. E. (2006) Before Folsom: The 12 Mile Creek site and the debate over the peopling of the Americas. *Plains Anthropologist* 51:141–156.

Hinsley, C. (1976) Amateurs and professionals in Washington anthropology, 1879 to 1903. In *American Anthropology: The Early Years*, ed. J. Murra, pp. 36–68. St. Paul: West Publishing.

Hinsley, C. (1981) *Savages and Scientists: The Smithsonian Institution and the Development of American Anthropology*. Washington DC: Smithsonian Institution Press.

Hinsley, C. M. (1985) From shell heaps to stellae: Early anthropology at the Peabody Museum. *History of Anthropology* 3:49–74.

Hinsley, C. M. (1992) The museum origins of Harvard anthropology, 1866–1915. In *Science at Harvard University: Historical Perspectives*, ed. C. Elliott and M. Rossiter, pp. 121–145. Bethlehem, PA: Lehigh University Press.

Hinsley, C. M. (1999) Frederic Ward Putnam, 1839–1915. In *Encyclopedia of Archaeology: The Great Archaeologists*, ed. T. Murray, pp. 141–154. Santa Barbara: ABC-CLIO.

Hinsley, C. M. (2003) Drab dove takes flight: The dilemmas of early Americanist archaeology in Philadelphia, 1889–1900. In *Archaeology and Archaeologists in Philadelphia*, ed. D. Fowler and D. Wilcox, pp. 1–20. Tuscaloosa: University of Alabama Press.

Holliday, V. T. (2000) The evolution of Paleoindian geochronology and typology on the Great Plains *Geoarchaeology* 15:227–290.

Holliday, V. T. and A. B. Anderson (1993) "Paleoindian," "Clovis" and "Folsom": A brief etymology. *Current Research in the Pleistocene* 10:79–81.

Holmes, F. (1984) Lavoisier and Krebs: The individual scientist in the near and deeper past. *Isis* 75:131–142.

Holmes, F. (1987) Scientific writing and scientific discovery. *Isis* 78:220–235.

Hopkins, D. M. (1967) Introduction. In *The Bering Land Bridge*, ed. D. Hopkins, pp. 1–6. Stanford: Stanford University Press.

Howells, W. W. (1992) Yesterday, today, and tomorrow. *Annual Review of Anthropology* 21:1–17.

Hull, D. (1979) In defense of presentism. *History and Theory* 18:1–15.

Hull, D. (1988) *Science as a Process: An Evolutionary Account of the Social and Conceptual Development of Science*. Chicago: University of Chicago Press.

Hull, D., P. Tessner, and A. Diamond (1978) Planck's principle. *Science* 202:717–723.

Hunter, R., D. Tvaryanas, D. Byers, and R. M. Stewart (2009) The Abbott Farm National Historic Landmark interpretive plan: Cultural resource technical document. Hunter Research Inc., Trenton, NJ. Available at http://www.state.nj.us/counties/mercer /about/community/openspace/abbott.html.

Jenkins, D. (1994) Object lessons and ethnographic displays: Museum exhibitions and the making of American anthropology. *Comparative Studies in Society and History* 36:242–270.

Joyce, A., W. Sandy, and S. Horan (1989) Dr. Charles Conrad Abbott and the question of human antiquity in the New World. *Bulletin of the Archaeological Society of New Jersey* 44:5970.

Judd, N. (1966) Frank H. H. Roberts, Jr. 1897–1966. *American Anthropologist* 68:1226–1232.

Judd, N. (1967) *The Bureau of American Ethnology: A Partial History*. Norman: University of Oklahoma Press.

Knight, D. (1987) Background and foreground: Getting things in context. *British Journal for the History of Science* 20:3–12.

Kohlstedt, S. G. (1980) *Science*: The struggle for survival, 1880–1894. *Science* 209:33–42.

Kohlstedt, S. G., M. Sokal, and B. Lewenstein (1999) *The Establishment of Science in America: 150 Years of the American Association for the Advancement of Science*. New Brunswick: Rutgers University Press.

Kraft, J. and R. Thomas (1976) Early man at Holly oak, Delaware. *Science* 192:756–761.

Krieger, A. D. (1961) Elias Howard Sellards, 1875–19 61. *American Antiquity* 27:225–28.

Kruger, J. and D. Dunning (1999) Unskilled and unaware of it: How difficulties in recognizing one's own incompetence lead to inflated self-asssessments. *Journal of Personality and Social Psychology* 77:1121–1134.

Kuhn, T. (1977) *The Essential Tension: Selected Studies in Scientific Tradition and Change*. Chicago: University of Chicago Press.

Lacey, M. (1979) The mysteries of earth making dissolve: A study of Washington's intellectual community and the origins of American environmentalism in the late 19th century. Unpublished PhD dissertation, Department of History, George Washington University.

Leech, M. (1941) *Reveille in Washington, 1860–1865*. New York: Harper & Row.

Leith, C. K. (1929) Chamberlin's work in Wisconsin. *Journal of Geology* 37:289–292.

Livingstone, D. (1987) *Nathaniel Southgate Shaler and the Culture of American Science*. Tuscaloosa: University of Alabama Press.

Livingstone, D. (2008) *Adam's Ancestors: Race, Religion, and the Politics of Human Origins*. Baltimore: Johns Hopkins University Press.

Lull, R. S. (1931) Memorial of Oliver Perry Hay. *Geological Society of America Bulletin* 42:30–48.

Lyman, R. L. and M. J. O'Brien (2006) *Measuring Time with Artifacts: A History of Methods in American Archaeology*. Lincoln: University of Nebraska Press.

Macdougall, D. (2004) *Frozen Earth: The Once and Future Story of Ice Ages*. Berkeley: University of California Press.

Madeira, P. (1964) *Men in Search of Man: The First Seventy-five Years of the University Museum of the University of Pennsylvania*. Philadelphia: University of Pennsylvania Press.

Mandlebaum, D., ed. (1949) *Selected Writings of Edward Sapir in Language, Culture, and Personality*. Berkeley: University of California Press.

Manning, T. G. (1967) *Government in Science: The U.S. Geological Survey 1867–1894*. Lexington: University of Kentucky Press.

Marcus, L. and R. Berger (1984) The significance of radiocarbon dates for Rancho La Brea. In *Quaternary Extinctions: A Prehistoric Revolution*, ed. P. S. Martin and R. G. Klein, pp. 159–183. Tucson: University of Arizona Press.

Mark, J. (1980) *Four Anthropologists: An American Science in Its Early Years*. New York: Science History Publications.

Mason, J. A. (1966) Pre-Folsom estimates of the age of man in America (with comments by Cotter). *American Anthropologist* 68:193–198.

Mayor, A. (2005) *Fossil Legends of the First Americans*. Princeton: Princeton University Press.

Mayr, E. (1982) *The Growth of Biological Thought*. Cambridge, MA: Harvard University Press.

McDonald, J. N. (1981) *North American Bison: Their Classification and Evolution*. Berkeley: University of California Press.

McMillan, R. B. (1976) Man and mastodon: A review of Koch's 1840 Pomme de Terre expeditions. In *Prehistoric Man and His Environments*, ed. R. Wood and R. MacMillan, pp. 81–96. New York: Academic.

McMillan, R. B. (2010) The discovery of fossil vertebrates on Missouri's western frontier. *Earth Sciences History* 29:26–51.

McMullin, E. (1987) Scientific controversy and its termination. In *Scientific Controversies: Case Studies in the Resolution and Closure of Disputes in Science and Technology*, ed. H. T. Engelhardt and A. L. Caplan, pp. 49–91. Cambridge, UK: Cambridge University Press.

Medawar, P. (1982) *Pluto's Republic: Incorporating the Art of the Soluble and Induction and Intuition in Scientific Thought*. Oxford: Oxford University Press.

Meltzer, D. J. (1983) The antiquity of man and the development of American archaeology. *Advances in Archaeological Method and Theory* 6:1–51.

Meltzer, D. J. (1985) North American archaeology and archaeologists, 1879–1934. *American Antiquity* 50:249–260.

Meltzer, D. J. (1989) A question of relevance. In *Tracing Archaeology's Past: The Historiography of Archaeology*, ed. A. Christenson, pp. 5–20. Carbondale: Southern Illinois University Press.

Meltzer, D. J. (1990) In search of a mammoth fraud. *New Scientist* 127(1725):51–55.

Meltzer, D. J. (1991) On "paradigms" and "paradigm bias" in controversies over human antiquity in America. In *The First Americans: Search and Research*, ed. T. Dillehay and D. Meltzer, pp. 13–49. Boca Raton: CRC Press.

Meltzer, D. J. (1994) The discovery of deep time: A history of views on the peopling of the Americas. In *Method and Theory for Investigating the Peopling of the Americas*, ed. R. Bonnichsen and D. G. Steele, pp. 7–26. Corvallis: Center for the Study of the First Americans.

Meltzer, D. J. (1998) Ephraim Squier, Edwin Davis, and the making of an American archaeological classic. In *Ancient Monuments of the Mississippi Valley, 150th Anniversary Edition*, pp. 1–97, by E. G. Squier and E. H. Davis. Washington, DC: Smithsonian Institution Press.

Meltzer, D. J. (2002) Starring the anthropologists in the *American Men of Science*. In *Anthropology, History, and American Indians: Essays in Honor of William C. Sturtevant*, ed. William Merrill and Ives Goddard. *Smithsonian Contributions to Anthropology* 44: 221–238.

Meltzer, D. J. (2003) In the heat of controversy: C. C. Abbott, the American Paleolithic, and the University Museum, 1889–1893. In *Archaeology and Archaeologists in Philadelphia*, ed. D. Fowler and D. Wilcox, pp. 48–87. Tuscaloosa: University of Alabama Press.

Meltzer, D. J. (2005) The seventy-year itch: Controversies over human antiquity and their resolution. *Journal of Anthropological Research* 61(4):433–468.

Meltzer, D. J. (2006a) *Folsom: New Archaeological Investigations of a Classic Paleoindian Bison Kill*. Berkeley: University of California Press.

Meltzer, D. J. (2006b) History of research on the Paleo–Indian. *Handbook of North American Indians*, vol. 3, pp. 110–128. D. H. Ubelaker, volume editor; W. C. Sturtevant, general editor. Washington, DC: Smithsonian Institution.

Meltzer, D. J. (2009) *First Peoples in a New World: Colonizing Ice Age America*. Berkeley: University of California Press.

Meltzer, D. J. (2010) When destiny takes a turn for the worse: William Henry Holmes and, incidentally, Franz Boas in Chicago, 1892–1897. *Histories of Anthropology Annual* 6:171–224.

Meltzer, D. J. and R. C. Dunnell (1992) Introduction. In *The Archaeology of William Henry Holmes*, ed. D. J. Meltzer and R. C. Dunnell, pp. vii–l. Washington, DC: Smithsonian Institution Press.

Meltzer, D. J. and W. C. Sturtevant (1983) The Holly Oak shell game: An historic archaeological fraud. In *Lulu Linear Punctated: Essays in Honor of G. Irving Quimby*, ed. R. C. Dunnell and D. K. Grayson. *Anthropological Papers* 72:325–352. Ann Arbor: Museum of Anthropology, University of Michigan.

Merrill, G. P. (1924) *The First One Hundred Years of American Geology*. New Haven: Yale University Press.

Merrill, G. P., ed. (1920) Contributions to a history of American state geological and natural history surveys. *United States National Museum Bulletin* 109.

Merton, R. K. (1968) The Matthew effect in science. *Science* 159:56–63.

Merton, R. K. (1988) The Matthew effect in science, II: Cumulative advantage and the symbolism of intellectual property. *Isis* 79:606–623.

Miller, H. S. (1970) *Dollars for Research: Science and Its Patrons in 19th Century America.* Seattle: University of Washington Press.

Morgan, M. (2010) A re-assessment of the human remains from the Upper Pecos Valley. In *Pecos Pueblo Revisited: The Biological and Social Context. Papers of the Peabody Museum of Archaeology and Ethnology* 85:27–41.

Morgan, M., ed. (2010) *Pecos Pueblo Revisited: The Biological and Social Context. Papers of the Peabody Museum of Archaeology and Ethnology* 85.

Morison, W. J. (1971) George Frederick Wright: In defense of Darwinism and Fundamentalism, 1838–1921. Unpublished PhD dissertation, Department of History, Vanderbilt University.

Moulton, F. R. (1929) Thomas Chrowder Chamberlin as a philosopher. *Journal of Geology* 37:368–379.

Mounier, R. (1972) The question of man's antiquity in the New World: 1840–1927. *Pennsylvania Archaeologist* 42:59–69.

Myster, S. and B. O'Connell (1997) Bioarchaeology of Iowa, Wisconsin and Minnesota. In *Bioarchaeology of the North Central United States*, ed. D. Owsley and J. Rose, pp. 147–239. Arkansas Archaeological Survey, Research Series 49.

Nelson, M. and J. Madsen (1979) The Hay-Romer camel debate: Fifty years later. *University of Wyoming Contributions to Geology* 18:47–50.

Noelke, V. (1974) The origin and early history of the Bureau of American Ethnology, 1879–1910. Unpublished PhD dissertation, Department of History, University of Texas, Austin.

Numbers, R. L. (1988) George Frederick Wright: From Christian Darwinist to Fundamentalist. *Isis* 79:624–645.

O'Brien, M. (2003) Nels Nelson and the measure of time. In *Picking the Lock of Time: Developing Chronology in American Archaeology*, ed. J. Truncer, pp. 64–87. Gainesville: University Press of Florida.

Olby, R. (1989) The dimensions of scientific controversy: The Biometric-Mendelian debate. *British Journal for the History of Science* 22:299–320.

Oldroyd, D. (1990) *The Highlands Controversy: Constructing Geological Knowledge through Fieldwork in Nineteenth–century Britain.* Chicago: University of Chicago Press.

Oldroyd, D. (2003) The earth sciences. In *From Natural Philosophy to the Sciences: Writing the History of Nineteenth-century Science*, ed. D. Cahan, pp. 88–128. Chicago: University of Chicago Press.

Parezo, N. and D. Fowler (2007) *Anthropology Goes to the Fair: The 1904 Louisiana Purchase Exposition.* Lincoln: University of Nebraska Press.

Penrose, R. A. F. (1929) The early days of the department of geology at the University of Chicago. *Journal of Geology* 37:320–327.

Pettit, M. (2006) "The Joy in Believing": The Cardiff Giant, commercial deceptions, and styles of observation in Gilded Age America. *Isis* 97:659–677.

Pinsky, V. (1992) Archaeology, politics, and boundary-formation: The Boas censure (1919) and the development of American archaeology during the inter-war years. In *Rediscovering Our Past: Essays on the History of American Archaeology*, ed. J. Reyman, pp. 161–189. Aldershot: Avebury.

Popper, K. (1963) *Conjectures and Refutations: The Growth of Scientific Knowledge.* New York: Basic Books.

Powell, J. (2005) *The First Americans: Race, Evolution, and the Origin of Native Americans.* Cambridge, UK: Cambridge University Press.

Purdy, B. A., K. Jones, J. J. Mecholsky, G. Bourne, R. C. Hulbert Jr., B. J. MacFadden, K. L. Church, M. W. Warren, T. F. Jorstad, D. J. Stanford, M. J. Wachowiak, and R. J. Speakman. (2011) Earliest art in the Americas: Incised image of a proboscidean on a mineralized extinct animal bone from Vero Beach, Florida. *Journal of Archaeological Science* 38:2908–2913.

Pyne, S. J. (1980) *Grove Karl Gilbert: A Great Engine of Research.* Austin: University of Texas Press.

Quimby, G. I. (1956) The locus of the Natchez pelvis find. *American Antiquity* 22: 77–79.

Rabbitt, M. (1979) *Minerals, Lands, and Geology for the Common Defense and General Welfare,* vol. 1: *Before 1879.* United States Geological Survey. Washington, DC: Government Printing Office.

Rabbitt, M. (1980) *Minerals, Lands, and Geology for the Common Defense and General Welfare,* vol. 2, *1879–1904.* United States Geological Survey. Washington, DC: Government Printing Office.

Regal, B. (2002) *Henry Fairfield Osborn: Race and the Search for the Origins of Man.* Hants, UK: Ashgate.

Reingold, N. (1976) Definitions and speculations: The professionalization of science in America in the nineteenth century. In *The Pursuit of Knowledge in the Early American Republic,* ed. A. Oleson and S. Brown, pp. 33–69. Baltimore: Johns Hopkins University Press.

Reybrouck, D. V. (2002) Boule's error: On the social context of scientific knowledge. *Antiquity* 76:158–164.

Rhees, W. J. (1901) *The Smithsonian Institution: Documents Relative to Its Origin and History, 1835–1899.* Smithsonian Miscellaneous Collections 42 and 43. Washington, DC: Government Printing Office.

Ridge, J. C. (2004) The Quaternary glaciation of western New England with correlations to surrounding areas. In *Quaternary Glaciations—Extent and Chronology.* Part 2: *North America,* ed. J. Ehlers and P. L. Gibbard, pp. 212–231. Developments in Quaternary Science, vol. 2b. Amsterdam: Elsevier.

Rogers, R. and L. Martin (1984) The 12 Mile Creek site: A reinvestigation. *American Antiquity* 49:757–764.

Rossiter, M. W. (1982) *Women Scientists in America: Struggles and Strategies to 1940.* Baltimore: Johns Hopkins University Press.

Rowe, J. (1975) Review of "A History of American Archaeology" by G. R. Willey and J. A. Sabloff. *Antiquity* 49:156–158.

Rowly-Conwy, P. (2006) The concept of prehistory and the invention of the terms "prehistoric" and "prehistorian": The Scandinavian origin, 1833–1850. *European Journal of Archaeology* 9:103–130.

Rudwick, M. J. S. (1971) Uniformity and progression: Reflections on the structure of geological theory in the age of Lyell. In *Perspectives in the History of Science and Technology,* ed. D. Roller, pp. 209–227. Norman: University of Oklahoma Press.

Rudwick, M. J. S. (1976) *The Meaning of Fossils: Episodes in the History of Paleontology.* New York: Science History Publications.

Rudwick, M. J. S. (1985) *The Great Devonian Controversy: The Shaping of Scientific Knowledge among Gentlemanly Specialists.* Chicago: University of Chicago Press.

Rudwick, M. J. S. (1997) *Georges Cuvier, Fossil Bones, and Geological Catastrophes: New Translations and Interpretations of the Primary Texts.* Chicago: University of Chicago Press.

Rudwick, M. J. S. (2008) *Worlds before Adam: The Reconstruction of Geohistory in the Age of Reform.* Chicago: University of Chicago Press.

Russett, C. E. (1976) *Darwin in America: The Intellectual Response 1865–1912.* San Francisco: W. H. Freeman.

Sackett, J. (2000) Human antiquity and the Old Stone Age: The Nineteenth Century background to paleoanthropology. *Evolutionary Anthropology* 37:36–49.

Schlereth, T. J. (1991) *Victorian America: Transformations in Everyday Life, 1876–1915.* New York: Harper Collins.

Schneer, C. (1978) The Great Taconic controversy. *Isis* 69:173–191.

Schultz, A. H. (1946) Biographical memoir of Aleš Hrdlička 1869–1943. *Biographical Memoirs of the National Academy of Sciences* 23:305–338.

Schultz, S. F. (1976) Thomas C. Chamberlin: An intellectual biography of a geologist and educator. Unpublished PhD dissertation, Department of History, University of Wisconsin.

Schultz, S. F. (1983) The debate over multiple glaciation in the United States: T. C. Chamberlin and G. F. Wright, 1889–1894. *Earth Sciences History* 2:122–129.

Secord, J. A. (1986) *Controversies in Victorian Geology: The Cambrian-Silurian Dispute.* Princeton: Princeton University Press.

Shor, E. (1971) *Fossils and Flies: The Life of a Compleat Scientist Samuel Wendell Williston (1851–1918).* Norman: University of Oklahoma Press.

Shor, E. (1974) *The Fossil Feud between E. D. Cope and O. C. Marsh.* Hicksville, NY: Exposition Press.

Simpson, G. G. (1952) Biographical memoir of Chester Stock 1892–1950. *Biographical Memoirs of the National Academy of Sciences* 27:335–362.

Sinclair, B. (1979) Americans abroad: Science and cultural nationalism in the early nineteenth century. In *The Sciences in the American Context: New Perspectives*, ed. N. Reingold, pp. 35–53. Washington, DC: Smithsonian Institution Press.

Skinner, M., S. Skinner, and R. Gooris (1977) Stratigraphy and biostratigraphy of late Cenozoic deposits in central Sioux County, western Nebraska. *American Museum of Natural History Bulletin* 158:271–367.

Sloan, P. (1976) The Buffon-Linnaeus controversy. *Isis* 67:356–375.

Snead, J. (2001) *Ruins and Rivals: The Making of Southwest Archaeology.* Tucson: University of Arizona Press.

Sommer, M. (2006) Mirror, mirror on the wall: Neanderthal as image and "distortion" in early 20th-century French science and press. *Social Studies of Science* 36: 207–240.

Spencer, F. (1979) Aleš Hrdlička, M.D., 1869–1943: A chronicle of the life and work of an American physical anthropologist. Unpublished PhD dissertation, Department of Anthropology, University of Michigan.

Spencer, F. (1988) Prologue to a scientific forgery: The British Eolithic movement from Abbeville to Piltdown. *History of Anthropology* 5:84–116.

Spencer, F. (1990) *Piltdown: A Scientific Forgery.* Oxford: Oxford University Press.

Spencer, F. (1997a) Hrdlička, Aleš (1869–1943). In *History of Physical Anthropology: An Encyclopedia*, vol. 1, ed. F. Spencer, pp. 503–505. New York: Garland.

Spencer, F. (1997b) Manouvier, Leonce-Pierre (1850–1927). In *History of Physical Anthropology: An Encyclopedia*, vol. 2, ed. F. Spencer, ed., pp. 642–643. New York: Garland.

Spencer, F. and F. Smith (1981) The significance of Ales Hrdlička's "Neanderthal Phase of Man": A historical and current assessment. *American Journal of Physical Anthropology* 56:435–459.

Stanton, W. (1960) *The Leopard's Spots: Scientific Attitudes toward Race in America, 1815–1859*. Chicago: University of Chicago Press.

Stegner, W. (1954) *Beyond the Hundredth Meridian: John Wesley Powell and the Second Opening of the West*. Boston: Houghton Mifflin.

Stewart, T. D. (1946) A re-examination of the fossil human skeletal remains from Melbourne, Florida, with further data on the Vero skull. *Smithsonian Miscellaneous Collections* 106:1–28.

Stewart, T. D. (1949) The development of the concept of morphological dating in connection with early man in America. *Southwestern Journal of Anthropology* 5:1–16.

Stewart, T. D. (1951) Antiquity of man in America demonstrated by the fluorine test. *Science* 113:391–392.

Stock, C. (1951) Biographical memoir of John C. Merriam 1869–1945. *Biographical Memoirs of the National Academy of Sciences* 26:209–232.

Stocking, G. (1960) Franz Boas and the founding of the American Anthropological Association. *American Anthropologist* 62:1–17.

Stocking, G. (1968) *Race, Culture, and Evolution: Essays in the History of Anthropology*. New York: Free Press.

Stocking, G. (1974) *A Franz Boas Reader: The Shaping of American Anthropology, 1883–1911*. Chicago: University of Chicago Press.

Stocking, G. (1976) Ideas and institutions in American anthropology: Thoughts toward a history of the interwar years. In *Selected Papers from the American Anthropologist, 1921–1945*, ed. George W. Stocking, pp. 1–53. Washington, DC: American Anthropological Association.

Stocking, G. W. (1987) *Victorian Anthropology*. New York: Free Press.

Swanton, J. (1935) Biographical memoir of William Henry Holmes, 1846–1933. *Biographical Memoirs of the National Academy of Sciences* 17:223–252.

Swanton, J. (1944) Notes regarding my adventures in anthropology and with anthropologists. Unpublished manuscript on file at the National Anthropological Archives, Smithsonian Institution, Washington, DC.

Taylor, R. E., L. Payen, and P. Slota (1992) The age of the Calaveras skull: Dating the "Piltdown Man" of the New World. *American Antiquity* 57:269–275.

Thomas, D. H. (2000) *Skull Wars: Kennewick Man, Archaeology, and the Battle for Native American Identity*. New York: Basic Books.

Thomson, K. (2008) *The Legacy of the Mastodon: The Golden Age of Fossils in America*. New Haven: Yale University Press.

Thoresen, T. H. (1977) Art, evolution, and history: A case study of paradigm change in anthropology. *Journal of the History of the Behavioral Sciences* 13:107–125.

Totten, S. M. and G. W. White (1985) Glacial geology and the North American craton. In *Geologists and Ideas: A History of North American Geology*, Geological Society of America, Centennial Special vol. 1:125–141.

Tozzer, A. M. (1933) Biographical memoir of Frederic Ward Putnam 1839–1915. *Biographical Memoirs of the National Academy of Sciences* 16:125–153.

Trigger, B. G. (1980) Archaeology and the image of the American Indian. *American Antiquity* 45:662–676.

Trigger, B. G. (1989) *A History of Archaeological Thought*. Cambridge, UK: Cambridge University Press.

Trigger, B. G. (2006) *A History of Archaeological Thought*, 2nd ed. Cambridge, UK: Cambridge University Press.

Truncer, J. (2003) Leslie Spier and the Middle Atlantic revolution that never happened. In *Picking the Lock of Time: Developing Chronology in American Archaeology*, ed. J. Truncer, pp. 88–103. Gainesville: University Press of Florida.

Van Riper, A. B. (1993) *Men among the Mammoths: Victorian Science and the Discovery of Human Prehistory*. Chicago: University of Chicago Press.

Vetter, J. (2008) Cowboys, scientists, and fossils: The field site and local collaboration in the American west. *Isis* 99:273–303.

Visher, S. (1947) *Scientists Starred, 1903–1943, in "American Men of Science": A Study of Collegiate and Doctoral Training, Birthplace, Distribution, Backgrounds, and Developmental Influences*. Baltimore: Johns Hopkins University Press.

Ward, C. (2003) The evolution of human origins. *American Anthropologist* 105:77–88.

Wedel, W. (1959) *An Introduction to Kansas Archaeology*. Bulletin 174. Washington, DC: Bureau of American Ethnology.

Wedel, W. (1981) Towards a history of Plains archaeology. *Great Plains Quarterly* 1:16–38.

Weigel, G. (1962) Fossil vertebrates of Vero, Florida. *Special Publication of the Florida Geological Survey* 10.

Weisensee, K. and R. Jantz (2010) Rethinking Hooton: A reexamination of the Pecos cranial and postcranial data using recent methods. In *Pecos Pueblo Revisited: The Biological and Social Context. Papers of the Peabody Museum of Archaeology and Ethnology* 85:43–55.

White, G. (1973) History of investigation and classification of Wisconsinan drift in north-central United States. In *The Wisconsin Stage*, ed. R. Black, R. Goldthwait, and H. Willman. *Geological Society of America Memoir* 136:3–34.

Wiebe, R. (1967) *The Search for Order 1877–1920*. New York: Hill and Wang.

Willey, G. (1967) Biographical memoir of Alfred Vincent Kidder, 1885–1936. *Biographical Memoirs of the National Academy of Sciences* 39:292–322.

Willey, G. and J. Sabloff (1974) *A History of American Archaeology*. London: Thames & Hudson.

Willey, G. and J. Sabloff (1980) *A History of American Archaeology*, 2nd ed. San Francisco: W. H. Freeman.

Willey, G. and J. Sabloff (1993) *A History of American Archaeology*, 3rd ed. San Francisco: W. H. Freeman.

Wilmsen, E. (1965) An outline of early man studies in the United States. *American Antiquity* 31:172–192.

Wilson, L. G. (1972) *Charles Lyell, the Years to 1841: The Revolution in Geology*. New Haven: Yale University Press.

Winegrad, D. P. (1993) *Through Time, across Continents: A Hundred Years of Archaeology and Anthropology at the University Museum*. Philadelphia: University of Pennsylvania Press.

Winnick, H. (1970) Science and morality in Thomas C. Chamberlin. *Journal of the History of Ideas* 31:441–456.

Wolf, J. and J. Mellett (1985) The role of "Nebraska man" in the creation-evolution debate. *Creation/Evolution* 16:31–43.

Wood, W. R. (1976) Archaeological investigations at the Pomme de Terre springs. In *Prehistoric Man and His Environments*, ed. R. Wood and R. MacMillan, pp. 97–107. New York: Academic.

Woodbury, R. B. (1993) *60 Years of Southwestern Archaeology: A History of the Pecos Conference*. Albuquerque: University of New Mexico Press.

ACKNOWLEDGMENTS

Having excavated several large, complex archaeological sites, and now having completed a historical study of an extensive, complex, multidisciplinary controversy, I can report that working at archaeology is not so very different from working at the history of science. Both require considerable digging (obviously, into the published literature and archives in the one case, the earth in the other) and routinely entail lengthy periods of "postexcavation" analysis and writing. It is not surprising that the duration of each is compounded when one undertakes both sorts of investigations simultaneously, as I have over the years, and it has meant sacrificing work on one to make time for another. Unfortunately, it was this historical study that often lost out to my archaeological projects (my thumb was not always on one side of the scale—occasionally I spent mornings in the field writing about the history of archaeology while my very able graduate students led the excavations on-site). As a result, this project has had a long gestation, one that began in the late twentieth century and finished in the twenty-first.

Still, starting this project when I did at least gave me the opportunity, even more treasured now than it was at the time, of communicating with a number of individuals who came of age as the human antiquity controversy ended, who knew many of the principals in that dispute, and who in some cases played key roles in developing the knowledge about North American Paleoindians that followed. They have all since passed away, yet though they are no longer here to be thanked I would like to record my gratitude all the same to Henry B. Collins, John L. Cotter, Glen Evans, James B. Griffin, Emil W. Haury, Jesse D. Jennings, Dorothy Cook Meade, George I. Quimby, C. Bertrand Schultz, T. Dale Stewart, and H. Marie Wormington.

The project's long incubation also provided a number of opportunities to test drive my ideas in talks and publications. I justified this to my colleagues—not disingenuously—as being of interest and significance in our own efforts to understand the Pleistocene peopling of the Americas (not to mention helping provide a broader context for our own disputes). They responded in kind: over the years I have benefited from the comments and advice of fellow archaeologists, as well as geologists, vertebrate paleontologists, and historians of science. For their thoughts, counsel, historical tips, and occasional loans of hard-to-find items from their personal libraries, I am grateful to Robert C. Dunnell, Jacob W. Gruber, Curtis M. Hinsley, Vance T. Holliday, Louis L. Jacobs, Bernard Means, George Milner, Ronald L. Numbers, Nathan Reingold, Jeremy A. Sabloff, Michael B. Schiffer, Frank Spencer, William C. Sturtevant, Bruce G. Trigger, and Gordon R. Willey. Two very thorough but anonymous reviewers for the University of Chicago Press read the manuscript and provided valuable comments. Amanda Thornton did excellent work checking the endnotes against the bibliography to ensure nothing was awry (no small task, that). And, not least, I am especially indebted to my longtime friend and colleague Donald K. Grayson, who helped in countless ways throughout this project—even to its inception. It was Don who, in the early 1980s and then working on his own historical study of the establishment of human antiquity in Europe, first got me curious about how the comparable process played out in the Americas—and thereby planted the idea that ultimately became this book.

My research into the archives of the participants and the archaeology of the sites in this controversy was made possible by grants from the History and Philosophy of Science Program at the National Science Foundation and the National Endowment for the Humanities, and the Potts and Sibley Foundation, and with additional in-kind support from the Smithsonian Institution—the latter through the good offices of Daniel Goodwin (then director of the Smithsonian Institution Press) and Bruce D. Smith.

The research trips to archives were eased in two ways: by family members Seymour and Maxine Garner, Jay and Gayle Halfond, Jeanne and Murray Halfond, Peter and Anne Meltzer, Stephen and Florence Meltzer, and Jean Siegel, all of whom kindly hosted my visits in various cities; and by the help I had at each stop from archivists, librarians, museum curators, and fellow archaeologists. Many of these individuals may no longer be at the institutions where they once were, but it would be uncharitable not to acknowledge their assistance. Unfortunately, I did not always catch a name (or, inexcusably, did not write it down). In the hope that those named can symbolize those I cannot, I would like to thank (in order by archive): Carol Spawn (Academy of Natural Sciences, Philadelphia, PA); Audrey Barnhart and Reid Miller (Agate Fossil Beds National Monument, Harrison, NE); Dorothy Cook and Grayson Meade (Agate Ranch, Agate, NB); Emmett Chisum (American Heritage Center, Laramie, WY); John Alexander, Belinda Kaye, and Lori Pendleton Thomas (American Museum of Natural History, New York, NY); Terry McNealy (Bucks County Historical Society, Doylestown, PA); Susan Vazquez (Carnegie Institution of Washington, Washington, DC); Bob Akerly and Kris Haglund (Denver Museum of Nature and Science, Denver, CO); Ann Van Arsdale and Jean Preston (Firestone Library, Princeton University, Princeton, NJ); Steve Nielsen and Hampton Smith (Minnesota Historical Society, St. Paul, MN); Vyrtis Thomas (National Anthropological Archives, Washington, DC); Roland Baumann (Oberlin College, Oberlin, OH); Sarah Polirer (Pusey Library, Harvard University, Cambridge, MA); Tom Burch and Emily Burch Hughes (Raton, NM); William Deiss and Libby Glenn (Smithsonian Institution Archives, Washington, DC); John Rick (Stanford University, Stanford, CA); Lynn Denton (Texas Memorial Museum, Austin, TX); Elizabeth O'Leary (University of Arizona, Tucson, AZ); Art Bettis and Earl Rogers (University of Iowa, Iowa City, IA); George Corner and Michael Voorhies (University of Nebraska, Lincoln, NE); Fred Burchsted (University of Texas, Austin, TX); and Lucy Fowler-Williams and Alessandro Pezzati (University Museum, University of Pennsylvania, Philadelphia, PA). I would also like to thank the several generations of Southern Methodist University (SMU) librarians who let me roam freely in the cordoned-off "Staff Only" basement of our Science and Engineering Library where most of nineteenth- and early twentieth-century journals and books were in deep storage (and well before they were available and searchable electronically). It was a privilege and trust I deeply appreciated.

As noted in chapter 1, I also visited many of the archaeological sites that had a part in this controversy (and excavated at several), and on a number of those occasions my visits were facilitated by knowledgeable guides. I would like to thank Jim Baum, then-mayor of Colorado City, TX, who got Vance Holliday and me close to the spot of the Lone Wolf Creek site, long since eroded away (chapter 8); Jim Concannon, who led me through heavy brush to show me the remnants of the fruit cellar dug by his grandfather and great uncle that yielded the Lansing skeletal remains (chapter 6); R. Michael

Stewart, who took me around the Trenton Gravel sites of Charles Abbott and others and explained what is now known of the archaeology of that locality (chapter 3); and even the unnamed homeowners who watched suspiciously but did not call the Washington, DC, police when I walked through their yards to get to William Henry Holmes's Piney Branch quarry site (chapter 5). Finally, I am especially grateful to Joseph and Ruth Cramer, who endowed the Quest Archaeological Research Program at SMU in 1996, which generously supports my archaeological investigations into North American Paleoindians. It is no coincidence that the first major project I undertook under Quest auspices was a multiseason excavation at the Folsom site, or that the subsequent analysis also incorporated the bison remains and artifacts that had been recovered from the site during the 1920s work (Meltzer 2006a). That project—at the very spot where the human antiquity controversy was resolved in September 1927 (chapter 8)—lay precisely at the crossroads of my archaeological and historical interests.

Much of the research and writing for this book was spread across multiple semester-length sabbaticals, for which I am grateful to a succession of deans of SMU's Dedman College: Hal Williams, Buddy Gray, James F. Jones, and most recently William Tsutsui, who commuted my sentence as department chair and provided a sabbatical over the 2012–2013 academic year. Part of the fall semester was spent writing at SMU's beautiful Taos campus, thanks to its director and my friend and colleague Mike Adler. In the spring semester, Ofer Bar-Yosef and Matthew Liebmann arranged an office and a visiting appointment for me in the Department of Anthropology at Harvard University, where they and other faculty—Rowan Flad, Dan Lieberman, Noreen Tuross, and Jason Ur (in Anthropology and Human Evolutionary Biology), and Jerry Mitrovica and Carl Wunsch (in Earth and Planetary Sciences)—were welcoming hosts.

It was at Harvard in the late spring of 2013 that I finished this book. I well appreciate the historical symmetry of doing so at the institution—and, indeed, in the very building—where Frederick Ward Putnam and Charles Abbott gave the American Paleolithic its formal debut nearly 140 years earlier.

David J. Meltzer
Dallas, Texas, September 2014

INDEX

Page numbers in *italics* refer to figures and tables.